1998 World Development Indicators

Photo credits: Curt Carnemark/World Bank and Still Pictures.

If you have questions and comments about this product, please contact:

Development Data Center,
The World Bank
1818 H Street, N.W., Room MC2-812, Washington, D.C. 20433, U.S.A.
Hotline: (800) 590 1906 or (202) 473 7824; fax (202) 522 1498.
Email: info@worldbank.org
Website: http://www.worldbank.org or http://www.worldbank.org/wdi

ISBN 0-8213-3701-4124-3

1998 World Development Indicators

The World Bank

Foreword

In my 1997 address to the Annual Meetings of the Board of Governors of the World Bank Group and the International Monetary Fund, I noted that our primary goal in development must be to reduce the disparities across and within countries—to bring more people into the economic mainstream and to promote equitable access to the fruits of development, regardless of nationality, race, or gender. I said then, and I write here, that the key development challenge of our time is the challenge of inclusion.

In the many international conferences of the 1990s, the world's leaders have committed themselves to reducing these disparities. And they have set clear targets for reducing poverty, improving child and maternal mortality, expanding education opportunities, closing the gender gap, and reversing environmental degradation. In 1996 the members of the OECD's Development Assistance Committee endorsed a set of related goals and committed themselves to supporting countries that want to achieve them through more effective development cooperation efforts.

The World Bank is equally committed to helping countries fight the scourges of poverty, illiteracy, disease, hunger, and environmental degradation. And together with our development partners we are working hard to meet the challenge of inclusion. The *World Development Indicators*, the product of a broad global knowledge partnership of national and international development institutions, is one contribution to this effort. We believe that by reporting regularly and systematically on progress toward the targets the international community has set for itself, we will focus attention on the task ahead and make those responsible for advancing the development agenda accountable for results.

This year's *World Development Indicators* begins with the first of a series of annual reports we plan to publish on progress toward the international development goals. The main—and not surprising—message is that these goals are difficult to attain. Countries that have succeeded in these areas have done so by sustaining economic growth, investing in their people, and implementing the right policies. But as the recent difficulties in East Asia warn, good and open governance that builds a social consensus is equally important. Without it, success can prove brittle.

In closing, I want to thank those of you who took me at my word last year and provided comments on the newly redesigned *World Development Indicators*. We have benefited greatly from your suggestions, which are reflected in some of the many further improvements I hope you will notice in this second edition. Please continue to review this product critically and to offer us your suggestions.

James D. Wolfensohn
President
The World Bank Group

Acknowledgments

This book and its companion volume, the *World Bank Atlas,* were prepared by a team led by K. Sarwar Lateef. The team consisted of Mehdi Akhlaghi, Aelim Chi, David Cieslikowski, Demet Kaya, Saeed Ordoubadi, Sulekha Patel, Eric Swanson, K. M. Vijayalakshmi, Amy Wilson, and Estela Zamora, working closely with other teams in the Development Economics Vice Presidency's Development Data Group. The CD-ROM development team included Mehdi Akhlaghi, Azita Amjadi, Elizabeth Crayford, Reza Farivari, Vasantha Hevaganinge, Yusri Harun, Angelo Kostopoulos, André Léger, Patricia McComas, and William Prince. The work was carried out under the management of Shaida Badiee.

The choice of indicators and textual content was shaped through close consultation with and substantial contributions from staff in the World Bank's four thematic networks—Environmentally and Socially Sustainable Development; Finance, Private Sector, and Infrastructure; Human Development; and Poverty Reduction and Economic Management—and staff of the International Finance Corporation and the Multilateral Investment Guarantee Agency. Most important, we received substantial help, guidance, and data from our external partners. For individual acknowledgments of contributions to the book's content, please see the *Credits* section. For a listing of our key partners, see the *Partners* section.

Bruce Ross-Larson was the principal editor, and Peter Grundy, the art director. The cover and page design and the layout and desktopping were done by the American Writing Division of Communications Development Incorporated, with Grundy & Northedge of London. The External Affairs Vice Presidency oversaw publication and dissemination of the book.

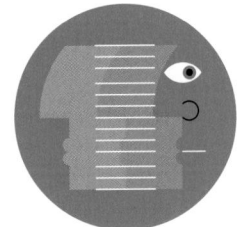

Now in its second year, the *World Development Indicators* continues to evolve. The response to last year's book and CD-ROM was encouraging, and we learned much from the comments of our readers and partners. This year's edition reflects many of your suggestions:

•   The introductions to each section focus on key development issues and trends. The first section, *World View*, reports on progress toward international development goals—goals that should be achieved early in the 21st century. This report will become a standard feature of the *World Development Indicators*. The introduction to the third section, *Environment,* reports on three big issues for development and the environment: genuine saving, trade in goods from polluting industries, and demand for transport fuels.

•   We have tried to provide more timely data. Coverage of most indicators extends through 1996, and the introduction to the fourth section, *Economy,* includes preliminary estimates of macroeconomic indicators for 1997 for 37 developing countries and projections of world economic growth through 2000. The sixth section, *Global Links,* includes recent estimates of capital flows.

•   Every section includes new indicators. *World View* contains two tables on long-term growth. *People* has new tables on employment, unemployment, and reproductive health. *Economy* includes the most recent data on relative prices from the International Comparison Programme. *States and Markets* has added a table on military expenditures and trade in arms. *Global Links* provides new coverage of interregional trade flows and tariff rates.

Although we have added new indicators and relocated some tables, we have maintained the book's general plan. Most tables show indicators for both a recent and an earlier year or period, and cover 148 economies with populations of more than one million. For users who need time series of primary data, the *World Development Indicators* CD-ROM contains time series from the *World Development Indicators* database, extended back to 1960 wherever data permits. And for those who want a brief overview of the world economy, the *World Bank Atlas* presents maps, tables, and figures for 43 indicators in 210 economies. In addition, the *Atlas* has been expanded to include *World View* and *Global Links* sections.

Producing the *World Development Indicators* has been stimulating. The process has strengthened the World Bank's partnerships with other international organizations, statistical offices, nongovernmental organizations, and notably with private suppliers and users of data. With the new international commitment to monitoring development progress, timely, accurate, and comprehensive indicators have become even more important. This obliges us to work harder to bring you the best information available. We hope we are succeeding in this, and we look forward to hearing from you. Please write us or send us email at info@worldbank.org. We would also appreciate your completing the user survey at the back of the book or at http://www.worldbank.org/wdi/.

Shaida Badiee
Director
Development Data Group

Contents

World View has been reorganized to provide better coverage of key international development goals and has been expanded with new tables showing long-term trends in development. Table 1.1 now includes rankings of country performance on total GNP, GNP growth, and GNP per capita. Data for small economies (populations under 1 million) and those with limited data availability are now shown in table 1.6.

Contents

People has new tables on employment, unemployment, and reproductive health. The introduction features a special presentation on the worldwide HIV/AIDS epidemic, drawing on recent work by UNAIDS and the World Bank.

Environment includes new tables on water pollution and sources of electricity generation. Coverage of air pollution has been expanded. Because of the importance of agriculture for rural welfare and environmental management, indicators on agricultural inputs and production have been moved here. The introduction reports on three big issues for development and the environment: genuine saving, trade in goods from polluting industries, and demand for transport fuels.

2 PEOPLE

3 ENVIRONMENT

Contents

The introduction to *Economy* provides estimates of 1997 values for macroeconomic indicators for 37 developing countries, drawn from the World Bank's operational database, and projections of regional economic growth through 2000. The table on the structure of consumption measured in purchasing power parity (PPP) prices has been updated with data from the latest round of PPP surveys by the International Comparison Programme, and a new table on relative prices has been added. Indicators for agricultural inputs and production have been moved to *Environment*.

The introduction to *States and Markets* includes information on government credibility and corruption adapted from the World Bank's *World Development Report 1997: The State in a Changing World*. New this year are data on defense spending and trade in arms and updated indicators for state-owned enterprises.

Global Links includes indicators on trade that appeared in Economy last year, and on global financial flows that appeared in States and Markets. New indicators include direction of world trade, exports by regional trading blocs, and the most recent data on average tariff rates and travel and tourism. The introduction looks at the problems of measuring global economic integration and includes estimates of private financial flows in 1997.

6 GLOBAL LINKS

BACK MATTER

Partners

Defining, gathering, and disseminating international statistics is a collective effort of many people and organizations. The indicators presented in the *World Development Indicators* are the fruit of decades of work at many levels, from the field workers who have administered censuses and house-hold surveys in every part of the world to the committees and working parties of the national and international statistical agencies that have developed the nomenclature, classifications, and stan-dards that are fundamental to an international statistical system. Nongovernmental organizations and the private sector have also made important contributions, both in gathering primary data and in organizing and publishing their results. And academic researchers have played a crucial role in developing statistical methods and carrying on a continuing dialogue about the quality and inter-pretation of statistical indicators. What all these contributors have in common is a strong belief that accurate data, readily available, will improve the quality of public and private decisionmaking.

The organizations listed here have made the *World Development Indicators* possible by shar-ing their data and their expertise with us. More important, their collaboration contributes to the success of the World Bank's development efforts and to those of many others to improve the quality of life of the world's people. These organizations, like the World Bank, draw on the efforts and resources of national statistical offices and other agencies that gather and maintain primary statistical databases but that are too numerous to name individually. Nevertheless, we acknowl-edge our debt and gratitude to all those whose efforts have helped to build a base of compre-hensive, quantitative information about the world and its people. For your convenience we have included URLs (Web addresses) for organizations that maintain websites. The addresses shown were active on 1 January 1998. For information about the World Bank, visit its website at www.worldbank.org.

International agencies

Food and Agriculture Organization

The Food and Agriculture Organization (FAO), a specialized agency of the United Nations, was founded in October 1945 with a mandate to raise nutrition levels and living standards, to increase agricultural productivity, and to better the condition of rural populations. Since its inception the FAO has worked to alleviate poverty and hunger by promoting agricultural devel-opment, improved nutrition, and the pursuit of food security—the access of all people at all times to the food they need for an active and healthy life. The organization provides direct development assistance; collects, analyzes, and disseminates information; offers policy and planning advice to governments; and serves as an international forum for debate on food and agricultural issues.

Statistical publications of the FAO include the *Production Yearbook, Trade Yearbook,* and *Fertilizer Yearbook*. The FAO makes much of its data available on diskette through its Agrostat PC system. FAO publications can be ordered from national sales agents or directly from the FAO Distribution and Sales Section, Viale delle Terme di Caracalla, 00100 Rome, Italy; website: www.fao.org/default.htm.

International Civil Aviation Organization

The International Civil Aviation Organization (ICAO), a specialized agency of the United Nations, was founded on 7 December 1944. It is responsible for establishing international standards and recommended practices and procedures for the technical, economic, and legal aspects of international civil aviation operations.

The ICAO promotes the adoption of safety measures, establishes visual and instrument flight rules for pilots and crews, develops aeronautical charts, coordinates aircraft radio frequencies, and sets uniform regulations for the operation of air services and customs procedures.

To obtain ICAO publications contact ICAO, Document Sales Unit, 999 University Street, Montreal, Quebec H3C 5H7, Canada; telephone: (514) 954 8022; fax: (514) 954 6769; email: sales_unit@icao.org; website: www.cam.org/~icao.

International Labour Organization

The International Labour Organization (ILO), a specialized agency of the United Nations, seeks the promotion of social justice and internationally recognized human and labor rights. Founded in 1919, it is the only surviving major creation of the Treaty of Versailles, which brought the League of Nations into being. It became the first specialized agency of the United Nations in 1946. The ILO has a structure that is unique within the United Nations system, a tripartite structure that has workers and employers participating as equal partners with governments in the work of its governing organs. As part of its mandate, the ILO maintains an extensive statistical publication program. *Yearbook of Labour Statistics* is its most comprehensive collection of labor force data.

Publications can be ordered from the International Labour Office, 4 route des Morillons, CH-1211 Geneva 22, Switzerland, or from sales agents and major booksellers throughout the world and ILO offices in many countries. Fax: (41 22) 798 86 85; website: www.ilo.org.

International Monetary Fund

The International Monetary Fund (IMF) was established at a conference in Bretton Woods, New Hampshire, U.S.A., on 1–22 July 1944. (The conference also established the World Bank.) The IMF came into official existence on 27 December 1945, and commenced financial operations on 1 March 1947. It currently has 181 member countries.

The statutory purposes of the IMF are to promote international monetary cooperation, facilitate the expansion and balanced growth of international trade, promote exchange rate stability, help establish a multilateral payments system, make the general resources of the IMF temporarily available to its members under adequate safeguards, and shorten the duration and lessen the degree of disequilibrium in the international balances of payments of members.

The IMF maintains an extensive program for the development and compilation of international statistics, and is responsible for collecting and reporting statistics on international financial transactions and the balance of payments. In April 1996 it undertook an important initiative aimed at improving the quality of international statistics, establishing the Special Data Dissemination Standard to guide members that have or seek access to international capital markets in providing economic and financial data to the public.

The IMF's major statistical publications include *International Financial Statistics, Balance of Payments Statistics Yearbook, Government Finance Statistics Yearbook,* and *Direction of Trade Statistics.*

For more information on IMF statistical publications contact the International Monetary Fund, Publications Services, Catalog Orders, 700 19th Street, N.W., Washington, D.C. 20431, U.S.A.; telephone: (202) 623 7430; fax: (202) 623 7201; telex: RCA 248331 IMF UR; email: pubweb@imf.org; website: www.imf.org; SDDS bulletin board: dsbb.imf.org.

International Telecommunication Union

Founded in Paris in 1865 as the International Telegraph Union, the International Telecommunication Union (ITU) took its current name in 1934 and became a specialized agency of the United Nations in 1947. The ITU is an intergovernmental organization within which the public and private sectors cooperate for the development of telecommunications. The ITU adopts international regulations and treaties governing all terrestrial and space uses of the frequency spectrum and the use of the geostationary-satellite orbit. It also develops standards for the interconnection of telecommunications systems worldwide. The ITU fosters the development of telecommunications in developing countries by establishing medium-term development policies and strategies in consultation with other partners in the sector and providing specialized technical assistance in management, telecommunications policy, human resource management, research and development, technology choice and transfer, network installation and maintenance, and investment financing and resource mobilization.

The *Telecommunications Yearbook* is the ITU's main statistical publication. Publications can be ordered from ITU Sales and Marketing Service, Place des Nations, CH-1211 Geneva 20, Switzerland; telephone: (41 22) 730 6141 (English), (41 22) 730 6142 (French), and (41 22) 730 6143 (Spanish); fax: (41 22) 730 5194; email: sales.online @itu.ch; telex: 421 000 uit ch; telegram: ITU GENEVE; website: www.itu.ch.

Organisation for Economic Co-operation and Development

The Organisation for Economic Co-operation and Development (OECD) was set up in 1948 as the Organisation for European Economic Co-operation (OEEC) to administer Marshall Plan funding in Europe. In 1960, when the Marshall Plan had completed its task, the OEEC's member countries agreed to bring in Canada and the United States to form an organization to coordinate policy among industrial countries.

The OECD is the international organization of the industrialized, market economy countries. Representatives of member countries meet at the OECD to exchange information and harmonize policy with a view to maximizing economic growth in member countries and helping nonmember countries develop more rapidly. The OECD has set up a number of specialized committees to further its aims. One of these is the Development Assistance Committee (DAC), whose members have agreed to coordinate their policies on assistance to developing and transition economies. Also associated with the OECD are several agencies or bodies that have their own governing statutes, including the International Energy Agency (IEA) and the Centre for Co-operation with Economies in Transition.

The OECD's main statistical publications include *Geographical Distribution of Financial Flows to Developing Countries, National Accounts of OECD Countries, Labour Force Statistics, Revenue Statistics of OECD Member Countries, International Direct Investment Statistics Yearbook, Basic Science and Technology Statistics, Industrial Structure Statistics,* and *Services: Statistics on International Transactions.*

For information on OECD publications contact OECD, 2, rue André-Pascal, 75775 Paris Cedex 16, France; telephone: (33 1) 45 24 82 00; fax: (33 1) 45 24 85 00; websites: www.oecd.org and www.oecdwash.org.

United Nations

The United Nations and its specialized agencies maintain a number of programs for the collection of international statistics, some of which are described elsewhere in this book. At United Nations headquarters the Statistics Division of the Department of Economic and Social Information and Policy Analysis provides a wide range of statistical outputs and services for producers and users of statistics worldwide. By increasing the global availability and use of official statistics, the division's work facilitates national and international policy formulation, implementation, and monitoring.

The Statistics Division publishes statistics on international trade, national accounts, demography and population, gender, industry, energy, environment, human settlements, and disability. Its major statistical publications include the *International Trade Statistics Yearbook, Yearbook of National Accounts,* and *Monthly Bulletin of Statistics,* along with general statistics compendiums such as the *Statistical Yearbook* and *World Statistics Pocketbook.*

For publications contact the United Nations Sales Section, DC2-0853, New York, N.Y. 10017, U.S.A.; fax: (212) 963 3489; email: statistics@un.org; website: www.un.org.

United Nations Children's Fund

The United Nations Children's Fund (UNICEF), the only organization of the United Nations dedicated exclusively to children, works with other United Nations bodies and with governments and nongovernmental organizations to improve children's lives in more than 140 developing countries through community-based services in primary health care, basic education, and safe water and sanitation.

UNICEF's major publications include *The State of the World's Children* and *The Progress of Nations.*

For information on UNICEF publications contact UNICEF House, 3 United Nations Plaza, New York, N.Y. 10017, U.S.A.; telephone: (212) 326 7000; fax: (212) 888 7465; telex: RCA-239521; website: www.unicef.org.

United Nations Conference on Trade and Development

The United Nations Conference on Trade and Development (UNCTAD) is the principal organ of the United Nations General Assembly in the field of trade and development. It was established as a permanent intergovernmental body in 1964 in Geneva with a view to accelerating economic growth and development, particularly in developing countries. UNCTAD discharges its mandate through policy analysis; intergovernmental deliberations, consensus building, and negotiation; monitoring, implementation, and follow-up; and technical cooperation.

UNCTAD produces a number of publications containing trade and economic statistics, including the *Handbook of International Trade and Development Statistics.*

For information contact UNCTAD, Palais des Nations, CH-1211 Geneva 10, Switzerland; telephone: (41 22) 907 12 34 or 917 12 34; fax: (41 22) 907 00 57; telex: 42962; website: www.unicc.org/unctad.

United Nations Educational, Scientific, and Cultural Organization

The United Nations Educational, Scientific, and Cultural Organization (UNESCO) is a specialized agency of the United Nations established in 1945 to promote "collaboration among nations through education, science, and culture in order to further universal respect for justice, for the rule of law, and for the human rights and fundamental freedoms . . . for the peoples of the world, without distinction of race, sex, language, or religion . . ."

UNESCO's principal statistical publications are the *Statistical Yearbook, World Education Report* (biennial), and *Basic Education and Literacy: World Statistical Indicators.*

For publications contact UNESCO Publishing, Promotion, and Sales Division, 1, rue Miollis F, 75732 Paris Cedex 15, France; fax: (33 1) 45 68 57 41; email: c.laje@unesco.org; website: www.unesco.org.

United Nations Environment Programme

The mandate of the United Nations Environment Programme (UNEP) is to provide leadership and encourage partnership in caring for the environment by inspiring, informing, and enabling nations and people to improve their quality of life without compromising that of future generations. UNEP publications include *Global Environment Outlook* and *Our Planet* (a bimonthly magazine). For information contact UNEP, P.O. Box 30552, Nairobi, Kenya; telephone: (254 2) 62 1234 or 3292; fax: (254 2) 62 3927 or 3692; website: www.unep.org.

United Nations Industrial Development Organization

The United Nations Industrial Development Organization (UNIDO) was established in 1966 to act as the central coordinating body for industrial activities and to promote industrial development and cooperation at the global, regional, national, and sectoral levels. In 1985 UNIDO became the sixteenth specialized agency of the United Nations, with a mandate to help develop scientific and technological plans and programs for industrialization in the public, cooperative, and private sectors.

UNIDO's databases and information services include the Industrial Statistics Database (INDSTAT), Commodity Balance Statistics Database (COMBAL), Industrial Development Abstracts (IDA), and the International Referral System on Sources of Information. Among its publications is the *International Yearbook of Industrial Statistics.*

For information contact UNIDO Public Information Section, Vienna International Centre, P.O. Box 300, A-1400 Vienna, Austria; telephone: (43 1) 211 31 5021 or 5022; fax: (43 1) 209 2669; email: unido-pinfo@unido.org; website: www.unido.org.

World Health Organization

The constitution of the World Health Organization (WHO) was adopted on 22 July 1946 by the International Health Conference, convened in New York by the Economic and Social Council. The objective of the WHO, a specialized agency of the United Nations, is the attainment by all people of the highest possible level of health. The WHO carries out a wide range of functions, including coordinating international health work; helping governments strengthen health services; providing technical assistance and emergency aid; working for the prevention and control of disease; promoting improved nutrition, housing, sanitation, recreation, and economic and working conditions; promoting and coordinating biomedical and health services research; promoting improved standards of teaching and training in health and medical professions; establishing international standards for biological, pharmaceutical, and similar products; and standardizing diagnostic procedures.

The WHO publishes the *World Health Statistics Annual* and many other technical and statistical publications.

For publications contact Distribution and Sales, Division of Publishing, Language, and Library Services, World Health Organization Headquarters, CH-1211 Geneva 27, Switzerland; telephone: (41 22) 791 2476 or 2477; fax: (41 22) 791 4857; email: publications@who.ch; website: www.who.ch.

The World Intellectual Property Organization

The World Intellectual Property Organization (WIPO) is a specialized agency of the United Nations based in Geneva, Switzerland. The objectives of WIPO are to promote the protection of intellectual property throughout the world through cooperation among states and, where appropriate, in collaboration with other international organizations and to ensure administrative cooperation among the intellectual property unions—that is, the "unions" created by the Paris and Berne Conventions and several subtreaties concluded by members of the Paris Union. WIPO is responsible for administering various multilateral treaties dealing with the legal and administrative aspects of intellectual property. A substantial part of WIPO's activities and resources is devoted to development cooperation with developing countries.

For information contact the World Intellectual Property Organization, 34, chemin des Colombettes, Geneva, Switzerland; mailing address: P.O. Box 18, CH-1211 Geneva 20, Switzerland; telephone: (41 22) 338 9111; fax: (41 22) 733 5428; telex: 412912 ompi ch; website: http://www.wipo.int.

World Trade Organization

The World Trade Organization (WTO), established on 1 January 1995, is the successor to the General Agreement on Tariffs and Trade (GATT). The WTO provides the legal and institutional foundation of the multilateral trading system and embodies the results of the Uruguay Round of trade negotiations, which ended with the Marrakesh Declaration of 15 April 1994. The WTO is mandated to administer and implement multilateral trade agreements, serving as a forum for multilateral trade negotiations, seeking to resolve trade disputes, overseeing national trade policies, and cooperating with other international institutions involved in global economic policymaking.

The WTO's Statistics and Information Systems Divisions compile statistics on world trade and maintain the Integrated Database, which contains the basic records of the outcome of the Uruguay Round. The *WTO Annual Report* includes a statistical appendix.

For publications contact World Trade Organization, Publications Services, Centre William Rappard, 154 rue de Lausanne, CH-1211 Geneva, Switzerland; telephone: (41 22) 739 5208 or 5308; fax: (41 22) 739 5458; email: publications@wto.org; website: www.wto.org.

Private and nongovernmental organizations

Currency Data & Intelligence, Inc.

Currency Data & Intelligence, Inc. is a research and publishing firm that produces currency-related products and undertakes research for international agencies and universities worldwide. Its flagship product, the *World Currency Yearbook,* is the most comprehensive source of information on currency. It includes official and unofficial exchange rates and discussions of economic, social, and political issues that affect the value of currencies in world markets. A second publication, the monthly *Global Currency Report,* covers devaluations and other critical developments in exchange rate restrictions and valuations and provides parallel market exchange rates.

For information contact Currency Data & Intelligence, Inc., 328 Flatbush Avenue, Suite 344, Brooklyn, N.Y. 11238, U.S.A.; telephone: (718) 230 7176; fax: (718) 230 1992; email: curncydata@aol.com.

Euromoney Publications PLC

Euromoney Publications PLC provides a wide range of financial, legal, and general business information. The monthly *Euromoney* magazine carries a semiannual rating of country creditworthiness.

For information contact Euromoney Publications PLC, Nestor House, Playhouse Yard, London EC4V 5EX, U.K.; telephone: (44 171) 779 8999; fax: (44 171) 779 8407; website: www.euromoney.com.

Institutional Investor, Inc.

Institutional Investor magazine is published monthly by Institutional Investor, Inc., which develops country credit ratings every six months based on information provided by leading international banks.

For information contact Institutional Investor, Inc., 488 Madison Avenue, New York, N.Y. 10022, U.S.A.; telephone: (212) 224 3300.

International Road Federation

The International Road Federation (IRF) is a not-for-profit, nonpolitical service organization representing the views and interests of road-related industries around the world. The IRF has more than 600 corporate and institutional members in approximately 100 countries—companies, associations, research institutes, and administrations concerned with developing and modernizing road infrastructure. To encourage better road and transport systems worldwide, the IRF assists in the transfer and application of technology and management practices that will produce maximum economic and social returns from national road investments, through its consultative status with the United Nations and the OECD and its advisory capacity with the European Union.

The IRF publishes *World Road Statistics*.

For information contact International Road Federation, 63 rue de Lausanne, CH-1202 Geneva, Switzerland; telephone: (41 22) 731 7150; fax: (41 22) 731 7158; email: IRD@dial.eunet.ch; website: http://is.eunet.ch/geneva-intl/gi/egi/egi149.html

Moody's Investor Service

Moody's Investors Service is a global credit analysis and financial opinion firm. It provides the international investment community with globally consistent credit ratings on debt and other securities issued by North American state and regional government entities, by corporations worldwide, and by some sovereign issuers. It also publishes extensive financial data in both print and electronic form. Clients of Moody's Investors Service include investment banks, brokerage firms, insurance companies, public utilities, research libraries, manufacturers, and government agencies and departments.

Moody's publishes *Sovereign, Subnational* and *Sovereign-Guaranteed Issuers*.

For information contact Moody's Investors Service, 99 Church Street, New York, N.Y. 10007, U.S.A.; website: www.moodys.com.

Political Risk Services

Political Risk Services is a global leader in political and economic risk forecasting and market analysis and has served international companies large and small for nearly 20 years. The data it contributed to this year's *World Development Indicators* come from the *International Country Risk Guide,* a monthly publication that monitors and rates political, financial, and economic risk in 130 countries. The guide's data series and commitment to independent and unbiased analysis make it the standard for any organization practicing effective risk management.

For information contact Political Risk Services, 6320 Fly Road, Suite 102, P.O. Box 248, East Syracuse, N.Y. 13057, U.S.A.; telephone: (315) 431 0511; fax: (315) 431 0200; email: custserv@polrisk.com; website: www.prsgroup.com

Price Waterhouse

Price Waterhouse is one of the world's largest international organizations of accountants and consultants. Founded in 1849, it now consists of a network of 27 individual practice firms in 119 countries and territories. Staffed with professionals committed to client service, it is equipped to advise on matters relating to international operations, not only in individual countries but also on a regional or global basis.

For information contact Price Waterhouse World Firm Services BV, Inc., 1251 Avenue of the Americas, New York, N.Y. 10020, U.S.A.; telephone: (212) 819 5000; fax: (212) 790 6620; telex: 362196; website: www.pricewaterhouse.com.

Partners

Standard and Poor's Rating Services

Standard and Poor's *Sovereign Ratings* provides issuer and local and foreign currency debt ratings for sovereign governments and for sovereign-supported and supranational issuers worldwide. Standard & Poor's Rating Services monitors the credit quality of $1.5 trillion worth of bonds and other financial instruments and offers investors global coverage of debt issuers. Standard & Poor's also has ratings on commercial paper, mutual funds, and the financial condition of insurance companies worldwide.

For information contact The McGraw-Hill Companies, Inc., Executive Offices, 1221 Avenue of the Americas, New York, N.Y. 10020, U.S.A.; subscriber services: (212) 208 1146; website: www.ratings.standardpoor.com.

World Conservation Monitoring Centre

WORLD CONSERVATION
MONITORING CENTRE

The World Conservation Monitoring Centre (WCMC) provides information on the conservation and sustainable use of the world's living resources and helps others to develop information systems of their own. It works in close collaboration with a wide range of organizations and people to increase access to the information needed for wise management of the world's living resources. Committed to the principle of data exchange with other centers and noncommercial users, the WCMC, whenever possible, places the data it manages in the public domain.

For information contact World Conservation Monitoring Centre, 219 Huntingdon Road, Cambridge CB3 0DL, U.K.; telephone: (44 12) 2327 7314; fax: (44 12) 2327 7136; website: www.wcmc.org.uk.

World Resources Institute

The World Resources Institute is an independent center for policy research and technical assistance on global environmental and development issues. The institute provides—and helps other institutions provide—objective information and practical proposals for policy and institutional change that will foster environmentally sound, socially equitable development. The institute's current areas of work include trade, forests, energy, economics, technology, biodiversity, human health, climate change, sustainable agriculture, resource and environmental information, and national strategies for environmental and resource management.

For information contact World Resources Institute, 1709 New York Avenue, N.W., Washington, D.C. 20006, U.S.A.; telephone: (202) 638 6300; fax: (202) 638 0036; telex 64414 WRIWASH; website: www.wri.org.

The World Bank Group

The World Bank Group is made up of five organizations: the International Bank for Reconstruction and Development (IBRD), the International Development Association (IDA), the International Finance Corporation (IFC), the Multilateral Investment Guarantee Agency (MIGA), and the International Centre for Settlement of Investment Disputes (ICSID).

Established in 1944 at a conference of world leaders in Bretton Woods, New Hampshire, the World Bank is a lending institution whose aim is to help integrate developing and transition countries with the global economy, and reduce poverty by promoting economic growth. The Bank lends for policy reforms and development projects and provides policy advice, technical assistance, and non-lending services to its 181 member countries.

Users guide

Principal sections

Are signposted by these icons:

Section 1 World view

Section 2 People

Section 3 Environment

Section 4 Economy

Section 5 States and markets

Section 6 Global links

The tables

Tables are numbered by section and display the identifying icons of each section. Countries and economies are listed alphabetically (except for Hong Kong, China, which appears after China). Data are shown for 148 economies with populations of more than 1 million people and for which data are regularly reported by the relevant authority, as well as Taiwan, China, in selected tables. Selected indicators for 62 other economies—small economies with populations between 30,000 and 1 million, smaller economies if they are members of the World Bank, and larger economies for which data are not regularly reported—are shown in table 1.6. The term *country*, used interchangeably with *economy*, does not imply political independence or official recognition by the World Bank, but refers to any territory for which authorities report separate social or economic statistics. When available, aggregate measures for income and regional groups appear at the end of each table.

Indicators

Indicators are shown for the most recent year or period for which data are available and, in most tables, for an earlier year or period. Time-series data are available in the *World Development Indicators* CD-ROM.

1

3.1 Land use and deforestation

| | Land area | Rural population density | Land use | | | | | | Forest area | Annual deforestation | |
	thousand sq. km 1995	people per sq. km 1995	Cropland % of land area 1980	1995	Permanent pasture % of land area 1980	1994	Other land % of land area 1980	1994	thousand sq. km 1995	sq. km 1990-95	average % change 1990-95
Albania	27	354	26	26	15	15	59	59	10	0	0.0
Algeria	2,382	165	3	3	15	13	82	83	19	234	1.2
Angola	1,247	248	3	3	43	43	54	54	222	2,370	1.0
Argentina	2,737	17	10	10	52	52	38	38	339	894	0.3
Armenia	28	198	..	25	..	24	..	54	3	−84	−2.7
Australia	7,682	6	6	6	57	54	37	40	409	−170	0.0
Azerbaijan	87	208	20	18	25	24	56	57	39	0	0.0
Bangladesh	130	1,157	70	67	5	5	25	52	10	0	0.0
Belarus	207	49	..	30	..	14	..	56	74	−688	−1.0
Belgium[a]	33	39	24	23	23	21	53	55	7	0	0.0
Benin	111	236	16	17	4	4	80	79	46	596	1.2
Bolivia	1,084	137	2	2	25	24	73	73	483	5,814	1.2
Bosnia and Herzegovina	51	516	..	13	..	24	..	61	27	0	0.0
Botswana	567	168	1	1	45	45	54	54	139	708	0.5
Brazil	8,457	65	6	8	20	22	74	71	5,511	25,544	0.5
Bulgaria	111	67	38	38	18	16	44	46	32	−6	0.0
Burkina Faso	274	255	10	13	22	22	68	66	43	320	0.7
Burundi	26	623	46	43	39	42	15	14	3	14	0.4
Cambodia	177	209	12	22	3	8	85	70	98	1,638	1.6
Cameroon	465	123	15	15	4	4	81	81	196	1,292	0.6
Canada	9,221	15	5	5	3	3	92	92	2,446	−1,764	−0.1
Central African Republic	623	103	3	3	5	5	92	92	299	1,282	0.4
Chad	1,259	154	3	3	36	36	62	62	110	942	0.8
Chile	749	57	6	6	17	17	77	77	79	292	0.4
China	9,326	913	11	10	36	43	53	47	1,333	866	0.1
Hong Kong, China	1	5,130	7	7	1	1	92	92	..	..	..
Colombia	1,039	419	5	6	39	39	56	55	530	2,622	0.5
Congo, Dem. Rep.	2,267	429	3	3	7	7	90	90	..	..	..
Congo, Rep.	342	755	0	0	29	29	70	70	195	416	0.2
Costa Rica	51	600	10	10	39	46	51	44	12	414	3.0
Côte d'Ivoire	318	272	10	13	41	41	49	46	55	308	0.6
Croatia	56	189	..	22	..	20	..	59	18	0	0.0
Cuba	110	71	30	41	24	20	46	39	18	236	1.2
Czech Republic	77	114	..	44	..	12	..	45	26	−2	0.0
Denmark	42	33	63	55	6	7	31	37	4	0	0.0
Dominican Republic	48	221	29	39	43	43	27	19	16	264	1.6
Ecuador	277	299	9	11	15	18	77	71	111	1,890	1.6
Egypt, Arab Rep.	995	1,144	2	3	..	..	..	..	0	0	0.0
El Salvador	21	572	35	37	29	28	36	35	1	38	3.3
Eritrea	101	673	..	5	..	69	..	26	3	0	0.0
Estonia	42	36	..	27	..	7	..	66	20	−196	−1.0
Ethiopia	1,000	421	..	12	..	20	..	69	136	624	0.5
Finland	305	74	..	..	1	0	..	..	200	166	0.1
France	550	80	34	35	23	19	42	45	150	−1,608	−1.1
Gabon	258	169	2	2	18	18	80	80	179	910	0.5
Gambia, The	10	452	..	..	19	20	..	..	1	8	0.9
Georgia	70	290	..	16	..	27	..	57	30	0	0.0
Germany	349	93	36	35	17	15	47	50	107	0	0.0
Ghana	228	391	16	20	37	37	47	43	90	1,172	1.3
Greece	129	178	30	27	41	41	29	32	65	−1,408	−2.3
Guatemala	108	479	16	18	12	24	72	58	38	824	2.1
Guinea	246	667	3	4	44	44	54	53	64	748	1.1
Guinea-Bissau	28	279	10	12	38	38	51	50	23	104	0.4
Haiti	28	873	32	33	18	18	49	49	0	8	3.4
Honduras	112	196	16	18	13	14	71	68	41	1,022	2.3

2

ENVIRONMENT

Land use and deforestation | 3.1

	Land area	Rural population density	Cropland % of land area		Permanent pasture % of land area		Other land % of land area		Forest area	Annual deforestation	
	thousand sq. km	people per sq. km	1980	1995	1980	1994	1980	1994	thousand sq. km	sq. km	average % change
	1995	1995							1995	1990-95	1990-95
Hungary	92	75	58	54	14	12	28	34	17	-88	-0.5
India	2,973	410	57	57	4	4	39	39	650	-72	0.0
Indonesia	1,812	732	14	17	7	7	79	77	1,098	10,844	1.0
Iran, Islamic Rep.	1,622	148	8	11	27	27	64	61	15	284	1.7
Iraq	437	96	12	13	9	9	78	78	1	0	0.0
Ireland	69	115	16	19	67	45	17	36	6	-140	-2.7
Israel	21	147	20	21	6	7	74	72	1	0	0.0
Italy	294	236	42	37	17	16	40	47	65	-58	-0.1
Jamaica	11	660	22	22	24	24	54	56	2	158	7.2
Japan	377	692	13	12	2	2	85	87	251	132	0.1
Jordan	89	375	4	5	9	9	87	87	0	12	2.5
Kazakhstan	2,671	21	..	12	..	70	..	17	105	-1,928	-1.9
Kenya	569	476	8	8	37	37	55	55	13	34	0.3
Korea, Dem. Rep.	120	504	16	17	0	0	84	83	62	0	0.0
Korea, Rep.	99	471	22	20	1	1	77	78	76	130	0.2
Kuwait	18	927	0	0	8	8	92	92	0	0	0.0
Kyrgyz Republic	192	336	..	7	..	47	..	46	7	0	0.0
Lao PDR	231	417	3	4	3	3	94	93	..	..	..
Latvia	62	40	..	28	..	13	..	59	29	-250	-0.9
Lebanon	10	236	30	30	1	1	69	69	1	52	7.8
Lesotho	30	470	..	..	66	66	..	..	0	0	0.0
Libya	1,760	41	1	1	7	8	91	91	4	0	0.0
Lithuania	65	35	..	46	..	8	..	46	20	-112	-0.6
Macedonia, FYR	25	130	..	26	..	25	..	49	10	2	0.0
Madagascar	582	379	5	5	41	41	54	53	151	1,300	0.8
Malawi	94	505	14	18	20	20	66	62	33	546	1.6
Malaysia	329	512	15	23	1	1	85	76	155	4,002	2.4
Mali	1,220	208	2	3	25	25	74	73	116	1,138	1.0
Mauritania	1,025	541	0	0	38	38	62	62	6	0	0.0
Mauritius	2	668	53	52	3	3	44	44	0	0	0.0
Mexico	1,909	95	13	14	39	42	48	45	554	5,080	0.9
Moldova	33	118	..	66	..	11	..	22	4	0	0.0
Mongolia	1,567	73	1	1	79	75	20	24	94	0	0.0
Morocco	446	148	18	21	47	47	35	32	38	118	0.3
Mozambique	784	391	4	4	56	56	40	40	169	1,162	0.7
Myanmar	658	351	15	15	1	1	84	84	272	3,874	1.4
Namibia	823	121	1	1	46	46	53	53	124	420	0.3
Nepal	143	659	16	21	13	12	71	69	48	548	1.1
Netherlands	34	193	24	27	35	31	41	42	3	0	0.0
New Zealand	268	32	13	12	53	51	34	38	79	-434	-0.6
Nicaragua	121	67	11	23	40	40	49	39	56	1,508	2.5
Niger	1,267	148	..	..	8	8	..	..	26	0	0.0
Nigeria	911	222	33	36	44	44	23	20	138	1,214	0.9
Norway	307	118	3	3	0	0	97	97	81	-180	-0.2
Oman	212	3,256	0	0	5	5	95	95	0	0	0.0
Pakistan	771	405	26	28	6	6	67	66	17	550	2.9
Panama	74	234	7	9	17	20	75	71	28	636	2.1
Papua New Guinea	453	6,023	1	1	0	0	99	99	369	1,332	0.4
Paraguay	397		4	6	40	55	56	40	115	3,266	2.6
Peru	1,280		3	3	21	21	76	76	676	2,168	0.3
Philippines	298	586	29	32	3	4	67	64	68	2,624	3.5
Poland	304	99	49	48	13	13	38	39	87	-120	-0.1
Portugal	92	278	34	33	9	10	57	57	29	-240	-0.9
Puerto Rico	9	3,022	11	9	38	26	51	65	3	24	0.9
Romania	230	107	46	43	19	21	35	36	62	12	0.0
Russian Federation	16,889	27	..	8	..	5	..	87	7,635	0	0.0

Statistics

Data are shown for economies as they were constituted in 1996, and historical data are revised to reflect current political arrangements. Exceptions are noted throughout the tables.

On 1 July 1997 China resumed its exercise of sovereignty over Hong Kong. Data for China do not include data for Hong Kong, China, or Taiwan, China, unless otherwise noted.

Data for the Democratic Republic of Congo (Congo, Dem. Rep. in the table listings) refer to the former Zaire. For clarity, this edition also uses the formal name of the Republic of Congo (Congo, Rep. in the table listings).

Data are shown whenever possible for the individual countries formed from the former Czechoslovakia—the Czech Republic and the Slovak Republic.

Data are shown for Eritrea whenever possible, but in most cases before 1992 Eritrea is included in the data for Ethiopia.

Data for Germany refer to the unified Germany unless otherwise noted.

Data for Jordan refer to the East Bank only unless otherwise noted.

In 1991 the Union of Soviet Socialist Republics (U.S.S.R.) was dissolved into 15 countries (Armenia, Azerbaijan, Belarus, Estonia, Georgia, Kazakhstan, Kyrgyz Republic, Latvia, Lithuania, Moldova, Russian Federation, Tajikistan, Turkmenistan, Ukraine, and Uzbekistan). Whenever possible, data are shown for the individual countries.

Data for the Republic of Yemen refer to that country from 1990 onward; data for previous years refer to aggregated data of the former People's Democratic Republic of Yemen and the former Yemen Arab Republic unless otherwise noted.

Whenever possible, data are shown for the individual countries formed from the former Yugoslavia—Bosnia and Herzegovina, Croatia, the former Yugoslav Republic of Macedonia, Slovenia, and the Federal Republic of Yugoslavia. All references to the Federal Republic of Yugoslavia in the tables are to the Federal Republic of Yugoslavia (Serbia/Montenegro).

Additional information about the data is provided in *Primary data documentation*. That section summarizes national and international efforts to improve basic data collection and gives information on primary sources, census years, fiscal years, and other background information. *Statistical methods* provides technical information on some of the general calculations and formulas used throughout the book.

Discrepancies in data presented in different editions of the *World Development Indicators* reflect updates by countries as well as revisions to historical series and changes in methodology. Thus readers are advised not to compare data series between editions of the *World Development Indicators* or between different World Bank publications. Consistent time-series data for 1960–96 are available on the *World Development Indicators* CD-ROM. Except where noted, growth rates are in real terms. (See *Statistical methods* for information on the methods used to calculate growth rates.) Data for some economic indicators for some economies are presented in fiscal years rather than calendar years; see *Primary data documentation*. All dollar figures are current U.S. dollars unless otherwise stated. The methods used for converting national currencies are described in *Statistical methods*.

The World Bank's classification of economies

For operational and analytical purposes the World Bank's main criterion for classifying economies is gross national product (GNP) per capita. Every economy is classified as low income, middle income (subdivided into lower middle and upper middle), or high income. For income classifications see the map on the inside front cover and the list on the front cover flap. Note that classification by income does not necessarily reflect development status. Because GNP per capita changes over time, the country composition of income groups may change from one *World Development Indicators* to the next. Once the classification is fixed for an edition, all historical data presented are based on the same country grouping using the most recent year for which GNP per capita data are available (1996 in this edition). Low-income economies are those with a GNP per capita of $785 or less in 1996. Middle-income economies are those with GNP per capita of more than $785 but less than $9,636. Lower-middle-income and upper-middle-income economies are separated at GNP per capita of $3,115. High-income economies are those with a GNP per capita of $9,636 or more.

Aggregate measures for income groups

The aggregate measures for income groups include 210 economies (economies presented in the main tables plus the economies listed in table 1.6) wherever data are available. To maintain consistency in the aggregate measures over time and between tables, missing data are imputed where possible. Most aggregates are totals (designated by a *t* if the aggregates include gap-filled estimates for missing data; otherwise totals are designated by an *s* for simple totals), median values (*m*), or weighted averages (*w*). Gap filling of amounts not allocated to countries may result in discrepancies between subgroup aggregates and overall totals. See *Statistical methods* for further discussion of aggregation methods.

Aggregate measures for regions

The aggregate measures for regions include only low- and middle-income economies (note that these measures include developing economies with populations of less than 1 million, including those listed in table 1.6). The country composition of regions is based on the World Bank's analytical regions and may differ from common geographic usage. For regional classifications see the map on the inside back cover and the list on the back cover flap. See *Statistical methods* for further discussion of aggregation methods.

3

3.1 Land use and deforestation

	Land area	Rural population density	Land use						Forest area	Annual deforestation	
	thousand sq. km 1995	people per sq. km 1995	Cropland % of land area 1980	1995	Permanent pasture % of land area 1980	1994	Other land % of land area 1980	1994	thousand sq. km 1995	sq. km 1990–95	average % change 1990–95
Rwanda	25	710	41	47	28	28	30	25	3	4	0.2
Saudi Arabia	2,150	88	1	2	40	56	60	42	2	18	0.8
Senegal	193	208	12	12	30	30	58	58	74	496	0.7
Sierra Leone	72	619	7	8	31	31	62	62	13	426	3.0
Singapore	1	0	13	2	..	..	..	..	0	0	0.0
Slovak Republic	48	148	..	33	..	17	..	49	20	–24	–0.1
Slovenia	20	414	..	14	..	25	..	61	11	0	0.0
South Africa	1,221	125	11	13	67	67	22	21	85	150	0.2
Spain	499	60	41	40	22	21	37	38	84	0	0.0
Sri Lanka	65	1,549	29	29	7	7	64	64	18	202	1.1
Sudan	2,376	142	5	5	41	46	54	48	416	3,526	0.8
Sweden	412	54	7	7	2	1	91	92	244	24	0.0
Switzerland	40	688	10	11	41	29	49	60	11	0	0.0
Syrian Arab Republic	184	134	31	32	46	45	23	22	2	52	2.2
Tajikistan	141	483	..	6	..	25	..	69	4	0	0.0
Tanzania	884	728	3	4	40	40	57	56	325	3,226	1.0
Thailand	511	278	36	40	1	2	63	58	116	3,294	2.6
Togo	54	138	43	45	4	4	53	52	12	186	1.4
Trinidad and Tobago	5	486	23	24	2	2	75	74	2	26	1.5
Tunisia	155	120	30	31	22	20	48	49	6	30	0.5
Turkey	770	77	37	35	13	16	50	48	89	0	0.0
Turkmenistan	470	177	..	3	..	64	..	33	38	0	0.0
Uganda	200	331	28	34	9	9	63	57	61	592	0.9
Ukraine	579	46	..	59	..	13	..	28	92	–54	–0.1
United Arab Emirates	84	1,172	0	1	2	3	97	96	1	0	0.0
United Kingdom	242	107	29	25	47	46	24	29	24	–128	–0.5
United States	9,159	34	21	21	26	26	53	53	2,125	–5,886	–0.3
Uruguay	175	25	8	7	78	77	14	15	8	4	0.0
Uzbekistan	414	327	..	11	..	50	..	39	91	–2,260	–2.7
Venezuela	882	115	4	4	20	21	76	75	440	5,034	1.1
Vietnam	325	1,082	20	21	1	1	79	79	91	1,352	1.4
West Bank and Gaza	..	..	..	..	..	..	..	..	..	..	..
Yemen, Rep.	528	702	3	3	30	30	67	67	0	0	0.0
Yugoslavia, FR (Serb./Mont.)	102	123	..	40	..	21	..	39	18	0	0.0
Zambia	743	97	7	7	40	40	53	53	314	2,644	0.8
Zimbabwe	387	244	7	8	44	44	49	48	87	500	0.6
World	**130,129 s**	**559 w**	**11 w**	**11 w**	**27 w**	**26 w**	**61 w**	**62 w**	**32,712 w**	**101,724 w**	**0.3 w**
Low income	39,294	634	12	13	31	32	56	54	6,227	38,690	0.6
Excl. China & India	26,994	515	8	9	32	32	59	59	4,243	37,896	0.9
Middle income	59,884	401	9	10	26	23	65	67	19,985	74,598	0.4
Lower middle income	39,310	462	10	11	21	18	68	71	12,884	37,888	0.3
Upper middle income	20,574	170	7	9	30	32	63	59	7,100	36,710	0.5
Low & middle income	99,178	583	10	11	28	27	61	62	26,211	113,288	0.4
East Asia & Pacific	15,869	841	11	12	30	34	59	54	3,756	29,826	0.8
Europe & Central Asia	23,864	124	..	13	..	16	..	71	8,590	–5,798	–0.1
Latin America & Carib.	20,064	230	7	8	28	29	65	63	9,064	57,766	0.6
Middle East & N. Africa	10,972	493	5	6	23	26	72	68	89	800	0.9
South Asia	4,781	509	44	45	11	10	46	45	744	1,316	0.2
Sub-Saharan Africa	23,628	356	6	7	34	34	58	57	3,969	29,378	0.7
High income	30,951	215	..	..	25	24	..	..	6,501	–11,564	–0.2

a. Includes Luxembourg.

120 | **1998** World Development Indicators

Footnotes

Known deviations from standard definitions or breaks in comparability over time or across countries are either footnoted in the tables or noted in *About the data*. When available data are deemed to be too weak to provide reliable measures of levels and trends or do not adequately adhere to international standards, the data are not shown.

Substantial improvements in social indicators have accompanied growth in average incomes. Infant mortality rates have fallen from 104 per 1,000 live births in 1970–75 to 59 in 1996. On average, life expectancy has risen by four months each year since 1970. Growth in food production has substantially outpaced that of population. Governments report rapid progress in primary school enrollment. Adult literacy has also risen, from 46 to 70 percent. And gender disparities have narrowed, with the average ratio of girls to boys in secondary schools rising from 70 to 100 in 1980 to 82 to 100 in 1993. The developing world today is healthier, wealthier, better fed, and better educated. (For a spirited accounting of development progress, see Fox 1995.)

But progress has been far from even (table 1b). Take mortality. All developing regions have seen infant and child mortality rates decline sharply. But South Asia's infant mortality rates today are about the same as East Asia's in the early 1970s, reflecting both poor progress in South Asia and the favorable initial

social conditions in East Asia that were responsible for its subsequent strong economic performance. And Sub-Saharan Africa's infant mortality rates are well above those in East Asia some 25 years ago. On average, 147 of every 1,000 African children die before the age of 5, and 91 in 1,000 before the age of 1. Ten African countries have under-five mortality rates in excess of 200 (Angola, Guinea, Guinea-Bissau, Malawi, Mali, Mozambique, Niger, Rwanda, Sierra Leone, and Zambia).

Gross primary school enrollment rates have risen in all regions. But Sub-Saharan Africa's rates, having risen from 50 percent of the eligible population to 80 percent by 1980, fell back to 72 in 1993, reflecting larger problems. Again, averages disguise wide country disparities. Six countries in Africa have fewer than half their children enrolled in primary school (Burkina Faso, Ethiopia, Guinea, Mali, Niger, and Sierra Leone).

Reducing gender disparities in education, as measured by female enrollment in secondary schools, is one area where Sub-Saharan Africa is doing relatively well. Progress is slower in South Asia, where the ratio of 66 girls to 100 boys is well below the developing world average.

Progress toward the 21st century

The uneven progress of development is worrying. The free flows of trade and capital that integrate the global economy may bring benefits to millions, but poverty and suffering persist. In an integrated world, disease, environmental degradation, civil strife, criminal activities, and illicit drugs are also global concerns. In response, international development agencies have begun to reexamine the way they do business. They are looking at impacts more than inputs by establishing performance targets. And they are seeking ways to enhance their accountability and transparency by measuring progress toward those targets.

In May 1996 the Development Assistance Committee of the OECD published *Shaping the 21st Century*. This policy paper, known as Strategy 21, spotlights the substantial achievements of

Table 1a

Population living on less than $1 a day in developing regions, 1987 and 1993

Region	Number (millions)		Share of population (%)	
	1987	1993	1987	1993
East Asia and the Pacific	464.0	445.8	28.8	26.0
Europe and Central Asia	2.2	14.5	0.6	3.5
Latin America and the Caribbean	91.2	109.6	22.0	23.5
Middle East and North Africa	10.3	10.7	4.7	4.1
South Asia	479.9	514.7	45.4	43.1
Sub-Saharan Africa	179.6	218.6	38.5	39.1
Total	1,227.2	1,313.9	30.1	29.4

Source: World Bank 1996e.

Table 1b

Progress in social indicators

Region	Infant mortality per 1,000 live births		Under-five mortality per 1,000		Gross primary school enrollment (% of relevant age group)		Gross secondary school enrollment (female as % of male)	
	1970	1996	1980	1996	1980	1995	1980	1993
East Asia and the Pacific	79	39	75	47	88	115	73	84
Europe and Central Asia	..	24	..	30	..	100	96	101
Latin America and the Caribbean	84	33	82	41	..	111	95	..
Middle East and North Africa	134	50	141	63	68	97	63	85
South Asia	139	73	174	93	67	99	50	66
Sub-Saharan Africa	137	91	193	147	80	75	50	82
All developing countries	107	59	133	80	79	103	70	83

Note: Numbers in italics are for 1994.

Source: World Bank staff estimates.

On the verge of the 21st century, dramatic political, social, and economic changes have overtaken the world. On this stage a new international dialogue on the future of development has begun. The focal point has been a succession of conferences sponsored by the United Nations, with governments coming together to chart development strategies for the next century—for children, for education, for environment, for population, for women, for social development. The debates, often heated, have renewed commitments to the eradication of poverty, the sustainability of the environment, the reduction of infant, child, and maternal mortality, and the elimination of gender differences in access to education, starting with universal primary education.

But such commitments are too often received with cynicism. After all, the world's poor are testament to the past lack of political will to achieve such goals. Can we make it different this time?

Perhaps, because new forces are at work. First, there is a new consensus on development. Gone with the Cold War are the sterile, ideological debates over the roles of the state and the market. In their place is a more pragmatic approach to effective and broad-based development strategies. Second, the success of some developing countries shows that the worst forms of poverty can be eradicated, and that investments in human capital and the poor can have high economic returns.

Advancing social development

Living standards have risen dramatically over the past 25 years. Despite an increase in population from 2.9 billion people in 1970 to 4.8 billion in 1996, per capita income growth in developing countries has averaged about 1.3 percent a year. While the number of people living in poverty continues to grow, hundreds of millions have had lifted from them the yoke of poverty and despair. As a result the proportion of the poor is holding steady at less than a third of the developing world's population (table 1a).

This global picture conceals large regional differences. The proportion in poverty is declining in Asia, where most of the poor live. But it is rising rapidly in Europe and Central Asia, and continuing to rise in Latin America and Sub-Saharan Africa. More than 4 in 10 households (over 500 million people) remain in poverty in South Asia, and the financial crisis in East Asia will slow the pace of poverty reduction there.

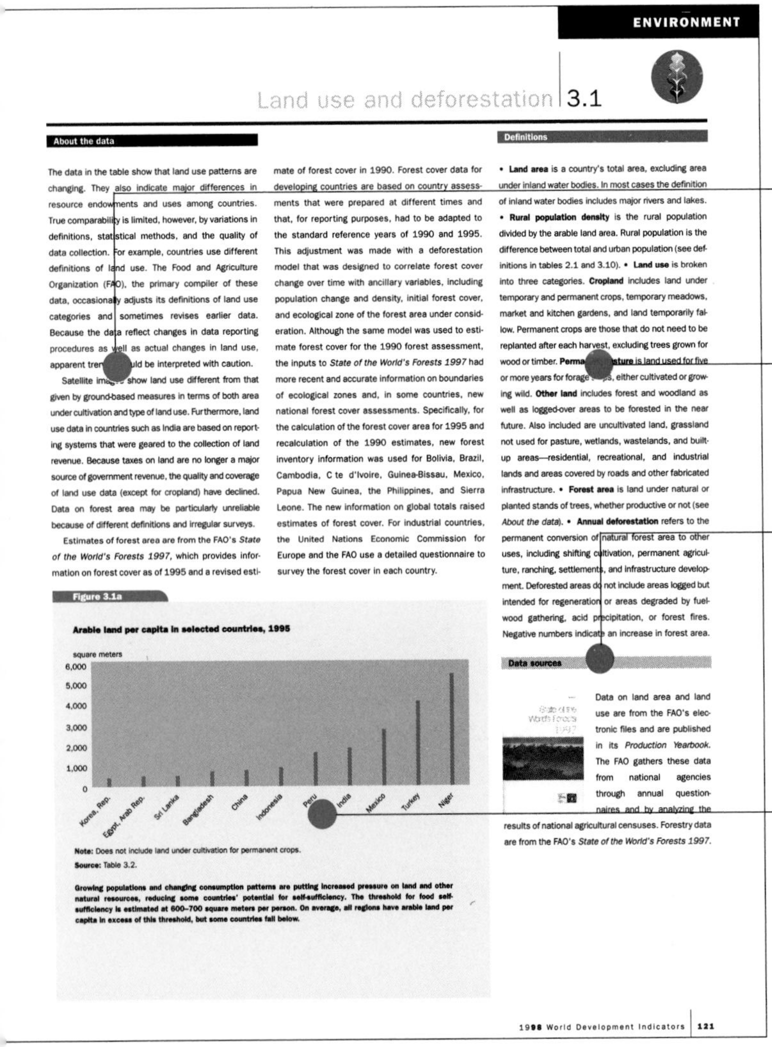

4

ENVIRONMENT

Land use and deforestation | 3.1

About the data

The data in the table show that land use patterns are changing. They also indicate major differences in resource endowments and uses among countries. True comparability is limited, however, by variations in definitions, statistical methods, and the quality of data collection. For example, countries use different definitions of land use. The Food and Agriculture Organization (FAO), the primary compiler of these data, occasionally adjusts its definitions of land use categories and sometimes revises earlier data. Because the data reflect changes in data reporting procedures as well as actual changes in land use, apparent trends should be interpreted with caution.

Satellite images show land use different from that given by ground-based measures in terms of both area under cultivation and type of land use. Furthermore, land use data in countries such as India are based on reporting systems that were geared to the collection of land revenue. Because taxes on land are no longer a major source of government revenue, the quality and coverage of land use data (except for cropland) have declined. Data on forest area may be particularly unreliable because of different definitions and irregular surveys.

Estimates of forest area are from the FAO's *State of the World's Forests 1997*, which provides information on forest cover as of 1995 and a revised esti-

mate of forest cover in 1990. Forest cover data for developing countries are based on country assessments that were prepared at different times and that, for reporting purposes, had to be adapted to the standard reference years of 1990 and 1995. This adjustment was made with a deforestation model that was designed to correlate forest cover change over time with ancillary variables, including population change and density, initial forest cover, and ecological zone of the forest area under consideration. Although the same model was used to estimate forest cover for the 1990 forest assessment, the inputs to *State of the World's Forests 1997* had more recent and accurate information on boundaries of ecological zones and, in some countries, new national forest cover assessments. Specifically, for the calculation of the forest cover area for 1995 and recalculation of the 1990 estimates, new forest inventory information was used for Bolivia, Brazil, Cambodia, C te d'Ivoire, Guinea-Bissau, Mexico, Papua New Guinea, the Philippines, and Sierra Leone. The new information on global totals raised estimates of forest cover. For industrial countries, the United Nations Economic Commission for Europe and the FAO use a detailed questionnaire to survey the forest cover in each country.

Definitions

• **Land area** is a country's total area, excluding area under inland water bodies. In most cases the definition of inland water bodies includes major rivers and lakes. • **Rural population density** is the rural population divided by the arable land area. Rural population is the difference between total and urban population (see definitions in tables 2.1 and 3.10). • **Land use** is broken into three categories. **Cropland** includes land under temporary and permanent crops, temporary meadows, market and kitchen gardens, and land temporarily fallow. Permanent crops are those that do not need to be replanted after each harvest, excluding trees grown for wood or timber. **Permanent pasture** is land used for five or more years for forage crops, either cultivated or growing wild. **Other land** includes forest and woodland as well as logged-over areas to be forested in the near future. Also included are uncultivated land, grassland not used for pasture, wetlands, wastelands, and built-up areas—residential, recreational, and industrial lands and areas covered by roads and other fabricated infrastructure. • **Forest area** is land under natural or planted stands of trees, whether productive or not (see *About the data*). • **Annual deforestation** refers to the permanent conversion of natural forest area to other uses, including shifting cultivation, permanent agriculture, ranching, settlements, and infrastructure development. Deforested areas do not include areas logged but intended for regeneration or areas degraded by fuelwood gathering, acid precipitation, or forest fires. Negative numbers indicate an increase in forest area.

Data sources

Data on land area and land use are from the FAO's electronic files and are published in its *Production Yearbook*. The FAO gathers these data from national agencies through annual questionnaires and by analyzing the results of national agricultural censuses. Forestry data are from the FAO's *State of the World's Forests 1997*.

Figure 3.1a

Arable land per capita in selected countries, 1995

square meters
6,000
5,000
4,000
3,000
2,000
1,000
0

Korea Rep. · Egypt, Arab Rep. · Sri Lanka · Bangladesh · China · Indonesia · Peru · India · Mexico · Turkey · Niger

Note: Does not include land under cultivation for permanent crops.
Source: Table 3.2.

Growing populations and changing consumption patterns are putting increased pressure on land and other natural resources, reducing some countries' potential for self-sufficiency. The threshold for food self-sufficiency is estimated at 600–700 square meters per person. On average, all regions have arable land per capita in excess of this threshold, but some countries fall below.

1998 World Development Indicators | 121

Notes about data

About the data provides a general discussion of international data standards, data collection methods, and sources of potential errors and inconsistencies. Readers are urged to read these notes to gain an understanding of the reliability and limitations of the data presented. For a full discussion of data collection methods and definitions readers should consult the technical documentation provided by the original compilers cited in *Data sources*.

Definitions

Definitions provide short descriptions of the main indicators in each table.

Sources

Partners are identified in the *Data sources* section following each table, and key publications of the partners drawn on for the table are identified. For a description of our partners and information on their data publications see the *Partners* section.

Figures

When appropriate, tables are accompanied by figures highlighting particular trends or issues.

Data presentation conventions and symbols

The cutoff date for data is 1 February 1998. The symbol .. means that data are not available or that aggregates cannot be calculated because of missing data in the year shown. A blank means not applicable or that an aggregate is not analytically meaningful. The numbers 0 and 0.0 mean zero or less than half the unit shown. Billion is 1,000 million. Trillion is 1,000 billion. The symbol / in dates, as in 1990/91, means that the period of time, usually 12 months, straddles two calendar years and refers to a crop year, a survey year, or a fiscal year. Figures in italics indicate data that are for years or periods other than those specified. Data for years that are more than three years from the range shown are footnoted. Dollars are current U.S. dollars unless otherwise noted.

the past 50 years as well as the large unfinished agenda, and calls for a global partnership to pursue a new development strategy that focuses on six key goals (distilled from the many set by various international conferences) for the start of the 21st century.

For economic well being
- Reducing by half the proportion of people in extreme poverty by 2015.

For social development
- Achieving universal primary education in all countries by 2015.
- Demonstrating progress toward gender equality and the empowerment of women by eliminating gender disparities in primary and secondary education by 2005.
- Reducing by two-thirds the mortality rates for infants and children under 5 and by three-fourths the mortality rates for mothers by 2015.
- Providing access to reproductive health services for all individuals of appropriate age no later than 2015.

For environmental sustainability and regeneration
- Implementing national strategies for sustainable development by 2005 to ensure that the current loss of environmental resources is reversed globally and nationally by 2015.

The goals are expressed in global terms but "must be pursued country by country through individual approaches that reflect local conditions and locally owned development strategies." Achieving them will also require building capacity for effective, democratic, and accountable governance, protection of human rights, and respect for the rule of law. Strategy 21 commits OECD countries to help countries that want to make a serious effort to attain these development goals.

The strong consensus emerging around this new development strategy provides an opportunity for the development community to galvanize support and political commitment for a set of people-centered goals that ordinary people in rich and poor countries can understand. The goals also emphasize account-

ability, and the World Bank will systematically monitor them in the countries it assists.

Reducing poverty

The common international poverty line of $1 a day suggests that some 60 percent of the world's poor live in India and China—and that 12 countries, each with more than 10 million in poverty, account for 80 percent of the world's poor (figure 1a). These countries require special attention. But this should not be at the expense of smaller, less populous countries where grinding poverty persists and governments are motivated to implement effective poverty reduction strategies.

The poverty goal calls for reducing by half the proportion of people in poverty by 2015. If nothing is done and the proportion in poverty stays at 30 percent, the number of people in poverty would rise from 1.3 billion to 1.9 billion by 2015, given the expected increase in population. A reduction to 15 percent by 2015 would reduce the number of people living in poverty to 900 million. (See the spread showing goals for 2015 that follows). Thus the goals for the 21st century call for lifting nearly 1 billion people out of poverty over the next two decades.

Is this feasible? A recent World Bank study (Demery and Walton 1997) explores the question. Income poverty is a function of growth and the extent to which the poor participate in growth. So, to answer the question, one must form a view of the prospects for growth and for inequality. Consider the growth in average real consumption per person required to halve the incidence of poverty using a $1 a day international poverty line and assuming no change in income distribution. Then compare these growth rates with past experience.

Plotting the growth in private consumption per capita required to reduce poverty based on the $1 poverty line against actual growth in 1990–95 shows that many countries have achieved the required rate of growth (figure 1b). Those countries above the diagonal experienced faster growth than required,

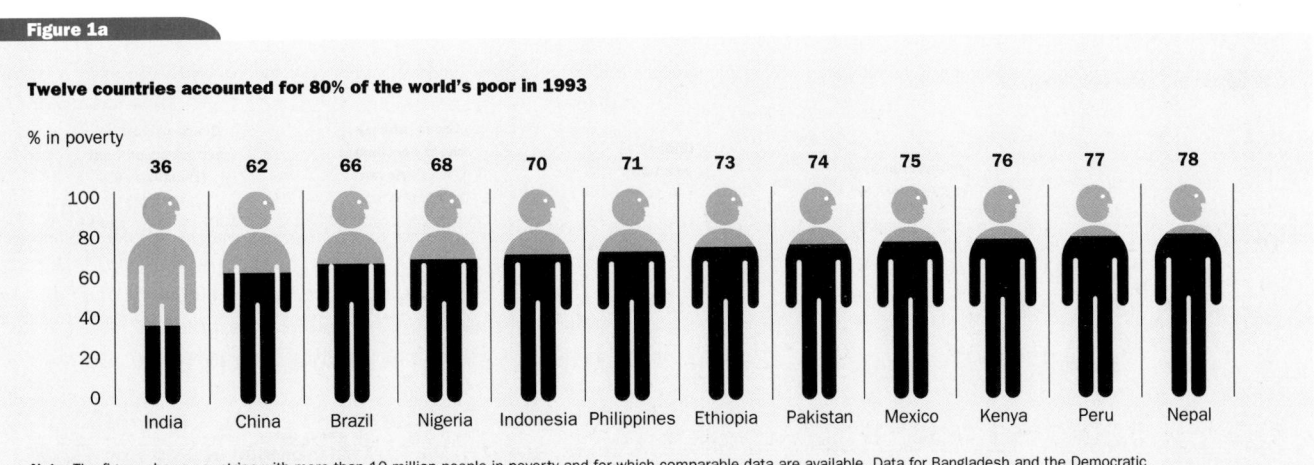

Figure 1a

Twelve countries accounted for 80% of the world's poor in 1993

% in poverty: India 36, China 62, Brazil 66, Nigeria 68, Indonesia 70, Philippines 71, Ethiopia 73, Pakistan 74, Mexico 75, Kenya 76, Peru 77, Nepal 78

Note: The figure shows countries with more than 10 million people in poverty and for which comparable data are available. Data for Bangladesh and the Democratic Republic of Congo are not available, but they are included in world estimates.

Source: World Bank staff estimates.

International comparisons of extreme poverty are based on a common international poverty line of $1 a person a day, expressed in 1985 international prices and adjusted to local currencies using purchasing power parity exchange rates. Most countries have their own poverty lines based on local views of minimum socially acceptable living standards (see *About the data* in table 2.7).

Global poverty targets will be met if China and India sustain their recent growth

Actual annual change in private consumption per capita, 1990–95 (%)

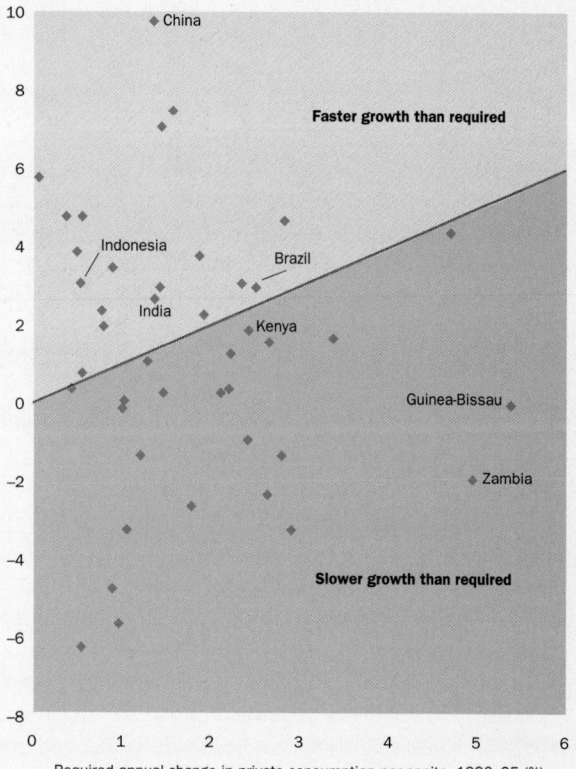

Source: World Bank staff estimates.

Required, actual, and projected regional growth in real consumption

%

Region	Annual growth required to reduce poverty by half by 2015	Real consumption per capita annual growth rate actual 1991–95	projected 1997–2000
East Asia and the Pacific	1.2	6.9	2.7
Europe and Central Asia	0.8	0.7	2.4
Latin America and the Caribbean	1.8	2.0	1.9
Middle East and North Africa	0.3	1.1	1.3
South Asia	1.3	1.9	3.5
Sub-Saharan Africa	1.9	–1.3	1.1

Source: Demery and Walton 1997; projections as of January 1998.

those below slower. Although many countries are not growing fast enough, the global poverty targets will be met if China and India sustain their recent growth.

Past trends may not be a good predictor of growth. Some regions may not be able to sustain rapid growth, and others are likely to improve their policies and performance. Table 1c compares the growth rates required to achieve a decline in poverty by one half against recent and projected growth scenarios for each developing region. Most regions should reach the goals by 2015—except Sub-Saharan Africa, where growth will fall short.

Forecasts tend to assume current policy choices and strategies. Demery and Walton show that policy changes that contribute to growth could make a big difference. Drawing on the work of Barro (1991) and Sachs and Warner (1995) to classify 36 developing countries by good and poor policies, they base their illustrative predictions of growth on the 1990 policy stance. If policies do not change, only half the countries are projected to achieve the growth required to meet the poverty target. If all 36 countries improve their policies, 28 should meet the poverty target.

Income distribution also matters. A highly unequal income distribution makes it harder to reduce poverty. Inequality is lower in South Asia and Eastern Europe and higher in Sub-Saharan Africa (particularly South Africa) and Latin America. While the distribution of income tends to be stable over time, there is some evidence of worsening inequality in East Asia. Reducing inequality will increase the numbers who benefit from the same average rate of growth. Conversely, higher inequality will increase the rate of growth needed to yield the same reduction in poverty. (See table 1.3 for estimates of the distribution-adjusted rate of growth in private consumption.)

Policies that promote growth are reasonably well understood. Policies that promote better income distribution are not. Policy areas deserving special attention from researchers and policy-makers in this regard include giving priority to rural development, assets redistribution, inclusive education systems, rapid growth in labor demand, and pro-poor tax and spending policies.

Reducing mortality rates and raising enrollment rates
And what of the social targets? If infant mortality rates were to remain at 1990 levels, the number of infant deaths would total some 8.8 million in 2015. Reaching the target of 35 deaths per 1,000 births would cause infant deaths to fall about 5.8 million a year, to slightly over 3 million. Similarly, attaining the primary school enrollment goal would require enrolling some 200 million more children in primary school in 2015 than are there today, an increase of 41 percent over current levels. These are ambitious targets.

How far are low- and lower-middle-income countries from these goals?

- 27 countries will need to reduce under-five mortality rates by more than 50 per 1,000 live births between now and 2015, with as many as 12 needing to reduce under-five mortality rates by 100 per 1,000.
- 18 of 26 countries reporting net primary enrollment data will need to increase net primary enrollment by more than 25 percentage points.

Countries with high infant mortality in 1970 continue to have high levels today

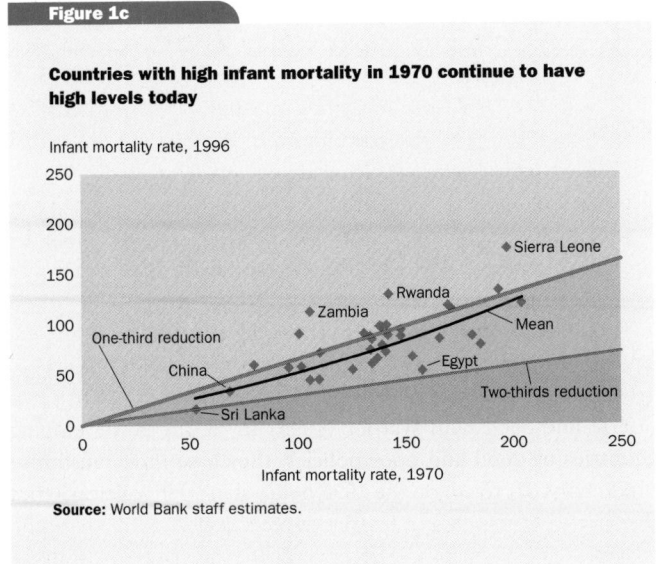

Infant mortality rate, 1996

Source: World Bank staff estimates.

Infant mortality rates, 1970 and 1996

Country	Infant mortality per 1,000 live births 1970	Infant mortality per 1,000 live births 1996	level in 1996	Ranking progress from 1970 to 1996
Bangladesh	140	77	16	14
Benin	148	87	22	20
Bolivia	153	67	12	7
Burkina Faso	141	98	30	31
Burundi	138	97	29	32
Cameroon	126	54	6	6
Central African Republic	139	96	28	30
Chad	171	115	32	26
China	69	33	2	10
Congo, Dem. Rep.	131	90	26	28
Congo, Rep.	101	90	25	35
Côte d'Ivoire	135	84	19	22
Egypt, Arab Rep.	158	53	5	2
Gambia, The	185	79	18	5
Ghana	111	71	13	24
Guyana	80	59	9	33
Haiti	141	72	14	12
Honduras	110	44	4	3
India	137	65	11	9
Kenya	102	57	8	16
Lesotho	134	74	15	15
Madagascar	181	88	23	11
Malawi	193	133	36	27
Mali	204	120	34	19
Mauritania	148	94	27	23
Nepal	166	85	20	13
Nicaragua	106	44	3	4
Niger	170	118	33	29
Nigeria	139	79	17	17
Pakistan	142	88	24	21
Rwanda	142	129	35	36
Senegal	135	60	10	8
Sierra Leone	197	174	37	34
Sri Lanka	53	15	1	1
Togo	134	87	21	25
Zambia	106	112	31	37
Zimbabwe	96	56	7	18

Source: World Bank staff estimates.

Reaching these targets will not be easy. Looking at past performance can provide some guidance for the future. Table 1d shows data on infant mortality rates in 1970 and 1996 for 37 low- and lower-middle-income countries. Aturupane, Glewwe, and Isenman (1994) used similar data for a larger set of developing countries to look at progress over time on infant mortality and primary school enrollment. Following their approach, the regression of the log of infant mortality in 1996 on the log of infant mortality in 1970 estimates the average reduction in infant mortality over the past 26 years. The slope of the regression line (0.58), shown in figure 1c, implies that the average reduction was 42 percent. The dashed lines represent reductions by one-third and two-thirds in infant mortality rates. Note that only one country—Sri Lanka—achieved a reduction of two-thirds. Countries with high levels of infant mortality a quarter century earlier continue to have high levels today (except Egypt, which has almost matched Sri Lanka's improvement). One other country, China, has achieved an infant mortality rate below the target level of 35.

Social indicators are closely related to GNP. Figure 1d shows the relationship between GNP per capita and under-five mortality and primary enrollment. As incomes increase, social indicators improve. Table 1e shows the ratio of the actual values of three social indicators to expected values based on GNP per capita for 43 low- and lower-middle-income countries. The United Nations Children's Fund calls the difference between these two values the "national performance gap" (UNICEF 1997). The results show that large differences persist even after accounting for income differences. The third column for each indicator shows the difference between the country's 1996 level and the development goal. Although some countries have been relatively successful—given the resources available to them—they remain far from the goals.

Demery and Walton (1997) provide a projection of child mortality rates, assuming past trends in mortality reduction and projected increases in per capita income and female education. They conclude that even with rapid growth, child mortality will be, on average, some 60 percent above the target for 2015. This is as much a comment on the past neglect of social development as on the magnitude of the task facing the international community.

This is not to say that the past must determine the future. With sufficient political will, improvements in female education, health programs, and incomes of the poor could bring the infant mortality target within reach. Otherwise, the cost in lost lives would be enormous:

- If infant mortality follows its current trend and declines by 2015 to 37 per 1,000 births, some 4.4 million infants will die each year.

	GNP per capita	Under-five mortality per 1,000			Net primary school enrollment			Gross female primary school enrollment		
	$ 1996	predicted on the basis of GNP per capita 1996	ratio of actual to predicted 1996	distance from 1996 value to goal[a]	predicted on the basis of GNP per capita 1996	ratio of actual to predicted 1996	distance from 1996 value to goal	predicted on the basis of GNP per capita 1996	ratio of actual to predicted 1996	distance from 1996 value to goal
Armenia	630	100	0.20	8				45	1.11	0
Bangladesh	260	152	0.74	74	61	1.14	30	42	1.06	5
Benin	350	135	1.04	95	63	0.84	47	43	0.77	16
Bolivia	830	53	1.94	67	70	1.30	9	47	1.00	3
Burkina Faso	230	155	1.02	113	61	0.52	68	42	0.92	11
Burundi	170	166	1.06	131	60	.84	49	42	1.08	5
Cameroon	610	91	1.12	67	..	..	..	46	1.02	3
Central African Republic	310	142	1.16	119	..	..	..	43	0.89	11
Chad	160	167	1.13	144	..	..	..	42	0.78	18
China	750	87	0.45	26	66	1.46	4	45	1.04	3
Congo, Dem Rep	130	167	0.86	99	60	0.90	46	42	1.02	7
Congo, Rep.	670	80	1.80	100	..	..	..	46	1.05	2
Côte d'Ivoire	660	82	1.83	105	68	0.76	48	46	0.93	7
Egypt, Arab Rep.	1,080	8	8.12	44	73	1.22	11	49	0.93	4
Ethiopia	100	176	1.01	132	59	0.35	79	..	..	..
Gambia, The	..[b]	..	..	..	63	0.88	45	43	0.96	9
Georgia	850	49	0.39	7	71	1.15	18	48	1.02	2
Haiti	310	150	0.87	86	62	0.42	74	43	1.13	2
Honduras	660	80	0.62	33	68	1.32	10	46	1.07	0
India	380	130	0.66	56	..	..	..	43	0.99	7
Kenya	320	140	0.64	59	..	..	..	43	1.15	1
Lesotho	660	82	1.38	75	68	0.96	35	46	1.15	−3
Madagascar	250	154	0.88	90	..	..	..	..	..	..
Malawi	180	167	1.30	172	60	1.53	8	42	1.13	3
Mali	240	152	1.45	175	62	0.37	77	43	0.92	11
Mozambique	80	179	1.19	169	59	0.69	59	41	1.02	8
Nepal	210	159	0.73	77	..	..	..	42	0.92	11
Nicaragua	380	130	0.44	40	63	1.24	21	44	1.15	0
Niger	200	..	..	..	61	0.41	75	42	0.89	12
Nigeria	240	154	0.85	86	..	..	..	42	1.04	6
Pakistan	480	113	1.09	81	..	..	..	44	0.69	19
Rwanda	190	162	1.26	160	61	1.17	29	42	1.18	0
Senegal	570	99	0.89	58	66	0.76	50	45	0.95	7
Sierra Leone	200	159	1.79	239	..	..	..	42	0.98	9
Sri Lanka	740	68	0.28	7	..	..	..	46	1.04	2
Tajikistan	340	138	0.27	25	..	..	..	43	1.12	1
Tanzania	170[c]	166	0.87	99	..	..	..	..	..	..
Togo	300	143	0.96	93	62	1.11	31	43	0.93	10
Uganda	300	143	0.98	96	..	..	..	43	1.03	6
Vietnam	290	145	0.33	32	..	..	..	..	..	..
Yemen, Rep.	380	130	1.00	86	..	..	..	44	0.64	22
Zambia	360	131	1.54	157	63	1.09	31	43	1.10	2
Zimbabwe	610	90	0.95	57	..	..	..	..	..	..

a. The goal for under-five mortality was determined to be the lesser of one-third of the 1996 level or 45. For countries where under-five mortality was less than 45 in 1996, the goal was assumed to be 12. b. Estimated to be low income (average per capita income of $785 or less). c. Refers to mainland Tanzania only.

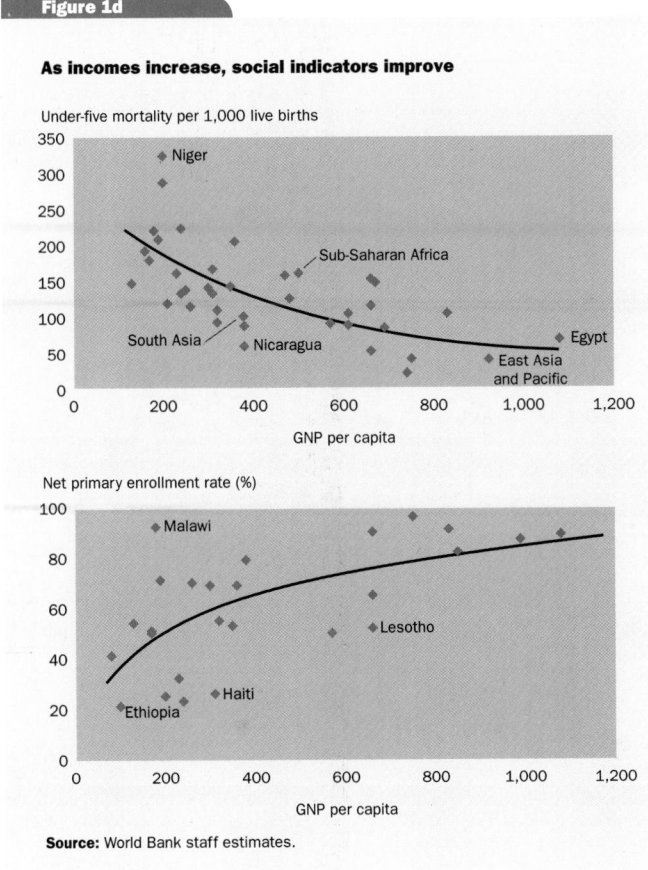

Figure 1d

As incomes increase, social indicators improve

Under-five mortality per 1,000 live births

Net primary enrollment rate (%)

Source: World Bank staff estimates.

- *Accelerating social development.* Social indicators will benefit from improvements in economic growth and income and wealth distribution, but there is still room for policies that target interventions that appear to have a large impact on health and educational outcomes. At the top of the list are female education, safe water and sanitation, and child immunization.

Second, donors and international development agencies must support countries that show a determination to take up the challenges of the goals for the 21st century. These goals are shared, and ownership of these goals must also be shared. This implies extending this dialogue to major international forums in the United Nations and Bretton Woods systems, and to the consultative groups and round tables that meet regularly to support development strategies in individual countries. Donors and their developing country partners need to surmount the obstacles to ownership and implementation of sound strategies in support of these international goals. The decline in aid flows will need to be reversed, particularly for countries committed to an accelerated development strategy and appropriate policies in support of these goals—and particularly for the current strong donor support for Strategy 21 to be seen as credible.

Third, international agencies must take the lead in strengthening the capacity of developing countries to monitor their progress on outcomes. This will involve ensuring that the statistical infrastructure in key countries is adequate to mount periodic surveys—and to collect and analyze data on the main outcome indicators and on the leading indicators that predict those outcomes. Data quality, a serious issue in many developing countries, will need to be urgently addressed if policy-makers are to take the monitoring of these goals seriously.

It will not be easy. But we must move forward. And as Strategy 21 notes, meeting goals as ambitious and important as these will take a significant effort by the world community. It calls for a substantial and serious commitment by developing nations—by their governments, their private sectors, and their civil societies. It calls for a renewed effort by international development agencies, both bilateral and multilateral, to be responsive to these development efforts in a coordinated partnership. And it calls for a new effort by advanced countries to achieve coherence in their aid and other policies that affect developing countries. As World Bank President James Wolfensohn (1997) told the World Bank Group's Board of Governors:

> *Without equity, we will not have global stability. Without a better sense of social justice, our cities will not be safe, and our societies will not be stable. Without inclusion, too many of us will be condemned to live separate, armed, frightened lives. Whether you broach it from the social or the economic or the moral perspective, this is a challenge that we cannot afford to ignore. There are not two worlds, there is one world. We share the same world, and we share the same challenge. The fight against poverty is the fight for peace, security, and growth for all of us.*

- An average two-thirds reduction to 20 per 1,000 for developing countries would reduce the number of infant deaths to 2.4 million—2 million fewer infant deaths each year.

The challenges

How do we move forward? First, developing countries must embark on strategies that help them attain these goals. In the areas of poverty and social development, this implies particular attention by policymakers to:

- *Accelerating economic growth.* Growth is the most powerful weapon in the fight for higher living standards. If Sub-Saharan Africa is to make a serious dent in its rising number of poor, it must improve its growth performance over the early 1990s by as much as 3 percentage points. Latin America and Europe and Central Asia require a more modest acceleration. Faster growth will require policies that encourage macroeconomic stability, shift resources to more efficient sectors, and accelerate integration with the global economy.
- *Improving the distribution of income and wealth.* The benefits of growth for the poor may be eroded if the distribution of income worsens, which might also undermine the incentives for growth-inducing economic reforms. Understanding what policies improve the distribution of income and wealth in a way that fosters incentives for growth should be high on the agenda of policymakers and researchers.

Goals for social development

Primary education
Achieving universal
primary education in
all countries by 2015

**Gender equality
in education**
Demonstrating progress
toward gender equality and
the empowerment of women by
eliminating gender disparities in primary
and secondary education by 2005

Infant, child, and maternal mortality
Reducing by two-thirds the mortality rates for
infants and children under five and by three-fourths
the mortality rates for mothers by 2015

**Infant and child
mortality**

**Maternal
mortality**

$\downarrow \dfrac{2}{3}$ $\downarrow \dfrac{3}{4}$

Reproductive health
Providing access to reproductive health
services for all individuals of appropriate
age by no later than 2015

Goals for economic well-being

Poverty
Reducing by half the proportion of people in extreme poverty by 2015

↓ 1/2

Goals for environmental sustainability and regeneration

Environment
Implementing national strategies for sustainable development by 2005 to ensure that the current loss of environmental resources is reversed globally and nationally by 2015

"We commit ourselves to the goal of eradicating poverty in the world through decisive national actions and international cooperation, as an ethical, social, political, and economic imperative of humankind"
—World Summit for Social Development, Copenhagen, March 1995

Meeting goals as ambitious and important as these will require significant efforts by the global community

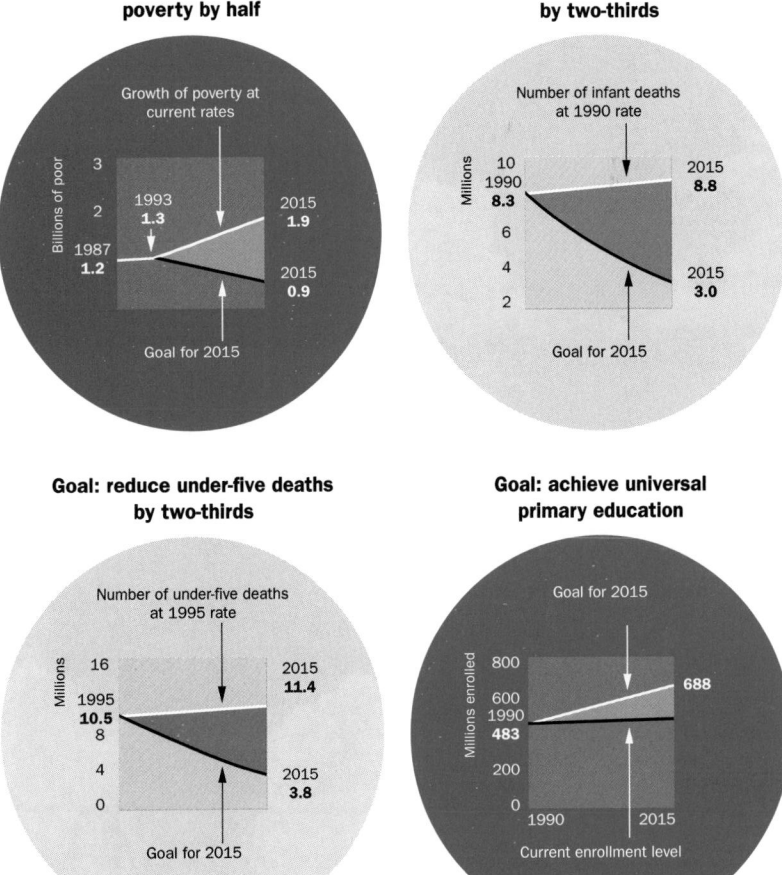

Goal: reduce number in poverty by half

Growth of poverty at current rates

Billions of poor

1993 **1.3**
1987 **1.2**
2015 **1.9**
2015 **0.9**

Goal for 2015

Goal: reduce infant deaths by two-thirds

Number of infant deaths at 1990 rate

Millions

1990 **8.3**
2015 **8.8**
2015 **3.0**

Goal for 2015

Goal: reduce under-five deaths by two-thirds

Number of under-five deaths at 1995 rate

Millions

1995 **10.5**
2015 **11.4**
2015 **3.8**

Goal for 2015

Goal: achieve universal primary education

Goal for 2015

Millions enrolled

1990 **483**
688

1990   2015

Current enrollment level

1.1 Size of the economy

	Population	Land area	Population density	GNP			GNP per capita			GNP PPP[a]		
	millions 1996	thousand sq. km 1995	people per sq. km 1996	$ billions 1996[b]	rank 1996	average annual growth % 1995–96	$ 1996[b]	rank 1996	average annual growth % 1995–96	$ billions 1996	per capita $ 1996	rank 1996
Albania	3	27	120	2.7	105	..	820	80	..	..	..	..
Algeria	29	2,382	10	43.7	48	4.1	1,520	63	1.8	132.7 c	4,620 c	56
Angola	11	1,247	9	3.0	104	1.3	270	113	−1.7	11.4	1,030	113
Argentina	35	2,737	10	295.1	18	4.0	8,380	27	2.7	335.6	9,530	32
Armenia	4	28	130	2.4	110	7.8	630	87	7.4	8.2	2,160	86
Australia	18	7,682	2	367.8	13	4.0	20,090	15	2.6	363.9	19,870	16
Austria	8	83	100	226.5	21	1.2	28,110	7	1.0	174.5	21,650	9
Azerbaijan	8	87	90	3.6	99	−0.4	480	94	−1.3	11.3	1,490	103
Bangladesh	122	130	930	31.2	51	5.5	260	114	3.8	122.9	1,010	116
Belarus	10	207	50	22.5	54	2.6	2,070	55	2.9	45.1	4,380	62
Belgium	10	33	310	268.6	19	1.6	26,440	9	1.4	227.5	22,390	7
Benin	6	111	50	2.0	112	6.2	350	104	3.2	6.9	1,230	109
Bolivia	8	1,084	7	6.3	79	5.0	830	79	2.6	21.7	2,860	77
Bosnia and Herzegovina	..	51	..	..	..	..	d	..	..	..	..	..
Botswana	1	567	3	..	..	..	e	..	..	10.9	7,390	40
Brazil	161	8,457	20	709.6	8	8.2	4,400	31	6.7	1,023.1	6,340	46
Bulgaria	8	111	80	9.9	68	−9.2	1,190	69	−8.8	35.8	4,280	64
Burkina Faso	11	274	40	2.4	109	6.2	230	119	3.3	10.1 c	950 c	117
Burundi	6	26	250	1.1	125	−8.9	170	125	−11.1	3.8	590	129
Cambodia	10	177	60	3.1	103	6.5	300	109	3.9	..	..	..
Cameroon	14	465	30	8.4	75	7.6	610	88	4.5	24.1	1,760	94
Canada	30	9,221	3	569.9	9	1.7	19,020	18	0.5	640.6	21,380	11
Central African Republic	3	623	5	1.0	127	−3.0	310	107	−5.0	4.8 c	1,430 c	105
Chad	7	1,259	5	1.0	126	3.0	160	127	0.5	5.8	880	121
Chile	14	749	20	70.1	38	10.1	4,860	29	8.5	168.7	11,700	28
China	1,215	9,326	130	906.1	7	10.0	750	81	8.9	4,047.3	3,330	72
Hong Kong, China[f]	6	1	6,370	153.3	26	4.7	24,290	13	2.2	153.1	24,260	4
Colombia	37	1,039	40	80.2	37	1.2	2,140	54	−0.5	251.7	6,720	43
Congo, Dem. Rep.	45	2,267	20	5.7	85	3.1	130	128	−0.1	35.7 c	790 c	124
Congo, Rep.	3	342	8	1.8	118	7.6	670	83	4.7	3.8	1,410	106
Costa Rica	3	51	70	9.1	72	−0.1	2,640	48	−2.0	22.3	6,470	45
Côte d'Ivoire	14	318	50	9.4	70	7.3	660	84	4.6	22.7	1,580	100
Croatia	5	56	90	18.1	60	4.6	3,800	36	4.7	20.5	4,290	63
Cuba	11	110	100	..	..	..	e	..	..	..	..	..
Czech Republic	10	77	130	48.9	46	4.4	4,740	30	4.6	112.1	10,870	29
Denmark	5	42	120	168.9	25	2.5	32,100	4	1.8	116.4	22,120	8
Dominican Republic	8	48	160	12.8	66	7.6	1,600	60	5.7	35.0	4,390	60
Ecuador	12	277	40	17.5	62	3.3	1,500	64	1.2	55.3	4,730	54
Egypt, Arab Rep.	59	995	60	64.3	40	5.4	1,080	73	3.5	169.5	2,860	78
El Salvador	6	21	280	9.9	69	2.6	1,700	58	0.0	16.2	2,790	80
Eritrea	4	101	40	..	..	..	d	..	..	..	..	..
Estonia	1	42	30	4.5	92	4.0	3,080	42	5.2	6.8	4,660	55
Ethiopia	58	1,000	60	6.0	81	10.7	100	129	7.2	29.1	500	131
Finland	5	305	20	119.1	31	3.8	23,240	14	3.5	93.6	18,260	18
France	58	550	110	1,533.6	4	1.4	26,270	10	1.0	1,255.6	21,510	10
Gabon	1	258	4	4.4	93	1.2	3,950	34	−1.2	7.1	6,300	47
Gambia, The	1	10	110	..	..	..	d	..	..	1.5 c	1,280 c	107
Georgia	5	70	80	4.6	91	..	850	78	..	9.8	1,810	91
Germany	82	349	230	2,364.6	3	1.3	28,870	6	0.9	1,729.2	21,110	12
Ghana	18	228	80	6.2	80	5.0	360	101	2.3	31.4 c	1,790 c	93
Greece	10	129	80	120.0	30	2.4	11,460	23	2.2	133.3	12,730	26
Guatemala	11	108	100	16.0	64	11.7	1,470	65	8.6	41.7	3,820	66
Guinea	7	246	30	3.8	98	4.4	560	92	1.8	11.6	1,720	96
Guinea-Bissau	1	28	40	0.3	130	6.1	250	115	3.7	1.1	1,030	114
Haiti	7	28	270	2.3	111	2.4	310	108	0.0	8.3 c	1,130 c	110
Honduras	6	112	50	4.0	97	2.7	660	85	−0.3	13.0	2,130	87

Size of the economy 1.1

	Population	Land area	Population density	GNP			GNP per capita			GNP PPP[a]		
						average annual growth			average annual growth		per capita	
	millions 1996	thousand sq. km 1995	people per sq. km 1996	$ billions 1996[b]	rank 1996	% 1995–96	$ 1996[b]	rank 1996	% 1995–96	$ billions 1996	$ 1996	rank 1996
Hungary	10	92	110	44.3	47	2.2	4,340	33	2.6	68.6	6,730	42
India	945	2,973	320	357.8	14	6.9	380	98	5.1	1,493.3	1,580	101
Indonesia	197	1,812	110	213.4	22	7.5	1,080	74	5.8	652.3	3,310	74
Iran, Islamic Rep.	63	1,622	40	..	..	2.8	..[e]	..	0.6	335.0	5,360	53
Iraq	21	437	50	..	..	..	..[e]	..	..	..	..	..
Ireland	4	69	50	62.0	42	9.9	17,110	19	8.7	60.7	16,750	21
Israel	6	21	280	90.3	34	..	15,870	20	..	103.0	18,100	19
Italy	57	294	200	1,140.5	6	1.0	19,880	16	0.7	1,141.3	19,890	15
Jamaica	3	11	240	4.1	96	–1.0	1,600	61	–1.9	8.8	3,450	71
Japan	126	377	330	5,149.2	2	3.9	40,940	2	3.6	2,945.3	23,420	5
Jordan	4	89	50	7.1	77	5.7	1,650	59	2.8	15.4	3,570	68
Kazakhstan	16	2,671	6	22.2	55	0.9	1,350	66	1.8	53.2	3,230	75
Kenya	27	569	50	8.7	73	5.7	320	106	3.1	30.9	1,130	111
Korea, Dem. Rep.	22	120	190	..	..	..	..[e]	..	..	..	..	..
Korea, Rep.	46	99	460	483.1	11	6.9	10,610	24	5.6	595.7	13,080	25
Kuwait	2	18	90	..	..	..	..[g]	..	..	..	..	..
Kyrgyz Republic	5	192	20	2.5	107	5.5	550	93	4.1	9.0	1,970	89
Lao PDR	5	231	20	1.9	115	6.8	400	97	4.0	5.9	1,250	108
Latvia	2	62	40	5.7	84	2.4	2,300	51	3.5	9.1	3,650	67
Lebanon	4	10	400	12.1	67	2.4	2,970	45	0.6	24.7	6,060	49
Lesotho	2	30	70	1.3	121	9.0	660	86	6.7	4.8 c	2,380 c	82
Libya	5	1,760	3	..	..	..	..[h]	..	..	..	..	..
Lithuania	4	65	60	8.5	74	2.6	2,280	52	2.7	16.3	4,390	61
Macedonia, FYR	2	25	80	2.0	114	1.3	990	76	0.6	..	..	..
Madagascar	14	582	20	3.4	101	3.5	250	116	0.5	12.3	900	119
Malawi	10	94	110	1.8	117	16.0	180	124	13.0	6.9	690	127
Malaysia	21	329	60	89.8	35	8.3	4,370	32	5.8	213.7	10,390	30
Mali	10	1,220	8	2.4	108	4.3	240	117	1.2	7.1	710	126
Mauritania	2	1,025	2	1.1	124	4.4	470	96	1.8	4.2	1,810	92
Mauritius	1	2	560	4.2	95	5.6	3,710	37	4.5	10.2	9,000	33
Mexico	93	1,909	50	341.7	16	6.6	3,670	38	4.7	713.8	7,660	37
Moldova	4	33	130	2.5	106	–10.0	590	90	–9.7	6.2	1,440	104
Mongolia	3	1,567	2	0.9	129	2.0	360	102	–0.1	4.6	1,820	90
Morocco	27	446	60	34.9	50	12.4	1,290	67	10.4	89.7	3,320	73
Mozambique	18	784	20	1.5	120	8.7	80	130	5.0	9.0 c	500 c	132
Myanmar	46	658	70	..	..	..	..[d]	..	..	..	..	..
Namibia	2	823	2	3.6	100	2.8	2,250	53	0.3	8.5 c	5,390 c	52
Nepal	22	143	150	4.7	90	4.6	210	120	1.8	24.0	1,090	112
Netherlands	16	34	460	402.6	12	4.2	25,940	11	3.9	323.5	20,850	13
New Zealand	4	268	10	57.1	45	0.6	15,720	21	–0.6	60.0	16,500	22
Nicaragua	5	121	40	1.7	119	7.3	380	99	4.2	7.9 c	1,760 c	95
Niger	9	1,267	7	1.9	116	3.3	200	121	–0.1	8.6 c	920 c	118
Nigeria	115	911	130	27.6	52	5.0	240	118	1.9	99.7	870	122
Norway	4	307	10	151.2	27	5.1	34,510	3	4.6	101.7	23,220	6
Oman	2	212	10	..	..	..	..[h]	..	..	18.9	8,680	34
Pakistan	134	771	170	63.6	41	3.1	480	95	0.3	213.6	1,600	99
Panama	3	74	40	8.2	76	5.8	3,080	43	4.1	18.9	7,060	41
Papua New Guinea	4	453	10	5.0	87	–0.1	1,150	72	–2.4	12.4 c	2,820 c	79
Paraguay	5	397	10	9.2	71	1.1	1,850	57	–1.5	17.2	3,480	70
Peru	24	1,280	20	58.7	44	2.0	2,420	49	0.0	107.1	4,410	59
Philippines	72	298	240	83.3	36	6.9	1,160	70	4.5	255.2	3,550	69
Poland	39	304	130	124.7	29	6.3	3,230	41	6.2	231.7	6,000	51
Portugal	10	92	110	100.9	32	2.4	10,160	25	2.4	133.6	13,450	24
Puerto Rico	4	9	430	..	..	..	..[h]	..	..	..	..	..
Romania	23	230	100	36.2	49	4.4	1,600	62	4.7	103.5	4,580	57
Russian Federation	148	16,889	9	356.0	15	–5.3	2,410	50	–5.0	619.0	4,190	65

1.1 Size of the economy

	Population	Land area	Population density	GNP			GNP per capita			GNP PPP[a]		
						average annual growth			average annual growth		per capita	
	millions 1996	thousand sq. km 1995	people per sq. km 1996	$ billions 1996[b]	rank 1996	% 1995–96	$ 1996[b]	rank 1996	% 1995–96	$ billions 1996	$ 1996	rank 1996
Rwanda	7	25	270	1.3	123	13.3	190	123	7.8	4.2	630	128
Saudi Arabia	19	2,150	9	..	..	..	[h]	..	..	188.3	9,700	31
Senegal	9	193	40	4.9	89	5.9	570	91	3.2	14.1	1,650	97
Sierra Leone	5	72	60	0.9	128	10.4	200	122	7.6	2.4	510	130
Singapore	3	1	4,990	93.0	33	7.6	30,550	5	5.6	81.9	26,910	2
Slovak Republic	5	48	110	18.2	59	6.6	3,410	40	6.3	39.9	7,460	38
Slovenia	2	20	100	18.4	58	3.2	9,240	26	3.2	24.1	12,110	27
South Africa	38	1,221	30	132.5	28	2.9	3,520	39	1.0	280.4 [c]	7,450 [c]	39
Spain	39	499	80	563.2	10	1.7	14,350	22	1.6	600.3	15,290	23
Sri Lanka	18	65	280	13.5	65	1.6	740	82	0.5	41.9	2,290	83
Sudan	27	2,376	10	..	..	..	[d]	..	..	..	..	..
Sweden	9	412	20	227.3	20	1.0	25,710	12	0.8	166.0	18,770	17
Switzerland	7	40	180	313.7	17	–0.8	44,350	1	–1.2	186.3	26,340	3
Syrian Arab Republic	15	184	80	16.8	63	3.4	1,160	71	0.6	43.8	3,020	76
Tajikistan	6	141	40	2.0	113	–7.0	340	105	–8.4	5.3	900	120
Tanzania	30	884	30	5.2	86	4.6	170 [i]	126	1.7	..	..	..
Thailand	60	511	120	177.5	24	5.4	2,960	46	4.4	402.0	6,700	44
Togo	4	54	80	1.3	122	7.4	300	110	4.3	7.0	1,650	98
Trinidad and Tobago	1	5	250	5.0	88	3.8	3,870	35	3.0	7.9	6,100	48
Tunisia	9	155	60	17.6	61	1.3	1,930	56	–0.4	41.5	4,550	58
Turkey	63	770	80	177.5	23	6.8	2,830	47	5.0	379.9	6,060	50
Turkmenistan	5	470	10	4.3	94	–2.4	940	77	–4.3	9.2	2,010	88
Uganda	20	200	100	5.8	83	9.4	300	111	6.2	20.3 [c]	1,030 [c]	115
Ukraine	51	579	90	60.9	43	–9.9	1,200	68	–8.5	113.1	2,230	84
United Arab Emirates	3	84	30	..	..	..	[g]	..	..	43.0 [c]	17,000 [c]	20
United Kingdom	59	242	240	1,152.1	5	2.6	19,600	17	2.3	1,173.3	19,960	14
United States	265	9,159	30	7,433.5	1	2.3	28,020	8	1.4	7,433.3	28,020	1
Uruguay	3	175	20	18.5	57	7.5	5,760	28	6.8	24.9	7,760	36
Uzbekistan	23	414	60	23.5	53	1.1	1,010	75	–0.8	56.9	2,450	81
Venezuela	22	882	30	67.3	39	–1.6	3,020	44	–3.7	181.4	8,130	35
Vietnam	75	325	230	21.9	56	9.3	290	112	7.3	118.3	1,570	102
West Bank and Gaza	2	..	..	..	..	..	[e]	..	..	..	..	..
Yemen, Rep.	16	528	30	6.0	82	–4.7	380	100	–7.8	12.5	790	125
Yugoslavia, FR (Serb./Mont.)	11	102	100	..	..	..	[e]	..	..	..	..	..
Zambia	9	743	10	3.4	102	6.1	360	103	3.4	7.9	860	123
Zimbabwe	11	387	30	6.8	78	8.1	610	89	5.8	24.7	2,200	85
World	**5,754 s**	**130,129 s**	**44 w**	**29,510 t**		**3.2 w**	**5,130 w**		**1.7 w**	**35,688 t**	**6,200 w**	
Low income	3,236	39,294	82	1,597		8.0	490		6.2	6,809	2,100	
Excl. China & India	1,076	26,994	40	333		5.1	310		2.5	1,268	1,180	
Middle income	1,599	59,884	27	4,141		3.7	2,590		2.4	8,305	5,200	
Lower middle income	1,125	39,310	29	1,963		1.9	1,740		0.6	4,699	4,180	
Upper middle income	473	20,574	23	2,178		6.4	4,600		5.0	3,606	7,620	
Low & middle income	4,835	99,178	49	5,738		5.2	1,190		3.6	15,114	3,130	
East Asia & Pacific	1,732	15,869	109	1,540		8.8	890		7.4	5,839	3,370	
Europe & Central Asia	478	23,864	20	1,050		–0.4	2,200		–0.4	2,059	4,310	
Latin America & Carib.	486	20,064	24	1,804		5.8	3,710		4.1	3,174	6,530	
Middle East & N. Africa	276	10,972	25	572			2,070		..	1,251	4,530	
South Asia	1,266	4,781	265	478		6.3	380		4.4	1,924	1,520	
Sub-Saharan Africa	596	23,628	25	295		4.8	490		1.9	867	1,450	
High income	919	30,951	30	23,772		2.7	25,870		2.0	20,574	22,390	

a. Purchasing power parity; see *Definitions*. b. Calculated using the World Bank Atlas method. c. The estimate is based on regression; others are extrapolated from the latest International Comparison Programme benchmark estimates. d. Estimated to be low income ($785 or less). e. Estimated to be lower middle income ($785 to $3,115). f. GNP data are GDP. g. Estimated to be high income ($9,636 or more). h. Estimated to be upper middle income ($3,116 to $9,635). i. Data refer to mainland Tanzania only.

Size of the economy | 1.1

Population, land area, and output are important measurements of economy size. They also provide a broad indication of actual and potential resources. Therefore, population, land area, and output—as measured by gross national product (GNP) or gross domestic product (GDP)—are used throughout the *World Development Indicators* to normalize other indicators.

Population estimates are generally based on extrapolations from the most recent national census. See *About the data* for tables 2.1 and 2.2 for further discussion on the measurement of population and population growth.

Land area is particularly important for understanding the agricultural capacity of an economy and the effects of human activity on the environment. See tables 3.1-3.4 for other measures of land area, rural population density, land use, and productivity. Land area differs from other measures of geographic size such as surface area, which includes inland bodies of water and some coastal waterways, and gross area which may include offshore territorial waters. Recent innovations in satellite mapping techniques and computer databases have resulted in more precise measurements of land and water areas.

GNP, the broadest measure of national income, measures the total domestic and foreign value added claimed by residents. GNP comprises GDP plus net receipts of primary income from nonresident sources. The World Bank uses GNP per capita in U.S. dollars to classify countries for analytical purposes

and to determine borrowing eligibility. See the *Users guide* for definitions of the income groups used in this book. Also see *About the data* for tables 4.1 and 4.2 for further discussion of the usefulness of national income as a measure of productivity or welfare.

When calculating GNP in U.S. dollars from GNP reported in national currencies, the World Bank follows its Atlas conversion method. This involves using a three-year average of exchange rates to smooth the effects of transitory exchange rate fluctuations. See Statistical methods for further discussion of the Atlas method. Note that growth rates are calculated from data in constant prices and national currency units, not from the Atlas estimates.

Because exchange rates do not always reflect international differences in relative prices, this table also shows GNP and GNP per capita estimates that are converted into international dollars using purchasing power parities (PPPs). PPPs provide a standard measure of real price levels between countries, just as conventional price indexes calculate real values over time. The PPP conversion factors used here are derived from the most recent round of price surveys—covering 118 countries—conducted by the International Comparison Programme (ICP). The surveys, completed in 1996, are based on a 1993 reference year. Estimates for countries not included in the survey are derived from statistical models using available data. See *About the data* for tables 4.10 and 4.11 for more information on the ICP and the calculation of PPPs.

• **Population** is based on the de facto definition of population, which counts all residents regardless of legal status or citizenship—except for refugees not permanently settled in the country of asylum, who are generally considered part of the population of the country of origin. The values shown are midyear estimates for 1996. See also table 2.1. • **Land area** is a country's total area, excluding areas under inland bodies of water. • **Population density** is midyear population divided by land area in square kilometers. • **Gross national product (GNP)** is the sum of value added by all resident producers plus any taxes (less subsidies) that are not included in the valuation of output plus net receipts of primary income (employee compensation and property income) from nonresident sources. Data are in current U.S. dollars converted using the World Bank Atlas method (see *Statistical methods*). Growth is calculated from constant price GNP in national currency units. • **GNP per capita** is gross national product divided by midyear population. GNP per capita in U.S. dollars is converted using the World Bank Atlas method. Growth is calculated from constant price GNP per capita in national currency units. • **GNP PPP** is gross national product converted to international dollars using purchasing power parity rates. An international dollar has the same purchasing power over GNP as the U.S. dollar in the United States. All ranks are calculated for economies reporting data.

Population estimates are prepared by World Bank staff from a variety of sources (see *Data sources* for table 2.1). Data on land area are from the Food and Agriculture Organization (see *Data sources* for table 3.1). GNP per capita is estimated by World Bank staff based on national accounts data collected by World Bank staff during economic missions or reported by national statistical offices to other international organizations such as the OECD. Data for high-income OECD economies come from the OECD. Purchasing power parity conversion factors are estimates by World Bank staff based on data collected by the International Comparison Programme.

Population and GNP in selected countries, 1996

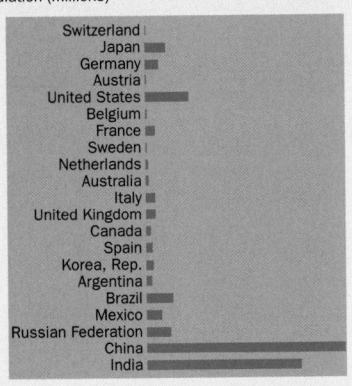

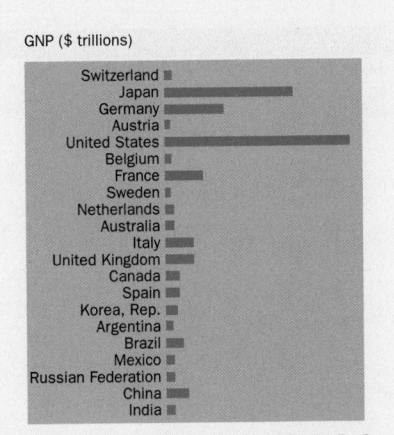

Note: Countries are ranked by GNP per capita.

Source: World Bank staff estimates.

1.2 | Quality of life

	Life expectancy at birth		Prevalence of child malnutrition	Sanitation	Safe water	Adult illiteracy rate		Commercial energy use
	Male years 1996	Female years 1996	% of children under 5 1990–96	% of population with access 1995	% of population with access 1995	% of people 15 and above Male 1995	% of people 15 and above Female 1995	kg of oil equivalent per capita 1995
Albania	69	75	..	..	..	..	..	314
Algeria	68	72	10	..	..	26	51	866
Angola	45	48	35	16	32	..	..	89
Argentina	69	77	2	89	64	4	4	1,525
Armenia	69	76	..	..	..	..	..	444
Australia	75	81	..	90	95	..	..	5,215
Austria	74	80	..	100	..	..	..	3,279
Azerbaijan	65	74	10	..	..	..	..	1,735
Bangladesh	57	59	68	35	79	51	74	67
Belarus	63	74	..	100	..	..	..	2,305
Belgium	73	80	..	100	..	..	..	5,167
Benin	52	57	24	20	50	51	74	20
Bolivia	59	63	16	44	60	10	24	396
Bosnia and Herzegovina	..	..	..	..	..	..	..	364
Botswana	50	53	27	55	70	20	40	383
Brazil	63	71	7	41	72	17	17	772
Bulgaria	67	75	..	99	..	..	..	2,724
Burkina Faso	45	47	33	18	78	71	91	16
Burundi	45	48	38	..	..	51	78	23
Cambodia	52	55	38	..	13	20	47	52
Cameroon	55	58	15	40	41	25	48	117
Canada	76	82	..	85	100	..	..	7,879
Central African Republic	46	51	23	..	18	32	48	29
Chad	47	50	..	21	24	38	65	16
Chile	72	78	1	83	..	5	5	1,065
China	68	71	16	21	90	10	27	707
Hong Kong, China	76	81	..	..	..	4	12	2,212
Colombia	67	73	8	63	76	9	9	655
Congo, Dem. Rep.	51	54	34	..	..	13	32	47
Congo, Rep.	49	54	24	9	47	17	33	139
Costa Rica	75	79	2	..	..	5	5	584
Côte d'Ivoire	53	55	24	54	72	50	70	97
Croatia	68	77	..	68	96	..	..	1,435
Cuba	74	78	8	66	93	4	5	949
Czech Republic	70	77	1	..	..	..	..	3,776
Denmark	73	78	..	100	100	..	..	3,918
Dominican Republic	69	73	6	78	71	18	18	486
Ecuador	67	73	17	64	70	8	12	553
Egypt, Arab Rep.	64	67	9	11	64	36	61	596
El Salvador	66	72	11	68	55	27	30	410
Eritrea	54	56	41	..	..	..	..	..
Estonia	63	76	..	..	..	..	..	3,454
Ethiopia	48	51	48	10	27	55	75	21
Finland	73	81	..	100	100	..	..	5,613
France	74	82	..	96	100	..	..	4,150
Gabon	53	57	15	76	67	26	47	587
Gambia, The	51	55	17	37	76	47	75	55
Georgia	69	77	..	..	..	..	..	342
Germany	73	80	..	100	..	..	..	4,156
Ghana	57	61	27	27	56	24	47	92
Greece	75	81	..	96	..	..	..	2,266
Guatemala	64	69	33	66	60	38	51	206
Guinea	46	47	24	70	62	50	78	64
Guinea-Bissau	42	45	23	20	23	32	58	37
Haiti	54	57	28	24	28	52	58	50
Honduras	65	69	18	62	65	27	27	236

Quality of life | 1.2

	Life expectancy at birth		Prevalence of child malnutrition	Sanitation	Safe water	Adult illiteracy rate		Commercial energy use
	Male years 1996	Female years 1996	% of children under 5 1990–96	% of population with access 1995	% of population with access 1995	% of people 15 and above Male 1995	% of people 15 and above Female 1995	kg of oil equivalent per capita 1995
Hungary	65	75	..	94	..	..	..	2,454
India	62	63	66	29	81	35	62	260
Indonesia	63	67	40	51	62	10	22	442
Iran, Islamic Rep.	69	70	16	..	..	22	34	1,374
Iraq	60	63	12	87	44	29	55	1,206
Ireland	74	79	..	100	..	..	..	3,196
Israel	75	79	..	70	99	..	..	3,003
Italy	75	81	..	100	..	..	..	2,821
Jamaica	72	77	10	74	70	19	11	1,191
Japan	77	83	3	85	..	..	..	3,964
Jordan	69	72	10	100	89	7	21	1,031
Kazakhstan	60	70	1	..	..	..	..	3,337
Kenya	57	60	23	77	53	14	30	109
Korea, Dem. Rep.	61	65	..	100	100	1	3	1,113
Korea, Rep.	69	76	..	100	89	1	3	3,225
Kuwait	74	79	6	100	..	18	25	9,381
Kyrgyz Republic	62	71	..	53	75	..	..	513
Lao PDR	52	54	40	19	39	31	56	40
Latvia	63	76	..	..	..	..	..	1,471
Lebanon	68	71	9	..	..	10	20	1,120
Lesotho	57	60	21	6	52	19	38	..
Libya	66	70	5	..	90	12	37	3,129
Lithuania	65	76	..	..	..	..	..	2,291
Macedonia, FYR	70	74	..	..	..	..	..	1,308
Madagascar	57	60	32	3	29	..	..	36
Malawi	43	43	28	53	45	28	58	38
Malaysia	70	74	23	91	88	11	22	1,655
Mali	48	52	31	31	37	61	77	21
Mauritania	52	55	48	..	..	50	74	102
Mauritius	68	75	15	100	98	13	21	388
Mexico	69	75	14	66	83	8	13	1,456
Moldova	64	71	..	50	..	..	..	963
Mongolia	64	67	12	..	..	..	..	1,045
Morocco	64	68	10	40	52	43	69	311
Mozambique	44	46	47	21	32	42	77	38
Myanmar	58	61	31	41	38	11	22	50
Namibia	55	57	26	34	..	..	..	..
Nepal	57	57	49	20	48	59	86	33
Netherlands	75	80	..	100	100	..	..	4,741
New Zealand	73	79	..	..	..	..	..	4,290
Nicaragua	65	70	24	31	61	35	33	265
Niger	44	49	43	15	53	79	93	37
Nigeria	51	55	35	36	39	33	53	165
Norway	75	81	..	100	..	..	..	5,439
Oman	69	73	14	79	..	..	..	1,880
Pakistan	62	65	40	30	60	50	76	243
Panama	72	76	7	87	83	9	10	678
Papua New Guinea	57	58	30	22	28	19	37	232
Paraguay	68	74	4	30	..	7	9	308
Peru	66	71	11	44	60	6	17	421
Philippines	64	68	30	..	..	5	6	307
Poland	68	77	..	100	..	..	..	2,448
Portugal	72	79	..	100	..	..	..	1,939
Puerto Rico	71	80	..	..	..	..	..	1,993
Romania	65	73	6	49	..	..	..	1,941
Russian Federation	60	73	3	..	..	..	..	4,079

	Life expectancy at birth		Prevalence of child malnutrition	Sanitation	Safe water	Adult illiteracy rate		Commercial energy use
	Male years 1996	Female years 1996	% of children under 5 1990–96	% of population with access 1995	% of population with access 1995	% of people 15 and above Male 1995	% of people 15 and above Female 1995	kg of oil equivalent per capita 1995
Rwanda	39	42	29	..	..	30	48	33
Saudi Arabia	69	71	..	86	93	29	50	4,360
Senegal	49	52	22	58	50	57	77	104
Sierra Leone	35	38	29	11	34	55	82	72
Singapore	74	79	14	97	100	4	14	7,162
Slovak Republic	69	77	..	51	..	..	..	3,272
Slovenia	71	78	..	90	..	..	..	2,806
South Africa	62	68	9	46	70	18	18	2,405
Spain	73	81	..	100	99	..	..	2,639
Sri Lanka	71	75	38	..	..	7	13	136
Sudan	53	56	34	22	50	42	65	65
Sweden	76	82	..	100	..	..	..	5,736
Switzerland	75	82	..	100	100	..	..	3,571
Syrian Arab Republic	66	71	..	78	85	14	44	1,001
Tajikistan	66	72	..	62	..	..	..	563
Tanzania	49	52	29	86	49	21	43	32
Thailand	67	72	13	70	81	4	8	878
Togo	49	52	25	22	..	33	63	45
Trinidad and Tobago	70	75	7	56	82	1	3	5,381
Tunisia	69	71	9	..	..	21	45	591
Turkey	66	71	10	94	92	8	28	1,009
Turkmenistan	62	69	..	60	85	..	..	3,047
Uganda	43	43	26	57	34	26	50	22
Ukraine	62	73	..	49	97	..	..	3,136
United Arab Emirates	74	76	7	95	98	21	20	11,567
United Kingdom	74	80	..	96	100	..	..	3,786
United States	74	80	..	85	90	..	..	7,905
Uruguay	70	77	4	82	83	3	2	639
Uzbekistan	66	72	4	18	..	..	..	2,043
Venezuela	70	76	5	58	79	8	10	2,158
Vietnam	66	70	45	21	36	4	9	104
West Bank and Gaza	..	..	..	..	..	..	..	..
Yemen, Rep.	54	54	30	51	52	..	..	192
Yugoslavia, FR (Serb./Mont.)	70	75	..	100	..	..	..	1,125
Zambia	44	45	29	23	43	14	29	145
Zimbabwe	55	57	16	58	74	10	20	424
World	**65 w**	**69 w**	..	**47 w**	**78 w**	**21 w**	**38 w**	**1,474 w**
Low income	62	64	..	28	76	24	45	393
Excl. China & India	55	58	..	36	51	36	55	132
Middle income	65	71	..	60	..	14	22	1,488
Lower middle income	64	70	..	58	..	14	25	1,426
Upper middle income	66	73	..	64	76	13	16	1,633
Low & middle income	63	67	..	37	76	21	39	762
East Asia & Pacific	67	70	..	29	84	9	24	657
Europe & Central Asia	64	73	..	..	..	..	..	2,690
Latin America & Carib.	66	73	..	57	73	12	15	969
Middle East & N. Africa	66	68	..	..	..	28	50	1,178
South Asia	61	63	..	30	78	38	64	231
Sub-Saharan Africa	51	54	..	37	45	34	53	238
High income	74	81	..	92	..	..[a]	..[a]	5,123

a. UNESCO estimates illiteracy to be less than 5 percent.

Quality of life | 1.2

About the data

The indicators in this table provide an overview of the conditions in which more than 5 billion of the world's people live. Although not perfectly correlated with income or consumption per capita, they tend to tell a common story: on average, the residents of poor countries enjoy fewer amenities, lack basic skills, and suffer higher rates of illness and, consequently, live shorter lives. These indicators complement those in table 1.3, which measure progress toward international goals for social and economic development.

Except for the adult illiteracy rate, all of the indicators shown here appear elsewhere in the *World Development Indicators*. For more information about them, see *About the data* for tables 2.14 (access to safe water and sanitation), 2.16 (child malnutrition), 2.17 (life expectancy), and 3.7 (commercial energy use).

Literacy is difficult to define and to measure. The definition here is based on the concept of functional literacy—the ability to use reading and writing skills effectively in the context of the society. To measure literacy using such a definition requires census or sample survey measurements under controlled conditions. In practice, many countries estimate the number of illiterate adults from self-reported data or from estimates of school completion. Because of these problems, comparisons across countries—and even over time within countries—should be made with caution.

Definitions

• **Life expectancy at birth** is the number of years a newborn infant would live if prevailing patterns of mortality at the time of its birth were to stay the same throughout its life. • **Prevalence of child malnutrition** is the percentage of children under 5 whose weight by age is less than minus two standard deviations from the median of the reference population. • **Access to sanitation** is the percentage of the population with excreta disposal facilities that can effectively prevent human, animal, and insect contact with excreta. Suitable facilities range from simple but protected pit latrines to flush toilets with sewerage. To be effective, all facilities must be correctly constructed and properly maintained. • **Access to safe water** is the percentage of the population with reasonable access to an adequate amount of safe water (including treated surface water and untreated but uncontaminated water, such as from springs, sanitary wells, and protected boreholes). In urban areas the source may be a public fountain or standpipe located not more than 200 meters away. In rural areas the definition implies that members of the household do not have to spend a disproportionate part of the day fetching water. An adequate amount of safe water is that needed to satisfy metabolic, hygienic, and domestic requirements–usually about 20 liters a person a day. The definition of safe water has changed over time. • **Adult illiteracy** rate is the percentage of adults aged 15 and above who cannot, with understanding, read and write a short, simple statement about their everyday life. • **Commercial energy use** is measured by indigenous energy production (from all commercial sources) plus imports and stock changes less exports and international marine bunkers, stated in kilograms of oil equivalents per capita.

Figure 1.2a

Women tend to live longer than men

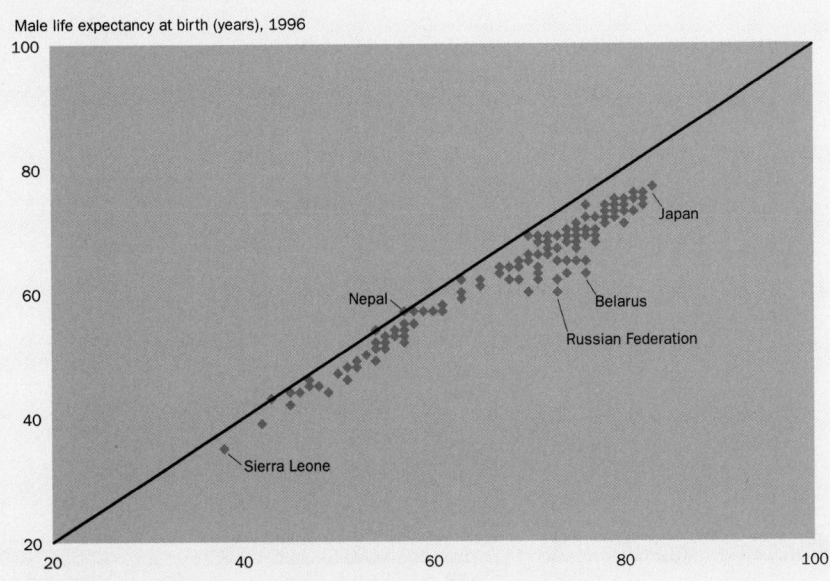

Source: World Bank staff estimates.

This figure plots male against female life expectancy for 148 economies. Most observations lie below the 45 degree line, reflecting women's longer life expectancies. Observations near the line may indicate that women in these countries lack access to adequate health care or receive less than an equitable share of other resources. In the Russian Federation and other states of the former Soviet Union, however, the large gap between men and women is the result of a recent drop in male life expectancy.

Data sources

The indicators here and throughout the rest of the book have been compiled by World Bank staff from primary and secondary sources. For most of the indicators shown in the tables in this section, the sources are cited in the notes to the tables referred to in *About the data*. Data on illiteracy are supplied by the United Nations Educational, Scientific, and Cultural Organization and published in its *Statistical Yearbook* (see *Data sources* for table 2.9).

1.3 | Development progress

	Private consumption per capita		Net primary enrollment ratio				Infant mortality rate		Under-5 mortality rate		Maternal mortality ratio	Health care
	annual average growth 1980–96	distribution	% of relevant age group Male		% of relevant age group Female		per 1,000 live births		per 1,000		per 100,000 live births	% of population with access
	uncorrected	corrected	1980	1995	1980	1995	1970	1996	1970	1996	1990–96	1993
Albania	..	..	..	95	..	97	66	37	..	40	28	..
Algeria	–1.9	–1.2	91	99	71	91	139	32	192	39	140	..
Angola	–7.4	..	..	..	..	..	178	124	..	209	1,500	24
Argentina	..	..	..	..	..	..	52	22	71	25	100	..
Armenia	–5.4	..	..	..	..	..	..	16	..	20	21	..
Australia	1.6	1.1	100	98	100	98	18	6	..	7	9	100
Austria	2.0	1.5	99	100	98	100	26	5	..	6	10	100
Azerbaijan	..	..	..	..	..	..	..	20	..	23	44	..
Bangladesh	0.0	0.0	..	66	..	58	140	77	237	112	850	74
Belarus	–4.5	–3.5	..	97	..	94	..	13	..	17	22	100
Belgium	1.7	1.3	97	98	98	98	21	7	..	7	10	100
Benin	–0.8	..	..	74	..	43	148	87	256	140	500	42
Bolivia	–0.7	–0.4	84	95	74	87	153	67	243	102	370	..
Bosnia and Herzegovina	..	..	..	..	..	..	59	19	..	..	..	..
Botswana	5.9	..	69	94	82	99	95	56	146	85	250	86
Brazil	0.0	0.0	..	..	..	..	95	36	135	42	160	..
Bulgaria	–0.7	–0.5	96	98	96	96	27	16	..	20	20	100
Burkina Faso	0.0	..	18	37	11	24	141	98	278	158	930	..
Burundi	–0.8	..	23	56	16	48	138	97	233	176	1,300	80
Cambodia	..	..	..	..	..	..	161	105	..	170	900	..
Cameroon	–2.5	..	..	..	..	..	126	54	215	102	550	15
Canada	1.3	0.9	..	96	..	94	19	6	..	7	6	99
Central African Republic	–2.4	..	73	65	41	43	139	96	238	164	700	13
Chad	–0.4	..	..	..	..	..	171	115	..	189	900	26
Chile	3.2	1.4	..	87	..	85	77	12	97	13	180	95
China	7.7	4.5	..	99	..	98	69	33	115	39	115	..
Hong Kong, China	5.3	..	95	90	96	92	19	4	23	6	7	..
Colombia	1.3	0.6	..	..	..	..	74	25	113	31	100	87
Congo, Dem. Rep.	–4.2	..	..	71	..	50	131	90	245	144	..	59
Congo, Rep.	–0.4	..	99	..	93	..	101	90	..	145	890	..
Costa Rica	0.7	0.4	89	86	90	87	62	12	85	15	55	97
Côte d'Ivoire	–2.6	–1.6	..	..	..	..	135	84	237	150	600	60
Croatia	..	..	..	83	..	82	..	9	..	10	12	..
Cuba	..	..	95	99	95	99	39	8	43	10	36	100
Czech Republic	..	..	..	98	..	98	21	6	..	7	7	..
Denmark	1.6	1.2	96	98	95	99	14	6	..	6	9	100
Dominican Republic	0.6	0.3	..	79	..	83	98	40	127	47	110	..
Ecuador	–0.2	–0.1	..	91	..	92	100	34	140	40	150	80
Egypt, Arab Rep.	2.0	1.3	..	95	..	82	158	53	235	66	170	99
El Salvador	2.8	1.4	..	78	..	80	103	34	161	40	300	..
Eritrea	..	..	..	33	..	30	..	64	..	120	1,400	..
Estonia	7.8	4.7	..	93	..	94	20	10	..	16	52	..
Ethiopia	–1.7	..	..	28	..	19	158	109	239	177	1,400	55
Finland	1.4	1.1	..	99	..	99	13	4	..	5	11	100
France	1.7	1.1	..	99	..	99	18	5	..	6	15	..
Gabon	–4.9	..	..	..	..	..	138	87	..	145	500	87
Gambia, The	0.5	..	66	64	34	46	185	79	..	107	1,100	..
Georgia	..	..	..	81	..	82	..	17	..	19	19	..
Germany	..	..	..	100	..	100	23	5	..	6	22	..
Ghana	0.1	0.1	..	..	..	..	111	71	187	110	740	25
Greece	1.9	..	103	98	103	98	30	8	..	9	10	..
Guatemala	–0.4	–0.1	..	..	..	..	106	41	168	56	190	60
Guinea	0.9	0.5	..	..	..	..	181	122	..	210	880	45
Guinea-Bissau	–1.0	–0.4	63	..	31	..	185	134	..	223	910	..
Haiti	–0.8	..	38	25	37	26	141	72	221	130	600	45
Honduras	–0.3	–0.1	78	89	78	91	110	44	170	50	220	62

Development progress 1.3

	Private consumption per capita		Net primary enrollment ratio				Infant mortality rate		Under-5 mortality rate		Maternal mortality ratio	Health care
	annual average growth 1980–96		% of relevant age group Male		% of relevant age group Female		per 1,000 live births		per 1,000		per 100,000 live births	% of population with access
		distribution										
	uncorrected	corrected	1980	1995	1980	1995	1970	1996	1970	1996	1990–96	1993
Hungary	1.4	1.0	94	92	95	94	36	11	..	13	14	..
India	2.3	1.6	..	..	..	..	137	65	202	85	437	..
Indonesia	4.3	2.8	93	99	83	95	118	49	172	60	390	43
Iran, Islamic Rep.	0.0	..	..	..	..	..	131	36	191	37	120	73
Iraq	..	..	100	83	94	74	102	101	123	136	310	98
Ireland	2.8	1.8	100	100	100	100	20	5	..	7	10	..
Israel	3.3	2.1	..	..	..	..	25	6	28	9	7	100
Italy	2.2	1.5	..	..	..	..	30	6	..	7	12	..
Jamaica	3.8	2.2	95	100	97	100	43	12	64	14	120	..
Japan	2.9	..	100	100	100	100	13	4	..	6	8	100
Jordan	–1.2	–0.7	94	89	91	89	..	30	107	35	150	90
Kazakhstan	..	..	..	..	..	..	..	25	..	30	53	..
Kenya	0.9	0.4	92	..	89	..	102	57	156	90	650	..
Korea, Dem. Rep.	..	..	..	..	..	..	51	56	..	45	..	100
Korea, Rep.	7.1	..	100	98	100	99	46	9	55	11	30	100
Kuwait	–5.5	..	89	65	80	65	48	11	54	14	18	100
Kyrgyz Republic	..	..	..	99	..	95	..	26	..	36	32	..
Lao PDR	..	..	..	75	..	61	146	101	..	140	650	..
Latvia	..	..	..	86	..	82	21	16	..	18	15	..
Lebanon	..	..	..	..	..	..	50	31	..	36	300	..
Lesotho	–2.8	–1.2	54	60	78	71	134	74	190	113	610	80
Libya	..	..	100	98	100	96	122	25	..	30	220	100
Lithuania	..	..	..	..	..	..	24	10	..	13	13	..
Macedonia, FYR	..	..	..	86	..	84	..	16	..	18	22	..
Madagascar	–2.7	–0.2	..	..	..	..	181	88	..	135	660	65
Malawi	–0.6	..	48	100	38	100	193	133	347	217	620	80
Malaysia	3.3	1.7	..	91	..	92	45	11	..	14	34	88
Mali	–1.1	..	..	30	..	19	204	120	..	220	580	..
Mauritania	–0.4	–0.2	..	64	..	55	148	94	..	155	800	..
Mauritius	5.4	..	80	96	79	96	60	17	83	20	112	99
Mexico	–0.3	–0.1	..	..	..	..	72	32	111	36	110	91
Moldova	..	..	..	..	..	..	..	20	..	24	33	..
Mongolia	..	..	..	78	..	81	102	53	..	71	65	100
Morocco	1.7	1.0	75	81	47	62	128	53	187	67	372	62
Mozambique	–1.7	..	39	45	33	35	171	123	281	214	1,500	30
Myanmar	..	..	..	..	..	..	128	80	179	109	580	..
Namibia	–0.6	..	..	..	..	..	118	61	..	92	220	..
Nepal	5.2	3.3	..	..	..	..	166	85	232	116	1,500	..
Netherlands	1.5	1.1	91	99	94	99	13	5	..	6	12	100
New Zealand	0.9	..	100	100	100	100	17	6	..	7	25	100
Nicaragua	–2.7	–1.3	96	82	100	85	106	44	165	57	160	..
Niger	–6.3	–4.0	..	32	..	18	170	118	..	..	593	30
Nigeria	–3.0	–1.7	..	..	..	..	139	78	..	130	1,000	67
Norway	1.5	1.1	98	99	98	99	13	4	..	6	6	100
Oman	..	..	54	72	31	70	119	18	..	20	..	89
Pakistan	1.5	1.1	..	..	..	..	142	88	183	123	340	85
Panama	1.9	0.8	88	91	89	92	47	22	68	25	55	82
Papua New Guinea	–0.4	–0.2	..	..	..	..	112	62	..	85	930	96
Paraguay	2.0	0.8	90	89	88	89	55	24	76	45	190	..
Peru	–0.9	–0.5	..	91	..	90	108	42	178	58	280	..
Philippines	0.8	0.4	95	..	92	..	66	37	82	44	208	..
Poland	0.6	0.4	98	97	98	96	33	12	..	15	10	100
Portugal	2.9	..	97	100	100	100	56	7	..	8	15	..
Puerto Rico	2.1	..	..	..	..	..	29	12	34	14	..	..
Romania	0.0	0.0	..	92	..	92	49	22	..	28	41	..
Russian Federation	..	..	..	100	..	100	..	17	..	25	53	..

1.3 Development progress

	Private consumption per capita		Net primary enrollment ratio				Infant mortality rate		Under-5 mortality rate		Maternal mortality ratio	Health care
	annual average growth 1980–96	distribution	% of relevant age group Male		% of relevant age group Female		per 1,000 live births		per 1,000		per 100,000 live births	% of population with access
	uncorrected	corrected	1980	1995	1980	1995	1970	1996	1970	1996	1990–96	1993
Rwanda	–1.8	–1.3	62	*76*	57	*76*	142	129	209	205	1,300	..
Saudi Arabia	..	..	60	63	37	61	119	22	..	28	18	*98*
Senegal	–1.0	–0.5	44	60	30	48	135	60	279	88	510	*40*
Sierra Leone	–2.4	–0.9	..	..	..	..	197	174	360	284	1,800	..
Singapore	4.9	..	100	..	99	..	20	4	25	5	10	*100*
Slovak Republic	–3.2	–2.5	..	..	..	..	25	11	..	13	8	..
Slovenia	..	..	..	100	..	99	24	5	..	6	5	..
South Africa	–0.1	0.0	..	*95*	..	*96*	79	49	..	66	230	..
Spain	2.3	1.6	100	*100*	100	*100*	28	5	..	6	7	..
Sri Lanka	2.6	1.8	..	..	..	..	53	15	100	19	30	*90*
Sudan	–1.9	..	..	..	..	..	118	74	176	116	370	*70*
Sweden	0.7	0.5	..	*100*	..	*100*	11	4	..	5	7	*100*
Switzerland	0.6	0.4	..	*100*	..	*100*	15	5	..	6	6	*100*
Syrian Arab Republic	0.4	..	99	95	80	87	96	31	128	36	179	*99*
Tajikistan	..	..	..	..	..	..	..	32	..	38	74	..
Tanzania	..	..	..	47	..	48	129	86	218	144	530	*93*
Thailand	5.6	3.0	..	..	..	..	73	34	102	38	200	*59*
Togo	–0.9	..	..	98	..	72	134	87	..	138	640	..
Trinidad and Tobago	–1.2	..	89	83	91	94	52	13	55	15	90	*99*
Tunisia	0.8	0.5	92	98	72	95	121	30	201	35	..	*90*
Turkey	–1.3	..	..	*98*	..	*94*	144	42	201	47	180	*100*
Turkmenistan	..	..	..	..	..	..	..	41	..	50	44	..
Uganda	1.7	1.0	..	..	..	..	109	99	..	141	550	*71*
Ukraine	..	..	..	..	..	..	22	14	..	17	30	*100*
United Arab Emirates	–0.5	..	72	*84*	75	*82*	87	15	83	17	..	*90*
United Kingdom	2.6	1.7	100	*100*	100	*100*	19	6	..	7	9	..
United States	1.8	1.1	95	*96*	96	*97*	20	7	..	8	12	..
Uruguay	3.1	..	..	95	..	95	46	18	56	22	85	..
Uzbekistan	..	..	..	..	..	..	..	24	..	35	24	..
Venezuela	–0.7	–0.4	..	*87*	..	90	53	22	62	28	200	..
Vietnam	..	..	..	..	..	..	104	40	..	48	105	..
West Bank and Gaza	..	..	..	..	..	..	..	..	..	..	..	..
Yemen, Rep.	..	..	..	..	..	..	186	98	..	130	1,400	..
Yugoslavia, FR (Serb./Mont.)	..	..	..	69	..	70	54	14	..	19	12	..
Zambia	–4.0	–2.1	81	*78*	73	*76*	106	112	181	202	230	*75*
Zimbabwe	0.6	..	100	..	100	..	96	56	137	86	280	..
World	**2.9 w**	**2.0 w**	**.. w**	**.. w**	**.. w**	**.. w**	**98 w**	**54 w**	**..**	**73 w**	**..**	**..**
Low income	3.6	2.6	..	..	..	..	113	68	..	94	..	..
Excl. China & India	–0.8	..	..	..	..	..	139	88	..	131	..	..
Middle income	1.2	..	..	..	..	..	94	37	..	45	..	..
Lower middle income	1.6	..	..	..	..	..	102	40	..	49	..	..
Upper middle income	0.3	0.1	..	..	..	..	76	30	..	36	..	..
Low & middle income	2.9	2.1	..	..	..	..	107	59	..	80	..	..
East Asia & Pacific	6.8	4.0	..	99	..	98	79	39	..	47	..	..
Europe & Central Asia	..	..	..	97	..	96	..	24	..	30	..	..
Latin America & Carib.	0.1	0.0	..	..	..	..	84	33	..	41	..	..
Middle East & N. Africa	0.6	..	..	..	..	..	134	50	..	63	..	..
South Asia	2.1	1.5	..	..	..	..	139	73	..	93	..	..
Sub-Saharan Africa	–1.8	..	..	..	..	..	137	91	..	147	..	..
High income	2.4	..	98	*98*	98	*98*	22	6	..	7	..	..

About the data

The indicators in this table are intended to measure progress toward the development goals for the 21st century proposed by the OECD's Development Assistance Committee and discussed in the introduction to this section. The net enrollment ratio, infant and child mortality rates, and the maternal mortality rate are included in the set of monitoring indicators identified in Strategy 21. For further discussion of the monitoring indicators, see the introduction to section 2 and *About the data* for the tables in which the indicators appear.

Estimates of the number of people living in poverty appear in table 2.7. The growth of private consumption per capita is included here as an indicator of the effect of economic development has on the welfare of individuals. Positive growth rates are generally associated with a reduction in poverty, but where the distribution of income or consumption is highly unequal, the poor may not share in the improvement. The relationship between the rate of poverty reduction and the distribution of income or consumption, as measured by an index such as the Gini index, is complicated. But Ravallion (1997) has found that the rate of poverty reduction is directly proportional to the "distribution-corrected rate of growth" of private consumption. The distribution-corrected rate of growth is calculated as $(1-G)r$, where G is the Gini index (0 = perfect equality, 1 = perfect inequality) and r is the rate of growth in mean private consumption. In empirical tests covering 23 developing countries, Ravallion estimated that factor of proportionality to be 4.4, implying a growth elasticity of poverty reduction of between 3.3 for a low Gini index of 0.25 and 1.8 for a high Gini index of 0.60.

Definitions

• **Growth of private consumption per capita** is the average annual rate of change in private consumption divided by the midyear population. See the definition of private consumption in table 4.9. • **Distribution-corrected growth of private consumption per capita** is 1 minus the Gini index multiplied by the annual rate of growth in private consumption. • **Net enrollment ratio** is the ratio of the number of children of official school age enrolled in school to the number of children of official school age in the population. • **Infant mortality rate** is the number of deaths of infants under one year of age during the indicated year per 1,000 live births in the same year. • **Under-5 mortality rate** is the probability of a child born in the indicated year dying before reaching the age of 5, if subject to current age-specific mortality rates. The probability is expressed as a rate per 1000. • **Maternal mortality ratio** is the number of women who die during pregnancy and childbirth, per 100,000 live births.

Data sources

The indicators here and throughout the rest of the book have been compiled by World Bank staff from primary and secondary sources. More information about the indicators and their sources can be found in the *About the data, Definitions,* and *Data sources* entries that accompany each table in subsequent sections.

1.4 Trends in long-term economic development

| | Gross national product | | Population | | Value added | | | Private consumption | Gross domestic fixed investment | Exports of goods and services |
| | average annual % growth | | average annual % growth | | average annual % growth Agriculture | average annual % growth Industry | average annual % growth Services | average annual % growth | average annual % growth | average annual % growth |
	Total 1965–96	per capita 1965–96	Total 1965–96	Labor force 1965–96	1965–96	1965–96	1965–96	1965–96	1965–96	1965–96
Albania	..	..	1.8	2.2	2.9	−6.1	−1.5	..	..	..
Algeria	3.9	0.9	2.7	3.5	5.0	2.6	5.0	5.5	2.7	2.2
Angola	..	..	2.4	1.9	..	..	..	..	..	..
Argentina	1.2	−0.3	1.4	1.4	1.3	1.0	2.4	..	..	4.8
Armenia	2.0	0.3	1.7	2.3	..	..	..	..	..	..
Australia	3.2	1.6	1.5	2.1	1.7	2.2	3.5	3.4	2.5	5.6
Austria	2.9	2.7	0.3	0.4	0.8	2.0	2.6	3.0	2.9	6.3
Azerbaijan	..	..	1.6	2.1	..	..	..	..	..	..
Bangladesh	3.5	1.0	2.3	3.7	1.9	4.1	5.2	2.9	4.4	7.2
Belarus	..	..	0.6	0.7	..	..	..	..	..	..
Belgium	2.5	2.3	0.2	0.5	2.7	..	..	2.6	1.7	5.2
Benin	3.1	0.1	2.8	2.2	3.7	3.1	3.1	2.6	..	3.7
Bolivia	1.8	−0.5	2.2	2.3	2.6	0.0	3.0	2.6	−3.1	0.8
Bosnia and Herzegovina	..	..	0.7	1.1	..	..	..	..	..	..
Botswana	13.0	9.2	3.1	2.8	3.6	14.9	10.8	..	..	..
Brazil	4.6	2.4	2.0	3.1	3.5	4.6	5.4	4.6	1.7	8.6
Bulgaria	−0.3	0.1	0.1	−0.1	−2.1	−0.4	1.5	−1.1	−4.9	−12.1
Burkina Faso	3.9	1.5	2.3	1.8	2.6	2.4	6.3	3.0	5.5	3.5
Burundi	4.0	1.6	2.2	2.0	2.9	5.2	4.0	3.3	2.6	2.6
Cambodia	..	..	1.6	1.5	..	..	..	..	..	..
Cameroon	4.3	1.4	2.7	2.2	3.2	7.2	3.8	3.4	1.0	7.1
Canada	3.3	2.0	1.3	2.3	1.2	2.4	4.1	3.5	4.3	6.0
Central African Republic	1.5	−0.8	2.2	1.5	1.7	2.3	1.0	1.7	2.1	1.0
Chad	1.6	−0.6	2.1	2.0	1.3	1.4	2.9	..	..	1.4
Chile	3.3	1.6	1.6	2.2	3.7	2.9	4.3	2.8	3.8	8.0
China	8.5	6.7	1.7	2.0	4.3	11.0	11.1	7.6	10.5	11.1
Hong Kong, China	7.5[a]	5.6[a]	1.7	2.7	..	..	..	7.6	8.2	11.3
Colombia	4.4	2.1	2.2	3.2	3.4	4.5	4.9	4.1	4.8	5.7
Congo, Dem. Rep.	−0.5	−3.5	3.0	2.5	1.9	−2.3	−1.4	0.2	0.1	2.9
Congo, Rep.	4.9	1.9	2.8	2.5	2.6	7.4	5.0	3.8	..	6.5
Costa Rica	4.0	1.2	2.6	3.4	3.2	4.8	4.1	3.2	4.8	6.9
Côte d'Ivoire	3.9	0.0	3.6	3.0	2.2	6.3	3.3	2.8	−0.4	5.3
Croatia	..	..	0.3	0.3	..	..	..	..	..	..
Cuba	..	..	1.1	2.1	..	..	..	..	..	..
Czech Republic	..	..	0.2	0.4	..	..	..	..	..	..
Denmark	2.1	1.8	0.3	0.8	2.3	1.9	2.3	1.7	0.4	4.5
Dominican Republic	4.7	2.3	2.3	3.2	3.1	5.8	5.2	4.7	6.5	5.9
Ecuador	4.9	2.2	2.6	3.0	3.5	6.5	4.9	4.4	3.2	7.4
Egypt, Arab Rep.	6.4	4.0	2.2	3.0	2.8	6.8	9.2	5.3	6.2	5.5
El Salvador	1.3	−0.6	2.1	2.6	0.7	0.5	2.0	1.7	2.3	0.7
Eritrea	..	..	..	..	..	..	..	..	..	..
Estonia	..	..	0.4	0.5	..	..	..	..	..	..
Ethiopia	..	..	2.6	2.4	..	..	..	..	..	..
Finland	2.8	2.4	0.4	0.6	0.3	3.0	3.3	2.8	1.1	4.8
France	2.7	2.1	0.6	0.7	1.9	0.9	2.7	2.9	2.0	5.6
Gabon	3.0	0.0	2.6	2.0	0.3	1.8	0.2	2.8	−3.1	5.7
Gambia, The	3.9	0.4	3.3	2.9	2.2	4.2	4.2	1.2	..	3.6
Georgia	..	..	0.6	0.8	..	..	..	..	..	..
Germany	..	..	0.2	0.4	..	..	..	..	..	..
Ghana	1.7	−0.9	2.5	2.5	1.2	0.4	3.6	1.3	0.6	−1.5
Greece	3.2	2.5	0.6	0.8	1.3	3.3	3.8	3.5	1.4	7.8
Guatemala	3.2	0.4	2.7	3.1	..	..	..	3.3	2.2	2.1
Guinea	..	..	2.0	1.6	..	..	..	..	..	..
Guinea-Bissau	2.8	0.0	2.3	1.9	1.2	2.5	9.2	0.8	..	2.2
Haiti	0.9	−0.9	1.8	1.1	..	..	..	1.7	..	3.6
Honduras	3.8	0.5	3.1	3.6	2.5	4.5	4.3	3.8	3.8	2.7

Trends in long-term economic development | 1.4

	Gross national product		Population		Value added			Private consumption	Gross domestic fixed investment	Exports of goods and services
	average annual % growth		average annual % growth		average annual % growth Agriculture	average annual % growth Industry	average annual % growth Services	average annual % growth	average annual % growth	average annual % growth
	Total 1965–96	per capita 1965–96	Total 1965–96	Labor force 1965–96	1965–96	1965–96	1965–96	1965–96	1965–96	1965–96
Hungary	0.9	1.1	0.0	–0.3	..	..	..	1.8	2.6	3.9
India	4.5	2.3	2.1	2.1	2.8	5.4	5.5	4.0	5.4	6.1
Indonesia	6.7	4.6	2.0	2.5	3.9	9.1	7.5	7.1	8.9	5.6
Iran, Islamic Rep.	1.1	–2.0	2.9	3.0	4.6	–0.2	0.3	3.6	–1.4	–2.2
Iraq	–0.3	–3.5	3.1	3.1	..	..	..	..	..	..
Ireland	3.6	2.7	0.7	0.8	..	..	..	3.1	2.7	8.4
Israel	3.8	1.3	2.5	2.9	..	..	..	5.5	3.1	7.2
Italy	2.8	2.6	0.3	0.6	0.9	..	..	3.4	1.6	5.5
Jamaica	0.7	–0.5	1.2	2.0	0.9	–0.2	2.2	2.0	–2.7	1.7
Japan	4.5	3.6	0.8	1.0	–0.1	4.6	4.8	4.2	4.7	7.7
Jordan	4.0	–0.3	4.2	4.3	7.0	5.9	4.3	5.2	5.4	8.6
Kazakhstan	..	..	1.0	1.6	..	..	..	..	..	..
Kenya	5.0	1.5	3.2	3.1	3.6	5.8	5.7	4.2	1.0	3.0
Korea, Dem. Rep.	..	..	1.9	2.4	..	..	..	..	..	..
Korea, Rep.	8.9	7.3	1.5	2.5	2.0	13.8	9.0	7.5	12.1	16.1
Kuwait	0.7	–3.4	3.8	4.7	9.8	–4.2	1.2	7.8	8.5	–3.0
Kyrgyz Republic	..	..	1.8	2.1	..	..	..	..	..	..
Lao PDR	..	..	2.1	1.8	..	..	..	..	..	..
Latvia	2.1	1.1	0.3	0.4	–3.6	–5.6	0.3	..	..	..
Lebanon	..	..	1.8	2.4	..	..	..	..	..	..
Lesotho	5.9	3.3	2.3	2.0	–1.7	13.2	6.9	3.8	11.2	5.5
Libya	1.2	–2.9	3.6	3.6	..	..	..	..	..	..
Lithuania	..	..	0.7	0.8	..	..	..	..	..	..
Macedonia, FYR	..	..	..	..	..	..	..	..	..	..
Madagascar	0.7	–2.0	2.6	2.4	1.5	–0.3	0.2	0.1	..	–0.7
Malawi	3.5	0.4	2.9	2.6	2.5	3.8	4.6	2.7	–2.7	3.6
Malaysia	6.8	4.1	2.4	2.9	3.7	8.5	7.0	6.1	9.9	9.5
Mali	2.9	0.5	2.3	2.0	3.2	2.9	2.2	2.6	6.2	7.1
Mauritania	1.9	–0.6	2.4	2.0	0.9	2.1	2.4	3.4	1.2	2.8
Mauritius	5.2	3.9	1.3	2.4	–0.3	7.6	7.0	5.0	4.1	5.6
Mexico	4.1	1.5	2.4	3.3	2.3	4.6	4.2	3.7	3.8	7.9
Moldova	..	..	0.8	0.8	..	..	..	..	..	..
Mongolia	..	..	2.6	2.7	..	..	..	..	..	..
Morocco	4.4	2.1	2.2	3.3	2.8	4.1	5.7	4.0	4.6	5.0
Mozambique	..	..	2.4	2.0	..	..	..	0.4	..	..
Myanmar	..	..	2.0	2.0	..	..	..	..	..	..
Namibia	3.3	0.5	2.5	2.1	1.0	1.2	1.4	2.1	1.6	2.3
Nepal	3.6	1.0	2.4	2.0	2.2	..	..	..	..	..
Netherlands	2.6	1.9	0.7	1.5	4.1	1.3	2.5	2.7	1.3	5.0
New Zealand	1.7	0.7	1.0	1.8	3.5	1.0	1.8	1.5	2.3	4.0
Nicaragua	–1.3	–4.2	2.9	3.4	0.2	0.1	–0.7	–1.0	0.2	–0.4
Niger	0.2	–2.8	2.9	2.7	–0.1	5.0	–0.4	0.3	..	–0.6
Nigeria	3.1	0.1	2.8	2.6	1.4	4.3	6.0	3.4	–2.5	2.2
Norway	3.5	3.0	0.5	1.2	1.2	3.6	2.5	2.8	1.9	5.3
Oman	9.5	5.0	3.9	3.6	..	..	..	..	..	..
Pakistan	5.9	2.7	2.9	3.4	4.0	6.8	6.3	5.2	4.5	6.4
Panama	3.1	0.8	2.2	2.8	2.2	2.1	2.2	3.9	2.8	2.6
Papua New Guinea	2.9	0.6	2.2	2.0	2.7	6.3	2.4	3.0	1.3	7.4
Paraguay	5.1	2.1	2.8	3.3	4.5	5.5	5.8	5.9	4.9	8.6
Peru	2.1	–0.4	2.3	2.9	..	..	..	1.9	2.0	1.7
Philippines	3.5	0.9	2.5	2.8	2.4	3.6	3.9	3.7	4.5	6.3
Poland	1.2	0.7	0.6	0.7	..	..	..	1.1	1.5	6.0
Portugal	3.5	3.1	0.3	1.1	..	..	..	3.4	2.9	5.4
Puerto Rico	2.5	1.2	1.2	2.0	1.7	4.3	3.2	2.8	1.4	4.4
Romania	0.3	0.0	0.5	0.0	..	..	..	..	..	..
Russian Federation	3.3	1.7	0.5	0.8	..	..	..	..	..	..

	Gross national product average annual % growth		Population average annual % growth		Value added			Private consumption average annual % growth	Gross domestic fixed investment average annual % growth	Exports of goods and services average annual % growth
	Total 1965–96	per capita 1965–96	Total 1965–96	Labor force 1965–96	average annual % growth Agriculture 1965–96	average annual % growth Industry 1965–96	average annual % growth Services 1965–96	average annual % growth 1965–96	average annual % growth 1965–96	average annual % growth 1965–96
Rwanda	2.9	0.1	2.3	2.7	2.7	2.1	4.4	3.2	6.9	3.5
Saudi Arabia	2.0	–3.0	4.4	4.9	7.6	3.4	4.9	..	..	..
Senegal	2.3	–0.5	2.7	2.4	1.3	3.9	2.4	2.2	3.0	3.2
Sierra Leone	0.7	–1.4	2.0	1.6	3.2	–0.7	–0.2	–0.2	–7.5	–5.0
Singapore	8.3	6.3	1.8	3.0	–1.4	8.6	8.3	6.6	9.6	12.2
Slovak Republic	..	..	0.6	1.3	..	..	..	..	..	..
Slovenia	..	..	0.6	0.8	..	..	..	..	..	..
South Africa	2.4	0.2	2.1	2.2	2.2	1.7	3.1	3.2	1.6	1.4
Spain	3.0	2.4	0.6	1.1	..	..	..	2.9	2.7	7.2
Sri Lanka	4.7	3.1	1.5	2.1	2.8	4.8	5.3	4.0	7.9	..
Sudan	2.4	–0.4	2.5	2.7	2.4	2.4	3.1	4.2	..	–3.0
Sweden	1.8	1.4	0.4	0.9	0.7	1.4	1.8	1.4	1.0	4.4
Switzerland	1.7	1.3	0.6	1.0	..	..	..	1.8	2.3	3.8
Syrian Arab Republic	5.8	2.4	3.1	3.3	4.4	8.4	6.7	3.8	–2.2	2.1
Tajikistan	..	..	2.7	2.6	..	..	..	..	..	..
Tanzania	..	..	3.0	2.8	..	..	..	..	..	..
Thailand	7.3	5.0	2.1	2.6	4.1	9.7	7.4	6.3	9.1	11.2
Togo	2.3	–0.7	3.0	2.4	3.1	2.9	1.5	2.8	–0.4	3.8
Trinidad and Tobago	1.5	0.3	1.2	1.8	–2.3	0.1	1.7	3.8	..	3.3
Tunisia	5.1	2.7	2.1	3.1	3.9	6.2	5.0	5.9	4.5	6.9
Turkey	3.8	1.5	2.2	2.0	1.3	5.6	4.9	4.5	..	11.0
Turkmenistan	..	..	2.8	3.0	..	..	..	..	..	..
Uganda	..	..	2.8	2.5	..	..	..	..	..	..
Ukraine	..	..	0.3	0.3	..	..	..	..	..	..
United Arab Emirates	3.8	–4.0	9.5	9.9	11.5	1.2	..	..	..	..
United Kingdom	2.1	1.9	0.2	0.5	..	..	..	2.5	1.7	4.0
United States	2.4	1.4	1.0	1.6	4.0	1.7	2.5	3.0	2.3	5.5
Uruguay	0.8	0.2	0.5	0.9	1.4	1.2	2.2	1.7	0.2	5.6
Uzbekistan	..	..	2.5	2.8	..	..	..	..	..	..
Venezuela	2.2	–0.8	2.8	3.6	2.8	1.4	2.7	2.6	1.8	1.2
Vietnam	..	..	..	..	..	..	..	..	..	..
West Bank and Gaza	..	..	..	..	..	..	..	..	..	..
Yemen, Rep.	..	..	3.1	3.4	..	..	..	..	..	..
Yugoslavia, FR (Serb./Mont.)	..	..	0.7	0.7	..	..	..	..	..	..
Zambia	1.0	–2.0	2.9	2.6	1.7	1.0	0.9	0.3	–7.4	–1.0
Zimbabwe	3.5	0.4	3.0	2.9	1.5	1.3	4.3	7.1	0.7	4.4
World	**3.1 w**	**1.2 w**	**1.7w**	**2.0 w**	**2.5 w**	**2.7 w**	**3.1 w**	**3.4 w**	**3.1 w**	**5.8 w**
Low income	5.3	3.1	2.0	2.2	3.1	7.5	6.1	4.6	6.9	5.9
Excl. China & India	3.1	0.4	2.5	2.6	..	..	..	..	..	3.0
Middle income	3.3	0.9	1.8	2.0	..	..	..	..	..	..
Lower middle income	3.4	0.8	1.7	2.0	..	..	..	..	..	..
Upper middle income	3.2	1.2	1.8	2.3	2.8	3.2	4.1	..	1.2	6.1
Low & middle income	3.8	1.6	1.9	2.1	3.1	..	3.9	..	..	..
East Asia & Pacific	7.4	5.5	1.8	2.1	4.1	9.7	8.3	6.7	9.4	8.8
Europe & Central Asia	–0.6	–1.3	0.8	0.9	..	..	..	..	..	..
Latin America & Carib.	3.3	1.1	2.1	2.8	2.7	3.3	4.0	4.0	2.0	5.2
Middle East & N. Africa	1.1	–1.8	2.7	3.2	4.4	0.0	2.0	..	..	..
South Asia	4.6	2.2	2.2	2.4	2.8	5.5	5.6	4.1	5.3	6.2
Sub-Saharan Africa	2.7	–0.2	2.7	2.5	1.7	2.6	3.3	2.9	–1.1	2.1
High income	3.0	2.2	0.8	1.2	2.0	2.6	3.0	3.3	2.9	5.9

a. Data are for GDP.

Trends in long-term economic development | 1.4

The long-term trends shown in this table provide a view of the relative rates of change of key social and economic indicators over the past 31 years. Like all averages, they reflect the general tendency but may disguise considerable year-to-year variation, especially in economic indicators. In viewing these growth rates, it may be helpful to keep in mind that a quantity growing at 2.3 percent a year will double in 30 years, while a quantity growing at 7 percent a year will double in 10 years.

All the indicators shown here appear elsewhere in the *World Development Indicators.* For more information about them, see *About the data* for tables 1.1 (gross national product and GNP per capita), 2.1 (population), 2.3 (labor force), 4.1 (value added by industrial origin), 4.8 (exports of goods and services), and 4.9 (private consumption).

Definitions

- **Average annual growth rates of gross national product, value added, private consumption, gross domestic fixed investment,** and **exports of goods** and **services** are calculated from data in 1987 constant prices using the least-squares method. See Statistical methods for more information on the calculation of growth rates. • **Gross national product** is the sum of value added by all resident producers plus any taxes (less subsidies) that are not included in the valuation of output plus net receipts of primary income (employee compensation and property income) from nonresident sources. Growth is calculated from constant price GNP in national currency units. • **GNP per capita** is gross national product divided by midyear population. • **Average annual growth of total population and labor force** is calculated using the exponential endpoint method. • **Labor force** comprises all people who meet the International Labour Organization's definition of the economically active population. • **Value added** is the net output of a sector after adding up all outputs and subtracting intermediate inputs. It is calculated without making deductions for depreciation of fabricated assets or depletion and degradation of natural resources. The industrial origin of value added is determined by the International Standard Industrial Classification (ISIC), revision 2. • **Agriculture** is the value added of ISIC major divisions 1–5. • **Industry** is the value added of ISIC division 10–15. • **Services** is the value added in ISIC divisions 15–37. • **Private consumption** is the market value of all goods and services, including durable products, purchased or received as income in kind by households and nonprofit institutions. It excludes purchases of dwellings but includes imputed rent for owner-occupied dwellings. • **Exports of goods and services** is the value of all goods and market services provided to the rest of the world.

Figure 1.4a

The world's 30 fastest-growing economies . . .

Average annual % growth of GNP per capita, 1965–96

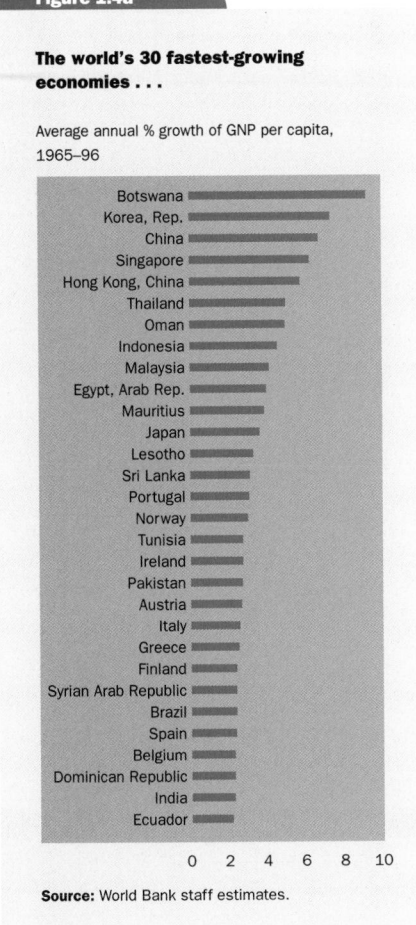

Source: World Bank staff estimates.

Figure 1.4b

. . . include many of the fastest-growing service sectors

Average annual % growth of value added in services, 1965–96

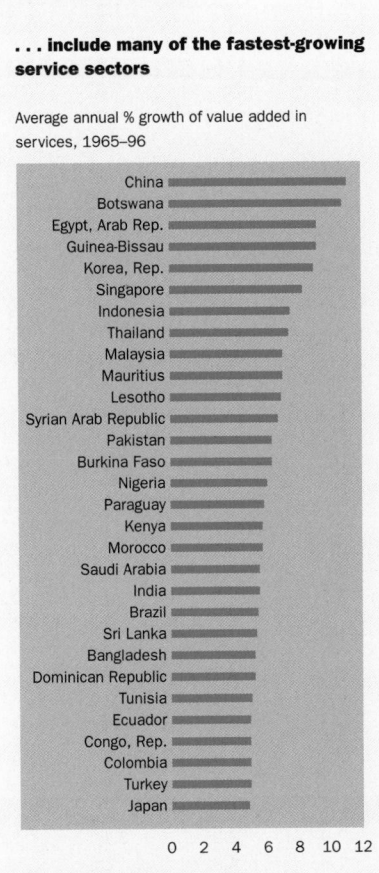

Source: World Bank staff estimates.

Data sources

The indicators here and throughout the rest of the book have been compiled by World Bank staff from primary and secondary sources. More information about the indicators and their sources can be found in the *About the data, Definitions,* and *Data sources* entries that accompany each table in subsequent sections.

1.5 Long-term structural change

	Agriculture value added		Labor force in agriculture		Urban population		Trade		Central government revenue		Money and quasi money	
	% of GDP		% of total labor force		% of total population		% of GDP		% of GDP		% of GDP	
	1970	1996	1970	1990	1970	1996	1970	1996	1970	1996	1970	1996
Albania	..	55	66	55	32	38	..	52	..	21	..	47
Algeria	11	13	47	26	40	56	51	56	..	..	51	22
Angola	..	7	78	75	15	32	..	118	..	..	..	..
Argentina	10	6	16	12	78	88	10	19	0	13	21	19
Armenia	..	44	27	18	59	69	..	86	..	..	..	6
Australia	6	4	8	6	85	85	29	42	21	25	44	63
Austria	..	2	15	8	65	64	60	78	28	36	53	91
Azerbaijan		23	35	31	50	56	..	62	..	..	..	9
Bangladesh	55	30	81	65	8	19	21	38	..	..	..	36
Belarus	..	16	35	20	44	72	..	96	..	..	..	13
Belgium	3	1	5	3	94	97	101	140	35	45	44	82
Benin	36	38	81	64	17	39	50	57	..	..	10	23
Bolivia	20	..	55	47	41	61	49	47	..	18	15	45
Bosnia and Herzegovina	..	..	50	11	27	42	..	..	..	..	..	..
Botswana	28	4	82	46	8	63	71	84	20	43	..	25
Brazil	12	14	45	23	56	79	14	15	0	..	17	26
Bulgaria	..	10	35	13	52	69	..	127	..	34	..	54
Burkina Faso	35	35	92	92	6	16	23	41	..	..	8	22
Burundi	71	57	94	92	2	8	22	19	..	13	9	18
Cambodia	..	51	79	74	12	21	14	69	..	..	..	10
Cameroon	31	40	85	70	20	46	51	32	..	13	14	13
Canada	..	..	8	3	76	77	43	73	19	21	37	62
Central African Republic	35	56	89	80	30	40	74	41	..	..	15	23
Chad	47	46	92	83	12	23	54	72	11	..	9	18
Chile	7	..	24	19	75	84	29	55	29	22	12	36
China	34	21	78	72	17	31	5	40	..	6	..	101
Hong Kong, China	..	0	4	1	88	95	181	285	..	..	..	166
Colombia	25	16	41	27	57	73	30	37	11	16	18	20
Congo, Dem. Rep.	15	64	75	68	30	29	35	68	11	5	8	2
Congo, Rep.	18	10	66	49	33	59	93	164	22	..	17	14
Costa Rica	23	16	43	26	40	50	63	91	15	26	19	36
Côte d'Ivoire	32	28	76	60	27	44	65	83	..	..	25	27
Croatia	..	12	50	16	40	56	..	95	..	45	..	29
Cuba	..	..	30	18	60	76	..	..	..	..	..	..
Czech Republic	..	6	17	11	52	66	..	117	..	36	..	75
Denmark	..	..	11	6	80	85	59	63	35	41	45	59
Dominican Republic	23	13	48	25	40	63	42	63	18	16	17	24
Ecuador	24	12	51	33	40	60	33	57	..	16	20	28
Egypt, Arab Rep.	29	17	52	40	42	45	33	46	..	37	34	75
El Salvador	40	13	57	36	39	45	49	54	10	11	20	37
Eritrea	..	10	86	80	..	17	..	117	..	..	..	..
Estonia	..	7	18	14	65	73	..	159	..	33	..	23
Ethiopia[a]	..	55	91	86	9	16	..	41	..	..	..	40
Finland	..	..	20	8	50	64	53	68	26	33	40	55
France	..	2	14	5	71	75	31	45	33	41	41	67
Gabon	19	7	79	52	25	51	88	96	..	..	15	13
Gambia, The	33	28	87	82	15	30	66	132	16	..	16	25
Georgia	..	35	37	26	48	59	..	44	..	..	..	..
Germany	..	1	9	4	80	87	..	46	..	31	..	64
Ghana	47	44	60	59	29	36	44	65	15	..	18	15
Greece	..	..	42	23	53	59	23	43	22	22	34	45
Guatemala	..	24	62	52	36	39	36	40	9	8	17	24
Guinea	..	26	92	87	14	30	..	41	..	..	..	9
Guinea-Bissau	47	54	89	85	15	22	34	42	..	..	..	0
Haiti	..	42	74	68	20	32	31	35	..	..	12	35
Honduras	32	22	65	41	29	44	62	100	12	..	19	27

Long-term structural change | 1.5

	Agriculture value added		Labor force in agriculture		Urban population		Trade		Central government revenue		Money and quasi money	
	% of GDP		% of total labor force		% of total population		% of GDP		% of GDP		% of GDP	
	1970	1996	1970	1990	1970	1996	1970	1996	1970	1996	1970	1996
Hungary	..	7	25	15	49	65	63	79	..	..	..	39
India	45	28	71	64	20	27	8	27	..	14	22	45
Indonesia	45	16	66	55	17	36	28	51	13	17	8	47
Iran, Islamic Rep.	..	25	44	39	42	60	..	36	..	24	..	35
Iraq	..	..	47	16	56	75	..	..	..	..	22	..
Ireland	..	..	26	14	52	58	79	134	30	37	54	50
Israel	..	..	10	4	84	91	70	69	31	39	44	67
Italy	8	3	19	9	64	67	33	51	..	44	73	60
Jamaica	7	8	33	25	42	54	71	123	..	..	30	45
Japan	6	2	20	7	71	78	20	17	11	..	69	111
Jordan	12	5	28	15	51	72	..	125	..	29	54	93
Kazakhstan	..	13	27	22	50	60	..	65	..	..	..	..
Kenya	33	29	86	80	10	30	60	70	17	24	27	41
Korea, Dem. Rep.	..	..	55	38	54	62	..	..	..	..	..	..
Korea, Rep.	25	6	49	18	41	82	37	69	15	21	29	43
Kuwait	0	0	2	1	78	97	84	104	42	..	36	89
Kyrgyz Republic	..	52	36	32	37	39	..	86	..	..	..	13
Lao PDR	..	52	81	78	10	21	..	65	..	30	..	22
Latvia	..	9	19	16	62	73	..	102	..	17	..	..
Lebanon	..	12	20	7	59	88	..	69	..	..	82	127
Lesotho	35	11	43	40	9	25	65	136	20	..	..	32
Libya	2	..	29	11	45	86	89	..	..	..	20	..
Lithuania	..	13	31	18	50	73	..	115	..	23	..	18
Macedonia, FYR	..	..	50	22	47	60	..	86	..	..	..	..
Madagascar	24	35	84	78	14	27	41	42	14	8	17	17
Malawi	44	40	91	87	6	14	63	49	16	..	18	15
Malaysia	29	13	54	27	34	54	80	183	20	25	30	85
Mali	66	48	93	86	14	27	33	56	..	..	14	21
Mauritania	29	25	84	55	14	53	74	115	..	..	9	17
Mauritius	16	10	34	17	42	41	85	126	..	19	35	73
Mexico	12	5	44	28	59	74	15	42	10	15	14	25
Moldova	..	50	54	33	32	52	..	118	..	..	..	16
Mongolia	..	31	48	32	45	61	..	89	..	24	..	21
Morocco	20	20	58	45	35	53	39	55	19	..	28	60
Mozambique	..	37	86	83	6	35	..	84	..	..	..	32
Myanmar	38	60	78	73	23	26	14	3	..	6	24	..
Namibia	..	14	64	49	19	37	..	107	..	..	..	39
Nepal	67	42	94	94	4	11	13	60	5	12	11	36
Netherlands	..	3	7	5	86	89	89	100	..	46	55	83
New Zealand	12	..	12	10	81	86	48	59	28	36	20	79
Nicaragua	25	34	50	28	47	63	55	106	12	25	14	35
Niger	65	39	93	90	9	19	29	37	..	..	5	13
Nigeria	41	43	71	43	20	40	20	28	10	..	9	17
Norway	..	2	12	6	65	73	74	72	32	41	49	53
Oman	16	..	57	45	11	77	93	89	38	32	..	31
Pakistan	37	26	59	52	25	35	22	37	..	19	41	41
Panama	..	8	42	26	48	56	..	185	..	26	22	65
Papua New Guinea	37	26	86	79	10	16	72	101	..	22	..	32
Paraguay	32	24	53	39	37	53	31	46	11	..	17	27
Peru	19	7	48	36	57	71	34	29	14	16	18	20
Philippines	30	21	58	46	33	55	43	94	13	18	23	49
Poland	..	6	39	27	52	64	..	49	..	40	..	33
Portugal	..	..	32	18	26	36	50	74	..	36	76	81
Puerto Rico	3	..	14	4	58	74	107	..	..	..	..	22
Romania	..	21	49	24	42	56	..	60	..	30	..	22
Russian Federation	..	7	19	14	63	76	..	42	..	18	..	14

	Agriculture value added		Labor force in agriculture		Urban population		Trade		Central government revenue		Money and quasi money	
	% of GDP		% of total labor force		% of total population		% of GDP		% of GDP		% of GDP	
	1970	1996	1970	1990	1970	1996	1970	1996	1970	1996	1970	1996
Rwanda	66	40	94	92	3	6	27	28	..	..	11	17
Saudi Arabia	4	..	64	19	49	83	89	72	..	..	13	50
Senegal	21	18	83	77	33	44	59	67	16	..	14	20
Sierra Leone	30	44	76	67	18	34	62	43	..	8	13	9
Singapore	2	0	3	0	100	100	232	356	21	26	62	81
Slovak Republic	..	5	17	12	41	59	..	126	..	..	..	66
Slovenia	..	5	50	6	37	52	..	111	..	..	..	35
South Africa	8	5	31	14	48	50	48	52	21	28	60	54
Spain	..	3	26	12	66	77	27	47	18	31	69	78
Sri Lanka	28	22	55	48	22	22	54	79	20	19	22	31
Sudan	43	..	77	69	16	32	31	..	18	..	17	20
Sweden	..	..	..	0	81	83	48	73	29	42	..	..
Switzerland	..	..	8	6	55	61	67	68	15	23	104	133
Syrian Arab Republic	20	..	50	33	43	53	39	..	25	23	35	51
Tajikistan	..	..	46	41	37	32	..	228	..	..	..	..
Tanzania	..	48b	90	84	7	25	..	58b	..	..	..	23
Thailand	26	11	80	64	13	20	34	83	12	19	27	75
Togo	34	35	74	66	13	31	88	69	..	..	17	27
Trinidad and Tobago	5	2	19	11	63	72	84	95	..	28	27	40
Tunisia	17	14	42	28	45	63	47	86	23	30	32	44
Turkey	40	17	71	53	38	70	10	49	14	18	20	27
Turkmenistan	..	..	38	37	48	45	..	..	..	..	..	7
Uganda	54	46	90	85	8	13	43	34	14	..	17	10
Ukraine	..	13	31	20	55	71	..	93	..	..	..	10
United Arab Emirates	..	..	9	8	57	84	..	139	..	2	..	54
United Kingdom	..	..	3	2	89	89	45	58	37	36	..	..
United States	..	..	4	3	74	76	11	24	18	21	63	60
Uruguay	20	9	19	14	82	91	40	38	23	31	21	36
Uzbekistan	..	26	44	35	37	41	..	69	..	..	..	..
Venezuela	6	4	26	12	72	86	38	61	17	21	19	17
Vietnam	..	27	77	71	18	19	..	97	..	..	..	18
West Bank and Gaza	..	..	..	0	..	..	..	..	..	..	..	..
Yemen, Rep.	..	18	70	61	13	34	..	91	..	20	..	39
Yugoslavia, FR (Serb./Mont.)	..	..	50	30	39	57	..	..	..	..	..	..
Zambia	11	18	79	75	30	43	90	84	22	18	25	16
Zimbabwe	15	14	77	68	17	33	..	82	..	..	..	26
World	.. w	.. w	58 w	49 w	35 w	46 w	25 w	43 w	19 w	26 w	..	..
Low income	41	27	78	69	17	29	16	43	..	10	..	..
Excl. China & India	42	34	79	67	15	29	39	57	..	..	..	..
Middle income	..	11	48	32	43	61	..	55	..	..	..	..
Lower middle income	..	12	50	36	39	56	..	60	..	20	..	..
Upper middle income	15	9	41	22	52	73	35	45	7	..	..	..
Low & middle income	..	15	69	58	26	40	..	52	..	19	..	..
East Asia & Pacific	39	20	79	69	18	32	17	58	..	11	..	..
Europe & Central Asia	..	11	39	23	48	66	..	64	..	25	..	..
Latin America & Carib.	17	10	44	26	53	74	24	33	6	..	..	..
Middle East & N. Africa	..	..	54	35	37	57	..	54	..	..	..	..
South Asia	44	28	73	64	18	27	14	30	..	14	..	..
Sub-Saharan Africa	30	24	80	68	16	32	48	56	17	..	..	..
High income	..	..	16	6	68	76	24	40	20	28	..	..

a. Data prior to 1992 include Eritrea. b. GNP and its components refer to mainland Tanzania only.

Long-term structural change | 1.5

About the data

Over a period of 25 years or longer, cumulative processes of change reshape an economy and the social order built on that economy. This table highlights some of the notable trends that have been at work for much of the 20th century: the shift of production from agriculture to manufacturing and services; the reduction of the agricultural labor force and the growth of urban centers; the expansion of trade; the increasing size of the central government in most countries—and the reversal of this trend in some; and the monetization of economies that have achieved stable macroeconomic management.

All the indicators shown here appear elsewhere in the *World Development Indicators*. For more information about them, see tables 2.5 (labor force in agriculture), 3.10 (urban population), 4.2 (agriculture value added), 4.12 (central government revenues), 4.15 (money and quasi money), and 6.1 (trade).

Figure 1.5a

Agriculture employs a large share of the workforce in many developing countries

% of total labor force

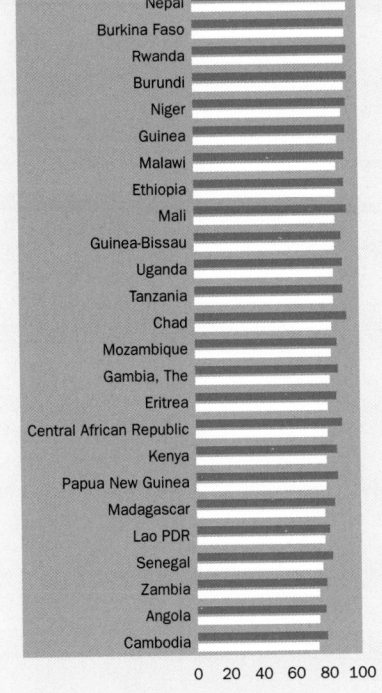

● 1970   ○ 1990

Source: International Labour Organization.

Definitions

• **Agriculture value added** is the sum of outputs of the agricultural sector (International Standard Industrial Classification major divisions 1–5) less the cost of intermediate inputs measured as a share of gross domestic product (GDP). • **Labor force in agriculture** is the percentage of the total labor force recorded as working in agriculture, hunting, forestry, and fishing (ISIC major divisions 1–5). • **Urban population** is the share of the total population living in areas defined as urban in each country. • **Trade** is the sum of exports and imports of goods and services measured as a share of GDP. • **Central government revenue** includes all revenue to the central government from taxes and nonrepayable receipts (other than grants) measured as a share of GDP. • **Money and quasi money** is the sum of currency outside banks and demand deposits other than those of the central government, plus the time, savings, and foreign currency deposits of resident sectors other than the central government. This measure of the money supply is commonly called M2.

Data sources

The indicators here and throughout the rest of the book have been compiled by World Bank staff from primary and secondary sources. More information about the indicators and their sources can be found in the *About the data*, *Definitions*, and *Data sources* entries that accompany each table in subsequent sections.

1.6 Key indicators for other economies

	Population	Land area	Population density	Gross national product						Life expectancy at birth	Adult illiteracy rate	Carbon dioxide emissions
						Per capita						
					average annual growth		average annual growth	PPP	PPP per capita		% of people 15	
	thousands	thousand sq. km	people per sq. km	$ millions	%	$	%	$ millions	$	years	and above	thousand tons
	1996	1995	1996	1996[a]	1995–96	1996[a]	1995-96	1996[b]	1996[b]	1996	1995	1995
Afghanistan	24,167	652.1	40	..	..	[c]	..	..	..	45	69	1,238
American Samoa	60	0.2	300	..	..	[d]	..	..	..	..	..	..
Andorra	71	0.2	160	..	..	[e]	..	..	..	..	..	..
Antigua and Barbuda	66	0.4	150	482	6.5	7,330	5.5	569	8,660	75	..	..
Aruba	77	0.2	400	..	..	[e]	..	..	..	..	..	..
Bahamas, The	284	10.0	30	..	..	[e]	..	2,891	10,180	73	2	1,707
Bahrain	599	0.7	870	..	..	[d]	..	8,368	13,970	73	15	14,832
Barbados	264	0.4	610	..	..	[d]	..	2,778	10,510	76	3	824
Belize	222	22.8	10	600	4.4	2,700	1.9	927	4,170	75	..	414
Bermuda	62	0.1	1,240	..	..	[e]	..	..	..	75	..	..
Bhutan	715	47.0	20	282	7.3	390	2.6	..	..	53	58	238
Brunei	290	5.3	50	..	..	[e]	..	..	..	75	12	8,233
Cape Verde	389	4.0	100	393	3.1	1,010	1.0	1,028 [f]	2,640 [f]	66	28	114
Cayman Islands	32	0.3	130	..	..	[e]	..	..	..	..	..	..
Channel Islands	148	..	..	..	..	[e]	..	..	..	78	..	..
Comoros	505	2.2	230	228	-4.7	450	-8.2	893 [f]	1,770 [f]	59	43	66
Cyprus	740	9.2	80	..	..	[e]	..	15,163 [f]	20,490 [f]	77	..	5,177
Djibouti	619	23.2	30	..	..	[g]	..	..	..	50	54	370
Dominica	74	0.8	100	228	3.3	3,090	2.3	323	4,390	74	..	..
Equatorial Guinea	410	28.1	10	217	24.1	530	20.5	1,104	2,690	50	..	132
Faeroe Islands	47	1.4	30	..	..	[e]	..	..	..	..	..	..
Fiji	803	18.3	40	1,983	6.8	2,470	5.6	3,268	4,070	72	8	737
French Guiana	153	88.2	2	..	..	[e]	..	..	..	..	..	872
French Polynesia	220	3.7	60	..	..	[e]	..	..	..	72	..	..
Greenland	58	341.7	0	..	..	[e]	..	..	..	68	..	..
Grenada	99	0.3	290	285	5.2	2,880	1.1	430	4,340	..	..	..
Guadeloupe	422	1.7	250	..	..	[d]	..	..	..	75	..	..
Guam	153	0.6	280	..	..	[e]	..	..	..	74	..	..
Guyana	839	196.9	4	582	13.5	690	11.3	1,912	2,280	64	2	934
Iceland	270	100.3	3	7,175	7.9	26,580	7.1	5,862	21,710	79	..	1,803
Isle of Man	72	0.6	120	..	..	[d]	..	..	..	..	..	..

About the data

This table shows data for 62 economies—small economies with populations between 30,000 and 1 million, smaller economies if they are members of the World Bank, and larger economies for which data are not regularly reported. Where data on GNP per capita are not available, the estimated range is given.

Definitions

• **Population** is based on the de facto definition of population, which counts all residents regardless of legal status or citizenship. Refugees not permanently settled in the country of asylum are generally considered part of the population of the country of origin. The values shown are midyear estimates for 1996. See also table 2.1. • **Land area** is a country's total area, excluding areas under inland bodies of water. • **Population density** is midyear population divided by land area in square kilometers. • **Gross national product** (GNP) is the sum of value added by all resident producers plus any taxes (less subsidies) that are not included in the valuation of output plus net receipts of primary income (employee compensation and property income) from nonresident

Key indicators for other economies | 1.6

	Population	Land area	Population density	Gross national product							Life expectancy at birth	Adult illiteracy rate	Carbon dioxide emissions
							Per capita						
					average annual growth		average annual growth		PPP	PPP per capita		% of people 15	thousand
	thousands	thousand sq. km	people per sq. km	$ millions	%	$	%	$ millions	$	years	and above	tons
	1996	1995	1996	1996[a]	1995–96	1996[a]	1995-96	1996[b]	1996[b]	1996	1995	1995
Kiribati	82	0.7	110	75	6.0	920	4.5	..	..	60	..	..
Liberia	2,810	96.3	30	..	..	[c]	..	..	..	49	62	319
Liechtenstein	31	0.2	190	..	..	[e]	..	..	..	..	..	..
Luxembourg	416	2.6	160	18,850	9.3	45,360	7.7	14,328	34,480	77	..	9,263
Macao	461	0.0	23,070	..	..	[e]	..	..	..	77	..	1,231
Maldives	256	0.3	850	277	10.4	1,080	6.9	802	3,140	64	7	183
Malta	373	0.3	1,170	..	..	[d]	..	5,174	13,870	77	..	1,726
Marshall Islands	57	0.2	290	108	4.3	1,890	2.7	..	..	..	..	..
Martinique	384	1.1	360	..	..	[e]	..	..	..	77	..	2,037
Mayotte	108	0.3	340	..	..	[d]	..	..	..	..	..	..
Micronesia, Fed. Sts.	109	0.7	160	225	4.3	2,070	2.5	..	..	66	..	..
Monaco	32	0.0	16,840	..	..	[e]	..	..	..	..	..	..
Netherlands Antilles	202	0.8	250	..	..	[e]	..	..	..	76	..	..
New Caledonia	197	18.6	10	..	..	[e]	..	..	..	74	..	1,715
Northern Mariana Islands	63	0.5	110	..	..	[e]	..	..	..	..	..	..
Palau	17	0.5	40	..	..	[d]	..	..	..	..	..	..
Qatar	658	11.0	60	..	..	[e]	..	10,745	16,330	72	21	29,019
Reunion	664	2.5	270	..	..	[e]	..	..	..	75	..	1,554
Samoa	172	2.8	60	200	6.9	1,170	6.4	..	..	69	..	132
São Tomé and Principe	135	1.0	140	45	6.9	330	3.1	..	..	64	25	77
Seychelles	77	0.5	170	526	3.3	6,850	1.3	..	..	71	21	..
Solomon Islands	389	28.0	10	349	2.2	900	0.0	876 [f]	2,250 [f]	63	..	161
Somalia	9,805	627.3	20	..	..	[c]	..	..	..	49	..	11
St. Kitts and Nevis	41	0.4	110	240	8.0	5,870	8.3	299	7,310	70	..	..
St. Lucia	158	0.6	260	553	4.6	3,500	3.2	777	4,920	70	..	..
St. Vincent and the Grenadines	112	0.4	290	264	4.7	2,370	4.4	465	4,160	73	..	..
Suriname	432	156.0	3	433	20.5	1,000	19.0	1,136	2,630	71	7	2,151
Swaziland	926	17.2	50	1,122	1.4	1,210	–1.6	3,075	3,320	57	23	454
Tonga	97	0.7	140	175	3.4	1,790	2.9	..	..	72	..	..
Vanuatu	173	12.2	10	224	8.7	1,290	5.7	522 [f]	3,020 [f]	64	..	62
Virgin Islands (U.S.)	98	0.3	290	..	..	[e]	..	..	..	76	..	..

a. Calculated using the World Bank Atlas method. b. Purchasing power parity, See the notes following these tables. c. Estimated to be low income ($785 or less). d. Estimated to be upper middle income $3,116 to $9,635). e. Estimated to be high income ($9,636 or more). f. The estimate is based on regression; others are extrapolated from the latest International Comparison Programme benchmark estimates. g.Estimated to be lower middle income ($786 to $3,115).

Data sources

sources. Data are in current U.S. dollars converted using the World Bank Atlas method (see Statistical methods). Rank is calculated for economies reporting data. Growth is calculated from constant price GNP in national currency units. • **GNP per capita** is gross national product divided by midyear population. GNP per capita in U.S. dollars is converted using the World Bank Atlas method. Rank is calculated for economies reporting data. Growth is calculated from constant price GNP per capita in national currency units. • **GNP in PPP terms** is gross national product converted to international dollars using purchasing power parity rates. An international dollar has the same purchasing power over GNP as the U.S. dollar in the United States. Rank is calculated for

economies reporting data. • **Life expectancy at birth** is the number of years a newborn infant would live if prevailing patterns of mortality at the time of its birth were to stay the same throughout its life. • **Adult illiteracy rate** is the percentage of adults aged 15 and above who cannot, with understanding, read and write a short, simple statement about their everyday life. • **Carbon dioxide emissions** are those stemming from the burning of fossil fuels and the manufacture of cement. They include carbon dioxide produced during consumption of solid fuels, liquid fuels, gas fuels, and gas flaring.

The indicators here and throughout the rest of the book have been compiled by World Bank staff from primary and secondary sources. More information about the indicators and their sources can be found in the *About the data*, *Definitions*, and *Data sources* entries that accompany each table in subsequent sections.

Development is about people and their well-being—about people developing their capabilities to provide for their families, to act as stewards of the environment, to form civil societies that are just and orderly. The international consensus emerging around a set of development goals for the 21st century (see the introduction to the first section, *World View*) captures many parts of well-being: current and future health status, educational attainment, and freedom from extreme deprivation. Here we look at the social indicators identified by a recent OECD–United Nations–World Bank conference (box 2a) and at the statistical systems that produce them. Are the indicators reliable? Do they accurately and adequately measure the outcomes they intend to track? Can good decisions be made based on the indicators? Too often the answer is no, but the alternative is to know nothing and do nothing.

Poverty

The 21st century goals call for reducing poverty by half by 2015. Inadequate income and consumption levels are not only undesirable in themselves, they can lead to such other problems as crime and violence, and the reduced capacity to enjoy the full benefits and opportunities offered by the community. But poverty is easier to define than to measure, and in many countries there is more than one definition of poverty and more than one way to measure it. The three indicators selected to measure progress in reducing poverty—the headcount index, the poverty gap index (see table 2.7 for definitions), and either the income or consumption share of the lowest quintile—reflect income (or consumption) poverty. The headcount index and the poverty gap are based on an international poverty line of $1 a day defined in constant prices and measured in purchasing power parity dollars. The advantage of a common poverty line is that it permits comparisons based explicitly on equivalent real baskets of goods and services. It also allows aggregations across countries to track regional poverty. The disadvantage is that it is not based on local development circumstances (and thus might not be adopted by a country) and that it varies widely from measures based on national poverty lines (see table 2.7). Not all countries will be able meet the poverty reduction goal, but poverty for large regions or even for developing countries as a group could be cut in half by 2015 through concerted effort.

All three indicators measure the economic dimensions of poverty. The headcount index provides a count of the people in poverty. The poverty gap measures the amount of additional income per capita, expressed as a proportion of the poverty line, that, if available to the poor, would lift them out of extreme poverty. The income consumption share of the lowest 20 percent measures the extent to which the poor share in economic growth. But being poor means much more than being poor in income. It means being poor in health, education, and access to goods and services and involves other sources of vulnerability. The three poverty indicators do not capture these noneconomic dimensions. Child malnutrition has been proposed as a cross-check on income poverty because the prevalence of malnourished children is an indication of poverty. It should be noted, however, that the absence of malnourished members does not mean that a household is not poor.

The reliability of the income poverty indicators depends on the quality of the income and consumption data, which are usually obtained from household surveys. In poor countries household income is difficult to measure because a number of activities, products, and services go unrecorded. In these instances calculations may be based on consumption (which tends to understate inequality and household income differentials). Many rural transactions are not conducted in cash, and a part of rural household consumption is obtained from what are called "common property resources" not usually recorded in household consumption aggregates. All this implies that estimates of household income and consumption have a high variance and may be understated. Estimates of income and consumption are also affected by the limitations of sample surveys: recall errors, short reference periods, and the exclusion from the sampling frame of people in remote areas and other marginal groups who are most likely to be poor.

The gathering of household income and consumption data therefore presents problems, particularly in subsistence societies. Some of the problems, particularly the practical ones relating to respondents' inability to remember, can be overcome by carefully questioning individual household members, paying frequent visits to the household, or applying consistency checks. But the sheer size of the task makes frequent surveys of this kind impractical for poor countries that lack sound statistical systems. This difficulty is reflected in coverage rates for country poverty indicators, which are well below 50 percent (table 2a). A concerted effort will have to be made to motivate and equip governments

Box 2a

Social goals and indicators for the 21st century

A recent OECD–United Nations–World Bank conference (held in Paris on February 16–17, 1998) identified 6 social goals and 16 complementary indicators to be monitored by the development community as part of a new international development strategy. (The table numbers in parentheses show where these indicators appear in the *World Development Indicators*.)

Reduce poverty by half
- Headcount index (table 2.7)
- Poverty gap index (table 2.7)
- Income inequality: share of income accruing to poorest 20 percent (table 2.8)
- Child malnutrition (table 2.16)

Provide universal primary education
- Net primary enrollment rate (table 2.10)
- Progression to grade 5 (table 2.11)
- Literacy rate of 15–24 year olds (table 1.2)[1]

Improve gender equality in education
- Gender differences in education and literacy (tables 1.3 and 2.12)

Reduce infant and child mortality
- Infant mortality rate (table 2.17)
- Under-5 mortality rate (table 2.17)

Reduce maternal mortality
- Maternal mortality ratio (table 2.15)
- Births attended by health staff (table 2.15)

Expand access to reproductive health services
- Contraceptive prevalence rate (table 2.15)
- Total fertility rate (table 2.15)
- HIV prevalence in pregnant 15–24 year olds (table 2.16)[1]

1. These data are not yet available, but the referenced tables show comparable indicators.

Table 2a

Coverage of poverty indicators by region, 1996

Region	Number of low- and middle-income countries	Number of countries for which data are available	% of population represented	Number of countries with two or more data sets
East Asia and the Pacific	21	5	88	5
Europe and Central Asia	27	18	72	18
Latin America and the Caribbean	34	15	84	10
Middle East and North Africa	15	5	47	3
South Asia	8	5	98	3
Sub-Saharan Africa	49	19	66	3
Total	154	67	84	42

Source: World Bank.

to undertake household surveys on a regular basis to monitor progress and aid policymaking.

The World Bank's program for improving the collection of data on poverty involves two main steps. First, an in-house review is being conducted to take stock of existing databases in all client countries. Second, based on the review results, strategies will be developed to increase awareness of the need to collect such information at the country level. Because the surveys needed for poverty data are the responsibility of client governments, capacity building at the country level will be key to a continued flow of reliable poverty data in developing countries.

Universal primary education

The formal education system is the principal means by which people acquire knowledge, skills, and shared values. The 21st century goals call for universal primary education by 2015. Through schooling, individual (and ultimately societal) ideas, aspirations, and behaviors change. Female autonomy, through reduced fertility and the ability to take advantage of opportunities that are an alternative to childbearing, is powerfully linked to education. And primary education is important because literacy and numeracy expand personal horizons and potential. It is also the entry point for future education. Universal primary education is a composite of three dimensions, each measured by a different indicator: access and participation are measured by net primary enrollment, retention by progression to grade 5, and achievement of basic literacy by the literacy rate of 15–24 year olds.

Net primary enrollment measures the percentage of the official primary school–age population that is enrolled in primary education. Data are typically collected during national

school censuses organized by the ministry of education at the beginning of each school year. Low enrollment rates signal inadequacies in providing universal access to primary education and may in turn identify factors that prevent children from enrolling or remaining in school. They do not, however, fully capture participation in the process, because participation rates require data on daily attendance by age, grade, and gender. Some statistical offices collect data on school attendance through household and sample surveys, but such periodic assessments serve more as a check on official enrollment statistics. In addition, coverage of net enrollment data is extremely limited (table 2b), and the rates themselves have been criticized for their unreliability (UNRISD 1993).

Progression to grade 5 is concerned with the retention of children in school and their eventual acquisition of basic literacy and numeracy. It has, as its starting point, the number of children enrolled. But again, enrollment does not mean attendance, and in the absence of detailed individual pupil records, which are costly to build and maintain, assumptions have to made about promotion, repetition, and attrition. So progression estimates are likely to be biased upward. More fundamentally, retention does not translate to acquisition of basic skills. Thus there is a need to identify actual learning outcomes through the formal (and nonformal) educational system that are universally accepted and can be applied by all countries.

The literacy rate of 15–24 year olds is an outcome measure that reflects skills acquired through both formal and nonformal training. Methods of measuring literacy vary within countries, and standards have changed over time. So changes in recorded literacy rates may not be a reliable measure of the success or failure of the education system. Still, levels of literacy are important ends in themselves because they represent a key element in the quality of life.

Literacy rates are usually derived from data on self-declared literacy in censuses or from updating census or survey estimates with current estimates of school enrollment, not criteria-based literacy tests. And although the United Nations Educational, Scientific, and Cultural Organization (UNESCO) has issued guidelines for estimating literacy levels, international comparability is affected by differences in methods (some countries test literacy in the official language, others in the mother tongue) and completeness of coverage.

UNESCO has implemented several initiatives to improve the quality and coverage of statistics, including an end of the decade education-for-all assessment, strengthening of national education statistical systems in Sub-Saharan Africa, and similar initiatives in other regions. It is also promoting the increased use of household and sample surveys to supplement national administrative files in order to monitor and test literacy status. Such surveys and the expansion of coverage of statistical information systems should also be used to provide process indicators to monitor the additional dimensions of universal primary education, such as the net intake of students at grade one, school attendance rates, and learning achievements.

Table 2b

Coverage of net primary enrollment indicators by continent, 1995

Continent	General data Total number of countries	Data by gender Number of reporting countries	% of countries represented	Number of reporting countries	% of countries represented
Africa	55	14	26	12	22
Americas	50	20	40	18	36
Asia and the Pacific	62	26	42	25	40
Total	167	60	36	55	33

Source: UNESCO.

Mortality reduction

The 21st century goals call for a two-thirds reduction in child mortality by 2015. Beyond its obvious relevance as a measure of health conditions, child mortality is one of the best indicators of overall socioeconomic development in a community. The two indicators selected to measure progress are the infant mortality rate and the under-five mortality rate. Both capture the threat to children, but each focuses on a different mix of risks: the infant mortality rate captures risks at the earliest stage of life closely related to the health of the mother and the socioeconomic circumstances of the family. The under-five mortality rate captures a range of influences on health that reflect communal development, and are most amenable to change.

For monitoring this goal in high-mortality countries, the infant mortality rate is less desirable than the under-five mortality rate (which includes infant mortality) for two reasons. First, less than 20 percent of under-five deaths are infant deaths. A large portion of these infant deaths are neonatal (occurring in the first month) and are more difficult to affect through policy interventions after the event. While it is possible to distinguish neonatal from post-neonatal deaths, such data are not easy to come by in the many countries where civil registration of deaths is incomplete and where infants dying during the first weeks of life, especially in remote rural areas, may not even have been recorded as being born.

Second, in low- and middle-income countries mortality between the ages of 1 and 5 can be quite high (in contrast to high-income countries, where less than 20 percent of under-five deaths are child deaths), reflecting the effects of malnutrition, incomplete immunization, the lack of adequate sanitation and safe water, and other basic preventive public health measures. For monitoring the effects of targeted interventions, the under-five mortality rate is therefore preferable.

All governments are committed to measuring these indicators, and the expertise to measure them exists in almost all countries. The principal sources are vital registration systems (covering at least 90 percent of the population) and direct or indirect estimates based on sample surveys and censuses. Unfortunately, effective vital registration systems are not common in developing countries. But most countries now have at least one estimate of infant or under-five mortality based on empirical data. Because building reliable vital registration systems is a lengthy process, the United Nations Children's Fund (UNICEF) recommends that countries use international survey programs such as the Demographic and Health Surveys (DHS) that contain questions to measure these indicators. It further recommends that survey measurements be done every three to five years.

Maternal mortality and access to reproductive health

The objective of reducing maternal mortality by three-fourths should be viewed in the context of the more comprehensive goal of providing access, through the primary health care system, to reproductive health services for all who need them. The two are interrelated: providing access to reproductive health care and thereby reducing maternal mortality is essential to improving women's health status. Improving women's social status and ensuring gender equity in health care are important strategies for achieving these goals. The importance of this goal for society cannot be emphasized enough. Providing reproductive health services to women—and men—will have a significant effect on their health and well-being, on the size of the future world population, and on the quality of life of future generations.

Measuring progress in providing access to reproductive health is a major challenge because the area broadly encompasses many health needs and behaviors, cultural and religious attitudes, and supply-side factors. The indicators cover several dimensions of reproductive health, and most are available in the majority of developing countries: contraceptive prevalence rate, total fertility rate, maternal mortality ratio, and births attended by trained and skilled health staff. Together, they indicate the degree to which reproductive health services are accessible and used. The use of contraception and lower fertility reduce the risk of dying during pregnancy and childbirth. Prenatal care and delivery attended by skilled personnel are essential for prompt identification, referral, and treatment of complications. An additional indicator, HIV prevalence in pregnant women age 15–24, has been proposed by the Joint United Nations Programme on HIV/AIDS (UNAIDS), although the methodology is still being finalized. Having this information is considered important, particularly where HIV prevalence is high (WHO 1997b).

A well-functioning referral system for emergency obstetric complications is an important dimension of access to reproductive health care and is essential to reducing maternal mortality. Key elements of referral systems include skilled birth attendants, emergency transport, and appropriately equipped and staffed referral centers. Measuring the quality, accessibility, and use of referral systems is difficult and should be augmented by maternal death audits, verbal autopsies, and documentation of the social and logistical factors related to referral, numbers of women referred, medical condition leading to referral and upon arrival at referral facility, and outcomes.

There has been much debate about the feasibility of using the maternal mortality ratio as a measure of the quality of care. Unless there is an excellent vital registration system in the country and medical attribution of cause of death, conducting a survey is the most practical way of getting estimates. But survey methods differ in several attributes, producing estimates that can be imprecise. Observed differences in the maternal mortality ratio may not reflect improved maternal health status. Instead they may be due to changes in the reporting system or to wide random fluctuation from a small number of events. The World Health Organization (WHO) and UNICEF have derived maternal mortality ratio estimates using a demographic model. But surveys cannot detect significant changes in maternal mortality over time. And model-based estimates cannot monitor trends in maternal mortality either.

Despite the problems with reliable data collection, maternal mortality ratio is now widely used. It may be combined with fertility to calculate a lifetime risk of dying during pregnancy and childbirth. According to WHO (1997b), any alternative outcome indicators for maternal health will have similar problems with reliable data collection.

Data issues are also likely to weaken the monitoring capability of the remaining indicators. Survey measures of total fertility and contraceptive prevalence are affected by recall errors and nonspecific reference periods. For contraceptive prevalence derived from program statistics, the accuracy of the assumptions is difficult to assess. International comparisons of births attended by health staff are limited by differing definitions of what constitutes trained and skilled. Nor does the measure reflect the content and quality of care provided, so that countries with similar levels could have large discrepancies in actual care provided. While there are no proposals to improve coverage of these indicators, international survey programs such as the DHS provide the most reliable information for their estimation.

The need for better data

Establishing a reliable system for monitoring living standards at disaggregated levels requires a well-functioning sample survey apparatus. Many countries already conduct excellent household surveys that generate sufficiently disaggregated information to facilitate planning, and international survey programs such as the DHS have helped to improve the quality and comparability of basic indicators of living standards. For other countries, however, the state of the data does not even allow an accurate estimate of basic indicators. For these countries the World Bank's Living Standards Measurement and Social Dimensions of Adjustment surveys, with bilateral and multilateral support, provide one way to build local capacity.

Beyond data generation, however, the organizational structure in which statistical work takes place is also important, including cultivation of demand for good data by governments. Development institutions such as the United Nations Development Programme and the World Bank can help nurture this demand within countries by emphasizing the need for well-informed policy decisions, based on reliable and timely information on economic and social progress.

AIDS has lowered life expectancy

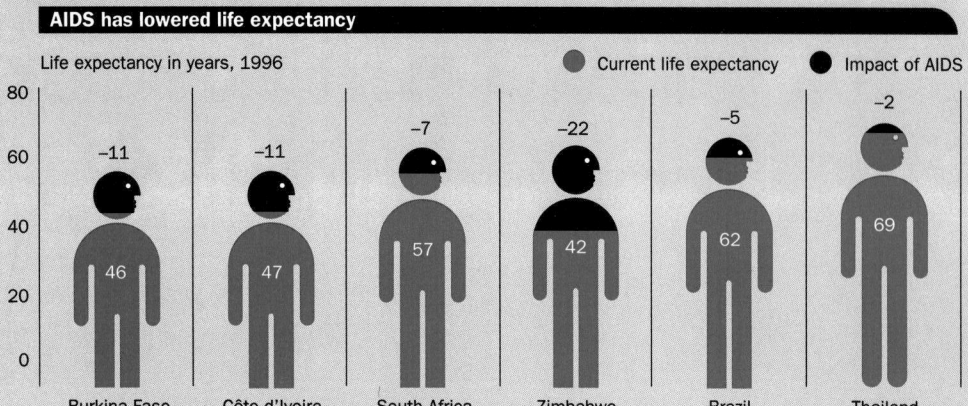

Life expectancy in years, 1996

● Current life expectancy   ● Impact of AIDS

Burkina Faso	−11 / 46
Côte d'Ivoire	−11 / 47
South Africa	−7 / 57
Zimbabwe	−22 / 42
Brazil	−5 / 62
Thailand	−2 / 69

The HIV/AIDS burden is especially large in Sub-Saharan Africa

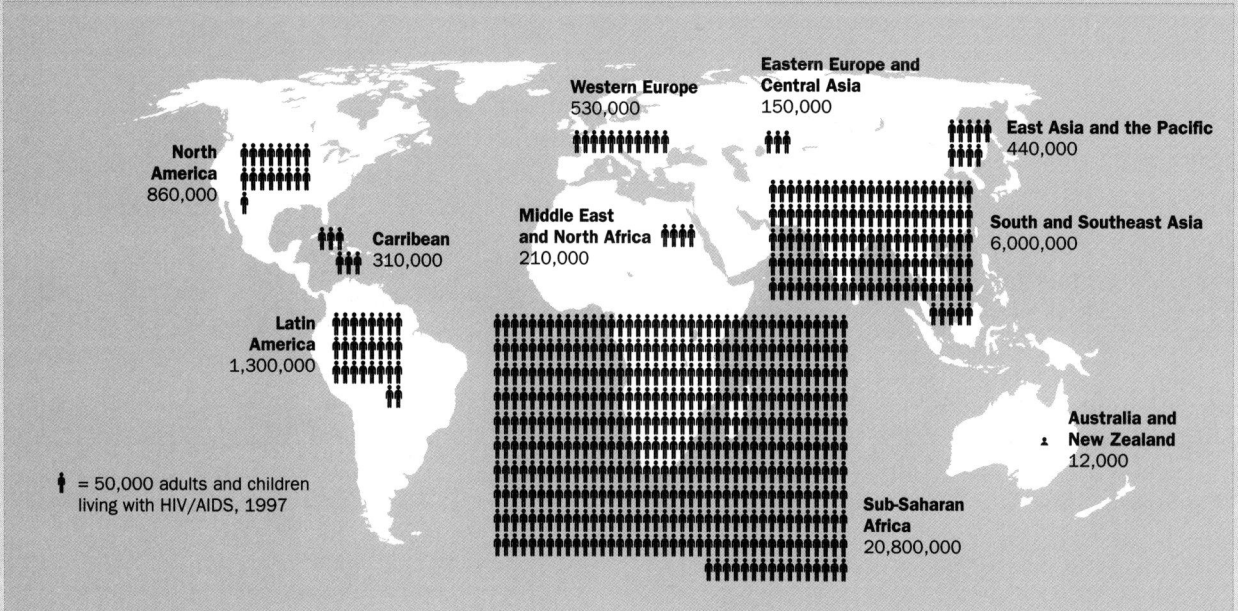

North America 860,000

Carribean 310,000

Latin America 1,300,000

Western Europe 530,000

Eastern Europe and Central Asia 150,000

East Asia and the Pacific 440,000

Middle East and North Africa 210,000

South and Southeast Asia 6,000,000

Australia and New Zealand 12,000

Sub-Saharan Africa 20,800,000

👤 = 50,000 adults and children living with HIV/AIDS, 1997

By 2020 HIV/AIDS will account for a sizable portion of deaths from infectious disease in developing countries

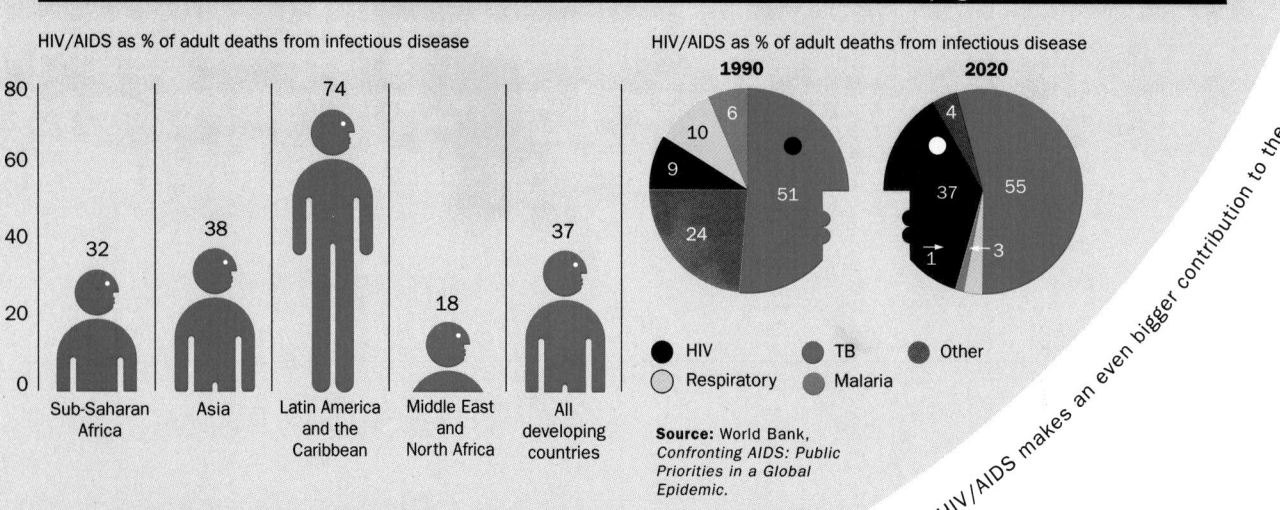

HIV/AIDS as % of adult deaths from infectious disease

Region	%
Sub-Saharan Africa	32
Asia	38
Latin America and the Caribbean	74
Middle East and North Africa	18
All developing countries	37

HIV/AIDS as % of adult deaths from infectious disease

1990 — HIV •, 51, 24, 9, 10, 6

2020 — 4, 37, 55, 1, 3

● HIV   ● TB   ● Other
○ Respiratory   ● Malaria

Source: World Bank, *Confronting AIDS: Public Priorities in a Global Epidemic.*

Life expectancy in several countries has returned to levels of more than 10 years ago

HIV/AIDS makes an even bigger contribution to the disease burden if the focus is on infectious diseases

AIDS is a large and growing problem that has already taken a terrible toll on people and their communities. Many countries still have an opportunity to avert a full-scale AIDS epidemic, but others—mostly in the developing world—are being forced to deal with the consequences of widespread HIV infection. In these countries, until prevention programs become more effective, life expectancy will fall, the number of orphans will increase, poverty will worsen, and health care systems will come under increasing strain.

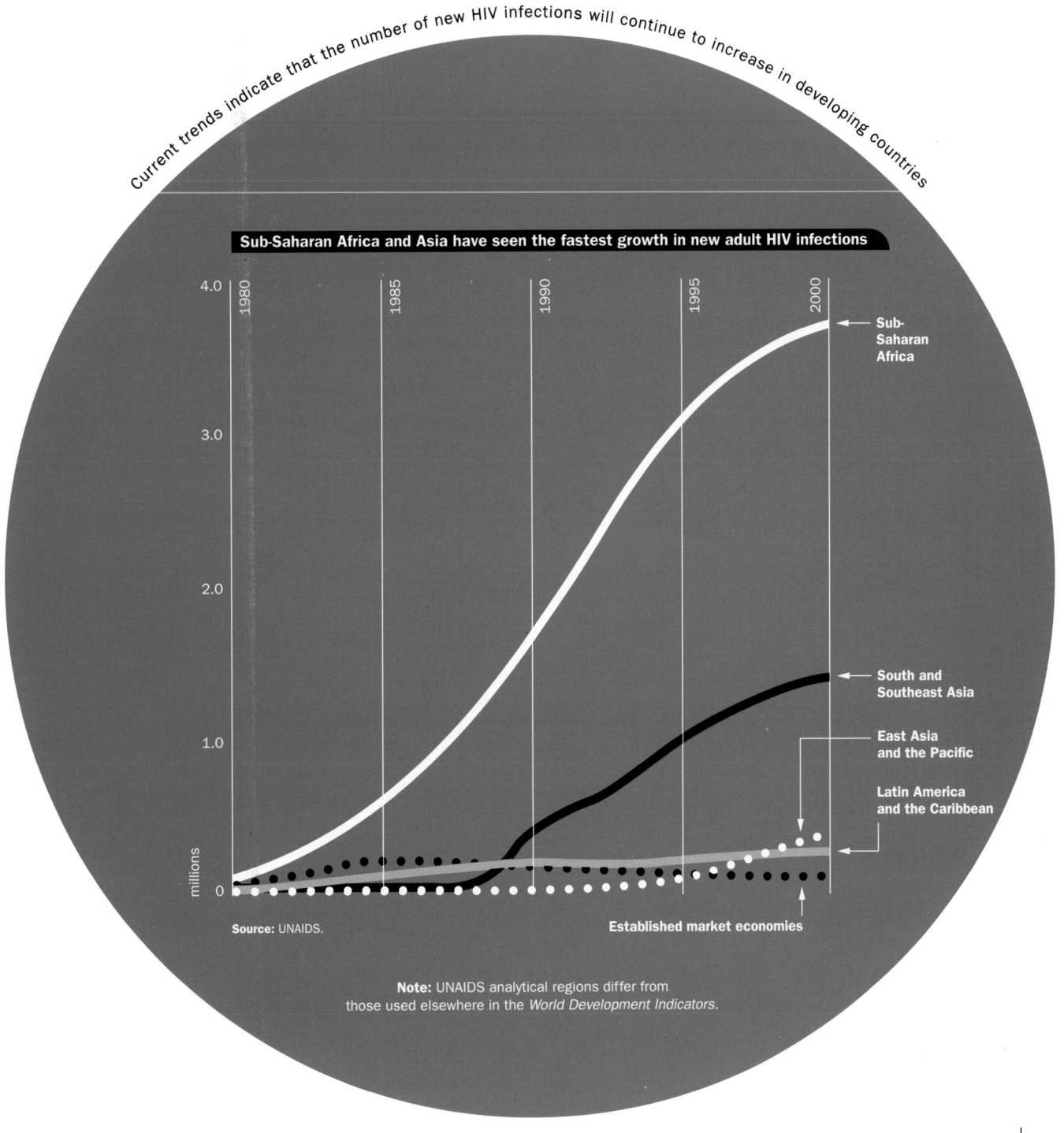

Current trends indicate that the number of new HIV infections will continue to increase in developing countries

Sub-Saharan Africa and Asia have seen the fastest growth in new adult HIV infections

millions

Sub-Saharan Africa

South and Southeast Asia

East Asia and the Pacific

Latin America and the Caribbean

Established market economies

Source: UNAIDS.

Note: UNAIDS analytical regions differ from those used elsewhere in the *World Development Indicators*.

2.1 | Population

	Total population (millions)			Average annual population growth rate (%)		Age dependency ratio (dependents as proportion of working-age population)		Population age 60 and above (% of total)		Women age 60 and above (per 100 men)	
	1980	**1996**	**2010**	**1980–96**	**1996–2010**	**1980**	**1996**	**1996**	**2010**	**1996**	**2010**
Albania	3	3	4	1.3	0.9	0.7	0.6	9.3	11.5	118	114
Algeria	19	29	37	2.7	1.9	1.0	0.7	5.8	6.5	112	113
Angola	7	11	16	2.9	2.7	0.9	1.0	4.6	4.3	121	119
Argentina	28	35	41	1.4	1.0	0.6	0.6	13.2	14.1	134	134
Armenia	3	4	4	1.2	0.4	0.6	0.5	11.7	14.0	138	143
Australia	15	18	20	1.4	0.7	0.5	0.5	15.6	19.3	122	115
Austria	8	8	8	0.4	0.0	0.6	0.5	19.4	23.3	154	129
Azerbaijan	6	8	8	1.3	0.5	0.7	0.6	9.5	10.2	145	154
Bangladesh	87	122	150	2.1	1.5	1.0	0.8	5.0	6.0	82	95
Belarus	10	10	10	0.4	−0.3	0.5	0.5	18.1	19.0	180	172
Belgium	10	10	10	0.2	0.0	0.5	0.5	21.2	23.9	137	130
Benin	3	6	8	3.0	2.7	1.0	1.0	4.4	4.6	124	126
Bolivia	5	8	10	2.2	2.1	0.9	0.8	6.0	6.4	120	123
Bosnia and Herzegovina	4	..	..	..	..	0.5	..	..	..	..	..
Botswana	1	1	2	3.1	1.4	1.0	0.8	3.7	3.5	169	152
Brazil	121	161	190	1.8	1.2	0.7	0.6	7.2	9.0	122	131
Bulgaria	9	8	7	−0.4	−0.9	0.5	0.5	20.7	24.4	127	138
Burkina Faso	7	11	15	2.7	2.4	1.0	1.0	4.7	3.8	111	140
Burundi	4	6	9	2.8	2.4	0.9	1.0	4.3	3.4	151	147
Cambodia	6	10	13	2.9	1.8	0.7	0.8	4.8	5.8	174	159
Cameroon	9	14	19	2.8	2.4	0.9	0.9	5.5	4.9	117	114
Canada	25	30	33	1.2	0.6	0.5	0.5	16.2	20.4	127	118
Central African Republic	2	3	4	2.3	1.9	0.8	0.9	6.1	5.2	132	134
Chad	4	7	9	2.4	2.3	0.8	0.8	5.7	5.2	123	121
Chile	11	14	17	1.6	1.1	0.6	0.6	9.7	12.6	134	130
China	981	1,215	1,349	1.3	0.7	0.7	0.5	9.8	11.9	101	100
Hong Kong, China	5	6	7	1.4	0.5	0.5	0.4	14.1	17.7	103	98
Colombia	28	37	45	1.8	1.3	0.8	0.6	7.8	9.2	109	125
Congo, Dem. Rep.	27	45	69	3.2	3.0	1.0	1.0	4.5	4.3	130	123
Congo, Rep.	2	3	4	3.0	2.4	0.9	1.0	5.6	4.1	145	144
Costa Rica	2	3	4	2.6	1.4	0.7	0.6	7.1	9.5	114	114
Côte d'Ivoire	8	14	19	3.5	2.1	1.0	0.9	4.6	4.8	94	90
Croatia	5	5	5	0.2	−0.3	0.5	0.5	21.3	23.3	155	143
Cuba	10	11	12	0.8	0.4	0.7	0.5	12.6	17.2	107	114
Czech Republic	10	10	10	0.1	−0.2	0.6	0.5	17.4	22.6	145	133
Denmark	5	5	5	0.2	0.0	0.5	0.5	19.5	23.0	132	121
Dominican Republic	6	8	10	2.1	1.4	0.8	0.6	6.3	7.9	102	106
Ecuador	8	12	15	2.4	1.6	0.9	0.7	6.5	7.6	113	119
Egypt, Arab Rep.	41	59	74	2.3	1.6	0.8	0.7	6.6	7.7	116	114
El Salvador	5	6	8	1.5	2.0	1.0	0.7	6.5	6.7	120	133
Eritrea	..	4	6	..	3.0	..	0.9	4.8	4.7	114	110
Estonia	1	1	1	−0.1	−1.0	0.5	0.5	18.6	23.9	187	191
Ethiopia	38	58	89	2.7	3.0	1.0	1.0	4.5	4.3	122	111
Finland	5	5	5	0.4	0.2	0.5	0.5	19.0	24.3	151	131
France	54	58	60	0.5	0.2	0.6	0.5	20.2	22.5	139	133
Gabon	1	1	2	3.0	2.1	0.7	0.8	8.9	8.0	118	122
Gambia, The	1	1	2	3.6	2.3	0.8	0.8	4.8	5.7	112	112
Georgia	5	5	5	0.4	0.0	0.5	0.5	16.7	18.7	155	157
Germany	78	82	81	0.3	−0.1	0.5	0.5	21.0	25.1	152	128
Ghana	11	18	24	3.1	2.3	0.9	0.9	4.8	5.1	118	118
Greece	10	10	11	0.5	0.1	0.6	0.5	22.4	25.5	121	125
Guatemala	7	11	15	2.9	2.3	1.0	0.9	5.3	5.3	108	117
Guinea	4	7	10	2.6	2.6	0.9	1.0	4.2	4.1	109	106
Guinea-Bissau	1	1	1	1.9	2.2	0.8	0.9	6.4	5.7	115	115
Haiti	5	7	9	2.0	1.6	0.8	0.8	5.9	5.8	119	129
Honduras	4	6	9	3.2	2.4	1.0	0.9	4.9	4.9	113	116

	Total population (millions)			Average annual population growth rate (%)		Age dependency ratio (dependents as proportion of working-age population)		Population age 60 and above (% of total)		Women age 60 and above (per 100 men)	
	1980	1996	2010	1980–96	1996–2010	1980	1996	1996	2010	1996	2010
Hungary	11	10	10	–0.3	–0.4	0.5	0.5	19.3	21.5	153	152
India	687	945	1,129	2.0	1.3	0.7	0.6	7.3	8.6	106	106
Indonesia	148	197	236	1.8	1.3	0.8	0.6	6.7	8.4	114	117
Iran, Islamic Rep.	39	63	81	2.9	1.9	0.9	0.8	6.4	6.5	79	84
Iraq	13	21	31	3.1	2.8	0.9	0.8	4.7	5.6	112	111
Ireland	3	4	4	0.4	0.5	0.7	0.5	15.1	17.4	126	120
Israel	4	6	7	2.4	1.7	0.7	0.6	10.9	11.7	122	116
Italy	56	57	55	0.1	–0.3	0.5	0.5	22.1	26.1	136	132
Jamaica	2	3	3	1.1	0.8	0.9	0.6	8.8	9.3	121	122
Japan	117	126	127	0.5	0.1	0.5	0.4	21.0	29.8	131	123
Jordan	2	4	6	4.3	2.6	1.1	0.8	4.6	5.6	78	96
Kazakhstan	15	16	17	0.6	0.1	0.6	0.6	10.2	11.8	174	159
Kenya	17	27	36	3.1	2.0	1.1	0.9	4.2	3.8	114	114
Korea, Dem. Rep.	18	22	26	1.5	0.9	0.8	0.5	7.4	9.6	168	119
Korea, Rep.	38	46	50	1.1	0.7	0.6	0.4	9.2	13.6	146	130
Kuwait	1	2	2	0.9	2.3	0.7	0.6	3.1	6.9	62	74
Kyrgyz Republic	4	5	5	1.4	1.1	0.8	0.7	8.5	7.8	156	146
Lao PDR	3	5	7	2.4	2.4	0.8	0.9	5.6	5.0	109	125
Latvia	3	2	2	–0.1	–0.7	0.5	0.5	19.1	23.3	194	191
Lebanon	3	4	5	1.9	1.4	0.8	0.6	8.3	8.3	115	125
Lesotho	1	2	3	2.4	2.0	0.9	0.8	6.0	6.6	126	119
Libya	3	5	7	3.3	2.3	1.0	0.8	4.9	6.3	83	87
Lithuania	3	4	4	0.5	–0.2	0.5	0.5	17.6	20.0	175	176
Macedonia, FYR	2	2	2	0.3	0.7	0.6	0.5	12.7	15.5	118	119
Madagascar	9	14	20	2.8	2.8	0.9	0.9	4.7	4.8	119	118
Malawi	6	10	14	3.1	2.3	1.0	1.0	4.2	4.0	118	107
Malaysia	14	21	26	2.5	1.6	0.8	0.7	6.0	7.9	117	114
Mali	7	10	15	2.6	2.8	1.0	1.0	4.2	3.9	131	140
Mauritania	2	2	3	2.6	2.3	0.9	0.8	5.1	5.2	124	117
Mauritius	1	1	1	1.0	1.0	0.6	0.5	8.5	11.0	130	131
Mexico	67	93	115	2.1	1.5	1.0	0.6	6.2	8.1	120	125
Moldova	4	4	4	0.5	0.0	0.5	0.5	13.6	14.8	157	152
Mongolia	2	3	3	2.6	1.9	0.9	0.7	5.8	5.9	123	111
Morocco	19	27	34	2.1	1.6	0.9	0.7	6.3	7.1	114	127
Mozambique	12	18	25	2.5	2.4	0.9	0.9	4.1	4.2	126	120
Myanmar	34	46	55	1.9	1.3	0.8	0.7	6.8	7.2	116	118
Namibia	1	2	2	2.7	2.1	0.9	0.8	5.7	5.5	120	115
Nepal	14	22	31	2.6	2.3	0.8	0.9	5.5	5.7	97	101
Netherlands	14	16	16	0.6	0.3	0.5	0.5	17.9	22.3	135	119
New Zealand	3	4	4	1.0	0.7	0.6	0.5	15.4	17.9	123	120
Nicaragua	3	5	6	3.0	2.4	1.0	0.9	4.5	5.1	117	116
Niger	6	9	14	3.3	3.0	1.0	1.0	3.9	3.7	123	130
Nigeria	71	115	166	3.0	2.6	0.9	0.9	4.1	4.3	130	127
Norway	4	4	5	0.4	0.3	0.6	0.5	20.1	22.0	130	119
Oman	1	2	4	4.2	3.8	0.9	1.0	3.9	4.8	99	..
Pakistan	83	134	190	3.0	2.5	0.9	0.9	4.9	5.5	96	97
Panama	2	3	3	2.0	1.3	0.8	0.6	7.6	9.6	103	105
Papua New Guinea	3	4	6	2.2	2.0	0.8	0.7	4.9	5.8	103	109
Paraguay	3	5	7	2.9	2.2	0.9	0.8	5.2	6.0	132	117
Peru	17	24	30	2.1	1.5	0.8	0.7	6.7	7.8	114	116
Philippines	48	72	92	2.5	1.8	0.8	0.7	5.4	6.8	115	116
Poland	36	39	40	0.5	0.2	0.5	0.5	15.8	18.2	151	149
Portugal	10	10	10	0.1	–0.2	0.6	0.5	21.0	21.2	149	147
Puerto Rico	3	4	4	1.0	0.7	0.7	0.5	13.7	16.6	125	152
Romania	22	23	22	0.1	–0.3	0.6	0.5	17.4	19.2	130	138
Russian Federation	139	148	143	0.4	–0.2	0.5	0.5	17.1	18.3	198	181

	Total population (millions)			Average annual population growth rate (%)		Age dependency ratio (dependents as proportion of working-age population)		Population age 60 and above (% of total)		Women age 60 and above (per 100 men)	
	1980	1996	2010	1980–96	1996–2010	1980	1996	1996	2010	1996	2010
Rwanda	5	7	11	1.7	3.5	1.0	1.0	3.6	3.1	122	125
Saudi Arabia	9	19	31	4.6	3.3	0.9	0.8	4.4	5.6	92	..
Senegal	6	9	12	2.7	2.3	0.9	0.9	4.6	3.9	111	111
Sierra Leone	3	5	6	2.2	2.2	0.9	1.0	4.4	4.2	130	131
Singapore	2	3	3	1.8	0.9	0.5	0.4	9.5	14.5	117	113
Slovak Republic	5	5	6	0.4	0.2	0.6	0.5	15.0	17.6	148	148
Slovenia	2	2	2	0.3	–0.1	0.5	0.4	17.8	22.3	160	138
South Africa	27	38	46	2.0	1.4	0.8	0.6	6.5	7.3	142	140
Spain	37	39	38	0.3	–0.1	0.6	0.5	20.6	22.8	134	141
Sri Lanka	15	18	21	1.4	1.0	0.7	0.5	8.9	11.8	106	120
Sudan	19	27	37	2.4	2.2	0.9	0.8	4.9	5.7	116	110
Sweden	8	9	9	0.4	0.0	0.6	0.6	21.9	26.0	129	119
Switzerland	6	7	7	0.7	0.3	0.5	0.5	19.3	23.6	137	126
Syrian Arab Republic	9	15	20	3.2	2.3	1.1	0.9	4.6	4.9	108	120
Tajikistan	4	6	7	2.5	1.3	0.9	0.8	6.5	6.4	131	124
Tanzania	19	30	42	3.1	2.2	1.0	0.9	4.1	3.9	120	114
Thailand	47	60	66	1.6	0.6	0.8	0.5	7.8	9.8	123	121
Togo	3	4	6	3.0	2.5	0.9	1.0	4.9	4.4	121	117
Trinidad and Tobago	1	1	1	1.1	0.9	0.7	0.6	9.0	11.1	102	101
Tunisia	6	9	11	2.2	1.4	0.8	0.6	7.1	7.8	102	117
Turkey	44	63	76	2.1	1.3	0.8	0.6	8.2	9.7	113	119
Turkmenistan	3	5	6	3.0	1.5	0.8	0.8	6.3	6.2	143	133
Uganda	13	20	27	2.7	2.4	1.0	1.0	3.6	2.5	116	99
Ukraine	50	51	47	0.1	–0.6	0.5	0.5	19.4	20.9	184	170
United Arab Emirates	1	3	3	5.5	1.9	0.4	0.4	3.2	10.2	45	..
United Kingdom	56	59	59	0.3	0.1	0.6	0.5	20.7	23.3	131	121
United States	227	265	294	1.0	0.7	0.5	0.5	16.4	18.8	134	126
Uruguay	3	3	3	0.6	0.6	0.6	0.6	17.0	17.2	133	141
Uzbekistan	16	23	29	2.3	1.6	0.9	0.8	6.6	6.5	141	130
Venezuela	15	22	28	2.5	1.6	0.8	0.7	6.2	8.4	116	116
Vietnam	54	75	94	2.1	1.6	0.9	0.7	7.2	6.8	136	142
West Bank and Gaza	1	2	4	4.0	3.5	0.9	0.9	4.4	4.3	103	119
Yemen, Rep.	9	16	25	3.8	3.3	1.1	1.0	3.9	3.4	126	148
Yugoslavia, FR (Serb./Mont.)	10	11	11	0.5	0.2	0.5	0.5	18.0	19.1	122	123
Zambia	6	9	12	3.0	1.9	1.1	1.0	3.7	3.3	101	106
Zimbabwe	7	11	14	3.0	1.5	1.0	0.8	4.7	4.3	112	103
World	4,427 t	5,754 t	6,788 t	1.6 w	1.2 w	0.7 w	0.6 w	9.6 w	10.8 w	121 w	117 w
Low income	2,375	3,236	3,948	1.9	1.4	0.8	0.7	7.4	8.5	105	105
Excl. China & India	706	1,076	1,471	2.6	2.2	0.9	0.9	5.1	5.2	114	115
Middle income	1,227	1,599	1,875	1.7	1.1	0.7	0.6	9.1	10.1	137	132
Lower middle income	867	1,125	1,313	1.6	1.1	0.7	0.6	9.2	10.1	140	133
Upper middle income	360	473	562	1.7	1.2	0.7	0.6	8.8	10.3	131	130
Low & middle income	3,602	4,835	5,824	1.8	1.3	0.8	0.6	8.0	9.0	116	114
East Asia & Pacific	1,359	1,732	1,975	1.5	0.9	0.7	0.5	8.7	10.6	106	104
Europe & Central Asia	428	478	490	0.7	0.2	0.6	0.5	14.6	16.0	163	156
Latin America & Carib.	358	486	588	1.9	1.4	0.8	0.6	7.5	9.1	120	127
Middle East & N. Africa	175	276	371	2.9	2.1	0.9	0.8	5.8	6.3	101	103
South Asia	902	1,266	1,555	2.1	1.5	0.8	0.7	6.7	7.9	103	104
Sub-Saharan Africa	379	596	844	2.8	2.5	0.9	0.9	4.6	4.5	123	120
High income	825	919	964	0.7	0.3	0.5	0.5	18.0	21.8	134	126

About the data

Knowing the size of a country's population, its growth rate, and its age distribution is important for evaluating the welfare of its citizens, assessing the productive capacity of its economy, and estimating the quantity of goods and services that will be needed to meet future needs. Thus governments, businesses, and anyone interested in analyzing economic performance must have accurate population estimates.

Population estimates are usually based on national censuses, but the frequency and quality of these censuses vary by country. Most countries conduct a complete enumeration no more than once a decade. Precensus and postcensus estimates are interpolations or extrapolations based on demographic models. Errors and undercounting occur even in high-income countries; in developing countries such errors may be substantial because of limits on transportation, communication, and resources required to conduct a full census. Moreover, the international comparability of population indicators is limited by differences in the concepts, definitions, data collection procedures, and estimation methods used by national statistical agencies and other organizations that collect population data.

Of the 148 economies listed in the table, 129 conducted a census between 1987 and 1997. The currentness of a census, along with the availability of complementary data from surveys or registration systems, is one of many objective ways to judge the quality of demographic data. In some European countries registration systems offer complete information on population in the absence of a census. See *Primary data documentation* for the most recent census or survey year and for registration completeness.

Current population estimates for developing countries that lack recent census-based population data, and precensus and postcensus estimates for countries with census data, are provided by national statistical offices or by the United Nations Population Division. The estimation methods require fertility, mortality, and net migration data, which are often collected from sample surveys, some of which may be small or have limited coverage. These estimates are the product of demographic modeling and so are also susceptible to biases and errors due to shortcomings of the model, as well as the data.

The quality and reliability of official demographic data are also affected by public trust in the government, the government's commitment to full and accurate enumeration, the confidentiality of and protection against misuse accorded to census data,

and the independence of census agencies from undue political influence.

Population projections are made using the cohort component method. This method compiles separate projections of future fertility, mortality, and net migration levels by age and gender, then applies them to the 1995 base year age and gender structure. Future fertility, mortality, and net migration levels are determined from demographic models that use current levels and trends as inputs. Countries where fertility has been falling are assumed to have further declines at the rate of the previous 10 years until fertility reaches the replacement level of about two children. In countries where fertility has remained high, the transition to smaller families is assumed to occur at the average rate of decline of countries that are currently making this transition. Countries where fertility is below two children per woman are assumed to remain at this level for another decade, after which fertility rates will gradually return to replacement level. Similarly, mortality changes are modeled by assuming that the rate of change in the previous decade will continue in the near future. Future mortality in countries with high levels of HIV infection is adjusted to reflect the lagged impact of the disease on mortality.

Figure 2.1a

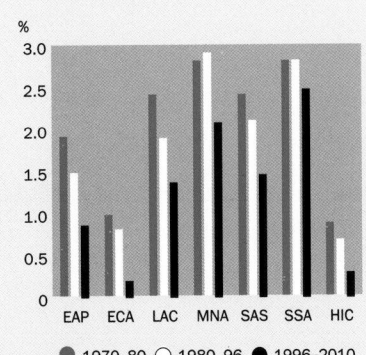

Population growth rates are declining

Source: World Bank staff estimates.

The world's population is expected to increase by more than 1 billion people over the next 14 years. Of this increase, 9 out of 10 people will be added in developing countries. While the highest growth rates will continue to be in Sub-Saharan Africa and the Middle East and North Africa, the variation in growth rates differs from the pattern of absolute increase in population. The largest population increases are expected in South Asia, East Asia, and Sub-Saharan Africa.

Definitions

• **Total population** of an economy includes all residents regardless of legal status or citizenship—except for refugees not permanently settled in the country of asylum, who are generally considered part of the population of their country of origin. The indicators shown are midyear estimates for 1980 and 1996 and projections for 2010. • **Average annual population growth rate** is the exponential change for the period indicated. See *Statistical methods* for more information. • **Age dependency ratio** is the ratio of dependents—people younger than 15 and older than 65—to the working-age population—those age 15–64. • **Population age 60 and above** is the percentage of the total population that is 60 or older. • **Women age 60 and above** is the ratio of women to men in that age group.

Data sources

The World Bank's population estimates are produced by the Human Development Network and the Development Data Group in consultation with the Bank's operational staff. Important inputs to the World Bank's demographic work come from the following sources: census reports and other statistical publications and electronic bulletins from country statistical offices; demographic and health surveys conducted by national sources; United Nations Department of Economic and Social Information and Policy Analysis, Statistics Division, *Population and Vital Statistics Report* (quarterly), and Population Division, *World Population Prospects: The 1996 Edition*; Eurostat, *Demographic Statistics* (various years); Council of Europe, *Recent Demographic Developments in Europe and North America 1996*; South Pacific Commission, *Pacific Island Populations Data Sheet 1997*; Centro Latinoamericano de Demografia, *Boletin Demografico* (various years); Economic and Social Commission for Western Asia, *Demographic and Related Socio-Economic Data Sheets 1995*; and U.S. Bureau of the Census, *World Population Profile 1996*. Projections are based on the methods discussed in Bos and others, *World Population Projections 1994–95*.

2.2 | Population dynamics

	Crude death rate per 1,000 people		Crude birth rate per 1,000 people		Projected additional population from 2000	Population momentum	Future population growth due to			Average annual population growth rates		
							Momentum	Above-replacement fertility	Mortality improvements	Age 0–14 %	Age 15–64 %	Age 65+ %
	1980	1996	1980	1996	millions	1996	millions	millions	millions	1990–96	1990–96	1990–96
Albania	6	6	29	20	2	1.4	1.2	0.1	0.5	–0.8	0.1	2.6
Algeria	12	5	42	26	31	1.6	18.4	4.8	8.1	0.4	3.5	3.3
Angola	23	19	50	48	37	1.5	6.4	23.7	6.5	3.3	2.8	2.7
Argentina	9	8	24	19	20	1.4	13.5	0.8	5.9	0.2	1.7	2.3
Armenia	6	7	23	13	1	1.2	0.9	–1.0	0.6	–0.5	1.1	6.4
Australia	7	7	15	14	2	1.2	3.2	–1.6	0.8	0.7	1.1	2.1
Austria	12	10	12	11	–2	1.0	0.0	–2.0	0.2	0.8	0.7	0.5
Azerbaijan	7	6	25	17	3	1.3	2.6	–1.8	1.7	0.0	1.0	4.6
Bangladesh	18	10	44	28	127	1.5	69.3	7.8	49.6	0.5	2.6	1.0
Belarus	10	13	16	9	–2	1.0	0.0	–3.2	1.7	–1.5	0.0	3.0
Belgium	12	11	13	11	–2	1.0	0.0	–1.9	0.4	–0.1	0.2	1.2
Benin	19	13	49	42	13	1.6	3.9	6.9	2.6	2.2	3.7	0.3
Bolivia	15	9	39	34	12	1.6	4.6	4.3	2.9	2.0	2.6	3.3
Bosnia and Herzegovina	7	..	19	..	..	..	..	..	..	..	..	..
Botswana	14	13	48	33	2	1.4	0.6	0.4	0.9	1.7	3.1	2.1
Brazil	9	7	31	21	107	1.4	61.6	0.3	44.6	–0.4	2.3	3.2
Bulgaria	11	14	15	9	–3	0.9	–0.6	–2.9	1.0	–2.9	–0.6	1.5
Burkina Faso	20	18	47	45	31	1.4	4.8	19.3	6.6	2.5	2.9	4.0
Burundi	18	17	46	43	18	1.4	3.1	11.1	3.8	3.0	2.4	0.7
Cambodia	27	13	40	34	14	1.4	5.0	5.1	4.3	3.7	2.0	3.2
Cameroon	15	11	47	40	29	1.5	8.0	12.6	8.1	2.7	3.1	2.7
Canada	7	7	15	13	1	1.1	3.7	–3.8	0.8	0.6	1.2	2.6
Central African Republic	19	17	43	38	6	1.4	1.5	2.4	1.7	2.0	2.4	1.9
Chad	22	17	44	42	15	1.5	3.4	8.5	3.0	2.6	2.5	2.2
Chile	7	5	24	19	8	1.4	6.2	0.0	1.5	1.0	1.6	3.1
China	6	7	18	17	348	1.2	276.7	–184.5	255.2	0.1	1.3	3.6
Hong Kong, China	5	5	17	10	–1	1.0	0.3	–2.1	0.5	–0.8	1.8	4.7
Colombia	7	6	30	23	28	1.4	17.5	0.4	9.8	0.8	2.1	5.0
Congo, Dem. Rep.	17	14	48	45	156	1.6	29.8	100.9	25.7	3.3	3.0	2.8
Congo, Rep.	16	15	46	43	8	1.5	1.5	4.5	1.4	3.1	2.6	3.8
Costa Rica	4	4	30	23	3	1.6	2.2	0.1	0.4	1.1	2.5	4.2
Côte d'Ivoire	16	12	51	37	24	1.5	8.4	9.1	6.9	2.2	3.6	3.8
Croatia	..	11	..	11	–1	0.9	–0.3	–1.0	0.6	–2.2	–0.4	4.9
Cuba	6	7	14	14	0	1.2	2.0	–3.0	1.0	–0.1	0.7	1.9
Czech Republic	13	11	15	9	–2	1.0	0.1	–3.6	1.1	–2.5	0.6	0.2
Denmark	11	12	11	13	0	1.0	–0.1	–0.5	0.4	0.9	0.3	–0.4
Dominican Republic	7	5	33	26	7	1.5	4.6	0.4	1.8	0.8	2.4	4.5
Ecuador	9	6	36	26	11	1.6	7.0	1.0	3.0	0.8	2.9	3.2
Egypt, Arab Rep.	13	8	39	26	58	1.5	31.3	9.0	17.7	0.9	2.6	3.3
El Salvador	11	6	39	31	7	1.6	4.1	1.2	1.7	0.5	3.6	4.2
Eritrea	..	13	..	40	11	1.5	2.1	6.1	2.3	2.8	2.6	3.3
Estonia	12	13	15	9	0	0.9	–0.1	–0.5	0.2	–3.1	–1.0	0.9
Ethiopia	20	17	47	48	216	1.5	34.8	148.5	33.0	2.5	1.9	1.8
Finland	9	10	13	12	0	0.9	–0.7	–0.4	1.0	0.1	0.3	1.5
France	10	9	15	13	–1	1.1	6.3	–8.0	0.8	–0.4	0.4	2.0
Gabon	18	14	33	36	2	1.4	0.5	1.0	0.4	3.3	2.2	2.5
Gambia, The	24	14	48	40	2	1.4	0.5	1.1	0.6	3.3	3.8	3.7
Georgia	9	7	18	11	0	1.1	0.6	–1.5	0.7	–1.2	–0.4	3.6
Germany	12	11	11	9	–23	0.9	–5.5	–19.9	2.8	0.4	0.4	1.0
Ghana	15	10	45	36	33	1.6	12.0	14.3	7.0	2.4	2.9	3.7
Greece	9	10	15	10	–2	1.0	0.2	–2.7	0.3	–2.0	0.6	3.4
Guatemala	11	7	43	35	20	1.7	8.2	7.5	4.0	2.3	3.2	4.3
Guinea	24	18	46	43	17	1.5	3.5	10.2	3.4	2.5	2.8	2.9
Guinea-Bissau	25	22	43	44	2	1.4	0.4	1.4	0.6	2.4	1.8	1.7
Haiti	15	12	37	32	9	1.4	3.3	2.1	3.5	2.0	2.1	1.1
Honduras	10	6	43	35	11	1.7	4.8	4.0	2.1	2.4	3.6	4.1

	Crude death rate		Crude birth rate		Projected additional population from 2000	Population momentum	Future population growth due to			Average annual population growth rates		
	per 1,000 people		per 1,000 people		millions		Momentum	Above-replacement fertility	Mortality improvements	Age 0–14 %	Age 15–64 %	Age 65+ %
	1980	1996	1980	1996		1996	millions	millions	millions	1990–96	1990–96	1990–96
Hungary	14	14	14	10	–2	1.0	–0.4	–2.5	1.5	–2.3	0.1	0.6
India	13	9	35	25	724	1.4	391.9	53.5	278.9	0.9	2.2	3.1
Indonesia	12	8	34	23	144	1.4	81.0	1.9	61.0	0.2	2.3	3.8
Iran, Islamic Rep.	11	5	44	26	66	1.6	37.1	13.1	16.1	0.6	3.8	5.3
Iraq	9	9	41	37	51	1.7	17.2	27.6	5.9	2.1	3.3	3.5
Ireland	10	9	22	14	1	1.3	0.9	–0.5	0.3	–2.1	1.4	0.5
Israel	7	6	24	21	4	1.4	2.7	0.5	0.7	2.0	4.2	1.5
Italy	10	10	11	9	–18	0.9	–3.0	–15.9	1.1	–1.8	0.1	2.2
Jamaica	7	6	28	21	2	1.5	1.3	–0.3	0.5	0.3	1.4	0.0
Japan	6	7	14	10	–30	0.8	–21.0	–28.8	19.4	–2.3	0.2	4.0
Jordan	..	5	..	31	7	1.7	3.4	2.3	1.3	4.0	6.0	5.7
Kazakhstan	8	10	24	15	6	1.2	3.3	–1.7	4.3	–1.5	0.0	2.7
Kenya	13	9	51	34	42	1.6	18.2	10.3	13.4	1.4	3.8	1.3
Korea, Dem. Rep.	6	9	22	22	10	1.2	5.0	–0.9	6.1	1.6	1.4	3.5
Korea, Rep.	6	6	22	15	6	1.2	7.5	–9.4	8.3	–1.0	1.4	3.5
Kuwait	4	2	37	22	1	1.5	0.9	0.3	0.2	–4.5	–5.2	1.4
Kyrgyz Republic	9	8	30	24	4	1.5	2.2	0.2	1.4	0.1	0.7	3.1
Lao PDR	20	14	45	40	10	1.5	2.7	4.7	2.2	2.7	2.3	6.2
Latvia	13	14	15	8	–1	1.0	–0.1	–0.9	0.3	–2.3	–1.2	0.8
Lebanon	9	7	30	24	3	1.5	2.1	0.1	1.0	1.4	2.1	3.2
Lesotho	15	11	41	32	3	1.5	1.1	1.0	0.8	1.2	2.8	1.9
Libya	12	5	46	28	9	1.6	3.6	2.9	1.9	1.1	3.6	5.8
Lithuania	10	12	16	11	0	1.0	0.1	–1.0	0.5	–1.1	–0.1	2.0
Macedonia, FYR	7	8	21	16	1	1.2	0.4	0.0	0.3	–0.7	0.9	3.5
Madagascar	16	11	46	41	38	1.6	9.9	21.6	6.8	2.1	3.1	4.7
Malawi	23	20	57	46	24	1.4	4.6	13.7	5.6	2.5	2.9	1.8
Malaysia	6	5	31	27	20	1.5	12.3	2.2	5.2	2.0	2.4	3.2
Mali	22	16	49	49	30	1.6	6.5	17.5	5.9	3.0	2.5	3.2
Mauritania	19	14	43	38	4	1.5	1.3	2.0	1.0	2.0	3.0	2.2
Mauritius	6	7	24	18	1	1.3	0.3	0.0	0.2	–0.3	1.6	3.1
Mexico	7	5	33	26	82	1.6	60.2	2.9	19.1	0.3	2.8	2.3
Moldova	10	12	20	12	1	1.1	0.5	–0.9	0.9	–1.4	0.2	1.9
Mongolia	11	7	38	28	3	1.6	1.6	0.4	0.8	0.7	3.1	1.5
Morocco	12	7	38	25	25	1.5	15.1	2.2	7.7	0.4	2.7	3.9
Mozambique	20	18	46	44	47	1.5	9.1	27.9	9.9	4.4	3.8	–0.9
Myanmar	14	10	36	27	39	1.4	21.1	2.7	15.3	0.8	2.2	2.7
Namibia	14	12	41	36	3	1.5	0.8	1.3	0.8	2.3	2.8	3.0
Nepal	17	11	43	37	42	1.5	12.6	20.4	9.3	2.6	2.7	2.5
Netherlands	8	9	13	12	–2	1.0	0.7	–2.8	0.6	0.8	0.5	1.3
New Zealand	9	8	16	16	1	1.2	0.8	–0.1	0.3	0.8	1.2	2.0
Nicaragua	11	6	45	33	7	1.7	3.7	1.9	1.5	1.9	3.9	4.6
Niger	23	18	51	51	33	1.5	5.6	21.7	5.7	3.5	3.0	2.9
Nigeria	18	13	50	41	280	1.5	68.3	152.7	59.0	2.8	3.1	1.4
Norway	10	10	12	14	0	1.0	0.1	–0.3	0.4	0.9	0.5	0.1
Oman	10	4	45	42	10	1.8	2.0	7.3	1.0	5.4	4.9	5.6
Pakistan	15	8	47	37	260	1.7	105.1	121.1	34.0	2.9	2.8	3.9
Panama	6	5	29	22	2	1.5	1.5	0.0	0.4	0.6	2.4	2.9
Papua New Guinea	14	10	37	32	7	1.4	2.0	2.7	2.0	1.8	2.4	5.6
Paraguay	7	5	36	30	7	1.7	3.6	1.8	1.3	2.3	3.0	2.0
Peru	11	6	35	25	22	1.5	13.0	1.5	7.1	0.7	2.7	3.8
Philippines	9	7	35	29	80	1.5	41.0	14.7	24.7	1.6	2.7	3.0
Poland	10	10	19	11	1	1.1	3.7	–8.6	5.5	–1.9	0.7	2.0
Portugal	10	11	16	11	–2	1.0	0.2	–2.7	0.9	–1.9	0.0	2.8
Puerto Rico	6	8	23	17	1	1.3	1.1	–0.3	0.5	–0.5	1.6	1.8
Romania	10	13	18	10	–4	1.0	0.3	–7.8	3.7	–3.2	0.0	2.0
Russian Federation	11	14	16	9	–20	1.0	–4.1	–42.6	26.5	–1.8	0.0	3.2

2.2 Population dynamics

	Crude death rate per 1,000 people		Crude birth rate per 1,000 people		Projected additional population from 2000	Population momentum	Future population growth due to			Average annual population growth rates		
							Momentum millions	Above-replacement fertility millions	Mortality improvements millions	Age 0–14 % 1990–96	Age 15–64 % 1990–96	Age 65+ % 1990–96
	1980	1996	1980	1996	millions	1996						
Rwanda	19	21	51	40	18	1.4	3.5	9.3	4.8	–0.3	–0.5	–3.4
Saudi Arabia	9	5	43	35	69	1.6	14.2	46.5	8.0	3.4	3.7	4.7
Senegal	20	14	46	40	18	1.4	3.8	8.9	5.2	2.2	2.8	2.6
Sierra Leone	29	27	49	48	10	1.4	1.8	5.1	2.7	3.3	1.8	–0.5
Singapore	5	5	17	16	0	1.1	0.4	–0.4	0.3	2.5	1.5	4.3
Slovak Republic	10	10	19	11	0	1.1	0.6	–1.5	0.8	–2.0	0.8	1.1
Slovenia	10	9	15	10	–1	0.9	–0.1	–0.6	0.3	–2.8	0.3	1.8
South Africa	12	8	36	27	32	1.5	18.5	1.5	12.3	0.0	2.8	1.1
Spain	8	9	15	9	–10	1.0	0.7	–12.9	1.8	–3.0	0.5	2.3
Sri Lanka	6	6	28	19	10	1.4	6.8	–0.4	3.5	–0.9	1.8	3.6
Sudan	17	12	45	34	48	1.5	13.8	21.6	13.0	0.9	3.0	3.0
Sweden	11	11	12	11	–1	0.9	–0.9	–1.4	1.4	1.3	0.4	0.1
Switzerland	9	9	12	12	–1	1.0	0.1	–1.1	0.1	1.5	0.7	1.0
Syrian Arab Republic	9	5	46	30	23	1.7	11.6	6.2	5.0	1.5	4.1	4.5
Tajikistan	8	5	37	22	6	1.6	3.9	0.0	1.8	1.0	2.3	4.1
Tanzania	15	14	47	41	66	1.5	16.3	32.8	17.2	2.7	3.1	3.2
Thailand	8	7	28	17	17	1.3	18.1	–14.8	13.2	–1.3	2.2	4.0
Togo	16	15	45	42	12	1.5	2.2	7.6	2.3	3.1	2.9	2.7
Trinidad and Tobago	7	7	29	16	1	1.3	0.4	0.0	0.2	–1.3	1.8	1.2
Tunisia	9	6	35	23	7	1.5	4.7	0.1	2.4	0.4	2.6	3.9
Turkey	10	7	32	22	43	1.4	27.5	0.7	14.8	–0.4	2.7	5.3
Turkmenistan	8	7	34	24	4	1.5	2.6	0.3	1.5	3.0	4.1	5.5
Uganda	18	19	49	49	50	1.4	8.4	27.2	14.2	3.3	3.0	1.7
Ukraine	11	15	15	9	–10	0.9	–2.6	–15.1	8.1	–1.7	–0.3	2.1
United Arab Emirates	5	3	30	19	2	1.2	0.6	0.6	0.3	3.8	5.8	6.8
United Kingdom	12	11	13	12	–3	1.1	3.5	–7.8	1.8	0.4	0.3	0.4
United States	9	8	16	15	70	1.2	49.1	11.4	9.9	1.1	1.0	1.2
Uruguay	10	10	19	17	1	1.2	0.7	0.0	0.4	–0.4	0.8	1.6
Uzbekistan	8	6	34	27	24	1.6	14.9	1.9	6.8	1.4	2.4	3.7
Venezuela	6	5	33	25	20	1.5	13.2	1.7	5.1	1.1	2.8	4.3
Vietnam	8	7	36	25	70	1.5	42.8	5.2	22.0	1.2	2.8	2.1
West Bank and Gaza	..	5	..	44	8	1.8	2.0	4.5	1.1	4.6	3.9	4.3
Yemen, Rep.	19	13	53	47	66	1.6	11.1	45.7	9.2	4.3	5.2	3.9
Yugoslavia, FR (Serb./Mont.)	9	11	18	13	1	1.0	0.4	–1.1	1.8	–1.4	–0.1	4.3
Zambia	15	18	50	43	15	1.4	4.0	5.4	5.7	1.9	3.6	1.8
Zimbabwe	13	10	49	31	13	1.5	5.6	1.5	5.6	1.6	2.9	4.2
World	**10 w**	**9 w**	**27 w**	**22 w**	**4,199 t**	**1.4 w**	**1,946.2 t**	**807.4 t**	**1,445.6 t**	**0.8 w**	**1.8 w**	**2.8 w**
Low income	11	9	31	26	3,129	1.5	1,278.8	857.8	992.1	1.2	2.0	3.2
Excl. China & India	16	12	45	37	2,057	1.4	610.2	988.8	458.0	2.1	2.7	2.5
Middle income	10	8	29	21	1,078	1.4	616.7	63.8	397.7	0.2	1.9	3.1
Lower middle income	10	9	29	21	721	1.4	418.0	16.5	286.7	0.2	1.9	3.4
Upper middle income	9	7	28	22	357	1.4	198.7	47.3	111.0	0.1	2.1	2.4
Low & middle income	11	9	30	24	4,207	1.4	1,895.5	921.6	1,389.8	0.9	1.9	3.2
East Asia & Pacific	8	7	22	19	761	1.3	509.3	–160.6	412.0	0.3	1.6	3.6
Europe & Central Asia	10	11	19	13	50	1.1	57.8	–97.6	89.7	–1.2	0.5	2.8
Latin America & Carib.	8	7	31	23	385	1.4	238.3	28.3	118.3	0.4	2.4	3.0
Middle East & N. Africa	11	7	41	29	433	1.6	173.8	172.3	86.4	1.5	3.5	4.1
South Asia	14	9	37	27	1,163	1.4	585.7	202.4	375.3	1.1	2.3	3.0
Sub-Saharan Africa	18	14	47	41	1,416	1.5	330.6	776.8	308.1	2.4	2.9	2.1
High income	9	9	15	12	–8	1.1	50.7	–114.2	55.8	0.4	1.1	2.1

About the data

The vital rates shown in the table are based on data derived from registration systems, censuses, and sample surveys conducted by national statistical offices. As with the basic demographic data in table 2.1, estimates for 1996 are based on projections from censuses or surveys from earlier years, and hence international comparisons are limited by differences in definitions and data collection and estimation methods.

Vital registers are the preferred source of these data, but in many developing countries systems for registering births and deaths do not exist or are incomplete because of deficiencies in geographic coverage or population coverage. For these countries, vital rates are estimated by applying various demographic methods to incomplete vital registration data or to data from surveys and censuses. The United Nations Department of Economic and Social Information and Policy Analysis has monitored vital registration systems for many years. Its quarterly publication, *Population and Vital Statistics Report*, shows that the proportion of countries with at least 90 percent complete vital registration increased from 46 percent in 1990 to 52 percent in 1997. Still, some of the most populous developing countries—China, India, Indonesia, Brazil, Pakistan, Nigeria, Bangladesh—do not have complete vital registration systems. As a result less than 25 percent of vital events worldwide are thought to be recorded.

In many countries fertility rates have fallen to near the two-child replacement level, and in some countries they have fallen well below that. But almost all these countries will continue to have growing populations over the next several decades as large cohorts born in previous years move through the reproductive ages, generating more births than are offset by deaths in the smaller, older cohorts. The reverse may happen in countries with aging populations and a history of low fertility rates. This phenomenon, called *population momentum*, is measured here as the ratio of the population when zero growth has been achieved to the population in 2000, assuming that fertility remains at replacement level from 2000 onward. A momentum ratio greater than one indicates that population will continue to grow even after replacement-level fertility has been achieved; a ratio of less than one indicates that population will decline.

Population will continue to grow in most countries for several reasons: fertility will remain above replacement level, increasing the size of each generation; population momentum in the age structure will lead to

more births than deaths (momentum greater than one); mortality will keep falling (the situation in most countries), with the greatest effect on population growth in countries where infant and child mortality are currently high; and net migration will be positive.

The table shows the contribution that each of these components makes to future population growth (mortality and migration are combined). For example, Algeria's population is projected to grow to 62 million before it stabilizes. Of the 31 million increase, about 18 million is the result of population momentum, 5 million is due to excess fertility, and 8 million is due to projected mortality decline. A negative value for any component indicates that current conditions are such that they would lead to population decline. A momentum indicator of less than one indicates that even a recovery to replacement-level fertility by 2000 will not prevent a decline in population.

Figure 2.2a

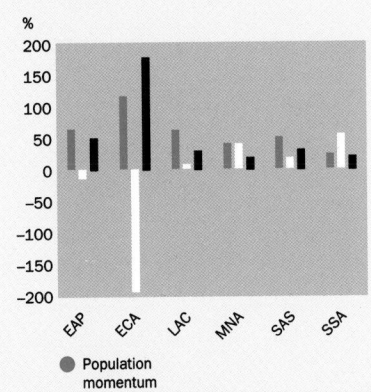

Population momentum tends to be the biggest contributor to population growth

- ● Population momentum
- ○ Above-replacement fertility
- ● Mortality improvements

Source: World Bank staff estimates.

Between 2000 and when stationary population is reached, the world's population will increase by about 4.2 billion people. This figure shows the contributions that population momentum, high fertility, and mortality improvements will make to future population growth. Population momentum is likely to be a major source of population growth in all regions except Sub-Saharan Africa. In Sub-Saharan Africa the persistence of fertility well above replacement level will account for more than half of the region's future population growth. In contrast, low fertility in Europe and Central Asia and East Asia will reduce overall population growth.

Definitions

- **Crude death rate** and **crude birth rate** are the number of deaths and the number of live births occurring during the year, per 1,000 midyear population. The difference between the crude birth rate and crude death rate is the rate of natural increase.
- **Projected additional population from 2000** is the projected increase in population between 2000 and the projected stationary population that is reached after fertility has been at replacement level for many decades. A negative number indicates a projected decline in population. • **Population momentum** is the ratio of the population when zero growth has been achieved to the population in year *t* (in this case the year 2000), given the assumption that fertility remains at replacement level from year *t* onward. • **Future population growth due to momentum** is the projected increase in population from 2000 onward that would occur if fertility were at replacement level. A negative number indicates that negative momentum has built up in the age structure as the result of fertility being below replacement level for several decades. • **Future population growth due to above-replacement fertility** is the projected change in population from 2000 onward that would occur if fertility were not at replacement level.
- **Future population growth due to mortality improvements** is the projected increase in population from 2000 onward due to projected changes in mortality and net migration. • **Average annual population growth rates** are calculated using the exponential end-point method (see *Statistical methods* for more information).

Data sources

The World Bank's population estimates are produced by the Human Development Network and the Development Data Group in consultation with the Bank's operational staff. Important inputs to the World Bank's demographic work come from the following sources: United Nations Department of Economic and Social Information and Policy Analysis, Statistics Division, *Population and Vital Statistics Report* (quarterly), and Population Division, *World Population Prospects: The 1996 Edition*; census reports and other statistical publications from country statistical offices; demographic and health surveys conducted by national sources; and Eurostat, *Demographic Statistics* (various years). Projections are based on the methods discussed in Bos and others, *World Population Projections 1994–95*.

2.3 Labor force structure

	Population age 15–64 (millions)		Labor force — Total millions			Average annual growth rate %		Female % of labor force		Children 10–14 % of age group	
	1980	1996	1980	1996	2010	1980–96	1996–2010	1980	1996	1980	1996
Albania	2	2	1	2	2	1.8	1.1	39	41	4	1
Algeria	9	17	5	9	15	3.7	3.6	21	25	7	1
Angola	4	5	3	5	8	2.3	2.7	47	46	30	27
Argentina	17	22	11	14	18	1.5	1.9	28	31	8	4
Armenia	2	2	1	2	2	1.2	0.9	48	48	0	0
Australia	10	12	7	9	10	1.8	0.7	37	43	0	0
Austria	5	5	3	4	4	0.5	–0.1	40	41	0	0
Azerbaijan	4	5	3	3	4	1.1	1.4	47	44	0	0
Bangladesh	44	66	41	61	81	2.4	1.9	42	42	35	30
Belarus	6	7	5	5	5	0.2	0.0	50	49	0	0
Belgium	6	7	4	4	4	0.3	–0.2	34	40	0	0
Benin	2	3	2	3	4	2.5	2.5	47	48	30	27
Bolivia	3	4	2	3	4	2.4	2.4	33	37	19	14
Bosnia and Herzegovina	3	..	2	..	..	..	..	33	1		
Botswana	0	1	0	1	1	2.9	1.7	50	46	26	16 ..
Brazil	70	103	48	72	87	2.4	1.2	28	35	19	16
Bulgaria	6	6	5	4	4	–0.5	–0.8	45	48	0	0
Burkina Faso	3	5	4	6	7	1.9	1.9	48	47	71	50
Burundi	2	3	2	3	5	2.5	2.6	50	49	50	49
Cambodia	4	6	3	5	7	2.5	2.3	56	53	27	24
Cameroon	5	7	4	6	8	2.5	2.5	37	38	34	25
Canada	17	20	12	16	17	1.5	0.5	40	45	0	0
Central African Republic	1	2	1	2	2	1.6	1.7	48	47	39	31
Chad	2	3	2	3	5	2.1	2.3	43	44	42	38
Chile	7	9	4	6	7	2.3	1.9	26	32	0	0
China	586	821	539	718	804	1.7	0.7	43	45	30	11
Hong Kong, China	3	4	2	3	4	1.6	0.7	34	37	6	0
Colombia	16	23	9	16	22	3.2	2.0	26	38	12	6
Congo, Dem. Rep.	14	22	12	19	29	2.7	2.8	45	44	33	29
Congo, Rep.	1	1	1	1	2	2.7	2.4	43	43	27	26
Costa Rica	1	2	1	1	2	3.1	1.9	21	30	10	5
Côte d'Ivoire	4	7	3	5	7	2.6	2.0	32	33	28	20
Croatia	3	3	2	2	2	0.2	–0.3	40	44	0	0
Cuba	6	8	4	5	6	1.9	0.7	31	38	0	0
Czech Republic	6	7	5	6	5	0.3	–0.3	47	47	0	0
Denmark	3	4	3	3	3	0.4	–0.3	44	46	0	0
Dominican Republic	3	5	2	3	5	2.8	2.2	25	29	25	16
Ecuador	4	7	3	4	6	3.2	2.5	20	27	9	5
Egypt, Arab Rep.	23	35	14	22	32	2.4	2.5	26	29	18	11
El Salvador	2	3	2	2	4	2.3	2.9	27	35	17	15
Eritrea	..	2	..	2	3	..	2.9	47	47	44	39
Estonia	1	1	1	1	1	–0.2	–0.7	51	49	0	0
Ethiopia	19	30	17	26	39	2.6	2.7	42	41	46	42
Finland	3	3	2	3	2	0.4	–0.3	46	48	0	0
France	34	38	24	26	27	0.5	0.3	40	44	0	0
Gabon	0	1	0	1	1	2.0	1.6	45	44	29	18
Gambia, The	0	1	0	1	1	3.1	2.4	45	45	44	36
Georgia	3	4	3	3	3	0.2	0.1	49	46	0	0
Germany	52	56	37	41	40	0.5	0.0	40	42	0	0
Ghana	6	9	5	8	12	2.8	2.4	51	51	16	13
Greece	6	7	4	4	5	1.0	0.3	28	37	5	0
Guatemala	4	6	2	4	6	2.9	3.1	22	27	19	16
Guinea	2	3	2	3	5	2.0	2.4	47	47	41	33
Guinea-Bissau	0	1	0	1	1	1.5	2.0	40	40	43	38
Haiti	3	4	3	3	4	1.4	1.5	45	43	33	25
Honduras	2	3	1	2	4	3.5	3.4	25	30	14	8

Labor force structure | 2.3

	Population age 15–64		Labor force								
	millions		Total millions			Average annual growth rate %		Female % of labor force		Children 10–14 % of age group	
	1980	1996	1980	1996	2010	1980–96	1996–2010	1980	1996	1980	1996
Hungary	7	7	5	5	4	–0.4	–0.4	43	44	0	0
India	395	574	300	408	519	1.8	1.6	34	32	21	14
Indonesia	83	124	59	91	124	2.6	2.0	35	40	13	9
Iran, Islamic Rep.	20	35	12	19	31	2.8	3.3	20	25	14	4
Iraq	7	12	4	6	9	2.8	3.3	17	18	11	3
Ireland	2	2	1	1	2	0.8	1.0	28	33	1	0
Israel	2	4	1	2	3	2.8	2.5	34	40	0	0
Italy	36	39	23	25	24	0.6	–0.2	33	38	2	0
Jamaica	1	2	1	1	2	1.9	1.3	46	46	0	0
Japan	79	87	57	66	66	0.9	0.0	38	41	0	0
Jordan	1	2	1	1	2	4.7	3.8	15	22	4	1
Kazakhstan	9	11	7	8	8	0.6	0.5	48	47	0	0
Kenya	8	14	8	13	18	3.1	2.2	46	46	45	41
Korea, Dem. Rep.	10	15	8	11	13	2.3	1.1	45	45	3	0
Korea, Rep.	24	32	16	22	26	2.1	1.2	39	41	0	0
Kuwait	1	1	0	1	1	2.6	1.7	13	29	0	0
Kyrgyz Republic	2	3	2	2	3	1.4	1.7	48	47	0	0
Lao PDR	2	2	2	2	3	1.9	2.5	45	47	31	27
Latvia	2	2	1	1	1	–0.3	–0.6	51	50	0	0
Lebanon	2	2	1	1	2	2.7	2.5	23	28	5	0
Lesotho	1	1	1	1	1	2.2	2.1	38	37	28	22
Libya	2	3	1	1	2	2.6	2.4	19	21	9	0
Lithuania	2	2	2	2	2	0.3	–0.1	50	48	0	0
Macedonia, FYR	1	1	1	1	1	0.7	0.8	36	41	1	0
Madagascar	4	7	4	7	10	2.5	2.8	45	45	40	35
Malawi	3	5	3	5	7	2.6	2.1	51	49	45	34
Malaysia	8	12	5	8	12	2.6	2.4	34	37	8	3
Mali	3	5	3	5	7	2.2	2.6	47	46	61	54
Mauritania	1	1	1	1	2	2.1	2.3	45	44	30	24
Mauritius	1	1	0	0	1	1.9	1.2	26	32	5	3
Mexico	34	57	22	37	52	3.0	2.3	27	31	9	6
Moldova	3	3	2	2	2	0.1	0.2	50	49	3	0
Mongolia	1	1	1	1	2	2.8	2.3	46	46	4	2
Morocco	10	16	7	11	15	2.4	2.4	34	35	21	5
Mozambique	6	9	7	9	13	1.8	2.5	49	48	39	34
Myanmar	19	28	17	23	29	1.8	1.5	44	43	28	24
Namibia	1	1	0	1	1	2.3	2.1	40	41	34	21
Nepal	8	12	7	10	15	2.3	2.4	39	40	56	45
Netherlands	9	11	6	7	7	1.4	0.0	31	40	0	0
New Zealand	2	2	1	2	2	1.7	0.9	34	44	0	0
Nicaragua	1	2	1	2	3	3.1	3.4	28	36	19	14
Niger	3	5	3	4	7	2.8	2.8	45	44	48	45
Nigeria	38	60	30	45	68	2.5	2.6	36	36	29	25
Norway	3	3	2	2	2	0.8	0.3	40	46	0	0
Oman	1	1	0	1	1	3.6	4.1	7	15	6	0
Pakistan	44	72	29	48	76	2.9	3.1	23	27	23	17
Panama	1	2	1	1	1	2.7	1.9	30	34	6	3
Papua New Guinea	2	3	2	2	3	2.0	2.1	42	42	28	19
Paraguay	2	3	1	2	3	2.7	2.7	27	29	15	7
Peru	9	15	5	9	13	2.9	2.4	24	29	4	2
Philippines	27	42	19	30	42	2.7	2.3	35	37	14	8
Poland	23	26	19	19	20	0.3	0.3	45	46	0	0
Portugal	6	7	5	5	5	0.4	0.0	39	43	8	2
Puerto Rico	2	2	1	1	2	1.7	1.3	32	36	0	0
Romania	14	15	11	11	10	–0.1	–0.1	46	44	0	0
Russian Federation	95	99	76	78	78	0.1	0.0	49	49	0	0

	Population age 15–64		Labor force								
	millions			Total millions		Average annual growth rate %		Female % of labor force		Children 10–14 % of age group	
	1980	1996	1980	1996	2010	1980–96	1996–2010	1980	1996	1980	1996
Rwanda	3	3	3	4	6	2.7	2.5	49	49	43	42
Saudi Arabia	5	11	3	6	10	5.0	3.2	8	14	5	0
Senegal	3	4	3	4	5	2.4	2.3	42	43	43	31
Sierra Leone	2	2	1	2	2	1.9	2.3	36	36	19	15
Singapore	2	2	1	1	2	2.0	0.8	35	38	2	0
Slovak Republic	3	4	2	3	3	0.8	0.2	45	48	0	0
Slovenia	1	1	1	1	1	0.2	–0.2	46	46	0	0
South Africa	15	23	10	15	19	2.2	1.6	35	37	1	0
Spain	23	27	14	17	17	1.1	0.2	28	36	0	0
Sri Lanka	9	12	5	8	10	2.0	1.6	27	35	4	2
Sudan	10	15	7	10	15	2.3	2.4	27	29	33	29
Sweden	5	6	4	5	5	0.6	–0.2	44	48	0	0
Switzerland	4	5	3	4	4	1.2	0.2	37	40	0	0
Syrian Arab Republic	4	8	2	4	7	2.9	3.1	23	26	14	5
Tajikistan	2	3	2	2	3	2.0	2.3	47	44	0	0
Tanzania	9	16	10	16	22	2.9	2.2	50	49	43	39
Thailand	26	40	24	35	39	2.1	0.8	47	46	25	15
Togo	1	2	1	2	3	2.5	2.6	39	40	36	28
Trinidad and Tobago	1	1	0	1	1	1.3	1.8	32	37	1	0
Tunisia	3	6	2	3	5	2.7	2.4	29	31	6	0
Turkey	25	40	19	29	38	2.5	1.8	35	36	21	23
Turkmenistan	2	3	1	2	3	2.6	2.5	47	45	0	0
Uganda	6	10	7	10	14	2.3	2.4	48	48	49	45
Ukraine	33	34	26	25	24	–0.2	–0.4	50	49	0	0
United Arab Emirates	1	2	1	1	2	4.3	1.8	5	14	0	0
United Kingdom	36	38	27	29	29	0.5	0.1	39	43	0	0
United States	151	174	110	134	150	1.2	0.7	42	46	0	0
Uruguay	2	2	1	1	2	1.3	0.9	31	41	4	2
Uzbekistan	9	13	6	9	14	2.3	2.4	48	46	0	0
Venezuela	8	13	5	9	13	3.2	2.4	27	33	4	1
Vietnam	28	44	26	38	48	2.3	1.6	48	49	22	8
West Bank and Gaza	..	..	..	..	..	..	..	..	..	..	..
Yemen, Rep.	4	8	2	5	9	3.9	3.8	33	29	26	20
Yugoslavia, FR (Serb./Mont.)	6	7	4	5	5	0.7	0.2	38	42	0	0
Zambia	3	5	2	4	5	2.8	2.3	45	45	19	16
Zimbabwe	3	6	3	5	7	2.9	1.8	44	44	37	29
World	**2,595 t**	**3,586 t**	**2,034 t**	**2,739 t**	**3,343 t**	**1.7 w**	**1.3 w**	**39 w**	**40 w**	**20 w**	**13 w**
Low income	1,352	1,973	1,153	1,604	2,000	1.9	1.5	40	40	28	18
Excl. China & India	371	578	315	478	678	2.4	2.3	40	40	31	27
Middle income	715	997	509	695	877	1.8	1.6	37	38	11	7
Lower middle income	506	700	367	494	626	1.7	1.6	39	39	11	7
Upper middle income	209	279	142	201	252	2.0	1.5	32	35	10	7
Low & middle income	2,067	2,970	1,663	2,299	2,878	1.9	1.5	39	39	23	14
East Asia & Pacific	796	1,140	704	966	1,127	1.9	1.0	42	44	27	11
Europe & Central Asia	276	313	215	234	250	0.5	0.4	47	46	3	4
Latin America & Carib.	200	300	130	201	266	2.6	1.9	28	33	13	9
Middle East & N. Africa	91	156	54	89	140	2.9	3.0	24	26	14	5
South Asia	508	749	389	546	716	2.0	1.8	34	33	23	17
Sub-Saharan Africa	196	312	171	263	379	2.5	2.4	42	42	35	30
High income	528	616	372	440	466	1.0	0.4	38	43	0	0

Labor force structure | 2.3

The labor force is the supply of labor in an economy. It includes people who are currently employed and people who are unemployed but seeking work. Not everyone who works is included, however. Unpaid workers, family workers, and students are usually omitted, and in some countries members of the military are also not counted. The size of the labor force tends to vary during the year as seasonal workers enter and leave the labor force.

Data on the labor force are compiled by the International Labour Organization (ILO) from census or labor force surveys. Despite the ILO's efforts to encourage the use of international standards, labor force data are not fully comparable because of differences among countries, and sometimes within countries, in definitions and methods of collection, classification, and tabulation. In some countries data on the labor force refer to people above a specific age, while in others there is no specific age provision. The reference period of the census or survey is another important source of differences: in some countries data refer to a person's status on the day of the census or survey or during a specific period before the inquiry date, while in others the data are recorded without reference to any period. In developing countries, where the household is often the basic unit of production and all members contribute to output, but some at low intensity or irregular intervals, the estimated labor force may significantly underestimate the numbers actually working (ILO 1990a, *Yearbook of Labour Statistics 1996*).

The population age 15–64 is often used to provide a rough estimate of the potential labor force. But in many developing countries children under 15 work full or part time. And in some high-income countries many workers postpone retirement past age 65. As a result labor force participation rates may systematically over- or underestimate actual rates.

The labor force estimates in the table were calculated by applying gender-specific activity rates from the ILO database to create a labor force series consistent with the World Bank's population estimates. This procedure sometimes results in estimates of the absolute size of the labor force that differ slightly from those published in the ILO's *Yearbook of Labour Statistics*.

Estimates of women in the labor force are not comparable internationally because in many countries large numbers of women assist on farms or in other family enterprises without pay, and countries differ in the criteria used to determine the extent to which such workers are to be counted as part of the labor force.

Reliable estimates of child labor are hard to obtain. In many countries child labor is officially presumed not to exist, and so is not included in surveys or covered in official data. Data are also subject to underreporting because they do not include children engaged in agricultural or household activities with their families.

- **Population age 15–64** is the number of people who could potentially be economically active, excluding children.
- **Total labor force** comprises people who meet the ILO definition of the economically active population: all people who supply labor for the production of goods and services during a specified period. It includes both the employed and the unemployed. While national practices vary in the treatment of such groups as the armed forces and seasonal or part-time workers, in general the labor force includes the armed forces, the unemployed, and first-time job-seekers, but excludes homemakers and other unpaid caregivers and workers in the informal sector. • **Average annual growth rate** of the labor force is calculated using the exponential end-point method (see *Statistical methods* for more information). • **Females as a percentage of the labor force** shows the extent to which women are active in the labor force. • **Children 10–14 in the labor force** is the share of that age group that is active in the labor force.

Population estimates are from the World Bank's population database. Labor force activity rates are from the ILO database, Estimates and Projections of the Economically Active Population, 1950–2010. The ILO publishes estimates of the economically active population in its *Yearbook of Labour Statistics*.

Figure 2.3a

Children work less as incomes rise

% of children who work, 1995

GNP per capita
(1987 $)

Source: ILO and World Bank estimates.

Child labor is a poverty issue. Children who work rather than attend school cannot fully develop their skills. And premature and extensive engagement in work can damage a child's health and social development, leading to lower earning power and reduced productivity over the longer term. Thus the cycle of poverty continues.

The incidence of child labor declines as per capita income rises. In countries where annual per capita income is $500 or less, the proportion of children age 10–14 who work is extremely high, at 30–50 percent (see table). But the rate falls to 10–30 percent in countries with annual incomes between $500 and $1,000. Many factors affect the prevalence of child labor, including culture and the structure of production. For instance, child labor tends to be more common in countries where agriculture accounts for a large share of GDP.

2.4 | Employment by occupation

	Employers and own-account workers				Employees				Contributing family workers			
	Male % of economically active male population		Female % of economically active female population		Male % of economically active male population		Female % of economically active female population		Male % of economically active male population		Female % of economically active female population	
	1980	1994	1980	1994	1980	1994	1980	1994	1980	1994	1980	1994
Albania	..	..	..	..	..	..	..	..	..	..	..	..
Algeria	..	..	..	..	..	..	..	..	..	..	..	..
Angola												
Argentina	..	..	..	..	..	..	..	..	..	..	..	..
Armenia		..		..								
Australia	15.8	16.6	11.2	10.5	78.0	73.2	79.1	79.4	0.3	0.7	0.5	1.3
Austria	15.6	10.9	19.1	8.0	84.4	87.7	80.9	86.9	1.4	1.4	9.1	5.1
Azerbaijan	..	..	..	..	..	..	..	..	..	..	..	..
Bangladesh	..	39.2	..	6.4	..	15.6	..	5.2		22.3		83.3
Belarus	..	..	..	..	..	..	..	..	..	..	..	..
Belgium	14.0	16.1	8.1	8.2	79.5	74.6	70.2	69.5	0.8	0.9	6.9	7.0
Benin	..	..	..	..	..	..	..	..	..	..	..	..
Bolivia	..	31.2	..	42.0	..	59.3	..	43.6		3.9		8.4
Bosnia and Herzegovina	..		..									
Botswana	..		..		..		..		..		..	
Brazil	..	28.7	..	21.9	..	61.2	..	64.4		6.4		10.2
Bulgaria	..	..	..	..	..	..	..	..	..	..	..	..
Burkina Faso	..	..	..	..	..	..	..	..	..	..	..	..
Burundi	..		..		..		..		..		..	
Cambodia	..		..		..		..		..		..	
Cameroon	61.2		58.7	..	21.3		3.5		9.2		32.7	
Canada	10.1	11.1	6.4	7.7	89.3	87.9	90.7	90.3	0.3	0.2	1.9	0.9
Central African Republic	..		..		..		..		..		..	
Chad	..	..	..	..	..	..	..	..	..	..	..	..
Chile	24.4	29.3	16.5	20.4	46.1	63.0	53.8	67.8	9.3	2.3	11.4	5.0
China	..		..		..		..		..		..	
Hong Kong, China	12.6	14.5	4.1	3.2	83.0	83.4	88.5	92.9	0.6	0.2	3.8	2.0
Colombia	..	32.9	..	24.1	..	66.3	..	62.2		0.6		2.2
Congo, Dem. Rep.	..	..	..	..	..	..	..	..	..	..	..	..
Congo, Rep.	..		..		..		..		..		..	
Costa Rica	22.6	26.1	10.9	19.1	70.9	70.3	82.8	76.4	5.0	3.2	3.2	3.5
Côte d'Ivoire	..	..	..	..	..	..	..	..	..	..	..	..
Croatia	..		..		..		..		..		..	
Cuba	..		..		..		..		..		..	
Czech Republic	..		..		..		..		..		..	
Denmark	17.2	12.7	3.0	3.2	81.1	86.8	89.1	93.1	0.1	0.2	5.5	3.4
Dominican Republic	..		..		..		..		..		..	
Ecuador	..		..		..		..		..		..	
Egypt, Arab Rep.	29.7	28.6	12.0	12.2	52.8	54.7	66.0	35.0	13.6	10.2	2.7	35.8
El Salvador	24.5	30.6	35.3	38.1	62.1	54.5	53.6	40.6	12.6	11.6	7.7	8.9
Eritrea	..		..		..		..		..		..	
Estonia	..		..		..		..		..		..	
Ethiopia	..		..		..		..		..		..	
Finland	11.2	16.2	9.0	8.8	85.3	79.9	86.3	87.0	1.6	0.8	2.4	0.4
France	..	..	..	..	..	..	..	..	..	..	..	..
Gabon	..		..		..		..		..		..	
Gambia, The	..	..	..	..	..	..	..	..	..	..	..	..
Georgia	..		..		..		..		..		..	
Germany	..	..	..	..	..	..	..	..	..	..	..	..
Ghana	..		..		..		..		..		..	
Greece	44.6	41.5	18.7	17.3	48.4	51.7	41.4	51.3	3.7	4.4	34.2	22.4
Guatemala	34.5	23.6	49.9	26.4	52.0	72.6	40.4	68.0	11.5	3.4	7.6	4.7
Guinea	..		..		..		..		..		..	
Guinea-Bissau	..		..		..		..		..		..	
Haiti	61.1	60.7	56.9	56.7	15.5	15.4	18.2	18.3	11.0	10.9	9.8	9.7
Honduras	..		..		..		..		..		..	

	Employers and own-account workers				Employees				Contributing family workers			
	Male % of economically active male population		Female % of economically active female population		Male % of economically active male population		Female % of economically active female population		Male % of economically active male population		Female % of economically active female population	
	1980	1994	1980	1994	1980	1994	1980	1994	1980	1994	1980	1994
Hungary	2.9	13.5	1.4	8.1	80.6	85.3	78.9	88.3	0.3	1.2	5.7	3.6
India	..	..	..	..	..	..	..	..	..	..	..	..
Indonesia	21.3	52.2	19.4	28.7	63.3	31.5	37.1	24.0	12.6	14.1	39.9	44.7
Iran, Islamic Rep.	..	..	..	..	..	..	..	..	..	..	..	..
Iraq	..	..	..	..	..	..	..	..	..	..	..	..
Ireland	23.5	23.6	6.3	6.8	68.7	72.7	84.4	88.6	2.2	1.2	4.7	1.7
Israel	23.7	19.0	11.6	8.7	71.2	74.4	78.1	79.9	0.9	0.3	4.3	1.5
Italy	24.4	25.9	13.9	13.8	68.5	62.3	63.5	63.7	2.4	2.5	9.7	6.5
Jamaica	..	..	..	..	..	..	..	..	..	..	..	..
Japan	19.0	14.1	13.4	8.9	75.5	81.0	62.0	75.5	3.2	1.8	22.5	12.4
Jordan	..	..	..	..	..	..	..	..	..	..	..	..
Kazakhstan	..	..	..	..	..	..	..	..	..	..	..	..
Kenya	..	..	..	..	..	..	..	..	..	..	..	..
Korea, Dem. Rep.	..	..	..	..	..	..	..	..	..	..	..	..
Korea, Rep.	38.0	33.7	22.4	18.4	49.0	61.2	38.0	56.3	6.8	2.0	36.1	23.0
Kuwait	..	..	..	..	..	..	..	..	..	..	..	..
Kyrgyz Republic	..	..	..	..	..	..	..	..	..	..	..	..
Lao PDR	..	..	..	..	..	..	..	..	..	..	..	..
Latvia	..	..	..	..	..	..	..	..	..	..	..	..
Lebanon	..	..	..	..	..	..	..	..	..	..	..	..
Lesotho	..	..	..	..	..	..	..	..	..	..	..	..
Libya	..	..	..	..	..	..	..	..	..	..	..	..
Lithuania	..	..	..	..	..	..	..	..	..	..	..	..
Macedonia, FYR	..	..	..	..	..	..	..	..	..	..	..	..
Madagascar	..	..	..	..	..	..	..	..	..	..	..	..
Malawi	..	..	..	..	..	..	..	..	..	..	..	..
Malaysia	..	25.0	..	13.7	..	71.4	..	71.5	..	3.6	..	14.8
Mali	..	..	..	..	..	..	..	..	..	..	..	..
Mauritania	..	..	..	..	..	..	..	..	..	..	..	..
Mauritius	..	..	..	..	..	..	..	..	..	..	..	..
Mexico	..	33.2	..	23.1	..	52.5	..	56.6	..	12.0	..	17.1
Moldova	..	..	..	..	..	..	..	..	..	..	..	..
Mongolia	..	..	..	..	..	..	..	..	..	..	..	..
Morocco	..	..	..	..	..	..	..	..	..	..	..	..
Mozambique	..	..	..	..	..	..	..	..	..	..	..	..
Myanmar	..	..	..	..	..	..	..	..	..	..	..	..
Namibia	..	..	..	..	..	..	..	..	..	..	..	..
Nepal	..	..	..	..	..	..	..	..	..	..	..	..
Netherlands	11.8	12.0	4.9	7.5	81.6	81.7	78.7	82.0	0.4	0.4	5.4	2.4
New Zealand	..	66.3	..	77.7	..	23.1	..	12.0	..	0.6	..	1.5
Nicaragua	..	..	..	..	..	..	..	..	..	..	..	..
Niger	..	..	..	..	..	..	..	..	..	..	..	..
Nigeria	51.6	..	65.3	..	33.7	..	15.7	..	7.6	..	11.7	..
Norway	13.6	11.0	4.2	4.8	83.5	82.1	87.6	89.1	1.4	0.7	5.5	1.2
Oman	..	..	..	..	..	..	..	..	..	..	..	..
Pakistan	..	45.8	..	13.0	..	34.0	..	22.6	..	15.6	..	47.6
Panama	33.6	33.9	12.1	11.2	57.7	58.5	78.8	80.1	6.2	4.4	2.7	2.0
Papua New Guinea	..	..	..	..	..	..	..	..	..	..	..	..
Paraguay	..	33.3	..	34.8	..	64.2	..	60.4	..	2.1	..	4.2
Peru	..	33.1	..	36.1	..	62.4	..	44.2	..	3.5	..	7.5
Philippines	40.7	39.5	24.7	30.4	40.5	42.3	38.9	40.8	15.4	10.4	28.0	19.3
Poland	..	24.6	..	18.8	..	67.9	..	69.7	..	4.7	..	7.5
Portugal	20.9	24.6	8.5	21.2	71.7	73.3	58.5	75.3	4.6	1.4	25.2	2.3
Puerto Rico	17.9	19.2	4.4	5.9	80.2	79.7	91.1	90.9	0.5	0.3	2.4	1.3
Romania	..	..	..	..	..	..	..	..	..	..	..	..
Russian Federation	..	..	..	..	..	..	..	..	..	..	..	..

2.4 | Employment by occupation

	Employers and own-account workers				Employees				Contributing family workers			
	Male % of economically active male population		Female % of economically active female population		Male % of economically active male population		Female % of economically active female population		Male % of economically active male population		Female % of economically active female population	
	1980	1994	1980	1994	1980	1994	1980	1994	1980	1994	1980	1994
Rwanda	..	..	..	..	..	..	..	..	..	..	..	..
Saudi Arabia	..	..	..	..	..	..	..	..	..	..	..	..
Senegal	..	..	..	..	..	..	..	..	..	..	..	..
Sierra Leone	..	..	..	..	..	..	..	..	..	..	..	..
Singapore	15.9	17.2	5.1	5.1	79.6	80.0	87.4	90.3	1.7	0.2	4.0	1.8
Slovak Republic	..	87.9	..	92.5	..	8.3	..	2.6	..	0.1	..	0.2
Slovenia	..	..	..	..	..	..	..	..	..	..	..	..
South Africa	..	..	..	..	..	..	..	..	..	..	..	..
Spain	21.7	20.1	14.1	12.0	71.8	72.9	60.7	67.8	3.2	2.2	17.5	6.0
Sri Lanka	28.2	29.1	11.8	14.9	54.7	56.4	53.7	51.8	6.4	4.6	13.6	12.0
Sudan	..	..	..	..	..	..	..	..	..	..	..	..
Sweden	10.1	13.8	3.9	5.3	88.0	76.6	92.9	87.4	0.2	0.4	0.9	0.5
Switzerland	..	15.4	..	9.4	..	83.3	..	85.7	..	1.3	..	4.9
Syrian Arab Republic	37.1	36.5	9.9	5.7	56.0	50.1	48.1	45.7	5.3	8.3	37.1	34.5
Tajikistan	..	..	..	..	..	..	..	..	..	..	..	..
Tanzania	..	..	..	..	..	..	..	..	..	..	..	..
Thailand	43.5	38.1	17.3	22.6	25.9	44.2	16.9	35.4	29.6	11.5	65.1	29.4
Togo	..	..	..	..	..	..	..	..	..	..	..	..
Trinidad and Tobago	14.6	21.5	9.7	12.8	81.3	75.2	80.8	81.1	2.8	1.7	6.4	5.4
Tunisia	..	..	..	..	..	..	..	..	..	..	..	..
Turkey	..	36.0	..	8.6	..	49.2	..	24.1	..	12.1	..	62.9
Turkmenistan	..	..	..	..	..	..	..	..	..	..	..	..
Uganda	..	..	..	..	..	..	..	..	..	..	..	..
Ukraine	..	..	..	..	..	..	..	..	..	..	..	..
United Arab Emirates	..	..	..	..	..	..	..	..	..	..	..	..
United Kingdom	..	14.9	..	6.5	..	71.0	..	84.1	..	0.3	..	0.9
United States	10.2	9.8	5.1	6.6	88.9	89.7	92.7	92.6	0.3	0.1	1.2	0.2
Uruguay	..	26.4	..	19.4	..	70.8	..	73.8	..	1.1	..	3.5
Uzbekistan	..	..	..	..	..	..	..	..	..	..	..	..
Venezuela	29.9	37.1	17.3	25.1	60.2	53.0	74.4	65.0	3.0	1.6	3.3	1.1
Vietnam	..	..	..	..	..	..	..	..	..	..	..	..
West Bank and Gaza	..	..	..	..	..	..	..	..	..	..	..	..
Yemen, Rep.	..	..	..	..	..	..	..	..	..	..	..	..
Yugoslavia, FR (Serb./Mont.)	..	..	..	..	..	..	..	..	..	..	..	..
Zambia	..	..	..	..	..	..	..	..	..	..	..	..
Zimbabwe	..	..	..	..	..	..	..	..	..	..	..	..

Employment by occupation | 2.4

This table shows the distribution of employment classified by occupational status according to the International Classification of Status in Employment (ICSE). ICSE classifications are based on the explicit or implicit employment contract workers have with other people or organizations. The basic criteria for defining classification groups are the type of economic risk and the type of authority over establishments and other workers that the job incumbent has or will have. Until 1993 the main ICSE groups were *employers*, *own-account workers*, *employees*, *members of producers cooperatives*, and *unpaid family workers*. In 1993 the group unpaid family workers was changed to *contributing family workers* and the group own-account workers was expanded to include people working in a family enterprise with the same degree of commitment as the head of the enterprise. These people, usually women, were formerly considered unpaid family workers.

Data on employment are drawn from labor force surveys, enterprise censuses and surveys, administrative records of social insurance schemes, and official national estimates. The concept of employment generally refers to people above a certain age who worked or who held a job during a reference period. Shares of occupational employment in the labor force are calculated using the International Labour Organization's (ILO) labor force estimates, which may differ from those based on the World Bank's population estimates as shown in table 2.3. Occupational categories should add up to 100 percent. Where they do not, the difference arises from people who are not classifiable by status.

Employment data include both full-time and part-time workers. There are, however, many differences in how countries define and measure employment status, particularly for part-time workers, students, members of the armed forces, and household workers. Because of these differences, the content of ICSE groups is not easily comparable across countries (ILO, *Yearbook of Labour Statistics 1996*, p. 64). In most countries managers and directors of incorporated enterprises are classified as employees, but in some they are classified as employers. Similarly, in most countries family members who receive regular remuneration in the form of wages, salaries, commissions, piece rates, or in-kind payments are classified as employees, but in some they are classified as contributing family workers. Some countries cannot accurately measure the number of contributing family workers. And many cannot distinguish between own-account workers and employers, so only the sum of the two groups is available.

Countries also take very different approaches to the treatment of unemployed people. In most countries unemployed people with previous job experience are classified according to their last job. In some countries, however, they and people seeking their first job are classified as persons not classifiable by status, and so are not included in the table.

• **Employers** operate, alone or with one or more partners, their own economic enterprise, or engage independently in a profession or trade, and hire one or more employees on a continuous basis. The definition of "a continuous basis" is determined by national circumstances. Partners may or may not be members of the same family or household. • **Own-account workers** operate, alone or with one or more partners, their own economic enterprise, or engage independently in a profession or trade, and hire no employees on a continuous basis. As with employers, partners may or may not be members of the same family or household. • **Employees** are people who work for a public or private employer and receive remuneration in the form of wages, salaries, commissions, tips, piece rates, or in-kind payments. • **Contributing family workers** (previously referred to as unpaid family workers) work without pay in an economic enterprise operated by a related person living in the same household and cannot be regarded as a partner because their commitment in terms of working time or other factors is not at a level comparable to that of the head of the enterprise. In countries where it is customary for young people to work without pay in an enterprise operated by a related person, the requirement of living in the same household is often eliminated.

Employment data are compiled by the World Bank's Development Data Group using an ILO database corresponding to table 2a in its *Yearbook of Labour Statistics*.

2.5 | Employment by economic activity

	Agriculture				Industry				Services			
	Male % of economically active male population		Female % of economically active female population		Male % of economically active male population		Female % of economically active female population		Male % of economically active male population		Female % of economically active female population	
	1980	1994	1980	1994	1980	1994	1980	1994	1980	1994	1980	1994
Albania	54	51	62	60	28	26	17	19	18	23	21	20
Algeria	27	18	69	57	33	38	6	7	40	44	25	36
Angola	67	65	87	86	13	14	1	2	20	21	11	13
Argentina	17	16	3	3	40	39	18	17	44	46	79	80
Armenia	21	24	21	11	48	47	38	39	31	29	41	51
Australia	7	6	4	3	34	32	15	11	44	58	67	80
Austria	9	6	13	8	51	48	24	19	41	45	62	72
Azerbaijan	28	27	42	36	36	35	20	21	36	38	38	43
Bangladesh	67	59	81	74	5	14	14	19	29	26	5	7
Belarus	29	26	23	13	44	45	33	36	28	29	44	51
Belgium	4	3	2	1	40	34	15	11	50	54	68	72
Benin	66	62	69	65	10	12	4	4	24	27	27	30
Bolivia	53	48	53	45	21	22	11	10	26	30	36	45
Bosnia and Herzegovina	26	9	37	16	45	54	24	37	30	37	39	48
Botswana	53	39	74	55	18	30	2	9	28	31	24	36
Brazil	41	28	26	14	28	28	14	13	31	43	60	74
Bulgaria	..	..	22	13	..	..	..	..	..	..	..	..
Burkina Faso	92	91	93	94	3	2	2	2	5	7	5	5
Burundi	88	86	98	98	4	4	1	1	9	10	1	1
Cambodia	70	69	80	78	7	7	7	8	23	24	14	14
Cameroon	65	62	87	83	11	12	2	3	23	26	11	14
Canada	6	4	3	3	34	32	14	11	52	63	74	85
Central African Republic	79	74	90	87	5	6	1	0	15	20	9	13
Chad	82	77	95	91	6	7	0	1	12	16	4	8
Chile	20	20	2	5	25	30	13	14	52	45	79	74
China	71	69	79	76	16	17	12	13	14	14	10	11
Hong Kong, China	1	1	1	1	46	39	56	33	52	60	43	66
Colombia	..	2	..	1	..	36	..	24	..	63	..	75
Congo, Dem. Rep.	62	58	84	81	18	20	4	5	20	23	12	14
Congo, Rep.	42	33	81	69	20	23	2	4	38	44	17	27
Costa Rica	43	34	5	6	23	27	20	26	34	39	75	68
Côte d'Ivoire	60	54	75	72	10	11	5	6	30	34	20	22
Croatia	23	17	28	15	38	38	27	28	39	45	45	57
Cuba	30	24	10	8	32	36	22	21	39	41	68	71
Czech Republic	14	13	11	9	67	54	44	36	19	33	45	55
Denmark	10	7	3	3	41	37	17	16	44	55	76	81
Dominican Republic	40	31	11	9	26	32	16	23	34	38	73	68
Ecuador	44	39	22	16	21	20	15	16	34	41	63	68
Egypt, Arab Rep.	43	32	8	43	20	23	10	9	32	38	56	31
El Salvador	56	50	8	7	20	22	18	19	24	29	73	74
Eritrea	79	77	88	85	7	8	2	2	14	16	11	13
Estonia	19	18	12	11	50	48	36	34	31	34	52	55
Ethiopia	90	86	89	86	2	2	2	2	8	11	10	12
Finland	12	10	9	5	42	38	21	14	38	49	64	76
France	9	6	7	4	43	38	22	17	48	56	71	78
Gabon	59	46	74	59	18	21	5	10	24	33	21	32
Gambia, The	77	74	93	92	10	12	2	3	13	14	5	6
Georgia	31	27	34	24	32	38	21	23	37	34	45	52
Germany	6	4	8	4	54	48	33	24	40	48	59	72
Ghana	66	64	57	55	12	12	14	14	22	25	29	31
Greece	25	19	39	23	34	32	18	17	40	47	39	51
Guatemala	64	64	17	16	17	16	27	23	19	21	56	61
Guinea	86	83	96	92	2	2	1	1	12	15	3	7
Guinea-Bissau	81	78	98	96	3	3	0	0	17	19	2	3
Haiti	79	76	61	57	8	9	8	8	13	15	31	35
Honduras	63	48	40	25	17	23	9	12	20	30	51	64

Employment by economic activity | 2.5

	Agriculture				Industry				Services			
	Male % of economically active male population		Female % of economically active female population		Male % of economically active male population		Female % of economically active female population		Male % of economically active male population		Female % of economically active female population	
	1980	1994	1980	1994	1980	1994	1980	1994	1980	1994	1980	1994
Hungary	21	19	15	11	47	42	38	32	32	39	47	57
India	63	59	83	74	15	17	9	15	22	24	8	11
Indonesia	59	54	56	56	12	14	12	13	29	31	32	31
Iran, Islamic Rep.	36	30	82	73	28	26	6	9	35	44	12	18
Iraq	21	12	62	39	24	19	11	9	55	69	27	52
Ireland	..	16	..	3	..	28	..	16	..	39	..	69
Israel	7	4	4	2	39	37	15	15	51	55	77	77
Italy	11	7	14	7	40	35	24	18	40	49	48	59
Jamaica	42	34	18	15	23	32	9	13	35	34	73	72
Japan	8	5	13	6	39	39	28	25	50	53	57	65
Jordan	11	10	58	41	27	28	3	4	62	63	39	55
Kazakhstan	28	28	20	15	38	37	25	25	34	35	55	60
Kenya	77	75	88	85	10	11	2	3	13	14	10	12
Korea, Dem. Rep.	39	35	52	42	37	38	20	23	24	27	28	35
Korea, Rep.	29	13	38	17	30	37	23	25	35	47	36	56
Kuwait	2	2	0	0	36	32	2	2	62	67	97	98
Kyrgyz Republic	34	36	33	28	34	30	23	23	32	34	44	50
Lao PDR	77	76	82	81	7	7	4	5	16	17	13	14
Latvia	18	19	14	12	49	47	35	33	32	34	50	55
Lebanon	13	6	20	10	29	34	21	22	58	59	59	68
Lesotho	26	29	64	59	52	41	5	5	22	30	31	36
Libya	16	7	63	28	29	27	3	5	55	66	34	68
Lithuania	26	23	29	13	47	47	30	34	27	30	41	53
Macedonia, FYR	30	21	47	23	38	40	23	41	32	39	30	36
Madagascar	72	70	93	88	9	10	2	3	19	20	5	9
Malawi	78	78	96	95	10	9	1	1	12	13	3	3
Malaysia	36	28	49	26	19	23	18	23	44	48	33	52
Mali	86	83	92	89	2	2	1	2	12	15	7	9
Mauritania	65	49	79	63	11	16	2	4	25	35	19	34
Mauritius	27	18	27	14	28	40	27	50	45	42	45	35
Mexico	43	35	19	12	30	25	28	20	28	40	53	69
Moldova	46	38	41	28	29	34	23	26	25	28	36	46
Mongolia	43	34	36	30	21	23	21	22	36	44	43	48
Morocco	48	35	72	63	23	28	14	19	29	37	14	18
Mozambique	72	70	97	96	14	14	1	1	14	15	2	3
Myanmar	72	70	80	78	9	11	7	9	19	19	12	14
Namibia	52	46	64	54	22	21	5	8	27	33	31	39
Nepal	91	91	98	98	1	0	0	0	8	9	2	2
Netherlands	6	5	2	2	36	30	12	9	50	58	74	77
New Zealand	13	13	7	7	39	33	21	14	48	54	72	79
Nicaragua	49	38	16	9	26	28	21	23	26	34	63	69
Niger	86	84	98	97	5	6	1	1	9	10	1	1
Nigeria	52	42	57	44	10	9	5	3	38	49	38	53
Norway	10	7	6	3	40	33	14	10	50	57	80	84
Oman	52	48	25	20	21	22	33	35	27	30	42	45
Pakistan	56	45	73	72	15	20	12	13	29	34	15	15
Panama	35	30	5	3	21	21	11	10	38	46	76	80
Papua New Guinea	76	72	92	89	8	9	2	3	16	18	6	8
Paraguay	58	51	9	8	20	23	22	20	23	26	70	72
Peru	45	41	25	22	20	20	14	12	35	39	61	66
Philippines	61	54	37	31	15	16	16	14	25	29	47	56
Poland	28	27	32	28	46	45	28	25	26	28	40	47
Portugal	21	10	31	12	41	40	24	24	32	49	37	62
Puerto Rico	8	6	0	0	30	29	26	22	24	27	73	78
Romania	26	21	45	28	50	53	29	40	24	26	25	32
Russian Federation	19	17	13	10	50	48	37	35	31	34	50	56

	Agriculture				Industry				Services			
	Male % of economically active male population		Female % of economically active female population		Male % of economically active male population		Female % of economically active female population		Male % of economically active male population		Female % of economically active female population	
	1980	1994	1980	1994	1980	1994	1980	1994	1980	1994	1980	1994
Rwanda	88	86	98	98	5	6	1	1	7	8	1	2
Saudi Arabia	45	20	25	12	17	21	5	6	39	59	70	82
Senegal	74	70	90	86	9	10	2	4	17	20	8	11
Sierra Leone	63	60	82	81	20	22	4	4	17	18	14	16
Singapore	1	0	1	0	32	35	39	30	63	62	57	68
Slovak Republic	15	14	13	9	37	35	34	31	48	50	54	60
Slovenia	14	5	17	6	49	52	37	39	37	43	46	54
South Africa	18	16	16	10	45	42	16	14	37	42	68	76
Spain	18	10	16	7	42	39	21	14	36	46	56	65
Sri Lanka	41	29	44	32	16	18	12	17	26	37	20	27
Sudan	66	64	88	84	9	10	4	5	24	26	8	11
Sweden	8	5	3	2	45	34	16	11	46	52	79	81
Switzerland	..	6	..	4	..	42	..	19	..	49	..	74
Syrian Arab Republic	27	22	78	69	35	30	7	6	39	49	15	25
Tajikistan	36	37	54	45	29	28	16	17	35	35	30	37
Tanzania	80	78	92	91	7	8	2	2	13	14	7	7
Thailand	..	42	..	38	..	26	..	19	..	26	..	31
Togo	70	66	67	65	12	12	7	7	19	22	26	29
Trinidad and Tobago	..	13	..	5	..	38	..	19	..	49	..	77
Tunisia	33	22	53	42	30	33	32	32	37	44	16	26
Turkey	45	38	88	82	22	24	5	7	33	38	7	11
Turkmenistan	33	34	46	41	32	30	16	14	35	36	38	44
Uganda	84	81	91	88	6	7	2	2	10	12	8	10
Ukraine	26	24	24	16	46	46	33	34	28	31	44	50
United Arab Emirates	5	9	0	0	40	30	7	2	55	61	93	97
United Kingdom	3	3	1	1	44	32	21	13	44	51	72	80
United States	5	4	2	2	39	34	19	13	52	62	78	84
Uruguay	22	21	4	4	31	31	23	21	47	48	74	75
Uzbekistan	35	34	46	35	34	30	19	19	32	35	36	45
Venezuela	19	19	3	2	32	25	19	13	48	48	76	76
Vietnam	71	70	75	73	16	17	10	11	13	13	15	16
West Bank and Gaza	..	..	..	..	..	..	..	..	..	..	..	..
Yemen, Rep.	60	50	98	88	19	22	1	6	21	29	1	7
Yugoslavia, FR (Serb./Mont.)	34	28	47	32	35	38	19	26	31	34	33	41
Zambia	69	68	85	83	13	13	3	3	19	19	13	14
Zimbabwe	63	58	85	81	19	13	4	2	18	29	12	17
World	50 w	48 w	56 w	52 w	24 w	23 w	15 w	15 w	26 w	29 w	28 w	32 w
Low income	69	66	80	76	14	15	10	12	17	19	10	12
Excl. China & India	69	65	79	75	11	13	7	8	20	23	14	17
Middle income	34	32	32	29	35	32	24	21	31	35	44	49
Lower middle income	35	35	35	35	35	31	24	21	30	33	40	42
Upper middle income	31	25	23	14	37	34	23	19	32	41	54	67
Low & middle income	59	56	66	62	20	20	14	15	21	23	20	23
East Asia & Pacific	69	67	75	72	15	16	12	13	15	17	13	15
Europe & Central Asia	25	23	27	22	45	43	31	30	30	33	42	48
Latin America & Carib.	..	29	..	12	..	28	..	16	..	42	..	71
Middle East & N. Africa	39	29	53	55	25	26	10	11	35	43	29	29
South Asia	64	59	82	75	13	16	10	15	23	25	8	10
Sub-Saharan Africa	69	65	80	75	11	11	4	4	20	23	17	21
High income	8	6	8	4	41	35	23	18	47	56	66	75

About the data

The International Labour Organization (ILO) classifies economic activity on the basis of the International Standard Industrial Classification (ISIC) of All Economic Activities. Because this classification is based on where work is performed (industry) rather than on the type of work performed (occupation), all of an enterprise's employees are classified under the same industry, regardless of their trade or occupation.

The ILO's *Yearbook of Labour Statistics* reports data by major divisions of the ISIC revision 2 or tabulation categories of the ISIC revision 3. In this table the reported divisions or categories are aggregated into three broad groups: agriculture, industry, and services. An increasing number of countries report economic activity according to the ISIC. Where data are supplied according to national classifications, however, industry definitions and descriptions may differ. Classification into broad groups also may obscure fundamental differences in countries' industrial patterns.

The distribution of economic activity by gender reveals some interesting patterns. Agriculture accounts for the largest share of female employment in much of Africa and Asia. Services account for much of the increase in women's economic participation in North Africa, Latin America and the Caribbean, and high-income economies. Worldwide, women are underrepresented in industry.

There are several explanations for the rising importance of service jobs for women. Many service jobs—such as nursing and social and clerical work—are considered "feminine" because of a perceived similarity with women's traditional roles. Moreover, women often do not receive training to take advantage of changing employment opportunities. Finally, the greater availability of part-time work in service jobs may lure more women, although it is not clear whether this is a cause or an effect (United Nations 1991).

Figure 2.5a

Women's labor force participation depends on how work is defined

% of labor force working in agriculture, 1994

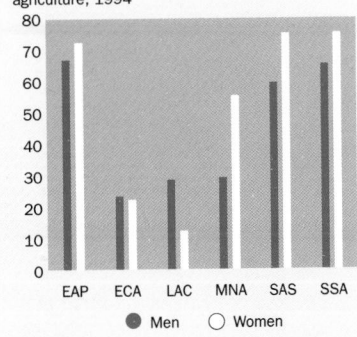

● Men   ○ Women

Source: ILO.

Although there are still significant differences between men's and women's work by sector, occupation, and type of work, women's overall labor force participation rates are moving closer to those of men. Women's labor force participation continues to be strongly influenced by gender differences in the definition of work in different countries. This is particularly evident in the informal sector and in agriculture, where it is sometimes difficult to distinguish between women's housework and their unpaid work in a family enterprise or in agricultural production.

Female labor force participation and women's share of the work force tend to be large in countries where women's contributions to family agriculture are defined as work. This is particularly evident in Africa, where several countries report more than 90 percent of the female work force in agriculture, resulting in high regional participation rates. In other countries, where the distinction between housework and a subsistence activity—such as tending a home garden—is less clear, the proportion of women active in agriculture can be substantially smaller than that of men. Thus women's work in agriculture and the informal sector warrants special attention in cross-country comparisons of women's share in the work force.

Definitions

• **Agriculture** includes hunting, forestry, and fishing, corresponding to major division 1 (ISIC revision 2) or tabulation categories A and B (ISIC revision 3).
• **Industry** includes mining and quarrying (including oil production), manufacturing, electricity, gas and water, and construction, corresponding to major divisions 2 through 5 (ISIC revision 2) or tabulation categories C through F (ISIC revision 3). • **Services** include wholesale and retail trade and restaurants and hotels; transport, storage, and communications; financing, insurance, real estate, and business services; and community, social, and personal services, corresponding to major divisions 6 through 9 (ISIC revision 2) or tabulation categories G through P (ISIC revision 3).

Data sources

Employment data are compiled by the World Bank's Development Data Group using an ILO database corresponding to table 2a in its *Yearbook of Labour Statistics*.

2.6 | Unemployment

	Male unemployment			Female unemployment			Total unemployment		
	% of male labor force			% of female labor force			% of total labor force		
	1980	1990	1996	1980	1990	1996	1980	1990	1996
Argentina	..	8.4	..	..	10.4	..	2.3	9.2	18.4
Australia	5.1	6.7	8.7	7.9	7.2	8.3	6.1	6.9	8.5
Austria	..	3.0	3.9	..	3.6	4.5	..	3.2	4.1
Belgium	..	4.5	10.5	..	11.4	18.2	..	7.2	13.8
Bolivia	..	6.9	3.7	..	7.8	4.5	..	7.3	4.2
Brazil	4.2	3.8	..	4.4	3.4	..	4.3	3.7	4.6
Bulgaria	..	..	14.1	..	..	14.2	..	1.7	14.2
Canada	..	8.1	9.8	..	8.1	9.4	..	8.1	9.7
Chile	10.6	5.7	..	10.0	5.7	..	10.4	5.6	6.4
Costa Rica	5.3	4.2	..	7.8	5.9	..	5.9	4.6	..
Czech Republic	..	..	3.5	..	..	4.1	..	0.3	3.1
Denmark	..	..	7.8	..	..	9.9	..	..	8.8
Dominican Republic	..	12.5	10.2	..	33.1	28.7	..	19.7	16.7
Finland	4.7	4.0	15.8	4.7	2.8	16.5	4.7	3.4	16.1
France	4.3	6.7	..	9.5	11.7	..	6.4	8.9	12.4
Germany	..	6.0	8.1	..	8.8	10.2	..	7.2	9.0
Greece	3.3	4.3	..	5.7	11.7	..	4.0	7.0	..
Hungary	..	..	10.7	..	..	7.6	..	0.8	11.0
Ireland	..	12.5	11.9	..	13.8	11.9	..	12.9	11.9
Israel	4.1	8.4	..	6.0	11.3	..	4.8	9.6	..
Italy	4.8	7.3	9.3	13.1	17.1	16.7	7.6	11.0	12.0
Jamaica	16.3	9.3	..	39.6	23.1	..	27.3	15.7	..
Japan	2.0	2.0	3.3	2.0	2.2	3.4	2.0	2.1	3.4
Korea, Rep.	..	2.9	2.3	..	1.8	1.6	..	2.4	2.0
Netherlands	6.3	5.4	7.0	13.4	10.7	8.3	7.9	7.5	7.6
New Zealand	..	8.1	6.1	..	7.2	6.1	..	7.8	6.1
Nicaragua	..	9.0	..	..	15.4	..	..	11.1	..
Norway	1.3	5.6	4.9	2.3	4.8	4.9	1.7	5.2	4.9
Panama	..	12.8	11.0	..	22.6	19.4	..	16.1	13.9
Paraguay	3.8	6.6	..	4.8	6.5	..	4.1	6.6	..
Philippines	3.2	7.1	..	7.5	9.8	..	4.8	8.1	..
Poland	..	..	9.8	..	..	13.7	..	6.1	14.0
Portugal	4.1	3.2	6.4	13.0	6.6	8.2	7.8	4.7	7.2
Puerto Rico	19.5	16.2	..	12.3	10.7	..	17.1	14.1	..
Romania	..	..	6.3	..	..	7.4	..	3.0	6.3
Russian Federation	..	..	9.6	..	..	9.0	..	..	9.3
Singapore	..	1.9	2.9	..	1.3	3.1	..	1.7	3.0
Slovak Republic	..	..	10.0	..	..	11.9	..	0.6	13.0
Spain	10.8	12.0	17.3	12.8	24.2	29.4	11.4	16.3	21.9
Sweden	1.7	1.7	8.5	2.3	1.6	7.5	2.0	1.6	8.0
Switzerland	..	..	4.4	..	..	5.1	..	..	4.7
Trinidad and Tobago	8.0	17.8	..	14.0	24.2	..	10.0	20.0	..
United Kingdom	..	6.9	9.2	..	6.5	6.4	..	6.7	8.0
United States	6.8	5.7	5.3	7.4	5.5	5.4	7.0	5.6	5.4
Venezuela	..	10.9	8.2	..	9.3	9.8	5.9	10.4	8.7

Unemployment | 2.6

The International Labour Organization (ILO) defines the unemployed as members of the economically active population who are without work but available for and seeking work, including people who have lost their jobs and those who have voluntarily left work. Some unemployment is unavoidable in all economies. At any time some workers are temporarily unemployed—between jobs as employers look for the right workers and workers search for better jobs. Such unemployment, often called frictional unemployment, results from the normal operation of labor markets. Changes in unemployment over time may reflect changes in the demand for and supply of labor, but they may also reflect changes in reporting practices. High and sustained unemployment, however, indicates serious inefficiencies in the allocation of resources.

The ILO definition of unemployment notwithstanding, reference periods and criteria for seeking work vary across countries in their treatment of people temporarily laid off and those seeking work for the first time. In many developing countries it is especially difficult to measure employment and unemployment in agriculture. The timing of a survey, for example, can maximize the seasonal effects of agricultural unemployment. And informal sector employment is difficult to quantify in the absence of regulation for registering and tracking such activities.

Data on unemployment are drawn from labor force sample surveys, employment office statistics, and administrative records of social insurance programs. Labor force surveys generally yield the most comprehensive data because they include groups—particularly people seeking work for the first time—not covered in other unemployment statistics. In addition, the quality and completeness of data obtained from social insurance programs and employment offices vary widely. The most common exclusion from these sources is discouraged workers who have given up their job search because they believe that no employment opportunities exist or do not register as unemployed after their benefits have been exhausted. Thus measured unemployment may be higher in economies that offer more or longer unemployment benefits. Economies for which unemployment data are not consistently available or were deemed unreliable have been omitted from the table.

• **Unemployment** is the share of the labor force that is without work but available for and seeking employment. Definitions of labor force and unemployment differ by country (see *About the data*).

Unemployment data are from an ILO database corresponding to table 3a in its *Yearbook of Labour Statistics*, the OECD's *Employment Outlook* (1997), and country statistical sources.

Unemployment continues to be high in transition economies

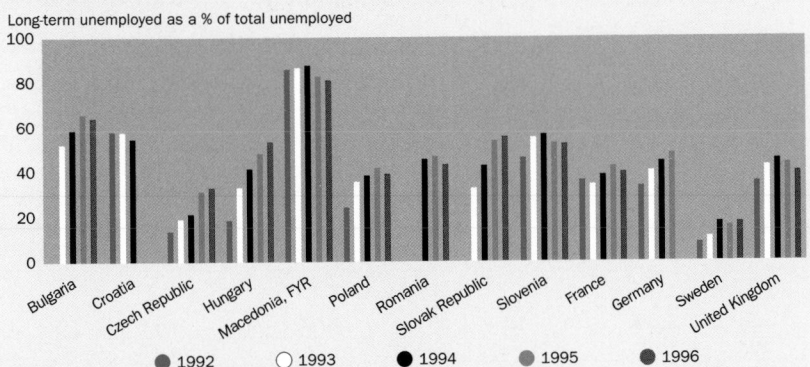

Long-term unemployed as a % of total unemployed

● 1992   ○ 1993   ● 1994   ● 1995   ● 1996

Source: OECD Centre for Co-operation with Economies in Transition Labour Market Database; World Bank Social Challenges in Transition Database; OECD 1997.

Growth in long-term unemployment has been one of the most troubling developments accompanying Central and Eastern Europe's transition from planned to market economies. Following an initial rise during 1989–93, unemployment growth in most countries has tapered off, and registered unemployment rates have stabilized or started to decline. But throughout the region, the long-term unemployed—individuals who have been out of work for more than a year—now make up the largest share of the unemployed. Although growth in long-term unemployment is not a phenomenon unique to transition economies, the situation in Central and Eastern Europe is serious because of gaps in social safety nets and a dearth of labor market programs targeted to the needs of the long-term unemployed.

The proportion of long-term unemployed grew steadily between 1992 and 1995 in all countries in the region except Croatia. In 1996 the share of long-term unemployed began to decline slightly in many countries, but it continued to increase in the Czech Republic, Hungary, and the Slovak Republic. Long-term unemployment in Central and Eastern Europe now resembles or even exceeds levels in Western Europe. In Bulgaria long-term unemployment accounts for more than 60 percent of total unemployment—a higher incidence than in Spain, which has experienced chronically high long-term unemployment. Rates approached or surpassed 40 percent in all countries except the Czech Republic, where it reached 33 percent. In FYR Macedonia it was 81 percent.

Long-term unemployment in Central and Eastern Europe is linked to job transition patterns. Individuals are more likely to be hired out of the public sector into the private sector, or between firms, than from the ranks of the unemployed or right out of school. As a result there is little movement out of unemployment, and the pool of unemployed has become increasingly homogeneous. Recent World Bank poverty assessments for Hungary and Romania show that demographic characteristics of the unemployed—age, ethnicity, education—are crucial risk factors for poverty in the region.

	National poverty line								International poverty line			
		Population below the poverty line				Population below the poverty line				Population below $1 a day	Poverty gap at $1 a day	Population below $2 a day	Poverty gap at $2 a day
	Survey year	Rural %	Urban %	National %	Survey year	Rural %	Urban %	National %	Survey year	%	%	%	%
Albania	1996	..	..	19.6	..	..	..	..		..	..	..	..
Algeria	1988	16.6	7.3	12.2	1995	30.3	14.7	22.6	1995	<2	..	17.6	4.4
Angola		..	..	..		..	..	..		..	..	..	..
Argentina	1991	..	..	25.5		..	..	..		..	..	..	..
Armenia		..	..	..		..	..	..		..	..	..	..
Australia		..	..	..		..	..	..		..	..	..	..
Austria		..	..	..		..	..	..		..	..	..	..
Azerbaijan		..	..	..		..	..	..		..	..	..	..
Bangladesh	1991–92	46.0	23.3	42.7	1995–96	39.8	14.3	35.6					
Belarus		..	..	..		..	..	..	1993	<2	..	6.4	0.8
Belgium		..	..	..		..	..	..		..	..	..	..
Benin	1995	..	..	33.0		..	..	..		..	..	..	..
Bolivia		..	..	..		..	..	..		..	..	..	..
Bosnia and Herzegovina		..	..	..		..	..	..		..	..	..	..
Botswana		..	..	..		..	..	..	1985–86	33.0	12.4	61.0	30.4
Brazil	1990	32.6	13.1	17.4		..	..	..	1995	23.6	10.7	43.5	22.4
Bulgaria		..	..	..		..	..	..	1992	2.6	0.8	23.5	6.0
Burkina Faso		..	..	..		..	..	..		..	..	..	..
Burundi	1990	..	..	36.2		..	..	..		..	..	..	..
Cambodia		..	..	..		..	..	..		..	..	..	..
Cameroon	1984	32.4	44.4	40.0		..	..	..		..	..	..	..
Canada		..	..	..		..	..	..		..	..	..	..
Central African Republic		..	..	..		..	..	..		..	..	..	..
Chad		..	..	..		..	..	..		..	..	..	..
Chile	1992	..	..	21.6	1994	..	..	20.5	1992	15.0	4.9	38.5	16.0
China	1994	11.8	<2	8.4	1995	9.2	<2	6.5	1995	22.2	6.9	57.8	24.1
Hong Kong, China		..	..	..		..	..	..		..	..	..	..
Colombia	1991	29.0	7.8	16.9	1992	31.2	8.0	17.7	1991	7.4	2.3	21.7	8.4
Congo, Dem. Rep.		..	..	..		..	..	..		..	..	..	..
Congo, Rep.		..	..	..		..	..	..		..	..	..	..
Costa Rica		..	..	..		..	..	..	1989	18.9	7.2	43.8	19.4
Côte d'Ivoire		..	..	..		..	..	..	1988	17.7	4.3	54.8	20.4
Croatia		..	..	..		..	..	..		..	..	..	..
Cuba		..	..	..		..	..	..		..	..	..	..
Czech Republic		..	..	..		..	..	..	1993	3.1	0.4	55.1	14.0
Denmark..		..	..	..		..	..	..		..	..	..	..
Dominican Republic	1989	27.4	23.3	24.5	1992	29.8	10.9	20.6	1989	19.9	6.0	47.7	20.2
Ecuador	1994	47.0	25.0	35.0	1995	..	..	..	1994	30.4	9.1	65.8	29.6..
Egypt, Arab Rep.		..	..	..		..	..	..	1990–91	7.6	1.1	51.9	15.3
El Salvador	1992	55.7	43.1	48.3		..	..	..		..	..	..	..
Eritrea..		..	..	..		..	..	..		..	..	..	..
Estonia	1994	14.7	6.8	8.9		..	..	..	1993	6.0	1.6	32.5	10.0
Ethiopia		..	..	..		..	..	..	1981–82	46.0	12.4	89.0	42.7
Finland		..	..	..		..	..	..		..	..	..	..
France..		..	..	..		..	..	..		..	..	..	..
Gabon..		..	..	..		..	..	..		..	..	..	..
Gambia, The	1992	..	..	64.0		..	..	..		..	..	..	..
Georgia..		..	..	..		..	..	..		..	..	..	..
Germany..		..	..	..		..	..	..		..	..	..	..
Ghana	1992	34.3	26.7	31.4		..	..	..		..	..	..	..
Greece..		..	..	..		..	..	..		..	..	..	..
Guatemala		..	..	..		..	..	..	1989	53.3	28.5	76.8	47.6
Guinea		..	..	..		..	..	..	1991	26.3	12.4	50.2	25.6
Guinea-Bissau	1991	60.9	24.1	48.8		..	..	..	1991	88.2	59.5	96.7	76.6
Haiti	1987	..	..	65.0		..	..	..		..	..	..	..
Honduras	1992	46.0	56.0	50.0		..	..	..	1992	46.9	20.4	75.7	41.9

Poverty | 2.7

	National poverty line								International poverty line				
		Population below the poverty line				Population below the poverty line				Population below $1 a day	Poverty gap at $1 a day	Population below $2 a day	Poverty gap at $2 a day
	Survey year	Rural %	Urban %	National %	Survey year	Rural %	Urban %	National %	Survey year	%	%	%	%
Hungary	1993	..	..	25.3		..	..	..	1993	<2	..	.10.7	2.1
India	1992	43.5	33.7	40.9	1994	36.7	30.5	35.0	1992	52.5	15.6	88.8	45.8
Indonesia	1987	16.4	20.1	17.4	1990	14.3	16.8	15.1	1995	11.8	1.8	58.7	19.3
Iran, Islamic Rep.		..	..	..		..	..	..		..	..	..	..
Iraq		..	..	..		..	..	..		..	..	..	..
Ireland		..	..	..		..	..	..		..	..	..	..
Israel		..	..	..		..	..	..		..	..	..	..
Italy		..	..	..		..	..	..		..	..	..	..
Jamaica	1992	..	..	34.2		..	..	..	1993	4.3	0.5	24.9	7.5
Japan		..	..	..		..	..	..		..	..	..	..
Jordan	1991	..	..	15.0		..	..	..	1992	2.5	0.5	23.5	6.3
Kazakhstan		..	..	..		..	..	..	1993	<2	..	12.1	2.5
Kenya	1992	46.4	29.3	42.0		..	..	..	1992	50.2	22.2	78.1	44.4
Korea, Dem. Rep.		..	..	..		..	..	..		..	..	..	..
Korea, Rep.		..	..	..		..	..	..		..	..	..	..
Kuwait		..	..	..		..	..	..		..	..	..	..
Kyrgyz Republic	1993	48.1	28.7	40.0		..	..	..	1993	18.9	5.0	55.3	21.4
Lao PDR	1993	53.0	24.0	46.1		..	..	..		..	..	..	..
Latvia		..	..	..		..	..	..		..	..	..	..
Lebanon		..	..	..		..	..	..		..	..	..	..
Lesotho	1993	53.9	27.8	49.2		..	..	..	1986–87	48.8	23.8	74.1	43.5
Libya		..	..	..		..	..	..		..	..	..	..
Lithuania		..	..	..		..	..	..	1993	<2	..	18.9	4.1
Macedonia, FYR		..	..	..		..	..	..		..	..	..	..
Madagascar		..	..	..		..	..	..	1993	72.3	33.2	93.2	59.6
Malawi	1990–91	..	..	54.0		..	..	..		..	..	..	..
Malaysia	1989	..	..	15.5		..	..	..	1989	5.6	0.9	26.6	8.5
Mali		..	..	..		..	..	..		..	..	..	..
Mauritania	1990	..	..	57.0		..	..	..	1988	31.4	15.2	68.4	33.0
Mauritius	1992	..	..	10.6		..	..	..		..	..	..	..
Mexico	1988	..	..	10.1		..	..	..	1992	14.9	3.8	40.0	15.9
Moldova		..	..	..		..	..	..	1992	6.8	1.2	30.6	9.7
Mongolia	1995	33.1	38.5	36.3		..	..	..		..	..	..	..
Morocco	1984–85	32.6	17.3	26.0	1990–91	18.0	7.6	13.1	1990–91	<2	..	19.6	4.6
Mozambique		..	..	..		..	..	..		..	..	..	..
Myanmar		..	..	..		..	..	..		..	..	..	..
Namibia		..	..	..		..	..	..		..	..	..	..
Nepal	1995–96	44.0	23.0	42.0		..	..	..	1995	50.3	16.2	86.7	44.6
Netherlands		..	..	..		..	..	..		..	..	..	..
New Zealand		..	..	..		..	..	..		..	..	..	..
Nicaragua	1993	76.1	31.9	50.3		..	..	..	1993	43.8	18.0	74.5	39.7
Niger		..	..	..		..	..	..	1992	61.5	22.2	92.0	51.8
Nigeria	1985	49.5	31.7	43.0	1992–93	36.4	30.4	34.1	1992–93	31.1	12.9	59.9	29.8
Norway		..	..	..		..	..	..		..	..	..	..
Oman		..	..	..		..	..	..		..	..	..	..
Pakistan	1991	36.9	28.0	34.0		..	..	..	1991	11.6	2.6	57.0	18.6
Panama		..	..	..		..	..	..	1989	25.6	12.6	46.2	24.5
Papua New Guinea		..	..	..		..	..	..		..	..	..	..
Paraguay	1991	28.5	19.7	21.8		..	..	..		..	..	..	..
Peru	1986	64.0	45.0	52.0	1991	68.0	50.3	54.0		..	..	..	..
Philippines	1985	58.0	42.0	52.0	1991	71.0	39.0	54.0	1991	28.6	7.7	64.5	28.2
Poland	1993	..	..	23.8		..	..	..	1993	6.8	4.7	15.1	7.7
Portugal		..	..	..		..	..	..		..	..	..	..
Puerto Rico		..	..	..		..	..	..		..	..	..	..
Romania	1994	27.9	20.4	21.5		..	..	..	1992	17.7	4.2	70.9	24.7
Russian Federation	1994	..	..	30.9		..	..	..	1993	<2	..	10.9	2.3

2.7 | Poverty

	National poverty line								International poverty line				
		Population below the poverty line				Population below the poverty line				Population below $1 a day %	Poverty gap at $1 a day %	Population below $2 a day %	Poverty gap at $2 a day %
	Survey year	Rural %	Urban %	National %	Survey year	Rural %	Urban %	National %	Survey year				
Rwanda	1993	..	..	51.2		..	..	..	1983–85	45.7	11.3	88.7	42.3
Saudi Arabia		..	..	..		..	..	..		..	..	..	..
Senegal								..	1991–92	54.0	25.5	79.6	47.2
Sierra Leone	1989	76.0	53.0	68.0		..	..	..		..	..	..	..
Singapore													
Slovak Republic									1992	12.8	2.2	85.1	27.5
Slovenia		..	..	..		..	..	..		..	..	..	..
South Africa		..	..	..		..	..	..	1993	23.7	6.6	50.2	22.5
Spain		..	..	..		..	..	..		..	..	..	..
Sri Lanka	1985–86	45.5	26.8	40.6	1990–91	38.1	28.4	35.3	1990	4.0	0.7	41.2	11.0
Sudan		..	..	..		..	..	..		..	..	..	..
Sweden		..	..	..		..	..	..		..	..	..	..
Switzerland		..	..	..		..	..	..		..	..	..	..
Syrian Arab Republic		..	..	..		..	..	..		..	..	..	..
Tajikistan		..	..	..		..	..	..		..	..	..	..
Tanzania	1991	..	..	51.1		..	..	..	1993	10.5	2.1	45.5	15.3
Thailand	1990	..	..	18.0	1992	15.5	10.2	13.1	1992	<2	..	23.5	5.4
Togo	1987–89	..	..	32.3		..	..	..		..	..	..	..
Trinidad and Tobago	1992	..	..	21.0		..	..	..		..	..	..	..
Tunisia	1985	29.2	12.0	19.9	1990	21.6	8.9	14.1	1990	3.9	0.9	22.7	6.8
Turkey		..	..	..		..	..	..		..	..	..	..
Turkmenistan		..	..	..		..	..	..	1993	4.9	0.5	25.8	7.6
Uganda	1993	..	..	55.0		..	..	..	1989–90	69.3	29.1	92.2	56.6
Ukraine	1995	..	..	31.7		..	..	..		..	..	..	..
United Arab Emirates		..	..	..		..	..	..		..	..	..	..
United Kingdom		..	..	..		..	..	..		..	..	..	..
United States		..	..	..		..	..	..		..	..	..	..
Uruguay		..	..	..		..	..	..		..	..	..	..
Uzbekistan		..	..	..		..	..	..		..	..	..	..
Venezuela	1989	..	..	31.3		..	..	..	1991	11.8	3.1	32.2	12.2
Vietnam	1993	57.2	25.9	50.9		..	..	..		..	..	..	..
West Bank and Gaza		..	..	..		..	..	..		..	..	..	..
Yemen, Rep.	1992	19.2	18.6	19.1		..	..	..		..	..	..	..
Yugoslavia, FR (Serb./Mont.)		..	..	..		..	..	..		..	..	..	..
Zambia	1991	88.0	46.0	68.0	1993	..	..	86.0	1993	84.6	53.8	98.1	73.4
Zimbabwe	1990–91	..	..	25.5		..	..	..	1990–91	41.0	14.3	68.2	35.5

About the data

International comparisons of poverty data entail both conceptual and practical problems. Different countries have different definitions of poverty, and consistent comparisons between countries can be difficult. Local poverty lines tend to have higher purchasing power in rich countries, where more generous standards are used than in poor countries.

Is it reasonable to treat two people with the same standard of living differently—in terms of their command over commodities—because one happens to live in a better-off country? Can we hold the real value of the poverty line constant between countries, just as we do when making comparisons over time?

Poverty measures based on an international poverty line attempt to do this. The commonly used $1 a day standard, measured in 1985 international prices and adjusted to local currency using purchasing power parities, was chosen for the World Bank's *World Development Report 1990: Poverty* because it is typical of the poverty lines in low-income countries. Purchasing power parity (PPP) exchange rates, such as those from the Penn World Tables, are used because they take into account the local prices of goods and services that are not traded internationally. But PPP rates were designed not for making international poverty comparisons, but for comparing aggregates from national accounts. As a result there is no certainty that an international poverty line measures the same degree of need or deprivation across countries.

Just as there are problems in comparing a poverty measure for one country with that for another, there can also be problems in comparing poverty measures within countries. For example, the cost of living is typically higher in urban than in rural areas. (Food staples, for example, tend to be more expensive in urban areas.) So the urban monetary poverty line should be higher than the rural poverty line. But it is not always clear that the actual difference between urban and rural poverty lines found in practice properly reflects the difference in the cost of living. For some countries the urban poverty line in common use has a higher real value—meaning that it allows poor people to buy more commodities for consumption—than does the rural poverty line. Sometimes the difference has been so large as to imply that the incidence of poverty is greater in urban than in rural areas, even though the reverse is found when adjustments are made only for differences in the cost of living. As with international comparisons, when the real value of the poverty line varies, it is not clear how meaningful such urban-rural comparisons are.

The problems of making poverty comparisons do not end there. Further issues arise in measuring household living standards. The choice between income and consumption as a welfare indicator is one issue. Incomes are generally more difficult to measure accurately, and consumption accords better with the idea of the standard of living than does income, which can vary over time even if the standard of living does not. But consumption data are not always available, and when they are not there is little choice but to use income. There are still other problems. Household survey questionnaires can differ widely, for example, in the number of distinct categories of consumer goods they identify. Survey quality varies, and even similar surveys may not be strictly comparable.

Comparisons across countries at different levels of development also pose a potential problem, because of differences in the relative importance of consumption of nonmarket goods. The local market value of all consumption in kind (including consumption from own production, particularly important in underdeveloped rural economies) should be included in the measure of total consumption expenditure. Similarly, the imputed profit from production of nonmarket goods should be included in income. This is not always done, though such omissions were a far bigger problem in surveys before the 1980s. Most survey data now include valuations for consumption or income from own production. Nonetheless, valuation methods vary—for example, some surveys use the price at the nearest market, while others use the average farmgate selling price.

The international poverty measures shown here are based on the most recent PPP estimates from the latest version of the Penn World Tables (PWT_5.6). It should be noted, however, that any revisions in the PPP of a country to incorporate better price indexes can produce dramatically different poverty lines in local currency.

Whenever possible, consumption has been used as the welfare indicator for deciding who is poor. When only household income is available, average income has been adjusted to accord with either a survey-based estimate of mean consumption (when available) or an estimate based on consumption data from national accounts. This procedure adjusts only the mean, however; nothing can be done to correct for the difference in Lorenz (income distribution) curves between consumption and income.

Empirical Lorenz curves were weighted by household size, so they are based on percentiles of population, not households. In all cases the measures of poverty have been calculated from primary data sources (tabulations or household data) rather than existing estimates. Estimation from tabulations requires an interpolation method; the method chosen was Lorenz curves with flexible functional forms, which have proved reliable in past work.

Definitions

• **Survey year** is the year in which the underlying data were collected. • **Rural poverty rate** is the percentage of the rural population living below the national rural poverty line. • **Urban poverty rate** is the percentage of the urban population living below the national urban poverty line. • **National poverty rate** is the percentage of the population living below the poverty line deemed appropriate for the country by its authorities. National estimates are based on population-weighted subgroup estimates from household surveys. • **Population below $1 a day** and **$2 a day** are the percentages of the population living on less than $1 a day and $2 a day at 1985 international prices, adjusted for purchasing power parity. • **Poverty gap** is the mean shortfall below the poverty line (counting the nonpoor as having zero shortfall) expressed as a percentage of the poverty line. This measure reflects the depth of poverty as well as its incidence.

Data sources

 Poverty measures are prepared by the World Bank's Development Research Group. National poverty lines are based on the Bank's country poverty assessments. International poverty lines are based on nationally representative primary household surveys conducted by national statistical offices or by private agencies under government or international agency supervision and obtained from government statistical offices and World Bank country departments.

The World Bank has prepared an annual review of poverty trends since 1993. The most recent is *Poverty Reduction and the World Bank: Progress in Fiscal 1996 and 1997*.

2.8 Distribution of income or consumption

| | Survey year | Gini index | \multicolumn{7}{c}{Percentage share of income or consumption} |
			Lowest 10%	Lowest 20%	Second 20%	Third 20%	Fourth 20%	Highest 20%	Highest 10%
Albania	..	..	..	..	..	..	..	..	..
Algeria	1995[a, b]	35.3	2.8	7.0	11.6	16.1	22.7	42.6	26.8
Angola	..	..	..	..	..	..	..	..	..
Argentina	..	..	..	..	..	..	..	..	..
Armenia	..	..	..	..	..	..	..	..	..
Australia	1989[c, d]	33.7	2.5	7.0	12.2	16.6	23.3	40.9	24.8
Austria	1987[c, d]	23.1	4.4	10.4	14.8	18.5	22.9	33.3	19.3
Azerbaijan	..	..	..	..	..	..	..	..	..
Bangladesh	1992[a, b]	28.3	4.1	9.4	13.5	17.2	22.0	37.9	23.7
Belarus	1993[c, d]	21.6	4.9	11.1	15.3	18.5	22.2	32.9	19.4
Belgium	1992[c, d]	25.0	3.7	9.5	14.6	18.4	23.0	34.5	20.2
Benin	..	..	..	..	..	..	..	..	..
Bolivia	1990[c, d]	42.0	2.3	5.6	9.7	14.5	22.0	48.2	31.7
Bosnia and Herzegovina	..	..	..	..	..	..	..	..	..
Botswana	..	..	..	..	..	..	..	..	..
Brazil	1995[c, d]	60.1	0.8	2.5	5.7	9.9	17.7	64.2	47.9
Bulgaria	1992[c, d]	30.8	3.3	8.3	13.0	17.0	22.3	39.3	24.7
Burkina Faso	..	..	..	..	..	..	..	..	..
Burundi	..	..	..	..	..	..	..	..	..
Cambodia	..	..	..	..	..	..	..	..	..
Cameroon	..	..	..	..	..	..	..	..	..
Canada	1994[c, d]	31.5	2.8	7.5	12.9	17.2	23.0	39.3	23.8
Central African Republic	..	..	..	..	..	..	..	..	..
Chad	..	..	..	..	..	..	..	..	..
Chile	1994[c, d]	56.5	1.4	3.5	6.6	10.9	18.1	61.0	46.1
China	1995[c, d]	41.5	2.2	5.5	9.8	14.9	22.3	47.5	30.9
Hong Kong, China	..	..	..	..	..	..	..	..	..
Colombia	1995[c, d]	57.2	1.0	3.1	6.8	10.9	17.6	61.5	46.9
Congo, Dem. Rep.	..	..	..	..	..	..	..	..	..
Congo, Rep.	..	..	..	..	..	..	..	..	..
Costa Rica	1996[c, d]	47.0	1.3	4.0	8.8	13.7	21.7	51.8	34.7
Côte d'Ivoire	1988[a, b]	36.9	2.8	6.8	11.2	15.8	22.2	44.1	28.5
Croatia	..	..	..	..	..	..	..	..	..
Cuba	..	..	..	..	..	..	..	..	..
Czech Republic	1993[c, d]	26.6	4.6	10.5	13.9	16.9	21.3	37.4	23.5
Denmark	1992[c, d]	24.7	3.6	9.6	14.9	18.3	22.7	34.5	20.5
Dominican Republic	1989[c, d]	50.5	1.6	4.2	7.9	12.5	19.7	55.7	39.6
Ecuador	1994[a, b]	46.6	2.3	5.4	8.9	13.2	19.9	52.6	37.6
Egypt, Arab Rep.	1991[a, b]	32.0	3.9	8.7	12.5	16.3	21.4	41.1	26.7
El Salvador	1995[c, d]	49.9	1.2	3.7	8.3	13.1	20.5	54.4	38.3
Eritrea	..	..	..	..	..	..	..	..	..
Estonia	1993[c, d]	39.5	2.4	6.6	10.7	15.1	21.4	46.3	31.3
Ethiopia	..	..	..	..	..	..	..	..	..
Finland	1991[c, d]	25.6	4.2	10.0	14.2	17.6	22.3	35.8	21.6
France	1989[c, d]	32.7	2.5	7.2	12.7	17.1	22.8	40.1	24.9
Gabon	..	..	..	..	..	..	..	..	..
Gambia, The	..	..	..	..	..	..	..	..	..
Georgia	..	..	..	..	..	..	..	..	..
Germany	1989[c, d]	28.1	3.7	9.0	13.5	17.5	22.9	37.1	22.6
Ghana	1992[a, b]	33.9	3.4	7.9	12.0	16.1	21.8	42.2	27.3
Greece	..	..	..	..	..	..	..	..	..
Guatemala	1989[c, d]	59.6	0.6	2.1	5.8	10.5	18.6	63.0	46.6
Guinea	1991[a, b]	46.8	0.9	3.0	8.3	14.6	23.9	50.2	31.7
Guinea-Bissau	1991[a, b]	56.2	0.5	2.1	6.5	12.0	20.6	58.9	42.4
Guyana	1993[a, b]	40.2	2.4	6.3	10.7	15.0	21.2	46.9	32.0
Haiti	..	..	..	..	..	..	..	..	..
Honduras	1996[c, d]	53.7	1.2	3.4	7.1	11.7	19.7	58.0	42.1

Distribution of income or consumption | 2.8

	Survey year	Gini index	Percentage share of income or consumption						
			Lowest 10%	Lowest 20%	Second 20%	Third 20%	Fourth 20%	Highest 20%	Highest 10%
Hungary	1993[c, d]	27.9	4.1	9.7	13.9	16.9	21.4	38.1	24.0
India	1994[a, b]	29.7	4.1	9.2	13.0	16.8	21.7	39.3	25.0
Indonesia	1995[a, b]	34.2	3.6	8.4	12.0	15.5	21.0	43.1	28.3
Iran, Islamic Rep.		..	..	..	..	..	..	..	..
Iraq		..	..	..	..	..	..	..	..
Ireland	1987[c, d]	35.9	2.5	6.7	11.6	16.4	22.4	42.9	27.4
Israel	1992[c, d]	35.5	2.8	6.9	11.4	16.3	22.9	42.5	26.9
Italy	1991[c, d]	31.2	2.9	7.6	12.9	17.3	23.2	38.9	23.7
Jamaica	1991[a, b]	41.1	2.4	5.8	10.2	14.9	21.6	47.5	31.9
Japan		..	..	..	..	..	..	..	..
Jordan	1991[a, b]	43.4	2.4	5.9	9.8	13.9	20.3	50.1	34.7
Kazakhstan	1993[c, d]	32.7	3.1	7.5	12.3	16.9	22.9	40.4	24.9
Kenya	1992[a, b]	57.5	1.2	3.4	6.7	10.7	17.0	62.1	47.7
Korea, Dem. Rep.		..	..	..	..	..	..	..	..
Korea, Rep.		..	..	..	..	..	..	..	..
Kuwait		..	..	..	..	..	..	..	..
Kyrgyz Republic	1993[c, d]	35.3	2.7	6.7	11.5	16.4	23.1	42.3	26.2
Lao PDR	1992[a, b]	30.4	4.2	9.6	12.9	16.3	21.0	40.2	26.4
Latvia	1993[c, d]	27.0	4.3	9.6	13.6	17.5	22.6	36.7	22.1
Lebanon		..	..	..	..	..	..	..	..
Lesotho	1986–87[a, b]	56.0	0.9	2.8	6.5	11.2	19.4	60.1	43.4
Libya		..	..	..	..	..	..	..	..
Lithuania	1993[c, d]	33.6	3.4	8.1	12.3	16.2	21.3	42.1	28.0
Luxembourg	1991[c, d]	26.9	4.2	9.5	13.6	17.7	22.4	36.7	22.3
Macedonia, FYR		..	..	..	..	..	..	..	..
Madagascar	1993[a, b]	43.4	2.3	5.8	9.9	14.0	20.3	50.0	34.9
Malawi		..	..	..	..	..	..	..	..
Malaysia	1989[c, b]	48.4	1.9	4.6	8.3	13.0	20.4	53.7	37.9
Mali		..	..	..	..	..	..	..	..
Mauritania	1988[a, b]	42.4	0.7	3.6	10.3	16.2	23.0	46.5	30.4
Mauritius		..	..	..	..	..	..	..	..
Mexico	1992[a, b]	50.3	1.6	4.1	7.8	12.5	20.2	55.3	39.2
Moldova	1992[c, d]	34.4	2.7	6.9	11.9	16.7	23.1	41.5	25.8
Mongolia	1995[a, b]	33.2	2.9	7.3	12.2	16.6	23.0	40.9	24.5
Morocco	1990–91[a, b]	39.2	2.8	6.6	10.5	15.0	21.7	46.3	30.5
Mozambique		..	..	..	..	..	..	..	..
Myanmar		..	..	..	..	..	..	..	..
Namibia		..	..	..	..	..	..	..	..
Nepal	1995–96[a, b]	36.7	3.2	7.6	11.5	15.1	21.0	44.8	29.8
Netherlands	1991[c, d]	31.5	2.9	8.0	13.0	16.7	22.5	39.9	24.7
New Zealand		..	..	..	..	..	..	..	..
Nicaragua	1993[a, b]	50.3	1.6	4.2	7.9	12.6	20.0	55.2	39.8
Niger	1992[a, b]	36.1	3.0	7.5	11.8	15.5	21.1	44.1	29.3
Nigeria	1992–93[a, b]	45.0	1.3	4.0	8.9	14.4	23.4	49.4	31.4
Norway	1991[c, d]	25.2	4.1	10.0	14.3	17.9	22.4	35.3	21.2
Oman		..	..	..	..	..	..	..	..
Pakistan	1991[a, b]	31.2	3.4	8.4	12.9	16.9	22.2	39.7	25.2
Panama	1991[c, d]	56.8	0.5	2.0	6.3	11.3	20.3	60.1	42.5
Papua New Guinea	1996[a, b]	50.9	1.7	4.5	7.9	11.9	19.2	56.5	40.5
Paraguay	1995[c, d]	59.1	0.7	2.3	5.9	10.7	18.7	62.4	46.6
Peru	1994[a, b]	44.9	1.9	4.9	9.2	14.1	21.4	50.4	34.3
Philippines	1994[a, b]	42.9	2.4	5.9	9.6	13.9	21.1	49.6	33.5
Poland	1992[a, b]	27.2	4.0	9.3	13.8	17.7	22.6	36.6	22.1
Portugal		..	..	..	..	..	..	..	..
Puerto Rico		..	..	..	..	..	..	..	..
Romania	1992[c, d]	25.5	3.8	9.2	14.4	18.4	23.2	34.8	20.2
Russian Federation	1993[c, d]	31.0	3.0	7.4	12.6	17.7	24.2	38.2	22.2

2.8 Distribution of income or consumption

	Survey year	Gini index	Percentage share of income or consumption						
			Lowest 10%	Lowest 20%	Second 20%	Third 20%	Fourth 20%	Highest 20%	Highest 10%
Rwanda	1983–85[a, b]	28.9	4.2	9.7	13.2	16.5	21.6	39.1	24.2
Saudi Arabia		..	..	..	..	..	..	..	..
Senegal	1991[a, b]	54.1	1.4	3.5	7.0	11.6	19.3	58.6	42.8
Sierra Leone	1989[a, b]	62.9	0.5	1.1	2.0	9.8	23.7	63.4	43.6
Singapore		..	..	..	..	..	..	..	..
Slovak Republic	1992[c, d]	19.5	5.1	11.9	15.8	18.8	22.2	31.4	18.2
Slovenia	1993[c, d]	29.2	4.0	9.3	13.3	16.9	21.9	38.6	24.5
South Africa	1993[a, b]	58.4	1.4	3.3	5.8	9.8	17.7	63.3	47.3
Spain	1990[c, d]	32.5	2.8	7.5	12.6	17.0	22.6	40.3	25.2
Sri Lanka	1990[a, b]	30.1	3.8	8.9	13.1	16.9	21.7	39.3	25.2
Sudan		..	..	..	..	..	..	..	..
Sweden	1992[c, d]	25.0	3.7	9.6	14.5	18.1	23.2	34.5	20.1
Switzerland	1982[c, d]	36.1	2.9	7.4	11.6	15.6	21.9	43.5	28.6
Syrian Arab Republic		..	..	..	..	..	..	..	..
Tajikistan		..	..	..	..	..	..	..	..
Tanzania	1993[a, b]	38.1	2.9	6.9	10.9	15.3	21.5	45.4	30.2
Thailand	1992[a, b]	46.2	2.5	5.6	8.7	13.0	20.0	52.7	37.1
Togo		..	..	..	..	..	..	..	..
Trinidad and Tobago		..	..	..	..	..	..	..	..
Tunisia	1990[a, b]	40.2	2.3	5.9	10.4	15.3	22.1	46.3	30.7
Turkey		..	..	..	..	..	..	..	..
Turkmenistan	1993[c, d]	35.8	2.7	6.7	11.4	16.3	22.8	42.8	26.9
Uganda	1992[a, b]	40.8	3.0	6.8	10.3	14.4	20.4	48.1	33.4
Ukraine	1992[c, d]	25.7	4.1	9.5	14.1	18.1	22.9	35.4	20.8
United Arab Emirates		..	..	..	..	..	..	..	..
United Kingdom	1986[c, d]	32.6	2.4	7.1	12.8	17.2	23.1	39.8	24.7
United States	1994[c, d]	40.1	1.5	4.8	10.5	16.0	23.5	45.2	28.5
Uruguay		..	..	..	..	..	..	..	..
Uzbekistan		..	..	..	..	..	..	..	..
Venezuela	1995[c, d]	46.8	1.5	4.3	8.8	13.8	21.3	51.8	35.6
Vietnam	1993[a, b]	35.7	3.5	7.8	11.4	15.4	21.4	44.0	29.0
West Bank and Gaza		..	..	..	..	..	..	..	..
Yemen, Rep.	1992[a, b]	39.5	2.3	6.1	10.9	15.3	21.6	46.1	30.8
Yugoslavia, FR (Serb./Mont.)		..	..	..	..	..	..	..	..
Zambia	1993[a, b]	46.2	1.5	3.9	8.0	13.8	23.8	50.4	31.3
Zimbabwe	1990[a, b]	56.8	1.8	4.0	6.3	10.0	17.4	62.3	46.9

a. Refers to expenditure shares by percentiles of population. b. Ranked by per capita expenditure. c. Refers to income shares by percentiles of population. d.Ranked by per capita income.

Distribution of income or consumption | 2.8

Inequality in the distribution of income is reflected in the percentage share of either income or consumption accruing to segments of the population ranked by income or consumption levels. The segments ranked lowest by personal income receive the smallest share of total income. The Gini index provides a convenient summary measure of the degree of inequality.

Data on personal or household income or consumption come from nationally representative household surveys. The data in the table refer to different years between 1985 and 1996. Footnotes to the survey year indicate whether the rankings are based on per capita income or consumption. For the first time, every distribution (including high-income economies) is based on percentiles of population—rather than households—with households ranked by income or expenditure per person. Where the original data from the household survey were available, they have been used to directly calculate the income (or consumption) shares by quintile. Otherwise, shares have been estimated from the best available grouped data.

The distribution indicators have been adjusted for household size, providing a more consistent measure of per capita income or consumption. No adjustment has been made for spatial differences in cost of living within countries, because the data needed for such calculations are generally unavailable. For further details on the estimation method for low- and middle-income economies, see Ravallion and Chen (1996).

Because the underlying household surveys differ in method and in the type of data collected, the distribution indicators are not strictly comparable across countries. These problems are diminishing as survey methods improve and become more standardized, but achieving strict comparability is still impossible (see the notes to table 2.7).

The following sources of noncomparability should be noted. First, the surveys can differ in many respects, including whether they use income or consumption expenditure as the living standard indicator. Income is typically more unequally distributed than consumption. In addition, the definitions of income used in surveys are usually very different from the economic definition of income (the maximum level of consumption consistent with keeping productive capacity unchanged). Consumption is usually a much better welfare indicator particularly in developing countries. Second, household units differ in size (number of members) and in extent of income sharing among members. Individuals differ in age and consumption needs. Differences between countries in these respects may bias distribution comparisons.

World Bank staff have made an effort to ensure that the data are as comparable as possible. Whenever possible, consumption has been used rather than income. Households have been ranked by consumption or income per capita in forming the percentiles, and the percentiles are of population, not households. The income distribution and Gini indexes for high-income countries are directly calculated from the Luxembourg Income Study database. The estimation method used here is consistent with that which is applied to developing countries.

• **Survey year** is the year in which the underlying data were collected. • **Gini index** measures the extent to which the distribution of income (or, in some cases, consumption expenditures) among individuals or households within an economy deviates from a perfectly equal distribution. A Lorenz curve plots the cumulative percentages of total income received against the cumulative number of recipients, starting with the poorest individual or household. The Gini index measures the area between the Lorenz curve and a hypothetical line of absolute equality, expressed as a percentage of the maximum area under the line. Thus a Gini index of zero represents perfect equality while an index of 100 implies perfect inequality. • **Percentage share of income or consumption** is the share that accrues to subgroups of population indicated by deciles or quintiles. Percentage shares by quintiles may not add up to 100 because of rounding.

Data on distribution are compiled by the World Bank's Development Research Group using primary household survey data obtained from government statistical agencies and World Bank country departments. Data for high-income economies are from national sources, supplemented by the Luxembourg Income Study database.

	Public expenditure on education		Expenditure per student						Expenditure on teaching materials		Primary pupil-teacher ratio	Duration of primary education
	% of GNP		Primary % of GNP per capita		Secondary % of GNP per capita		Tertiary % of GNP per capita		Primary % of total for level	Secondary % of total for level	pupils per teacher	years
	1980	1995ᵃ	1980	1994	1980	1995ᵃ	1980	1995ᵃ	1994	1994	1995ᵃ	1995ᵃ
Albania	..	3.4	..	..	..	23.0	..	36.0	..	5.4	18	8
Algeria	7.8	..	8.9	10.6	..	..	..	..	1.1	0.0	27	6
Angola	..	..	..	..	..	..	..	..	..	..	..	4
Argentina	2.7	4.5	6.5	16.2	..	12.0	10.4	17.0	..	..	..	7
Armenia	..	..	..	..	..	..	..	19.0	..	..	22	4
Australia	5.5	5.6	..	..	..	..	29.6	30.0	..	..	16	6
Austria	5.6	5.5	16.1	18.8	..	25.0	37.9	32.0	..	..	12	4
Azerbaijan	..	3.0	..	..	..	..	..	13.0	..	0.3	20	4
Bangladesh	1.5	2.3	4.8	..	..	23.0	46.8	30.0	..	..	..	5
Belarus	5.2	5.6	19.6	37.4	..	..	32.8	20.0	..	..	20	4
Belgium	6.1	5.7	17.8	..	..	25.0	34.8	35.0	0.2	..	12	6
Benin	..	3.1	..	..	..	22.0	..	240.0	..	..	49	6
Bolivia	4.4	6.6	13.7	..	..	18.0	..	67.0	..	..	..	8
Bosnia and Herzegovina	..	..	..	..	..	..	..	..	..	..	..	..
Botswana	..	9.6	13.6	..	..	..	..	665.5	..	..	26	7
Brazil	3.6	..	8.7	..	11.0	..	0.1	..	..	..	23	8
Bulgaria	4.5	4.2	17.5	28.9	..	..	..	21.0	..	..	17	4
Burkina Faso	2.6	3.6	26.5	..	..	..	3,371.1	..	0.8	..	58	6
Burundi	..	2.8	24.2	14.2	..	69.0	..	941.0	1.4	2.5	65	6
Cambodia	..	..	..	..	..	..	..	..	..	..	45	5
Cameroon	3.2	..	10.0	..	..	..	362.8	..	..	..	46	6
Canada	6.9	7.3	..	..	..	..	27.9	36.0	..	4.0	16	6
Central African Republic	..	..	22.1	..	..	..	..	..	..	..	..	6
Chad	..	2.2	..	12.3	..	33.0	..	234.0	..	0.9	62	6
Chile	4.6	2.9	9.6	8.5	..	9.0	..	21.0	0.0	..	27	8
China	2.5	2.3	3.8	5.6	..	14.0	..	81.0	..	..	24	5
Hong Kong, China	..	2.8	..	..	..	12.0	..	52.0	0.3	..	24	6
Colombia	1.9	3.5	5.2	10.5	..	11.0	41.1	29.0	..	..	25	5
Congo, Dem. Rep.	2.6	..	..	..	..	..	748.9	..	..	..	45	6
Congo, Rep.	7.0	5.9	10.1	..	..	..	..	224.0	0.1	..	70	6
Costa Rica	7.8	4.5	13.1	10.6	..	19.0	76.1	44.0	0.4	..	31	6
Côte d'Ivoire	7.2	..	22.5	..	113.0	..	..	..	..	..	45	6
Croatia	..	5.3	..	..	..	..	..	..	..	..	20	4
Cuba	7.2	..	10.4	..	..	..	..	28.5	5.7	..	14	6
Czech Republic	..	6.1	..	41.2	..	25.0	..	41.0	..	36.1	20	4
Denmark	6.9	8.3	38.4	..	..	..	..	55.0	4.3	..	10	6
Dominican Republic	2.2	1.9	3.1	2.9	..	5.0	..	5.0	..	..	35	8
Ecuador	5.6	3.4	5.6	3.9	..	15.0	22.3	34.0	..	..	26	6
Egypt, Arab Rep.	5.7	5.6	..	..	..	..	..	108.0	..	1.0	24	5
El Salvador	3.9	2.2	12.4	..	..	5.0	103.5	8.0	..	..	28	9
Eritrea	..	..	..	..	..	..	..	..	..	..	41	5
Estonia	..	6.6	..	..	..	..	..	40.0	..	3.4	17	5
Ethiopia	..	4.7	19.4	56.9	..	62.0	..	592.0	2.5	..	33	6
Finland	5.3	7.6	20.7	24.0	..	30.0	27.8	46.0	6.2	4.7	..	6
France	5.0	5.9	12.0	15.9	..	26.0	21.8	24.0	..	0.3	19	5
Gabon	2.7	..	..	5.5	..	..	..	..	..	0.6	52	6
Gambia, The	3.3	5.5	21.1	..	..	28.0	..	235.0	..	..	30	6
Georgia	..	5.2	..	..	..	..	..	28.0	..	..	16	4
Germany	..	4.7	..	..	..	..	..	35.0	..	..	18	4
Ghana	3.1	..	3.9	..	..	..	..	..	1.6	..	28	6
Greece	..	3.7	8.3	..	..	19.0	27.0	29.0	2.4	..	16	6
Guatemala	..	1.7	4.9	6.2	..	5.0	..	33.0	..	..	34	6
Guinea	..	..	..	10.4	..	38.0	..	498.0	..	..	49	6
Guinea-Bissau	..	..	32.7	..	..	..	..	..	..	..	..	6
Haiti	1.5	..	5.9	..	..	..	65.3	..	1.5	..	..	6
Honduras	3.2	3.9	10.9	..	..	22.0	72.1	59.0	3.6	..	35	6

Education policy and infrastructure | 2.9

	Public expenditure on education		Expenditure per student						Expenditure on teaching materials		Primary pupil-teacher ratio	Duration of primary education
	% of GNP		Primary % of GNP per capita		Secondary % of GNP per capita		Tertiary % of GNP per capita		Primary % of total for level	Secondary % of total for level	pupils per teacher	years
	1980	1995[a]	1980	1994	1980	1995[a]	1980	1995[a]	1994	1994	1995[a]	1995[a]
Hungary	4.7	6.0	14.0	26.0	..	28.0	75.3	73.0	..	..	11	8
India	2.8	3.5	9.4	11.9	..	13.0	..	78.0	..	..	63	5
Indonesia	1.7	..	..	..	..	..	..	..	..	..	23	6
Iran, Islamic Rep.	7.5	4.0	22.4	8.2	..	12.0	..	62.0	..	..	32	5
Iraq	3.0	..	7.0	..	..	..	..	..	..	..	22	6
Ireland	..	6.3	11.5	14.9	..	23.0	38.8	38.0	0.3	0.5	23	6
Israel	7.9	6.6	15.4	..	..	29.0	52.2	31.0	9.9	..	16	8
Italy	..	4.9	..	19.9	..	26.0	..	23.0	..	1.4	11	5
Jamaica	7.0	8.2	14.0	14.7	..	25.0	166.6	193.0	1.7	..	37	6
Japan	5.8	3.8	14.8	..	..	19.0	21.1	16.0	4.8	..	18	6
Jordan	..	6.3	..	..	..	..	..	111.0	..	3.0	21	10
Kazakhstan	..	4.5	..	..	..	..	..	20.0	..	..	20	4
Kenya	6.8	7.4	15.6	17.7	..	47.0	808.2	540.0	..	..	31	8
Korea, Dem. Rep.	..	..	..	..	..	..	..	..	..	..	..	4
Korea, Rep.	3.7	3.7	10.4	14.7	..	12.0	7.1	6.0	2.2	0.1	32	6
Kuwait	2.4	5.6	6.1	..	..	..	27.9	..	6.9	..	15	4
Kyrgyz Republic	7.2	6.8	..	..	..	..	..	49.0	..	0.6	20	4
Lao PDR	..	2.4	..	4.7	..	25.0	..	55.0	..	0.3	30	5
Latvia	3.3	6.3	..	..	..	..	..	45.0	..	..	14	4
Lebanon	..	2.0	..	..	..	..	..	..	..	..	12	6
Lesotho	5.1	5.9	8.8	12.6	..	51.0	642.3	399.0	0.4	..	49	7
Libya	3.4	..	..	..	..	..	..	..	..	..	..	9
Lithuania	5.5	6.1	..	..	..	..	..	51.0	..	0.4	17	4
Macedonia, FYR	..	5.5	..	20.6	..	..	..	..	..	0.1	20	8
Madagascar	4.4	..	7.8	..	35.6	..	..	..	1.1	..	40	5
Malawi	3.4	5.7	7.5	9.6	..	145.0	1,136.7	979.0	6.0	8.1	62	8
Malaysia	6.0	5.3	12.0	10.9	..	22.0	148.6	77.0	2.2	8.8	20	6
Mali	3.8	2.2	32.9	17.5	..	35.0	..	522.0	2.2	..	66	6
Mauritania	..	5.0	30.4	12.7	..	59.0	..	157.0	..	4.8	52	6
Mauritius	5.3	4.3	15.6	..	..	..	163.0	..	0.0	..	22	6
Mexico	4.7	5.3	4.3	7.8	..	20.0	..	61.0	1.3	0.0	29	6
Moldova	..	6.1	..	..	..	..	..	..	..	..	23	4
Mongolia	..	5.6	..	..	..	34.0	..	74.0	..	..	25	3
Morocco	6.1	5.6	15.5	15.5	..	51.0	..	74.0	0.2	..	28	6
Mozambique	4.4	..	..	..	..	..	..	..	..	..	58	5
Myanmar	1.7	1.3	..	..	..	10.0	..	21.0	..	..	48	5
Namibia	1.5	9.4	..	..	..	44.0	..	86.0	..	..	32	7
Nepal	1.8	2.9	14.6	8.4	..	11.9	271.9	156.0	7.2	..	39	5
Netherlands	7.6	5.3	13.8	..	..	20.0	53.7	44.0	3.1	..	19	6
New Zealand	5.8	6.7	15.0	16.9	..	23.0	33.3	39.0	7.9	..	18	6
Nicaragua	3.4	..	7.8	13.1	..	85.9	..	..	0.9	0.3	38	6
Niger	3.1	..	25.4	..	..	1,492.6	..	..	..	..	37	6
Nigeria	6.4	..	4.5	..	..	344.6	..	..	..	..	37	6
Norway	7.2	8.3	30.0	38.2	..	..	28.7	50.0	2.0	5.1	9	6
Oman	2.1	4.6	..	16.8	..	23.0	..	..	..	2.2	26	6
Pakistan	2.0	..	8.7	..	..	..	235.6	..	..	..	..	5
Panama	4.8	5.2	12.0	11.7	..	13.0	29.1	47.0	1.8	..		6
Papua New Guinea	..	..	..	..	..	..	..	..	..	..	33	6
Paraguay	1.5	2.9	..	7.9	..	11.0	..	52.0	..	..	24	6
Peru	3.1	..	7.2	..	..	..	5.1	..	0.7	..	28	6
Philippines	1.7	2.2	6.1	..	..	..	..	..	..	..	35	6
Poland	..	4.6	8.2	14.7	..	19.0	..	42.0	..	..	16	8
Portugal	3.8	5.4	13.5	17.2	..	20.0	..	25.0	0.2	..	12	6
Puerto Rico	..	..	..	..	..	..	..	..	..	..	..	8
Romania	3.3	3.2	..	21.7	..	7.0	..	40.0	..	..	20	4
Russian Federation	3.5	4.1	..	..	..	..	..	..	..	..	20	3

	Public expenditure on education % of GNP		Expenditure per student Primary % of GNP per capita		Secondary % of GNP per capita		Tertiary % of GNP per capita		Expenditure on teaching materials Primary % of total for level	Secondary % of total for level	Primary pupil-teacher ratio pupils per teacher	Duration of primary education years
	1980	1995a	1980	1994	1980	1995a	1980	1995a	1994	1994	1995a	1995a
Rwanda	2.7	..	11.1	..	..	..	..	..	3.5	..	..	7
Saudi Arabia	4.1	5.5	..	40.8	..	..	..	63.0	..	..	13	6
Senegal	..	3.6	24.6	..	..	..	..	..	4.0	..	58	6
Sierra Leone	3.8	..	..	..	..	..	..	..	..	..	..	7
Singapore	2.8	3.0	6.8	..	..	13.0	30.9	32.0	0.0	..	..	6
Slovak Republic	..	4.4	..	22.1	..	4.0	..	39.0	..	..	24	4
Slovenia	..	5.8	..	23.0	..	24.0	..	38.0	..	6.9	14	4
South Africa	..	6.8	..	32.3	..	..	..	59.0	..	..	37	7
Spain	..	5.0	..	14.1	..	21.0	..	18.0	..	..	18	5
Sri Lanka	2.7	3.1	..	..	..	..	62.2	64.0	..	..	28	5
Sudan	4.8	..	26.9	..	..	..	440.6	..	..	..	36	8
Sweden	9.0	8.0	43.0	45.2	..	25.6	76.0	..	3.9	6.8	11	6
Switzerland	5.0	5.5	..	..	..	..	55.7	..	..	..	12	6
Syrian Arab Republic	4.6	..	8.0	..	..	17.0	..	..	1.9	7.9	24	6
Tajikistan	8.2	8.6	..	..	..	..	29.7	39.0	..	..	23	4
Tanzania	4.4	..	11.1	..	..	..	..	2,195.3	..	..	37	7
Thailand	3.4	4.2	8.8	..	..	11.0	..	25.0	1.0	..	20	6
Togo	5.6	5.6	8.3	11.9	..	42.0	891.5	521.0	0.2	2.3	51	6
Trinidad and Tobago	4.0	4.5	9.2	..	..	17.0	55.1	77.0	5.7	..	25	7
Tunisia	5.4	6.8	11.8	13.5	..	23.0	193.9	89.0	2.0	..	25	6
Turkey	2.8	3.4	8.0	13.2	..	9.0	107.7	51.0	0.1	0.1	28	5
Turkmenistan	..	..	..	..	..	..	..	..	..	..	..	4
Uganda	1.2	..	3.7	..	..	..	..	..	..	..	35	7
Ukraine	5.6	7.7	21.2	42.9	..	..	38.5	20.0	..	..	20	4
United Arab Emirates	1.3	1.8	..	..	..	..	..	..	..	..	17	6
United Kingdom	5.6	5.5	16.0	..	..	22.0	79.7	44.0	2.9	..	19	6
United States	6.7	5.3	27.1	..	..	24.0	48.3	23.0	..	..	16	6
Uruguay	2.3	2.8	9.3	8.3	..	8.0	..	28.4	5.8	..	20	6
Uzbekistan	6.4	9.5	..	..	..	..	..	28.0	..	0.3	21	4
Venezuela	4.4	5.2	3.0	..	..	..	56.8	..	1.3	..	23	9
Vietnam	..	2.7	..	..	..	..	..	..	..	..	34	5
West Bank and Gaza	..	..	..	..	..	..	..	..	..	..	42	6
Yemen, Rep.	..	7.5	..	..	..	..	..	..	..	..	..	9
Yugoslavia, FR (Serb./Mont.)	..	..	..	..	..	..	..	..	..	9.3	22	4
Zambia	4.5	1.8	10.6	..	..	9.0	762.3	160.0	2.8	..	39	7
Zimbabwe	6.6	8.5	24.2	18.9	..	39.0	259.8	234.0	0.1	4.1	39	7
World	**4.4 m**	**5.2 m**	**11.6 m**	**14.7 m**	..	**22.0 m**	**55.7 m**	**44.5 m**	..		**32 w**	**6 m**
Low income	3.4	3.6	11.0	12.4	..	33.5	362.8	158.5	..	..	41	6
Excl. China & India	3.4	3.9	11.1	12.6	..	34.5	362.8	192.0	..	..	..	6
Middle income	4.4	5.2	9.3	14.1	..	17.0	55.9	41.5	..	..	24	6
Lower middle income	4.5	5.2	10.4	12.4	..	13.0	49.8	40.0	..	..	24	6
Upper middle income	4.0	5.0	9.2	16.5	..	19.5	65.2	42.0	..	..	24	7
Low & middle income	3.9	4.6	10.4	12.9	..	22.0	107.7	57.0	..	..	35	6
East Asia & Pacific	2.1	2.6	7.5	5.6	..	18.0	148.6	64.5	..	..	25	5
Europe & Central Asia	5.0	5.6	15.7	23.0	..	21.0	38.5	39.0	..	..	21	4
Latin America & Carib.	3.9	3.9	8.7	8.5	..	12.5	55.1	39.0	..	..	26	5
Middle East & N. Africa	5.0	5.6	10.3	14.5	..	23.0	193.9	81.5	..	..	26	6
South Asia	2.0	3.0	9.1	10.2	..	13.0	148.9	71.0	..	..	62	5
Sub-Saharan Africa	4.1	5.3	15.6	12.7	..	42.0	748.9	240.0	..	..	41	6
High income	5.6	5.5	15.0	..	..	22.5	30.2	33.5	..	..	17	6

a. Data are from UNESCO's forthcoming *World Education Report 1998*. They are not yet available in time series.

Education policy and infrastructure | 2.9

About the data

Data on education are compiled by the United Nations Educational, Scientific, and Cultural Organization (UNESCO) from official responses to surveys and from reports provided by education authorities in each country. Because coverage, definitions, and data collection methods vary across countries and over time within countries, data on education should be interpreted with caution. Although exceptions are noted in the table, readers seeking greater detail should consult the country- and indicator-specific notes in the source cited below. In addition, Behrman and Rosenzweig (1994) contains a general discussion of the reliability of data on education.

The data on education spending refer solely to public spending—that is, spending on public education plus subsidies for private education. Unless specified, the data exclude foreign aid for education. They also may exclude spending by religious schools, which play a significant role in many developing countries. Data for some countries and for some years refer to spending by the ministry of education of the central government only (excluding education expenditures by other ministries and departments, local authorities, and so on). Data for a few countries include private spending, although national practices vary with respect to whether parents or schools pay for books, uniforms, and other supplies.

In most cases the percentage of GNP devoted to education spending has little or no correlation with cross-national indicators of educational attainment. This percentage can be expected to be reflected in education indicators only when comparing countries that have the same national income per capita. Otherwise, this percentage reflects effort rather than achievement.

The comparability of pupil-teacher ratios is affected by whether both full- and part-time teachers are included, whether teachers are assigned nonteaching duties, and by differences in class size by grade and in number of hours taught. Moreover, the underlying enrollment levels are subject to a variety of reporting errors. (See *About the data* in table 2.10 for further discussion of enrollment data.) While the pupil-teacher ratio is often used to compare the quality of schooling across countries, it is not strongly related to the value added of schooling systems (Behrman and Rosenzweig 1994).

In many countries the duration of primary education changed between 1980 and 1995 (see table 2.10 for definitions of primary, secondary, and tertiary education). As a result the relative size of public spending on education by level and primary pupil-teacher ratios also may have changed. These changes may affect the comparability of enrollment ratios over time and across countries.

Definitions

• **Public expenditure on education** is the percentage of GNP accounted for by public spending on public education plus subsidies to private education at the primary, secondary, and tertiary levels. • **Expenditure on teaching materials** is the percentage of public spending on teaching materials (textbooks, books, and other scholastic supplies) to total public spending on primary or secondary education. • **Primary pupil-teacher ratio** is the number of pupils enrolled in primary school divided by the number of primary school teachers (regardless of their teaching assignment). • **Duration of primary education** is the minimum number of grades (years) a child is expected to cover in primary schooling.

Data sources

International data on education are compiled by UNESCO's Division of Statistics in cooperation with national commissions for UNESCO and national statistical services. The data in the table were compiled using a UNESCO electronic database corresponding to various tables in its *Statistical Yearbook 1996*.

2.10 Access to education

| | Gross enrollment ratio | | | | | | | | Net enrollment ratio | | |
| | Preprimary % of relevant age group | Primary % of relevant age group | | Secondary % of relevant age group | | Tertiary % of relevant age group | | Primary % of relevant age group | | Secondary % of relevant age group | |
	1995	1980	1995	1980	1995	1980	1995	1980	1995	1980	1995
Albania	38	113	87	67	35	8	10	..	96	..	..
Algeria	2	94	107	33	62	6	11	81	95	31	56
Angola	68	174	88	21	14	0	1	70	..	..	..
Argentina	50	106	108	56	72	22	38	..	..	..	59
Armenia	22	..	82	..	79	30	49	..	..	..	..
Australia	73	112	108	71	147a	25	72	100	98	70	89
Austria	76	99	101	99	104	26	45	99	100	..	90
Azerbaijan	20	115	104	93	74	24	20	..	..	..	..
Bangladesh	..	61	92	18	..	3	..	..	..	..	..
Belarus	80	104	97	98	..	39	..	..	95	..	..
Belgium	116	104	103	91	144a	26	49	97	98	..	98
Benin	3	67	72	16	16	1	3	..	59	..	..
Bolivia	..	87	..	37	..	16	..	79	..	16	..
Bosnia and Herzegovina	..	..	..	..	..	..	..	..	..	..	..
Botswana	..	91	115	19	56	1	4	75	96	14	45
Brazil	56	98	112	33	45	11	11	80	90	14	19
Bulgaria	62	98	94	84	78	16	39	96	97	73	75
Burkina Faso	..	17	38	3	8	0	1	15	31	..	7
Burundi	..	26	70	3	7	1	1	20	52	..	5
Cambodia	5	..	122	32	27	1	2	..	..	..	..
Cameroon	11	98	88	18	27	2	..	..	..	15	..
Canada	63	99	102	88	106	57	103	..	95	..	92
Central African Republic	..	71	58	14	10	1	1	56	..	..	..
Chad	1	..	55	..	9	..	1	..	..	..	..
Chile	96	109	99	53	69	12	28	..	86	..	55
China	29	113	118	46	67	2	5	..	99	..	..
Hong Kong, China	84	107	96	64	75	10	..	95	91	61	71
Colombia	28	124	114	41	67	9	17	..	85	..	50
Congo, Dem. Rep.	1	80	72	24	26	1	2	..	61	..	23
Congo, Rep.	1	141	114	74	53	5	..	96	..	..	..
Costa Rica	70	105	107	48	50	21	32	89	92	39	43
Côte d'Ivoire	2	75	69	19	23	3	4	..	..	..	..
Croatia	31	..	86	..	82	19	28	..	82	..	66
Cuba	89	106	105	81	80	17	14	95	99	..	59
Czech Republic	91	..	96	..	96	18	21	..	98	..	88
Denmark	81	96	99	105	118	28	45	96	99	88	86
Dominican Republic	20	118	103	42	41	..	..	..	81	..	22
Ecuador	49	117	109	53	50	35	..	..	92	..	..
Egypt, Arab Rep.	8	73	100	50	74	16	18	..	89	..	65
El Salvador	31	74	88	25	32	13	18	..	79	..	21
Eritrea	4	..	57	..	19	..	1	..	31	..	15
Estonia	56	98	91	..	86	25	38	..	94	..	77
Ethiopia	1	36	31	9	11	0	1	..	24	..	..
Finland	39	96	100	100	116	32	67	..	99	..	93
France	84	111	106	85	111	25	50	100	99	79	88
Gabon	..	..	142	..	..	..	..	..	..	..	..
Gambia, The	24	53	73	11	22	..	2	50	55	..	18
Georgia	32	..	82	..	73	30	38	..	82	..	71
Germany	84	..	102	98	103	34	43	..	100	..	88
Ghana	..	79	76	41	37	2	..	..	..	..	..
Greece	61	103	..	81	95	17	38	103	..	..	85
Guatemala	32	71	84	18	25	8	8	58	..	13	..
Guinea	9	36	48	17	12	5	..	..	37	..	..
Guinea-Bissau	..	68	64	6	..	..	..	47	..	3	..
Haiti	..	76	..	14	..	1	..	38	..	..	..
Honduras	14	98	111	30	32	8	10	78	90	..	21

	Gross enrollment ratio							Net enrollment ratio			
	Preprimary % of relevant age group	Primary % of relevant age group		Secondary % of relevant age group		Tertiary % of relevant age group		Primary % of relevant age group		Secondary % of relevant age group	
	1995	1980	1995	1980	1995	1980	1995	1980	1995	1980	1995
Hungary	86	96	97	70	81	14	19	95	93	..	73
India	5	83	100	30	49	5	6	..	..	..	..
Indonesia	19	107	114	29	48	..	15	88	97	..	42
Iran, Islamic Rep.	7	87	99	42	69	..	15	..	..	..	..
Iraq	8	113	90	57	44	9	..	99	79	47	37
Ireland	107	100	104	90	114	18	37	100	100	78	85
Israel	71	95	99	73	89	29	41	..	..	..	..
Italy	96	100	98	72	74	27	41	..	97	..	..
Jamaica	81	103	109	67	66	7	6	96	100	64	64
Japan	49	101	102	93	99	31	40	100	100	93	96
Jordan	25	104	94	75	..	27	..	93	..	68	..
Kazakhstan	29	84	96	93	83	34	33	..	..	..	..
Kenya	36	115	85	20	24	1	..	91	..	..	..
Korea, Dem. Rep.	..	..	..	..	..	..	..	..	..	..	..
Korea, Rep.	85	110	101	78	101	15	52	100	99	70	96
Kuwait	52	102	73	80	64	11	25	85	65	..	54
Kyrgyz Republic	8	116	107	110	81	16	14	..	97	..	..
Lao PDR	7	113	107	21	25	0	2	..	68	..	18
Latvia	44	78	89	100	85	24	26	..	84	..	78
Lebanon	74	111	109	59	81	30	27	..	..	..	..
Lesotho	..	102	99	18	28	1	2	66	65	13	16
Libya	..	125	110	76	97	8	16	100	97	62	..
Lithuania	36	79	96	114	84	35	28	..	..	..	80
Macedonia, FYR	24	100	89	61	57	28	18	..	85	..	51
Madagascar	..	133	72	..	14	3	3	..	..	..	..
Malawi	..	60	135	3	98	1	2	43	100	..	66
Malaysia	..	93	91	48	61	4	11	..	91	..	..
Mali	3	26	34	8	9	1	..	20	25	..	..
Mauritania	0	37	78	11	15	..	4	..	60	..	..
Mauritius	85	93	107	50	62	1	6	79	96	..	..
Mexico	71	120	115	49	58	14	14	..	100	..	..
Moldova	45	83	94	78	80	30	25	..	..	..	..
Mongolia	23	107	88	91	59	..	15	..	80	..	57
Morocco	63	83	83	26	39	6	11	62	72	20	..
Mozambique	..	99	60	5	7	0	1	36	40	..	6
Myanmar	..	91	100	22	32	5	5	..	..	..	..
Namibia	11	..	133	..	62	..	8	..	92	..	36
Nepal	..	86	110	22	38	3	5	..	..	..	..
Netherlands	100	100	107	93	139[a]	29	49	93	99	81	..
New Zealand	77	111	104	83	117	27	58	100	100	81	93
Nicaragua	20	98	110	43	47	13	9	98	83	23	27
Niger	1	25	29	5	7	0	..	21	..	4	..
Nigeria	..	105	89	16	30	2	4	..	..	..	..
Norway	98	100	99	94	92	26	55	98	99	84	94
Oman	3	51	80	12	66	..	5	43	71	10	56
Pakistan	..	39	74	14	26	..	3	..	..	..	..
Panama	76	106	106	61	68	21	30	89	..	46	..
Papua New Guinea	1	59	80	12	14	2	3	..	..	..	..
Paraguay	38	106	109	27	38	9	10	89	89	..	33
Peru	36	114	123	59	70	17	31	86	91	..	53
Philippines	13	112	116	64	79	24	27	94	100	45	60
Poland	45	100	98	77	96	18	27	98	97	70	83
Portugal	58	123	128	37	102[a]	11	34	98	100	..	78
Puerto Rico	..	..	..	..	..	48	..	..	..	..	..
Romania	53	102	100	71	66	12	18	..	92	..	73
Russian Federation	63	102	108	96	87	46	43	..	100	..	..

	Gross enrollment ratio							Net enrollment ratio			
	Preprimary % of relevant age group	Primary % of relevant age group		Secondary % of relevant age group		Tertiary % of relevant age group		Primary % of relevant age group		Secondary % of relevant age group	
	1995	1980	1995	1980	1995	1980	1995	1980	1995	1980	1995
Rwanda	..	63	82	3	11	0	..	59	76	..	8
Saudi Arabia	8	61	78	29	58	7	15	49	62	21	48
Senegal	2	46	65	11	16	3	3	37	54	..	..
Sierra Leone	..	52	..	14	..	1	..	..	..	..	..
Singapore	..	108	104	58	62	8	34	99	..	..	..
Slovak Republic	71	..	97	..	91	..	20	..	..	..	..
Slovenia	66	..	98	..	91	..	32	..	100	..	..
South Africa	28	85	117	..	84	..	17	..	96	..	52
Spain	69	109	105	87	118	23	46	100	100	74	94
Sri Lanka	..	103	113	55	75	3	5	..	..	..	..
Sudan	37	50	54	16	13	2	..	..	..	..	..
Sweden	60	97	105	88	132a	31	43	..	100	..	96
Switzerland	94	..	107	..	91	18	32	..	100	..	..
Syrian Arab Republic	7	100	101	46	44	17	18	89	91	39	39
Tajikistan	10	..	89	..	82	24	20	..	..	..	..
Tanzania	..	93	67	3	5	..	1	68	48	..	..
Thailand	58	99	87	29	55	15	20	..	..	..	..
Togo	3	118	118	33	27	2	3	..	85	..	..
Trinidad and Tobago	10	99	96	70	72	4	8	90	88	..	64
Tunisia	..	102	116	27	61	5	13	82	97	23	..
Turkey	6	96	105	35	56	5	18	..	96	..	50
Turkmenistan	..	..	..	..	..	23	..	..	..	..	..
Uganda	..	50	73	5	12	1	2	39	..	..	..
Ukraine	54	102	87	94	91	42	41	..	..	..	..
United Arab Emirates	57	89	95	52	78	3	9	74	83	..	71
United Kingdom	29	103	115	83	134a	19	48	100	100	79	92
United States	68	99	102	91	97	56	81	95	96	..	89
Uruguay	33	107	111	62	82	17	27	..	95	..	..
Uzbekistan	54	81	77	105	93	29	32	..	..	..	..
Venezuela	43	93	94	21	35	21	29	82	88	14	20
Vietnam	35	109	114	42	47	2	4	95	..	..	..
West Bank and Gaza	..	..	..	..	..	..	..	..	..	..	..
Yemen, Rep.	1	..	79	..	23	..	4	..	..	..	..
Yugoslavia, FR (Serb./Mont.)	31	29	72	..	65	..	21	..	..	..	..
Zambia	..	90	89	16	28	2	3	77	77	..	16
Zimbabwe	..	85	116	8	47	1	7	..	..	..	..
World	**33** w	**97** w	**103** w	**49** w	**62** w	**14** w	.. w	.. w	.. w	.. w	.. w
Low income	19	93	107	34	56	3	6	..	..	..	..
Excl. China & India	..	75	82	21	..	3	..	..	..	..	..
Middle income	38	100	105	54	60	19	19	..	93	..	..
Lower middle income	32	99	104	57	60	21	22	..	94	..	..
Upper middle income	56	101	107	47	62	14	14	..	91	..	..
Low & middle income	24	95	103	41	53	8	..	..	..	..	..
East Asia & Pacific	30	111	115	43	65	3	6	..	99	..	..
Europe & Central Asia	48	97	100	84	81	31	32	..	96	..	..
Latin America & Carib.	56	106	111	42	53	14	15	..	91	..	..
Middle East & N. Africa	14	87	97	42	64	11	15	..	..	..	..
South Asia	5	76	99	27	49	5	6	..	..	..	..
Sub-Saharan Africa	..	78	75	14	27	1	..	..	..	..	..
High income	69	102	103	87	104	35	57	..	98	..	..

a. Includes training for the unemployed.

About the data

School enrollment data are important indicators of the size and capacity of a country's education system and may be useful measures of education outcomes, but they are notoriously rife with errors. The indicators in the table are reported to the United Nations Educational, Scientific, and Cultural Organization (UNESCO) by national education authorities on the basis of annual enrollment surveys, typically conducted at the beginning of the school year. They do not reflect actual rates of attendance or dropouts during the school year. Furthermore, school administrators may have incentives to exaggerate enrollments. Behrman and Rosenzweig (1994), comparing official school enrollment data for Malaysia in 1988 with gross school attendance rates from a household survey, found that the official statistics systematically overstated enrollment.

Overage or underage enrollments may occur, particularly when parents prefer for cultural or economic reasons to have children start school at other than the official age. Children's age at enrollment may also be inaccurately estimated or misstated, especially in communities where registration of births is not strictly enforced. Parents who want to enroll their underage children in primary school may do so by overstating the age of the child. And in some education systems ages for children repeating a grade may be deliberately or inadvertently underreported.

As an international indicator, the gross primary enrollment ratio has an inherent weakness: the length of primary education differs significantly across countries (see table 2.9), so a short duration increases the ratio, and a long duration decreases it (partly because of more dropouts among older children). Other problems affecting cross-country comparisons of enrollment data stem from errors in estimates of school-age populations. Age-gender structures from censuses or vital registration systems, the primary sources of data on school-age populations, are commonly subject to underenumeration (especially of young children) in order to circumvent laws or regulations; errors are also introduced when parents round up children's ages. While census data are often adjusted for age bias, adjustments are rarely made for inadequate vital registration systems. Compounding these problems, pre- and post-census estimates of school-age children are interpolations or projections (see the discussion of demographic data in the notes to table 2.1) based on models that may miss important demographic events.

In using enrollment data, it is also important to consider repetition rates, which are quite high in some

developing countries, leading to a substantial number of overage children enrolled in each grade and raising the gross enrollment ratio. Thus gross enrollment ratios provide an indication of the capacity of each level of the education system, but a high ratio does not necessarily indicate a successful education system. Net enrollment ratios provide a better indicator of a school system's efficiency, but neither indicator measures the quality of the education provided.

Figure 2.10a

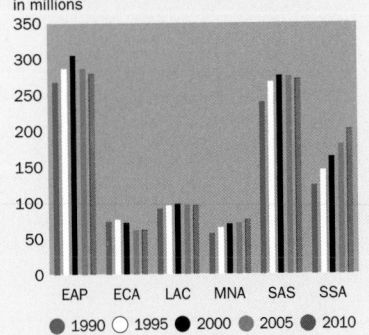

Enrollments are improving, but the school-age population is growing

Projected growth of school-age population, in millions

● 1990 ○ 1995 ● 2000 ● 2005 ● 2010

Source: World Bank staff estimates.

Enrollment ratios have improved considerably in all regions and at all education levels, particularly the primary level. Still, a substantial portion of children of school age continue to be out of school. The challenge for developing countries is to create an environment, both in school and out of school, that is conducive to bringing out-of-school children to schools and retaining them. In many countries this challenge will be exacerbated by sizable projected growth in the population age 6–14—growth that will put increasing pressure on the physical and financial resources of education systems.

Definitions

• **Gross enrollment ratio** is the ratio of total enrollment, regardless of age, to the population of the age group that officially corresponds to the level of education shown. Estimates are based on UNESCO's classification of education levels, as follows. • **Preprimary** provides education for children not old enough to enter school at the primary level. • **Primary** provides the basic elements of education at elementary or primary schools (see table 2.9 for the duration of primary school). • **Secondary** provides general or specialized instruction at middle, secondary, or high schools, teacher training schools, and vocational or technical schools; this level of education is based on at least four years of instruction at the primary level. • **Tertiary** requires, as a minimum condition of admission, the successful completion of education at the secondary level or evidence of attainment of an equivalent level of knowledge and is provided at universities, teachers colleges, and higher-level professional schools. • **Net enrollment ratio** is the ratio of the number of children of official school age (as defined by the education system) enrolled in school to the number of children of official school age in the population.

Data sources

Enrollment ratios are from UNESCO's *Statistical Yearbook 1997*.

2.11 | Educational attainment

	Percentage of cohort reaching grade 4				Progression to secondary school (general)				Average years of schooling			
	Male %		Female %		Male %		Female %		Male		Female	
	1980	1991	1980	1991	1980	1991	1980	1991	1980	1992	1980	1992
Albania	..	..	..	..	..	..	..	..	..	..	..	..
Algeria	92	97	91	96	55	78	62	83	9	11	6	9
Angola	..	..	..	..	..	..	..	..	8	..	7	..
Argentina	..	..	..	..	..	..	..	..	..	13	..	14
Armenia	..	..	..	..	..	..	..	..	..	..	..	..
Australia	..	..	..	..	..	..	..	..	12	13	12	14
Austria	..	..	..	..	..	..	..	..	11	15	11	14
Azerbaijan	..	..	..	..	..	..	..	..	..	..	..	..
Bangladesh	..	..	..	..	..	56	..	47	..	..	..	..
Belarus	..	..	..	..	..	..	..	..	..	..	..	..
Belgium	78	..	81	..	..	..	..	..	14	14	13	14
Benin	..	64	..	62	41	..	39	..	..	..	..	..
Bolivia	..	..	..	..	..	92	..	90	9	11	8	9
Bosnia and Herzegovina	..	..	..	..	..	..	..	..	..	..	..	..
Botswana	..	91	..	95	..	74	..	77	7	10	8	11
Brazil	..	..	..	..	..	..	..	..	9	..	9	..
Bulgaria	..	93	..	90	..	40	..	88	11	11	11	12
Burkina Faso	79	81	79	82	..	..	..	..	2	3	1	2
Burundi	83	78	83	76	8	12	7	11	3	5	2	4
Cambodia	..	..	..	..	..	..	..	..	..	..	..	..
Cameroon	81	..	81	..	24	32	19	30	8	..	6	..
Canada	..	..	..	..	..	..	..	..	15	17	15	18
Central African Republic	..	85	..	81	38	..	35	..	..	..	..	..
Chad	..	74	..	65	..	36	..	35	..	..	..	..
Chile	..	..	..	..	79	..	80	..	..	12	..	12
China	..	..	..	..	..	..	..	..	..	..	..	..
Hong Kong, China	100	..	100	..	87	..	93	..	12	..	12	..
Colombia	..	72	..	74	..	..	..	..	..	..	..	..
Congo, Dem. Rep.	77	..	70	..	..	31	..	34	..	7	..	4
Congo, Rep.	91	88	91	89	86	..	80	..	..	..	..	..
Costa Rica	80	90	84	91	..	66	..	67	10	10	10	9
Côte d'Ivoire	94	85	91	83	25	..	21	..	..	..	..	..
Croatia	..	..	..	..	..	..	..	..	..	11	..	11
Cuba	..	..	..	..	..	..	..	..	..	12	..	13
Czech Republic	..	..	..	..	..	..	..	..	..	..	..	..
Denmark	..	98	..	98	..	..	..	..	14	15	14	15
Dominican Republic	..	..	..	..	..	..	..	..	..	10	..	10
Ecuador	..	..	..	..	..	..	..	..	..	..	..	..
Egypt, Arab Rep.	95	..	65	..	..	..	..	..	..	11	..	9
El Salvador	..	..	..	..	28	21	26	18	..	9	..	9
Eritrea	..	..	..	..	..	..	..	..	..	..	..	..
Estonia	..	..	..	..	..	..	..	..	..	12	..	13
Ethiopia	..	..	..	..	..	..	..	..	..	..	..	..
Finland	..	100	..	100	..	..	..	..	..	..	..	..
France	..	..	..	..	..	..	..	..	13	14	13	15
Gabon	82	..	79	..	..	..	..	..	..	..	..	..
Gambia, The	..	..	..	..	41	..	42	..	..	5	6	3
Georgia	..	..	..	..	..	..	..	..	..	..	..	..
Germany	..	..	..	..	..	..	..	..	..	15	..	14
Ghana	87	..	82	..	..	..	..	..	..	..	..	..
Greece	98	..	98	..	..	..	..	..	12	13	12	13
Guatemala	..	..	..	..	..	..	..	..	..	..	..	..
Guinea	..	80	..	73	..	49	..	44	..	4	..	2
Guinea-Bissau	63	..	46	..	71	..	46	..	6	..	3	..
Haiti	..	60	..	60	38	80	45	92	..	..	..	..
Honduras	..	..	..	..	..	..	..	..	..	..	..	..

Educational attainment | 2.11

	Percentage of cohort reaching grade 4				Progression to secondary school (general)				Average years of schooling			
	Male %		Female %		Male %		Female %		Male		Female	
	1980	1991	1980	1991	1980	1991	1980	1991	1980	1992	1980	1992
Hungary	96	*97*	96	*97*	..	..	..	..	9	12	10	12
India	..	..	..	..	..	..	..	..	..	..	..	..
Indonesia	..	..	..	..	..	..	..	..		10		9
Iran, Islamic Rep.	..	94	..	93	..	83	..	82		10		8
Iraq	..	..	..	..	*46*	..	*51*	..	12	9	9	7
Ireland	..	..	..	..	..	..	..	..	11	13	11	13
Israel	..	..	..	..	..	..	..	..	..	..	..	..
Italy	100	100	100	100	98	99	98	100	..	..	..	..
Jamaica	..	*98*	..	*100*	*100*	98	89	95	10	11	11	11
Japan	100	100	100	100	..	..	..	..	13		12	
Jordan	95	*100*	95	*97*	..	..	88	70	12	11	12	12
Kazakhstan	..	..	..	..	..	..	..	..	..	..	..	..
Kenya	..	..	..	..	..	..	..	..	..	..	..	..
Korea, Dem. Rep.	..	..	..	..	..	..	..	..				
Korea, Rep.	96	100	96	100	99	..	96	..	12	14	11	13
Kuwait	..	..	..	..	98	..	98	..	11		11	
Kyrgyz Republic	..	..	..	..	..	..	..	..	..	..		..
Lao PDR	..	..	..	..	..	65	..	60		8		6
Latvia	..	..	..	..	..	..	..	..	..	..	..	..
Lebanon	..	..	..	..	..	..	..	..	..	..	..	..
Lesotho	61	74	77	84	..	..	..	..	7	8	10	10
Libya	..	..	..	..	..	..	..	..	..	..	..	..
Lithuania	..	..	..	..	..	..	..	..	..	..	..	..
Macedonia, FYR	..	..	..	..	..	..	..	..	..	..	..	..
Madagascar	..	63	..	64	*44*	..	*41*	..	..	..	..	..
Malawi	62	*73*	55	*68*	..	..	..	..		6		5
Malaysia	..	98	..	99	..	..	..	..	..	..	..	..
Mali	..	..	..	..	41	61	36	61		2		1
Mauritania	..	82	..	83	..	37	..	29	..	..	..	..
Mauritius	..	*99*	..	*99*	47	45	47	48	..	..	..	..
Mexico	..	..	..	..	..	..	..	..	..	..	..	..
Moldova	..	..	..	..	..	..	..	..	..	..	..	..
Mongolia	..	..	..	..	..	..	..	..	..	..	..	..
Morocco	90	85	89	85	..	79	..	84	8	8	5	6
Mozambique	..	66	..	60	*25*	39	*23*	39	5	4	4	3
Myanmar	..	..	..	..	..	..	..	..	..	..	..	..
Namibia	..	..	..	..	..	76	..	72		12		13
Nepal	..	..	..	..	..	79	..	77	..	..	..	..
Netherlands	97	..	100	..	65	..	75	..	14	16	13	15
New Zealand	..	97	..	97	..	..	..	..	14	15	13	16
Nicaragua	51	..	55	..	..	..	..	..	8	8	9	9
Niger	82	..	79	..	..	*42*	..	*37*		3		1
Nigeria	..	..	..	..	..	..	..	..	..	..	..	..
Norway	99	..	100	..	..	..	..	..	13	15	13	16
Oman	..	..	..	..	*74*	84	*83*	88	5	8	2	7
Pakistan	..	..	..	..	..	..	..	..	..	..	..	..
Panama	87	*85*	88	88	..	..	..	..	11	11	11	11
Papua New Guinea	..	68	..	67	..	..	..	..	..	..	..	..
Paraguay	..	79	..	81	..	..	..	..		9		8
Peru	85	..	83	..	81	..	78	..	11	..	10	..
Philippines	..	..	..	..	..	..	..	..	10	11	11	11
Poland	..	..	..	..	..	..	..	..	12	12	12	12
Portugal	..	..	..	..	..	*67*	..	*78*	..	..	..	..
Puerto Rico	..	..	..	..	..	..	..	..	..	..	..	..
Romania	..	..	..	..	..	..	..	..	..	..	11	11
Russian Federation	..	..	..	..	..	..	..	..	..	..	..	..

	Percentage of cohort reaching grade 4				Progression to secondary school (general)				Average years of schooling			
	Male %		Female %		Male %		Female %		Male		Female	
	1980	1991	1980	1991	1980	1991	1980	1991	1980	1992	1980	1992
Rwanda	83	72	84	75	5	..	2	..	..	6	..	6
Saudi Arabia	91	..	90	..	85	96	94	94	7	9	5	8
Senegal	93	94	90	90	..	..	..	..	..	6	..	4
Sierra Leone	..	..	..	..	..	..	..	..	..	..	..	..
Singapore	..	..	..	..	82	..	84	..	11	..	11	..
Slovak Republic	..	..	..	..	..	..	..	..	..	..	..	..
Slovenia	..	..	..	..	..	..	..	..	..	..	..	..
South Africa	..	..	..	..	..	87	..	91	..	12	..	12
Spain	95	97	95	98	..	..	91	..	13	14	12	15
Sri Lanka	..	97	..	98	..	88	..	92	..	..	..	..
Sudan	78	..	78	..	..	..	..	..	..	..	..	..
Sweden	99	..	100	..	..	..	..	..	12	14	13	14
Switzerland	92	..	94	..	42	46	42	48	14	15	13	14
Syrian Arab Republic	94	95	91	95	76	68	76	61	11	10	8	9
Tajikistan	..	..	..	..	..	..	..	..	..	..	..	..
Tanzania	..	89	..	90	..	..	..	..	..	..	..	..
Thailand	..	..	..	..	..	..	..	..	..	..	..	..
Togo	90	84	84	79	39	40	34	35	..	11	..	6
Trinidad and Tobago	..	..	..	..	..	..	..	..	11	11	11	11
Tunisia	94	93	90	93	31	60	31	60	10	11	7	10
Turkey	..	99	..	98	47	62	33	44	..	..	..	..
Turkmenistan	..	..	..	..	..	..	..	..	..	..	..	..
Uganda	..	..	..	..	..	..	..	..	..	..	..	..
Ukraine	..	..	..	..	..	..	..	..	..	..	..	..
United Arab Emirates	..	94	..	93	91	92	93	96	8	11	7	12
United Kingdom	..	..	..	..	..	..	..	..	13	15	13	15
United States	..	..	..	..	..	..	..	..	14	16	15	16
Uruguay	93	99	99	99	..	..	..	..	..	..	..	..
Uzbekistan	..	..	..	..	..	..	..	..	..	..	..	..
Venezuela	..	..	..	..	69	70	70	75	..	10	..	11
Vietnam	..	..	..	..	..	..	..	..	..	..	..	..
West Bank and Gaza	..	..	..	..	..	..	..	..	..	..	..	..
Yemen, Rep.	..	..	..	..	..	..	..	..	..	..	..	..
Yugoslavia, FR (Serb./Mont.)	..	..	..	..	..	..	..	..	..	..	..	..
Zambia	..	..	..	..	..	..	..	..	..	..	..	..
Zimbabwe	..	81	..	80	..	..	..	..	..	..	..	..

Indicators of students' progress through school provide a measure of an education system's success in maintaining a flow of students from one grade to the next and thus in imparting a particular level of education. Although school attendance is mandatory in most countries, at least through the primary level, students drop out of school for a variety of reasons—including discouragement over poor performance, the cost of schooling, and the opportunity cost of time spent in school. In addition, students' progress to higher grades may be limited by the availability of teachers, classrooms, and educational materials.

The rate of progression, or persistence, is measured by the proportion of a single-year cohort of students that eventually reaches a particular grade or level of schooling. Because tracking data for individual students are not available, aggregate student flows from one grade to the next are estimated using data on average promotion, repetition, and dropout rates. Other flows caused by new entrants, reentrants, grade skipping, migration, or school transfers during the school year are not considered. This procedure, called the reconstructed cohort method, makes three simplifying assumptions: dropouts never return to school; promotion, repetition, and dropout rates remain constant over the entire period in which the cohort is enrolled in school; and the same rates apply to all pupils enrolled in a given grade, regardless of whether they previously repeated a grade.

Because data from the United Nations Educational, Scientific, and Cultural Organization (UNESCO) do not include dropouts or dropout rates, the number of dropouts is estimated as the difference between enrollments in successive grades in successive years, after netting out repeaters. The remaining students are assumed to be promoted. Repeated application of the same calculations leads to an estimate of the number of students entering each successive grade (Fredricksen 1991).

The percentage of the cohort reaching grade 4, rather than some other grade, is shown for two reasons. First, four grades are the minimum needed to acquire literacy (United Nations 1993b). Second, using grade 4 minimizes the effect of repetition at or close to the final grade of primary education.

Progression to secondary school measures the percentage of students in the final grade of primary school who enter the first year of the general secondary system. The comparability of this indicator across time and between countries may be affected by changes in the definition of the primary and secondary levels, rules governing repetition and promotion, and the availability of special programs and other alternatives to the general secondary education system.

The average years of schooling measures educational attainment for men and women.

• **Percentage of cohort reaching grade 4** is the share of children enrolled in primary school in 1980 and 1991 who reached grade 4 in 1983 and 1994, respectively. The estimate is based on the reconstructed cohort method (see *About the data*). • **Progression to secondary school** (**general**) is the number of new entrants in the first grade of secondary school (general) divided by the number of children enrolled in the final grade of primary school in the previous year (according to the country's duration of primary education, as shown in table 2.9). • **Average years of schooling** is the average number of years of formal schooling received.

Estimates of the percentage of cohort reaching grade 4 and progression to secondary school were compiled using UNESCO's database on enrollment by level, grade or field, and gender.

2.12 Gender and education

	Primary education				Secondary general				Secondary vocational			
	Teachers % female		Pupils % female		Teachers % female		Pupils % female		Teachers % female		Pupils % female	
	1980	1995a	1980	1994	1980	1995a	1980	1994	1980	1994	1980	1994
Albania	50	60	47	48	46	51	59	54	32	50	41	31
Algeria	37	44	42	46	..	45	39	47	..	20	21	34
Angola	..	..	47	..	..	33	..	..	..	..	21	..
Argentina	92	..	49	..	75	..	64	..	..	..	47	..
Armenia	..	97	..	50	..	85	..	..	..	..	..	..
Australia	70	76	49	49	45	52	50	49	..	..	..	..
Austria	75	84	49	49	54	61	49	49	36	44	41	43
Azerbaijan	..	83	..	47	..	..	..	48	..	..	32	38
Bangladesh	8	..	37	..	7	..	24	..	5	..	2	..
Belarus	..	..	..	48	..	..	..	50	..	..	..	..
Belgium	59	72	49	49	..	..	..	..	..	..	..	..
Benin	23	24	32	..	..	..	26	..	..	..	..	..
Bolivia	48	..	47	..	..	..	..	..	..	..	..	..
Bosnia and Herzegovina	..	..	..	..	..	..	..	..	..	..	..	..
Botswana	72	76	55	50	35	44	56	53	45	32	25	23
Brazil	85	..	49	..	..	..	..	..	..	..	..	..
Bulgaria	72	89	49	48	64	75	68	67	49	59	40	38
Burkina Faso	20	24	37	39	..	18	33	34	..	21	40	49
Burundi	47	47	39	45	18	..	25	38	10	14	18	39
Cambodia	..	37	..	44	..	28	..	38	..	..	..	..
Cameroon	20	32	45	47	18	25	34	40	24	28	39	41
Canada	66	67	49	48	44	67	49	49	..	..	..	..
Central African Republic	25	..	37	..	12	..	25	..	25	25	49	..
Chad	..	8	..	32	..	4	..	17	..	..	..	..
Chile	..	72	49	49	..	..	55	54	..	..	47	47
China	37	47	45	47	25	36	40	44	25	25	34	46
Hong Kong, China	73	76	48	..	..	51	51	..	..	..	32	..
Colombia	79	80	50	50	41	..	50	..	42	42	45	..
Congo, Dem. Rep.	..	22	42	43	..	..	..	30	..	..	..	..
Congo, Rep.	25	36	48	48	8	15	40	41	..	..	54	..
Costa Rica	79	..	49	49	57	..	54	52	50	..	50	48
Côte d'Ivoire	15	18	40	43	..	..	28	34	..	..	49	..
Croatia	73	89	49	49	..	67	..	65	..	58	..	46
Cuba	75	81	48	49	50	61	51	54	25	34	46	48
Czech Republic	..	93	..	49	..	72	..	52	..	55	..	41
Denmark	..	58	49	49	..	49	51	52	..	..	41	46
Dominican Republic	..	71	40	50	..	50	..	57	..	45	75	57
Ecuador	65	68	49	49	38	44	48	47	37	..	60	55
Egypt, Arab Rep.	47	53	40	45	35	41	36	44	21	21	38	45
El Salvador	65	..	49	49	24	..	43	48	32	..	48	53
Eritrea	..	35	..	44	..	13	..	42	..	3	..	11
Estonia	..	89	49	48	..	83	..	53	..	64	..	47
Ethiopia	22	27	35	38	10	10	36	46	..	..	..	18
Finland	..	..	49	49	..	..	53	53	42	42	47	54
France	68	78	48	48	58	..	49	51	42	..	68	45
Gabon	27	44	49	50	28	19	42	..	17	..	28	..
Gambia, The	34	34	35	41	27	..	30	..	20	..	19	..
Georgia	..	94	..	..	..	..	..	..	..	..	..	..
Germany	..	85	..	49	..	48	..	50	..	35	..	44
Ghana	42	..	44	46	21	..	38	..	21	..	25	..
Greece	48	55	48	48	55	56	50	50	24	44	20	34
Guatemala	62	..	45	46	..	..	43	..	..	..	39	..
Guinea	14	25	33	33	10	12	28	24	4	..	34	25
Guinea-Bissau	24	..	32	..	20	..	22	..	3	22	14	..
Haiti	49	..	46	..	11	..	47	..	..	..	..	..
Honduras	74	73	50	50	..	..	50	..	..	..	49	..

Gender and education | 2.12

	Primary education				Secondary general				Secondary vocational			
	Teachers % female		Pupils % female		Teachers % female		Pupils % female		Teachers % female		Pupils % female	
	1980	1995ᵃ	1980	1994	1980	1995ᵃ	1980	1994	1980	1994	1980	1994
Hungary	80	84	49	49	61	67	65	63	..	..	39	45
India	26	32	39	43	30	35	32	38	..	..	32	13
Indonesia	33	52	46	48	..	39	36	46	..	27	27	40
Iran, Islamic Rep.	57	55	40	47	32	46	39	45	10	17	16	24
Iraq	48	68	46	45	42	55	32	39	24	52	29	26
Ireland	74	78	49	49	..	..	51	50	..	..	72	49
Israel	..	83	49	49	..	..	56	53	..	..	46	45
Italy	87	93	49	49	64	71	48	50	45	45	41	43
Jamaica	87	89	50	49	67	..	52	..	56	56	65	..
Japan	57	60	49	49	..	34	50	50	..	28	47	45
Jordan	59	61	48	49	44	48	46	55	28	37	30	35
Kazakhstan	..	97	..	49	..	74	..	52	..	..	..	..
Kenya	31	40	47	49	..	33	42	44	..	..	..	..
Korea, Dem. Rep.	..	..	..	..	..	..	..	..	..	..	..	..
Korea, Rep.	37	59	49	48	28	41	46	47	20	25	44	53
Kuwait	56	71	48	49	50	54	46	49	..	25	..	34
Kyrgyz Republic	88	83	49	50	58	71	49	51	..	38	50	50
Lao PDR	30	42	45	43	26	39	38	39	..	26	28	31
Latvia	..	97	49	48	..	81	..	52	..	..	..	45
Lebanon	..	..	48	48	..	..	51	53	8	35	40	47
Lesotho	75	79	59	53	48	51	60	59	47	..	56	46
Libya	47	..	47	49	24	..	39	..	12	..	25	..
Lithuania	97	98	49	48	85	82	..	52	..	..	..	42
Macedonia, FYR	..	53	..	48	..	52	..	60	..	..	..	44
Madagascar	..	56	49	49	..	..	..	50	..	..	11	34
Malawi	32	38	41	47	..	..	29	39	..	..	..	4
Malaysia	44	59	49	49	46	54	48	51	22	34	29	27
Mali	20	23	36	39	..	18	29	34	..	..	8	34
Mauritania	9	20	35	45	8	10	21	36	..	4	7	23
Mauritius	43	50	49	49	39	44	48	51	..	..	22	..
Mexico	..	..	49	48	..	..	43	48	..	..	66	59
Moldova	96	97	49	49	..	..	51	51	..	..	..	43
Mongolia	..	91	49	51	..	67	52	58	44	56	63	48
Morocco	30	38	37	42	..	32	38	42	..	26	23	40
Mozambique	22	23	43	42	27	19	29	40	15	24	17	25
Myanmar	54	67	48	48	61	74	45	..	..	..	..	..
Namibia	..	65	..	50	..	46	..	55	..	20	..	39
Nepal	10	16	28	39	..	..	..	..	..	..	..	..
Netherlands	46	54	49	50	26	30	52	52	..	31	41	41
New Zealand	66	81	49	49	41	56	49	50	..	49	82	47
Nicaragua	78	84	51	50	..	56	52	53	..	..	56	49
Niger	30	34	35	38	22	23	29	33	15	12	8	13
Nigeria	33	46	43	44	8	..	36	..	38	..	17	..
Norway	56	..	49	49	..	..	51	51	..	..	47	41
Oman	34	50	34	48	27	48	25	48	6	..	7	17
Pakistan	32	..	33	31	30	..	26	..	20	20	17	33
Panama	80	..	48	..	55	..	51	..	47	47	54	..
Papua New Guinea	27	37	41	45	34	35	32	41	31	31	..	31
Paraguay	..	55	48	48	..	65	49	51	..	..	..	43
Peru	60	58	48	..	46	39	46	..	..	..	40	..
Philippines	80	..	49	50	..	..	53	..	..	..	..	..
Poland	..	..	49	48	..	..	71	68	..	..	44	41
Portugal	..	..	48	..	59	..	48	..	..	..	..	..
Puerto Rico	..	..	..	..	..	..	..	..	..	..	..	..
Romania	70	84	49	49	53	65	65	53	41	54	45	42
Russian Federation	98	98	49	49	76	79	51	52	..	..	..	..

	Primary education				Secondary general				Secondary vocational			
	Teachers % female		Pupils % female		Teachers % female		Pupils % female		Teachers % female		Pupils % female	
	1980	1995a	1980	1994	1980	1995a	1980	1994	1980	1994	1980	1994
Rwanda	38	..	48	50	..	..	28	..	..	..	55	..
Saudi Arabia	39	51	39	47	33	49	37	44	..	15	..	12
Senegal	24	26	40	43	15	15	34	35	..	..	25	35
Sierra Leone	22	..	42	..	21	..	30	..	..	..	..	..
Singapore	66	..	48	..	56	..	51	..	24	..	23	..
Slovak Republic	..	91	..	49	..	75	..	51	..	63	..	48
Slovenia	..	92	..	49	..	76	..	52	..	58	..	44
South Africa	..	58	..	49	..	64	..	53	..	..	..	..
Spain	67	71	49	48	43	..	51	51	31	..	46	52
Sri Lanka	..	83	48	48	..	62	51	51	..	..	..	..
Sudan	31	60	40	44	26	..	37	..	..	..	21	..
Sweden	..	72	49	49	..	64	51	51	..	..	52	50
Switzerland	..	69	49	49	..	..	49	50	..	..	39	41
Syrian Arab Republic	54	65	43	47	22	44	37	44	15	34	..	..
Tajikistan	..	51	..	49	..	34	..	47	..	..	..	..
Tanzania	37	43	47	49	28	25	33	43	..	..	..	..
Thailand	49	..	48	49	57	..	46	50	..	..	..	..
Togo	21	14	38	40	13	11	24	26	..	..	..	..
Trinidad and Tobago	66	74	50	49	52	56	50	51	..	..	..	..
Tunisia	29	49	42	47	..	36	39	47	..	..	..	..
Turkey	41	44	45	47	36	40	35	39	34	39	..	..
Turkmenistan	..	..	..	..	..	..	..	..	..	..	..	..
Uganda	30	32	43	44	20	..	29	38	..	..	..	..
Ukraine	97	98	49	49	..	..	..	51	..	..	..	..
United Arab Emirates	54	69	48	48	48	55	45	51	..	..	..	..
United Kingdom	78	82	49	49	49	57	49	49	..	..	57	52
United States	..	86	49	49	..	56	49	49	..	..	..	..
Uruguay	..	..	49	49	..	..	58	55	..	..	..	..
Uzbekistan	78	82	49	49	48	49	46	48	..	..	..	..
Venezuela	83	75	50	50	..	..	..	58	..	..	..	..
Vietnam	65	..	47	..	58	..	47	..	..	..	..	..
West Bank and Gaza	..	48	..	..	..	41	..	..	..	..	..	..
Yemen, Rep.	11	..	..	28	..	..	..	..	..	..	..	..
Yugoslavia, FR (Serb./Mont.)	..	75	..	49	..	..	..	51	..	..	..	..
Zambia	40	43	47	48	..	..	35	38	3	..	..	..
Zimbabwe	38	41	48	48	..	32	42	44	..	..	..	..
World	**45 w**	**53 w**	**45 w**	**46 w**	**.. w**	**.. w**	**..**	**.. w**	**..**	**..**	**34 w**	**.. w**
Low income	32	39	42	44	26	33	..	41	..	..	30	31
Excl. China & India	31	..	42	..	..	..	..	..	..	..	..	..
Middle income	..	..	47	48	..	..	..	..	..	..	..	..
Lower middle income	65	68	47	48	..	..	..	..	..	..	..	..
Upper middle income	..	..	49	..	..	..	..	..	..	..	..	..
Low & middle income	42	47	44	45	31	..	..	..	..	..	32	..
East Asia & Pacific	41	46	45	47	28	35	..	44	..	..	33	45
Europe & Central Asia	84	85	48	48	..	67	..	..	..	..	..	..
Latin America & Carib.	..	..	49	..	..	..	..	..	..	..	..	..
Middle East & N. Africa	44	51	42	45	35	44	..	..	..	..	26	30
South Asia	24	31	38	41	27	34	..	38	..	..	27	15
Sub-Saharan Africa	29	38	43	46	..	..	..	..	..	..	..	..
High income	..	76	49	49	..	49	..	49	..	..	..	..

a. Data are from UNESCO's forthcoming *World Education Report 1998*. They are not yet available in time series.

About the data

Although data on female enrollment suffer from the same problems affecting data on general enrollment discussed in the notes to table 2.10, female enrollment as a share of total enrollment is a relatively simple indicator that does not raise serious problems of cross-country comparability. Most countries could achieve gender parity in primary and secondary schools, especially if education resources kept pace with the population of children. Yet disparities remain, and female enrollment rates tend to be positively correlated with other indicators of development (UNRISD 1977).

Girls' enrollments have caught up with boys' in most high-income countries, as well as in Latin America and the Caribbean. But they lag behind in Sub-Saharan Africa, South Asia, and the Middle East. In low- and lower-middle-income countries dropout rates at the primary level are higher for girls than for boys, indicating that the gender gap in these countries is wider than is reflected by enrollment rates. One reason for this is early child-bearing in many of these countries, which is clearly incompatible with schooling.

The economic incentives for educating girls lie in the opportunities women have to work. Teaching has always been one of the first professions open to women, making the number of female teachers a revealing indicator of employment opportunities. In addition, female teachers are important role models for girls, particularly in societies where female education is not encouraged or male teaching of females is forbidden. Over the past decade the proportion of female primary school teachers has increased everywhere. But data on teachers may not reflect the functions they perform. Schools may employ teachers in many capacities outside the classroom, and the responsibilities assigned to male and female teachers may differ systematically.

Figure 2.12a

Gender disparities in education do not respond to changes in GNP per capita

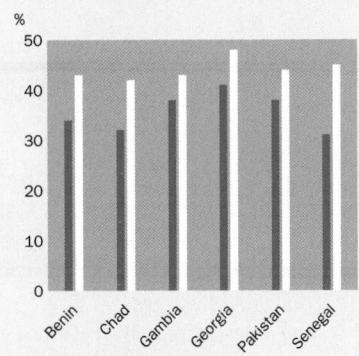

● Female pupils as % of total primary enrollment, 1994
○ Predicted % of female pupils based on GNP per capita, 1994

Source: UNESCO; World Bank staff estimates.

Although the disparity between the enrollment of boys and girls has narrowed, the percentage of enrolled girls continues to lag behind that of boys in many parts of the developing world. The obstacles to female education stem from many factors: national education policies that affect boys and girls differently; uneven distribution of primary schools, especially in rural areas; lack of schools for girls in systems segregated by sex; perceived irrelevance of primary school curricula to women's employment possibilities; and demand for the household labor of girls.

Gender disparities in educational enrollment are not correlated with an overall standard of living such as GNP per capita, so gender disparity is not something that economies "grow out of" (Filmer, King, and Pritchett 1998). Thus any strategy to improve female enrollment should aim at establishing supportive national policies, providing access to schools with adequate infrastructure, and reducing the direct and opportunity cost of girls' attendance.

Definitions

• **Female teachers as a percentage of total teachers** includes full-time and part-time teachers. • **Female pupils as a percentage of total pupils** includes enrollments in public and private schools but may exclude specialized schools and training programs.

Data sources

The estimates in this table were compiled using the United Nations Educational, Scientific, and Cultural Organization's electronic database on institutions, teachers, and pupils.

	Health expenditure			Health expenditure per capita		Physicians		Hospital beds		Inpatient admission rate	Average length of stay	Outpatient visits per capita
	Public % of GDP 1990–95a	Private % of GDP 1990–95a	Total % of GDP 1990–95a,b	PPP $ 1990–95	$ 1990–95	per 1,000 people 1980	per 1,000 people 1994	per 1,000 people 1980	per 1,000 people 1994	% of population 1990–97	days 1990–97	1990–97
Albania	2.7	..	..	..	..	0.9	1.3	..	3.0	..	11	2
Algeria	3.3	1.3	4.6	247	109	..	0.8	..	2.1	..	..	..
Angola	4.0	..	..	..	..	..	0.0	..	1.3	..	..	..
Argentina	4.3	6.3	10.6	932	877	..	2.7	..	4.6	..	..	..
Armenia	3.1	4.7	7.8	140	10	3.5	3.1	8.4	7.8	8	15	3
Australia	6.0	3.0	8.9	1,728	1,578	1.8	2.2	..	8.9	14	14	11
Austria	5.9	1.9	7.9	1,720	1,926	2.3	2.6	11.2	9.4	25	11	6
Azerbaijan	1.4	6.1	7.5	96	3	3.4	3.8	9.7	10.0	6	18	1
Bangladesh	1.2	1.3	2.4	35	5	0.1	0.2	0.2	0.3	..	..	..
Belarus	5.3	1.1	6.4	280	245	3.4	4.1	12.5	12.4	26	18	11
Belgium	7.0	1.0	8.0	1,784	2,082	2.5	3.7	9.4	7.6	20	12	8
Benin	1.7	..	..	..	..	0.1	0.1	1.5	0.2	..	..	..
Bolivia	2.7	2.4	5.0	138	38	0.5	0.4	..	1.4	..	..	..
Bosnia and Herzegovina	..	..	..	..	..	..	0.6	..	2.0	..	15	..
Botswana	1.9	1.4	3.1	171	109	0.1	0.2	2.4	1.6	..	..	..
Brazil	2.7	4.7	7.4	428	261	0.8	1.4	..	3.0	0c	..	2
Bulgaria	4.0	1.4	6.9	296	197	2.5	3.3	11.1	10.2	18	14	6
Burkina Faso	2.3	3.2	5.5	43	22	0.0	..	..	0.3	..	..	..
Burundi	0.9	..	..	..	..	..	0.1	..	0.7	..	..	..
Cambodia	0.7	6.5	7.2	..	18	0.1	0.1	..	2.1	..	..	..
Cameroon	1.0	0.4	1.4	33	7	..	0.1	..	2.6	..	..	..
Canada	6.8	2.7	9.6	2,238	1,835	1.8	2.2	..	5.4	13	12	7
Central African Republic	1.9	..	..	..	..	0.0	0.0	1.6	0.9	..	..	..
Chad	3.4	0.1	3.5	26	6	..	..	..	0.7	..	..	..
Chile	2.5	4.0	6.5	652	241	..	1.1	3.4	3.2	..	..	..
China	2.1	1.8	3.8	100	23	0.9	1.6	2.0	2.4	4	15	..
Hong Kong, China	1.9	2.5	4.3	1,036	944	0.8	1.3	4.0	..	2	..	1
Colombia	3.0	4.4	7.4	487	138	..	0.9	1.6	1.4	..	..	..
Congo, Dem. Rep.	0.2	..	..	..	..	..	0.1	..	1.4	..	..	..
Congo, Rep.	1.8	3.2	6.3	170	102	0.1	0.3	..	3.3	..	..	..
Costa Rica	6.3	2.2	8.5	536	214	..	0.9	3.3	2.5	..	..	..
Côte d'Ivoire	1.4	2.0	3.4	71	22	..	0.1	..	0.8	..	..	..
Croatia	8.5	1.6	10.1	..	302	..	2.0	..	5.9	14	..	..
Cuba	7.9	..	..	..	..	1.4	3.6	..	5.4	..	..	..
Czech Republic	7.7	1.9	9.6	970	383	..	2.9	..	7.4	19	13	16
Denmark	5.3	1.1	6.4	1,508	1,849	2.4	2.9	..	5.0	25	8	5
Dominican Republic	2.0	3.3	5.3	220	71	..	1.1	..	2.0	..	..	..
Ecuador	2.0	3.3	5.3	253	78	..	1.5	1.9	1.6	..	..	..
Egypt, Arab Rep.	1.6d	2.1d	3.7d	..	..	1.1	1.8	2.0	2.1	3	8	4
El Salvador	1.2	3.8	5.0	132	74	0.3	0.7	..	1.5	..	..	..
Eritrea	1.1	0.9	2.0	..	..	..	..	..	..	..	..	..
Estonia	6.3	..	..	..	..	4.2	3.1	12.4	8.4	18	12	6
Ethiopia	1.7	..	..	..	..	0.0	0.0	0.3	0.2	..	..	..
Finland	5.7	1.9	7.7	1,521	1,526	1.9	2.7	15.5	10.1	23	12	4
France	8.0	1.9	9.9	2,156	2,576	2.2	2.8	..	9.0	21	11	6
Gabon	0.6	..	..	..	..	0.5	0.5	..	3.2	..	..	..
Gambia, The	1.9	..	..	..	..	..	..	..	0.6	..	..	..
Georgia	0.8	..	..	..	..	4.8	4.2	10.7	8.2	5	13	2
Germany	8.2	2.3	10.4	2,123	2,578	2.2	3.3	..	9.7	21	14	13
Ghana	1.3	0.1	1.4	30	4	..	..	..	1.5	..	..	..
Greece	5.5	1.8	7.3	706	488	2.4	4.0	6.2	5.0	14	8	..
Guatemala	0.9	1.7	2.7	92	33	..	0.3	..	1.1	..	..	..
Guinea	1.2	..	..	..	..	0.0	0.2	..	0.6	..	..	..
Guinea-Bissau	1.1	..	..	..	..	0.1	..	1.8	1.5	..	..	..
Haiti	1.3	2.3	3.6	35	8	0.1	0.1	0.7	0.8	..	..	..
Honduras	2.8	2.8	5.6	121	34	0.3	0.4	1.3	1.0	..	..	..

Health expenditure, services, and use | 2.13

	Health expenditure			Health expenditure per capita		Physicians		Hospital beds		Inpatient admission rate	Average length of stay	Outpatient visits per capita
	Public % of GDP 1990–95[a]	Private % of GDP 1990–95[a]	Total % of GDP 1990–95[a,b]	PPP $ 1990–95	$ 1990–95	per 1,000 people 1980	1994	per 1,000 people 1980	1994	% of population 1990–97	days 1990–97	1990–97
Hungary	6.8	0.5	7.3	496	295	2.5	3.6	9.1	9.6	24	10	14
India	0.7	4.4	5.6	68	24	0.4	0.4	0.8	0.8	..	..	..
Indonesia	0.7	1.1	1.8	76	17	0.1	0.2	..	0.7	..	6	..
Iran, Islamic Rep.	2.8	2.0	4.8	239	1,343	0.3	0.3	1.5	1.4	..	..	..
Iraq	..	..	..	..	..	0.6	0.6	1.9	1.7	..	..	..
Ireland	5.4	1.3	6.7	1,451	1,151	1.3	2.0	9.7	5.0	15	7	..
Israel	2.1	2.1	4.1	560	825	2.5	..	5.1	6.0	..	..	..
Italy	5.4	2.4	7.7	1,605	1,404	1.3	1.7	..	6.5	16	11	..
Jamaica	3.0	2.3	5.4	212	91	0.4	0.5	..	2.1	..	..	..
Japan	5.7	1.6	7.2	1,587	2,580	1.4	1.8	11.3	16.2	9	46	16
Jordan	3.7	4.2	7.9	347	118	0.8	1.6	1.3	1.6	11	3	3
Kazakhstan	2.2	..	..	..	..	3.2	3.8	13.1	12.2	16	17	1
Kenya	1.9	1.0	2.5	34	13	0.1	0.0	..	1.7	..	..	..
Korea, Dem. Rep.	..	..	..	..	..	2.5	..	..	..	..	..	..
Korea, Rep.	1.8	3.6	5.4	518	420	0.6	1.2	1.7	4.1	6	19	2
Kuwait	3.6	..	..	..	..	1.7	0.2	4.1	..	..	..	..
Kyrgyz Republic	3.7	..	..	..	..	2.9	3.1	12.0	9.9	16	15	1
Lao PDR	1.3	1.3	2.6	..	8	..	0.2	..	2.6	..	..	4
Latvia	4.4	..	..	..	..	4.1	3.0	13.7	11.9	21	14	4
Lebanon	2.1	3.3	5.3	..	..	1.7	1.9	..	3.1	..	..	..
Lesotho	3.5	..	..	..	..	..	0.0	..	..	..	..	..
Libya	..	..	..	..	..	1.3	1.1	4.8	4.2	..	..	..
Lithuania	5.1	..	..	..	..	3.9	4.0	12.1	11.1	20	14	7
Macedonia, FYR	7.3	0.9	8.3	..	..	..	2.3	..	5.5	9	15	3
Madagascar	1.1	..	..	..	..	0.1	0.1	..	0.9	..	..	0.1
Malawi	2.3	..	..	..	..	0.0	0.0	..	1.6	..	..	..
Malaysia	1.4	1.0	2.5	220	85	0.3	0.4	2.3	2.0	..	..	..
Mali	2.0	1.3	2.9	15	11	0.0	0.1	..	..	..	..	..
Mauritania	1.8	4.1	5.2	75	35	..	0.1	..	0.7	..	..	..
Mauritius	2.2	1.7	3.4	408	109	0.5	0.8	3.1	3.1	0[c]	..	4
Mexico	2.8	3.0	5.3	365	223	0.9	1.3	..	1.2	6	4	2
Moldova	4.9	..	..	..	..	3.1	3.6	12.0	12.2	19	18	8
Mongolia	4.8	0.7	6.7	174	158	9.9	2.7	11.2	11.5	23	..	5
Morocco	1.6	1.6	3.4	126	36	0.1	0.4	1.2	1.1	..	..	..
Mozambique	4.6	..	..	..	..	0.0	..	1.1	0.9	..	..	..
Myanmar	0.4	..	..	..	..	0.2	0.1	0.9	0.6	..	..	..
Namibia	3.7	3.7	7.6	303	153	..	0.2	..	..	..	..	..
Nepal	1.2	3.8	5.0	60	9	0.0	0.1	0.2	0.2	..	..	..
Netherlands	6.7	2.0	8.8	1,813	2,198	2.1	2.5	12.5	11.3	6	33	6
New Zealand	5.7	1.8	7.4	1,260	1,018	1.6	2.1	..	7.3	14	7	..
Nicaragua	4.3	3.5	7.8	..	34	0.4	0.7	..	1.8	..	..	..
Niger	1.6	..	..	..	..	..	0.0	..	..	..	..	..
Nigeria	0.3	1.0	1.4	18	5	0.1	0.2	0.9	1.7	..	..	..
Norway	6.6	1.4	8.0	2,080	2,274	1.9	3.3	15.0	13.5	15	10	4
Oman	2.5	..	..	..	..	0.5	0.9	1.6	2.1	1	5	4
Pakistan	0.8	2.7	3.5	70	17	0.3	0.5	0.6	0.7	..	..	3
Panama	5.4	2.0	7.5	485	201	1.0	1.8	..	2.5	..	..	..
Papua New Guinea	2.8	..	..	..	..	0.1	0.1	5.5	4.0	..	..	..
Paraguay	1.0	3.3	4.3	161	72	0.6	0.3	..	0.6	..	..	..
Peru	2.6	2.3	4.9	199	106	0.7	1.0	..	1.4	0[c]	..	2
Philippines	1.3	1.0	2.4	60	22	0.1	0.1	1.7	1.1	..	..	..
Poland	4.8	1.1	6.0	283	226	1.8	2.3	5.6	6.3	14	11	6
Portugal	4.5	3.6	8.1	1,058	797	2.0	2.9	..	4.3	11	10	3
Puerto Rico	..	6.0	..	..	..	..	..	..	..	..	..	..
Romania	3.6	..	..	..	..	1.5	1.8	8.8	7.7	18	10	5
Russian Federation	4.1	0.6	4.8	225	96	4.0	3.8	13.0	11.8	20	17	8

	Health expenditure			Health expenditure per capita		Physicians		Hospital beds		Inpatient admission rate	Average length of stay	Outpatient visits per capita
	Public % of GDP 1990–95[a]	Private % of GDP 1990–95[a]	Total % of GDP 1990–95[a,b]	PPP $ 1990–95	$ 1990–95	per 1,000 people 1980	1994	per 1,000 people 1980	1994	% of population 1990–97	days 1990–97	1990–97
Rwanda	1.9	..	..	..	..	0.0	0.0	1.5	1.7	..	..	..
Saudi Arabia	3.1	..	..	202	159	0.5	1.3	1.5	2.5	..	..	..
Senegal	2.5	..	..	..	..	0.1	0.1	..	0.7	..	..	..
Sierra Leone	1.6	2.0	3.6	22	18	0.1	..	1.2	..	..	..	..
Singapore	1.3	2.3	3.6	845	987	0.9	1.4	4.2	3.6	12	..	..
Slovak Republic	6.0	..	..	..	..	..	2.8	..	7.1	20	11	12
Slovenia	7.4	..	..	..	..	..	2.2	7.0	5.8	16	11	..
South Africa	3.6	4.3	7.9	396	257	..	..	..	..	..	..	..
Spain	6.0	1.7	7.6	1,166	1,043	2.8	4.1	..	4.0	10	11	..
Sri Lanka	1.4	0.4	1.9	61	12	0.1	0.1	2.9	2.7	..	..	..
Sudan	..	2.7	0.3	..	29	0.1	..	0.9	1.1	..	..	..
Sweden	6.0	1.3	7.3	1,523	1,724	2.2	3.0	14.8	6.5	19	8	3
Switzerland	7.2	2.8	10.0	2,395	3,533	..	3.1	..	20.8	15	..	11
Syrian Arab Republic	..	..	..	..	..	0.4	0.8	1.1	1.1	..	..	..
Tajikistan	6.4	..	..	..	..	2.4	2.1	10.0	8.8	16	15	..
Tanzania	3.0	..	..	..	..	..	..	1.4	0.9	..	..	..
Thailand	1.4	3.9	5.3	336	111	0.1	0.2	1.5	1.7	..	..	..
Togo	1.7	2.2	3.4	40	20	0.1	0.1	..	1.5	..	..	..
Trinidad and Tobago	2.6	1.3	3.9	381	151	0.7	0.7	..	3.2	..	..	..
Tunisia	3.0	2.9	5.9	..	104	0.3	0.6	2.1	1.8	8	..	..
Turkey	2.7	1.5	4.2	239	100	0.6	1.1	2.2	2.5	6	6	1
Turkmenistan	2.8	..	..	..	..	2.9	3.2	10.6	11.5	17	15	..
Uganda	1.6	2.2	3.9	61	10	0.0	..	1.5	0.9	..	..	..
Ukraine	5.0	..	..	..	..	3.7	4.4	12.5	12.2	23	17	10
United Arab Emirates	2.0	0.5	2.5	378	379	1.1	0.8	2.8	3.1	11	5	..
United Kingdom	5.8	1.1	6.9	1,373	1,208	1.6	1.5	9.3	4.9	23	10	6
United States	6.6	7.7	14.2	3,801	3,667	1.8	2.5	5.9	4.2	12	8	6
Uruguay	2.0	6.5	8.5	642	439	2.0	3.2	..	4.5	..	..	..
Uzbekistan	3.5	..	..	..	..	2.9	3.3	11.5	8.7	19	14	..
Venezuela	2.3	4.8	7.1	602	202	0.8	1.6	0.3	2.6	..	..	..
Vietnam	1.1	4.1	5.2	..	..	0.2	0.4	3.5	3.8	7	8	3
West Bank and Gaza	..	..	..	..	..	..	..	..	..	..	..	..
Yemen, Rep.	1.2	1.5	2.6	..	39	0.1	0.1	..	0.8	..	..	..
Yugoslavia, FR (Serb./Mont.)	..	..	..	..	..	3.4	2.0	13.6	5.4	8	12	2
Zambia	2.4	0.7	3.3	31	362	0.1	0.1	3.5	..	..	..	..
Zimbabwe	2.0	4.2	6.5	122	86	0.2	0.1	3.1	0.5	..	..	..
World	**3.2 w**	**2.8 w**	**5.4 w**	**532 w**	**505 w**	**1.0 w**	**1.4 w**	**3.5 w**	**3.8 w**	..	..	..
Low income	1.5	2.7	4.2	78	22	0.6	1.0	1.5	1.6	..	..	..
Excl. China & India	1.1	2.0	3.1	47	18	..	0.4	1.5	1.5	..	..	..
Middle income	4.3	2.4	5.1	264	209	1.4	1.6	..	4.6	..	..	..
Lower middle income	..	..	..	..	..	1.6	1.8	6.9	5.4	..	..	..
Upper middle income	..	..	..	..	..	1.0	1.6	..	3.3	..	..	..
Low & middle income	2.4	2.6	4.5	139	83	0.9	1.2	2.7	2.7	..	..	..
East Asia & Pacific	1.7	1.9	3.6	106	27	0.8	1.4	2.0	2.1	..	..	..
Europe & Central Asia	4.4	1.1	5.4	315	138	3.0	3.1	10.4	9.1	..	..	..
Latin America & Carib.	2.9	3.9	6.7	425	248	0.8	1.4	..	..	..	..	..
Middle East & N. Africa	2.4	2.2	4.5	211	433	0.7	..	1.7	1.8	..	..	..
South Asia	1.2	3.8	5.0	64	21	0.3	0.4	0.7	0.7	..	..	..
Sub-Saharan Africa	1.6	1.6	2.9	87	55	..	..	..	1.2	..	..	..
High income	6.9	3.7	9.6	2,227	2,404	1.8	2.5	..	7.4	..	..	..

a. Data are for most recent year available. b. Data may not sum to totals because of rounding. c. Less than 0.5. d. Data are for 1997.

About the data

Most industrial countries have national health accounting systems that track and compare public and private health care expenditures. Data on private and public health expenditures are required for the public sector to rationalize its spending and to devise policies that are both efficient and equitable. Few developing countries, however, have national health accounts. As a result cross-country comparisons of health financing data are difficult, especially because records of private out-of-pocket expenditures are often lacking. Compiling estimates of public health expenditures is also complicated in countries where state, provincial, and local governments are involved in health care financing because such data are not regularly reported and are often of poor quality. Furthermore, in some countries health services are considered social services, and so are excluded from health sector expenditures. The data on health expenditures shown here were collected by the World Bank as part of its health, nutrition, and population strategy. No estimates were made for countries with incomplete data.

Health services indicators (physicians and hospital beds per 1,000 people) and health utilization indicators (inpatient admission rates, average length of stay, and outpatient visits) come from a variety of sources (see below). Data are lacking for many countries, and for others comparability is limited by differences in definitions. For example, some countries incorrectly include retired physicians or those working outside the health sector in estimates of health personnel. Moreover, it is important to recognize that these indicators show the availability and use of health services but do not reflect their quality—that is, how well trained physicians are or how well equipped hospitals are.

Average length of stay in hospitals is one indicator of the efficiency of resource use. Longer stays may reflect a waste of resources if patients are kept in hospitals beyond the time medically required, inflating demand for hospital beds and increasing hospital costs. Aside from differences in cases and financing methods, cross-country variations in average length of stay may result from differences in the role of hospitals. Many developing countries do not have separate extended facilities, so hospitals become the source of long-term as well as acute care. Data for some countries may not include all public and private hospitals. Admission rates may be overstated in some countries if outpatient surgeries are counted as hospital admissions. And in many countries outpatient visits, especially emergency visits, may result in double counting if a patient receives treatment in more than one department.

Figure 2.13a

Low-income countries devote relatively less public spending to health . . .

Log public health expenditure, 1990–95

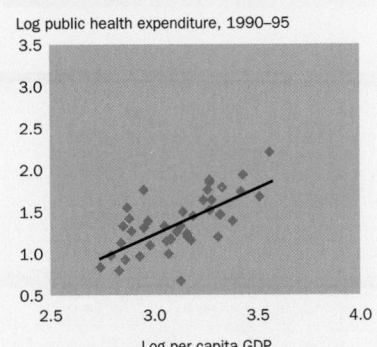

Log per capita GDP

. . . than do high-income countries

Log public health expenditure, 1990–95

Note: Public health expenditure and per capita GDP have been adjusted for purchasing power parity.

Source: See *Data sources*.

The income elasticity of health spending, defined as the percentage change in health spending resulting from a percentage change in income, can provide a useful measure of how differences in income translate into differences in health expenditures. Globally, every 1 percent increase in per capita income causes total health expenditures to increase by 1.24 percent. Income elasticities are 1.08 for low-income countries, 1.10 for middle-income countries, and 1.96 for high-income countries. Thus countries with higher incomes tend to spend a larger share of their income on health. Public health expenditures increase by 1.33 percent for every 1 percent increase in per capita income. Private health expenditures are less responsive to income changes (income elasticity of 0.99).

Definitions

• **Public health expenditure** consists of recurrent and capital spending from government (central and local) budgets, external borrowings and grants (including donations from international agencies and nongovernmental organizations), and social (or compulsory) health insurance funds. • **Private health expenditure** includes direct household (out-of-pocket) spending, private insurance, charitable donations, and direct service payments by private corporations. • **Total health expenditure** is the sum of public and private health expenditures. It covers the provision of health services (preventive and curative), family planning activities, nutrition activities, and emergency aid designated for health but does not include provision of water and sanitation. • **Physicians** are defined as graduates of any faculty or school of medicine who are working in the country in any medical field (practice, teaching, research). • **Hospital beds** include inpatient beds available in public, private, general, and specialized hospitals and rehabilitation centers. In most cases beds for both acute and chronic care are included. • **Inpatient admission rate** is the percentage of the population admitted to hospitals during a year. • **Average length of stay** is the average duration of inpatient hospital admissions. • **Outpatient visits per capita** is the number of visits to health care facilities per capita, including repeat visits.

Data sources

Health expenditure estimates come from country sources, supplemented by information from international agencies and World Bank country and sector studies; including the Human Development Network's *Sector Strategy: Health, Nutrition, and Population.* Data were also drawn from World Bank public expenditure reviews, the International Monetary Fund's government finance data files, and other studies. Data for private expenditure are largely from household surveys and World Bank poverty assessments and sector studies. The Organisation for Economic Co-operation and Development (OECD) provided data on public and private health expenditures and health services and use data for member countries. Data for physicians and beds are from the World Health Organization (WHO), supplemented by country data.

2.14 | Access to health services

	Health care		Safe water		Sanitation		Child immunization			
	% of population with access		% of population with access		% of population with access		Measles % of children under 12 months		DPT % of children under 12 months	
	1980	1993	1980	1995	1980	1995	1980	1995	1980	1995
Albania	100	..	92	..	..	..	90	91	94	97
Algeria	..	..	77	..	..	..	17	69	33	75
Angola	70	24	..	32	..	16	26	32	6	21
Argentina	..	..	..	64	..	89	58	76	41	82
Armenia	..	..	..	..	..	..	..	95	..	83
Australia	99	100	99	95	99	90	68	..	..	86
Austria	..	100	100	..	85	100	25	60	90	90
Azerbaijan	..	..	..	..	..	..	..	91	..	90
Bangladesh	..	74	..	79	..	35	0	96	0	91
Belarus	..	100	..	..	50	100	..	96	..	89
Belgium	..	100	..	..	99	100	50	70	95	97
Benin	..	42	..	50	..	20	..	81	..	87
Bolivia	..	..	..	60	..	44	13	83	11	88
Bosnia and Herzegovina	..	..	..	..	..	..	..	57	..	67
Botswana	..	86	..	70	..	55	63	68	64	76
Brazil	..	..	..	72	..	41	56	78	40	83
Bulgaria	..	100	96	..	..	99	98	93	97	100
Burkina Faso	..	..	35	78	5	18	23	55	2	47
Burundi	..	80	..	..	..	..	30	44	38	57
Cambodia	..	..	..	13	..	..	..	75	..	79
Cameroon	20	15	..	41	..	40	16	51	5	48
Canada	..	99	97	100	60	85	..	98	80	93
Central African Republic	..	13	16	18	..	..	12	70	13	40
Chad	..	26	..	24	..	21	..	24	1	18
Chile	..	95	..	..	..	83	87	93	94	98
China	..	..	..	90	..	21	78	89	58	92
Hong Kong, China	..	..	..	..	..	..	74	72	73	83
Colombia	88	87	..	76	..	63	14	77	15	94
Congo, Dem. Rep.	80	59	..	..	..	..	18	41	18	35
Congo, Rep.	..	..	..	47	..	9	49	39	42	50
Costa Rica	..	97	..	..	..	..	60	94	86	85
Côte d'Ivoire	..	60	20	72	17	54	..	57	..	40
Croatia	..	..	..	96	..	68	..	90	..	87
Cuba	..	100	61	93	31	66	48	100	67	100
Czech Republic	..	..	..	..	..	..	..	96	..	96
Denmark	..	100	100	100	100	100	..	88	85	89
Dominican Republic	..	..	..	71	..	78	29	100	35	83
Ecuador	..	80	..	70	..	64	24	100	10	74
Egypt, Arab Rep.	100	99	90	64	70	11	78	82	84	91
El Salvador	..	..	..	55	..	68	45	94	43	100
Eritrea	..	..	..	..	..	..	..	29	..	35
Estonia	..	..	..	..	..	..	74	81	84	84
Ethiopia	..	55	4	27	..	10	4	38	3	47
Finland	..	100	..	100	100	100	70	98	92	100
France	..	..	..	100	85	96	0	78	79	89
Gabon	..	87	..	67	..	76	..	50	14	48
Gambia, The	..	..	42	76	..	37	71	68	80	78
Georgia	..	..	..	..	..	..	..	63	..	..
Germany	..	..	..	..	..	100	35	75	..	80
Ghana	..	25	..	56	..	27	16	54	7	51
Greece	..	..	..	..	..	96	..	70	72	78
Guatemala	..	60	..	60	..	66	23	84	43	80
Guinea	..	45	..	62	12	70	..	69	..	73
Guinea-Bissau	30	..	24	23	..	20	..	68	9	74
Haiti	..	45	..	28	..	24	..	24	3	30
Honduras	..	62	..	65	..	62	35	90	31	96

Access to health services | 2.14

	Health care		Safe water		Sanitation		Child immunization			
	% of population with access		% of population with access		% of population with access		Measles % of children under 12 months		DPT % of children under 12 months	
	1980	1993	1980	1995	1980	1995	1980	1995	1980	1995
Hungary	..	..	..	..	..	94	99	100	99	100
India	50	..	..	81	..	29	0	84	31	86
Indonesia	..	43	..	62	..	51	0	89	0	92
Iran, Islamic Rep.	50	73	50	..	60	..	39	88	32	99
Iraq	..	98	74	44	..	87	35	88	13	91
Ireland	..	..	..	..	..	100	10	..	34	65
Israel	..	100	..	99	..	70	69	94	84	92
Italy	..	..	99	..	99	100	5	50	..	50
Jamaica	..	..	..	70	..	74	12	82	34	90
Japan	..	100	..	..	..	85	69	68	60	85
Jordan	..	90	89	89	76	100	29	92	30	100
Kazakhstan	..	..	..	..	..	..	..	72	..	80
Kenya	..	..	..	53	..	77	..	35	..	40
Korea, Dem. Rep.	..	100	..	100	..	100	29	98	50	96
Korea, Rep.	..	100	..	89	..	100	4	92	70	100
Kuwait	100	100	100	..	100	100	48	93	67	100
Kyrgyz Republic	..	..	..	75	..	53	..	89	..	83
Lao PDR	..	..	..	39	..	19	..	65	..	54
Latvia	..	..	..	..	..	..	..	85	..	65
Lebanon	..	..	92	..	59	..	..	65	4	94
Lesotho	..	80	18	52	12	6	49	82	56	58
Libya	100	100	90	90	70	..	65	..	60	96
Lithuania	..	..	..	..	..	..	..	94	..	96
Macedonia, FYR	..	..	..	..	..	..	..	85	..	87
Madagascar	..	65	..	29	..	3	..	59	48	67
Malawi	40	80	..	45	..	53	49	99	58	98
Malaysia	..	88	..	88	75	91	11	81	58	90
Mali	20	..	..	37	..	31	..	49	..	46
Mauritania	..	..	..	..	..	..	45	53	18	50
Mauritius	100	99	..	98	..	100	34	85	87	89
Mexico	51	91	..	83	..	66	35	90	41	92
Moldova	..	..	..	..	..	50	..	98	..	86
Mongolia	90	100	..	..	..	..	..	85	76	88
Morocco	..	62	32	52	50	40	17	87	43	90
Mozambique	..	30	9	32	10	21	32	71	56	57
Myanmar	30	..	20	38	20	41	..	66	4	84
Namibia	..	..	..	..	..	34	..	57	..	61
Nepal	10	..	11	48	0	20	2	78	8	65
Netherlands	..	100	100	100	100	100	91	95	96	97
New Zealand	..	100	87	..	..	..	80	87	76	89
Nicaragua	..	..	..	61	..	31	15	81	15	85
Niger	..	30	..	53	..	15	19	38	6	19
Nigeria	40	67	..	39	..	36	55	50	..	44
Norway	..	100	..	..	100	100	80	93	90	92
Oman	75	89	15	..	..	79	22	98	18	100
Pakistan	65	85	38	60	16	30	1	53	2	58
Panama	..	82	..	83	..	87	47	84	47	86
Papua New Guinea	..	96	..	28	..	22	1	55	32	50
Paraguay	..	..	..	..	..	30	19	76	17	79
Peru	..	..	..	60	..	44	21	97	14	94
Philippines	..	..	..	..	..	..	9	86	47	86
Poland	100	100	67	..	50	100	92	96	96	95
Portugal	..	..	57	..	..	100	54	94	73	93
Puerto Rico	..	..	..	..	..	..	..	..	..	..
Romania	..	..	77	..	50	49	83	93	..	98
Russian Federation	..	..	..	..	..	..	..	91	..	72

| | Health care | | Safe water | | Sanitation | | Child immunization | | | |
| | % of population with access | | % of population with access | | % of population with access | | Measles % of children under 12 months | | DPT % of children under 12 months | |
	1980	1993	1980	1995	1980	1995	1980	1995	1980	1995
Rwanda	..	..	..	..	..	..	42	74	17	83
Saudi Arabia	85	98	91	93	76	86	8	94	41	96
Senegal	..	40	..	50	..	58	..	80	..	80
Sierra Leone	26	..	..	34	13	11	36	44	13	41
Singapore	..	100	100	100	..	97	47	88	84	95
Slovak Republic	..	..	..	..	43	51	..	99	..	99
Slovenia	..	..	..	..	..	90	..	91	..	98
South Africa	..	..	..	70	..	46	74	77	74	81
Spain	..	..	98	99	95	100	8	90	..	88
Sri Lanka	90	90	..	..	..	..	..	88	46	91
Sudan	..	70	..	50	..	22	1	74	1	77
Sweden	..	100	..	..	85	100	56	96	99	99
Switzerland	..	100	..	100	85	100	..	..	..	89
Syrian Arab Republic	..	99	71	85	45	78	13	98	13	92
Tajikistan	..	..	..	..	..	62	..	90	..	95
Tanzania	72	93	..	49	..	86	45	75	59	79
Thailand	30	59	..	81	..	70	..	86	49	94
Togo	..	..	..	..	..	22	..	65	..	73
Trinidad and Tobago	..	99	..	82	..	56	..	87	24	89
Tunisia	95	90	72	..	46	..	65	89	36	92
Turkey	..	100	67	92	..	94	27	65	42	66
Turkmenistan	..	..	..	85	..	60	..	90	..	87
Uganda	..	71	..	34	..	57	22	79	9	79
Ukraine	..	100	..	97	50	49	..	96	53	94
United Arab Emirates	96	90	100	98	75	95	34	90	11	90
United Kingdom	..	..	..	100	..	96	52	92	44	92
United States	..	..	..	90	98	85	86	89	96	94
Uruguay	..	..	..	83	..	82	50	80	53	87
Uzbekistan	..	..	..	..	..	18	..	71	..	65
Venezuela	..	..	..	79	..	58	50	94	56	68
Vietnam	75	..	..	36	..	21	1	95	4	94
West Bank and Gaza	..	..	..	..	..	..	..	..	..	..
Yemen, Rep.	16	..	..	52	..	51	2	49	1	53
Yugoslavia, FR (Serb./Mont.)	..	..	..	..	58	100	95	81	90	92
Zambia	..	75	..	43	..	23	..	78	83	76
Zimbabwe	55	..	..	74	5	58	56	77	39	80
World	..	..	..	78 w	..	47 w	42 w	82 w	46 w	84 w
Low income	..	..	..	76	..	28	40	80	39	81
Excl. China & India	..	..	..	51	..	36	18	63	13	63
Middle income	..	..	..	..	..	60	38	86	41	87
Lower middle income	..	..	..	..	..	58	..	86	35	86
Upper middle income	..	..	..	76	..	64	56	84	54	88
Low & middle income	..	..	..	76	..	37	39	82	40	83
East Asia & Pacific	..	..	..	84	..	29	64	88	49	91
Europe & Central Asia	..	..	..	..	..	..	..	87	..	81
Latin America & Carib.	..	..	..	73	..	57	44	84	38	86
Middle East & N. Africa	..	..	..	..	..	..	40	83	42	90
South Asia	..	..	..	78	..	30	0	82	25	83
Sub-Saharan Africa	..	..	..	45	..	37	35	56	..	55
High income	..	..	..	..	..	92	53	82	79	88

Access to health services | 2.14

The indicators in the table are provided to the World Health Organization (WHO) by member states as part of their efforts to monitor and evaluate progress in implementing national health-for-all strategies. Because reliable, observation-based statistical data for these indicators do not exist in many developing countries, in most cases the data are estimates. In some cases these estimates may be skewed by a country's desire to show progress or to establish a need for international assistance.

Access indicators measure the supply of services but reveal little about benefits or rate of use. For example, data on access to health care provide no information on the quality of care or on how the consumption of services differs among groups within countries, regions, or communities. Moreover, unless these indicators are based on survey statistics, they may not fully reflect the situation. In many developing countries services by nongovernmental organizations and private charities play an increasingly important role for the poor and for many rural residents, widening the gap between official statistics and the actual production and consumption of many essential services. It is not known, however, whether such services truly replace publicly provided services, and if so, how they differ in quantity and quality from public services. In addition, health care facilities tend to be concentrated in urban areas. Separate data for rural areas (not shown here) indicate much lower coverage and access.

People's health is also influenced by the environment in which they live. A lack of clean water and basic sanitation is the main reason diseases transmitted by feces are so common in developing countries. Drinking water contaminated by feces deposited near homes and an inadequate supply of water cause diseases that account for 10 percent of the total disease burden in developing countries (World Bank 1993c). To date, however, efforts to improve the provision of water, sanitation, and drainage have been disappointing. At the end of the 1980s—which had been declared the International Drinking Water Supply and Sanitation Decade by a coalition of international aid agencies—most people in poor regions still lacked adequate sanitation.

Governments in developing countries usually finance immunization against measles and DPT (diphtheria, pertussis or whooping cough, and tetanus) as part of the basic public health package, but personnel with limited training are often used to provide the vaccines. According to the World Bank's *World Development Report 1993: Investing in Health,*

these diseases account for about 10 percent of the disease burden among children under 5, compared with an expected 23 percent had vaccination coverage remained at the 1970s level. In many developing countries, however, data recording practices make immunization difficult to measure (WHO 1996a).

• **Percentage of population with access to health care** is the share of the population that can expect treatment for common diseases and injuries, including essential drugs on the national list, within one hour's walk or travel. • **Percentage of population with access to safe water** is the share of the population with reasonable access to an adequate amount of safe water (including treated surface water and untreated but uncontaminated water, such as from springs, sanitary wells, and protected boreholes). In urban areas the source may be a public fountain or standpipe located not more than 200 meters away. In rural areas the definition implies that members of the household do not have to spend a disproportionate part of the day fetching water. An adequate amount of safe water is that needed to satisfy metabolic, hygienic, and domestic requirements—usually about 20 liters a person a day. The definition of safe water has changed over time. • **Percentage of population with access to sanitation** is the share of the population with at least adequate excreta disposal facilities that can effectively prevent human, animal, and insect contact with excreta. Suitable facilities range from simple but protected pit latrines to flush toilets with sewerage. To be effective, all facilities must be correctly constructed and properly maintained. • **Child immunization** is the rate of vaccination coverage of children under one year of age for four diseases—measles and DPT (diphtheria, pertussis or whooping cough, and tetanus). A child is considered adequately immunized against measles after receiving one dose of vaccine, and against DPT after receiving two or three doses of vaccine, depending on the immunization scheme.

The table was produced using information provided to the WHO by countries as part of their responsibility for monitoring progress toward "health for all" and reported in the WHO's *World Health Report 1996* and *1997*; the WHO's Expanded Programme of Immunization Information System; and WHO, the United Nations Children's Fund (UNICEF), and the Water Supply and Sanitation Collaborative Council's *Water Supply* and *Sanitation Sector Monitoring Report 1996.*

2.15 | Reproductive health

	Total fertility rate		Adolescent fertility rate	Unwanted fertility rate	Contraceptive prevalence rate	Births attended by trained health staff		Maternal mortality ratio
	births per woman		births per 1,000 women age 15–19	births per 1,000 women age 15–49	% of women age 15–49	% of total		per 100,000 live births
	1980	**1996**	**1995**	**1990–97**	**1990–96**	**1985**	**1992**	**1990–96**
Albania	3.6	2.6	26	..	..	99	..	28[a]
Algeria	6.7	3.4	17	..	51	..	..	140[a]
Angola	6.9	6.8	218	..	..	15	16	1,500[b]
Argentina	3.3	2.7	62	..	..	..	..	100[b]
Armenia	2.3	1.6	50	..	..	..	..	21[a]
Australia	1.9	1.8	31	..	..	99	..	9[b]
Austria	1.6	1.4	23	..	..	..	..	10[b]
Azerbaijan	3.2	2.1	33	..	..	..	..	44[a]
Bangladesh	6.1	3.4	116	1.3	45	..	7	850[b]
Belarus	2.0	1.3	39	..	..	100	..	22[a]
Belgium	1.7	1.5	11	..	..	100	..	10[b]
Benin	7.0	5.9	127	..	17	34	34	500[c]
Bolivia	5.5	4.4	82	1.9	45	36	29	370[c]
Bosnia and Herzegovina	2.1	..	28	..	..	..	..	
Botswana	6.7	4.3	106	..	..	52	..	250[b]
Brazil	3.9	2.4	37	..	77	73	..	160[c]
Bulgaria	2.1	1.2	60	..	..	100	..	20[a]
Burkina Faso	7.5	6.7	149	0.9	8	..	33	930[b]
Burundi	6.8	6.4	66	..	..	..	26	1,300[b]
Cambodia	4.7	4.6	108	..	..	..	..	900[b]
Cameroon	6.5	5.5	136	0.6	16	..	25	550[b]
Canada	1.7	1.7	25	..	..	99	100	6[b]
Central African Republic	5.8	5.0	145	..	14	..	..	700[c]
Chad	5.9	5.6	183	..	..	..	21	900[c]
Chile	2.8	2.3	48	..	..	95	..	180[a]
China	2.5	1.9	17	..	85	..	51	115[d]
Hong Kong, China	2.0	1.2	13	..	..	..	..	7[b]
Colombia	3.8	2.7	80	0.8	72	51	..	100[b]
Congo, Dem. Rep.	6.6	6.3	221	..	..	..	..	
Congo, Rep.	6.2	6.0	140	..	..	..	..	890[b]
Costa Rica	3.7	2.7	67	..	..	..	..	55[b]
Côte d'Ivoire	7.4	5.1	136	1.0	11	..	..	600[c]
Croatia	..	1.6	28	..	..	..	..	12[a]
Cuba	2.0	1.6	68	..	..	99	100	36[a]
Czech Republic	2.1	1.2	34	..	69	..	..	7[a]
Denmark	1.5	1.8	18	..	..	..	..	9[b]
Dominican Republic	4.2	3.1	53	..	64	98	44	110[b]
Ecuador	5.0	3.1	68	..	57	27	..	150[b]
Egypt, Arab Rep.	5.1	3.3	56	1.0	48	..	24	170[b]
El Salvador	5.3	3.5	91	..	53	35	..	300[b]
Eritrea	..	5.9	125	..	8	..	..	1,400[b]
Estonia	2.0	1.3	36	..	..	..	..	52[a]
Ethiopia	6.6	7.0	164	..	4	58	..	1,400[b]
Finland	1.6	1.8	20	..	..	..	..	11[b]
France	1.9	1.7	17	..	..	..	..	15[b]
Gabon	4.5	5.0	150	..	..	92	..	500[b]
Gambia, The	6.5	5.3	167	..	..	54	65	1,100[b]
Georgia	2.3	1.5	40	..	..	..	..	19[a]
Germany	1.4	1.3	14	..	..	..	..	22[b]
Ghana	6.5	5.0	109	..	20	73	42	740[b]
Greece	2.2	1.4	19	..	..	..	..	10[b]
Guatemala	6.2	4.6	106	1.1	32	..	22	190[c]
Guinea	6.1	5.7	213	..	2	..	76	880[c]
Guinea-Bissau	6.0	6.0	186	..	..	..	16	910[b]
Haiti	5.9	4.3	70	1.8	18	20	..	600[c]
Honduras	6.5	4.5	112	..	47	50	63	220[b]

	Total fertility rate		Adolescent fertility rate	Unwanted fertility rate	Contraceptive prevalence rate	Births attended by trained health staff		Maternal mortality ratio
	births per woman		births per 1,000 women age 15–19	births per 1,000 women age 15–49	% of women age 15–49	% of total		per 100,000 live births
	1980	1996	1995	1990–97	1990–96	1985	1992	1990–96
Hungary	1.9	1.5	31	..	..	..	99	14[a]
India	5.0	3.1	81	0.8	43	33	75	437[c]
Indonesia	4.3	2.6	57	0.5	55	31	..	390[c]
Iran, Islamic Rep.	6.1	3.8	80	..	52	..	70	120[b]
Iraq	6.4	5.3	61	..	..	24	74	310[b]
Ireland	3.2	1.9	23	..	60	..	..	10[b]
Israel	3.2	2.6	28	..	..	99	..	7[b]
Italy	1.6	1.2	14	..	..	100	..	12[b]
Jamaica	3.7	2.3	67	..	..	89	88	120[b]
Japan	1.8	1.4	6	..	..	100	100	8[a]
Jordan	6.8	4.4	43	..	..	75	86	150[b]
Kazakhstan	2.9	2.1	40	..	..	..	..	53[a]
Kenya	7.8	4.6	95	2.0	..	..	..	650[b]
Korea, Dem. Rep.	3.0	2.1	30	..	..	100	..	..
Korea, Rep.	2.6	1.7	8	..	..	65	95	30[a]
Kuwait	5.3	2.9	45	..	..	99	98	18[a]
Kyrgyz Republic	4.1	3.0	44	..	..	..	..	32[a]
Lao PDR	6.7	5.7	59	..	..	..	..	650[b]
Latvia	2.0	1.2	34	..	..	..	..	15[a]
Lebanon	4.0	2.7	43	..	..	45	..	300[b]
Lesotho	5.6	4.6	55	..	23	28	..	610[b]
Libya	7.3	4.0	106	..	45	76	68	220[b]
Lithuania	2.0	1.4	34	..	..	..	..	13[a]
Macedonia, FYR	2.5	2.1	38	..	..	..	..	22[a]
Madagascar	6.5	5.7	145	0.9	17	62	71	660[c]
Malawi	7.6	6.5	151	1.0	22	59	41	620[c]
Malaysia	4.2	3.4	30	..	..	82	92	43[a]
Mali	7.1	6.7	190	0.7	7	27	..	580[c]
Mauritania	6.3	5.1	123	..	..	23	..	800[b]
Mauritius	2.7	2.1	42	..	75	84	91	112[a]
Mexico	4.5	2.9	57	..	..	..	45	110[b]
Moldova	2.4	1.9	46	..	..	..	..	33[a]
Mongolia	5.4	3.3	45	..	..	100	100	65[b]
Morocco	5.4	3.3	38	1.4	50	26	..	372[e]
Mozambique	6.5	6.1	122	..	..	28	29	1,500[b]
Myanmar	5.1	3.4	30	..	..	25	97	580[b]
Namibia	5.9	4.9	130	..	29	..	71	220[c]
Nepal	6.1	5.0	82	..	..	10	..	1,500[b]
Netherlands	1.6	1.5	8	..	..	..	..	12[b]
New Zealand	2.0	2.0	43	..	..	99	100	25[b]
Nicaragua	6.2	4.0	136	..	44	..	42	160[b]
Niger	7.4	7.4	222	0.3	4	47	21	593[c]
Nigeria	6.9	5.4	120	1.0	6	..	45	1,000[b]
Norway	1.7	1.9	22	..	..	..	100	6[b]
Oman	9.9	7.0	123	..	..	60	90	..
Pakistan	7.0	5.1	107	1.2	14	24	70	340[b]
Panama	3.7	2.6	61	..	..	83	85	55[b]
Papua New Guinea	5.7	4.7	44	..	..	34	20	930[b]
Paraguay	4.8	3.9	72	..	51	22	..	190[c]
Peru	4.5	3.1	52	1.5	55	44	..	280[b]
Philippines	4.8	3.6	47	1.2	48	..	76	208[c]
Poland	2.3	1.6	28	..	..	..	..	10[a]
Portugal	2.2	1.4	23	..	..	..	..	15[b]
Puerto Rico	2.6	1.9	48	..	78	..	..	..
Romania	2.4	1.3	34	..	57	99	..	41[a]
Russian Federation	1.9	1.3	31	..	34	..	..	53[a]

2.15 Reproductive health

	Total fertility rate		Adolescent fertility rate	Unwanted fertility rate	Contraceptive prevalence rate	Births attended by trained health staff		Maternal mortality ratio
	births per woman		births per 1,000 women age 15–19	births per 1,000 women age 15–49	% of women age 15–49	% of total		per 100,000 live births
	1980	1996	1995	1990–97	1990–96	1985	1992	1990–96
Rwanda	8.3	6.1	65	2.0	21	..	28	1,300[b]
Saudi Arabia	7.3	6.2	61	..	..	79	..	18[a]
Senegal	6.7	5.7	118	0.9	7	..	..	510[c]
Sierra Leone	6.5	6.5	203	..	..	25	..	1,800[b]
Singapore	1.7	1.7	13	..	..	100	100	10[b]
Slovak Republic	2.3	1.5	35	..	..	..	..	8[a]
Slovenia	2.1	1.3	19	..	..	..	..	5[a]
South Africa	4.6	2.9	68	..	69	..	..	230[b]
Spain	2.2	1.2	11	..	..	96	..	7[b]
Sri Lanka	3.5	2.3	33	..	..	87	85	30[a]
Sudan	6.5	4.7	84	..	10	20	..	370[a]
Sweden	1.7	1.7	20	..	..	100	..	7[b]
Switzerland	1.6	1.5	7	..	..	..	100	6[b]
Syrian Arab Republic	7.4	4.0	89	..	40	37	80	179[a]
Tajikistan	5.6	3.7	48	..	..	..	..	74[a]
Tanzania	6.7	5.6	123	0.7	18	74	..	530[c]
Thailand	3.5	1.8	18	..	..	59	71	200[b]
Togo	6.6	6.2	124	..	..	..	..	640[b]
Trinidad and Tobago	3.3	2.1	46	..	..	90	..	90[b]
Tunisia	5.2	2.8	32	..	60	60	50	..
Turkey	4.3	2.6	44	0.9	..	78	..	180[b]
Turkmenistan	4.9	3.3	26	..	..	..	..	44[a]
Uganda	7.2	6.7	193	1.3	15	..	..	550[e]
Ukraine	2.0	1.3	48	..	..	100	..	30[a]
United Arab Emirates	5.4	3.5	58	..	..	96	94	..
United Kingdom	1.9	1.7	30	..	..	98	..	9[a]
United States	1.8	2.1	46	..	..	99	100	12[b]
Uruguay	2.7	2.2	47	..	..	..	100	85[b]
Uzbekistan	4.8	3.4	43	..	..	..	..	24[a]
Venezuela	4.1	3.0	60	..	..	..	82	200[a]
Vietnam	5.0	3.0	42	..	..	..	..	105[a]
West Bank and Gaza	..	..		..	..	..	..	..
Yemen, Rep.	7.9	7.2	141	1.7	..	..	..	1,400[b]
Yugoslavia, FR (Serb./Mont.)	2.3	1.9	41	..	..	..	..	12[a]
Zambia	7.0	5.8	122	..	26	..	43	230[c]
Zimbabwe	6.8	3.9	68	0.8	58	67	49	280[c]
World	**3.7 w**	**2.8 w**	**67 w**			**.. w**	**.. w**	
Low income	4.3	3.2	81			..	..	
Excl. China & India	6.3	4.9	115			..	..	
Middle income	3.8	2.6	54			..	..	
Lower middle income	3.8	2.6	51			..	..	
Upper middle income	3.7	2.6	63			..	..	
Low & middle income	4.1	3.0	72			..	..	
East Asia & Pacific	3.1	2.2	25			..	..	
Europe & Central Asia	2.5	1.8	40			..	..	
Latin America & Carib.	4.1	2.8	72			..	..	
Middle East & N. Africa	6.1	4.0	53			..	..	
South Asia	5.3	3.4	106			..	..	
Sub-Saharan Africa	6.6	5.6	136			..	..	
High income	1.9	1.7	27			..	..	

a. Official estimate. b. UNICEF–WHO estimate based on statistical modeling. c. Indirect estimate based on sample survey. d. Based on a survey covering 30 provinces. e. Based on sample survey.

The number of women and men in need of reproductive health services is expected to nearly double over the next two decades (Conly and Epp 1997). Thus any action taken now to expand reproductive choices—including improved access to safe and reliable contraception—is likely to have a significant effect on the health, well-being, eventual size, and quality of life of a country's population.

Reproductive health behavior is complex and is influenced by a broad range of relevant interventions. Fertility outcomes, maternal mortality, births attended by skilled providers, and contraceptive prevalence are complex measures and indicate the demand for, access to, and use of reproductive health services.

Total and adolescent fertility rates are based on data from vital registration systems or, in their absence, from censuses or sample surveys. Provided that the surveys are fairly recent, the estimated rates are generally considered reliable. In cases where no empirical information on age-specific fertility rates is available, a model is used to estimate the proportion of all births that are teenage births. As with other basic demographic data (see *About the data* in table 2.1), international comparisons of fertility rates are limited by differences in data definitions and collection and estimation methods. Fertility rates for 1996 are generally based on projections from censuses or surveys from earlier years.

The unwanted fertility rate is based on survey responses by women of reproductive age and so is affected by response bias. In many developing countries fertility is not seen as within the control of an individual; women do not report a numerical ideal family size, and hence no birth is reported as unwanted. In such cases women are assumed to want all their births.

Contraceptive prevalence reflects all methods—ineffective traditional methods as well as highly effective modern methods. Unmarried women are often excluded from surveys, which may bias the estimate. Contraceptive prevalence rates are obtained mainly from demographic and health surveys and contraceptive prevalence surveys (see *Primary data documentation* for the most recent survey year).

Births attended by health staff is an indicator of a health system's ability to provide adequate care for pregnant women. Good health care improves maternal health and reduces mortality. However, data may not reflect this because health information systems are often weak, maternal deaths are underreported, and rates of maternal mortality are difficult to measure. The data in the table are from the World Health Organization (WHO), supplemented by data from the United Nations Children's Fund (UNICEF). They are based on national sources, derived from official community and hospital records; some reflect only births in hospitals and other medical institutions. In some cases smaller private and rural hospitals are excluded, and sometimes even primitive local facilities are included. Thus the coverage is not always comprehensive, and cross-country comparisons should be made with extreme caution.

Civil registers in many developing countries provide extremely unreliable mortality statistics, especially for maternal mortality. Classifying a death as maternal requires a cause of death attribution, which depends on the information available at the time of death. In many developing countries causes of death are assigned by nonphysicians and often attributed to "ill-defined causes." Even when causes are assigned by medically qualified staff with the aid of diagnostic information, some doubts remain about the diagnosis in the absence of autopsies and the assignment of appropriate International Classification of Diseases (ICD) codes. Maternal deaths are also likely to go unrecorded if they occur in remote and rural areas. Differences in definitions also may affect the comparability of estimates over time and across countries. The maternal mortality ratios shown here are official estimates from administrative records, survey-based indirect estimates, or derived from a demographic model developed by UNICEF and the WHO. Official or survey-based estimates are shown wherever they are available. In all cases the standard errors of maternal mortality ratios are large, which makes the ratio particularly unsuitable for monitoring changes over a short period.

• **Total fertility rate** is the number of children that would be born to a woman if she were to live to the end of her childbearing years and bear children in accordance with current age-specific fertility rates. • **Adolescent fertility rate** is the number of births to women age 15–19 per 1,000 women in the same age group. • **Unwanted fertility** rate is the difference between the total fertility rate and the wanted fertility rate. Unwanted births are defined as those that exceed the number considered ideal or wanted by women of reproductive age. • **Contraceptive prevalence rate** is the percentage of women who are practicing, or whose sexual partners are practicing, any form of contraception. It is usually measured for married women age 15-49 only. • **Births attended by trained health staff** is the percentage of deliveries attended by personnel trained to give the necessary supervision, care, and advice to women during pregnancy, labor, and the postpartum period, to conduct deliveries on their own, and to care for newborns. • **Maternal mortality ratio** is the number of women who die during pregnancy and childbirth, per 100,000 live births.

Data on reproductive health come from demographic and health surveys and from WHO and UNICEF, *Revised 1990 Estimates of Maternal Mortality: A New Approach.*

	Prevalence of anemia	Low-birthweight babies		Prevalence of child malnutrition	Smoking prevalence		Incidence of tuberculosis	Adult HIV-1 seroprevalence			
									% infected		Women attending urban antenatal clinic
	% of pregnant women 1985–95	% of births		% of children under 5 1990–96	% of adult male 1985–95	female 1985–95	per 100,000 population 1995	Survey year	Urban high-risk group	Survey year	
		1980	1989–95								
Albania	..	..	7	..	50	8	40	..	..	..	..
Algeria	42	..	9	10	53	10	53	1981–89	0.0 ᵃ	..	..
Angola	29	..	19	35	..	..	225	1988	24.7 ᵃ,ᵇ	1995	1.0 ᵇ
Argentina	26	..	7	2	40	23	50	1996	41.4 ᶜ	1995	2.8 ᵈ
Armenia	..	..	..	..	..	..	40	..	..	..	..
Australia	..	..	..	..	29	21	6	..	..	..	..
Austria	..	..	6	..	42	27	20	..	..	..	..
Azerbaijan	..	..	..	10	..	..	47	1995	0.0 ᵃ	..	..
Bangladesh	53	..	34	68	60	15	220	1996	0.6 ᵉ,ᶠ	..	..
Belarus	..	..	5	..	..	..	50	..	..	..	..
Belgium	..	..	6	..	31	19	16	..	..	..	..
Benin	41	..	10	24	..	..	135	1993–94	53.3 ᵇ,ᵉ	1993	0.4 ᵍ
Bolivia	51	10	10	16	50	21	335	1988	5.1 ᶠ,ᵍ,ʰ	..	..
Bosnia and Herzegovina	..	..	..	..	..	..	80	..	..	..	..
Botswana	..	..	8	27	21	..	400	1995	42.8 ᵃ,ᵍ	1995	34.2 ᵍ
Brazil	33	..	11	7	40	25	80	1994–95	40.4 ᶜ,ᵍ	1995	1.7 ᵍ,ⁱ
Bulgaria	..	..	6	..	49	17	40	1993	0.0 ᵃ	1993	0.0
Burkina Faso	24	21	21	33	..	..	289	1994	60.4 ᵇ,ᵉ	1995	12.0
Burundi	68	..	..	38	..	..	367	1986	18.5 ᵃ,ᶠ	1993	17.2
Cambodia	..	..	..	38	..	..	235	1996	43.0 ᵉ,ᵍ	1996	3.2
Cameroon	44	..	13	15	..	..	194	1994	21.2 ᵇ,ᵉ	1996	1.9
Canada	..	6	6	..	31	29	8	..	..	..	..
Central African Republic	67	23	15	23	..	..	139	1994–95	34.0 ᵃ	1993	10.0 ᵍ
Chad	37	11	..	..	..	..	167	..	..	1992	4.5 ᵍ
Chile	13	..	7	1	38	25	67	1994	0.7 ᵃ,ᵍ	1994	0.1
China	52	..	6	16	61	7	85	1994	66.5 ᶜ,ʲ	1993	0.0 ʲ
Hong Kong, China	..	..	..	..	..	..	140	..	..	..	..
Colombia	24	3	9	8	35	19	67	1994	26.2 ᵍ,ʰ	1994	0.5 ᵍ
Congo, Dem. Rep.	76	13	15	34	..	..	333	1995	30.3 ᵉ	1993	4.6
Congo, Rep.	..	15	16	24	..	..	250	1987	49.2 ᵉ,ᶠ,ᵍ	1994	7.1
Costa Rica	28	..	7	2	35	20	15	1994	4.9 ʰ	1992	0.0
Côte d'Ivoire	..	14	14	24	..	..	196	1994–95	67.6 ᵇ,ᵉ	1995–96	11.6 ᵇ,ᵍ
Croatia	..	..	8	..	37	38	65	..	..	..	..
Cuba	47	..	8	8	39	25	20	1993	0.0 ᵃ	1996	0.0 ⁱ
Czech Republic	23	..	6	1	43	31	25	1995	10.3 ʰ	1995	0.0
Denmark	..	..	5	..	37	37	12	..	..	..	..
Dominican Republic	..	..	16	6	66	14	110	1994	7.7 ʰ	1995	2.8 ᵍ
Ecuador	17	..	13	17	..	..	166	1988	28.8 ᶠ,ᵍ,ʰ	1992	0.3
Egypt, Arab Rep.	24	7	12	9	40	1	78	1994	7.6 ᶜ	1992	0.0
El Salvador	14	9	11	11	38	12	110	1995–96	6.0 ᵃ	1994–95	0.0 ⁱ
Eritrea	..	..	..	41	..	..	155	1989	5.8 ᵉ	..	..
Estonia	..	..	..	..	52	24	60	1996	0.0 ᵃ	..	..
Ethiopia	42	..	16	48	..	..	155	1991	67.5 ᵉ,ᵍ	1991	4.9 ᵍ
Finland	..	4	5	..	27	19	15	..	..	..	..
France	..	5	5	..	40	27	20	..	..	..	..
Gabon	..	..	10	15	..	..	100	1988	4.2 ᵃ,ᵇ	1994	1.7
Gambia, The	80	35	10	17	..	..	166	1993	34.7 ᵇ,ᵉ	1993–95	1.7 ᵇ
Georgia	..	..	..	..	..	..	70	..	..	..	..
Germany	..	..	..	..	37	22	18	..	..	..	..
Ghana	..	..	17	27	..	..	222	1986–87	30.8 ᵉ,ᵍ	1995	2.2 ᵇ,ᵍ
Greece	..	6	9	..	46	28	12	..	..	..	..
Guatemala	39	10	14	33	38	25	110	1990–93	5.3 ᵃ,ᵍ	1990–91	0.0
Guinea	..	18	21	24	40	2	166	1994	36.6 ᵉ	1990–91	0.7 ᵇ
Guinea-Bissau	74	13	20	23	..	..	220	1987	36.7 ᵇ,ᵉ,ᶠ	1995	6.9 ᵇ
Haiti	38	..	15	28	..	..	333	1989	41.9 ᵉ	1993	8.4
Honduras	14	9	9	18	36	11	133	1992	30.0 ᵈ,ʰ	1996	1.0 ᵈ

Health: risk factors and future challenges | 2.16

	Prevalence of anemia	Low-birthweight babies		Prevalence of child malnutrition	Smoking prevalence		Incidence of tuberculosis	Adult HIV-1 seroprevalence			
									% infected		Women attending urban antenatal clinic
	% of pregnant women 1985–95	% of births 1980	% of births 1989–95	% of children under 5 1990–96	% of adult male 1985–95	% of adult female 1985–95	per 100,000 population 1995	Survey year	Urban high-risk group	Survey year	
Hungary	..	..	9	..	40	27	50	..	..	..	..
India	88	..	33	66	40	3	220	1994	51.0 d,e	1995	0.3 k
Indonesia	64	..	14	40	53	4	220	1994	0.3 e	1986–87	0.0
Iran, Islamic Rep.	..	4	12	16	..	..	50	..	..	..	..
Iraq	18	6	15	12	40	5	150	..	..	..	..
Ireland	..	..	4	..	29	28	18	..	..	..	..
Israel	..	..	..	..	45	30	12	..	..	..	..
Italy	..	7	7	..	38	26	25	..	..	..	..
Jamaica	40	10	11	10	43	13	10	1994–95	24.6 e	1996	0.7
Japan	..	..	6	3	59	15	42	..	..	..	..
Jordan	..	..	7	10	43	5	14	..	..	..	..
Kazakhstan	11	..	..	1	..	..	77	..	..	..	..
Kenya	35	18	16	23	52	7	140	1992	85.5 e	1995	13.7 g
Korea, Dem. Rep.	..	..	..	..	..	..	162	..	..	..	..
Korea, Rep.	..	9	4	..	68	7	162	1988	0.1 e	..	..
Kuwait	40	7	7	6	52	12	40	..	..	..	..
Kyrgyz Republic	..	..	..	..	..	..	68	..	..	..	..
Lao PDR	..	..	18	40	..	..	235	1990–93	1.2 e	..	..
Latvia	..	..	..	..	67	12	70	..	..	..	..
Lebanon	..	..	..	9	..	..	35	..	..	..	..
Lesotho	7	8	11	21	38	1	250	1993	15.2 a,g	1993	6.1
Libya	..	5	5	5	..	..	12	..	..	..	..
Lithuania	..	..	..	..	52	10	82	1995	0.0 a	1995	0.0
Macedonia, FYR	..	..	..	..	..	..	60	..	..	..	..
Madagascar	..	..	10	32	29	28	310	1995	0.3 a	1995	0.1
Malawi	55	22	20	28	..	..	173	1994	78.0 e,g	1996	32.8
Malaysia	56	10	8	23	41	4	67	1992	29.5 c	..	..
Mali	58	13	17	31	..	..	289	1995	55.5 b,e	1994	3.5 g
Mauritania	..	..	11	48	..	..	220	1993–94	0.9 a	1993–94	0.5 b
Mauritius	29	..	8	15	47	4	50	1988–91	0.8 a	1986	0.0
Mexico	14	..	12	14	38	14	60	1994	32.7 h	1996	0.0
Moldova	50	..	..	..	..	..	70	..	..	1995	0.0
Mongolia	45	..	10	12	40	7	100	1987–93	0.0 a,e	1987–93	0.0
Morocco	45	9	9	10	40	9	125	1990	7.1 e,f	1993	0.2
Mozambique	58	16	20	47	..	..	189	1994	24.0 a,g	1994	10.5 g
Myanmar	58	..	16	31	..	..	189	1995	56.5 c	1995	1.3
Namibia	16	..	12	26	..	..	400	1992	7.2 a,d	1996	17.6
Nepal	65	..	26	49	..	..	167	1993	0.9 e	1992	0.0 d,i
Netherlands	..	4	4	..	36	29	13	..	..	..	..
New Zealand	..	5	6	..	24	22	10	..	..	..	..
Nicaragua	36	..	15	24	..	..	110	1990–91	1.6 e,f	..	..
Niger	41	..	15	43	..	..	144	1993	15.4 b,e	1993	1.3 b
Nigeria	55	18	16	35	24	7	222	1993–94	22.5 e,g	1993–94	3.8 g
Norway	..	4	5	..	36	36	8	..	..	..	..
Oman	54	..	10	14	..	..	20	..	..	..	..
Pakistan	37	..	25	40	27	4	150	1995	11.5 c	1995	0.2 d,g
Panama	..	8	10	7	56	20	90	1984–86	3.1 h	1994	0.3 i
Papua New Guinea	13	..	23	30	46	28	275	1992	0.1 a,g	1992	0.0
Paraguay	29	7	8	4	24	6	166	1987–90	8.8 h	1992	0.0
Peru	53	9	11	11	41	13	250	1989–90	41.0 h	..	..
Philippines	48	..	..	30	43	8	400	1993	0.6 e,g	..	..
Poland	16	8	8	..	51	29	50	1995	4.7 c,i	..	..
Portugal	..	8	5	..	38	15	60	..	..	..	..
Puerto Rico	..	..	..	..	..	..	8	..	..	..	..
Romania	31	..	..	6	..	..	120	..	..	..	..
Russian Federation	30	..	..	3	67	30	99	1995	0.5 h	1995	0.0

	Prevalence of anemia	Low-birthweight babies		Prevalence of child malnutrition	Smoking prevalence		Incidence of tuberculosis	Adult HIV-1 seroprevalence			
									% infected		Women attending urban antenatal clinic
	% of pregnant women 1985–95	% of births 1980	% of births 1989–95	% of children under 5 1990–96	% of adult male 1985–95	% of adult female 1985–95	per 100,000 population 1995	Survey year	Urban high-risk group	Survey year	
Rwanda	..	..	17	29	..	..	260	1984	87.9 e,f	1995	25.3 g
Saudi Arabia	..	..	..	..	53	..	22	..	..	..	..
Senegal	26	..	11	22	48	35	166	1994	22.1 e,g	1994	1.1 b,g
Sierra Leone	..	..	17	29	..	..	167	1995	26.7 e	1990	0.8 b,d
Singapore	..	8	7	14	32	3	82	..	..	..	..
Slovak Republic	..	..	6	..	43	26	40	1995	0.0 a,c	1992	0.0
Slovenia	..	..	6	..	35	23	35	1995	0.0 c	1995	0.0
South Africa	37	..	..	9	52	17	222	1994	20.1 a	1995	10.4 m
Spain	..	1	1	..	48	25	250	..	..	..	..
Sri Lanka	39	25	17	38	55	1	49	1993	0.5 d,e	..	..
Sudan	36	17	15	34	..	..	211	1989	19.1 a,g	1995	3.0 d
Sweden	..	..	5	..	22	24	7	..	..	..	..
Switzerland	..	..	5	..	36	26	18	..	..	..	..
Syrian Arab Republic	..	10	8	..	..	..	58	..	..	..	..
Tajikistan	50	..	..	..	..	..	133	..	..	..	..
Tanzania	..	..	14	29	..	..	187	1993	49.5 e	1995–96	13.9 g
Thailand	57	12	13	13	49	4	173	1995	34.4 c,g	1995	2.4 g,i
Togo	48	..	20	25	..	..	244	1993	7.3 a,b,d	..	..
Trinidad and Tobago	53	..	10	7	42	8	20	1983–84	40.0 h	1990	0.3 i
Tunisia	..	7	10	9	58	6	55	1987	0.0 e	1991	0.0
Turkey	..	8	8	10	63	24	57	1992	0.1 a	1987–88	0.0
Turkmenistan	..	..	..	..	27	1	72	..	..	..	..
Uganda	30	..	..	26	10	0	300	1987	86.0 e	1994–95	21.2
Ukraine	..	6	5	..	..	..	50	1995	13.0 c,n	1995	0.0
United Arab Emirates	46	8	8	7	..	..	30	..	..	..	..
United Kingdom	..	..	..	..	28	26	12	..	..	..	..
United States	..	7	7	..	28	23	10	..	..	..	..
Uruguay	20	..	8	4	41	27	20	1996	13.0 b,c	1991	0.0
Uzbekistan	..	..	..	4	40	1	55	..	..	..	..
Venezuela	29	..	9	5	..	..	44	1994	25.0 d,h	..	..
Vietnam	52	..	17	45	73	4	166	1995	7.5 c,f	1995	0.0 g
West Bank and Gaza	..	..	..	..	..	..	..	..	..	..	..
Yemen, Rep.	..	..	19	30	..	..	96	1992	0.0 a,f	..	..
Yugoslavia, FR (Serb./Mont.)	..	..	..	..	52	31	50	..	..	..	..
Zambia	34	..	13	29	39	7	345	1992–93	58.0 a	1994	27.9
Zimbabwe	..	15	14	16	36	15	207	1994–95	86.0 b,e	1995	35.2 g
World					**48 w**	**12 w**	**129 w**				
Low income					51	6	160				
Excl. China & India					..	..	195				
Middle income					48	16	118				
Lower middle income					52	13	136				
Upper middle income					42	22	74				
Low & middle income					50	9	146				
East Asia & Pacific					59	6	125				
Europe & Central Asia					58	26	72				
Latin America & Carib.					40	20	89				
Middle East & N. Africa					..	..	71				
South Asia					41	4	209				
Sub-Saharan Africa					..	..	220				
High income					39	22	37				

a. Patients with sexually transmitted diseases. b. HIV-1 and/or HIV-2. c. Injecting drug users. d. Sample size unknown. e. Sex workers. f. Data are best available but not reliable because of small sample size. g. Data averaged. h. Homosexual or bisexual men. i. Not specifically urban. j. For Yunnan Province. k. For Tamil Nadu State. l. UNAIDS data. m. National data. n. Data are from UNAIDS 1996.

Health: risk factors and future challenges | 2.16

About the data

The limited availability of data on health status is a major constraint to assessing the health situation in developing countries. Surveillance data are lacking for a number of major public health concerns. Estimates of prevalence and incidence are available for some diseases but are often unreliable and variable. National health authorities differ widely in their capability and willingness to collect or report information. Even when intentions are good, reporting is based on definitions that may vary widely across countries or over time. To compensate for the paucity of data and ensure a reasonable degree of reliability and international comparability, the World Health Organization (WHO) prepares estimates in accordance with epidemiological and statistical procedures.

Adequate quantities of micronutrients (vitamins and minerals) are essential for healthy growth and development. Studies indicate that more people are deficient in iron (anemic) than any other micronutrient, and most of those suffering are women of reproductive age. Anemia during pregnancy can harm both the mother and the fetus, causing loss of the baby, premature birth, or low birthweight. Estimates of the prevalence of anemia among pregnant women are generally drawn from clinical data, which suffer from two weaknesses: one, the sample is not random, but based on those who seek care; and two, private clinics or hospitals may not be part of the reporting network.

Low birthweight, which is associated with maternal malnutrition, raises the risk of infant mortality and stunts growth in infancy and childhood, increasing the incidence of other forms of retarded development. Estimates of low-birthweight infants are drawn mostly from hospital records. But many births in developing countries take place at home without assistance from formal medical practitioners and are seldom recorded. How this factor skews the data is uncertain. A hospital birth may indicate higher income and therefore better nutrition, or it could indicate a higher-risk birth, possibly skewing the data toward lower birthweight. Changes in this indicator are more likely to reflect changes in reporting practices than improvements or deterioration. The data should be treated with caution and no comparisons within or across countries should be attempted.

Estimates of child malnutrition are from national survey data on weight for age. Weight for age is a composite indicator of both weight for height (wasting) and height for age (stunting). The disadvantage of this indicator is that it cannot indicate whether malnutrition is due to wasting or stunting. Still, weight for age is useful for comparisons with earlier surveys because it was the first anthropometric measure in general use. Assessment methods vary, but the indicator used here reflects weight less than minus two standard deviations from the median weight for age of the U.S. National Center of Health Statistics reference population age 0–59 months. This reference population, adopted by the WHO in 1983, is based on children from the United States, who are assumed to be well nourished.

Data on smoking are obtained through surveys and should be interpreted with caution because a one-time estimate of the prevalence of smoking does not give any information on its duration (usually longer for males).

Tuberculosis (TB) has reemerged as a global health problem. From an economic point of view this epidemic is about wasted lives and lost productivity. From a health perspective it is about the need to efficiently organize and finance the health sector to serve the needs of the population. And from a social perspective it is about the need to provide equitable access to appropriate health services because TB is most likely to be contracted by the poor. Data on case notifications and treatment outcomes are reported to the WHO by national TB control offices. WHO checks these data for inconsistencies and adjusts them where necessary. The data in the table show the overall incidence of TB rather than just smear-positive incidence.

Adult HIV-1 seroprevalence rates reflect the rate of HIV-1 infection for each country's adult population. The global HIV pandemic currently involves two HIV viruses: HIV-1 and HIV-2. HIV-1 is the dominant type worldwide. HIV-2 appears to be less easily transmitted than HIV-1, and the progression from HIV-2 infection to AIDS appears to be slower than that for HIV-1. AIDS is late-stage infection characterized by a severely weakened immune system that can no longer ward off life-threatening opportunistic infections and cancers. This table uses only seroprevalence surveys measuring HIV-1, except where otherwise noted. Estimates of HIV seroprevalence are not based on national samples. Most HIV data originate from diagnostic centers or screening programs and so are subject to selection (usually high-risk groups) and participation bias. The results from high-risk groups should not be considered indicative of prevalence in the general, low-risk population (World Bank 1997a).

Definitions

• **Prevalence of anemia,** or iron deficiency, is defined as hemoglobin levels less than 11 grams per deciliter among pregnant women. • **Low-birthweight babies** are newborns weighing less than 2,500 grams, with the measurement taken within the first hours of life, before significant postnatal weight loss has occurred. • **Prevalence of child malnutrition** is the percentage of children under 5 whose weight for age is less than minus two standard deviations from the median of the reference population (see *About the data*). • **Smoking prevalence** is the percentage of men and women over 15 who smoke tobacco products. • **Incidence of tuberculosis** is the estimated number of new tuberculosis cases (all forms). • **Adult HIV-1 seroprevalence** is the estimated percentage of people over 15 who are HIV-1 positive.

Data sources

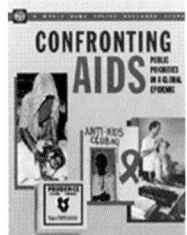

The data presented here are drawn from a variety of sources, including the United Nations *Update on the Nutrition Situation*; the WHO's *World Health Statistics Annual*, *Global Tuberculosis Control Report 1997*, and *Tobacco or Health: A Global Status Report 1997*; the World Bank's *Confronting AIDS: Public Priorities in a Global Epidemic;* the WHO-EC Collaborating Centre on AIDS' *European HIV Prevalence Database;* and the U.S. Bureau of Census' HIV/AIDS Surveillance Database.

	Life expectancy at birth		Infant mortality rate		Under-five mortality rate		Child mortality rate		Adult mortality rate			
			per 1,000 live births		per 1,000		Male per 1,000	Female per 1,000	Male per 1,000		Female per 1,000	
	years											
	1980	1996	1980	1996	1980	1996	1988–97	1988–97	1980	1995	1980	1995
Albania	69	72	47	37	..	40	15	15	140	122	82	65
Algeria	59	70	98	32	139	39	..	..	226	177	197	133
Angola	41	46	153	124	..	209	..	..	569	493	458	406
Argentina	70	73	35	22	38	25	..	..	205	176	102	84
Armenia	73	73	26	16	..	20	..	..	158	209	85	108
Australia	74	78	11	6	..	7	..	..	167	110	85	60
Austria	73	77	14	5	..	6	2	1	197	148	92	64
Azerbaijan	68	69	30	20	..	23	..	..	262	231	127	91
Bangladesh	48	58	132	77	207	112	47	62	383	314	388	292
Belarus	71	69	16	13	..	17	..	..	255	301	95	100
Belgium	73	77	12	7	..	7	2	1	173	135	90	68
Benin	49	55	120	87	205	140	89	90	486	472	397	399
Bolivia	52	61	118	67	171	102	53	47	357	292	273	237
Bosnia and Herzegovina	70	..	31	..	..	..	..	..	181	..	108	..
Botswana	58	51	69	56	80	85	18	16	341	212	278	153
Brazil	63	67	67	36	86	42	8	9	221	181	161	123
Bulgaria	71	71	20	16	..	20	..	..	190	213	106	106
Burkina Faso	44	46	121	98	241	158	107	110	467	426	362	340
Burundi	47	47	121	97	195	176	101	114	489	481	400	403
Cambodia	40	53	201	105	..	170	..	..	473	370	355	298
Cameroon	50	56	94	54	172	102	64	75	489	413	415	341
Canada	75	79	10	6	..	7	2	1	161	125	85	65
Central African Republic	46	49	117	96	193	164	63	64	540	505	424	406
Chad	42	48	147	115	206	189	..	..	556	470	449	385
Chile	69	75	32	12	37	13	3	2	218	155	120	82
China	67	70	42	33	60	39	10	11	185	155	148	130
Hong Kong, China	74	79	11	4	12	6	..	..	150	109	87	57
Colombia	66	70	45	25	58	31	7	7	237	214	162	118
Congo, Dem. Rep.	49	53	111	90	..	..	..	..	..	..	..	..
Congo, Rep.	50	51	89	90	..	145	..	..	408	405	298	313
Costa Rica	73	77	20	12	29	15	..	..	159	115	100	68
Côte d'Ivoire	51	54	108	84	157	150	..	..	421	392	346	333
Croatia	70	72	21	9	..	10	..	..	233	176	106	78
Cuba	74	76	20	8	22	10	..	..	135	122	94	78
Czech Republic	70	74	16	6	..	10	2	2	225	195	102	83
Denmark	74	75	8	6	..	6	1	1	163	145	102	92
Dominican Republic	64	71	74	40	92	47	18	20	183	155	138	100
Ecuador	63	70	67	34	98	40	12	9	229	179	176	110
Egypt, Arab Rep.	56	65	120	53	175	66	22	28	257	278	204	238
El Salvador	57	69	81	34	125	40	17	20	410	229	178	154
Eritrea	46	55	..	64	..	120	82	69	..	429	..	342
Estonia	69	69	17	10	..	16	..	..	291	284	110	95
Ethiopia	41	49	155	109	213	177	..	..	491	442	401	352
Finland	73	77	8	4	..	5	1	1	206	150	74	64
France	74	78	10	5	..	6	2	2	190	155	85	58
Gabon	48	55	116	87	..	145	..	..	474	386	387	322
Gambia, The	40	53	159	79	..	107	83	79	584	511	466	419
Georgia	71	72	25	17	..	19	..	..	210	189	94	77
Germany	73	76	12	5	..	6	2	1	177	145	90	70
Ghana	53	59	100	71	157	110	63	62	400	320	334	253
Greece	74	78	18	8	..	9	..	..	134	113	86	61
Guatemala	58	66	81	41	140	56	22	24	336	245	266	166
Guinea	40	46	185	122	..	210	122	112	589	498	507	497
Guinea-Bissau	39	44	168	134	..	223	..	..	535	584	517	572
Haiti	52	55	123	72	200	130	59	58	348	391	275	329
Honduras	60	67	70	44	101	50	..	..	306	166	237	111

Mortality | 2.17

	Life expectancy at birth		Infant mortality rate		Under-five mortality rate		Child mortality rate		Adult mortality rate			
	years		per 1,000 live births		per 1,000		Male per 1,000 1988–97	Female per 1,000 1988–97	Male per 1,000		Female per 1,000	
	1980	1996	1980	1996	1980	1996	1988–97	1988–97	1980	1995	1980	1995
Hungary	70	70	23	11	..	13	2	2	270	330	130	138
India	54	63	116	65	173	85	29	42	261	229	279	219
Indonesia	55	65	90	49	124	60	30	27	368	262	308	205
Iran, Islamic Rep.	60	70	92	36	130	37	..	..	221	158	190	149
Iraq	62	62	80	101	93	136	..	..	207	182	191	143
Ireland	73	76	11	5	..	7	2	1	175	125	103	72
Israel	73	77	15	6	19	9	2	2	138	105	85	65
Italy	74	78	15	6	..	7	1	1	163	125	80	57
Jamaica	71	74	21	12	34	14	..	..	186	144	121	90
Japan	76	80	8	4	..	6	1	1	129	101	70	47
Jordan	..	71	41	30	64	35	6	6	..	171	..	120
Kazakhstan	67	65	33	25	..	30	8	7	312	296	140	120
Kenya	55	58	72	57	115	90	33	33	417	362	339	295
Korea, Dem. Rep.	67	63	32	56	..	..	..	..	270	215	156	102
Korea, Rep.	67	72	26	9	18	11	..	..	270	230	156	96
Kuwait	71	77	27	11	33	14	6	5	172	126	116	68
Kyrgyz Republic	65	67	43	26	..	36	..	..	296	276	131	120
Lao PDR	45	53	127	101	..	140	..	..	531	444	439	375
Latvia	69	69	20	16	..	18	..	..	281	328	106	102
Lebanon	65	70	48	31	..	36	..	..	241	191	181	135
Lesotho	53	58	108	74	..	113	..	..	371	347	279	258
Libya	57	68	79	25	..	30	6	5	276	215	218	166
Lithuania	71	70	20	10	..	13	..	..	243	304	92	97
Macedonia, FYR	..	72	54	16	..	18	..	..	..	144	..	92
Madagascar	51	58	138	88	175	135	85	82	353	445	278	384
Malawi	44	43	169	133	271	217	126	114	429	553	349	487
Malaysia	67	72	30	11	..	14	4	4	230	182	149	110
Mali	42	50	184	120	291	220	136	138	454	412	362	326
Mauritania	47	53	120	94	..	155	..	..	505	467	416	396
Mauritius	66	71	32	17	38	20	..	..	277	222	181	116
Mexico	67	72	51	32	76	36	15	17	205	162	121	89
Moldova	66	67	35	20	..	24	..	..	289	275	173	128
Mongolia	58	65	82	53	..	71	..	..	320	221	273	182
Morocco	58	66	99	53	147	67	21	19	264	213	207	163
Mozambique	44	45	155	123	285	214	..	..	468	431	361	339
Myanmar	52	60	109	80	134	109	..	..	384	308	313	252
Namibia	53	56	90	61	108	92	30	34	427	356	366	304
Nepal	48	57	132	85	179	116	..	..	376	327	395	354
Netherlands	76	77	9	5	..	6	2	1	133	110	74	65
New Zealand	73	76	13	6	..	7	..	..	177	137	91	70
Nicaragua	59	68	90	44	120	57	..	..	277	177	189	130
Niger	42	47	150	118	300	..	212	232	562	510	453	401
Nigeria	46	53	99	78	196	130	118	102	535	450	453	377
Norway	76	78	8	4	..	6	2	1	144	118	71	60
Oman	60	71	41	18	..	20	13	17	389	201	326	134
Pakistan	55	63	124	88	161	123	22	37	283	208	291	228
Panama	70	74	32	22	47	25	..	..	172	139	117	88
Papua New Guinea	51	58	67	62	..	85	..	..	514	371	478	339
Paraguay	67	71	50	24	59	45	10	12	198	158	144	108
Peru	60	68	81	42	126	58	29	31	287	211	229	157
Philippines	61	66	52	37	69	44	28	25	323	254	259	189
Poland	71	72	21	12	..	15	2	2	253	179	105	92
Portugal	71	75	24	7	..	8	..	..	199	163	95	76
Puerto Rico	74	75	19	12	22	14	..	..	159	147	78	61
Romania	69	69	29	22	..	28	7	5	216	270	116	119
Russian Federation	67	66	22	17	..	25	3	2	341	472	120	172

	Life expectancy at birth		Infant mortality rate		Under-five mortality rate		Child mortality rate		Adult mortality rate			
	years		per 1,000 live births		per 1,000		Male per 1,000	Female per 1,000	Male per 1,000		Female per 1,000	
	1980	**1996**	**1980**	**1996**	**1980**	**1996**	**1988–97**	**1988–97**	**1980**	**1995**	**1980**	**1995**
Rwanda	46	41	128	129	218	205	87	73	503	542	409	461
Saudi Arabia	61	70	65	22	..	28	..	..	283	181	241	149
Senegal	45	50	91	60	218	88	96	80	586	561	516	496
Sierra Leone	35	37	190	174	335	284	..	..	540	589	527	470
Singapore	71	76	12	4	13	5	1	1	199	130	115	75
Slovak Republic	70	73	21	11	..	13	..	..	226	221	105	93
Slovenia	70	74	15	5	..	6	..	..	250	188	105	81
South Africa	57	65	67	49	..	66	..	..	..	..	..	..
Spain	76	77	12	5	..	6	2	2	144	140	69	57
Sri Lanka	68	73	34	15	48	19	10	9	210	172	152	108
Sudan	48	54	94	74	132	116	62	63	537	445	462	378
Sweden	76	79	7	4	..	5	1	1	142	102	76	60
Switzerland	76	78	9	5	..	6	1	1	145	115	70	58
Syrian Arab Republic	62	69	56	31	74	36	..	..	..	217	..	154
Tajikistan	66	69	58	32	..	38	..	..	190	200	129	197
Tanzania	50	50	108	86	176	144	59	52	451	485	370	417
Thailand	63	69	49	34	58	38	11	11	280	199	210	119
Togo	49	50	110	87	175	138	75	90	457	377	375	311
Trinidad and Tobago	68	73	35	13	39	15	4	3	234	170	166	130
Tunisia	62	70	69	30	100	35	19	19	227	171	224	148
Turkey	61	69	109	42	133	47	12	14	153	158	98	111
Turkmenistan	64	66	54	41	..	50	..	..	263	250	154	122
Uganda	48	43	116	99	180	141	82	72	463	622	395	558
Ukraine	69	67	17	14	..	17	..	..	282	294	112	112
United Arab Emirates	68	75	55	15	..	17	6	5	153	122	106	92
United Kingdom	74	77	12	6	..	7	1	1	160	120	96	69
United States	74	77	13	7	..	8	2	2	194	160	102	85
Uruguay	70	74	37	18	43	22	..	..	176	174	91	83
Uzbekistan	67	69	47	24	..	35	..	..	219	209	116	101
Venezuela	68	73	36	22	42	28	..	..	219	173	123	94
Vietnam	63	68	57	40	60	48	..	..	262	206	204	136
West Bank and Gaza	..	68	..	38	..	..	10	7	..	..	..	..
Yemen, Rep.	49	54	141	98	198	130	41	47	382	384	304	331
Yugoslavia, FR (Serb./Mont.)	70	72	33	14	..	19	..	..	164	170	106	99
Zambia	51	44	90	112	149	202	96	93	482	534	413	494
Zimbabwe	55	56	82	56	107	86	26	26	389	391	321	393
World	**63 w**	**67 w**	**80 w**	**54 w**	**132 w**	**73 w**	**3 w**	**3 w**	**247 w**	**217 w**	**189 w**	**164 w**
Low income	58	63	98	68	145	94	..	..	263	231	241	206
Excl. China & India	51	56	116	88	175	131	..	..	402	354	346	304
Middle income	63	68	65	37	..	45	6	6	268	238	168	141
Lower middle income	63	67	69	40	..	49	7	7	285	260	180	155
Upper middle income	66	70	53	30	..	36	2	2	226	181	136	107
Low & middle income	60	65	87	59	133	80	6	6	265	233	215	184
East Asia & Pacific	65	68	56	39	75	47	..	..	222	180	180	145
Europe & Central Asia	68	68	41	24	..	30	..	..	269	303	114	128
Latin America & Caribbean	65	70	59	33	82	41	3	2	225	182	151	114
Middle East & North Africa	59	67	96	50	141	63	..	..	249	211	208	177
South Asia	54	62	120	73	174	93	..	..	278	239	292	230
Sub-Saharan Africa	48	52	115	91	193	147	..	..	487	448	404	376
High income	74	77	13	6	..	7	2	2	174	142	91	70

Mortality 2.17

About the data

Age-specific mortality data such as infant and child mortality rates, along with life expectancy at birth, are probably the best general indicators of a community's current health status and are often cited as overall measures of a population's welfare or quality of life. They may be used nationally to identify populations in need, or internationally to compare levels of socioeconomic development. Despite variations in the quality of these data, discussed below, there is general agreement that age-specific mortality rates, especially child mortality rates, are a key indicator in any health monitoring system.

The main sources of mortality data are vital registration systems and direct or indirect estimates based on sample surveys or censuses. However, civil registers with relatively complete vital registration systems—that is, systems covering at least 90 percent of the population—are fairly uncommon in developing countries. Thus estimates must be obtained from sample surveys or derived by applying indirect estimation techniques to registration, census, or survey data. Survey data are subject to recall error and require large samples, especially if disaggregation is required. Indirect estimates rely on estimated actuarial ("life") tables that may be inappropriate to the population concerned. The life expectancy at birth that is estimated using this method would be accurate only if current mortality conditions were to remain the same for the entire life of the birth cohort.

Life expectancy at birth and age-specific mortality rates for 1996 are generally estimates based on the most recently available census or survey (see *Primary data documentation*). Extrapolations based on dated surveys may not be reliable for monitoring changes in health status.

Figure 2.17a

Child mortality rates show gender discrimination

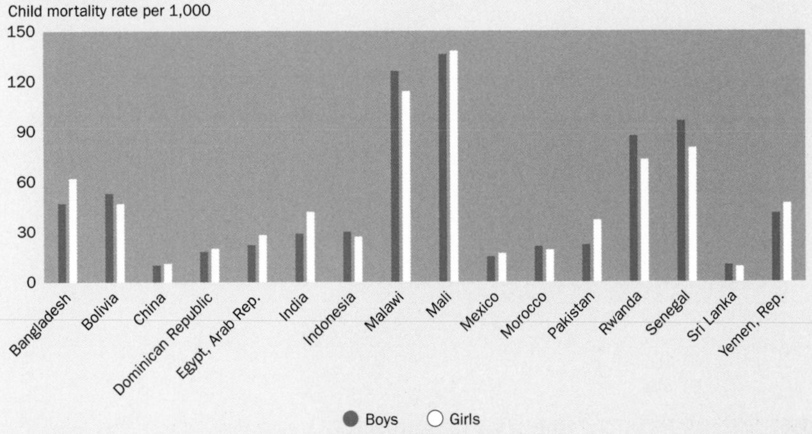
Child mortality rate per 1,000
150
120
90
60
30
0
Bangladesh
Bolivia
China
Dominican Republic
Egypt, Arab Rep.
India
Indonesia
Malawi
Mali
Mexico
Morocco
Pakistan
Rwanda
Senegal
Sri Lanka
Yemen, Rep.
Boys
Girls

Source: Demographic and health surveys and World Bank staff estimates.

In many countries parents have either a preference for sons or a preference for a certain sex distribution of their children. Son preference is most prominent in North Africa, South Asia, and East Asia, and has been documented by demographic and health surveys in these regions. No consistent pattern for gender preference has been found in Sub-Saharan Africa, while in some countries in Latin America there is a weak preference for daughters.

Preferences for boys lead to discrimination in how parents treat their sons and daughters. For example, boys receive clear preference with respect to school attendance. Findings are less clear and consistent in other areas, although in some South Asian countries one-year-old boys have higher immunization rates than one-year-old girls. The effect of gender preferences on mortality is often difficult to ascertain because infant mortality is higher for boys than for girls in all countries. Child mortality (between ages 1 and 5) better captures the effect of gender discrimination, as malnutrition and medical interventions are often more important for this age group. When female child mortality is higher, there is good reason to believe that girls are discriminated against.

The data provide only indirect evidence of discrimination by parents, but an alternative explanation exists. One consequence of son preference is that girls tend to grow up in larger families than boys, as parents attempt to achieve their desired sex distribution of children through continued childbearing. The higher number of siblings reduces the amount of resources per child even if there is no discrimination in allocation of household resources. It has been estimated that, if gender preferences were absent, pregnancy rates would decrease by 9–21 percent in countries that have high son preference.

Definitions

• **Life expectancy at birth** is the number of years a newborn infant would live if prevailing patterns of mortality at the time of its birth were to stay the same throughout its life. • **Infant mortality rate** is the number of infants who die before reaching one year of age, per 1,000 live births in a given year. • **Under-five mortality rate** is the probability that a newborn baby will die before reaching age 5, if subject to current age-specific mortality rates. • **Child mortality rate** is the probability of dying between the ages of 1 and 5, if subject to current age-specific mortality rates. • **Adult mortality rate** is the probability of dying between the ages of 15 and 60—that is, the population of 15-year olds who will die before their 60th birthday.

Data sources

POPULATION
AND
VITAL STATISTICS
REPORT

United Nations Department of Economic and Social Information and Policy Analysis, *Population and Vital Statistics Report;* demographic and health surveys from national sources; and United Nations Children's Fund (UNICEF), *The State of the World's Children 1998.*

ENVIRONMENT

Where developing countries are making economic progress, they risk repeating the mistakes of the past by putting growth before the environment—because growth can be a two-edged sword. Although economic growth raises living standards and gives people the means to enjoy their environment, it is often accompanied by urbanization, more motor vehicles, and increased energy consumption. And because unbridled growth can lead to congestion, overloaded infrastructure, and dangerous declines in air and water quality, growth at the expense of the environment is likely to be unsustainable.

Economic and social change is putting increasing pressure on the world's environmental resources. Much of the world's biological diversity is in developing nations and it is estimated to be disappearing at 50 to 100 times natural rates. Wetlands and forests are being lost at 0.3 to 1 percent a year. Greenhouse gas emissions are growing strongly with increasing economic activity. Reversing these trends will require actions by both developed and developing countries.

Many governments are adopting policies for sustainable development—that is, development that preserves the opportunities for well-being of both current and future generations. Economic growth and better environmental management can be complementary, because growth provides the resources to improve the environment. Striking a better balance between the costs and benefits of economic development will require reliable information to guide policy design and track progress toward sustainable development.

Measuring the environment

Understanding the environment and its links to economic activity requires a sound base of data and indicators. Some indicators deal with environmental "goods" such as protected areas or biodiversity (table 3.4). Others measure "bads" (deforestation, soil loss, air and water pollution). Still others monitor the effects of environmental degradation—waterborne disease, species loss, and numbers of threatened species. Such indicators are important because the links between the environment and the economy are often direct and immediate.

Many relevant indicators are not available because of weaknesses in country coverage and concerns about the quality and comparability of data. Moreover, some environmental indicators are not meaningful at a national level.

Although the world is divided into nation-states, air and water pollution do not respect national boundaries, and many other environmental problems are highly localized and location-specific. Thus a comprehensive set of environmental indicators must embrace local, national, regional, and global aspects of environmental problems.

The main indicators presented here cover important themes for which national information is available: land use, deforestation, biodiversity, protected areas, freshwater and water pollution, energy production and use, energy efficiency and net trade, sources of electricity generation, carbon dioxide emissions, urbanization, traffic and congestion, air pollution, and government commitment. There are important innovations in environmental indicators, however, for cases where limitations in data and coverage do not permit comprehensive national-level tabulations. Three such indicators are highlighted here.

"Green GNP" is one indicator gaining currency. There is widespread concern that standard national accounts indicators do not reflect environmental depletion and degradation and so may send false policy signals for nations aiming for environmentally sustainable development. While a greener measure of gross national product would have some policy uses, a related measure, genuine saving (described below), gets directly to the question of whether a country is on a sustainable path.

Trade and the environment is another issue high on the international agenda, particularly since the formation of the World Trade Organization in 1995. Whether trade liberalization is good or bad for the environment is hotly debated, and questions of environmental protection and competitiveness are of great concern to developing countries. A significant issue is whether polluting industries and firms will move to countries where environmental legislation is poor or weakly enforced. Indicators comparing exports to imports of pollution-intensive sectors can speak to these questions.

Finally, while many aspects of growth are beneficial to the environment (rising income means increased willingness and ability to pay for environmental protection), certain concomitants of growth are harmful. Indicators relating growth in incomes to the demand for polluting transport fuels provide an important link between growth and the environment.

Genuine saving

Achieving sustainable development is a process of creating and maintaining wealth. For this to be a satisfactory definition, however, it is essential that wealth be broadly conceived to include human capital, natural resources, and the natural environment. The rate at which this expanded notion of wealth is being created (or destroyed) is measured by an indicator of *genuine saving*. This is a comprehensive measure of a country's rate of saving after accounting for investments in human capital, depreciation of produced assets, and depletion and degradation of the environment.

Genuine saving departs from standard national accounting conventions in several ways, notably by expanding the range of

assets being valued. Natural resource extraction (which includes the economic rent associated with the scarcity of the resource being exploited) is explicitly treated in genuine saving by deducting the value of depletion of the underlying resource. (Where forests, water resources, and other renewable assets are sustainably managed, there is no net depletion). Deducting pollution damages, including lost welfare in the form of human sickness and death, is also necessary if it is assumed that society is aiming to maximize welfare. Finally, genuine saving estimates consider current education spending (on books, teacher salaries, and the like) as a component of saving (rather than consumption, as in the traditional national accounts), since education spending is an investment in human capital.

Table 3a provides genuine saving estimates for selected countries in Latin America and the Caribbean. *Extended domestic investment* is measured as gross domestic investment plus current education spending. While this adjustment does not have a large effect for some countries (Bolivia), it more than doubles the rate of domestic investment for others (Haiti). The next step in the accounting is to deduct net foreign borrowing, add net official transfers, and subtract depreciation of produced assets to arrive at an extended measure of net saving.

Next the value of resource depletion is deducted from extended net saving to arrive at *genuine saving I*. The natural resources included are bauxite, copper, gold, iron ore, lead, nickel, silver, tin, coal, crude oil, natural gas, phosphate rock, and timber. (Several assets—including water, fish, and soil—are not included because of difficulties in valuation.) The depletion of metals and minerals is measured as the difference between extraction values at world prices and the total cost of production (including depreciation of fixed assets and return on capital). (For technical reasons this approach probably overstates depletion. More detailed estimates could embody rising scarcity rents and account for the reserve life by applying a discount rate.) The difference between the rental value of roundwood harvest and the corresponding value of natural growth both in forests and plantations gives the measure of timber depletion. *Genuine saving II* equals genuine saving I less pollution damage. Because much pollution damage is localized (and difficult to estimate without location-specific data), table 3a includes only global damage from carbon dioxide emissions.

Figure 3a

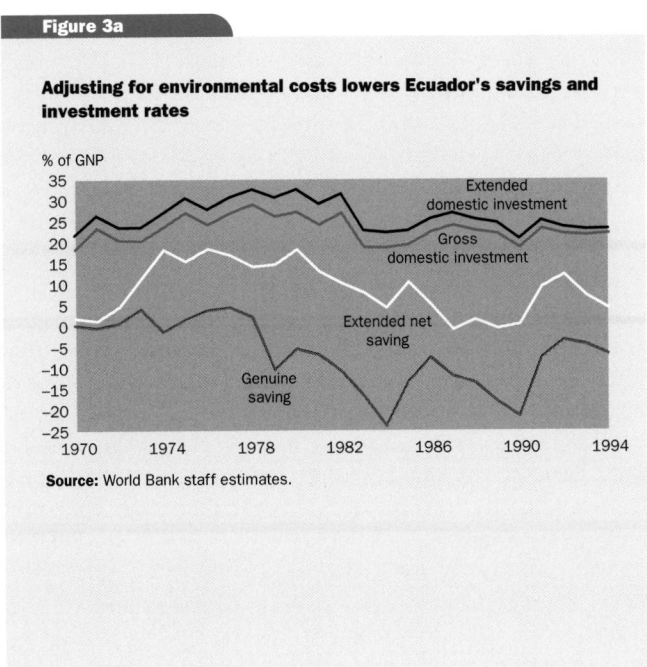

Adjusting for environmental costs lowers Ecuador's savings and investment rates

Source: World Bank staff estimates.

Table 3a

Genuine savings rates in selected countries in Latin America and the Caribbean

% of GNP	Average 1970s	Genuine saving II Average 1980s	Average 1990–94	Gross domestic investment 1994	Extended domestic investment 1994	Extended net saving 1994	Genuine saving I 1994	Genuine saving II 1994
Antigua and Barbuda[a]	..	..	11.0	26.7	29.8	13.4	13.4	12.9
Argentina	17.6	3.8	5.0	20.2	20.7	6.7	6.4	6.1
Barbados[a]	7.8	12.2	11.4	14.0	19.6	12.6	12.3	11.9
Belize	..	15.9	17.9	26.3	30.0	14.5	14.5	14.2
Bolivia	–3.8	–35.6	–14.0	15.2	15.5	–5.0	–9.4	–9.9
Brazil	12.6	9.4	11.3	21.3	25.4	14.2	12.9	12.7
Chile	–1.8	–3.4	10.2	27.4	30.3	20.3	13.2	12.9
Colombia	6.7	4.2	5.2	20.3	23.7	9.9	6.0	5.5
Costa Rica	13.0	12.2	..	28.6	31.1	..	..	..
Dominican Republic	13.2	9.7	7.2	20.5	21.9	11.4	11.0	10.6
Ecuador	0.7	–12.6	–8.5	22.3	23.4	4.5	–5.9	–6.4
El Salvador	11.4	1.8	3.1	18.7	19.7	8.9	7.1	6.9
Grenada[a]	..	22.0	14.8	30.7	33.7	10.2	10.2	9.9
Guatemala	9.2	–0.1	0.1	16.8	18.8	3.9	1.7	1.5
Haiti	0.3	–2.0	–11.9	1.7	3.8	–4.4	–15.8	–16.0
Jamaica	–0.6	–9.4	0.3	23.8	28.3	12.1	5.5	5.0
Mexico	9.1	–3.0	2.3	24.2	25.6	6.9	3.3	2.8
Paraguay	14.9	13.2	3.7	23.2	24.6	–0.1	–0.1	–0.3
Peru	5.8	–0.8	6.7	24.1	27.3	9.7	8.2	7.9
Trinidad and Tobago	–5.8	–20.6	–11.6	15.0	21.3	13.7	–1.4	–3.6
Uruguay	13.2	4.1	4.0	13.3	16.8	3.8	3.4	3.2
Venezuela	1.9	–17.6	–17.9	13.7	19.9	10.0	–11.1	–12.3

Note: *Extended domestic investment* equals gross domestic investment plus education spending. *Extended net saving* equals extended domestic investment minus net foreign borrowing plus net official transfers minus depreciation of produced assets. *Genuine saving I* equals extended net saving minus depletion of natural resources. *Genuine saving II* equals genuine saving I minus damage from carbon dioxide emissions. Only countries with adequate data are included.

a. Data are for 1993 and the associated averages for 1990–93.

Source: World Bank staff estimates.

Figure 3a shows genuine saving in Ecuador from 1970 to 1994 and carries some important messages. First, Ecuador's genuine saving rate was near zero or negative for much of the period of oil exploitation. Second, investment in human capital as a share of GNP shrank for the last decade. Finally, negative genuine saving implies that total wealth is in decline. Policies resulting in persistently negative genuine saving lead to unsustainability.

As well as serving as an indicator of sustainability, genuine saving has other advantages as a policy indicator. It presents resource and environmental issues within a framework that finance and development planning ministries can understand. It reinforces the need to boost domestic saving, and hence the need for sound macroeconomic policies. It highlights the fiscal aspects of environment and resource management, since collecting resource royalties and charging pollution taxes both raise development finance and ensure efficient use of the environment. Measuring genuine saving also makes the growth-environment tradeoff more explicit, because countries planning to grow today and protect the environment tomorrow will have depressed rates of genuine saving.

Trade and the environment

Are developing countries net exporters and developed countries net importers of pollution-intensive goods? The export-import ratio for polluting goods can shed some light on this issue.

The export-import ratio compares the total value of exports to the total value of imports of the products of each country's six most polluting industrial sectors. These sectors

Table 3b

Export-import ratios for selected countries, 1986 and 1995

	1986	1995		1986	1995
Algeria	1.01	0.95	Malaysia	0.39	0.36
Argentina	0.60	0.58	Mexico	0.82	0.71
Australia	1.00	1.11	Morocco	0.82	0.66
Austria	1.29	1.21	Netherlands	1.91	1.33
Belgium[a]	2.62	2.04	New Zealand	0.46	0.80
Bolivia	1.24	0.51	Norway	1.26	1.19
Brazil	1.57	1.02	Oman	0.11	0.27
Canada	1.91	2.05	Pakistan	0.06	0.02
Chile	2.75	2.52	Panama	0.04	0.07
Colombia	0.35	0.34	Peru	0.92	1.01
Egypt, Arab Rep.	0.20	0.35	Philippines	0.44	0.20
El Salvador	0.19	0.20	Poland	0.95	0.98
Finland	2.42	2.81	Portugal	0.69	0.60
Germany	1.14	1.18	Senegal	0.92	1.16
Greece	0.55	0.48	Singapore	1.63	0.65
Guatemala	0.09	0.15	Spain	1.00	0.77
Honduras	0.03	0.03	Sweden	1.60	1.65
India	0.13	0.37	Switzerland[b]	0.82	1.01
Indonesia	0.48	0.43	Thailand	0.16	0.17
Ireland	0.69	1.08	Tunisia	0.80	0.67
Israel	0.69	0.57	Turkey	0.55	0.41
Italy	0.77	0.71	United Kingdom	0.89	0.85
Jamaica	0.66	1.05	Uruguay	0.22	0.24
Japan	1.26	1.19	United States	0.51	0.89
Jordan	0.32	0.41	Venezuela	2.61	0.95
Korea, Rep.	0.65	0.68	Zimbabwe	0.89	0.56
Madagascar	0.06	0.05			

a. Includes Luxembourg. b. Includes Liechtenstein.

Source: World Bank staff estimates.

were identified, first, by ranking pollution control spending per unit of output for industries in the United States and other OECD economies, and then by ranking emission intensities (in terms of air pollutants, water pollutants, and heavy metals) for U.S. industries. The six most polluting sectors are iron and steel, nonferrous metals, industrial chemicals, petroleum refineries, nonmetallic mineral products, and pulp and paper products. (Some highly polluting sectors such as low-technology coal-fired thermal power stations that are basically domestic in orientation are not included.).

Table 3b shows the calculated export-import ratio for selected countries in 1986 and 1995. A ratio greater than one indicates that the country is a net exporter of polluting products. Contrary to a common perception, the results show that with few exceptions developing countries tend not to specialize in heavily polluting industries—instead, exports are lower than imports for the polluting sectors and the export-import ratio is less than one. Figure 3b shows that lower-income countries tend to have lower export-import ratios. Most high-income countries have ratios near or greater than one. These countries, particularly those with large resource sectors, appear to be the source of polluting goods.

Has trade liberalization in 1986–95 influenced this pattern? Whereas the average export-import ratio for middle-income countries has fallen, those for high- and low-income countries increased (see figure 3b). For high-income countries the ratio increased by 29 percent, to 1.32. For the United States the increase was 75 percent. For low-income countries the increase was 71 percent, possibly the result of the rapid export growth typical of the early stages of industrialization. Mexico, which has swiftly lowered trade barriers, had a lower export-import ratio in 1995 (0.71) than in 1986 (0.82), signaling a shift away from pollution-intensive goods. But Sub-Saharan Africa and Asia tended to increase their ratios during this period (figure 3c).

There may be several explanations for these results. Environmental protection costs may be lower than wage and capital costs, so that specialization is driven largely by entrenched technologies and by an economy's relative abundance of labor and physical capital. Thus countries with a large labor supply tend to specialize in relatively "clean" labor-intensive sectors, whereas physical- and human capital–intensive countries specialize in more polluting sectors. (World Bank research finds that the five most pollution-intensive sectors are about three times as energy intensive, twice as physical capital intensive, and 2.5 times less labor intensive than the five cleanest sectors; Mani and Wheeler 1997.) The tendency of countries to specialize in sectors in which they are relatively well endowed with factor inputs is reinforced by lower trade barriers. Given that the most capital-intensive economies are in the OECD, this implies that pollution-intensive production increasingly takes place in countries with relatively stringent regulation. As environmental regulations become more strict, however, comparative advantage may shift. Moreover, some of these industries tend to be relatively immobile, given their heavy dependence on a natural resource as a main factor of production. This cannot explain changes in the export-import ratio

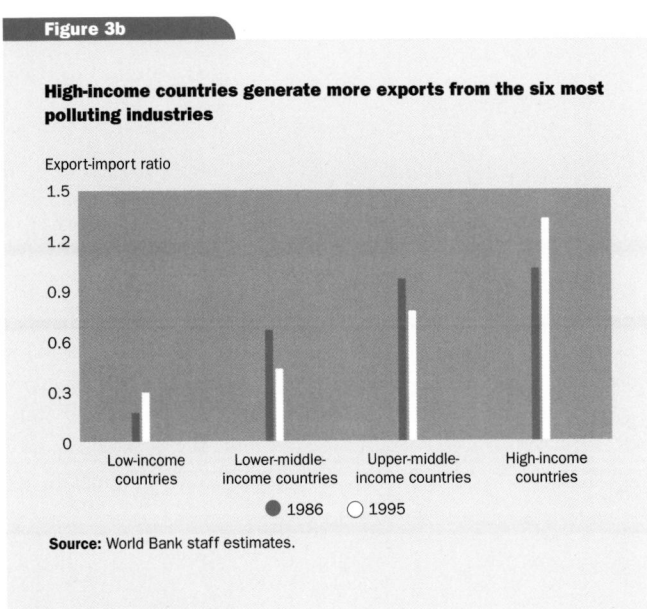

Figure 3b

High-income countries generate more exports from the six most polluting industries

Export-import ratio

Source: World Bank staff estimates.

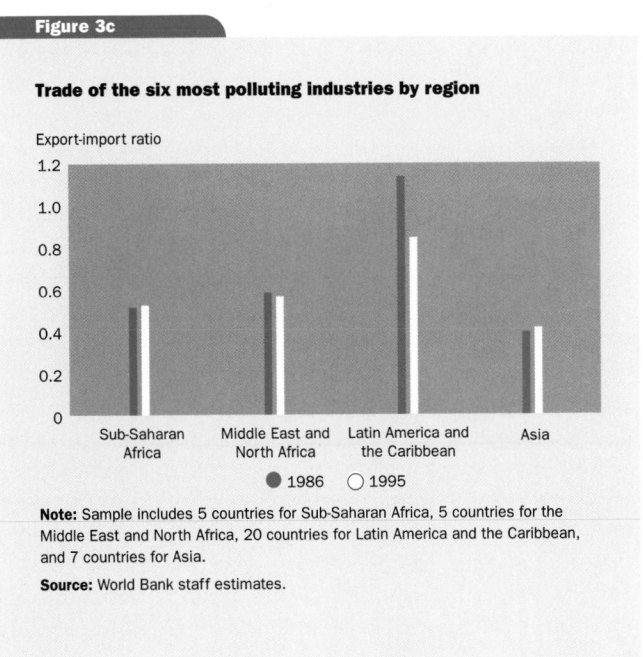

Figure 3c

Trade of the six most polluting industries by region

Export-import ratio

Note: Sample includes 5 countries for Sub-Saharan Africa, 5 countries for the Middle East and North Africa, 20 countries for Latin America and the Caribbean, and 7 countries for Asia.

Source: World Bank staff estimates.

over time, however. And some resource-dependent sectors, such as petroleum refineries, tend to be close to the market rather than to the source of the input.

Growth and the environment

Many of the indicators in this section relate to energy production and use and to the emissions of carbon dioxide associated with fossil fuel combustion. This is because energy use is both pervasive in economic activities and pollution-intensive. While global pollutants such as carbon dioxide are emphasized in the following indicator tables, energy production and use is also a major source of local pollutants such as acid rain and particulates suspended in air. Excess mortality and morbidity are strongly linked to high concentrations of particulates.

Income elasticity of demand for motive fuels in selected Asian economies, 1973–90

	Gasoline	Diesel
Bangladesh	0.82	1.28
Hong Kong, China	0.89	0.56
India	1.31	1.46
Indonesia	1.63	1.59
Korea, Rep.	1.22	..
Malaysia	1.62	1.53
Nepal	1.91	3.16
Philippines	..	3.16
Thailand	0.85	2.42

Source: World Bank staff estimates.

Consumption of motive fuel (diesel and gasoline) is particularly important in the urban environment because pollutants are emitted at ground level, where there is the greatest exposure to humans. Historical data on energy production and consumption tell much about pressures on the environment, but the rate of growth of energy demand can help indicate the likely *future* state of the environment. Extrapolating past growth rates is one method of analysis, but greater insight is gained by estimating one of the key determinants of energy use—the income elasticity of demand for motive fuels. This is economists' jargon for a simple ratio: the percentage change in demand for gasoline (for example) divided by the percentage change in income. Does demand for motive fuels rise proportionately at a greater or lesser rate than income? The answer to this question has profound implications for the relationship between growth and the environment (table 3c).

The pattern is clear. Except for Hong Kong and (for gasoline) Thailand and Bangladesh, income elasticities for motive fuels are greater than one and in many cases sharply greater. Hong Kong may be an outlier because of its limited land area, high vehicle taxes, and well-developed public transit system. Setting aside the outliers, a 1 percent increase in income leads to a 1.2–1.9 percent increase in demand for gasoline and a 1.3–3.2 percent increase in demand for diesel. Although only income elasticities are shown, the underlying analysis embodies a more complete specification of demand, including the effects of own-price and cross-price change (that is, how an increase in the diesel price would affect the demand for gasoline).

Past behavior may or may not be a good guide to the future, but the pressures that growth in incomes could place on the urban environment in developing countries are evident. If economic growth rates of 6–8 percent a year are typical of countries that have made the transition to industrialization and urbanization, growth rates in motive fuel demand of 10–15 percent a year are possible. Without policies to curb pollution emissions, especially of particulates, serious health consequences could follow.

Industrial countries accounted for most global carbon dioxide emissions in 1995

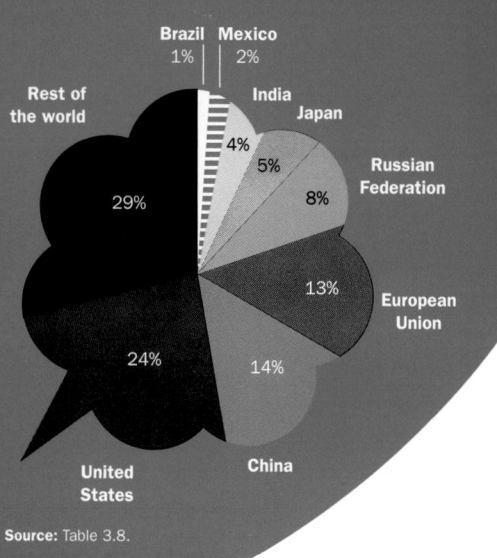

Brazil 1%
Mexico 2%
Rest of the world 29%
India 4%
Japan 5%
Russian Federation 8%
European Union 13%
China 14%
United States 24%

Source: Table 3.8.

The average American produces 21 metric tons of carbon dioxide a year, three times as much as the average Indian—twice as much as the average German—and the average Chinese, about 3 metric tons a year,

Energy sources are expected to shift

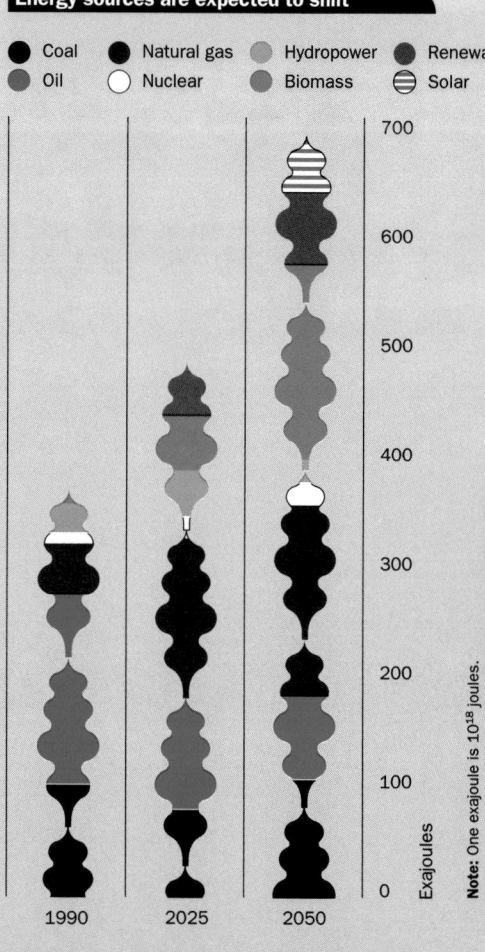

● Coal　● Natural gas　Hydropower　Renewables
Oil　○ Nuclear　Biomass　Solar

1990　2025　2050

700
600
500
400
300
200
100
0
Exajoules

Note: One exajoule is 10^{18} joules.
Source: Intergovernmental Panel on Climate Change 1996.

Carbon dioxide emissions varied considerably in 1995

Metric tons per capita

0　5　10　15　20　25

United States
Norway
Canada
Russian Federation
Germany
United Kingdom
Japan
Ukraine
Korea, Rep.
Italy
France
Mexico
China
Brazil
India

Source: Table 3.8.

Wealthy countries consume a disproportionate share of the world's energy

Energy use per capita, 1995
Kilograms of oil equivalent

Low-income countries excl. China and India
Low-income countries
Middle-income countries
High-income countries

0　1,000　2,000　3,000　4,000　5,000　6,000

20% of the world's population uses about 60% of its commercial energy. . .

Population 20%　=　Energy use 60%

. . . while 80% of the world's population uses about 40% of its energy

Population 80%　=　Energy use 40%

Source: Tables 2.1 and 3.7.

The United States, European Union, and Japan contain 13 percent of the world's people—but account for 42 percent of global carbon dioxide emissions

Carbon dioxide emissions have more than tripled since 1950—to 23 billion metric tons a year

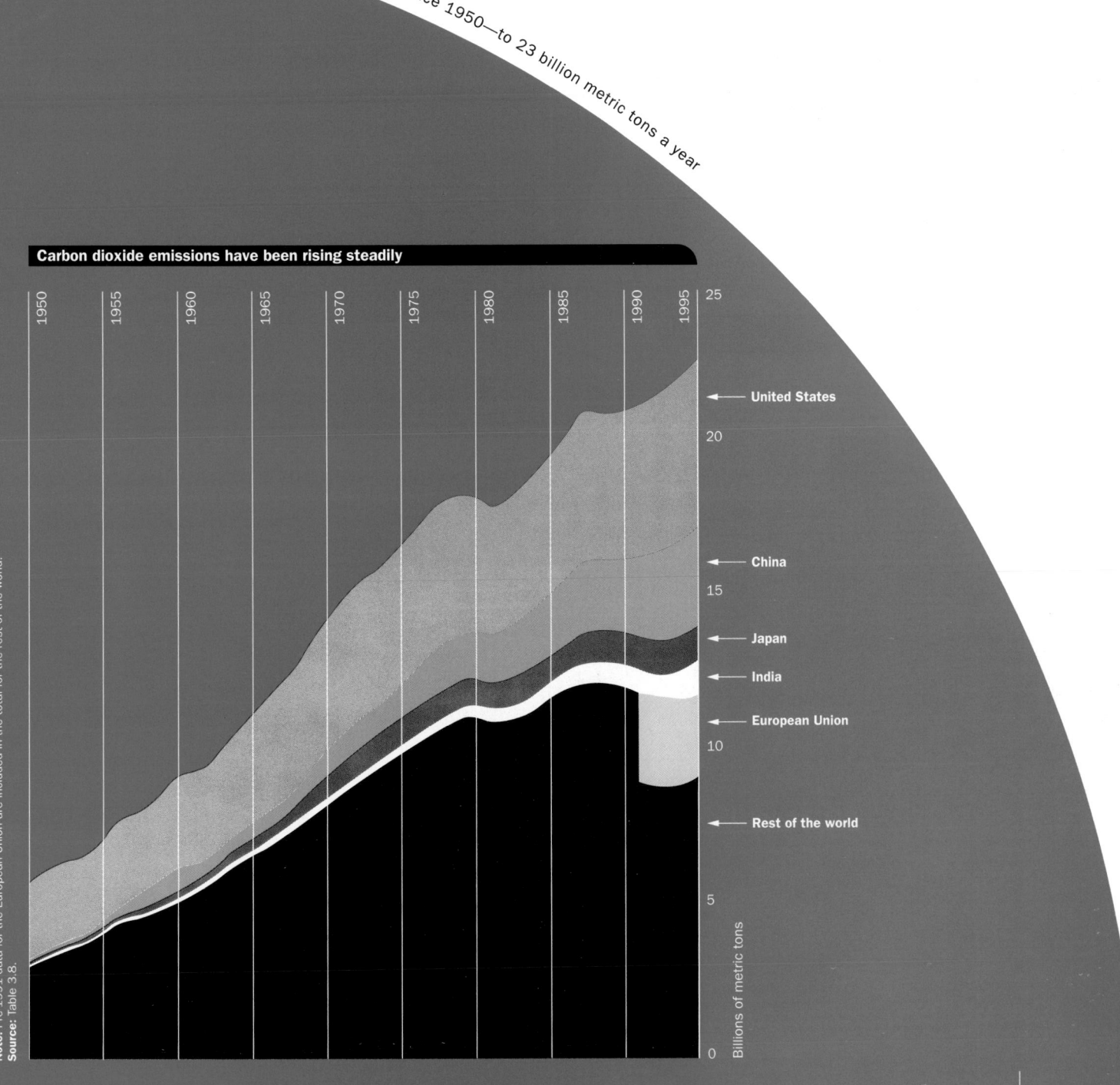

Carbon dioxide emissions have been rising steadily

1950 1955 1960 1965 1970 1975 1980 1985 1990 1995

25

← United States

20

← China

15

← Japan

← India

← European Union

10

← Rest of the world

5

0

Billions of metric tons

Note: Pre-1991 data for the European Union are included in the total for the rest of the world.
Source: Table 3.8.

3.1 Land use and deforestation

	Land area	Rural population density	Land use						Forest area	Annual deforestation	
			Cropland % of land area		Permanent pasture % of land area		Other land % of land area				average
	thousand sq. km	people per sq. km							thousand sq. km	sq. km	% change
	1995	**1995**	**1980**	**1995**	**1980**	**1994**	**1980**	**1994**	**1995**	**1990–95**	**1990–95**
Albania	27	354	26	26	15	15	59	59	10	0	0.0
Algeria	2,382	165	3	3	15	13	82	83	19	234	1.2
Angola	1,247	248	3	3	43	43	54	54	222	2,370	1.0
Argentina	2,737	17	10	10	52	52	38	38	339	894	0.3
Armenia	28	198	..	25	..	24	..	54	3	−84	−2.7
Australia	7,682	6	6	6	57	54	37	40	409	−170	0.0
Austria	83	202	20	18	25	24	56	57	39	0	0.0
Azerbaijan	87	208	..	23	..	25	..	52	10	0	0.0
Bangladesh	130	1,157	70	67	5	5	25	29	10	88	0.8
Belarus	207	49	..	30	..	14	..	56	74	−688	−1.0
Belgium[a]	33	39	24	23	23	21	53	55	7	0	0.0
Benin	111	236	16	17	4	4	80	79	46	596	1.2
Bolivia	1,084	137	2	2	25	24	73	73	483	5,814	1.2
Bosnia and Herzegovina	51	516	..	13	..	24	..	61	27	0	0.0
Botswana	567	168	1	1	45	45	54	54	139	708	0.5
Brazil	8,457	65	6	8	20	22	74	71	5,511	25,544	0.5
Bulgaria	111	67	38	38	18	16	44	46	32	−6	0.0
Burkina Faso	274	255	10	13	22	22	68	66	43	320	0.7
Burundi	26	623	46	43	39	42	15	14	3	14	0.4
Cambodia	177	209	12	22	3	8	85	70	98	1,638	1.6
Cameroon	465	123	15	15	4	4	81	81	196	1,292	0.6
Canada	9,221	15	5	5	3	3	92	92	2,446	−1,764	−0.1
Central African Republic	623	103	3	3	5	5	92	92	299	1,282	0.4
Chad	1,259	154	3	3	36	36	62	62	110	942	0.8
Chile	749	57	6	6	17	17	77	77	79	292	0.4
China	9,326	913	11	10	36	43	53	47	1,333	866	0.1
Hong Kong, China	1	5,130	7	7	1	1	92	92	..	..	..
Colombia	1,039	419	5	6	39	39	56	55	530	2,622	0.5
Congo, Dem. Rep.	2,267	429	3	3	7	7	90	90	..	..	..
Congo, Rep.	342	755	0	0	29	29	70	70	195	416	0.2
Costa Rica	51	600	10	10	39	46	51	44	12	414	3.0
Côte d'Ivoire	318	272	10	13	41	41	49	46	55	308	0.6
Croatia	56	189	..	22	..	20	..	59	18	0	0.0
Cuba	110	71	30	41	24	20	46	39	18	236	1.2
Czech Republic	77	114	..	44	..	12	..	45	26	−2	0.0
Denmark	42	33	63	55	6	7	31	37	4	0	0.0
Dominican Republic	48	221	29	39	43	43	27	19	16	264	1.6
Ecuador	277	299	9	11	15	18	77	71	111	1,890	1.6
Egypt, Arab Rep.	995	1,144	2	3	..	..	..	..	0	0	0.0
El Salvador	21	572	35	37	29	28	36	35	1	38	3.3
Eritrea	101	673	..	5	..	69	..	26	3	0	0.0
Estonia	42	36	..	27	..	7	..	66	20	−196	−1.0
Ethiopia	1,000	421	..	12	..	20	..	69	136	624	0.5
Finland	305	74	..	..	1	0	..	..	200	166	0.1
France	550	80	34	35	23	19	42	45	150	−1,608	−1.1
Gabon	258	169	2	2	18	18	80	80	179	910	0.5
Gambia, The	10	452	..	..	19	20	..	..	1	8	0.9
Georgia	70	290	..	16	..	27	..	57	30	0	0.0
Germany	349	93	36	35	17	15	47	50	107	0	0.0
Ghana	228	391	16	20	37	37	47	43	90	1,172	1.3
Greece	129	178	30	27	41	41	29	32	65	−1,408	−2.3
Guatemala	108	479	16	18	12	24	72	58	38	824	2.1
Guinea	246	667	3	4	44	44	54	53	64	748	1.1
Guinea-Bissau	28	279	10	12	38	38	51	50	23	104	0.4
Haiti	28	873	32	33	18	18	49	49	0	8	3.4
Honduras	112	196	16	18	13	14	71	68	41	1,022	2.3

Land use and deforestation | 3.1

	Land area	Rural population density	Land use						Forest area	Annual deforestation	
			Cropland % of land area		Permanent pasture % of land area		Other land % of land area				average
	thousand sq. km 1995	people per sq. km 1995	1980	1995	1980	1994	1980	1994	thousand sq. km 1995	sq. km 1990–95	% change 1990–95
Hungary	92	75	58	54	14	12	28	34	17	−88	−0.5
India	2,973	410	57	57	4	4	39	39	650	−72	0.0
Indonesia	1,812	732	14	17	7	7	79	77	1,098	10,844	1.0
Iran, Islamic Rep.	1,622	148	8	11	27	27	64	61	15	284	1.7
Iraq	437	96	12	13	9	9	78	78	1	0	0.0
Ireland	69	115	16	19	67	45	17	36	6	−140	−2.7
Israel	21	147	20	21	6	7	74	72	1	0	0.0
Italy	294	236	42	37	17	16	40	47	65	−58	−0.1
Jamaica	11	660	22	22	24	24	54	56	2	158	7.2
Japan	377	692	13	12	2	2	85	87	251	132	0.1
Jordan	89	375	4	5	9	9	87	87	0	12	2.5
Kazakhstan	2,671	21	..	12	..	70	..	17	105	−1,928	−1.9
Kenya	569	476	8	8	37	37	55	55	13	34	0.3
Korea, Dem. Rep.	120	504	16	17	0	0	84	83	62	0	0.0
Korea, Rep.	99	471	22	20	1	1	77	78	76	130	0.2
Kuwait	18	927	0	0	8	8	92	92	0	0	0.0
Kyrgyz Republic	192	336	..	7	..	47	..	46	7	0	0.0
Lao PDR	231	417	3	4	3	3	94	93	..	..	..
Latvia	62	40	..	28	..	13	..	59	29	−250	−0.9
Lebanon	10	236	30	30	1	1	69	69	1	52	7.8
Lesotho	30	470	..	..	66	66	..	..	0	0	0.0
Libya	1,760	41	1	1	7	8	91	91	4	0	0.0
Lithuania	65	35	..	46	..	8	..	46	20	−112	−0.6
Macedonia, FYR	25	130	..	26	..	25	..	49	10	2	0.0
Madagascar	582	379	5	5	41	41	54	53	151	1,300	0.8
Malawi	94	505	14	18	20	20	66	62	33	546	1.6
Malaysia	329	512	15	23	1	1	85	76	155	4,002	2.4
Mali	1,220	208	2	3	25	25	74	73	116	1,138	1.0
Mauritania	1,025	541	0	0	38	38	62	62	6	0	0.0
Mauritius	2	668	53	52	3	3	44	44	0	0	0.0
Mexico	1,909	95	13	14	39	42	48	45	554	5,080	0.9
Moldova	33	118	..	66	..	11	..	22	4	0	0.0
Mongolia	1,567	73	1	1	79	75	20	24	94	0	0.0
Morocco	446	148	18	21	47	47	35	32	38	118	0.3
Mozambique	784	391	4	4	56	56	40	40	169	1,162	0.7
Myanmar	658	351	15	15	1	1	84	84	272	3,874	1.4
Namibia	823	121	1	1	46	46	53	53	124	420	0.3
Nepal	143	659	16	21	13	12	71	69	48	548	1.1
Netherlands	34	193	24	27	35	31	41	42	3	0	0.0
New Zealand	268	32	13	12	53	51	34	38	79	−434	−0.6
Nicaragua	121	67	11	23	40	40	49	39	56	1,508	2.5
Niger	1,267	148	..	..	8	8	..	..	26	0	0.0
Nigeria	911	222	33	36	44	44	23	20	138	1,214	0.9
Norway	307	118	3	3	0	0	97	97	81	−180	−0.2
Oman	212	3,256	0	0	5	5	95	95	0	0	0.0
Pakistan	771	405	26	28	6	6	67	66	17	550	2.9
Panama	74	234	7	9	17	20	75	71	28	636	2.1
Papua New Guinea	453	6,023	1	1	0	0	99	99	369	1,332	0.4
Paraguay	397	105	4	6	40	55	56	40	115	3,266	2.6
Peru	1,280	182	3	3	21	21	76	76	676	2,168	0.3
Philippines	298	586	29	32	3	4	67	64	68	2,624	3.5
Poland	304	99	49	48	13	13	38	39	87	−120	−0.1
Portugal	92	278	34	33	9	10	57	57	29	−240	−0.9
Puerto Rico	9	3,022	11	9	38	26	51	65	3	24	0.9
Romania	230	107	46	43	19	21	35	36	62	12	0.0
Russian Federation	16,889	27	..	8	..	5	..	87	7,635	0	0.0

3.1 Land use and deforestation

	Land area	Rural population density	Land use						Forest area	Annual deforestation	
	thousand sq. km 1995	people per sq. km 1995	Cropland % of land area 1980	1995	Permanent pasture % of land area 1980	1994	Other land % of land area 1980	1994	thousand sq. km 1995	sq. km 1990–95	average % change 1990–95
Rwanda	25	710	41	47	28	28	30	25	3	4	0.2
Saudi Arabia	2,150	88	1	2	40	56	60	42	2	18	0.8
Senegal	193	208	12	12	30	30	58	58	74	496	0.7
Sierra Leone	72	619	7	8	31	31	62	62	13	426	3.0
Singapore	1	0	13	2	..	..	..	..	0	0	0.0
Slovak Republic	48	148	..	33	..	17	..	49	20	–24	–0.1
Slovenia	20	414	..	14	..	25	..	61	11	0	0.0
South Africa	1,221	125	11	13	67	67	22	21	85	150	0.2
Spain	499	60	41	40	22	21	37	38	84	0	0.0
Sri Lanka	65	1,549	29	29	7	7	64	64	18	202	1.1
Sudan	2,376	142	5	5	41	46	54	48	416	3,526	0.8
Sweden	412	54	7	7	2	1	91	92	244	24	0.0
Switzerland	40	688	10	11	41	29	49	60	11	0	0.0
Syrian Arab Republic	184	134	31	32	46	45	23	22	2	52	2.2
Tajikistan	141	483	..	6	..	25	..	69	4	0	0.0
Tanzania	884	728	3	4	40	40	57	56	325	3,226	1.0
Thailand	511	278	36	40	1	2	63	58	116	3,294	2.6
Togo	54	138	43	45	4	4	53	52	12	186	1.4
Trinidad and Tobago	5	486	23	24	2	2	75	74	2	26	1.5
Tunisia	155	120	30	31	22	20	48	49	6	30	0.5
Turkey	770	77	37	35	13	16	50	48	89	0	0.0
Turkmenistan	470	177	..	3	..	64	..	33	38	0	0.0
Uganda	200	331	28	34	9	9	63	57	61	592	0.9
Ukraine	579	46	..	59	..	13	..	28	92	–54	–0.1
United Arab Emirates	84	1,172	0	1	2	3	97	96	1	0	0.0
United Kingdom	242	107	29	25	47	46	24	29	24	–128	–0.5
United States	9,159	34	21	21	26	26	53	53	2,125	–5,886	–0.3
Uruguay	175	25	8	7	78	77	14	15	8	4	0.0
Uzbekistan	414	327	..	11	..	50	..	39	91	–2,260	–2.7
Venezuela	882	115	4	4	20	21	76	75	440	5,034	1.1
Vietnam	325	1,082	20	21	1	1	79	78	91	1,352	1.4
West Bank and Gaza	..	..	..	..	..	..	..	..	..	..	..
Yemen, Rep.	528	702	3	3	30	30	67	67	0	0	0.0
Yugoslavia, FR (Serb./Mont.)	102	123	..	40	..	21	..	39	18	0	0.0
Zambia	743	97	7	7	40	40	53	53	314	2,644	0.8
Zimbabwe	387	244	7	8	44	44	49	48	87	500	0.6
World	130,129 s	559 w	11 w	11 w	27 w	26 w	61 w	62 w	32,712 s	101,724 s	0.3 w
Low income	39,294	634	12	13	31	32	56	54	6,227	38,690	0.6
Excl. China & India	26,994	515	8	9	32	32	59	59	4,243	37,896	0.9
Middle income	59,884	401	9	10	26	23	65	67	19,985	74,598	0.4
Lower middle income	39,310	462	10	11	21	18	68	71	12,884	37,888	0.3
Upper middle income	20,574	170	7	9	30	32	63	59	7,100	36,710	0.5
Low & middle income	99,178	583	10	11	28	27	61	62	26,211	113,288	0.4
East Asia & Pacific	15,869	841	11	12	30	34	59	54	3,756	29,826	0.8
Europe & Central Asia	23,864	124	..	13	..	16	..	71	8,590	–5,798	–0.1
Latin America & Carib.	20,064	230	7	8	28	29	65	63	9,064	57,766	0.6
Middle East & N. Africa	10,972	493	5	6	23	26	72	68	89	800	0.9
South Asia	4,781	509	44	45	11	10	46	45	744	1,316	0.2
Sub-Saharan Africa	23,628	356	6	7	34	34	58	57	3,969	29,378	0.7
High income	30,951	215	..	..	25	24	..	..	6,501	–11,564	–0.2

a. Includes Luxembourg.

About the data

The data in the table show that land use patterns are changing. They also indicate major differences in resource endowments and uses among countries. True comparability is limited, however, by variations in definitions, statistical methods, and the quality of data collection. For example, countries use different definitions of land use. The Food and Agriculture Organization (FAO), the primary compiler of these data, occasionally adjusts its definitions of land use categories and sometimes revises earlier data. Because the data reflect changes in data reporting procedures as well as actual changes in land use, apparent trends should be interpreted with caution.

Satellite images show land use different from that given by ground-based measures in terms of both area under cultivation and type of land use. Furthermore, land use data in countries such as India are based on reporting systems that were geared to the collection of land revenue. Because taxes on land are no longer a major source of government revenue, the quality and coverage of land use data (except for cropland) have declined. Data on forest area may be particularly unreliable because of different definitions and irregular surveys.

Estimates of forest area are from the FAO's *State of the World's Forests 1997,* which provides information on forest cover as of 1995 and a revised esti-

mate of forest cover in 1990. Forest cover data for developing countries are based on country assessments that were prepared at different times and that, for reporting purposes, had to be adapted to the standard reference years of 1990 and 1995. This adjustment was made with a deforestation model that was designed to correlate forest cover change over time with ancillary variables, including population change and density, initial forest cover, and ecological zone of the forest area under consideration. Although the same model was used to estimate forest cover for the 1990 forest assessment, the inputs to *State of the World's Forests 1997* had more recent and accurate information on boundaries of ecological zones and, in some countries, new national forest cover assessments. Specifically, for the calculation of the forest cover area for 1995 and recalculation of the 1990 estimates, new forest inventory information was used for Bolivia, Brazil, Cambodia, Côte d'Ivoire, Guinea-Bissau, Mexico, Papua New Guinea, the Philippines, and Sierra Leone. The new information on global totals raised estimates of forest cover. For industrial countries, the United Nations Economic Commission for Europe and the FAO use a detailed questionnaire to survey the forest cover in each country.

Definitions

- **Land area** is a country's total area, excluding area under inland water bodies. In most cases the definition of inland water bodies includes major rivers and lakes.
- **Rural population density** is the rural population divided by the arable land area. Rural population is the difference between total and urban population (see definitions in tables 2.1 and 3.10). • **Land use** is broken into three categories. **Cropland** includes land under temporary and permanent crops, temporary meadows, market and kitchen gardens, and land temporarily fallow. Permanent crops are those that do not need to be replanted after each harvest, excluding trees grown for wood or timber. **Permanent pasture** is land used for five or more years for forage crops, either cultivated or growing wild. **Other land** includes forest and woodland as well as logged-over areas to be forested in the near future. Also included are uncultivated land, grassland not used for pasture, wetlands, wastelands, and built-up areas—residential, recreational, and industrial lands and areas covered by roads and other fabricated infrastructure. • **Forest area** is land under natural or planted stands of trees, whether productive or not (see *About the data*). • **Annual deforestation** refers to the permanent conversion of natural forest area to other uses, including shifting cultivation, permanent agriculture, ranching, settlements, and infrastructure development. Deforested areas do not include areas logged but intended for regeneration or areas degraded by fuelwood gathering, acid precipitation, or forest fires. Negative numbers indicate an increase in forest area.

Data sources

Data on land area and land use are from the FAO's electronic files and are published in its *Production Yearbook.* The FAO gathers these data from national agencies through annual questionnaires and by analyzing the results of national agricultural censuses. Forestry data are from the FAO's *State of the World's Forests 1997.*

Figure 3.1a

Arable land per capita in selected countries, 1995

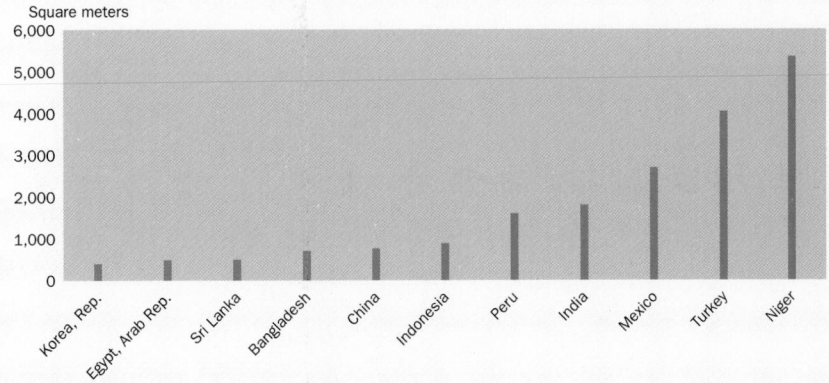

Note: Does not include land under cultivation for permanent crops.

Source: Table 3.2.

Growing populations and changing consumption patterns are putting increased pressure on land and other natural resources, reducing some countries' potential for self-sufficiency. The threshold for food self-sufficiency is estimated at 600–700 square meters per person. On average, all regions have arable land per capita in excess of this threshold, but some countries fall below.

3.2 | Agricultural inputs

	Arable land		Irrigated land		Area under cereal production		Agricultural land per worker		Fertilizer consumption		Pesticide consumption	
	hectares per capita		% of cropland		thousand hectares		hectares		Hundreds of grams per hectare of arable land		Grams per hectare of arable land	
	1979–81	1994–96	1979–81	1994–96	1979–81	1994–96	1979–81	1992–94	1979–81	1994–96	1984–86	1994–96
Albania	0.22	0.18	53.0	48.4	367	246	1.61	1.29	1,556	267	..	391
Algeria	0.37	0.27	3.4	6.9	2,968	2,488	23.6	20.0	277	145	..	836
Angola	0.41	0.28	2.2	2.1	705	915	21.4	16.5	49	33	..	..
Argentina	0.89	0.72	5.8	6.3	11,099	9,342	123.0	113.7	46	208	..	1,139
Armenia	..	0.15	..	44.7	..	179	..	4.5	..	129	..	..
Australia	2.97	2.65	3.5	4.9	15,986	14,626	1119.0	1074.4	269	358	..	..
Austria	0.20	0.18	0.2	0.3	1,062	820	11.1	13.6	2,615	1,736	..	..
Azerbaijan	..	0.21	..	50.0	..	613	..	4.3	..	244	..	..
Bangladesh	0.10	0.07	17.1	37.3	10,823	10,713	0.3	0.3	458	1,316	..	..
Belarus	..	0.59	..	1.9	..	2,528	..	9.4	..	849	..	..
Belgium[a]	..	..	0.1	0.1	426	336	12.5	13.6	5,323	4,190	11,052	..
Benin	0.39	0.26	0.3	0.5	525	681	2.0	1.6	12	146	61	..
Bolivia	0.35	0.29	6.6	3.7	559	698	27.2	22.4	23	32	..	..
Bosnia and Herzegovina	..	0.13	..	0.3	..	296	..	11.8	..	0	..	..
Botswana	0.44	0.27	0.5	0.3	153	197	103.0	102.4	33	26	..	..
Brazil	0.32	0.32	3.3	4.9	20,612	19,799	12.6	17.4	915	894	..	811
Bulgaria	0.43	0.48	28.3	19.0	2,110	2,057	6.8	11.8	2,334	468	..	..
Burkina Faso	0.40	0.33	0.4	0.7	2,026	2,998	2.4	1.9	26	69	..	1
Burundi	0.24	0.15	0.7	1.3	203	197	1.0	0.8	10	33	160	123
Cambodia	0.30	0.39	4.9	4.5	1,241	1,832	1.1	1.5	45	26	..	..
Cameroon	0.68	0.46	0.2	0.3	1,021	881	3.3	2.6	56	50	737	253
Canada	1.86	1.54	1.3	1.6	19,561	19,113	92.2	173.9	415	519	..	643
Central African Republic	0.81	0.60	..	..	194	136	4.8	4.2	5	6	..	12
Chad	0.70	0.51	0.2	0.4	907	1,514	24.3	20.3	6	22	51	..
Chile	0.36	0.28	29.6	29.9	820	602	21.6	18.2	321	1,000	..	2,701
China	0.10	0.08	45.1	51.8	..	..	1.1	1.0	1,493	3,539	..	..
Hong Kong, China	0.00	0.00	43.5	28.6	0	0	0.3	0.3	..	..	..	9,750
Colombia	0.13	0.07	7.7	16.6	1,361	1,346	12.7	12.9	812	2,383	4,351	..
Congo, Dem. Rep.	0.26	0.17	0.1	0.1	..	..	2.6	1.9	11	15	..	..
Congo, Rep.	0.08	0.06	0.7	0.6	19	29	24.7	21.6	28	134	76	..
Costa Rica	0.12	0.09	12.1	23.8	136	61	9.2	9.6	2,650	4,407	..	18,726
Côte d'Ivoire	0.24	0.21	1.4	1.7	1,008	1,589	7.5	6.2	261	224	..	..
Croatia	..	0.23	..	0.2	..	623	..	7.3	..	1,700	..	3,060
Cuba	0.27	0.34	23.0	20.2	224	166	6.7	7.7	2,030	532	..	..
Czech Republic	..	0.30	..	0.7	..	1,606	..	7.1	..	1,073	..	1,154
Denmark	0.52	0.45	14.5	20.1	1,818	1,480	15.6	19.1	2,453	1,883	95	1,923
Dominican Republic	0.19	0.17	11.7	13.7	149	137	5.2	5.6	572	727	..	..
Ecuador	0.20	0.14	19.4	8.1	419	1,013	6.5	6.6	471	895	..	1,672
Egypt, Arab Rep.	0.06	0.05	100.0	100.0	2,007	2,659	0.3	0.4	2,864	3,493	14,494	..
El Salvador	0.12	0.10	14.8	15.8	422	443	2.0	2.0	1,330	1,386	..	..
Eritrea	..	0.12	..	5.4	..	301	..	6.1	..	..	..	..
Estonia	..	0.76	..	..	..	303	..	12.6	..	370	..	138
Ethiopia	..	0.20	..	1.7	4,890	..	..	1.5	..	139	..	21
Finland	0.54	0.50	..	..	1,190	999	9.5	13.7	1,868	1,421	779	450
France	0.32	0.32	4.6	8.0	9,804	8,429	16.5	25.6	3,260	2,628	..	..
Gabon	0.42	0.29	0.9	0.8	6	15	21.1	22.3	20	13	..	..
Gambia, The	0.26	0.16	..	..	54	90	1.3	0.9	132	49	..	29
Georgia	..	0.15	..	42.0	..	295	..	4.3	..	394	..	..
Germany	0.15	0.14	3.7	3.9	7,692	6,528	7.3	12.6	4,249	2,421	..	..
Ghana	0.18	0.17	0.2	0.1	902	1,243	3.8	2.9	104	43	..	16
Greece	0.30	0.23	24.2	38.0	1,600	1,326	7.9	9.6	1,927	2,239	2,540	..
Guatemala	0.18	0.13	5.0	6.5	716	644	2.4	2.5	726	1,355	..	..
Guinea	0.13	0.10	12.8	10.9	708	673	5.4	4.1	19	16	..	147
Guinea-Bissau	0.32	0.28	6.0	5.0	142	135	3.9	3.4	24	20	..	..
Haiti	0.10	0.08	7.9	9.6	416	421	0.8	0.7	62	88	..	19
Honduras	0.43	0.29	4.1	3.6	421	482	4.7	5.1	163	340	..	..

Agricultural inputs | 3.2

	Arable land		Irrigated land		Area under cereal production		Agricultural land per worker		Fertilizer consumption		Pesticide consumption	
	hectares per capita		% of cropland		thousand hectares		hectares		Hundreds of grams per hectare of arable land		Grams per hectare of arable land	
	1979–81	1994–96	1979–81	1994–96	1979–81	1994–96	1979–81	1992–94	1979–81	1994–96	1984–86	1994–96
Hungary	0.47	0.47	3.6	4.2	2,878	2,820	7.1	9.2	2,906	683	6,271	3,294
India	0.24	0.18	22.8	29.5	104,350	99,891	0.9	0.8	342	826	..	..
Indonesia	0.12	0.09	16.2	15.2	11,825	14,682	1.1	0.9	645	1,439	..	..
Iran, Islamic Rep.	0.36	0.28	35.5	39.3	8,062	9,284	10.7	8.7	430	598	..	..
Iraq	0.40	0.27	32.1	61.3	2,159	3,202	9.4	14.7	172	650	..	..
Ireland	0.33	0.37	..	..	425	276	24.7	25.0	5,373	5,647	..	..
Israel	0.08	0.06	49.3	44.6	129	106	6.1	7.6	2,384	2,963	..	..
Italy	0.17	0.14	19.3	24.7	5,082	4,196	6.3	8.6	2,295	2,297	19,876	..
Jamaica	0.08	0.07	13.6	14.3	4	3	1.7	1.6	923	1,601	..	..
Japan	0.04	0.03	62.6	61.8	2,724	2,378	0.9	1.3	4,687	4,267	..	..
Jordan	0.14	0.08	11.0	18.2	158	79	12.1	7.5	404	503	2,362	..
Kazakhstan	..	2.00	..	7.0	..	18,839	..	128.7	..	36	..	..
Kenya	0.23	0.15	0.9	1.5	1,692	1,796	3.9	2.7	160	267	..	..
Korea, Dem. Rep.	0.09	0.08	58.9	73.0	1,625	1,447	0.6	0.5	4,688	4,421	..	..
Korea, Rep.	0.05	0.04	59.6	66.5	1,689	1,183	0.4	0.7	3,920	5,337	..	..
Kuwait	0.00	0.00	..	..	0	0	14.3	15.3	4,500	2,000	..	..
Kyrgyz Republic	..	0.25	..	77.6	..	590	..	17.6	..	270	..	..
Lao PDR	0.21	0.19	15.4	18.4	751	580	1.1	1.0	40	46	..	..
Latvia	..	0.68	..	..	..	447	..	12.1	..	550	..	190
Lebanon	0.07	0.05	28.1	28.7	34	39	3.0	6.5	1,653	1,837	..	..
Lesotho	0.22	0.16	..	..	203	123	9.8	7.5	150	188	..	..
Libya	0.58	0.36	10.7	21.7	538	464	64.1	138.7	357	475	..	..
Lithuania	..	0.79	..	..	..	1,070	..	10.1	..	283	..	..
Macedonia, FYR	..	0.31	..	9.9	..	235	..	7.2	..	814	..	..
Madagascar	0.29	0.20	21.5	35.0	1,309	1,330	7.3	5.3	31	38	..	29
Malawi	0.21	0.17	1.3	1.6	1,155	1,336	1.2	0.9	235	183	..	..
Malaysia	0.07	0.09	6.7	4.5	729	707	2.4	4.2	4,273	6,735	..	..
Mali	0.31	0.33	2.9	2.6	1,346	2,866	10.1	7.6	60	83	..	..
Mauritania	0.12	0.09	25.1	23.6	125	292	..	..	59	195	..	..
Mauritius	0.10	0.09	15.0	17.0	0	0	1.2	1.8	2,547	3,037	..	..
Mexico	0.35	0.27	20.3	23.5	9,547	10,424	12.2	12.0	570	562	..	..
Moldova	..	0.41	..	14.1	..	708	..	3.8	..	651	..	..
Mongolia	0.71	0.54	3.0	6.1	559	359	404.4	350.0	83	9	28	..
Morocco	0.39	0.33	15.0	13.5	4,414	5,374	7.3	7.4	269	344	..	..
Mozambique	0.24	0.17	2.1	3.4	1,077	1,640	8.3	6.9	109	27	..	..
Myanmar	0.28	0.21	10.4	14.3	5,133	6,645	0.8	0.6	111	130	9	..
Namibia	0.64	0.51	0.6	0.8	195	329	152.8	136.1	..	..	..	..
Nepal	0.16	0.13	22.5	31.0	2,251	3,122	0.6	0.5	98	328	..	21
Netherlands	0.06	0.06	58.5	61.5	225	194	6.4	6.4	8,620	5,849	..	..
New Zealand	0.80	0.44	5.2	9.1	193	153	120.8	98.9	1,965	4,212	..	2,370
Nicaragua	0.41	0.55	6.0	3.3	266	357	16.5	21.0	392	141	..	353
Niger	0.63	0.53	..	..	3,872	6,789	5.0	3.9	10	17	..	..
Nigeria	0.39	0.28	0.7	0.7	6,048	17,990	4.4	4.4	59	82	..	..
Norway	0.20	0.22	..	..	311	340	5.9	8.6	3,146	2,187	..	941
Oman	0.01	0.01	92.7	98.4	2	3	6.3	4.6	840	6,875	15,133	24,125
Pakistan	0.24	0.16	72.7	79.8	10,693	12,254	1.4	1.2	525	1,085	..	330
Panama	0.22	0.19	5.0	4.8	166	168	9.4	8.9	692	700	..	..
Papua New Guinea	0.01	0.01	..	..	2	2	0.4	0.3	3,827	2,383	17,833	2,635
Paraguay	0.52	0.46	3.4	3.0	304	545	33.7	38.4	44	105	..	..
Peru	0.19	0.16	33.0	41.2	732	816	13.9	11.3	381	445	..	..
Philippines	0.11	0.08	14.0	16.7	6,790	6,621	1.0	0.9	639	1,091	..	..
Poland	0.41	0.37	0.7	0.7	7,875	8,575	3.5	3.7	2,393	1,031	995	500
Portugal	0.25	0.23	20.1	20.9	1,099	679	3.3	5.3	1,113	1,112	..	..
Puerto Rico	0.02	0.01	39.0	51.3	1	0	8.2	8.1	..	..	..	..
Romania	0.44	0.41	21.9	31.3	6,340	6,177	4.0	6.6	1,448	508	..	..
Russian Federation	..	0.88	..	4.0	..	52,901	..	21.9	..	123	..	..

	Arable land		Irrigated land		Area under cereal production		Agricultural land per worker		Fertilizer consumption		Pesticide consumption	
	hectares per capita		% of cropland		thousand hectares		hectares		Hundreds of grams per hectare of arable land		Grams per hectare of arable land	
	1979–81	1994–96	1979–81	1994–96	1979–81	1994–96	1979–81	1992–94	1979–81	1994–96	1984–86	1994–96
Rwanda	0.15	0.13	0.4	0.3	239	108	0.7	0.7	3	0	..	..
Saudi Arabia	0.20	0.20	28.9	38.7	388	906	69.8	137.7	228	929	..	..
Senegal	0.42	0.28	2.6	3.1	1,216	1,235	3.9	3.0	104	78	..	214
Sierra Leone	0.14	0.11	4.1	5.4	434	348	3.1	2.7	58	62	..	..
Singapore	0.00	0.00	..	..	..	..	0.5	0.2	22,333	50,010	..	..
Slovak Republic	..	0.28	..	18.6	..	868	..	..	..	675	..	3,469
Slovenia	..	0.12	..	0.7	..	110	..	18.2	..	3,538	..	6,248
South Africa	0.46	0.40	8.4	8.1	6,645	6,389	50.3	51.0	874	506	..	57
Spain	0.42	0.39	14.8	17.8	7,391	6,612	12.2	18.7	1,012	1,211	..	..
Sri Lanka	0.06	0.05	28.4	29.2	864	903	0.8	0.7	1,757	2,270	..	..
Sudan	0.66	0.49	14.5	15.0	4,407	8,637	22.1	19.1	51	43	..	90
Sweden	0.36	0.31			1,505	1,130	14.4	17.4	1,654	1,147	..	..
Switzerland	0.06	0.06	6.2	5.8	172	206	10.7	8.0	4,623	3,547	..	4,632
Syrian Arab Republic	0.60	0.37	9.6	18.1	2,642	3,509	14.6	11.5	250	672	..	..
Tajikistan	..	0.14	..	83.5	..	260	..	5.5	..	854	..	..
Tanzania	0.12	0.11	4.1	4.9	2,835	3,168	4.6	3.2	143	98	..	..
Thailand	0.35	0.29	16.4	23.5	10,625	10,584	1.1	1.0	177	816	1,492	..
Togo	0.76	0.51	0.3	0.3	416	694	3.2	2.6	13	59	..	97
Trinidad and Tobago	0.06	0.06	17.8	18.0	4	5	2.8	2.6	1,064	867	28,486	..
Tunisia	0.51	0.32	4.9	7.4	1,416	1,091	9.3	10.5	212	296	..	..
Turkey	0.57	0.40	9.6	15.3	13,499	13,958	3.4	2.9	529	650	1,591	..
Turkmenistan	..	0.31	..	87.8	..	442	..	58.6	..	914	..	..
Uganda	0.32	0.27	0.1	0.1	752	1,319	1.3	1.1	1	5	..	15
Ukraine	..	0.64	..	7.5	..	12,225	..	8.5	..	310	..	..
United Arab Emirates	0.01	0.02	237.7	86.8	0	1	8.7	3.8	2,250	9,508	..	..
United Kingdom	0.12	0.10	2.0	1.8	3,930	3,193	26.2	28.2	3,185	3,790	..	..
United States	0.83	0.71	10.8	11.4	72,630	62,929	111.0	118.2	1,092	1,061	1,983	..
Uruguay	0.48	0.40	5.4	10.7	614	591	78.5	76.0	564	765	..	..
Uzbekistan	..	0.18	..	88.9	..	1,538	..	8.8	..	1,155	..	..
Venezuela	0.19	0.13	3.6	5.2	814	815	28.1	24.0	711	986	..	..
Vietnam	0.11	0.08	24.1	29.6	5,964	7,430	0.4	0.3	302	2,488	..	..
West Bank and Gaza	..	..	..	..	..	..	..	..	..	..	..	..
Yemen, Rep.	0.16	0.10	19.9	31.3	865	724	9.9	7.2	93	85	..	..
Yugoslavia, FR (Serb./Mont.)	..	0.35	..	1.6	..	..	..	..	..	225	..	..
Zambia	0.89	0.59	0.4	0.9	..	..	19.1	14.7	145	104	..	171
Zimbabwe	0.36	0.27	3.1	4.5	1,633	1,942	8.4	6.0	609	615	911	578
World	**0.27 w**	**0.24 w**	**16.6 w**	**17.6 w**	**492,369 s**	**597,529 s**	**3.9 w**	**3.8 w**	**864 w**	**935 w**	**..**	**.. w**
Low income	0.19	0.15	24.5	28.3	190,761	214,825	1.9	1.7	530	1,142	..	..
Excl. China & India	0.28	0.21	14.7	17.1	86,411	114,934	4.5	3.6	168	328	..	..
Middle income	0.29	0.34	13.4	12.9	145,335	245,730	8.0	9.7	783	548	..	..
Lower middle income	0.24	0.34	17.9	14.5	83,573	181,878	4.4	6.8	670	455	..	..
Upper middle income	0.38	0.33	8.1	9.4	61,762	63,852	18.6	22.3	912	776	..	..
Low & middle income	0.22	0.21	19.8	19.9	336,096	460,555	2.9	3.0	628	826	..	..
East Asia & Pacific	0.12	0.09	..	..	45,260	50,899	1.3	1.1	1,119	2,588	..	..
Europe & Central Asia	..	0.61	..	9.8	33,071	130,512	..	13.6	..	321	..	147
Latin America & Carib.	0.33	0.28	9.8	11.1	49,979	49,080	15.8	17.0	586	672	..	..
Middle East & N. Africa	0.29	0.21	23.5	31.2	25,654	29,822	10.7	11.9	421	651	..	..
South Asia	0.23	0.17	27.8	35.1	132,128	129,316	0.9	0.8	357	842	..	35
Sub-Saharan Africa	0.36	0.26	3.7	4.0	50,003	70,926	7.9	6.0	160	132	..	..
High income	0.46	0.41	..	..	156,273	136,974	36.2	51.5	1,321	1,231	..	..

a. Includes Luxembourg.

Agricultural inputs | 3.2

About the data

Agricultural activities provide developing countries with food and revenue, but they also can degrade natural resources. Poor farming practices can cause soil erosion and the loss of fertility. Efforts to increase productivity through the use of chemical fertilizers, pesticides, and intensive irrigation methods also have environmental costs and health impacts. Excessive chemical fertilizers can alter the chemistry of soil. Pesticide poisoning is common in developing countries. And salinization of irrigated land diminishes soil fertility. Thus the appropriate use of inputs for agricultural production has far-reaching effects.

This table provides indicators of major inputs to agricultural production: land, labor, fertilizers, and pesticides. There is no single correct mix of inputs; appropriate levels and application rates vary by country and over time, depending on the type of crops, the climate and soils, and the production process used. Most of the data shown here and in table 3.3 are collected by the Food and Agriculture Organization (FAO) through annual questionnaires. The FAO tries to impose standard definitions and reporting methods, but exact consistency across countries and over time is not possible.

The calculation of agricultural land per worker is based on estimates by the International Labour Organization of the number of persons employed in agriculture (see table 2.5). Data on agricultural employment should be used with caution. In many countries much agricultural employment is informal and unrecorded, including substantial work performed by women and children.

Fertilizer consumption measures the quantity of plant nutrients in the form of nitrogen, potassium, and phosphorous compounds available for direct application. Consumption is calculated as production plus imports minus exports. Traditional nutrients—animal and plant manures—are not included. Because some chemical compounds used for fertilizers have other industrial applications, the consumption data may overstate the quantity available for crops.

Data on pesticides refer to major groups of pesticides, herbicides, seed treatments, and plant growth regulators used in or sold to the agricultural sector. They are usually shown in terms of active ingredients, but some countries report weights including fillers. Country coverage and time series are incomplete because of the limited information provided by countries to the FAO. The FAO is currently revising its pesticides database.

Figure 3.2a

Arable land is no longer expanding

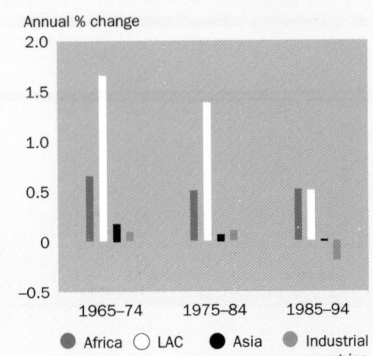

Source: FAO.

Over the past ten years growth in arable land area—which historically has been the main source of agricultural growth—fell to near zero on a global basis. A small increase in arable land in developing countries was offset by a decrease in industrial countries. Arable land area is still expanding in Africa and to some extent in Latin America, but in Asia the land frontier has been reached. With increasing demand for diversified crop and livestock products, the area under traditional food staples (mainly cereals) has begun to decline in Asia. The world is now almost entirely dependent on increased yields to expand agricultural supply.

Figure 3.2b

The distribution of fertilizer consumption has changed

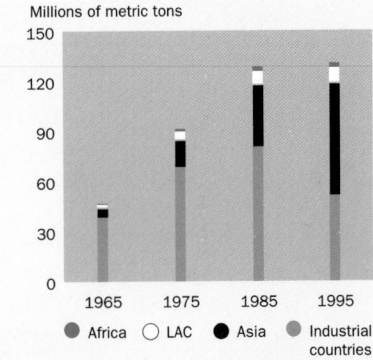

Source: FAO.

Fertilizer consumption grew rapidly after World War II but has leveled off in recent years. Fertilizer use has declined sharply in industrial countries because of concerns about its environmental effects. But consumption has expanded rapidly in developing countries—especially in Asia, which now accounts for more than half of world fertilizer consumption, up from less than 10 percent in 1965. Still, growth in fertilizer use is slowing in many countries, because higher fertilizer doses offer low marginal returns. Fertilizer use is lowest in Sub-Saharan Africa (at 13 kilograms of nutrients per hectare of arable land, compared with 85 in South Asia), and over the past 10 years its use has stagnated.

Definitions

• **Arable land** includes land defined by the FAO as land under temporary crops (double-cropped areas are counted once), temporary meadows for mowing or for pasture, land under market or kitchen gardens, and land temporarily fallow. Land abandoned as a result of shifting cultivation is not included. • **Irrigated land** refers to areas purposely provided with water, including land irrigated by controlled flooding. Cropland refers to arable land and land used for permanent crops (see table 3.1) • **Land under cereal production** refers to harvested areas, although some countries report only sown or cultivated areas. • **Agricultural land per worker** is the ratio of land used as cropland and permanent pasture (see table 3.1) to the number of workers in agriculture (see table 2.5), expressed in hectares. • **Fertilizer consumption** measures the quantity of plant nutrients used per unit of arable land. Fertilizer products cover nitrogenous, potash, and phosphate fertilizers (including ground rock phosphate). The crop year (July through June) is the time reference for fertilizer consumption. • **Pesticide consumption** refers to use or sale to the agricultural sector of substances that reduce or eliminate unwanted plants or animals, especially insects. They include major groups of pesticides such as insecticides, mineral oils, herbicides, plant growth regulators, bacteria and seed treatments, and other active ingredients.

Data sources

Data on arable land, irrigated land, and land under cereal production are from the FAO's electronic files and are published in its *Production Yearbook*. Data on agricultural employment, fertilizers, and pesticides are from electronic files that the FAO makes available to the World Bank.

3.3 Agricultural output and productivity

	Crop production index		Food production index		Livestock production index		Cereal yield		Agricultural productivity			
							kilograms per hectare		Agriculture value added per worker 1987 $		Agriculture value added per hectare of agricultural land 1987 $	
	1989–91 = 100		1989–91 = 100		1989–91 = 100							
	1979–81	1994–96	1979–81	1994–96	1979–81	1994–96	1979–81	1994–96	1979–81	1994–96	1979–81	1992–94
Albania	..	..	..	..	..	..	2,500	2,537	908	1,161	565	752
Algeria	80	116	71	115	52	110	656	953	2,713	3,612	109	180
Angola	102	143	92	126	88	102	526	405	..	149	..	9
Argentina	84	125	95	116	107	104	2,183	2,811	6,248	7,028	51	62
Armenia	..	108	..	78	..	62	..	1,431	..	1,275	..	261
Australia	80	123	92	118	91	108	1,321	1,740	17,222	22,256	16	21
Austria	93	95	92	101	95	105	4,131	5,384	10,695	15,659	956	1,088
Azerbaijan	..	56	..	55	..	64	..	1,599	..	..	..	..
Bangladesh	80	100	79	103	81	125	1,938	2,602	187	226	587	863
Belarus	..	99	..	68	..	63	..	2,161	..	3,023	..	380
Belgium[a]	85	122	88	114	89	113	4,861	7,369	..	..	..	..
Benin	54	147	63	126	69	108	698	1,008	374	563	188	321
Bolivia	71	128	71	120	76	111	1,183	1,561	1,135	..	42	..
Bosnia and Herzegovina	..	..	..	..	..	..	..	2,957	..	..	..	..
Botswana	90	95	88	102	88	103	203	300	392	483	4	5
Brazil	75	113	70	117	68	120	1,496	2,383	1,217	2,384	93	119
Bulgaria	108	69	105	68	96	57	3,853	2,615	4,446	6,240	650	513
Burkina Faso	59	121	63	121	60	117	575	808	155	182	64	93
Burundi	80	93	80	94	88	96	1,081	1,298	218	177	212	270
Cambodia	55	115	51	116	32	121	1,025	1,638	..	131	..	86
Cameroon	91	113	83	114	61	108	849	1,313	861	827	252	313
Canada	78	115	80	111	88	111	2,173	2,702	12,317	30,202	131	154
Central African Republic	103	111	80	111	49	110	529	790	456	516	96	119
Chad	67	117	91	117	120	109	587	639	148	198	6	10
Chile	71	120	72	125	76	131	2,124	4,409	1,729	3,042	79	150
China	67	125	61	144	45	181	..	..	113	193	106	184
Hong Kong, China	134	77	97	52	189	41	1,712	..	..	..	..	..
Colombia	84	102	76	109	73	114	2,452	2,623	1,579	2,172	123	165
Congo, Dem. Rep.	72	105	72	106	77	105	..	..	218	219	83	113
Congo, Rep.	82	111	80	112	80	115	825	936	544	629	21	28
Costa Rica	71	122	73	123	77	115	2,498	3,426	2,544	3,790	280	373
Côte d'Ivoire	74	110	71	118	75	119	867	1,111	1,527	1,354	195	212
Croatia	..	77	..	57	..	45	..	4,340	..	..	..	..
Cuba	87	57	90	62	91	66	2,458	1,877	..	..	..	..
Czech Republic	..	84	..	82	..	79	..	4,167	..	..	..	..
Denmark	65	89	83	102	95	112	4,040	5,934	18,790	38,131	1,166	1,684
Dominican Republic	96	91	85	104	69	123	3,004	4,034	1,325	1,587	251	262
Ecuador	78	135	77	131	73	128	1,633	2,066	1,267	1,790	194	259
Egypt, Arab Rep.	76	117	68	118	68	116	4,053	6,082	757	1,331	2,691	2,990
El Salvador	120	107	91	107	89	113	1,702	1,887	1,417	1,300	733	674
Eritrea	..	128	..	106	..	91	..	699	..	..	..	..
Estonia	..	67	..	56	..	46	..	1,756	..	6,266	..	526
Ethiopia	92	..	90	..	89	..	1,186	..	..	181	..	116
Finland	75	93	93	92	107	92	2,511	3,478	20,171	31,457	2,100	2,072
France	88	101	94	101	98	104	4,700	6,698	13,699	30,035	838	1,113
Gabon	78	111	80	107	87	105	1,718	1,779	1,412	1,516	67	74
Gambia, The	79	81	83	84	93	107	1,284	1,106	215	167	162	199
Georgia	..	65	..	71	..	75	..	1,881	..	..	..	..
Germany	90	100	91	89	98	84	4,166	6,049	..	..	..	..
Ghana	72	152	73	143	80	104	807	1,399	813	684	215	227
Greece	87	108	91	102	99	97	3,090	3,590	5,595	7,726	685	766
Guatemala	90	104	70	111	77	114	1,578	1,915	..	1,240	..	503
Guinea	90	125	97	126	116	124	958	1,219	..	225	..	54
Guinea-Bissau	66	111	69	111	78	111	711	1,389	186	292	54	78
Haiti	103	85	106	91	112	112	1,009	929	..	..	..	..
Honduras	90	107	88	104	81	112	1,170	1,436	959	1,490	200	268

Agricultural output and productivity | 3.3

	Crop production index		Food production index		Livestock production index		Cereal yield		Agricultural productivity			
							kilograms per hectare		Agriculture value added per worker 1987 $		Agriculture value added per hectare of agricultural land 1987 $	
	1989–91 = 100		1989–91 = 100		1989–91 = 100							
	1979–81	1994–96	1979–81	1994–96	1979–81	1994–96	1979–81	1994–96	1979–81	1994–96	1979–81	1992–94
Hungary	93	77	91	73	94	68	4,519	3,910	..	4,679	..	485
India	71	114	68	115	63	119	1,324	2,136	304	404	338	520
Indonesia	67	116	64	119	50	138	2,837	3,895	422	481	376	519
Iran, Islamic Rep.	58	134	61	135	68	130	1,108	1,826	4,415	6,157	404	696
Iraq	74	105	78	93	82	57	832	717	..	..	..	..
Ireland	94	106	83	105	83	106	4,733	6,687	..	..	..	..
Israel	98	103	86	108	79	114	1,840	1,424	..	..	..	..
Italy	106	100	101	102	93	104	3,548	4,716	10,516	17,876	1,650	1,964
Jamaica	99	130	86	116	74	100	1,667	1,374	711	1,045	433	591
Japan	108	96	94	98	85	98	5,252	6,119	9,832	16,712	11,279	12,445
Jordan	59	142	61	148	51	162	521	1,209	3,129	2,769	224	461
Kazakhstan	..	59	..	70	..	73	..	650	..	..	..	..
Kenya	75	109	68	101	60	97	1,364	1,822	268	240	68	90
Korea, Dem. Rep.	..	..	..	..	..	..	3,405	3,472	..	..	..	..
Korea, Rep.	88	102	78	115	53	143	4,986	5,813	1,950	5,302	5,229	6,961
Kuwait	41	106	99	129	109	134	3,124	4,937	4,564	..	288	224
Kyrgyz Republic	..	87	..	81	..	78	..	1,923	..	69	..	4
Lao PDR	73	111	71	115	64	127	1,402	2,561	..	..	..	..
Latvia	..	77	..	57	..	47	..	1,854	..	3,870	..	349
Lebanon	52	112	58	117	88	128	1,307	1,969	..	..	..	..
Lesotho	95	117	89	109	88	113	977	1,560	291	194	35	24
Libya	78	89	82	95	69	89	430	679	..	..	..	..
Lithuania	..	72	..	65	..	57	..	2,024	..	..	..	..
Macedonia, FYR	..	94	..	96	..	101	..	2,719	..	..	..	..
Madagascar	83	101	82	104	90	104	1,664	2,010	190	178	26	34
Malawi	84	107	91	102	79	98	1,161	1,194	162	156	145	153
Malaysia	75	106	55	122	41	141	2,828	3,052	2,235	4,052	941	942
Mali	59	120	80	114	95	113	804	809	251	259	24	33
Mauritania	60	137	86	100	89	95	384	750	..	..	5	7
Mauritius	92	96	89	104	66	135	2,536	4,339	1,764	3,762	1,607	1,902
Mexico	88	107	85	117	85	124	2,152	2,506	1,372	1,518	109	123
Moldova	..	71	..	63	..	49	..	2,711	..	..	..	..
Mongolia	45	37	88	80	93	85	573	734	..	..	..	..
Morocco	55	100	56	101	60	98	811	1,236	565	919	78	111
Mozambique	109	109	99	106	83	96	603	647	..	92	..	12
Myanmar	89	142	88	139	86	117	2,521	3,015	..	..	..	..
Namibia	81	99	108	107	116	108	377	264	1,295	1,458	8	9
Nepal	63	109	65	109	75	109	1,615	1,819	173	198	271	406
Netherlands	80	110	87	104	89	102	5,696	7,752	23,131	41,245	3,489	5,932
New Zealand	75	126	91	117	95	111	4,089	5,356	10,693	13,373	86	132
Nicaragua	123	110	118	120	140	116	1,475	1,687	3,268	3,697	212	155
Niger	94	123	101	120	110	114	440	338	292	256	57	63
Nigeria	52	134	58	132	82	125	1,269	1,172	479	684	111	150
Norway	91	95	92	99	95	102	3,634	3,807	19,593	34,809	3,172	3,403
Oman	60	110	63	88	62	96	982	2,180	1,041	..	155	328
Pakistan	66	111	66	125	60	131	1,608	1,943	323	466	227	382
Panama	97	96	86	102	71	113	1,524	2,078	1,954	2,320	208	246
Papua New Guinea	86	106	86	106	85	105	2,087	1,698	671	752	1,756	2,186
Paraguay	58	99	61	113	62	113	1,511	2,137	1,698	2,204	49	54
Peru	81	125	78	123	78	119	1,944	2,915	..	..	..	..
Philippines	88	109	86	116	74	138	1,611	2,283	777	780	782	835
Poland	85	86	88	83	98	79	2,345	2,854	..	1,359	..	366
Portugal	85	90	72	97	72	106	1,102	2,310	..	..	..	715
Puerto Rico	131	72	100	86	90	91	8,925	5,733	6,379	..	817	..
Romania	114	96	111	97	110	96	2,854	2,812	..	3,007	..	393
Russian Federation	..	77	..	71	..	67	..	1,313	..	..	..	..

	Crop production index		Food production index		Livestock production index		Cereal yield		Agricultural productivity			
									Agriculture value added per worker 1987 $		Agriculture value added per hectare of agricultural land 1987 $	
	1989–91 = 100		1989–91 = 100		1989–91 = 100		kilograms per hectare					
	1979–81	1994–96	1979–81	1994–96	1979–81	1994–96	1979–81	1994–96	1979–81	1994–96	1979–81	1992–94
Rwanda	89	68	90	72	81	86	1,134	1,491	306	206	445	378
Saudi Arabia	27	105	31	95	33	111	820	4,143	1,641	..	23	..
Senegal	78	97	75	106	64	131	690	824	328	375	92	118
Sierra Leone	80	94	85	95	84	107	1,249	1,192	365	344	117	123
Singapore	595	57	154	42	174	40	..	..	8,791	20,215	18,956	72,942
Slovak Republic	..	89	..	76	..	67	..	4,298	..	..	..	497
Slovenia	..	112	..	96	..	97	..	5,026	..	..	..	..
South Africa	97	101	93	98	90	92	2,117	1,918	2,361	2,870	45	49
Spain	83	90	82	95	84	107	1,986	2,478	..	8,699	..	496
Sri Lanka	99	105	98	108	93	133	2,462	2,568	489	561	592	801
Sudan	122	129	104	125	91	119	659	527	889	..	42	..
Sweden	92	86	100	96	104	102	3,595	4,399	18,485	28,590	1,263	1,577
Switzerland	95	94	96	97	99	97	4,883	6,362	..	..	..	..
Syrian Arab Republic	101	147	94	134	73	103	1,156	1,660	3,426	..	212	..
Tajikistan	..	61	..	70	..	64	..	1,109	..	..	..	..
Tanzania	82	94	77	98	69	109	1,063	1,310	..	..	..	..
Thailand	80	111	80	108	65	116	1,911	2,434	375	554	338	488
Togo	70	116	77	117	52	119	729	762	404	461	119	189
Trinidad and Tobago	120	102	102	105	84	99	3,167	3,649	4,822	3,586	1,801	1,245
Tunisia	69	93	68	99	64	123	828	1,164	1,384	2,286	142	232
Turkey	77	106	76	105	81	102	1,869	2,019	1,208	1,168	354	404
Turkmenistan	..	87	..	121	..	119	..	2,570	..	..	..	..
Uganda	68	110	71	107	85	113	1,555	1,552	..	592	..	515
Ukraine	..	69	..	70	..	64	..	2,410	..	..	..	..
United Arab Emirates	39	192	47	169	42	137	5,608	7,528	8,928	..	970	2,076
United Kingdom	79	100	92	101	98	101	4,792	6,909	..	..	..	..
United States	99	114	95	113	89	112	4,151	5,136	17,719	..	156	261
Uruguay	86	129	87	123	86	117	1,644	3,016	5,379	6,535	65	80
Uzbekistan	..	86	..	108	..	110	..	1,762	..	1,228	..	150
Venezuela	77	108	78	120	82	122	1,904	2,719	3,103	3,270	110	139
Vietnam	67	127	64	127	54	130	2,049	3,504	..	801	..	2,640
West Bank and Gaza	..	..	..	..	..	..	..	..	..	..	..	..
Yemen, Rep.	82	109	75	113	69	118	1,038	1,046	..	..	..	..
Yugoslavia, FR (Serb./Mont.)	..	83	..	94	..	104	..	..	..	..	..	..
Zambia	66	93	74	97	86	102	..	..	116	100	6	7
Zimbabwe	78	100	82	92	85	94	1,359	1,163	294	266	34	41
World	**80 w**	**113 w**	**80 w**	**116 w**	**81 w**	**118 w**	**2,230 w**	**2,561 w**	**.. w**	**.. w**	**..w**	**236 w**
Low income	70	121	67	130	58	153	1,349	1,846	209	293	128	206
Excl. China & India	75	116	76	118	75	118	1,380	1,595	..	393	79	135
Middle income	80	109	79	110	82	108	1,997	2,120	..	..	..	202
Lower middle income	79	110	77	111	77	110	1,957	1,919	..	..	..	320
Upper middle income	82	107	81	109	85	107	2,050	2,694	..	..	..	114
Low & middle income	74	116	72	122	70	131	1,629	1,993	..	459	..	206
East Asia & Pacific	70	124	65	139	50	174	2,216	3,067	..	..	..	..
Europe & Central Asia	..	..	..	..	..	..	..	1,774	..	..	..	..
Latin America & Carib.	82	112	80	115	82	116	1,840	2,494	1,586	2,292	90	116
Middle East & N. Africa	69	119	67	118	65	113	1,173	1,926	1,918	..	185	..
South Asia	73	113	70	115	64	122	1,410	2,136	290	383	337	519
Sub-Saharan Africa	77	115	79	113	84	106	1,100	1,041	458	392	53	68
High income	93	107	92	106	92	106	3,522	4,470	..	..	..	..

a. Includes Luxembourg.

Agricultural output and productivity | 3.3

About the data

The agricultural production indexes in the table are prepared by the Food and Agriculture Organization (FAO). The FAO obtains data from official and semi-official reports of crop yields, area under production, and livestock numbers. If data are not available, the FAO makes estimates. The indexes are calculated using the Laspeyres formula: production quantities of each commodity are weighted by average international commodity prices in the base period and summed for each year. Because the FAO's indexes are based on the concept of agriculture as a single enterprise, estimates of the amounts retained for seed and feed are subtracted from the production data to avoid double counting. The resulting aggregate represents production available for any use except as seed and feed. The FAO's indexes may differ from other sources because of differences in coverage, weights, concepts, time periods, calculation methods, and use of international prices.

To ease cross-country comparisons, the FAO uses international commodity prices to value production. These prices, expressed in international dollars (equivalent in purchasing power to the U.S. dollar), are derived using a Geary-Khamis formula for agriculture (see Inter-Secretariat Working Group on National Accounts 1993, sections 16–93). This method assigns a single price to each commodity so that, for example, one metric ton of wheat has the same price regardless of where it was produced. The use of international prices eliminates fluctuations in the value of output due to transitory movements of nominal exchange rates unrelated to the purchasing power of the domestic currency. Unlike the International Comparison Programme (ICP), the FAO calculates international prices only for agricultural products. Substantial differences may arise between the implicit exchange rate derived by the ICP and that of the FAO. For further discussion of the FAO's methods see FAO (1986). (See tables 4.10 and 4.11 for a discussion of the ICP.)

Data on cereal yields may be affected by a variety of reporting and timing differences. The FAO allocates production data to the calendar year in which the bulk of the harvest took place. But most of a crop harvested near the end of a year will be used in the following year. In general, cereal crops harvested for hay or harvested green for food, feed, or silage and those used for grazing are excluded. But millet and sorghum, which are grown as feed for livestock and poultry in Europe and North America, are used as food in Africa, Asia, and countries of the former Soviet Union.

Agricultural productivity is measured by value added per unit of input. (See tables 4.1 and 4.2 for further discussion on the calculation of value added in national accounts.) Agricultural value added includes that from forestry and fishing. Thus interpretations of land productivity should be made with caution.

To smooth annual fluctuations in agricultural activity, the indicators in the table have been averaged over three years.

Figure 3.3a

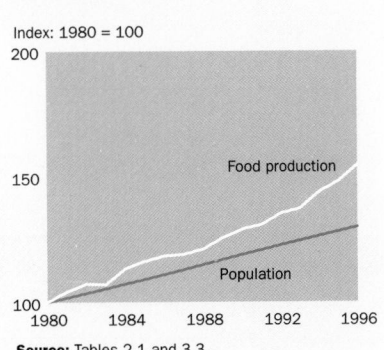

Food production has outpaced population growth

Index: 1980 = 100

Source: Tables 2.1 and 3.3.

Food production has increased by more than 50 percent since 1980, outstripping population growth of about 30 percent in the same period. Growth in food consumption has not been evenly distributed, however, and the gain in production may have reduced crop diversity and natural habitats—and increased chemical contamination (see also tables 3.2 and 3.4).

Definitions

• **Crop production index** shows agricultural production for each year relative to the base period 1989–91. It includes all crops except fodder crops. Regional and income group aggregates for the FAO's production indexes are calculated from the underlying values in international dollars, normalized to the base period 1989–91. Data in this table are three-year averages. However, missing observations have not been estimated or imputed. • **Food production index** covers food crops that are considered edible and that contain nutrients. Coffee and tea are excluded because, although edible, they have no nutritive value. • **Livestock production index** includes meat and milk from all sources, dairy products such as cheese and eggs, honey, raw silk, wool, and hides and skins. • **Cereal yield,** measured as kilograms per hectare of harvested land, includes wheat, rice, maize, barley, oats, rye, millet, sorghum, buckwheat, and mixed grains. Production data on cereals refer to crops harvested for dry grain only. Cereal crops harvested for hay or harvested green for food, feed, or silage and those used for grazing are excluded. • **Agricultural productivity** refers to value added per agricultural worker and valued added per hectare of agricultural land (measured as the sum of arable land, permanent cropland, and permanent pasture) measured in constant 1987 U.S. dollars.

Data sources

Agricultural production indexes are prepared by the FAO and published annually in its *Production Yearbook.* The FAO makes these data and data on cereal yields and agricultural employment available to the World Bank in electronic files that may contain more recent information than the published versions.

3.4 | Biodiversity and protected areas

	Nationally protected areas		Mammals		Birds		Higher plants[a]	
	thousand sq. km 1994[b]	% of total land area 1994[b]	Species 1994[b]	Threatened species 1994[b]	Species 1994[b]	Threatened species 1994[b]	Species 1994[b]	Threatened species 1994[b]
Albania	0.3	1.2	68	3	306	5	2,965	50
Algeria	119.2	5.0	92	11	375	7	3,100	145
Angola	26.4	2.1	276	16	909	13	5,000	25
Argentina	43.7	1.6	320	20	976	40	9,000	170
Armenia	2.1	7.6	..	1	..	5	..	..
Australia	940.8	12.2	252	43	751	51	15,000	1,597
Austria	20.8	24.2	83	3	414	3	2,950	22
Azerbaijan	1.9	2.2	..	3	..	6	..	1
Bangladesh	1.0	0.7	109	16	684	28	5,000	24
Belarus	2.7	1.2	..	5	..	4	..	..
Belgium	0.8	..	58	2	429	3	1,400	3
Benin	7.8	7.0	188	7	423	1	2,000	3
Bolivia	92.3	8.5	316	21	1,274	27	16,500	49
Bosnia and Herzegovina	0.3	0.5	..	..	..	2	..	..
Botswana	106.6	18.8	164	8	550	5	..	4
Brazil	321.9	3.8	394	45	1,635	103	55,000	463
Bulgaria	3.7	3.3	81	1	374	11	3,505	94
Burkina Faso	26.6	9.7	147	6	453	1	1,100	..
Burundi	0.9	3.5	107	6	596	5	2,500	1
Cambodia	30.0	17.0	123	19	429	16	..	7
Cameroon	20.5	4.4	297	21	874	14	8,000	74
Canada	823.6	9.0	193	6	578	5	2,920	649
Central African Republic	61.1	9.8	209	9	662	2	3,600	..
Chad	114.9	9.1	134	13	532	3	1,600	12
Chile	137.3	18.3	91	11	448	15	5,125	292
China	580.8	6.2	499	94	1,186	183	30,000	1,009
Hong Kong, China	..	..	..	..	..	..	..	..
Colombia	93.8	9.0	359	24	1,695	62	50,000	376
Congo, Dem. Rep.	99.2	4.4	415	23	1,096	26	11,000	7
Congo, Rep.	11.8	3.4	200	13	569	3	4,350	3
Costa Rica	6.5	12.5	205	8	850	10	11,000	456
Côte d'Ivoire	19.9	6.3	230	16	694	11	3,517	66
Croatia	3.9	6.9	..	..	..	4	..	..
Cuba	11.5	8.1	31	10	342	13	6,004	811
Czech Republic	10.7	13.8	..	3	..	5	..	..
Denmark	13.9	32.7	43	1	439	2	1,200	6
Dominican Republic	10.5	21.7	20	3	254	10	5,000	73
Ecuador	111.1	40.1	302	20	1,559	50	18,250	375
Egypt, Arab Rep.	7.9	0.8	98	7	439	10	2,066	84
El Salvador	0.1	0.2	135	2	420	..	2,500	35
Eritrea	..	..	112	3	537	3	..	..
Estonia	4.1	10.4	65	5	330	2	1,630	2
Ethiopia	60.2	6.0	255	21	813	17	6,500	153
Finland	27.4	9.0	60	3	425	4	1,040	11
France	56.0	10.2	93	5	506	5	4,500	117
Gabon	10.5	4.1	190	12	629	4	6,500	78
Gambia, The	0.2	2.3	108	3	504	1	966	..
Georgia	1.9	2.7	..	3	..	5	..	1
Germany	91.9	26.3	..	..	..	..	..	..
Ghana	11.0	4.9	222	12	725	7	3,600	32
Greece	2.2	1.7	95	5	398	9	4,900	539
Guatemala	13.3	7.7	250	5	669	4	8,000	315
Guinea	1.6	0.7	190	13	552	11	3,000	35
Guinea-Bissau	..	..	108	5	319	1	1,000	..
Haiti	0.1	0.4	3	3	220	10	4,685	28
Honduras	8.6	7.7	173	5	684	4	5,000	55

Why does biodiversity matter?

It is said that species are going extinct at 50–100 times the natural rate. Why should we be concerned? Biodiversity conservation, and the protection of ecosystems where biodiversity is greatest, are sometimes dismissed as being an elitist concern for a few charismatic species. In fact, biodiversity and the ecosystems that contain it generate a wide range of benefits to human society. Diverse ecosystems often contain a variety of economically useful products that can be harvested or used as inputs for production. They also provide economically valuable services, such as:

• Improving water quality and quantity for agriculture, industry, and human consumption.

• Reducing sedimentation in reservoirs and waterways.

• Minimizing floods, landslides, coastal erosion, and droughts.

• Providing recreational opportunities.

• Filtering excess nutrients.

• Providing essential habitats for species containing genetic material that can be used to develop useful products such as pharmaceuticals and improved crops. Moreover, many people value species and ecosystems for aesthetic, moral, or spiritual reasons, even if they do not use them.

Although all these benefits are real, many of them do not enter markets. This is one of the reasons biodiversity tends to be undervalued. With normal market transactions, buyers know what they are getting for their money—a kilogram of rice, a pair of shoes, a movie ticket. With biodiversity, however, there is much less certainty about the value, and even the quantity, of what is being "bought."

Habitat destruction—the main cause of biodiversity loss

Although there are cases where hunting has caused species to go extinct (for example, the passenger pigeon in the United States in the 20th century), the main cause of biodiversity loss is usually habitat destruction, driven by activities such as logging, conversion to agriculture, infrastructure development, or human settlement. Agriculture has played a major role in this process because crop and livestock production affect the largest portion of the earth's surface and are the world's biggest users of land and freshwater. In many countries the conversion of land to agriculture is closely related to logging because many logged areas are later cultivated, and roads built for logging facilitate new settlement.

Habitat conversion can lead directly to the extinction of species. Even if species survive the conversion of

	Nationally protected areas		Mammals		Birds		Higher plants[a]	
	thousand sq. km 1994[b]	% of total land area 1994[b]	Species 1994[b]	Threatened species 1994[b]	Species 1994[b]	Threatened species 1994[b]	Species 1994[b]	Threatened species 1994[b]
Hungary	5.7	6.2	72	2	363	7	2,148	24
India	143.4	4.8	316	40	1,219	71	15,000	1,256
Indonesia	185.6	10.2	436	57	1,531	104	27,500	281
Iran, Islamic Rep.	83.0	5.1	140	9	502	12	..	1
Iraq	..	..	81	4	381	11	..	2
Ireland	0.5	0.7	25	..	417	1	892	9
Israel	3.1	14.9	92	7	500	8	..	38
Italy	22.8	7.7	90	4	490	6	5,463	273
Jamaica	0.0	0.2	24	2	262	7	2,746	371
Japan	27.6	7.3	132	17	583	31	4,700	704
Jordan	2.9	3.3	71	8	361	4	2,200	10
Kazakhstan	9.9	0.3	..	9	..	14	..	..
Kenya	35.0	6.2	359	16	1,068	22	6,000	158
Korea, Dem. Rep.	0.6	0.5	..	7	390	16	2,898	7
Korea, Rep.	6.9	7.0	49	6	372	19	2,898	69
Kuwait	0.3	1.5	21	2	321	3	234	..
Kyrgyz Republic	2.8	1.5	..	4	..	5	..	1
Lao PDR	24.4	10.6	172	25	651	23	..	5
Latvia	7.8	12.5	83	4	325	5	1,153	..
Lebanon	0.0	0.4	54	5	329	5	..	4
Lesotho	0.1	0.2	33	2	281	3	1,576	7
Libya	1.7	0.1	76	8	323	2	1,800	57
Lithuania	6.3	9.8	68	4	305	4	1,200	..
Macedonia, FYR	2.2	8.5	..	..	..	..	..	..
Madagascar	11.2	1.9	105	33	253	28	9,000	189
Malawi	10.6	11.3	195	6	645	9	3,600	61
Malaysia	14.8	4.5	286	20	736	31	15,000	510
Mali	40.1	3.3	137	12	622	5	1,741	14
Mauritania	17.5	1.7	61	10	541	3	1,100	3
Mauritius	0.0	2.0	4	3	81	9	700	222
Mexico	98.5	5.1	450	24	1,026	34	25,000	1,048
Moldova	0.1	0.2	68	1	270	6	..	1
Mongolia	61.7	3.9	134	8	390	11	2,272	1
Morocco	3.7	0.8	105	7	416	11	3,600	195
Mozambique	0.0	0.0	179	9	678	13	5,500	92
Myanmar	1.7	0.3	251	20	999	43	7,000	29
Namibia	102.2	12.4	154	12	609	6	3,128	23
Nepal	11.1	8.1	167	23	824	23	6,500	21
Netherlands	4.3	11.5	55	2	456	3	1,170	1
New Zealand	60.7	22.9	10	3	287	45	2,160	236
Nicaragua	9.0	7.4	200	6	750	3	7,000	78
Niger	84.2	6.6	131	10	482	2	1,170	..
Nigeria	29.7	3.3	274	22	862	8	4,614	9
Norway	55.4	18.0	54	3	453	3	1,650	20
Oman	9.9	17.6	56	5	430	5	1,018	4
Pakistan	37.2	4.8	151	10	671	22	4,929	12
Panama	13.3	17.8	218	11	929	9	9,000	561
Papua New Guinea	0.8	0.2	214	33	708	31	10,000	95
Paraguay	15.0	3.7	305	8	600	22	7,500	12
Peru	41.8	3.3	344	29	1,678	60	17,121	377
Philippines	6.1	2.0	153	22	556	86	8,000	371
Poland	30.7	10.1	79	4	421	5	2,300	27
Portugal	5.8	6.3	63	6	441	7	2,500	240
Puerto Rico	..	..	..	..	..	..	..	..
Romania	10.7	4.7	84	3	368	11	3,175	122
Russian Federation	705.4	3.9	..	17	..	35	..	127

part of their habitat, their long-term survival may be threatened by fragmentation and disturbance of the rest of the ecosystem. As habitats become smaller, the number of species they can support per unit area falls, and the populations of wide-ranging or more aggressive species (often introduced species) may expand at the expense of species with more specialized habitat requirements. In Hawaii, for example, the natural ecosystem has been under continuous attack by introduced plant and animal species. Some birds unique to Hawaii build their nests on the ground and are now threatened by pigs and mongoose, both introduced species. Species are also threatened by agricultural chemicals and by changes in water sources caused by human use.

Protected areas are on the rise

Protection of endangered ecosystems is one way to ensure that biodiversity and its many benefits are preserved for current and future generations. Since the first national park (Yellowstone National Park in the United States) was created 125 years ago, the number of established protected areas and the total area protected have grown dramatically (figure 3.4a). Recent growth of protected areas has been especially rapid in low- and middle-income countries. Almost 7 percent of the world's land area is now protected, still short of the 10–15 percent recommended by some experts.

The real question, however, is not just how many hectares of land are protected, but which land is protected and how effective the protection is. The extent of physical area covered does not necessarily mean that the most vulnerable species, genes, or ecosystems are protected. For this reason the selection and establishment of protected areas must be based on a partnership between experts on biodiversity, other potential users, and government. Experience has also demonstrated the need to work with surrounding communities rather than rely on a strict enforcement approach.

Knowledge of national biological resources and the extent to which they are endangered varies greatly. In most countries more is known about the number of birds and mammals (and the number endangered) than about the number of higher plants. Plant species, however, are often a more important source of potentially usable biological material than birds or mammals. Many poor or very poor countries are among those with the largest number of plant species (the so-called megadiversity countries). For example, Brazil and Colombia are estimated to have more than 50,000 species of higher plants apiece, and China, Indonesia, and Mexico each have more

3.4 | Biodiversity and protected areas

	Nationally protected areas		Mammals		Birds		Higher plants[a]	
	thousand sq. km 1994[b]	% of total land area 1994[b]	Species 1994[b]	Threatened species 1994[b]	Species 1994[b]	Threatened species 1994[b]	Species 1994[b]	Threatened species 1994[b]
Rwanda	3.3	13.3	151	14	666	6	2,288	..
Saudi Arabia	62.0	2.9	77	6	413	10	1,729	6
Senegal	21.8	11.3	155	9	610	5	2,062	32
Sierra Leone	0.8	1.1	147	12	622	12	2,090	12
Singapore	0.0	4.9	45	3	295	6	2,000	14
Slovak Republic	10.2	21.1	..	3	..	4	..	..
Slovenia	1.1	5.4	69	3	361	3	..	11
South Africa	69.7	5.7	247	25	790	16	23,000	953
Spain	42.5	8.5	82	7	506	10	..	896
Sri Lanka	8.0	12.3	88	4	428	11	3,000	436
Sudan	93.8	3.9	267	16	937	9	3,132	8
Sweden	29.8	7.3	60	3	463	4	4,916	19
Switzerland	7.3	18.5	75	2	400	3	1,650	9
Syrian Arab Republic	..	..	..	4	..	6	..	10
Tajikistan	0.9	0.6	..	6	..	9	..	..
Tanzania	139.4	15.7	322	16	1,005	30	10,000	406
Thailand	70.2	13.7	265	22	915	44	11,000	382
Togo	6.5	11.9	196	8	558	1	2,000	..
Trinidad and Tobago	0.2	3.1	100	1	433	2	1,982	16
Tunisia	0.4	0.3	78	5	356	6	2,150	24
Turkey	10.7	1.1	116	4	418	13	8,472	1,827
Turkmenistan	11.1	2.4	..	8	..	9	..	1
Uganda	19.1	9.6	338	15	992	10	5,000	6
Ukraine	4.9	0.9	..	4	..	10	2,927	16
United Arab Emirates	..	..	25	2	360	4	..	..
United Kingdom	51.1	21.2	50	1	219	2	1,550	28
United States	1,302.1	11.4	428	22	768	46	16,302	1,845
Uruguay	0.3	0.2	81	4	365	9	2,184	11
Uzbekistan	2.4	0.6	..	7	..	11	..	5
Venezuela	263.2	29.8	305	12	1,296	22	20,000	107
Vietnam	13.3	4.1	213	25	761	45	..	350
West Bank and Gaza	..	..	..	..	..	..	..	..
Yemen, Rep.	..	..	66	4	366	12	..	149
Yugoslavia, FR (Serb./Mont.)	3.5	3.4	..	..	..	..	..	..
Zambia	63.6	8.6	229	7	736	10	4,600	9
Zimbabwe	30.7	7.9	270	9	648	7	4,200	94

World	8,603.2 s	6.7 w
Low income	1,999.0	5.2
Excl. China & India	1,274.8	4.9
Middle income	2,994.1	5.1
Lower middle income	2,160.9	5.6
Upper middle income	833.3	4.1
Low & middle income	4,993.2	5.1
East Asia & Pacific	966.3	6.2
Europe & Central Asia	857.8	3.6
Latin America & Carib.	1,303.4	6.5
Middle East & N. Africa	290.8	3.0
South Asia	212.4	4.4
Sub-Saharan Africa	1,362.5	5.8
High income	3,610.1	11.9

a. Flowering plants only. b. Data may refer to earlier years. They are the most recent reported by the World Conservation Monitoring Center in 1994.

than 25,000 species. Many high-income countries have fewer than 5,000 species of higher plants. Therein lies a paradox: species diversity and the potential benefits of unexplored genetic material are concentrated in precisely those countries that can least afford to protect ecosystems and where protected areas and biodiversity tend to be under the greatest threat.

Conserving biodiversity while reaping its economic benefits

Habitat conservation provides only a partial answer to the challenge of conserving biodiversity. To also reap the benefits of biodiversity, complementarities must be sought between biodiversity protection and economic activities. Such complementarities are particularly important for agriculture, which depends on many services provided by the environment, such as crop pollination and genes for developing improved crop varieties and livestock breeds. Exploiting biodiversity could substantially boost agricultural production. At the same time, damage to biodiversity often hurts agriculture. Reconciling biodiversity conservation with increased production to meet the needs of growing human populations will be a major challenge.

Some countries have been quite successful in linking economic benefits and biodiversity conservation. Costa Rica, for example, has a rich mix of biodiversity packed into a small country. It now advertises itself as "the Natural Country" and has developed a major ecotourism industry focused on activities associated with protected areas. Foreign exchange earnings from tourism now rival those of traditional exports like coffee and bananas. In addition, tourism revenues help generate political support for continued management of Costa Rica's extensive protected area system, which covers almost 13 percent of the country.

Political and popular support are essential if conservation of protected areas and the biodiversity they contain are to be sustained over the long term. Information—on the value of biodiversity and on the direct economic benefits flowing from protected areas and their wise management—is probably the best way to build this consensus.

Biodiversity and protected areas │ 3.4

About the data

Habitat conservation is vital for stemming the decline in biodiversity. Conservation efforts traditionally have focused on protected areas, which have grown substantially in recent decades. Measures of species richness are one of the most straightforward ways to indicate how important an area is for biodiversity. The number of small plants and animals is usually estimated by sampling of plots. It is also important to know which aspects are under most immediate threat. This, however, requires a large amount of data and time-consuming analysis. For this reason, global analyses of threatened species status have been carried out for only a few groups of organisms. Birds are the only species for which the status of all members has been assessed. Mammals approach birds in this respect, but an estimated 45 percent of mammal species remain to be assessed.

The table shows information on protected areas, numbers of certain species, and numbers of those species under threat. The World Conservation Monitoring Centre (WCMC)—a joint venture of the United Nations Environment Programme (UNEP), World Wide Fund for Nature (WWF), and World Conservation Union (IUCN)—compiles these data from a variety of sources. Because of differences in definitions and reporting practices, cross-country comparability is limited. Compounding these problems, available data cover different periods.

Nationally protected areas are areas of at least 1,000 hectares that fall into one of five management categories defined by the WCMC:
• Scientific reserves and strict nature reserves with limited public access.
• National parks of national or international significance (not materially affected by human activity).
• Natural monuments and natural landscapes with unique aspects.
• Managed nature reserves and wildlife sanctuaries.
• Protected landscapes and seascapes (which may include cultural landscapes).

The first three categories, referred to as "totally protected," are areas maintained in a natural state and closed to extractive uses. The last two categories, referred to as "partially protected," are areas that may be managed for specific uses, such as recreation or tourism, or that provide optimal conditions for certain species or communities of wildlife. Some natural resource extraction is allowed within these areas. Designating land as a protected area does not necessarily mean, however, that protection is in force. For

small countries that may only have protected areas smaller than 1,000 hectares, this limit will result in an underestimate of the extent and number of protected areas.

Threatened species are defined according to the IUCN's classification categories: endangered (in danger of extinction and survival unlikely if causal factors continue operating), vulnerable (likely to move into the endangered category in the near future if causal factors continue operating), rare (not endangered or vulnerable, but at risk), indeterminate (known to be endangered, vulnerable, or rare but not enough information is available to say which), out of danger (formerly included in one of the above categories but now considered relatively secure because appropriate conservation measures are in effect), and insufficiently known (suspected but not definitely known to belong to one of the above categories).

Figures on species are not necessarily comparable across countries because taxonomic concepts and coverage vary. And while the number of mammals and birds is fairly well known, it is difficult to make an accurate account of plants. Although the data in the table should be interpreted with caution, especially for numbers of threatened species (where our knowledge is very incomplete), they do identify countries that are major sources of global biodiversity and show national commitments to habitat protection. Until a practical indicator of the effectiveness of protection is available, these data are the best available on the distribution and potential protection of global biodiversity resources.

Definitions

• **Nationally protected areas** are totally or partially protected areas of at least 1,000 hectares that are designed as national parks, natural monuments, nature reserves or wildlife sanctuaries, protected landscapes and seascapes, or scientific reserves with limited public access. The data do not include sites protected under local or provincial law. Total land area is used to calculate the percentage of total area protected (see table 3.1). • **Mammals** exclude whales and porpoises. • **Birds** are listed for countries included within their breeding or wintering ranges. • **Higher plants** refer to native vascular plant species. • **Threatened species** are the number of species classified by the IUCN as endangered, vulnerable, rare, indeterminate, out of danger, or insufficiently known.

Data sources

Data on protected areas are from the WCMC's Protected Areas Data Unit. Data on species are from the WCMC's *Biodiversity Data Sourcebook*, the WCMC's *Global Biodiversity: Status of the Earth's Living Resources*, and the IUCN's 1996 *Red List of Threatened Animals*.

Figure 3.4a

Worldwide, the number and coverage of protected areas have increased dramatically

Number of protected areas / Area protected (millions of hectares)

Source: Dixon and Sherman 1990 and IUCN.

3.5 | Freshwater

	Freshwater resources	Annual freshwater withdrawals					Access to safe water			
	cubic meters per capita **1996**	billion cu. m[a]	% of total resources[a]	% for agriculture[b]	% for industry[b]	% for domestic[b]	Urban % of population **1980**	**1995**	Rural % of population **1980**	**1995**
Albania	13,542	0.2[c]	0.4	76	18	6	..	..	..	..
Algeria	483	4.5	32.4	60[d]	15[d]	25[d]	100	..	70	..
Angola	16,577	0.5	0.3	76[d]	10[d]	14[d]	..	69	..	15
Argentina	19,705	27.6[c]	4.0	73	18	9	..	73	..	17
Armenia	2,411	3.8	41.8	72[d]	15[d]	13[d]	..	..	..	..
Australia	18,731	14.6[c]	4.3	33	2	65	..	..	..	..
Austria	6,986	2.4	4.2	9[d]	58[d]	33[d]	..	..	..	..
Azerbaijan	1,068	15.8	195.1	74[d]	22[d]	4[d]	..	..	..	..
Bangladesh	11,153	22.5	1.7	96	1	3	24	42	40	80
Belarus	3,612	3.0	8.1	19	49	32	..	..	..	..
Belgium	827	9.0	107.5	4	85	11	..	..	..	..
Benin	1,829	0.2	1.5	67[d]	10[d]	23[d]	..	41	..	53
Bolivia	39,536	1.2	0.4	85	5	10	..	75	..	27
Bosnia and Herzegovina	..	..	..	..	..	..	..	..	..	..
Botswana	1,959	0.1	3.8	48[d]	20[d]	32[d]	..	100	..	53
Brazil	32,163	36.5	0.7	59	19	22	..	85	..	31
Bulgaria	2,154	13.9	77.2	22	76	3	..	..	..	..
Burkina Faso	1,640	0.4	2.2	81[d]	0[d]	19[d]	..	..	..	..
Burundi	561	0.1	2.8	64[d]	0[d]	36[d]	..	..	..	..
Cambodia	8,574	0.5	0.6	94	1	5	..	20	..	12
Cameroon	19,596	0.4	0.1	35[d]	19[d]	46[d]	..	71	..	24
Canada	95,097	45.1	1.6	12	70	18	..	..	..	..
Central African Republic	42,166	0.1	0.0	74[d]	5[d]	21[d]	..	18	5	18
Chad	2,269	0.2	1.2	82[d]	2[d]	16[d]	27	48	30	17
Chile	32,458	16.8[c]	3.6	89	5	6	..	..	..	..
China	2,304	460.0	16.4	87	7	6	..	93	..	89
Hong Kong, China	..	..	..	..	..	..	..	..	..	..
Colombia	28,571	5.3	0.5	43	16	41	..	88	..	48
Congo, Dem. Rep.	20,670	0.4	0.0	23[d]	16[d]	61[d]	..	..	..	..
Congo, Rep.	345,619	0.0	0.0	11[d]	27[d]	62[d]	..	..	..	..
Costa Rica	27,600	1.4[c]	1.4	89	7	4	..	..	..	..
Côte d'Ivoire	5,346	0.7	0.9	67[d]	11[d]	22[d]	30	59	10	81
Croatia	12,870	..	..	..	..	..	..	98	..	80
Cuba	3,131	8.1[c]	23.5	89	2	9	..	96	..	85
Czech Republic	5,642	2.7	4.7	2[d]	57[d]	41[d]	..	..	..	..
Denmark	2,090	1.2	10.9	43	27	30	..	100	..	100
Dominican Republic	2,511	3.0	14.9	89	6	5	..	74	..	67
Ecuador	26,842	5.6	1.8	90	3	7	..	82	..	55
Egypt, Arab Rep.	47	55.1	1,967.9	86[d]	8[d]	6[d]	93	82	61	50
El Salvador	3,270	1.0[c]	5.3	89	4	7	..	78	..	37
Eritrea	757	..	..	..	..	..	..	..	..	..
Estonia	8,663	3.3	26.0	3[d]	92[d]	5[d]	..	..	..	..
Ethiopia	1,889	2.2	2.0	86[d]	3[d]	11[d]	..	90	..	20
Finland	21,463	2.2	2.0	3	85	12	..	100	..	100
France	3,084	37.7	21.0	15	69	16	..	100	..	100
Gabon	145,778	0.1	0.0	6[d]	22[d]	72[d]	75	80	34	30
Gambia, The	2,616	0.0	0.7	91[d]	2[d]	7[d]	100	..	27	..
Georgia	10,737	4.0	6.9	42[d]	37[d]	21[d]	..	..	..	..
Germany	1,172	46.3	48.2	20[d]	70[d]	11[d]	..	..	..	..
Ghana	1,729	0.3[c]	1.0	52[d]	13[d]	35[d]	..	70	..	49
Greece	4,310	5.0	11.2	63	29	8	..	..	..	..
Guatemala	10,615	0.7[c]	0.6	74	17	9	..	91	..	43
Guinea	33,436	0.7	0.3	87[d]	3[d]	10[d]	..	61	20	62
Guinea-Bissau	14,628	0.0	0.1	36[d]	4[d]	60[d]	..	18	..	27
Haiti	1,499	0.0	0.4	68	8	24	..	37	..	23
Honduras	9,084	1.5	2.7	91	5	4	..	81	..	53

Freshwater | 3.5

	Freshwater resources	Annual freshwater withdrawals					Access to safe water			
	cubic meters per capita **1996**	billion cu. m[a]	% of total resources[a]	% for agriculture[b]	% for industry[b]	% for domestic[b]	Urban % of population **1980**	**1995**	Rural % of population **1980**	**1995**
Hungary	589	6.8	113.5	36	55	9	..	..	..	..
India	1,957	380.0[c]	20.5	93	4	3	72	85	..	79
Indonesia	12,839	16.6	0.7	76	11	13	..	78	..	54
Iran, Islamic Rep.	2,051	70.0[c]	54.6	92[d]	2[d]	6[d]	70	..	33	..
Iraq	1,647	42.8[c]	121.6	92[d]	5[d]	3[d]	92	..	22	..
Ireland	12,962	0.8[c]	1.7	10	74	16	..	..	..	..
Israel	299	1.9	108.8	79[d]	5[d]	16[d]	..	..	..	..
Italy	2,778	56.2	35.3	59	27	14	..	..	..	..
Jamaica	3,259	0.3[c]	3.9	86	7	7	..	92	..	48
Japan	4,350	90.8	16.6	50	33	17	..	..	..	..
Jordan	158	0.5[c]	66.2	75[d]	3[d]	22[d]	100	..	65	..
Kazakhstan	4,579	37.9	50.3	79[d]	17[d]	4[d]	..	..	..	..
Kenya	738	2.1	10.1	76[d]	4[d]	20[d]	..	67	..	49
Korea, Dem. Rep.	2,984	14.2	21.1	73	16	11	..	100	..	100
Korea, Rep.	1,451	27.6	41.8	46	35	19	..	..	..	..
Kuwait	0	0.5	..	60[d]	2[d]	38[d]	100	..	100	..
Kyrgyz Republic	10,315	11.0	23.4	95[d]	3[d]	2[d]	..	..	..	..
Lao PDR	9,840	1.0	2.1	82	10	8	..	40	..	39
Latvia	6,707	0.7	4.2	14[d]	44[d]	42[d]	..	..	..	..
Lebanon	1,030	1.3[c]	30.7	68[d]	4[d]	28[d]	95	..	85	..
Lesotho	2,571	0.1	1.0	56[d]	22[d]	22[d]	37	14	14	64
Libya	116	4.6	766.7	87[d]	2[d]	11[d]	..	..	..	..
Lithuania	4,206	4.4	28.2	3	90	7	..	..	..	..
Macedonia, FYR	..	..	..	..	..	..	..	..	..	..
Madagascar	24,590	16.3	4.8	99[d]	0[d]	1[d]	..	83	..	10
Malawi	1,747	0.9	5.1	86[d]	3[d]	10[d]	60	52	53	44
Malaysia	22,174	9.4[c]	2.1	47	30	23	..	100	..	74
Mali	6,001	1.4	2.3	97[d]	1[d]	2[d]	58	36	20	38
Mauritania	171	1.6[c]	407.5	92[d]	2[d]	6[d]	..	..	..	..
Mauritius	1,940	0.4[c]	16.4	77[d]	7[d]	16[d]	100	95	98	100
Mexico	3,836	77.6[c]	21.7	86	8	6	..	91	..	62
Moldova	231	3.7	370.0	23	70	7	..	..	..	..
Mongolia	9,776	0.6	2.2	62	27	11	..	..	..	..
Morocco	1,110	10.9	36.2	92[d]	3[d]	5[d]	63	98	2	14
Mozambique	5,547	0.6	0.6	89	2[d]	9[d]	82	17	2	40
Myanmar	23,582	4.0	0.4	90	3	7	..	36	..	39
Namibia	3,913	0.3	4.0	68[d]	3[d]	29[d]	..	..	..	..
Nepal	7,714	2.7	1.6	95	1	4	75	64	6	49
Netherlands	644	7.8	78.1	34	61	5	..	100	..	100
New Zealand	89,959	2.0	0.6	44	10	46	..	..	..	..
Nicaragua	38,862	0.9[c]	0.5	54	21	25	..	81	..	27
Niger	375	0.5	14.3	82[d]	2[d]	16[d]	..	46	40	55
Nigeria	1,929	3.6	1.6	54[d]	15[d]	31[d]	..	63	..	26
Norway	87,651	2.0	0.5	8	72	20	..	..	..	..
Oman	456	1.2	123.2	93[d]	2[d]	5[d]	70	..	10	..
Pakistan	1,858	155.6[c]	62.7	96[d]	2[d]	2[d]	77	77	22	52
Panama	53,852	1.3	0.9	77	11	12	..	..	..	..
Papua New Guinea	181,993	0.1	0.0	49	22	29	..	84	..	17
Paraguay	18,971	0.4	0.5	78	7	15	..	..	..	17
Peru	1,647	6.1	15.3	72	9	19	..	74	..	24
Philippines	4,492	29.5[c]	9.1	61	21	18	..	..	..	..
Poland	1,279	12.3	24.9	11	76	13	..	..	..	..
Portugal	3,827	7.3	19.2	48	37	15	..	..	..	..
Puerto Rico	..	..	..	..	..	..	..	..	..	..
Romania	1,637	26.0	70.3	59	33	8	..	..	..	..
Russian Federation	29,191	117.0	2.7	23[d]	60[d]	17[d]	..	..	..	..

3.5 | Freshwater

	Freshwater resources	Annual freshwater withdrawals					Access to safe water			
							Urban % of population		Rural % of population	
	cubic meters per capita 1996	billion cu. m[a]	% of total resources[a]	% for agriculture[b]	% for industry[b]	% for domestic[b]	1980	1995	1980	1995
Rwanda	937	0.8	12.2	94[d]	2[d]	5[d]	..	..	..	..
Saudi Arabia	124	17.0[c]	709.2	90[d]	1[d]	9[d]	92	..	87	..
Senegal	3,093	1.4	5.2	92[d]	3[d]	5[d]	..	82	..	28
Sierra Leone	34,557	0.4	0.2	89[d]	4[d]	7[d]	50	58	3	21
Singapore	197	0.2[c]	31.7	4	51	45	100	100	..	..
Slovak Republic	5,765	1.8	5.8	..	..	..	..	..	..	..
Slovenia	..	..	..	..	..	..	..	..	..	..
South Africa	1,190	13.3	29.7	72[d]	11[d]	17[d]	..	..	..	..
Spain	2,809	30.8	27.9	62	26	12	..	..	..	..
Sri Lanka	2,361	6.3[c]	14.6	96	2	2	..	..	..	..
Sudan	1,283	17.8	50.9	94[d]	1[d]	4[d]	49	66	45	45
Sweden	19,903	2.9	1.7	9	55	36	..	..	..	..
Switzerland	6,008	1.2	2.8	4	73	23	..	100	..	100
Syrian Arab Republic	483	14.4	205.9	94[d]	2[d]	4[d]	77	92	65	78
Tajikistan	11,186	12.6	19.0	88[d]	7[d]	5[d]	..	..	..	..
Tanzania	2,623	1.2	1.5	89[d]	2[d]	9[d]	88	65	40	45
Thailand	1,833	31.9	29.0	90	6	4	..	89	..	72
Togo	2,719	0.1	0.8	25[d]	13[d]	62[d]	..	..	..	..
Trinidad and Tobago	3,932	0.2[c]	2.9	35	38	27	..	83	..	80
Tunisia	385	3.1	87.2	89[d]	3[d]	9[d]	96	..	29	..
Turkey	3,126	31.6	16.1	72[d]	11[d]	16[d]	..	98	62	85
Turkmenistan	217	22.8	2,280.0	91	8	1	..	..	..	..
Uganda	1,976	0.2	0.5	60	8	32	..	47	..	32
Ukraine	1,047	34.7	65.3	30	54	16	..	..	..	..
United Arab Emirates	59	2.1	1,406.7	92[d]	1[d]	7[d]	100	98	100	98
United Kingdom	1,208	11.8	16.6	3	77	20	..	100	..	100
United States	9,270	467.3	19.0	42[d]	45[d]	13[d]	..	..	..	..
Uruguay	18,420	0.7[c]	1.1	91	3	6	..	..	..	..
Uzbekistan	702	82.2	504.3	84[d]	12[d]	4[d]	..	..	..	..
Venezuela	38,367	4.1[c]	0.5	46	11	43	..	80	..	75
Vietnam	4,990	28.9	7.7	78	9	13	..	53	..	32
West Bank and Gaza	..	..	..	..	..	..	..	..	..	..
Yemen, Rep.	260	2.9	71.5	92[d]	1[d]	7[d]	..	88	..	17
Yugoslavia, FR (Serb./Mont.)	..	..	..	..	..	..	..	..	..	..
Zambia	8,703	1.7	2.1	77[d]	7[d]	16[d]	..	64	..	27
Zimbabwe	1,254	1.2	8.7	79[d]	7[d]	14[d]	..	99	..	65
World	7,342 w	..	..	68 w	22 w	10 w	.. w	.. w	.. w	.. w
Low income	4,089	..	..	90	5	4	..	..	..	..
Excl. China & India	8,295	..	..	92	4	4	..	..	..	..
Middle income	12,719	..	..	66	23	11	..	..	..	..
Lower middle income	11,154	..	..	66	24	10	..	..	..	..
Upper middle income	16,447	..	..	68	17	14	..	..	..	..
Low & middle income	6,961	..	..	80	13	7	..	..	..	..
East Asia & Pacific	5,072	..	..	84	8	7	..	89	..	82
Europe & Central Asia	11,411	..	..	52	37	11	..	..	..	..
Latin America & Carib.	22,011	..	..	77	11	12	..	..	..	..
Middle East & N. Africa	854	..	..	84	8	8	84	..	45	..
South Asia	3,017	..	..	95	3	2	70	83	..	74
Sub-Saharan Africa	7,821	..	..	85	4	10	..	..	..	..
High income	9,378	..	..	40	45	15	..	..	..	..

a. Data refer to any year from 1980 to 1996 unless otherwise noted. b. Unless otherwise noted, sectoral withdrawal shares are estimated for 1987. Data may not sum to 100 percent because of rounding. c. Data refer to estimates for years before 1980 (see *Primary data documentation*). d. Data refer to years other than 1987 (see *Primary data documentation*).

Freshwater | 3.5

About the data

Data on freshwater resources are based on estimates of runoff into rivers and recharge of groundwater. These estimates are based on different sources and refer to different years, so cross-country comparisons of data on freshwater resources should be made with caution. Because they are collected intermittently, the data may hide significant variations in total renewable water resources from one year to the next. The data also fail to distinguish between seasonal and geographic variations in water availability within countries. Data for small countries and countries in arid and semiarid zones are less reliable than those for larger countries and countries with higher rainfall. Finally, caution is needed in comparing data on annual freshwater withdrawal, which are subject to variations in collection and estimation methods.

While information on access to safe water is widely used, it is extremely subjective, and such terms as "adequate amount" and "safe" may have very different meanings in different countries despite official World Health Organization definitions (see *Definitions* for table 2.14). Even in industrial countries treated water may not always be safe to drink. While access to safe water is equated with connection to a public supply system, this does not take account of variations in the quality and cost (broadly defined) of the service once connected. Thus cross-country comparisons must be made cautiously. Changes over time within countries may result from changes in definitions or measurements.

Figure 3.5a

Annual renewable freshwater resources are coming under increasing strain

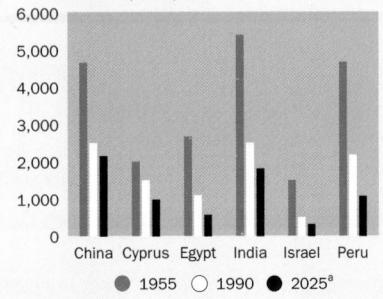

Cubic meters per capita

● 1955 ○ 1990 ● 2025ª

a. Projected.
Source: Serageldin 1995.

Global per capita water supplies are a third lower than they were 25 years ago. The reason? The world's population has increased by some 2 billion people since then. By 2025 further increases are expected to boost demand for water by more than 650 percent, leaving many countries subject to periodic water stress (defined as annual per capita availability of less than 1,700 cubic meters). Today 25 countries have renewable water resources of less than 1,000 cubic meters per capita, and another 27 have less than 2,000. By 2025, 52 countries inhabited by some 3 billion people are expected to suffer from periodic water stress or chronic water scarcity.

Figure 3.5b

Agriculture drinks up a lot of the world's freshwater

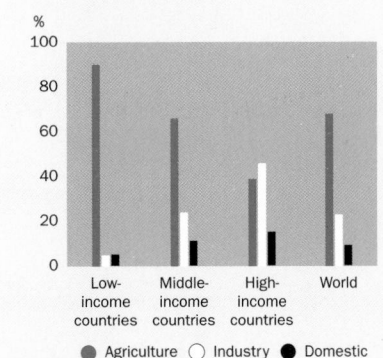

%

● Agriculture ○ Industry ● Domestic

Source: Table 3.5.

Agriculture uses two-thirds of the world's freshwater. Although irrigated agriculture has accounted for much of the dramatic increase in world food supplies over the past 20 years, constraints on water supplies will limit its capacity to do so in the future. In low-income countries, where an enormous share of water goes to agriculture, efforts must be made to increase efficiency.

Definitions

• **Freshwater resources** refer to internal renewable resources, which include flows of rivers and groundwater from rainfall in the country but not river flows from other countries. Freshwater resources per capita are calculated using the World Bank's population estimates (see table 2.1). • **Annual freshwater withdrawals** refer to total water withdrawal, not counting evaporation losses from storage basins. Withdrawals also include water from desalination plants in countries where they are a significant source of water withdrawals. Withdrawal data are for single years between 1980 and 1996 unless otherwise indicated. Withdrawals can exceed 100 percent of renewable supplies when extraction from nonrenewable aquifers or desalination plants is considerable, when river flows from other countries are used substantially, or if there is significant water reuse. Withdrawals for agriculture and industry are total withdrawals for irrigation and livestock production and for direct industrial use (including withdrawals for cooling thermoelectric plants). Withdrawals for domestic uses include drinking water, municipal use or supply, and use for public services, commercial establishments, and homes. For most countries sectoral withdrawal data are estimated for 1987–95. • **Access to safe water** refers to the percentage of people with reasonable access to an adequate amount of safe drinking water in a dwelling or within a convenient distance of their dwelling (see *About the data*).

Data sources

Data are compiled by the World Resources Institute from various sources and published in its *World Resources 1998–99*. The Département Hydrogéologie in Orléans, France, compiles data on water resources and withdrawals from published documents, including national, United Nations, and professional literature. The Institute of Geography at the National Academy of Sciences in Moscow also compiles global water data on the basis of published work and, where necessary, estimates water resources and consumption from models that use other data, such as area under irrigation, livestock populations, and precipitation.

3.6 | Water pollution

	Emissions of organic water pollutants				Industry shares of emissions of organic water pollutants							
	kilograms per day		kilograms per day per worker		Primary metals %	Paper and pulp %	Chemicals %	Food and beverages %	Stone, ceramics, and glass %	Textiles %	Wood %	Other %
	1980	1993	1980	1993	1993	1993	1993	1993	1993	1993	1993	1993
Albania	..	2,431	..	0.06	..	91.8	..	..	7.5	..	..	0.7
Algeria	60,290	..	0.19	..	..	..	..	..	..	..	..	..
Angola	..	..	..	..	..	..	..	..	..	..	..	..
Argentina	244,711	179,432	0.18	0.20	7.7	11.8	7.7	57.4	0.3	8.6	1.5	5.1
Armenia	..	34,982	..	0.11	6.4	3.9	5.7	35.1	0.5	31.4	1.4	15.5
Australia	204,333	173,490	0.18	0.19	12.4	22.7	6.8	43.4	0.2	5.3	2.8	6.4
Austria	108,416	86,692	0.16	0.15	14.6	19.9	9.4	34.3	0.3	6.9	3.9	10.7
Azerbaijan	..	..	..	..	..	..	..	..	..	..	..	..
Bangladesh	66,713	171,087	0.16	0.17	2.9	7.2	4.1	36.9	0.1	47.1	0.7	1.1
Belarus	..	..	..	..	..	..	..	..	..	..	..	..
Belgium	136,452	113,460	0.16	0.16	14.4	17.7	11.6	36.8	0.2	8.8	2.0	8.4
Benin	1,646	..	0.28	..	..	..	..	..	..	..	..	..
Bolivia	9,343	5,724	0.22	0.24	5.8	11.8	7.3	62.6	0.3	8.6	2.7	0.9
Bosnia and Herzegovina	..	..	..	..	..	..	..	..	..	..	..	..
Botswana	1,106	2,598	0.31	0.25	..	..	..	99.6	..	..	..	0.4
Brazil	866,790	855,432	0.16	0.17	10.4	13.5	9.1	45.8	0.3	11.5	3.0	6.4
Bulgaria	151,016	111,310	0.14	0.14	11.5	6.2	16.2	42.1	0.3	14.8	2.0	6.9
Burkina Faso	2,347	..	0.29	..	..	..	..	..	..	..	..	..
Burundi	756	1,617	0.23	0.26	..	8.4	4.8	68.9	0.1	17.0	..	0.8
Cambodia	..	11,881	..	0.17	..	3.5	3.3	60.1	0.6	25.1	5.9	1.5
Cameroon	14,280	13,029	0.29	0.30	3.8	5.9	1.6	79.4	0.0	5.1	3.6	0.5
Canada	330,241	300,071	0.18	0.18	10.1	30.1	8.7	34.5	0.1	5.9	3.3	7.3
Central African Republic	760	749	0.31	0.19	..	8.2	6.5	66.6	..	..	18.1	0.6
Chad	..	..	..	..	..	..	..	..	..	..	..	..
Chile	44,371	82,825	0.21	0.24	6.6	10.1	6.4	65.0	0.1	7.4	1.9	2.5
China	3,358,203	5,339,072	0.14	0.15	22.0	10.0	14.0	33.3	0.4	11.5	0.4	8.3
Hong Kong, China	102,002	86,140	0.11	0.13	1.1	28.5	7.0	14.8	0.0	39.9	0.5	8.3
Colombia	96,055	97,024	0.19	0.19	4.0	13.2	11.3	52.1	0.3	14.6	1.0	3.5
Congo, Dem. Rep.	..	..	..	..	..	..	..	..	..	..	..	..
Congo, Rep.	848	..	0.37	..	..	..	..	..	..	..	..	..
Costa Rica	..	27,624	..	0.20	0.4	10.1	8.6	58.6	0.1	18.6	1.6	2.0
Côte d'Ivoire	13,898	..	0.24	..	..	..	..	..	..	..	..	..
Croatia	..	55,440	..	0.16	9.3	13.8	8.9	43.1	0.3	14.2	3.7	6.8
Cuba	114,708	..	0.28	..	..	..	..	..	..	..	..	..
Czech Republic	..	171,227	..	0.13	24.5	9.9	6.7	31.3	0.4	12.5	2.3	12.4
Denmark	65,465	87,244	0.17	0.18	2.3	28.0	7.3	47.7	0.1	3.7	2.7	8.1
Dominican Republic	54,935	..	0.38	..	..	..	..	..	..	..	..	..
Ecuador	25,297	28,053	0.23	0.22	2.8	11.9	10.6	62.1	0.2	8.4	1.7	2.4
Egypt, Arab Rep.	169,146	198,373	0.19	0.19	11.7	7.1	9.1	50.5	0.3	17.5	0.5	3.5
El Salvador	9,390	7,663	0.24	0.22	7.5	12.0	9.9	49.6	0.1	19.4	0.5	1.1
Eritrea	..	..	..	..	..	..	..	..	..	..	..	..
Estonia	..	..	..	..	..	..	..	..	..	..	..	..
Ethiopia	..	18,593	..	0.23	2.1	9.5	2.4	59.0	0.1	24.9	1.5	0.4
Finland	92,275	68,255	0.17	0.19	7.8	41.3	6.5	31.1	0.2	3.3	3.3	6.6
France	716,285	609,940	0.14	0.15	11.9	20.7	11.0	37.0	0.2	6.7	1.8	10.8
Gabon	2,661	..	0.15	..	..	..	..	..	..	..	..	..
Gambia, The	549	..	0.30	..	..	..	..	..	..	..	..	..
Georgia	..	..	..	..	..	..	..	..	..	..	..	..
Germany	..	1,046,176	..	0.12	15.6	15.3	15.1	27.9	0.2	6.4	2.0	17.6
Ghana	15,868	..	0.20	..	..	..	..	..	..	..	..	..
Greece	65,304	59,701	0.17	0.19	6.0	12.1	8.0	51.8	0.3	16.6	1.5	3.8
Guatemala	20,856	22,606	0.25	0.24	2.1	8.7	8.7	67.3	0.2	10.5	1.4	1.1
Guinea	..	..	..	..	..	..	..	..	..	..	..	..
Guinea-Bissau	..	..	..	..	..	..	..	..	..	..	..	..
Haiti	4,734	..	0.19	..	..	..	..	..	..	..	..	..
Honduras	12,395	27,565	0.23	0.22	0.7	7.3	6.7	69.8	0.1	5.5	8.6	1.2

Water pollution | 3.6

	Emissions of organic water pollutants				Industry shares of emissions of organic water pollutants							
	kilograms per day		kilograms per day per worker		Primary metals %	Paper and pulp %	Chemicals %	Food and beverages %	Stone, ceramics, and glass %	Textiles %	Wood %	Other %
	1980	1993	1980	1993	1993	1993	1993	1993	1993	1993	1993	1993
Hungary	201,888	151,311	0.15	0.18	9.9	7.6	8.1	54.9	0.2	10.8	1.8	6.8
India	1,457,474	1,441,293	0.21	0.20	15.6	8.1	7.3	50.9	0.2	12.9	0.3	4.8
Indonesia	214,010	537,142	0.22	0.19	..	7.8	10.4	58.9	0.2	15.4	4.8	2.6
Iran, Islamic Rep.	72,334	101,763	0.15	0.16	21.7	7.8	7.9	38.2	0.6	17.6	0.8	5.5
Iraq	31,805	17,882	0.18	0.15	..	15.4	16.6	43.2	0.8	18.3	0.4	5.2
Ireland	43,544	33,417	0.19	0.17	1.6	17.3	9.6	54.5	0.2	7.5	1.5	7.7
Israel	39,113	50,030	0.15	0.16	4.1	19.3	8.4	44.3	0.2	12.3	2.1	9.3
Italy	442,712	353,906	0.13	0.13	17.0	16.1	10.5	25.8	0.3	16.1	2.1	12.1
Jamaica	11,123	17,752	0.25	0.27	0.7	7.3	4.6	75.4	0.1	10.0	1.1	0.8
Japan	1,456,016	1,548,021	0.14	0.14	9.9	22.0	8.8	36.5	0.2	7.9	1.9	12.8
Jordan	4,146	11,166	0.17	0.17	4.1	15.3	15.9	49.8	0.7	7.6	3.4	3.3
Kazakhstan	..	..	..	..	..	..	..	..	..	..	..	..
Kenya	26,150	44,065	0.19	0.23	..	11.5	5.6	68.6	0.1	9.1	1.9	3.2
Korea, Dem. Rep.	..	..	..	..	..	..	..	..	..	..	..	..
Korea, Rep.	281,900	358,610	0.14	0.13	12.8	15.4	11.2	25.8	0.3	20.8	1.5	12.2
Kuwait	6,921	9,052	0.16	0.16	2.5	16.1	11.4	47.9	0.4	12.4	3.7	5.5
Kyrgyz Republic	..	25,426	..	0.19	14.2	3.0	1.1	53.5	0.5	26.1	1.5	..
Lao PDR	..	..	..	..	..	..	..	..	..	..	..	..
Latvia	..	42,866	..	0.15	2.4	7.9	5.3	57.2	0.3	14.5	3.8	8.5
Lebanon	13,137	..	0.24	..	..	..	..	..	..	..	..	..
Lesotho	190	87	0.11	0.09	..	69.5	27.6	..	2.0	..	..	0.9
Libya	3,532	..	0.21	..	..	..	..	..	..	..	..	..
Lithuania	..	..	..	..	..	..	..	..	..	..	..	..
Macedonia, FYR	..	29,054	..	0.17	16.6	8.4	6.0	37.7	0.1	24.5	2.0	4.7
Madagascar	9,196	..	0.23	..	..	..	..	..	..	..	..	..
Malawi	12,224	..	0.32	..	..	..	..	..	..	..	..	..
Malaysia	77,215	136,055	0.15	0.12	6.8	14.3	15.2	31.8	0.2	11.1	7.6	13.1
Mali	1,774	..	0.30	..	..	..	..	..	..	..	..	..
Mauritania	..	..	..	..	..	..	..	..	..	..	..	..
Mauritius	8,949	18,229	0.21	0.15	1.5	4.2	2.3	35.4	0.1	54.6	0.9	1.0
Mexico	130,993	167,335	0.22	0.18	11.0	9.8	12.6	51.9	0.3	7.6	0.5	6.4
Moldova	..	54,263	..	0.17	2.1	3.2	1.5	69.0	0.3	15.0	1.7	7.1
Mongolia	2,376	..	0.15	..	..	..	..	..	..	..	..	..
Morocco	26,598	33,752	0.15	0.14	..	..	..	36.3	..	57.5	..	6.2
Mozambique	..	0	..	0.00	..	..	..	..	..	..	..	..
Myanmar	..	..	..	..	..	..	..	..	..	..	..	..
Namibia	..	..	..	..	..	..	..	..	..	..	..	..
Nepal	18,692	28,860	0.25	0.14	1.9	5.5	3.4	46.5	1.6	39.0	1.5	0.7
Netherlands	165,416	136,071	0.18	0.18	..	28.1	12.2	46.4	0.0	2.1	1.5	9.7
New Zealand	59,012	45,849	0.21	0.22	4.6	20.4	5.3	56.5	0.1	6.4	2.5	4.3
Nicaragua	9,647	..	0.28	..	..	..	..	..	..	..	..	..
Niger	372	..	0.19	..	..	..	..	..	..	..	..	..
Nigeria	72,082	..	0.17	..	..	..	..	..	..	..	..	..
Norway	67,897	47,494	0.19	0.20	11.5	33.1	5.5	38.3	0.1	1.8	2.4	7.3
Oman	..	..	..	0.00	..	..	..	..	..	..	..	..
Pakistan	75,125	..	0.17	..	..	..	..	..	..	..	..	..
Panama	8,207	10,821	0.26	0.28	1.1	8.2	5.5	76.4	0.2	6.7	1.2	0.7
Papua New Guinea	4,222	..	0.22	..	..	..	..	..	..	..	..	..
Paraguay	..	3,250	..	0.28	2.3	9.9	6.0	73.6	0.3	6.7	0.3	0.9
Peru	50,367	..	0.18	..	..	..	..	..	..	..	..	..
Philippines	182,052	181,714	0.19	0.19	4.1	8.1	6.9	52.9	0.1	21.4	3.7	2.7
Poland	580,869	365,580	0.14	0.16	13.4	6.6	7.7	48.4	0.3	12.5	2.4	8.7
Portugal	105,441	77,451	0.16	0.16	6.0	8.7	0.4	44.7	0.4	31.4	3.8	4.6
Puerto Rico	23,224	23,466	0.15	0.15	..	10.2	15.2	46.8	0.2	19.6	0.8	7.3
Romania	352,368	146,154	0.12	0.08	..	16.1	2.9	1.8	..	41.4	11.3	26.6
Russian Federation	..	..	..	..	..	..	..	..	..	..	..	..

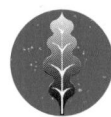

3.6 | Water pollution

	Emissions of organic water pollutants				Industry shares of emissions of organic water pollutants							
	kilograms per day		kilograms per day per worker		Primary metals %	Paper and pulp %	Chemicals %	Food and beverages %	Stone, ceramics, and glass %	Textiles %	Wood %	Other %
	1980	1993	1980	1993	1993	1993	1993	1993	1993	1993	1993	1993
Rwanda	..	..	..	..	..	..	..	..	..	..	..	..
Saudi Arabia	18,181	..	0.12	0.00	..	..	..	..	..	..	..	..
Senegal	9,865	..	0.31	..	..	..	..	..	..	..	..	..
Sierra Leone	..	..	..	..	..	..	..	..	..	..	..	..
Singapore	28,120	32,410	0.10	0.09	2.7	26.0	10.8	19.9	0.1	10.9	1.8	27.7
Slovak Republic	..	..	..	..	..	..	..	..	..	..	..	..
Slovenia	..	39,846	..	0.15	17.7	17.1	8.4	25.3	..	17.6	4.1	9.8
South Africa	237,599	251,142	0.17	0.17	12.7	16.7	9.5	41.3	0.2	11.0	2.7	5.9
Spain	376,253	318,506	0.16	0.17	10.7	15.4	9.3	45.6	0.3	8.7	2.8	7.2
Sri Lanka	30,086	51,328	0.18	0.19	1.4	8.0	6.4	52.0	0.2	29.9	0.7	1.2
Sudan	..	..	..	..	..	..	..	..	..	..	..	..
Sweden	130,439	102,341	0.15	0.16	11.6	36.4	7.4	27.9	0.1	1.5	3.1	11.9
Switzerland	..	38,161	..	0.15	..	76.9	..	..	..	8.0	..	15.1
Syrian Arab Republic	36,262	23,754	0.19	0.19	2.8	1.4	7.4	65.6	0.5	16.0	5.4	0.9
Tajikistan	..	..	..	..	..	..	..	..	..	..	..	..
Tanzania	20,648	..	0.21	..	..	..	..	..	..	..	..	..
Thailand	214,426	256,930	0.22	0.16	6.3	7.6	6.8	46.4	0.3	26.4	1.9	4.3
Togo	963	..	0.27	..	..	..	..	..	..	..	..	..
Trinidad and Tobago	6,737	7,384	0.19	0.25	..	13.9	6.1	70.1	0.2	6.2	1.3	2.2
Tunisia	20,294	25,610	0.16	0.19	5.6	5.6	5.1	62.7	0.8	17.4	0.7	2.1
Turkey	160,173	168,548	0.20	0.19	15.8	8.0	7.0	46.6	0.3	17.0	0.7	4.5
Turkmenistan	..	..	..	..	..	..	..	..	..	..	..	..
Uganda	..	..	..	..	..	..	..	..	..	..	..	..
Ukraine	..	666,233	..	0.14	18.3	3.7	7.3	46.9	0.5	10.1	2.0	11.3
United Arab Emirates	4,524	..	0.15	..	..	..	..	..	..	..	..	..
United Kingdom	964,510	680,865	0.15	0.16	8.9	24.7	10.1	37.1	0.2	7.2	1.7	10.0
United States	2,742,993	2,477,830	0.14	0.15	8.3	32.7	9.5	28.2	0.1	7.8	2.4	10.8
Uruguay	34,270	37,825	0.21	0.23	1.4	11.1	5.4	65.1	0.2	13.9	1.0	1.9
Uzbekistan	..	..	..	..	..	..	..	..	..	..	..	..
Venezuela	84,797	103,961	0.20	0.21	14.3	13.2	10.4	48.2	0.2	8.7	1.4	3.5
Vietnam	..	..	..	..	..	..	..	..	..	..	..	..
West Bank and Gaza	..	..	..	..	..	..	..	..	..	..	..	..
Yemen, Rep.	..	..	..	..	..	..	..	..	..	..	..	..
Yugoslavia, FR (Serb./Mont.)	..	..	..	..	..	..	..	..	..	..	..	..
Zambia	13,605	13,453	0.23	0.23	3.5	9.4	7.5	63.4	0.2	12.4	1.6	2.1
Zimbabwe	32,681	35,535	0.20	0.21	13.5	12.5	4.7	48.7	0.1	15.8	1.9	2.8
World	**18,745,247 s**	..	..	..	..	..	..	..	..	..	..	..
Low income	5,286,072	..	..	..	..	..	..	..	..	..	..	..
Excl. China & India	470,396	..	..	..	..	..	..	..	..	..	..	..
Middle income	4,461,968	..	..	..	..	..	..	..	..	..	..	..
Lower middle income	1,992,457	..	..	..	..	..	..	..	..	..	..	..
Upper middle income	2,469,512	..	..	..	..	..	..	..	..	..	..	..
Low & middle income	9,748,041	..	..	..	..	..	..	..	..	..	..	..
East Asia & Pacific	3,843,665	..	..	..	..	..	..	..	..	..	..	..
Europe & Central Asia	1,450,659	..	..	..	..	..	..	..	..	..	..	..
Latin America & Carib.	1,879,176	..	..	..	..	..	..	..	..	..	..	..
Middle East & N. Africa	437,544	..	..	..	..	..	..	..	..	..	..	..
South Asia	1,635,990	..	..	..	..	..	..	..	..	..	..	..
Sub-Saharan Africa	502,005	..	..	..	..	..	..	..	..	..	..	..
High income	8,997,206	..	..	..	..	..	..	..	..	..	..	..

Note: Industry shares of emissions may not sum to 100 percent because of rounding.

Water pollution | 3.6

About the data

Emissions of organic pollutants from industrial activities are a major source of water quality degradation. Water quality standards and pollution levels are generally measured in terms of concentration or load—the rate of occurrence of a substance in an aqueous solution. These substances include organic matter, metals, minerals, sediment, bacteria, and toxic chemicals. Because water pollution tends to be sensitive to local conditions, it is not meaningful to show national data for most pollutants. This table, however, focuses on organic water pollution resulting from industrial activities in a number of countries.

The data in the table come from an international study of industrial emissions that may be the first to include data from developing countries (Hettige, Mani, and Wheeler 1998). Unlike the estimates from engineering or economic models used in previous studies, these estimates are based on actual measurements of plant-level water pollution. The focus is on organic water pollution—measured by biochemical oxygen demand, or BOD—because it provides the most plentiful and reliable source of comparable cross-country emissions data. BOD measures the strength of an organic waste in terms of the amount of oxygen consumed in breaking it down. A sewage overload in natural waters exhausts the water's dissolved oxygen content. Wastewater treatment, by contrast, reduces BOD.

Data on water pollution are more readily available than other emissions data because most industrial pollution control programs start by regulating organic water emissions. Such data are fairly reliable because sampling techniques for measuring water pollution are more widely understood and much less expensive than those for air pollution.

In their study Hettige, Mani, and Wheeler (1998) used plant- and sector-level information on emissions and employment from 13 national environmental protection agencies and sector-level information on output and employment from the United Nations Industrial Development Organization (UNIDO). Their econometric analysis found that the ratio of BOD to employment in each industrial sector is about the same across countries. This finding allowed the authors to estimate BOD loads across countries and over time. The estimated BOD intensities per unit of employment were multiplied by sectoral employment numbers from UNIDO's industry database for 1975–93. The sectoral emissions estimates were then totaled to get daily BOD emissions in kilograms per day for each country and year.

Definitions

• **Emissions of organic water pollutants** are measured by biochemical oxygen demand, which refers to the amount of oxygen that bacteria in water will consume in breaking down waste. This is a standard water-treatment test for the presence of organic pollutants. • **Emissions per worker** are total emissions divided by the number of industrial workers. • **Industry shares of emissions of organic water pollutants** refer to emissions from manufacturing activities as defined by two-digit divisions of the International Standard Industrial Classification (ISIC), revision 2: *primary metals* (ISIC division 37); *paper and pulp* (34), *chemicals* (35), *food and beverages* (31), *stone, ceramics, and glass* (36), *textiles* (32), *wood* (33), and *other* (38 and 39).

Data sources

The indicators are from a 1998 study by Hemamala Hettige, Muthukumara Mani, and David Wheeler, "Industrial Pollution in Economic Development: Kuznets Revisited" (available as "New Ideas in Pollution Regulation" on the World Bank website at http://www.worldbank.org/NIPR). Sectoral employment numbers are from UNIDO's industry database.

3.7 Energy production and use

	Commercial energy production		Commercial energy use			Commercial energy use per capita			Net energy imports	
	thousand metric tons of oil equivalent		thousand metric tons of oil equivalent		average annual % growth	kg of oil equivalent		average annual % growth	% of commercial energy use	
	1980	1995	1980	1995	1980–95	1980	1995	1980–95	1980	1995
Albania	3,053	940	2,674	1,020	–6.4	1,001	314	–7.7	–14	8
Algeria	66,730	109,257	12,078	24,346	4.2	647	866	1.4	–452	–349
Angola	7,700	26,189	937	959	0.5	133	89	–2.3	–722	–2,631
Argentina	36,661	66,055	39,716	53,016	1.9	1,413	1,525	0.5	8	–25
Armenia	1,263	244	1,070	1,671	–1.8	346	444	–3.1	–18	85
Australia	86,096	186,625	70,372	94,200	2.2	4,790	5,215	0.7	–22	–98
Austria	7,654	8,481	23,449	26,383	1.3	3,105	3,279	0.9	67	68
Azerbaijan	14,821	14,719	15,001	13,033	–3.9	2,433	1,735	–5.1	1	–13
Bangladesh	1,113	5,962	2,809	8,061	7.4	32	67	5.1	60	26
Belarus	2,566	2,793	2,385	23,808	10.3	247	2,305	9.7	–8	88
Belgium	7,986	11,628	46,100	52,378	1.6	4,682	5,167	1.3	83	78
Benin	..	232	149	107	–3.3	43	20	–6.2	100	–117
Bolivia	3,553	4,478	1,599	2,939	3.2	299	396	1.0	–122	–52
Bosnia and Herzegovina	..	470	..	1,595	..	..	364	..	..	71
Botswana	260	250	384	555	2.5	426	383	–0.8	32	55
Brazil	25,777	73,172	73,041	122,928	4.2	602	772	2.3	65	40
Bulgaria	7,541	9,810	28,476	22,878	–2.5	3,213	2,724	–2.1	74	57
Burkina Faso	0	0	144	162	1.1	21	16	–1.5	100	100
Burundi	1	5	58	144	6.4	14	23	3.5	98	97
Cambodia	13	22	393	517	2.1	60	52	–1.0	97	96
Cameroon	2,855	5,380	774	1,556	3.3	89	117	0.4	–269	–246
Canada	207,359	350,629	192,942	233,328	1.6	7,845	7,879	0.3	–7	–50
Central African Republic	17	24	59	94	2.6	26	29	0.2	71	74
Chad	0	0	93	101	0.6	21	16	–1.8	100	100
Chile	3,871	4,361	7,732	15,131	5.4	694	1,065	3.6	50	71
China	428,693	866,556	413,176	850,521	5.1	421	707	3.7	–4	–2
Hong Kong, China	0	0	5,628	13,615	6.2	1,117	2,212	5.0	100	100
Colombia	13,047	54,361	13,962	24,120	3.5	501	655	1.6	7	–125
Congo, Dem. Rep.	1,478	1,948	1,487	2,058	2.2	55	47	–1.1	1	5
Congo, Rep.	3,387	9,031	262	367	2.6	157	139	–0.5	–1,193	–2,361
Costa Rica	181	380	949	1,971	6.0	415	584	3.3	81	81
Côte d'Ivoire	192	435	1,435	1,362	1.2	175	97	–2.4	87	68
Croatia	..	3,917	..	6,852	..	..	1,435	..	..	43
Cuba	293	1,223	9,645	10,437	0.1	992	949	–0.9	97	88
Czech Republic	40,002	30,448	45,766	39,013	–1.2	4,473	3,776	–1.2	13	22
Denmark	896	15,497	19,734	20,481	0.7	3,852	3,918	0.6	95	24
Dominican Republic	50	171	2,211	3,801	4.3	388	486	2.1	98	96
Ecuador	10,774	20,967	4,209	6,343	2.6	529	553	0.1	–156	–231
Egypt, Arab Rep.	33,374	59,287	15,176	34,678	5.4	371	596	2.9	–120	–71
El Salvador	407	703	1,004	2,322	5.7	221	410	4.2	59	70
Eritrea	..	..	..	..	..	..	..	..	..	..
Estonia	..	3,117	..	5,126	..	..	3,454	..	..	39
Ethiopia	55	158	624	1,178	4.9	17	21	2.0	91	87
Finland	6,912	12,911	25,022	28,670	1.5	5,235	5,613	1.1	72	55
France	46,829	126,866	190,109	241,322	2.1	3,528	4,150	1.6	75	47
Gabon	9,090	18,703	831	644	–4.3	1,203	587	–7.2	–994	–2,804
Gambia, The	0	0	53	61	0.9	83	55	–2.9	100	100
Georgia	4,706	478	4,474	1,850	–3.3	882	342	–3.7	–5	74
Germany	184,238	142,712	358,995	339,287	–0.2	4,585	4,156	–0.5	49	58
Ghana	554	526	1,303	1,564	2.7	121	92	–0.5	57	66
Greece	3,696	9,053	15,960	23,698	3.2	1,655	2,266	2.7	77	62
Guatemala	230	589	1,443	2,191	3.6	209	206	0.6	84	73
Guinea	38	58	356	422	1.3	80	64	–1.4	89	86
Guinea-Bissau	0	0	31	40	2.1	38	37	0.3	100	100
Haiti	19	32	241	357	0.1	45	50	–1.8	92	91
Honduras	67	235	636	1,401	5.1	174	236	1.9	89	83

Energy production and use | 3.7

	Commercial energy production		Commercial energy use			Commercial energy use per capita			Net energy imports	
	thousand metric tons of oil equivalent		thousand metric tons of oil equivalent		average annual % growth	kg of oil equivalent		average annual % growth	% of commercial energy use	
	1980	1995	1980	1995	1980–95	1980	1995	1980–95	1980	1995
Hungary	14,442	13,295	28,556	25,103	–1.0	2,667	2,454	–0.7	49	47
India	73,760	196,941	93,897	241,291	6.5	137	260	4.4	21	18
Indonesia	94,717	169,325	25,904	85,785	8.9	175	442	7.0	–266	–97
Iran, Islamic Rep.	83,430	216,406	38,347	84,069	6.3	980	1,374	3.2	–118	–157
Iraq	136,616	31,100	12,003	25,061	4.1	923	1,206	0.8	–1,038	–24
Ireland	1,894	3,601	8,484	11,461	2.2	2,495	3,196	2.0	78	69
Israel	151	562	8,607	16,650	5.0	2,219	3,003	2.6	98	97
Italy	19,644	28,645	138,629	161,360	1.4	2,456	2,821	1.3	86	82
Jamaica	10	10	2,164	3,003	2.7	1,015	1,191	1.6	100	100
Japan	43,247	99,468	346,567	497,231	2.8	2,968	3,964	2.3	88	80
Jordan	0	192	1,713	4,323	5.2	785	1,031	0.7	100	96
Kazakhstan	76,799	64,354	76,799	55,432	–3.1	5,153	3,337	–3.8	0	–16
Kenya	91	518	1,991	2,907	3.5	120	109	0.3	95	82
Korea, Dem. Rep.	28,275	21,538	30,932	24,600	–1.2	1,751	1,113	–2.6	9	12
Korea, Rep.	9,644	20,570	41,426	145,099	9.6	1,087	3,225	8.4	77	86
Kuwait	94,084	111,227	9,561	14,494	0.3	6,953	9,381	0.2	–884	–667
Kyrgyz Republic	2,190	1,377	1,938	2,315	5.0	534	513	3.4	–13	41
Lao PDR	236	220	107	184	2.6	33	40	0.1	–121	–20
Latvia	261	322	566	3,702	22.9	222	1,471	22.9	54	91
Lebanon	73	69	2,376	4,486	3.2	791	1,120	1.2	97	98
Lesotho	0	0	..	..	..	..	..	..	..	..
Libya	96,537	77,825	7,048	15,781	4.5	2,316	3,129	1.1	–1,270	–393
Lithuania	186	3,316	11,353	8,510	–3.2	3,326	2,291	–3.8	98	61
Macedonia, FYR	..	1,621	..	2,572	..	..	1,308	..	..	37
Madagascar	38	84	391	484	1.6	45	36	–1.2	90	83
Malawi	99	154	334	374	1.6	54	38	–1.6	70	59
Malaysia	15,049	62,385	9,522	33,252	9.8	692	1,655	7.0	–58	–88
Mali	21	42	164	207	1.7	25	21	–0.9	87	80
Mauritania	0	0	214	231	0.5	138	102	–2.0	100	100
Mauritius	21	34	339	435	2.6	351	388	1.7	94	92
Mexico	149,365	201,957	98,904	133,371	2.2	1,486	1,456	0.0	–51	–51
Moldova	35	24	..	4,177	..	..	963	..	..	99
Mongolia	1,195	2,190	1,943	2,576	1.8	1,168	1,045	–0.9	38	15
Morocco	617	440	4,518	8,253	4.4	233	311	2.2	86	95
Mozambique	1,293	160	1,123	662	–1.6	93	38	–3.5	–15	76
Myanmar	1,940	2,167	1,858	2,234	0.2	55	50	–1.7	–4	3
Namibia	0	0	..	..	..	..	..	..	..	..
Nepal	15	97	174	700	9.3	12	33	6.5	91	86
Netherlands	71,830	65,705	65,000	73,292	1.4	4,594	4,741	0.8	–11	10
New Zealand	5,592	12,436	9,190	15,409	3.9	2,952	4,290	3.1	39	19
Nicaragua	44	302	696	1,159	3.4	248	265	0.3	94	74
Niger	14	56	210	330	2.0	38	37	–1.2	93	83
Nigeria	105,512	104,475	9,879	18,393	3.4	139	165	0.4	–968	–468
Norway	55,743	182,428	18,819	23,715	1.8	4,600	5,439	1.4	–196	–669
Oman	14,756	45,403	1,010	4,013	9.2	917	1,880	4.6	–1,361	–1,031
Pakistan	6,970	18,612	11,451	31,536	7.0	139	243	3.8	39	41
Panama	83	202	1,419	1,783	1.6	725	678	–0.4	94	89
Papua New Guinea	80	2,500	705	1,000	2.4	228	232	0.2	89	–150
Paraguay	58	3,578	544	1,487	7.1	173	308	4.1	89	–141
Peru	11,188	8,388	8,233	10,035	0.6	476	421	–1.5	–36	16
Philippines	2,789	6,006	13,357	21,542	3.6	276	307	0.9	79	72
Poland	120,774	94,666	124,557	94,472	–2.0	3,501	2,448	–2.5	3	0
Portugal	1,481	1,870	10,291	19,245	4.6	1,054	1,939	4.6	86	90
Puerto Rico	35	42	8,042	7,444	0.9	2,508	1,993	–0.1	100	99
Romania	51,631	30,008	63,751	44,026	–2.9	2,872	1,941	–3.1	19	32
Russian Federation	749,289	928,870	764,349	604,461	–3.0	5,499	4,079	–3.4	2	–54

3.7 Energy production and use

	Commercial energy production		Commercial energy use			Commercial energy use per capita			Net energy Imports	
	thousand metric tons of oil equivalent		thousand metric tons of oil equivalent		average annual % growth	kg of oil equivalent		average annual % growth	% of commercial energy use	
	1980	1995	1980	1995	1980–95	1980	1995	1980–95	1980	1995
Rwanda	29	46	190	211	−0.7	37	33	−2.6	85	78
Saudi Arabia	533,071	469,820	35,355	82,742	5.2	3,772	4,360	0.3	−1,408	−468
Senegal	0	46	875	866	−0.3	158	104	−3.0	100	95
Sierra Leone	0	0	310	326	0.5	96	72	−1.7	100	100
Singapore	0	0	6,049	21,389	10.0	2,651	7,162	8.1	100	100
Slovak Republic	3,251	4,846	20,646	17,447	−1.3	4,142	3,272	−1.7	84	72
Slovenia	1,623	2,578	4,269	5,583	0.7	2,245	2,806	0.4	62	54
South Africa	66,740	116,160	59,051	88,882	1.8	2,175	2,405	−0.2	−13	−31
Spain	15,781	31,422	68,583	103,491	3.2	1,834	2,639	2.9	77	70
Sri Lanka	127	383	1,411	2,469	2.7	96	136	1.3	91	84
Sudan	58	81	1,140	1,745	3.3	61	65	0.9	95	95
Sweden	16,133	31,549	40,984	50,658	1.3	4,932	5,736	0.9	61	38
Switzerland	7,030	10,961	20,814	25,142	1.7	3,294	3,571	0.9	66	56
Syrian Arab Republic	9,495	34,287	5,343	14,121	5.9	614	1,001	2.5	−78	−143
Tajikistan	1,986	1,325	1,650	3,283	8.9	416	563	6.1	−20	60
Tanzania	86	135	1,023	947	0.8	55	32	−2.3	92	86
Thailand	535	19,430	12,093	52,125	11.1	259	878	9.4	96	63
Togo	1	0	195	185	0.9	75	45	−2.1	99	100
Trinidad and Tobago	13,127	12,991	3,860	6,925	4.0	3,567	5,381	2.8	−240	−88
Tunisia	6,149	4,579	3,083	5,314	4.0	483	591	1.7	−99	14
Turkey	17,190	26,079	31,314	62,187	4.9	704	1,009	2.6	45	58
Turkmenistan	8,034	32,589	7,948	13,737	..	2,778	3,047	−9.8	−1	−137
Uganda	153	185	320	430	2.8	25	22	0.0	52	57
Ukraine	109,708	80,700	97,893	161,586	2.1	1,956	3,136	1.9	−12	50
United Arab Emirates	93,915	138,821	8,576	28,454	7.5	8,222	11,567	1.6	−995	−388
United Kingdom	197,738	254,967	201,168	221,911	1.0	3,571	3,786	0.7	2	−15
United States	1,546,307	1,655,644	1,801,406	2,078,265	1.3	7,928	7,905	0.3	14	20
Uruguay	233	477	2,206	2,035	0.7	757	639	0.1	89	77
Uzbekistan	4,615	49,135	4,821	46,543	11.6	302	2,043	8.9	4	−6
Venezuela	132,919	187,498	35,011	47,140	1.7	2,354	2,158	−0.9	−280	−298
Vietnam	2,728	13,808	4,024	7,694	4.1	75	104	1.8	32	−79
West Bank and Gaza	..	..	..	..	..	..	..	..	..	..
Yemen, Rep.	..	17,394	1,364	2,933	5.3	160	192	1.2	..	−493
Yugoslavia, FR (Serb./Mont.)	..	11,295	..	11,865	..	..	1,125	..	..	5
Zambia	1,146	898	1,685	1,302	−2.1	294	145	−5.0	32	31
Zimbabwe	2,024	3,567	2,797	4,673	4.4	399	424	1.3	28	24
World	6,273,572 t	8,385,643 t	6,325,980 t	8,244,516 t	3.2 t	1,456 w	1,474 w	1.1 w	0 w	0 w
Low income	666,864	1,301,090	587,166	1,227,330	5.5	252	393	3.3	−14	−6
Excl. China & India	164,411	237,593	80,093	135,518	5.8	118	132	2.8	−105	−75
Middle income	2,821,534	3,509,935	1,935,029	2,342,066	5.8	1,604	1,488	2.6	−46	−50
Lower middle income	1,671,934	2,203,208	1,359,924	1,579,612	9.0	1,598	1,426	5.1	−23	−39
Upper middle income	1,149,600	1,306,727	575,105	762,454	1.9	1,618	1,633	0.0	−100	−71
Low & middle income	3,488,398	4,811,025	2,522,195	3,569,396	5.6	706	751	3.1	−38	−35
East Asia & Pacific	576,250	1,166,252	514,939	1,082,697	5.3	391	657	3.7	−12	−8
Europe & Central Asia	1,235,966	1,413,336	1,340,682	1,284,686	8.8	3,333	2,690	−2.7	8	−10
Latin America & Carib.	402,279	642,539	319,888	463,321	2.7	893	969	0.7	−26	−39
Middle East & N. Africa	985,969	1,073,532	142,738	315,726	5.2	822	1,178	2.2	−591	−240
South Asia	84,738	225,491	110,649	286,730	6.6	123	231	4.3	23	21
Sub-Saharan Africa	203,196	289,875	93,323	136,236	2.0	248	238	−0.8	−118	−113
High income	2,785,174	3,574,618	3,803,785	4,675,120	1.7	4,611	5,123	1.1	27	24

In developing countries growth in commercial energy use is closely related to growth in the modern sectors—industry, motorized transport, and urban areas. This connection is less robust in more developed countries. Thus commercial energy use per capita reflects the size of the modern sector as well as climatic, geographic, and economic factors (such as the relative price of energy). Because commercial energy is widely traded, it is necessary to distinguish between its production and use. Net energy imports show the extent to which an economy's use exceeds its domestic production.

Energy data are compiled by the International Energy Agency (IEA) and the United Nations Statistical Division (UNSD). IEA data for non-OECD countries is based on national energy data that have been adjusted to conform with annual questionnaires completed by OECD member governments. UNSD data are primarily from responses to questionnaires sent to national governments, supplemented by official national statistical publications and by data from intergovernmental organizations. When official data are not available, the UNSD prepares estimates based on the professional and commercial literature. The variety of sources affects the cross-country comparability of data.

Commercial energy use refers to domestic primary energy use before transformation to other end-use fuels (such as electricity and refined petroleum products). The use of firewood, dried animal manure, and other traditional fuels is not included. All forms of commercial energy—primary energy and primary electricity—are converted into oil equivalents. To convert nuclear electricity into oil equivalents, a notional thermal efficiency of 33 percent is assumed; for hydroelectric power, 100 percent efficiency is assumed.

Figure 3.7a

Since 1980 low- and middle-income countries have seen rapid growth in commercial energy use

Annual % change, 1980–95

Source: Table 3.7.

Although high-income countries use nearly four times as much commercial energy as low-income countries, since 1980 growth in energy use has been greatest in rapidly industrializing middle-income economies. At its current pace, energy use in low- and middle-income countries will double every 13 years. Slower growth in high-income countries reflects the generally lower economic growth in these economies, as well as their greater energy efficiency.

Figure 3.7b

Low-income countries are exporting less energy

Net energy imports/commercial energy use (%)

● 1980 ○ 1995

Source: Table 3.7.

High-income countries depend on imports for roughly a quarter of their energy use, a ratio that was remarkably constant from 1980 to 1995. Middle-income countries have been the main net exporters of energy, with net exports equal to roughly half of domestic use. This ratio was also nearly constant from 1980 to 1995. Low-income countries, however, have seen a substantial drop in net energy exports, particularly when China and India are excluded.

• **Commercial energy production** refers to commercial forms of primary energy—petroleum (crude oil, natural gas liquids, and oil from nonconventional sources), natural gas, and solid fuels (coal, lignite, and other derived fuels)—and primary electricity, all converted into oil equivalents (see *About the data*). • **Commercial energy use** refers to apparent consumption, which is equal to indigenous production plus imports and stock changes, minus exports and fuels supplied to ships and aircraft engaged in international transportation (see *About the data*). • **Net energy imports** are calculated as energy use less production, both measured in oil equivalents. A minus sign indicates that the country is a net exporter.

Data on commercial energy production and use are primarily from the IEA's electronic files that are also published in its annual publications, *Energy Statistics and Balances of Non-OECD Countries*, *Energy Statistics of OECD Countries*, and *Energy Balances of OECD Countries*, and the United Nations *Energy Statistics Yearbook*.

	GDP per unit of energy use		Traditional fuel use		Carbon dioxide emissions					
	1987 $ per kg oil equivalent		% of total energy use		Total million metric tons		Per capita metric tons		kg per 1987 $ of GDP	
	1980	1995	1980	1995	1980	1995	1980	1995	1980	1995
Albania	0.7	1.8	12.3	8.6	4.8	1.8	1.8	0.6	2.6	1.0
Algeria	4.1	2.7	2.7	2.0	66.2	91.3	3.5	3.2	1.3	1.4
Angola	..	7.7	47.1	59.5	5.3	4.6	0.8	0.4	..	0.6
Argentina	2.8	2.5	6.5	4.0	107.5	129.5	3.8	3.7	1.0	1.0
Armenia	2.1	0.6				3.6	..	1.0	..	3.4
Australia	2.4	2.8	2.1	3.8	202.8	289.8	13.8	16.0	1.2	1.1
Austria	4.6	5.5	1.4	2.8	52.2	59.3	6.9	7.4	0.5	0.4
Azerbaijan		0.2	..	..		42.6	..	5.7		14.6
Bangladesh	4.5	3.0	67.8	49.9	7.6	20.9	0.1	0.2	0.6	0.9
Belarus	..	0.7	..	0.8	..	59.3	..	5.7		3.4
Belgium	2.8	3.2	0.2	0.9	127.2	103.8	12.9	10.2	1.0	0.6
Benin	7.9	18.4	84.9	92.5	0.5	0.6	0.1	0.1	0.4	0.3
Bolivia	2.9	2.0	19.8	12.8	4.5	10.5	0.8	1.4	1.0	1.8
Bosnia and Herzegovina	..	..	..	9.9	..	1.8	..	0.4	..	..
Botswana	2.3	5.1	35.7	..	1.0	2.2	1.1	1.5	1.1	0.8
Brazil	3.4	2.7	41.2	27.5	183.4	249.2	1.5	1.6	0.7	0.8
Bulgaria	0.7	1.0	0.7	0.8	75.3	56.7	8.5	6.7	3.6	2.5
Burkina Faso	11.2	16.4	91.4	93.3	0.4	1.0	0.1	0.1	0.3	0.4
Burundi	13.9	7.7	92.7	88.8	0.1	0.2	0.0	0.0	0.1	0.2
Cambodia	..	2.6	71.2	75.3	0.3	0.5	0.0	0.0	..	0.4
Cameroon	9.7	6.1	69.4	77.3	3.9	4.1	0.4	0.3	0.5	0.4
Canada	1.7	2.0	0.6	0.6	420.9	435.7	17.1	14.7	1.3	0.9
Central African Republic	18.2	13.6	90.8	89.0	0.1	0.2	0.0	0.1	0.1	0.2
Chad	6.2	10.7	87.4	90.2	0.2	0.2	0.0	0.0	0.4	0.1
Chile	2.3	2.4	14.5	13.3	27.9	44.1	2.5	3.1	1.6	1.2
China	0.3	0.7	8.0	5.6	1,476.8	3,192.5	1.5	2.7	10.9	5.5
Hong Kong, China	5.3	5.4	0.9	0.3	16.4	31.0	3.3	5.0	0.5	0.4
Colombia	2.1	2.1	21.4	21.1	39.8	67.5	1.4	1.8	1.4	1.3
Congo, Dem. Rep.	4.4	2.3	79.5	83.9	3.5	2.1	0.1	0.0	0.5	0.4
Congo, Rep.	5.7	6.6	55.9	61.0	0.4	1.3	0.2	0.5	0.3	0.5
Costa Rica	4.2	3.3	40.4	12.7	2.5	5.2	1.1	1.6	0.6	0.8
Côte d'Ivoire	6.7	8.4	53.5	67.2	4.7	10.4	0.6	0.7	0.5	0.9
Croatia	..	..	..	3.0	..	17.0	..	3.6	..	..
Cuba	..	..	28.1	19.7	30.7	29.1	3.2	2.6	..	..
Czech Republic	..	0.8	..	0.5	..	112.0	..	10.8	..	3.4
Denmark	4.4	5.7	0.3	3.3	62.9	54.9	12.3	10.5	0.7	0.5
Dominican Republic	2.1	1.9	28.3	12.1	6.4	11.8	1.1	1.5	1.4	1.6
Ecuador	2.3	2.2	26.5	14.8	13.4	22.6	1.7	2.0	1.4	1.6
Egypt, Arab Rep.	1.8	1.6	5.0	3.3	45.2	91.7	1.1	1.6	1.6	1.6
El Salvador	4.4	2.5	50.3	42.9	2.1	5.2	0.5	0.9	0.5	0.9
Eritrea	..	..	..	..	..	..	..	..	..	..
Estonia	..	0.8	..	2.3	..	16.4	..	11.1	..	4.3
Ethiopia	..	7.4	92.4	90.1	1.8	3.5	0.0	0.1	..	0.4
Finland	2.9	3.3	3.8	5.1	54.9	51.0	11.5	10.0	0.8	0.5
France	4.1	4.3	1.3	1.0	482.7	340.1	9.0	5.8	0.6	0.3
Gabon	5.1	7.9	33.6	51.8	4.8	3.5	6.9	3.2	1.1	0.7
Gambia, The	3.2	4.4	79.7	81.2	0.2	0.2	0.2	0.2	0.9	0.8
Georgia	..	..	..	1.3	..	7.7	..	1.4	..	..
Germany	..	..	..	0.7	..	835.1	..	10.2		
Ghana	3.6	4.6	68.2	79.0	2.4	4.0	0.2	0.2	0.5	0.6
Greece	2.8	2.2	2.8	1.5	51.7	76.3	5.4	7.3	1.2	1.5
Guatemala	5.0	4.4	53.1	59.9	4.5	7.2	0.6	0.7	0.6	0.7
Guinea	..	6.7	68.4	69.9	0.9	1.1	0.2	0.2	..	0.4
Guinea-Bissau	3.9	5.8	76.1	70.5	0.1	0.2	0.2	0.2	1.1	1.0
Haiti	6.8	3.3	82.4	80.3	0.8	0.6	0.1	0.1	0.5	0.5
Honduras	5.6	3.8	61.2	49.3	2.1	3.9	0.6	0.7	0.6	0.7

Energy efficiency and emissions | 3.8

	GDP per unit of energy use		Traditional fuel use		Carbon dioxide emissions					
	1987 $ per kg oil equivalent		% of total energy use		Total million metric tons		Per capita metric tons		kg per 1987 $ of GDP	
	1980	1995	1980	1995	1980	1995	1980	1995	1980	1995
Hungary	0.8	1.0	2.1	1.8	82.5	55.9	7.7	5.5	3.7	2.3
India	1.9	1.7	34.7	23.3	347.3	908.7	0.5	1.0	1.9	2.2
Indonesia	2.0	1.6	51.6	29.9	94.6	296.1	0.6	1.5	1.8	2.1
Iran, Islamic Rep.	3.0	2.2	1.6	0.8	116.1	263.8	3.0	4.3	1.0	1.4
Iraq	7.2	..	0.2	0.1	44.0	99.0	3.4	4.8	0.5	..
Ireland	3.1	4.1	0.1	0.2	25.2	32.2	7.4	9.0	0.9	0.7
Israel	3.4	3.5	..	..	21.1	46.3	5.4	8.4	0.7	0.8
Italy	4.8	5.5	0.7	1.9	371.9	410.0	6.6	7.2	0.6	0.5
Jamaica	1.3	1.2	6.2	8.0	8.4	9.1	4.0	3.6	3.1	2.4
Japan	5.5	6.1	0.1	0.5	907.4	1,126.8	7.8	9.0	0.5	0.4
Jordan	2.7	1.9	..	..	4.7	13.3	2.2	3.2	1.0	1.7
Kazakhstan	..	0.3	..	0.1	..	221.5	..	13.3	..	13.8
Kenya	3.1	3.4	75.4	76.1	6.2	6.7	0.4	0.3	1.0	0.7
Korea, Dem. Rep.	..	..	2.7	3.9	124.9	257.0	7.1	11.6	..	..
Korea, Rep.	1.8	1.8	5.7	0.7	125.2	373.6	3.3	8.3	1.7	1.5
Kuwait	2.7	2.0	..	..	24.7	48.7	18.0	31.5	1.0	1.7
Kyrgyz Republic	..	0.5	..	..	..	5.5	..	1.2	..	4.9
Lao PDR	..	9.6	86.6	85.1	0.2	0.3	0.1	0.1	..	0.2
Latvia	12.1	1.3	..	18.0	..	9.3	..	3.7	..	1.9
Lebanon	0.0	1.3	4.3	2.6	6.2	13.3	2.1	3.3	..	2.4
Lesotho	..	..	..	..	..	..	..	..	..	..
Libya	5.7	..	1.7	0.8	26.9	39.4	8.8	7.8	0.7	..
Lithuania	..	0.8	..	5.6	..	14.8	..	4.0	..	2.1
Macedonia, FYR	..	..	..	6.9	..	..	..	..	..	..
Madagascar	6.9	5.8	77.1	83.8	1.6	1.1	0.2	0.1	0.6	0.4
Malawi	3.1	3.8	89.1	86.8	0.7	0.7	0.1	0.1	0.7	0.5
Malaysia	2.4	1.9	14.4	6.6	28.0	106.6	2.0	5.3	1.2	1.7
Mali	10.8	12.1	85.2	87.4	0.4	0.5	0.1	0.0	0.2	0.2
Mauritania	3.8	5.0	0.7	..	0.6	3.1	0.4	1.3	0.8	2.7
Mauritius	3.7	6.6	44.1	41.6	0.6	1.5	0.6	1.3	0.5	0.5
Mexico	1.3	1.3	4.4	4.4	255.0	357.8	3.8	3.9	2.0	2.1
Moldova	..	..	..	0.5	..	10.8	..	2.5	..	..
Mongolia	..	..	14.0	3.6	6.8	8.5	4.1	3.4	..	..
Morocco	3.4	2.8	5.4	4.7	15.9	29.3	0.8	1.1	1.1	1.3
Mozambique	1.2	3.4	72.6	86.0	3.2	1.0	0.3	0.1	2.3	0.4
Myanmar	..	..	66.5	69.4	4.8	7.0	0.1	0.2	..	..
Namibia	..	..	..	..	..	..	..	..	..	..
Nepal	12.6	6.4	94.8	88.9	0.5	1.5	0.0	0.1	0.2	0.3
Netherlands	3.0	3.7	0.0	0.5	152.6	135.9	10.8	8.8	0.8	0.5
New Zealand	3.4	2.8	0.2	..	17.6	27.4	5.6	7.6	0.6	0.6
Nicaragua	5.5	3.1	50.4	45.8	2.0	2.7	0.7	0.6	0.5	0.7
Niger	12.1	7.5	78.0	79.6	0.6	1.1	0.1	0.1	0.2	0.5
Nigeria	2.6	1.9	63.7	56.6	68.1	90.7	1.0	0.8	2.6	2.7
Norway	3.9	4.7	0.8	1.1	90.4	72.5	22.1	16.6	1.2	0.7
Oman	3.9	3.1	..	..	5.9	11.4	5.3	5.3	1.5	0.9
Pakistan	1.9	1.6	27.2	20.2	31.6	85.4	0.4	0.7	1.5	1.7
Panama	3.3	3.9	26.4	19.4	3.5	6.9	1.8	2.6	0.7	1.0
Papua New Guinea	3.9	4.6	64.1	58.9	1.8	2.5	0.6	0.6	0.7	0.5
Paraguay	6.0	3.4	66.1	51.5	1.5	3.8	0.5	0.8	0.4	0.7
Peru	2.5	2.4	18.7	22.9	23.5	30.6	1.4	1.3	1.2	1.3
Philippines	2.5	2.0	35.8	30.5	36.5	61.2	0.8	0.9	1.1	1.4
Poland	0.5	0.7	0.4	1.1	456.2	338.0	12.8	8.8	7.6	5.1
Portugal	3.5	2.7	1.1	0.7	27.1	51.9	2.8	5.2	0.7	1.0
Puerto Rico	2.4	4.2	..	..	14.0	15.5	4.4	4.2	0.7	0.4
Romania	0.5	0.7	1.5	21.5	191.8	121.1	8.6	5.3	5.7	3.9
Russian Federation	0.5	0.5	..	1.1	..	1,818.0	..	12.3	..	6.1

	GDP per unit of energy use		Traditional fuel use		Carbon dioxide emissions					
	1987 $ per kg oil equivalent		% of total energy use		Total million metric tons		Per capita metric tons		kg per 1987 $ of GDP	
	1980	**1995**	**1980**	**1995**	**1980**	**1995**	**1980**	**1995**	**1980**	**1995**
Rwanda	9.2	6.3	84.8	85.7	0.3	0.5	0.1	0.1	0.2	0.4
Saudi Arabia	2.7	1.2	..	..	130.7	254.3	14.0	13.4	1.4	2.6
Senegal	4.2	6.1	48.6	55.9	2.8	3.1	0.5	0.4	0.8	0.6
Sierra Leone	2.7	2.2	63.5	69.4	0.6	0.4	0.2	0.1	0.7	0.6
Singapore	2.3	2.0	0.0	0.0	30.1	63.7	13.2	21.3	2.2	1.5
Slovak Republic	..	0.9	..	0.5	..	38.0	..	7.1	..	2.3
Slovenia	..	..	..	0.8	..	11.7	..	5.9	..	..
South Africa	1.3	1.0	4.5	3.9	211.3	305.8	7.8	8.3	2.8	3.4
Spain	3.6	3.5	0.5	0.6	200.0	231.6	5.3	5.9	0.8	0.6
Sri Lanka	3.4	3.8	54.3	48.4	3.4	5.9	0.2	0.3	0.7	0.6
Sudan	12.9	*12.1*	76.4	76.4	3.3	3.5	0.2	0.1	0.2	*0.2*
Sweden	3.4	3.4	3.9	2.5	71.4	44.6	8.6	5.0	0.5	0.3
Switzerland	7.3	7.5	1.1	2.1	40.9	38.9	6.5	5.5	0.3	0.2
Syrian Arab Republic	1.9	1.3	0.1	0.0	19.3	46.0	2.2	3.3	1.9	2.6
Tajikistan	..	0.5	..	..	..	3.7	..	0.6	..	2.5
Tanzania	..	..	83.7	89.6	1.9	2.4	0.1	0.1	..	..
Thailand	2.8	2.1	48.3	32.7	40.1	175.0	0.9	2.9	1.2	1.6
Togo	6.4	7.1	38.3	73.1	0.6	0.7	0.2	0.2	0.5	0.6
Trinidad and Tobago	1.5	0.7	1.8	1.0	16.7	17.1	15.4	13.3	3.0	3.5
Tunisia	2.4	2.4	15.4	12.9	9.4	15.3	1.5	1.7	1.3	1.2
Turkey	1.9	1.8	18.0	3.1	76.3	165.9	1.7	2.7	1.3	1.5
Turkmenistan	..	..	..	..	..	28.3	..	6.3	..	..
Uganda	..	24.8	87.2	89.2	0.6	1.0	0.1	0.1	..	0.1
Ukraine	..	0.2	..	0.4	..	438.2	..	8.5	..	..
United Arab Emirates	3.6	..	..	..	36.3	68.3	34.8	27.8	1.2	..
United Kingdom	2.8	3.5	0.0	1.1	585.1	542.1	10.4	9.3	1.0	0.7
United States	2.1	2.6	1.2	4.2	4,515.3	5,468.6	19.9	20.8	1.2	1.0
Uruguay	3.4	4.4	20.4	26.7	5.8	5.4	2.0	1.7	0.8	0.6
Uzbekistan	..	0.3	..	..	..	98.9	..	4.3	..	7.3
Venezuela	1.3	1.2	1.0	1.2	89.6	180.2	6.0	8.3	2.0	3.1
Vietnam	..	7.8	53.5	49.1	16.8	31.7	0.3	0.4	..	0.5
West Bank and Gaza	..	..	..	..	..	..	..	..	..	..
Yemen, Rep.	..	..	..	..	1.2	..	0.1	..	..	..
Yugoslavia, FR (Serb./Mont.)	..	..	..	..	..	33.0	..	3.1	..	..
Zambia	1.3	1.7	54.6	71.2	3.5	2.4	0.6	0.3	1.6	1.1
Zimbabwe	1.6	1.4	33.6	27.4	9.6	9.7	1.4	0.9	2.2	1.5
World	**2.2 w**	**2.4 w**	**7.1 w**	**6.8 w**	**13,585.7 t**	**22,700.2 t**	**3.4 w**	**4.1 w**	**1.1 w**	**1.2 w**
Low income	0.9	1.1	25.4	19.0	2,037.8	4,503.7	0.9	1.4	4.2	3.4
Excl. China & India	3.4	2.7	64.0	56.1	213.7	402.5	0.3	0.4	1.1	1.2
Middle income	1.2	1.1	14.0	7.7	2,775.7	7,073.8	2.9	4.5	1.7	2.6
Lower middle income	1.0	1.0	16.8	7.5	1,209.0	4,942.5	2.0	4.5	..	3.2
Upper middle income	1.7	1.5	11.7	8.2	1,566.7	2,131.3	4.6	4.6	1.8	1.9
Low & middle income	1.1	1.1	18.8	12.0	4,813.5	11,577.5	1.5	2.5	2.3	2.9
East Asia & Pacific	..	0.9	15.8	11.6	1,832.7	4,140.0	1.4	2.5	6.0	3.9
Europe & Central Asia	..	0.6	3.5	2.1	887.9	3,733.7	..	7.9	..	5.1
Latin America & Carib.	2.2	2.0	20.3	15.6	850.5	1,219.8	2.4	2.6	1.2	1.4
Middle East & N. Africa	3.3	1.8	2.0	1.2	499.5	982.9	2.9	3.9	1.1	1.7
South Asia	2.0	1.7	37.5	25.6	392.4	1,024.1	0.4	0.8	1.8	2.1
Sub-Saharan Africa	2.1	1.9	46.6	47.4	350.5	477.1	0.9	0.8	1.8	1.9
High income	2.9	3.4	1.1	2.5	8,772.1	11,122.7	12.0	12.5	0.9	0.7

About the data

The ratio of real GDP to energy use provides a measure of energy efficiency. Differences in this ratio over time and across countries are influenced by structural changes in the economy, changes in the energy efficiency of particular sectors of the economy, and differences in fuel mixes.

For traditional fuels, fuelwood and charcoal consumption estimates are calculated by the Food and Agriculture Organization (FAO) based on population data and country-specific per capita consumption figures. Estimates of bagasse consumption are based on sugar production data.

Carbon dioxide (CO_2) emissions, largely a byproduct of energy production and use (see table 3.7), are the largest source of greenhouse gases, which are associated with global warming. Anthropogenic CO_2 emissions result primarily from fossil fuel combustion and cement manufacturing. Combustion of different fossil fuels releases different amounts of CO_2 for the same level of energy production. Burning oil releases about 50 percent more CO_2 than burning natural gas, and burning coal releases about twice as much. During cement manufacturing about 0.5 metric ton of CO_2 is released for each ton of cement produced.

The Carbon Dioxide Information Analysis Center (CDIAC), sponsored by the U.S. Department of Energy, calculates annual anthropogenic emissions of CO_2. These calculations are derived from data on fossil fuel consumption, based on the World Energy Data Set maintained by the United Nations Statistical Division, and from data on world cement manufacturing, based on the Cement Manufacturing Data Set maintained by the U.S. Bureau of Mines.

Although the estimates of global CO_2 emissions are probably within 10 percent of actual emissions (as calculated from global average fuel chemistry and use), country estimates may have larger error bounds. Trends estimated from a consistent time series tend to be more accurate than individual values. Each year the CDIAC recalculates the entire time series from 1950 to the present, incorporating its most recent findings and the latest corrections to its database. Estimates do not include fuels supplied to ships and aircraft engaged in international transportation because of the difficulty of apportioning these fuels among the countries benefiting from that transport.

Figure 3.8a

Carbon dioxide emissions are increasing everywhere

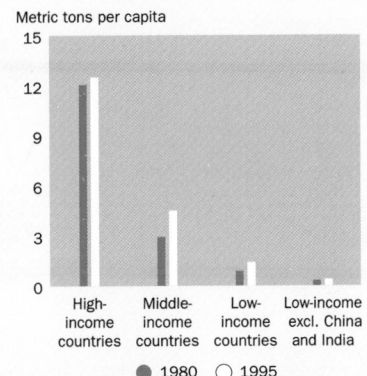

Metric tons per capita

● 1980 ○ 1995

Source: Table 3.8.

The main sources of carbon dioxide, the principal greenhouse gas, are the burning of fossil fuels and the manufacturing of cement. Decoupling economic growth and carbon dioxide emissions in both industrial and developing countries will be essential to preventing global warming. On average, high-income economies emitted 5 times as much carbon dioxide per capita as low- and middle-income countries in 1995, and 32 times as much per capita as low-income countries excluding China and India. Although per capita emissions declined slightly in high-income countries in the early 1990s, since 1980 emissions have grown in every income group—especially middle-income countries.

Figure 3.8b

In 1995 high-income countries emitted the most carbon dioxide

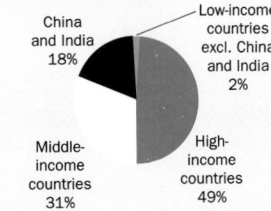

China and India 18%
Low-income countries excl. China and India 2%
Middle-income countries 31%
High-income countries 49%

Source: Table 3.8.

Carbon dioxide emissions per capita are highest in high-income countries, but large populations in low- and middle-income countries meant that they produced more than half of the world's carbon dioxide emissions in 1995. Middle-income countries, China, and India accounted for most of the developing world's emissions; the remaining low-income countries contributed a mere 2 percent.

Definitions

- **GDP per unit of energy use** is the U.S. dollar estimate of real GDP (at 1987 prices) per kilogram of oil equivalent of commercial energy use (see table 3.7).
- **Traditional fuel use** includes estimates of the consumption of fuelwood, charcoal, bagasse, and animal and vegetable wastes. Total energy use comprises commercial energy use (see table 3.7) and traditional fuel use. • **Carbon dioxide emissions** are those stemming from the burning of fossil fuels and the manufacture of cement. They include carbon dioxide produced during consumption of solid fuels, liquid fuels, gas fuels, and gas flaring.

Data sources

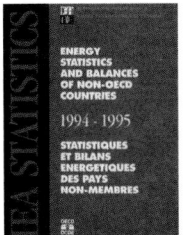

Underlying data on commercial energy production and use are from International Energy Agency (IEA) and United Nations sources. Data on CO_2 emissions are based on several sources as reported by the World Resources Institute. The main source is the Carbon Dioxide Information Analysis Center, Environmental Sciences Division, Oak Ridge National Laboratory, in the state of Tennessee in the United States. Traditional fuel data are from the World Resources Institute's *World Resources,* the United Nations *Energy Statistics Yearbook,* and FAO electronic files.

3.9 | Sources of electricity

	Electricity production		Sources of electricity										
	billion kwh		Hydropower %		Coal %		Oil %		Gas %		Nuclear power %	
	1980	1995	1980	1995	1980	1995	1980	1995	1980	1995	1980	1995
Albania	3.7	4.4	79.4	95.2	0.0	0.0	20.6	4.8	0.0	0.0	..	..
Algeria	7.1	19.7	3.6	1.0	..	..	12.2	3.3	84.1	95.7	..	..
Angola	0.7	1.0	88.1	93.8	..	..	11.9	6.3	..	..	..	..
Argentina	39.7	64.6	38.1	41.3	2.5	2.5	31.9	4.8	21.0	40.0	5.9	10.9
Armenia	13.0	5.6	12.0	34.5			54.8	39.3	0.0	20.8	33.2	5.5
Australia	95.2	173.4	13.6	9.2	73.3	77.0	5.4	1.7	7.3	10.3	..	..
Austria	41.6	55.1	69.1	67.3	7.0	10.6	14.0	3.1	9.2	15.5	..	..
Azerbaijan	15.0	17.0	7.3	9.1	..	..	92.7	90.9	0.0	0.0	..	..
Bangladesh	2.4	10.9	24.8	3.4	..	..	26.6	16.3	48.6	80.3	..	..
Belarus	34.1	24.9	0.1	0.1	0.0	0.0	99.9	28.5	0.0	71.5	..	..
Belgium	53.1	73.6	0.5	0.5	29.4	26.2	34.7	1.8	11.2	13.8	23.6	56.2
Benin	0.0	0.0	..	..	..	..	100.0	100.0	..	..	..	..
Bolivia	1.6	3.0	68.8	53.2	..	..	10.3	7.6	18.4	36.7	..	..
Bosnia and Herzegovina	..	2.2	..	64.5	..	35.5	..	0.0	..	..	..	..
Botswana	..	..	..	..	..	..	..	..	..	..	..	..
Brazil	139.4	275.4	92.5	92.2	2.0	1.5	3.8	2.7	0.0	0.2	0.0	0.9
Bulgaria	34.8	40.7	10.7	3.1	49.2	43.1	22.5	3.6	0.0	7.9	17.7	42.4
Burkina Faso	..	..	..	..	..	..	..	..	..	..	..	..
Burundi	..	..	..	..	..	..	..	..	..	..	..	..
Cambodia	..	..	..	..	..	..	..	..	..	..	..	..
Cameroon	1.5	2.7	93.9	96.9	..	..	6.1	3.1	..	..	..	..
Canada	373.3	551.4	67.3	60.5	16.0	15.1	3.7	1.9	2.5	3.9	10.2	17.7
Central African Republic	..	..	..	..	..	..	..	..	..	..	..	..
Chad	..	..	..	..	..	..	..	..	..	..	..	..
Chile	11.8	28.0	62.5	65.7	13.8	24.7	21.5	8.2	1.2	0.9	..	..
China	300.6	1,007.7	19.4	18.9	59.3	73.4	21.1	6.1	0.2	0.2	0.0	1.3
Hong Kong, China	12.6	27.9	..	..	0.0	97.7	100.0	2.3	..	..	..	..
Colombia	20.6	45.4	70.1	70.4	9.7	11.4	2.1	1.0	17.6	16.7	..	..
Congo, Dem. Rep.	4.4	6.2	95.5	96.2	..	..	4.5	3.8	..	..	..	..
Congo, Rep.	0.2	0.4	64.1	98.6	..	..	34.0	0.7	1.9	0.7	..	..
Costa Rica	2.2	5.1	94.7	86.1	..	..	5.3	13.7	..	..	..	..
Côte d'Ivoire	1.7	2.3	77.6	42.1	..	..	22.4	57.9	..	..	..	..
Croatia	..	8.9	..	59.4	..	2.7	..	27.7	..	10.1	..	..
Cuba	9.9	11.2	1.0	0.8	..	..	99.0	90.9	0.0	0.1	..	..
Czech Republic	52.7	60.6	4.6	3.3	84.8	74.0	9.6	1.0	1.1	0.8	0.0	20.2
Denmark	26.8	36.8	0.1	0.1	81.8	75.2	18.0	9.4	0.0	9.7	..	..
Dominican Republic	3.3	6.5	17.7	30.6	0.0	4.5	82.3	64.4	..	..	..	..
Ecuador	3.4	8.9	25.9	62.0	..	..	74.1	38.0	..	..	..	..
Egypt, Arab Rep.	18.9	54.8	51.8	19.7	..	..	27.7	37.3	20.5	43.0	..	..
El Salvador	1.6	3.4	68.3	42.2	..	..	8.6	36.3	..	..	..	..
Eritrea	..	..	..	..	..	..	..	..	..	..	..	..
Estonia	18.9	8.7	0.0	0.0	0.0	96.5	100.0	1.2	0.0	2.2	..	..
Ethiopia	0.7	1.3	70.2	86.8	..	..	27.6	7.9	..	..	..	..
Finland	40.7	63.9	25.1	20.2	42.6	26.7	10.8	2.3	4.2	10.4	17.2	30.1
France	256.9	489.3	26.9	14.6	27.2	5.4	18.9	1.6	2.7	0.8	23.8	77.1
Gabon	0.5	0.9	49.1	77.1	..	..	50.9	11.8	0.0	11.1	..	..
Gambia, The	..	..	..	..	..	..	..	..	..	..	..	..
Georgia	14.7	6.8	43.8	69.3	..	..	56.2	2.8	0.0	27.9	..	..
Germany	466.3	532.6	4.1	3.7	62.9	55.8	5.7	1.7	14.2	8.1	11.9	28.9
Ghana	5.3	6.2	99.4	99.3	..	..	0.6	0.7	..	..	..	..
Greece	22.7	41.2	15.0	8.6	44.8	69.6	40.1	21.5	0.0	0.2	..	..
Guatemala	1.7	3.2	16.7	67.2	..	..	83.3	26.1	..	..	..	..
Guinea	..	..	..	..	..	..	..	..	..	..	..	..
Guinea-Bissau	..	..	..	..	..	..	..	..	..	..	..	..
Haiti	0.3	0.5	70.1	75.4	..	..	29.9	20.6	..	..	..	..
Honduras	0.9	2.7	84.0	99.6	..	..	16.0	0.4	..	..	..	..

Sources of electricity | 3.9

	Electricity production		Sources of electricity										
			Hydropower %		Coal %		Oil %		Gas %		Nuclear power %	
	billion kwh											
	1980	1995	1980	1995	1980	1995	1980	1995	1980	1995	1980	1995
Hungary	23.9	34.0	0.5	0.5	46.9	29.5	24.6	16.3	28.0	11.9	0.0	41.2
India	119.3	414.6	39.0	20.1	49.9	69.4	7.7	2.9	0.8	5.9	2.5	1.7
Indonesia	8.4	61.2	16.0	14.2	0.0	23.5	84.0	26.8	0.0	32.0	..	..
Iran, Islamic Rep.	22.4	85.4	25.1	8.5	..	..	50.1	35.8	24.8	55.7	..	..
Iraq	11.4	29.0	6.1	2.0	..	..	93.9	98.0	..	..	..	..
Ireland	10.6	17.6	7.9	4.0	16.4	51.3	60.4	15.2	15.2	29.3	..	..
Israel	12.5	30.3	0.0	0.2	0.0	62.1	100.0	37.7	0.0	0.0	..	..
Italy	183.5	237.4	24.7	15.9	9.9	11.6	57.0	50.9	5.0	19.8	1.2	0.0
Jamaica	1.5	5.8	8.3	2.1	..	..	87.9	93.2	..	..	..	..
Japan	572.5	980.9	15.4	8.4	9.6	17.8	46.2	22.9	14.2	19.5	14.4	29.7
Jordan	1.1	5.6	0.0	0.3	..	..	100.0	86.4	0.0	13.3	..	..
Kazakhstan	..	66.7	..	12.5	..	72.0	..	7.3	..	8.2	..	0.0
Kenya	1.5	3.7	71.1	83.3	..	..	28.9	8.9	..	..	..	..
Korea, Dem. Rep.	35.0	36.0	64.3	63.9	35.7	36.1	..	..	..	..	..	..
Korea, Rep.	37.2	184.7	5.3	3.0	6.7	26.4	78.7	22.8	0.0	11.5	9.3	36.3
Kuwait	9.4	23.7	..	..	..	..	37.2	21.7	62.8	78.3	..	..
Kyrgyz Republic	9.2	12.3	53.1	90.0	0.0	6.0	46.9	3.9	0.0	0.0	..	..
Lao PDR	..	..	..	..	..	..	..	..	..	..	..	..
Latvia	4.7	4.0	64.9	73.8	0.0	2.5	35.1	10.5	0.0	13.2	..	..
Lebanon	2.8	5.3	30.9	13.8	..	..	69.1	86.2	..	..	..	..
Lesotho	..	..	..	..	..	..	..	..	..	..	..	..
Libya	4.8	18.0	..	..	..	..	100.0	100.0	..	..	..	..
Lithuania	11.7	13.5	4.0	2.8	..	..	96.0	8.3	0.0	1.5	0.0	87.4
Macedonia, FYR	..	6.1	..	13.1	..	86.3	..	0.6	..	..	..	..
Madagascar	..	..	..	..	..	..	..	..	..	..	..	..
Malawi	..	..	..	..	..	..	..	..	..	..	..	..
Malaysia	10.0	45.5	13.9	13.7	0.0	8.6	84.7	39.5	1.3	38.3	..	..
Mali	..	..	..	..	..	..	..	..	..	..	..	..
Mauritania	..	..	..	..	..	..	..	..	..	..	..	..
Mauritius	..	..	..	..	..	..	..	..	..	..	..	..
Mexico	67.0	152.5	25.2	18.0	0.0	9.4	57.9	51.4	15.5	11.9	0.0	5.5
Moldova	15.4	8.4	2.6	3.3	0.0	38.3	97.4	11.1	0.0	47.2	..	..
Mongolia	..	..	..	..	..	..	..	..	..	..	..	..
Morocco	5.2	12.0	28.9	5.0	19.5	46.3	51.6	48.7	..	..	..	..
Mozambique	14.0	0.6	96.8	8.9	0.0	13.0	3.2	63.6	0.0	14.6	..	..
Myanmar	1.5	3.8	53.5	40.4	2.0	0.1	31.3	14.3	13.2	45.2	..	..
Namibia	..	..	..	..	..	..	..	..	..	..	..	..
Nepal	0.2	1.2	82.3	96.9	..	..	17.7	3.1	..	..	..	..
Netherlands	64.8	81.1	0.0	0.1	13.7	35.6	38.4	4.8	39.8	51.8	6.5	5.0
New Zealand	22.3	36.2	85.0	76.1	1.9	2.0	0.1	0.1	7.7	13.6	..	..
Nicaragua	1.1	1.8	47.8	22.8	..	..	52.2	57.4	..	..	..	..
Niger	..	..	..	..	..	..	..	..	..	..	..	..
Nigeria	7.1	14.5	39.0	38.0	0.4	0.0	45.1	24.1	15.5	38.0	..	..
Norway	83.8	122.0	99.8	99.4	0.0	0.2	0.1	0.0	0.0	0.2	..	..
Oman	0.8	6.5	..	..	..	..	0.0	19.3	100.0	80.7	..	..
Pakistan	15.0	53.6	58.2	42.7	0.2	0.1	1.1	29.4	40.5	26.9	0.0	1.0
Panama	2.0	3.5	49.4	66.9	..	..	48.4	31.1	..	..	..	..
Papua New Guinea	..	..	..	..	..	..	..	..	..	..	..	..
Paraguay	0.8	41.6	88.6	99.9	..	..	2.5	0.0	..	..	..	..
Peru	10.0	16.0	69.8	85.9	..	..	27.3	11.9	1.7	1.4	..	..
Philippines	18.0	29.7	19.6	10.9	1.0	6.8	67.9	62.8	..	..	..	..
Poland	121.9	136.7	2.7	1.4	89.4	97.3	7.8	1.1	0.1	0.2	..	..
Portugal	15.2	33.2	52.7	25.2	2.3	40.6	42.9	31.1	..	..	..	..
Puerto Rico	..	..	..	..	..	..	..	..	..	..	..	..
Romania	67.5	59.3	18.7	28.2	31.4	35.1	9.6	9.8	38.2	26.9	..	..
Russian Federation	804.9	859.0	16.1	20.5	0.0	18.3	77.2	9.2	0.0	40.1	6.7	11.6

3.9 | Sources of electricity

	Electricity production		Sources of electricity									
	billion kwh		Hydropower %		Coal %		Oil %		Gas %		Nuclear power %	
	1980	1995	1980	1995	1980	1995	1980	1995	1980	1995	1980	1995
Rwanda	..	..	..	..			..	..	..	..	..	..
Saudi Arabia	20.5	93.9	..	..			58.5	55.3	41.5	44.7	..	..
Senegal	0.6	0.9	..	..			100.0	100.0	..	..	..	..
Sierra Leone	..	..	..	..			..	..	..	..	..	..
Singapore	7.0	22.1	..	..			100.0	82.8	0.0	17.2	..	..
Slovak Republic	20.0	25.6	11.3	19.1	37.9	22.6	17.9	4.7	10.2	9.0	22.7	44.6
Slovenia	8.0	12.6	42.3	25.6	51.6	34.3	3.9	2.3	2.2	0.0	0.0	37.8
South Africa	99.0	186.8	1.0	0.3	99.0	93.5			..	..	0.0	6.0
Spain	109.2	165.6	27.1	14.0	30.0	40.5	35.2	8.8	2.7	2.3	4.7	33.5
Sri Lanka	1.7	4.8	88.7	92.7	..	..	11.3	7.3	..	..	..	..
Sudan	1.1	1.3	61.7	71.0	..	..	27.7	10.2	..	..	..	..
Sweden	96.3	147.0	61.1	45.6	0.2	2.2	10.4	2.4	0.0	0.5	27.5	47.6
Switzerland	48.2	62.3	68.1	56.5	0.1	0.0	1.0	0.6	0.6	1.3	29.8	40.0
Syrian Arab Republic	4.0	15.3	64.7	45.4	..	..	31.9	32.1	3.4	22.5	..	..
Tajikistan	13.6	14.8	93.4	98.8	..	..	6.6	1.2	..	..	..	..
Tanzania	1.0	1.8	100.0	85.9	..	..	0.0	14.1	..	..	..	..
Thailand	14.4	80.1	8.8	8.4	9.8	18.5	81.4	30.5	0.0	42.3	..	..
Togo	..	..	..	..			..	..	..	..	..	..
Trinidad and Tobago	2.0	4.3	..	..	..	..	2.3	0.0	96.0	99.2	..	..
Tunisia	2.9	7.3	0.8	0.5	..	..	68.3	65.5	30.9	34.0	..	..
Turkey	23.1	86.2	49.0	41.2	25.8	32.5	25.2	6.7	0.0	19.2	..	..
Turkmenistan	6.7	9.8	0.1	0.0	..	..	100.0	0.0	0.0	100.0	..	..
Uganda	..	..	..	..			..	..	..	..	..	..
Ukraine	236.0	194.0	5.7	5.2	0.0	32.9	88.3	4.3	0.0	21.3	6.0	36.3
United Arab Emirates	6.3	19.1	..	..	..	..	44.8	18.8	55.2	81.2	..	..
United Kingdom	284.1	332.9	1.4	1.6	73.2	43.0	11.7	10.7	0.7	17.5	13.0	26.7
United States	2,427.3	3,558.4	11.5	8.8	51.2	51.5	10.8	2.5	15.3	14.9	11.0	20.1
Uruguay	3.4	6.3	67.9	87.8	..	..	30.7	11.0	..	..	..	..
Uzbekistan	33.9	47.2	14.6	15.0	0.0	6.5	85.4	2.4	0.0	76.1	..	..
Venezuela	36.9	73.5	39.6	70.0	..	..	14.5	5.4	45.9	24.5	..	..
Vietnam	3.6	14.4	41.8	77.9	39.9	7.4	18.3	8.2	0.0	6.5	..	..
West Bank and Gaza	..	..	..	..			..	..	..	..	..	..
Yemen, Rep.	0.5	2.4	..	..			100.0	100.0	..	..	..	..
Yugoslavia, FR (Serb./Mont.)	..	37.2	..	30.2	..	63.7	..	2.4	..	3.7	..	..
Zambia	9.5	7.8	98.8	99.7	0.7	0.3	0.5	0.0	..	..	..	..
Zimbabwe	4.3	7.3	87.6	32.3	11.8	67.7	0.6	0.0	..	..	..	..
World	**8226.9 s**	**13180.7 s**	**20.8 w**	**18.9 w**	**33.1 w**	**37.5 w**	**28.3 w**	**9.9 w**	**8.8 w**	**14.7 w**	**8.7 w**	**17.7 w**
Low income	566.8	1636.7	32.8	23.8	42.4	63.5	21.8	7.6	1.8	3.9	1.3	1.3
Excl. China & India	146.9	214.4	55.1	53.8	1.4	5.1	34.6	23.2	5.8	17.0	2.9	0.4
Middle income	2229.2	3304.9	21.2	26.1	15.3	25.3	54.8	15.4	4.7	24.2	3.6	8.2
Lower middle income	1601.7	2137.3	18.0	23.7	3.8	20.2	69.4	14.7	3.9	31.4	4.6	9.3
Upper middle income	627.5	1167.6	29.4	30.6	44.6	34.6	17.6	16.6	6.6	10.8	1.1	6.1
Low & middle income	2796.0	4941.6	23.6	25.4	20.8	37.9	48.1	12.8	4.1	17.4	3.2	5.9
East Asia & Pacific	391.6	1278.3	23.1	19.6	49.5	61.7	26.6	11.0	0.3	5.9	0.0	1.0
Europe & Central Asia	1648.0	1808.9	13.7	18.1	13.4	32.0	65.7	8.2	2.1	28.1	5.0	13.4
Latin America & Carib.	360.8	763.5	59.9	64.7	2.0	4.2	24.6	16.6	11.6	10.0	0.6	2.4
Middle East & N. Africa	104.0	360.0	20.5	7.6	1.0	1.5	52.2	49.6	26.4	41.3	..	..
South Asia	138.5	485.1	41.5	23.1	43.0	59.3	7.4	6.2	5.9	9.8	2.2	1.5
Sub-Saharan Africa	153.1	245.8	30.6	16.6	64.4	73.1	4.2	3.0	0.7	2.3	0.0	4.6
High income	5430.9	8239.1	19.4	15.0	39.4	37.3	18.1	8.1	11.2	13.0	11.5	24.8

Sources of electricity | 3.9

Use of energy in general, and access to electricity in particular, are major factors in improving people's standard of living. Electricity generation, however, has the potential to damage the environment. Whether such damage occurs largely depends on how electricity is generated. For example, burning coal as an energy input emits twice as much carbon dioxide—a major contributor to global warming—as does burning an equivalent amount of natural gas (see table 3.8). Nuclear energy does not generate any carbon dioxide emissions, but it is not as safe as solar energy. The table shows how "clean" generated electricity is.

The International Energy Agency (IEA) compiles data on energy inputs used to generate electricity. IEA data for non-OECD countries are based on national energy data that have been adjusted to conform with annual questionnaires completed by OECD member governments. In addition, estimates are sometimes made to complete major aggregates from which key data are missing, and adjustments are made to compensate for differences in definitions. The IEA makes these estimates in consultation with national statistical offices, oil companies, electricity utilities, and national energy experts.

The IEA occasionally revises its time series to reflect political changes. For example, since 1990 estimates of energy balances have been constructed for countries of the former Soviet Union. Energy statistics for some countries undergo continuous changes in coverage or methodology. For example, in recent years more detailed energy accounts have become available for some countries. Thus breaks in series are unavoidable.

Figure 3.9a

Sources of global electricity generation are shifting

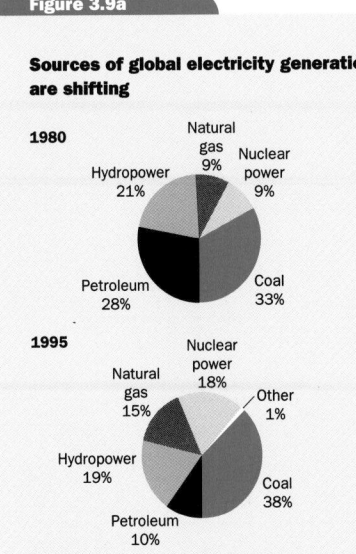

1980
Natural gas 9%
Nuclear power 9%
Hydropower 21%
Petroleum 28%
Coal 33%

1995
Nuclear power 18%
Natural gas 15%
Other 1%
Hydropower 19%
Coal 38%
Petroleum 10%

Source: Table 3.9.

Electricity generation has increased more than 60 percent since 1980—twice as much as population growth (see tables 2.1 and 3.9). Because fossil fuels are the main sources of generated electricity, this increase has been associated with various environmental impacts, including increased carbon dioxide emissions and depleted nonrenewable natural resources. As these figures show, since 1980 there has been a move toward "cleaner" sources of energy. Still, sources with the least environmental impact, such as solar energy, account for a small fraction of generated electricity.

• **Electricity production** is measured at the terminals of all alternator sets in a station. In addition to electricity generated by hydropower, coal, oil, gas, and nuclear power, it covers that generated by geothermal, solar, wind, and tide and wave energy, as well as that from combustible renewables and waste. Production includes the output of electricity plants that are designed to produce electricity only as well as that of combined heat and power plants.
• **Sources of electricity** refer to the inputs used to generate electricity: hydropower, coal, oil, gas, and nuclear power. *Hydropower* refers to electricity produced by hydroelectric power plants, *oil* refers to crude oil and petroleum products, *gas* refers to natural gas but excludes natural gas liquids, and *nuclear* refers to electricity produced by nuclear power plants. Shares may not sum to 100 percent because other sources of generated electricity (such as geothermal, solar, and wind) are not shown.

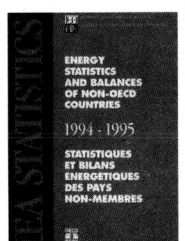

Data on electricity production are from the IEA's electronic files and its annual publications, *Energy Statistics and Balances of Non-OECD Countries, Energy Statistics of OECD Countries,* and *Energy Balances of OECD Countries.*

3.10 | Urbanization

	Urban population millions		Urban population % of total population		Population in urban agglomerations of more than one million % of total population			Population in the largest city % of urban population		Access to sanitation in urban areas % of urban population	
	1980	1996	1980	1996	1980	1996	2015	1980	1995	1980	1995
Albania	0.9	1.2	34	38	..	..	..	..	..	..	..
Algeria	8.1	16.2	43	56	11	13	16	25	24	95	..
Angola	1.5	3.5	21	32	13	20	30	63	62	..	34
Argentina	23.3	31.1	83	88	35	41	36	43	39	..	100
Armenia	2.0	2.6	66	69	34	34	41	51	50	..	..
Australia	12.6	15.5	86	85	47	57	59	26	23	..	..
Austria	4.9	5.2	65	64	27	26	27	42	40	..	100
Azerbaijan	3.3	4.2	53	56	26	25	29	48	44	..	..
Bangladesh	9.8	23.0	11	19	5	10	15	33	39	20	77
Belarus	5.4	7.4	57	72	14	17	20	24	24	..	..
Belgium	9.4	9.9	95	97	12	11	11	13	11	..	100
Benin	0.9	2.2	27	39	..	..	..	..	..	..	54
Bolivia	2.4	4.7	46	61	14	17	19	30	28	..	64
Bosnia and Herzegovina	1.5	..	36	42	..	..	..	..	..	..	..
Botswana	0.1	0.9	15	63	..	..	..	..	..	..	91
Brazil	80.3	127.3	66	79	27	33	34	16	13	..	55
Bulgaria	5.4	5.7	61	69	12	14	20	20	21	..	100
Burkina Faso	0.6	1.8	9	16	..	..	..	44	50	..	42
Burundi	0.2	0.5	4	8	..	..	..	..	..	..	..
Cambodia	0.8	2.2	12	21	..	..	..	..	..	..	..
Cameroon	2.7	6.2	31	46	6	19	14	19	22	..	73
Canada	18.6	23.0	76	77	29	35	35	16	19	..	..
Central African Republic	0.8	1.3	35	40	..	..	..	..	..	..	..
Chad	0.8	1.5	19	23	..	..	..	40	58	..	73
Chile	9.0	12.1	81	84	33	34	36	41	41	..	100
China	192.3	377.0	20	31	8	11	14	6	4	..	58
Hong Kong, China	4.6	6.0	92	95	91	93	86	100	99	..	..
Colombia	17.8	27.4	64	73	22	34	30	20	23	..	76
Congo, Dem. Rep.	7.8	13.1	29	29	..	10	..	28	34	8	..
Congo, Rep.	0.7	1.6	41	59	..	39	..	67	65	17	11
Costa Rica	1.0	1.7	43	50	..	..	..	61	55	..	..
Côte d'Ivoire	2.9	6.3	35	44	15	20	31	44	46	13	59
Croatia	2.3	2.7	50	56	..	..	..	28	37	..	72
Cuba	6.6	8.4	68	76	20	20	22	29	27	..	71
Czech Republic	6.5	6.8	64	66	12	12	13	18	18	..	..
Denmark	4.3	4.5	84	85	27	25	25	32	30	..	100
Dominican Republic	2.9	5.0	51	63	25	58	36	50	65	..	76
Ecuador	3.7	7.0	47	60	14	28	31	30	27	..	87
Egypt, Arab Rep.	17.9	26.6	44	45	23	25	25	38	37	95	20
El Salvador	1.9	2.6	42	45	..	22	..	40	48	..	78
Eritrea	..	0.6	..	17	..	..	..	..	..	..	..
Estonia	1.0	1.1	70	73	..	..	..	..	..	..	..
Ethiopia	4.0	9.2	11	16	3	4	6	30	28	..	..
Finland	2.9	3.3	60	64	..	21	..	22	33	100	100
France	39.5	43.7	73	75	21	20	20	23	22	..	100
Gabon	0.2	0.6	34	51	..	..	..	..	..	..	79
Gambia, The	0.1	0.3	20	30	..	..	..	..	..	..	83
Georgia	2.6	3.2	52	59	22	25	31	42	43	..	..
Germany	64.7	71.0	83	87	38	41	42	10	9	..	100
Ghana	3.4	6.4	31	36	9	10	14	30	27	..	50
Greece	5.6	6.2	58	59	31	30	37	54	50	..	100
Guatemala	2.6	4.3	37	39	..	21	..	29	53	..	78
Guinea	0.9	2.0	19	30	12	24	35	65	81	54	..
Guinea-Bissau	0.1	0.2	17	22	..	..	..	..	..	21	32
Haiti	1.3	2.4	24	32	13	21	27	55	64	..	42
Honduras	1.3	2.7	35	44	..	..	..	32	38	..	89

	Motor vehicles per 1,000 people		Motor vehicles per kilometer of road		Passenger cars per 1,000 people		Two-wheelers per 1,000 people		Road traffic (million vehicle kilometers)		Traffic accidents (people injured or killed per 1,000 vehicles)	
	1980	1996	1980	1996	1980	1996	1980	1996	1980	1996	1980	1996
Hungary	108	273	13	17	95	239	..	15	..	..	22	5
India	2	7	1	3	..	4	..	24	..	..	..	65
Indonesia	8	22	8	11	..	11	..	48	..	..	59	..
Iran, Islamic Rep.	..	38	..	15	..	29	..	..	..	..	..	15
Iraq	..	14	..	6	..	1	..	..	..	..	..	32
Ireland	236	307	9	12	216	272	8	6	14,917	28,390	11	10
Israel	123	263	114	98	107	208	..	12	10,442	31,060	38	32
Italy	334	674	65	122	303	571	114	44	226,569	453,160	12	7
Jamaica	..	50	..	7	..	41	..	..	..	..	..	..
Japan	323	552	34	60	203	374	102	120	389,052	689,800	16	14
Jordan	56	68	25	44	41	50	2	0	623	2,154	63	54
Kazakhstan	..	80	..	10	..	61	..	..	..	8,617	..	15
Kenya	8	13	3	6	7	10	1	1	..	6,200	74	75
Korea, Dem. Rep.	..	..	..	..	..	..	..	..	..	..	..	..
Korea, Rep.	14	195	11	106	7	151	6	54	8,728	56,940	212	42
Kuwait	390	404	..	156	..	338	..	10	12,189	..	7	..
Kyrgyz Republic	..	32	..	8	..	32	..	1	..	2,562	..	26
Lao PDR	..	4	..	1	..	3	..	49	..	..	..	..
Latvia	..	189	..	7	..	153	..	8	..	..	..	5
Lebanon	..	320	..	205	..	298	..	13	..	..	..	..
Lesotho	10	19	3	8	3	6	0	0	..	445	85	77
Libya	..	138	..	9	..	87	..	..	..	198	..	7
Lithuania	..	238	..	14	..	212	..	5	..	4,247	..	11
Macedonia, FYR	..	142	..	35	..	139	..	..	..	1	..	11
Madagascar	..	6	3	2	..	5	..	..	..	..	..	..
Malawi	5	6	3	2	2	3	2	1	..	..	..	..
Malaysia	..	152	..	33	52	131	101	164	..	3	..	17
Mali	..	4	..	3	..	3	..	..	..	..	..	22
Mauritania	..	13	..	4	..	8	..	..	..	..	..	..
Mauritius	44	88	23	54	27	63	27	85	46	..	..	4
Mexico	..	140	..	52	61	92	..	3	..	910	..	15
Moldova	..	54	..	19	..	39	..	25	..	..	..	20
Mongolia	..	26	..	2	..	12	..	10	..	..	..	44
Morocco	..	50	..	22	..	40	..	1	..	18	..	..
Mozambique	..	1	..	0	..	0	..	..	..	..	..	..
Myanmar	..	2	..	2	..	1	..	..	..	..	..	5
Namibia	..	83	..	2	..	40	..	1	..	2,149	..	..
Nepal	..	..	..	..	..	..	..	..	..	1	..	..
Netherlands	343	400	..	49	322	363	61	27	70,825	108,100	12	2
New Zealand	492	562	17	22	420	461	43	13	16,545	..	11	9
Nicaragua	..	30	..	8	9	16	..	5	..	150	..	29
Niger	6	6	..	5	5	4	..	..	..	240	63	..
Nigeria	4	12	3	7	3	7	4	3	..	..	123	..
Norway	342	470	17	22	302	379	36	46	..	25,386	..	6
Oman	..	134	..	9	..	97	..	2	..	..	..	23
Pakistan	2	7	5	4	2	5	3	12	..	31,950	71	28
Panama	..	99	..	24	..	76	..	..	..	..	..	33
Papua New Guinea	..	26	..	6	..	7	..	..	..	..	..	..
Paraguay	..	24	..	4	..	14	..	..	..	..	..	..
Peru	..	121	..	40	..	58	..	..	..	193	..	..
Philippines	..	13	..	5	6	9	4	11	..	..	..	..
Poland	86	248	10	26	67	209	..	43	44,597	118,530	..	8
Portugal	145	370	26	50	..	277	..	79	283	84,590	31	19
Puerto Rico	..	285	..	74	..	232	..	..	..	33,531	..	..
Romania	..	124	..	18	..	107	..	15	..	..	..	4
Russian Federation	..	158	..	24	..	92	..	..	..	..	..	9

	Motor vehicles				Passenger cars		Two-wheelers		Road traffic		Traffic accidents	
	per 1,000 people		per kilometer of road		per 1,000 people		per 1,000 people		million vehicle kilometers		people injured or killed per 1,000 vehicles	
	1980	1996	1980	1996	1980	1996	1980	1996	1980	1996	1980	1996
Rwanda	2	4	2	2	1	2	1	0	..	..	..	..
Saudi Arabia	163	149	26	18	67	90	..	0	..	*94,141*	..	*13*
Senegal	19	14	8	8	..	10	..	0	..	..	56	86
Sierra Leone	..	6	..	2	..	4	..	2	..	*29*	..	..
Singapore	..	167	..	168	*71*	120	*55*	44	..	..	..	14
Slovak Republic	..	217	..	32	..	198	..	38	..	*651*	..	11
Slovenia	..	387	..	52	..	365	..	4	..	8,037	..	11
South Africa	133	134	18	*16*	86	106	7	7	52,939	..	25	26
Spain	239	455	120	52	202	376	33	33	70,489	134,541	13	7
Sri Lanka	..	14	..	3	*8*	6	6	28	..	8,950	..	75
Sudan	..	12	..	28	..	10	..	..	..	..	..	..
Sweden	370	450	24	29	347	414	2	13	35,000	65,440	7	6
Switzerland	383	501	38	50	356	462	128	101	..	50,650	14	8
Syrian Arab Republic	..	28	..	10	..	10	..	..	..	..	..	24
Tajikistan	..	1	..	1	..	0	..	..	..	..	..	..
Tanzania	3	5	1	2	2	1	1	1	..	..	..	*104*
Thailand	13	106	13	97	9	28	19	171	16,824	99,900	29	*13*
Togo	..	27	1	15	..	19	..	14	..	*386*	..	..
Trinidad and Tobago	..	113	..	18	..	94	..	..	..	..	..	..
Tunisia	38	64	10	25	20	29	2	1	..	..	45	..
Turkey	23	70	4	11	..	55	..	14	14,785	41,015	26	25
Turkmenistan	..	..	..	..	..	..	..	..	..	..	..	..
Uganda	1	4	1	..	1	2	0	2	479	..	..	*130*
Ukraine	..	92	..	27	..	93	..	59	..	60,168	..	5
United Arab Emirates	..	99	..	52	..	79	..	..	..	..	..	..
United Kingdom	303	399	50	63	268	359	24	10	245,900	436,470	19	*14*
United States	..	767	25	32	536	521	30	14	2,418,620	2,577,600	..	17
Uruguay	..	166	..	63	..	151	..	100	..	..	..	4
Uzbekistan	..	..	..	..	..	..	..	..	..	..	..	..
Venezuela	112	88	27	23	92	68	41	..	56,900	..	32	..
Vietnam	..	..	..	..	..	..	..	45	..	..	..	..
West Bank and Gaza	..	..	..	..	..	..	..	..	..	..	..	..
Yemen, Rep.	..	34	..	8	8	15	3	..	1,251	11,476	..	15
Yugoslavia, FR (Serb./Mont.)	118	163	23	35	..	150	..	..	..	..	59	12
Zambia	..	26	..	6	..	17	..	..	..	..	..	..
Zimbabwe	..	32	..	19	..	29	..	32	..	..	..	54
World	**72 w**	**121 w**	..	..	..	**91 w**	**.. w**	**25 w**	..	..	..	***14 w***
Low income	2	8	..	..	..	4	..	13	..	..	..	..
Excl. China & India	..	10	..	..	..	6	..	..	..	..	..	..
Middle income	..	91	..	..	..	65	..	..	..	..	..	12
Lower middle income	..	70	..	..	..	46	..	..	..	..	..	13
Upper middle income	101	139	..	..	70	*111*	..	..	..	..	..	12
Low & middle income	14	36	..	..	..	23	..	19	..	..	..	14
East Asia & Pacific	3	15	..	..	..	7	..	22	..	..	..	18
Europe & Central Asia	..	142	..	..	..	109	..	..	..	..	..	10
Latin America & Carib.	..	92	..	..	62	72	..	..	..	..	..	..
Middle East & N. Africa	..	53	..	..	..	35	..	..	..	..	..	21
South Asia	2	6	..	..	..	4	..	20	..	..	..	61
Sub-Saharan Africa	..	20	..	..	..	14	..	..	..	..	..	36
High income	321	559	..	..	338	427	..	52	..	..	..	14

Traffic and congestion | 3.11

About the data

Traffic congestion in urban areas constrains economic productivity, damages people's health, and worsens their quality of life. In recent years ownership of passenger cars has increased, and the expansion of economic activity has contributed to the transport by road of more goods and services over greater distances. These developments have increased demand for roads and vehicles, adding to urban congestion, air pollution, health hazards, traffic accidents, and injuries.

Congestion, the most visible cost of expanding vehicle ownership, is reflected in the indicators in the table. Other relevant indicators, such as average vehicle speed in major cities or cost of traffic congestion, exact a heavy toll on economic productivity but are not included because data are incomplete or difficult to compare. Motor vehicles also emit particulate air pollution—the dust and soot from their exhaust—which is proving to be far more damaging to human health than was once believed. (See table

3.12 for information on suspended particulates and other air pollutants.)

The data in the table are compiled by the International Road Federation (IRF) through questionnaires sent to national organizations. The IRF uses a hierarchy of sources to gather as much information as possible. The primary sources are national road associations. In the absence of such an association, or in cases of nonresponse, other agencies are contacted, including road directorates, ministries of transport or public works, and central statistical offices. As a result the compiled data are of uneven quality. In addition, the coverage of each indicator may differ across countries because of differences in definitions. Moreover, comparability is limited when time-series data are reported. The data do not capture the quality or age of vehicles or the condition or width of roads. Thus comparisons over time and between countries should be made with caution.

Definitions

• **Motor vehicles per 1,000 people** include cars, buses, and freight vehicles but do not include two-wheelers. Population refers to midyear population in the year for which data are available. • **Motor vehicles per kilometer of road** include cars, buses, and freight vehicles but do not include two-wheelers. • **Passenger cars** refer to individual four-wheel transport. • **Two-wheelers** refer to mopeds and motorcycles. • **Road traffic** is the number of vehicles multiplied by the average distances they travel. • **Traffic accidents** refer to accident-related injuries reported to the authorities and to deaths resulting from accidents that occur within 30 days of the accident.

Data sources

The data in the table are from the IRF's annual *World Road Statistics*.

Figure 3.11a

Global production of motor vehicles continues to rise

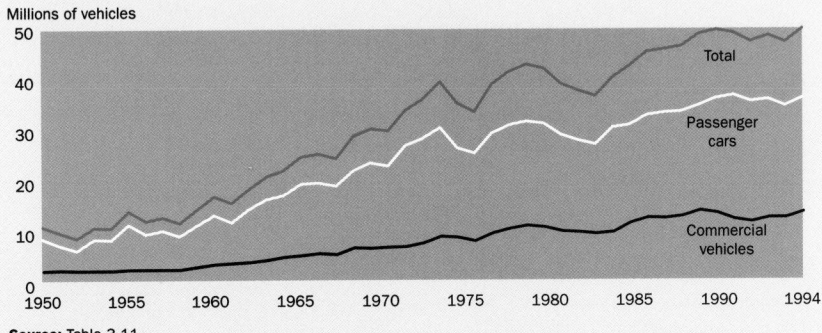

Source: Table 3.11.

In recent decades, along with household income, ownership of passenger cars has increased substantially. In 1946 about 46 million vehicles were registered worldwide. By 1994 that number exceeded 600 million. Growth in the number of private vehicles has substantially exceeded that of total vehicles, leading to relatively less use of public and mass transportation and thus to more pollution. The total number of vehicles (excluding motorized two- and three-wheel vehicles) is expected to reach 820 million by 2010.

3.12 | Air pollution

Country	City	City population	Total suspended particulates	Sulfur dioxide	Nitrogen dioxide
		thousands **1995**	micrograms per cu. m **1995[a]**	micrograms per cu. m **1995[a]**	micrograms per cu. m **1995[a]**
Argentina	Córdoba City	1,294	97	..	97
Australia	Sydney	3,590	54	28	..
	Melbourne	3,094	35	0	30
	Perth	1,220	45	5	19
Austria	Vienna	2,060	47	14	42
Belgium	Brussels	1,122	78	20	48
Brazil	São Paulo	16,533	86	43	83
	Rio de Janeiro	10,181	139	129	..
Bulgaria	Sofia	1,188	195	39	122
Canada	Toronto	4,319	36	17	43
	Montreal	3,320	34	10	42
	Vancouver	1,823	29	14	37
Chile	Santiago	4,891	..	29	81
China	Shanghai	13,584	246	53	73
	Beijing	11,299	377	90	122
	Tianjin	9,415	306	82	50
Colombia	Bogotá	6,079	120	..	..
Croatia	Zagreb	981	71	31	..
Cuba	Havana	2,241	..	1	5
Czech Republic	Prague	1,225	59	32	23
Denmark	Copenhagen	1,326	61	7	54
Ecuador	Guayaquil	1,831	127	15	..
	Quito	1,298	175	31	..
Egypt, Arab Rep.	Cairo	9,690	..	69	..
Finland	Helsinki	1,059	40	4	35
France	Paris	9,523	14	14	57
Germany	Frankfurt	3,606	36	11	45
	Berlin	3,317	50	18	26
	Munich	2,238	45	8	53
Ghana	Accra	1,673	137	..	..
Greece	Athens	3,093	178	34	64
Hungary	Budapest	2,017	63	39	51
Iceland	Reykjavik	100	24	5	42
India	Bombay	15,138	240	33	39
	Calcutta	11,923	375	49	34
	Delhi	9,948	415	24	41
Indonesia	Jakarta	8,621	271	..	..
Iran, Islamic Rep.	Tehran	6,836	248	209	..
Ireland	Dublin	911	..	20	..
Italy	Milan	4,251	77	31	248
	Rome	2,931	73	..	..
	Torino	1,294	151	..	..
Japan	Tokyo	26,959	49	18	68
	Osaka	10,609	43	19	63
	Yokohama	3,178	..	100	13
Kenya	Nairobi	1,810	69	..	..
Korea, Rep.	Seoul	11,609	84	44	60
	Pusan	4,082	94	60	51
	Taegu	2,432	72	81	62
Malaysia	Kuala Lumpur	1,238	85	24	..
Mexico	Mexico City	16,562	279	74	130
Netherlands	Amsterdam	1,108	40	10	58
New Zealand	Auckland	945	26	3	20
Norway	Oslo	477	15	8	43
Philippines	Manila	9,286	200	33	..
Poland	Katowice	3,552	..	83	79
	Warsaw	2,219	..	16	32

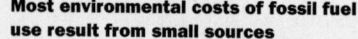

Figure 3.12a

Most environmental costs of fossil fuel use result from small sources

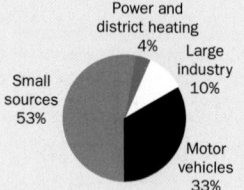

Power and district heating 4%
Large industry 10%
Small sources 53%
Motor vehicles 33%

Note: Data are an average for five cities: Krakow, Poland; Manila, Philippines; Mumbai, India; Santiago, Chile; and Shanghai, China.

Source: Lovei 1997.

Fuel combustion accounts for most air pollution emissions. Although the composition of pollution sources and contribution of different fuels vary by city, motor vehicles and small, dispersed sources—refuse incinerators, small industrial, commercial, and household boilers and stoves—usually do the most damage to air quality.

Country	City	City population	Total suspended particulates	Sulfur dioxide	Nitrogen dioxide
		thousands 1995	micrograms per cu. m 1995[a]	micrograms per cu. m 1995[a]	micrograms per cu. m 1995[a]
	Lodz	1,063	..	21	43
Portugal	Lisbon	1,863	61	8	52
Romania	Bucharest	2,100	82	10	71
Russian Federation	Moscow	9,269	100	109	..
	Omsk	1,199	100	9	30
Singapore	Singapore	2,848	..	20	30
Slovak Republic	Bratislava	651	62	21	27
South Africa	Capetown	2,671	..	21	72
	Johannesburg	1,849	..	19	31
	Durban	1,149	..	31	..
Spain	Madrid	4,072	42	11	25
	Barcelona	2,819	117	11	43
Sweden	Stockholm	1,545	9	5	29
Switzerland	Zurich	897	31	11	39
Thailand	Bangkok	6,547	223	11	23
Turkey	Istanbul	7,911	..	120	..
	Ankara	2,826	57	55	46
	Izmir	2,031	..	..	..
Ukraine	Kiev	2,809	100	14	51
United Kingdom	London	7,640	..	25	77
	Manchester	2,434	..	26	49
	Birmingham	2,271	..	9	45
United States	New York	16,332	..	26	79
	Los Angeles	12,410	..	9	74
	Chicago	6,844	..	14	57
Venezuela	Caracas	3,007	53	33	57

a. Data are most recent available for 1990–95. Most are for 1995.

Where coal is the primary fuel for power plants, steel mills, industrial boilers, and domestic heating, the result is usually high levels of urban air pollution—especially particulates and sometimes sulfur dioxide—and widespread acid deposition if the sulfur content of the coal is high. Countries such as China, India, Poland, and Turkey fit this pattern today, as many high-income countries once did. Where coal is not an important primary fuel or is used by plants with effective dust controls, the worst air pollutant emissions are caused by the combustion of petroleum products—diesel oil, heating oil, and heavy fuel oil. Industrial plants and motor vehicles—especially those with two-stroke engines, which do not fully process their fuel—are usually the worst offenders.

Data on air pollution are based on reports from urban monitoring sites. Annual means (measured in micrograms per cubic meter) are average concentrations observed at these sites. Coverage is not comprehensive because, due to lack of resources or different priorities, not all cities have monitoring systems. For example, data are reported for just 3 cities in Africa but for more than 87 cities in China. Pollutant concentrations are sensitive to local conditions, and even in the same city different monitoring sites may register different concentrations. Thus these data should be considered only a general indication of air quality in each city, and cross-country comparisons should be made with caution. World Health Organization (WHO) annual mean guidelines for air quality standards are 90 micrograms per cubic meter for total suspended particulates, 50 micrograms per cubic meter for sulfur dioxide, and 50 micrograms per cubic meter for nitrogen dioxide.

- **City population** is the number of residents of the city as defined by national authorities and reported to the United Nations. • **Total suspended particulates** refer to smoke, soot, dust, and liquid droplets from combustion that are in the air. Particulate levels indicate the quality of the air people are breathing and the state of a country's technology and pollution controls. • **Sulfur dioxide** (SO_2) is an air pollutant produced when fossil fuels containing sulfur are burned. It contributes to acid rain and can damage human health, particularly that of the young and the elderly. • **Nitrogen dioxide** (NO_2) is a poisonous, pungent gas formed when nitric oxide combines with hydrocarbons and sunlight, producing a photochemical reaction. These conditions occur in both natural and anthropogenic activities. NO_2 is emitted by bacteria, nitrogenous fertilizers, aerobic decomposition of organic matter in oceans and soils, combustion of fuels and biomass, and motor vehicles and industrial activities.

Data sources

Copyright ©1997
World Health Organization
All rights reserved

Healthy Cities
Air Management

Information System
AMIS 1.0, 1997
WHO, Geneva

The data in the table are from WHO's Healthy Cities Air Management Information System and the World Resources Institute, which relies on various national sources as well as, among others, the United Nations Environment Programme (UNEP) and WHO's *Urban Air Pollution in Megacities of the World*, the OECD's *Compendiums of Environmental Data*, the U.S. Environmental Protection Agency's *National Air Quality and Emissions Trends Report 1995* and AIRS Executive International database, the *China Environmental Yearbook 1996*, and the *Korea Statistical Yearbook 1995*.

3.13 | Government commitment

	Environmental strategy or action plan	Country environmental profile	Biodiversity assessment, strategy, or action plan	Frequency of reporting on trade in endangered species	Participation in treaties[a]			
				% of years reported[b]	Climate change	Ozone layer	CFC control	Law of the Sea[c]
Albania	1993	..	..		1995			
Algeria	..	..		50	1994	1993	1993	1996
Angola		..						1994
Argentina	1992	..		82	1994	1990	1990	1996
Armenia	..	..			1994			
Australia	1992	..	1994	88	1994	1988	1989	1995
Austria		..		100	1994	1988	1989	1995
Azerbaijan	..	..			1995			1996
Bangladesh	1991	1989	1990	80	1994	1990	1990	
Belarus	..	..				1988	1989	
Belgium	..	..		100	1996	1989	1989	
Benin	1993	..		0	1994	1993	1993	
Bolivia	1994	1986	1988	62	1995	1995	1995	1995
Bosnia and Herzegovina		..	..		1992			1994
Botswana	1990	1986	1991	86	1994	1992	1992	1994
Brazil		..	1988	41	1994	1990	1990	1994
Bulgaria		..	1994	0	1995	1991	1991	1996
Burkina Faso	1993	1994	..	0	1994	1989	1989	
Burundi	1994	1981	1989	0				
Cambodia	1997	1994			1996			
Cameroon	..	1989	1989	92	1995	1989	1989	1994
Canada	1990	..	1994	100	1994	1988	1989	
Central African Republic	..	..	..	50	1995		1993	
Chad	1990	1982	..	0	1994	1989	1994	
Chile		1987	1993	65	1995		1990	
China	1994	..	1994	100	1994	1989	1991	1996
Hong Kong, China		..	..					
Colombia		1990	1988	64	1995	1990	1994	
Congo, Dem. Rep.		1986	1990	69	1995	1995		1994
Congo, Rep.	..	..	1990	100	1997	1995	1995	
Costa Rica	1990	1987	1992	76	1994	1991	1991	1994
Côte d'Ivoire	1994	..	1991		1995	1993	1993	1994
Croatia	..	..	..		1996	1992	1991	1994
Cuba	..	..		50	1994	1992	1992	1994
Czech Republic	1991	..		100	1994	1993	1993	1996
Denmark	1994	..		100	1994	1988	1989	
Dominican Republic	..	1984	1995	80		1993	1993	
Ecuador	1993	1987	1995	76	1994	1990	1990	
Egypt, Arab Rep.	1992	1992	1988	0	1995	1988	1989	1994
El Salvador	1994	1985	1988	20	1996	1993	1992	
Eritrea	1995	..	..		1995			
Estonia	..	..			1994	1997	1997	
Ethiopia	1994	..	1991	75	1994	1995	1995	
Finland	1995	..	..	75	1994	1988	1989	1996
France	1990	..	..	100	1994	1988	1989	1996
Gabon	..	..	1990	67		1994	1994	
Gambia, The	1992	1981	1989	20	1994	1990	1990	1994
Georgia	..	..			1994	1996	1996	1996
Germany	..	..		100	1994	1988	1989	1994
Ghana	1992	1985	1988	81	1995	1989	1989	1994
Greece	..	..			1994	1989	1989	1995
Guatemala	1994	1984	1988	83	1996	1988	1990	
Guinea	1994	1983	1988	45	1994	1992	1992	1994
Guinea-Bissau	1993	..	1991	0	1996			1994
Haiti	..	1985	..		1996			1996
Honduras	1993	1989	..	29	1996	1994	1994	1994

Table 3.13a

Status of national environmental action plans

Completed

Albania	Grenada	Nepal
Azerbaijan	Guinea	Nicaragua
Bangladesh	Guinea-Bissau	Nigeria
Belarus	Guyana	Pakistan
Benin	Honduras	Papua New
Bhutan	Hungary	Guinea
Botswana	India	Bulgaria
Indonesia	Philippines	Poland
Burkina Faso	Kenya	Romania
Burundi	Lao PDR	Rwanda
Cambodia	Latvia	São Tomé and
Cape Verde	Lebanon	Principe
China	Lesotho	Senegal
Comoros	Lithuania	Seychelles
Congo, Rep.	Macedonia, FYR	Sierra Leone
Costa Rica	Madagascar	Sri Lanka
Côte d'Ivoire	Maldives	St. Kitts and
El Salvador	Mexico	Nevis
Estonia	Moldova	Uganda
Ethiopia	Mongolia	Ukraine
Gambia, The	Mozambique	Zambia
Ghana		

Being prepared

Armenia	Haiti	St. Lucia
Djibouti	Kazakhstan	St. Vincent &
Dominican Rep.	Korea, Rep.	Grenadines
Ecuador	Malaysia	Togo
Equatorial Guinea	Mali	Uzbekistan
Gabon	Niger	Vietnam
Georgia	Paraguay	Zimbabwe

Source: World Resources Institute, International Institute for Environment and Development, and IUCN 1996; World Bank.

Government commitment | 3.13

	Environmental strategy or action plan	Country environmental profile	Biodiversity assessment, strategy, or action plan	Frequency of reporting on trade in endangered species	Participation in treaties[a]			
				% of years reported[b]	Climate change	Ozone layer	CFC control	Law of the Sea[c]
Hungary	1995	..	..	57	1994	1988	1989	
India	1993	1989	1994	100	1994	1991	1992	1995
Indonesia	1992	1994	1993	92	1994	1992	1992	1994
Iran, Islamic Rep.	..	..	..	31	1996	1991	1991	
Iraq								1994
Ireland	..	..	..		1994	1988	1989	
Israel	..	..	..	25	1994	1992	1992	
Italy	..	..	..	86	1994	1988	1989	1995
Jamaica	1994	1987			1995		1993	1994
Japan	..	..	..	92	1994	1988	1989	1996
Jordan	1991	1979		31	1994	1989	1989	1995
Kazakhstan					1995			
Kenya	1994	1989	1992	54	1994	1989	1989	1994
Korea, Dem. Rep.	..	..	..		1995	1995	1995	
Korea, Rep.	..	..	..		1994	1992	1992	1996
Kuwait	..	..	..		1995	1993	1993	1994
Kyrgyz Republic	..	..	..					
Lao PDR	1995	..	..		1995			
Latvia	..	..	..		1995	1995	1995	
Lebanon	..	..	..		1995	1993	1993	1995
Lesotho	1989	1982			1995	1994	1994	
Libya	..	..	..			1990	1990	
Lithuania	..	..	..		1995	1995	1995	
Macedonia, FYR	..	..	..		1994	1994	1994	
Madagascar	1988	..	1991	82	1996	1997	1997	
Malawi	1994	1982	..	70	1994	1991	1991	
Malaysia	1991	1979	1988	86	1994	1989	1989	1997
Mali	..	1991	1989		1995	1995	1995	1994
Mauritania	1988	1984	..		1994	1994	1994	1996
Mauritius	1990	..	..	88	1994	1992	1992	1994
Mexico	..	..	1988	100	1994	1988	1989	1994
Moldova	..	..	..		1995	1997	1997	
Mongolia	1995	..	..		1994	1996	1996	
Morocco	..	1980	1988	44	1996	1996	1996	
Mozambique	1994	..	..	73	1995	1994	1994	
Myanmar	..	1982	1989		1995	1994	1994	1996
Namibia	1992	..	..	0	1995	1993	1993	1994
Nepal	1993	1983	..	76	1994	1994	1994	
Netherlands	1994	..	..	100	1994	1988	1989	1996
New Zealand	1994	..	..	67	1994	1988	1989	1996
Nicaragua	1994	1981	..	80	1996	1993	1993	
Niger	..	1985	1991	41	1995		1993	
Nigeria	1990	..	1992	18	1994	1989	1989	1994
Norway	..	..	1994	100	1994	1988	1989	1996
Oman	..	1981	..		1995			1994
Pakistan	1994	1994	1991	94	1994	1993	1993	
Panama	1990	1980	..	86	1995	1989		1996
Papua New Guinea	1992	1994	1993	75	1994	1993	1993	
Paraguay	..	1985	..	60	1994	1993	1993	1994
Peru	..	1988	1988	59	1994	1989	1993	
Philippines	1989	1992	1989	82	1994	1991	1991	1994
Poland	1993	..	1991	0	1994	1990	1990	
Portugal	1995	..	..	55	1994	1989	1989	
Puerto Rico	..	..	..					
Romania	..	..	..		1994	1993	1993	1997
Russian Federation	..	..	1994		1995	1988	1989	

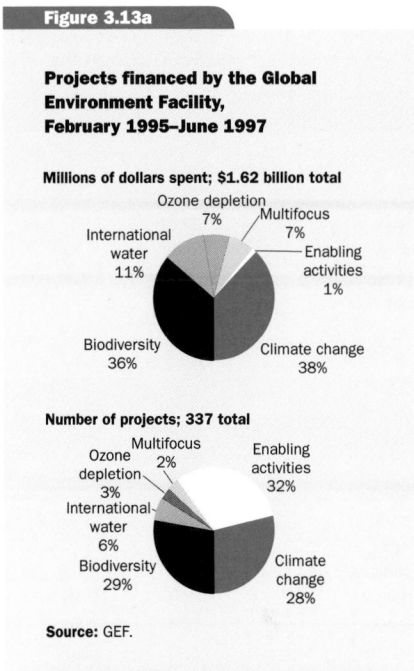

Figure 3.13a

Projects financed by the Global Environment Facility, February 1995–June 1997

Millions of dollars spent; $1.62 billion total

Ozone depletion 7%
Multifocus 7%
International water 11%
Enabling activities 1%
Biodiversity 36%
Climate change 38%

Number of projects; 337 total

Multifocus 2%
Ozone depletion 3%
International water 6%
Enabling activities 32%
Biodiversity 29%
Climate change 28%

Source: GEF.

Through its implementing agencies—the United Nations Development Programme, United Nations Environment Programme, and World Bank—the Global Environment Facility provides funding to developing and transition economies for projects and activities targeting global benefits in one or more of four focus areas: biodiversity, climate change, international waters, and ozone depletion. Activities involving desertification and deforestation, as they relate to the focus areas, are also eligible for funding. Enabling activities lay the foundations for designing and implementing measures to achieve the overall goals in the focus areas. Between February 1995 and June 1997 the facility provided $1.62 billion for a total of 337 projects.

	Environmental strategy or action plan	Country environmental profile	Biodiversity assessment, strategy, or action plan	Frequency of reporting on trade in endangered species	Participation in treaties[a]			
				% of years reported[b]	Climate change	Ozone layer	CFC control	Law of the Sea[c]
Rwanda	1991	1987	..	27				
Saudi Arabia	..	..	..		1995	1993	1993	
Senegal	1984	1990	1991	80	1995	1993	1993	1994
Sierra Leone	1994	..	..		1995			1995
Singapore	1993	1988	1995	100		1989	1989	1994
Slovak Republic	..	..	..		1994	1993	1993	1996
Slovenia	..	..	..		1996		1992	1994
South Africa	1993	..	..	94		1990	1990	1994
Spain	..	..	..	100	1994	1988	1989	
Sri Lanka	1994	1983	1991	54	1994	1990	1990	1994
Sudan	..	1989	..	44	1994		1993	1994
Sweden	..	..	..	94	1994	1988	1989	1996
Switzerland	..	..	..	100	1994	1988	1989	
Syrian Arab Republic	..	1981	..		1996	1990	1990	
Tajikistan	..	..	..					
Tanzania	1994	1989	1988	75		1993	1993	1994
Thailand	..	1992	..	56	1995	1989	1989	
Togo	1991	..	..	69	1995	1991	1991	1994
Trinidad and Tobago	..	..	..	67	1994	1989	1989	1994
Tunisia	1994	1980	1988	100	1994	1989	1989	1994
Turkey	..	1982	..			1991	1991	
Turkmenistan	..	..	..		1995	1994	1994	
Uganda	1994	1982	1988	0	1994	1988	1989	1994
Ukraine	..	..	..			1988	1989	
United Arab Emirates	..	..	..	0	1996	1990	1990	
United Kingdom	1995	..	1994	100		1988	1989	
United States	1995	..	1995	88	1994	1988	1989	
Uruguay	..	..	..	59	1994	1989	1991	1994
Uzbekistan	..	..	..		1994	1993	1993	
Venezuela	..	..	..	79	1995	1988	1989	
Vietnam	..	..	1993		1995	1994	1994	1994
West Bank and Gaza	..	..	..					
Yemen, Rep.	..	1990	1992		1996	1996	1996	1994
Yugoslavia, FR (Serb./Mont.)	..	..	..					
Zambia	1994	1988	..	45	1994	1990	1990	1994
Zimbabwe	1987	1982	..	83	1994	1993	1993	1994

a. The year the treaty entered into force in that country. b. Includes all trade reported by members of the Convention on International Trade in Endangered Species of Wild Flora and Fauna (CITES) as of May 1993. c. Convention became effective November 16, 1994.

The Kyoto Protocol on climate change

The Kyoto Protocol was adopted at the third conference of the parties to the United Nations Framework Convention on Climate Change, held in Kyoto, Japan in December 1997.

When the protocol enters into force, it will impose binding emission limitations in industrial and transition economies known as "annex 1" countries. These countries agreed to ensure that their greenhouse gas emissions do not exceed their assigned amounts, with a view to reducing overall emissions of such gases by at least 5.2 percent over 1990 levels by 2008–12.

The protocol will:

- Allow quantification of emissions to take into account carbon sinks (such as forests).
- Provide for possible carbon emissions trading among annex 1 countries.
- Allow for joint project implementation activities among annex 1 countries.
- Create a "clean development mechanism" that enables annex 1 countries to undertake joint implementation projects in developing countries.
- Commit all countries to cooperate in the development and transfer of climate-friendly technologies, commit industrial countries to help developing countries create a private sector–enabling environment, and commit annex 1 countries to eliminating relevant market distortions.
- Enter into force after ratification by 55 countries that account for at least 55 percent of carbon dioxide (CO_2) emissions by annex 1 countries. (The United States accounts for 38 percent of CO_2 emissions by annex 1 countries.)

Government commitment | 3.13

Unlike most other tables in this book, this table presents qualitative rather than quantitative indicators. National environmental strategies and participation in international treaties on environmental issues provide some evidence of government commitment to sound environmental management. But the signing of these treaties does not always imply ratification. Nor does it guarantee that governments will comply with treaty obligations.

In many countries efforts to halt environmental degradation have failed, primarily because governments have neglected to make this issue a priority, a reflection of competing claims on scarce resources. To address this problem, many countries are preparing national environmental strategies—some focusing narrowly on environmental issues, others dealing with the integration of environmental, economic, and social concerns. Among such initiatives are conservation strategies and environmental action plans. Some countries have also prepared country environmental profiles and biological diversity strategies and profiles.

National conservation strategies—promoted by the World Conservation Union (IUCN)—provide a comprehensive, cross-sectoral analysis of conservation and resource management issues to help integrate environmental concerns with the development process. Such strategies discuss a country's current and future needs, institutional capabilities, prevailing technical conditions, and the status of natural resources.

National environmental action plans (NEAPs), supported by the World Bank and other development agencies, describe a country's main environmental concerns, identify the principal causes of environmental problems, and formulate policies and actions to deal with them (table 3.13a). The NEAP is a continuing process in which governments develop comprehensive environmental policies, recommend specific actions, and outline the investment strategies, legislation, and institutional arrangements required to implement them.

Country environmental profiles identify how national economic and other activities can stay within the constraints imposed by the need to conserve natural resources. Some profiles consider issues of equity, justice, and fairness. Biodiversity profiles—prepared by the World Conservation Monitoring Centre and the IUCN—provide basic background on species diversity, protected areas, major ecosystems and habitat types, and legislative and administrative support. In an effort to establish a scientific base-

line for measuring progress on biodiversity conservation, the United Nations Environment Programme (UNEP) coordinates global biodiversity assessment.

To address global issues, many governments have also signed international treaties and agreements, launched in the wake of the 1972 United Nations Conference on Human Environment in Stockholm and the 1992 United Nations Conference on Environment and Development in Rio de Janeiro:

• The Convention on Climate Change aims to stabilize atmospheric concentrations of greenhouse gases at levels that will prevent human activities from interfering dangerously with the global climate.

• The Vienna Convention for the Protection of the Ozone Layer aims to protect human health and the environment by promoting research on the effects of changes in the ozone layer and on alternative substances (such as substitutes for chlorofluorocarbons) and technologies, monitoring the ozone layer, and taking measures to control the activities that produce adverse effects.

• The Montreal Protocol for CFC Control requires that countries help protect the earth from excessive ultraviolet radiation by cutting chlorofluorocarbon consumption by 20 percent over their 1986 level by 1994 and by 50 percent over their 1986 level by 1999, with allowances for increases in consumption by developing countries.

• The United Nations Convention on the Law of the Sea, which became effective in November 1994, establishes a comprehensive legal regime for seas and oceans, establishes rules for environmental standards and enforcement provisions, and develops international rules and national legislation to prevent and control marine pollution.

• The Convention on International Trade in Endangered Species of Wild Flora and Fauna (CITES) prohibits commercial trade in endangered species and requires signatories to closely monitor trade in species that may become depleted by trade.

To help developing countries comply with their obligations under these agreements, the Global Environment Facility (GEF) was created to focus on global improvement in biodiversity, climate change, international waters, and ozone layer depletion. The UNEP, United Nations Development Programme, and World Bank manage the GEF according to the policies of its governing body of country representatives. The World Bank is responsible for the Global Environmental Trust Fund and is chair of the GEF.

• **Environmental strategies and action plans** provide a comprehensive, cross-sectoral analysis of conservation and resource management issues to help integrate environmental concerns with the development process. They include national conservation strategies, national environmental action plans, national environmental management strategies, and national sustainable development strategies. The years shown refer to the year in which a strategy or action plan was adopted. • **Country environmental profiles** identify how national economic and other activities can stay within the constraints imposed by the need to conserve natural resources. The years shown refer to the year in which a profile was completed. • **Biodiversity assessments, strategies, or action plans** covers biodiversity assessments, country strategies or action plans, and biodiversity profiles (see *About the data*). • **Frequency of reporting on trade in endangered species** refers to the percentage of years for which a country has submitted an annual report to the CITES Secretariat since it became a party to the Convention on International Trade in Endangered Species. • **Participation in treaties** covers four international treaties (see *About the data*). • **Climate change** refers to the Convention on Climate Change (signed in New York in 1992). • **Ozone layer** refers to the Vienna Convention for the Protection of the Ozone Layer (1985). • **CFC control** refers to the Montreal Protocol for CFC Control (formally, the Protocol on Substances that Deplete the Ozone Layer, signed in 1987). • **Law of the Sea** refers to the United Nations Convention on the Law of the Sea (signed in Montego Bay, Jamaica, in 1982). The years shown refer to the year in which a treaty entered into force in a country.

 Data are from the World Resources Institute, UNEP, and UNDP's *World Resources 1994–95;* the World Resources Institute, International Institute for Environment and Development, and IUCN's 1996 *World Directory of Country Environmental Studies;* and the World Bank Environment Department's 1996 *National Environmental Strategies: Learning from Experience.*

Developing countries set the pace for world economic growth during the past two decades, but the gap between leaders and laggards grew wider. While some of the most heavily populated countries saw incomes rise, others experienced further deterioration in living standards, often linked to political and social upheavals.

Low- and middle-income economies with the strongest growth rates were generally characterized by:

- Commitment to improved policy environments in terms of domestic macroeconomic management and economic liberalization.
- Open door policies toward foreign trade and foreign investment, which contributed to closer integration with the global economy.
- Strong growth of the service sector and expanding trade in services.
- A decreasing share of central government expenditure in GDP.

The indicators presented in this section document the patterns of economic growth over 1980–96. (Long-term changes, back to 1965, are shown in tables 1.4 and 1.5.) Tables 4a and 4b provide current estimates and projections of major macroeconomic aggregates through 2000. Tables 4c and 4d provide recent, highly preliminary estimates of major macroeconomic aggregates for selected developing countries.

Recent economic developments

Starting from a 30-year low of 2 percent in the early 1990s, the pace of world economic expansion quickened to closer to 3 percent a year in 1996–97, similar to growth rates achieved during the 1980s. Developing countries, excluding Eastern Europe and the former Soviet Union, achieved GDP increases of more than 5 percent. Low- and middle-income economies in East Asia (including China) made the largest gains, averaging more than 10 percent a year, though progress slackened in the mid-1990s because of flagging exports and, more recently, the financial crisis that struck several of the fastest-growing Asian economies. In South Asia growth accelerated to more than 6 percent in 1996 from 4.5 percent in the previous five years, but growth slowed in India and Pakistan in 1997 while picking up in Bangladesh and Sri Lanka.

Foreign trade expansion enabled Latin America and the Caribbean to maintain growth rates of more than 3 percent a year through the 1990s, almost twice the increase achieved during the 1980s. Argentina, Mexico, and Peru all showed increasing economic growth in 1997 and, particularly Mexico, large increases in exports. Chile's growth declined slightly but remained above average.

Merchandise trade in the Middle East and North Africa has helped to underpin GDP growth of about 3 percent a year since 1991, compared with increases below 1 percent a year in the 1980s. However, the region's growth rate barely kept up with population increases and remains far below the annual gains recorded 20 and 30 years ago. Preliminary 1997 data show mixed results, with Egypt and Lebanon posting above-average growth but Morocco declining.

Prospects for the long-underperforming Sub-Saharan Africa region brightened when GDP growth of almost 4 percent was achieved in 1996, resulting in a real increase in per capita incomes. Preliminary 1997 data for Ghana, Nigeria, and South Africa show positive but modest growth rates.

Eastern Europe and the former Soviet Union experienced a sharp decline in growth during the first half of the 1990s—more than 20 percent a year in some cases. By 1996 the downtrend had slowed markedly, particularly in those countries that allowed the market economy to replace failing state enterprises and inflexible trading regimes. Growth was still negative for the region as a whole, but the annual rate of decline had fallen to 2 percent, from an average of almost 8 percent in the preceding five years. In Russia preliminary estimates for 1997 show no real decrease in the economy and a modest increase in exports.

Some aspects of growth in developing economies stand out:

•   *Service sectors expanded rapidly and increased their share of total output.* The growth of service activities in East Asia and the Pacific was three times the world average in the 1990s. South Asia experienced similar growth. In Latin America and the Caribbean services account for more than half of total output and continue to grow faster than the rest of the economy (tables 4.1 and 4.2). Since 1980 low- and middle-income countries, particularly in Europe and Central Asia, have increased their share of fast-growing world exports of services from 15 to almost 20 percent (tables 4.6 and 4.7).

•   *Trade grew faster than GDP, but the pace slackened.* Since the 1980s export growth has exceeded the expansion of world output. In 1996 weakness in industrial country import demand, the recent dollar appreciation (which affected the competitiveness of exporting countries with currencies linked to the dollar), and lower prices for certain commodities and semimanufactured goods all played a part in slowing world trade growth to 4 percent. Developing countries did better than the rest of the world: in East Asia merchandise trade expanded at an average annual rate of 17 percent during the 1990s. Exports of both goods and services grew to the equivalent of 30 percent of GDP, about twice the 1980 ratio. South Asian export earnings also grew faster than average. Latin America and the Caribbean managed double-digit increases after a decade of stagnation (tables 4.4 and 6.2).

•   *Increases in domestic savings helped to boost investment.* Here again the East Asia region, particularly China, stood out for its strong savings and investment performance. Private sector failures caused by unsound banking practices and overinvested property sectors led to serious liquidity problems in some of the region's "tiger" economies in 1997, with worldwide financial repercussions. Investment spend-

ing in Sub-Saharan Africa grew by more than 3 percent after a decade of retrenchment in the 1980s, but domestic savings rates remained below world averages (tables 4.8 and 4.9).

•   *Inflation was kept low.* Rates of inflation accelerated worldwide in the 1990s compared with the 1980s. Against this background, developing countries that managed to keep inflationary pressures under control mostly achieved better-than-average GDP growth rates (tables 4.1 and 4.15).

A global service economy

Service industries increased their share of the world economy during the past two decades, while the relative shares of agriculture and industry shrank in most developing regions. Compared with about 55 percent in 1980, this sector contributed almost two-thirds of global GDP in 1996, and one-fifth of the value of world exports. Developing countries' exports of services grew at an average annual rate of 12 percent in the 1990s, twice as fast as those from industrial regions (World Bank 1997b, p. 11). Services also absorb an increasing share of household consumption expenditures in developing countries. New estimates of purchasing power parities (PPPs) show that services typically account for between one-third and one-half of household consumption. Within individual countries low-income groups spend as large a share of their budgets on services as do the better off, but generally pay less for the services that they buy (tables 4.10 and 4.11).

While production by multinational firms is becoming increasingly important in manufacturing, the impact of global service industries in developing countries is still small, concentrated in finance, communications, and leisure industries. Even so, deregulation and the opening of sectors such as telecommunications and utilities to foreign investors are speeding the process of globalization (World Bank 1997b, p. 39). Recent experiences in low- and middle-income countries of the former Soviet Union have shown that the process can put considerable pressure on domestic prices. As subsidies and regulatory controls are eliminated, public services are required to become more efficient and cost-effective, particularly in capital-intensive sectors like housing, utilities, and transportation. In 17 of 22 of these countries, including Russia, the rise in consumer prices for services far exceeded the overall rate of inflation during 1996. The increases were attributable partly to the process of convergence with the prices prevailing in market economy trading partners, and partly to the direct impact of prices set by subsidiaries of multinational companies for services ranging from fast food to oil prospecting expertise. Adjustments in domestic prices reflecting these external influences have pushed up overall inflation rates, particularly in the early stages of transition from central control to market economy systems (IMF 1997b, p. 109).

Prospects for growth in 1998–2000

The buoyant growth of developing countries in the past few years—reflected in tables 4c and 4d and in many of the tables throughout the *World Development Indicators*—was interrupted by the outbreak of the financial crisis in East Asia during the middle of 1997. By the end of the year four developing Asian economies had been severely affected—Indonesia, Malaysia, the Philippines, and Thailand—

along with the Republic of Korea, which only recently graduated to the high-income category. The immediate effects of the crisis were a severe devaluation of currencies, a drop in stock market values, and a sharp decline in gross capital flows. In varying degrees, the crisis in Asia has affected confidence in other developing countries—especially where investors perceive weaknesses in banking systems or risks in growing short-term debt and overvalued currencies. The result is a reduction in expected short-term growth rates for developing countries, and a much greater degree of uncertainty over the course of the next three years.

Nevertheless, the economic environment facing developing countries remains generally favorable (see table 4a). G-7 inflation and international interest rates are expected to remain relatively low. The strong growth of world trade characterizing the growth of the mid-1990s is likely to continue, although the prices of many commodities, especially oil, are expected to fall in the near term, causing a decline in the terms of trade for many developing countries. Continuing growth in the United States and Europe will stimulate demand for exports, especially from countries that have undergone real exchange rate depreciation. Japan's stalled recovery, and its effect on other Asian economies, is the major uncertainty in the G-7 outlook.

If the reform programs and financial restructurings undertaken in the wake of the Asian financial crisis are successful, growth of the developing countries should rebound after a period of adjustment in 1998 (see table 4b). The regions most affected—East Asia (excluding China) and Latin America—will see sharply reduced growth in 1998, but will substantially recover by 1999. Capital flows will resume as confidence is restored. The prospects for other regions are tied less to events in Asia and depend more on continuing progress in macroeconomic management, trade liberalization, and political stability. In South Asia current account deficits have generally been small, real exchange rates have been stable, and the banking system is less exposed to short-term debt. With continuing economic reform, though politically difficult, growth is expected to pick up to an average of 6 percent in 1998–2000. The Middle East and North Africa will be more acutely affected by a decline in oil prices. Growth at near the levels achieved in 1996 depends in large part on the success of diversified exporters such as Egypt, Jordan, Lebanon, Morocco, Tunisia, and Syria. Despite the improved performance of many countries and increased attractiveness to private investors, the region faces many challenges and is subject to significant downside risk. In Sub-Saharan Africa growth is expected to improve despite declining terms of trade, uncertain capital flows, and possible extreme weather patterns. In Europe and Central Asia the effects of the Asian crisis appear to have been short-lived, but large or rising current account deficits and fragile domestic financial systems expose a number of countries to further risk. Growth in the region is strongly tied to recovery in Western Europe and expanding trade.

Note: Preliminary 1997 macroeconomic aggregates are estimates prepared by World Bank staff and are subject to revision. The projections for 1998–2000 were prepared by the staff of the Development Prospects Group as part of their work on *Global Economic Prospects and the Developing Countries 1998* (forthcoming).

Table 4a

External environment for developing country growth

Annual % change, except for LIBOR

	Current estimate		Projection		
	1996	1997	1998	1999	2000
GDP growth in G-7 economies[a]	2.3	2.7	2.2	2.5	2.5
Inflation in G-7 economies[a,b]	2.1	1.8	1.8	2.1	2.3
World export volume	5.1	7.8	6.3	6.3	6.3
Six-month LIBOR (% per year)[c,d]	5.7	5.8	5.7	5.8	6.1
Price indexes[d]					
Nonfuel commodity prices	−5.8	2.2	−9.8	−1.5	2.0
Petroleum prices	18.9	−6.1	−11.5	2.9	0.0
G-5 export unit value of manufactures[e]	−4.1	−4.0	2.5	3.1	2.9

a. Canada, France, Germany, Italy, Japan, the United Kingdom, and the United States.
b. Consumer price index in local currency aggregated using 1988–90 GDP weights.
c. London interbank offered rate.
d. Measured in U.S. dollars.
e. Unit value index of manufactures exports from G-5 to developing countries, expressed in U.S. dollars.
Source: World Bank staff estimates.

Table 4b

Growth of world GDP, 1996–2000

Annual % change

	Current estimate		Projection		
	1996	1997	1998	1999	2000
World	2.9	3.2	2.6	3.1	3.2
High-income economies	2.5	2.8	2.4	2.6	2.7
Low- and middle-income economies	4.5	4.8	4.0	4.8	5.2
excl. Eastern Europe and former Soviet Union	..	5.3	4.2	4.9	5.1
East Asia and the Pacific	8.6	7.8	5.7	6.3	6.7
excl. China	..	4.0	0.7	3.1	4.3
Europe and Central Asia	−0.3	2.3	3.0	4.0	5.1
Latin America and the Caribbean	3.4	4.8	2.7	3.7	3.8
Middle East and North Africa	4.1	3.1	2.7	3.2	3.5
South Asia	6.5	5.6	5.8	6.1	6.3
Sub-Saharan Africa	3.8	3.4	3.4	4.5	4.5

Source: World Bank staff estimates.

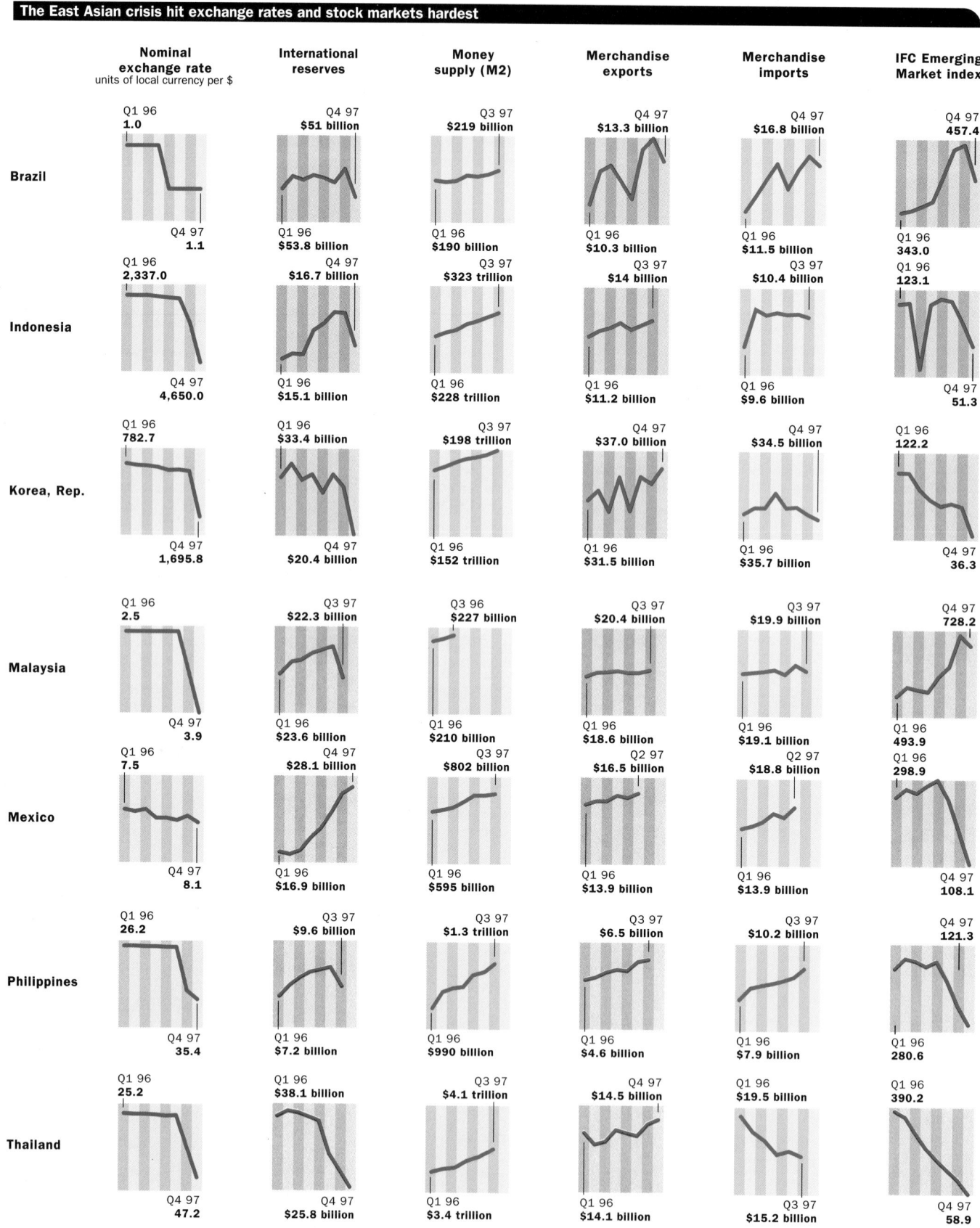

	Nominal exchange rate units of local currency per $	International reserves	Money supply (M2)	Merchandise exports	Merchandise imports	IFC Emerging Market index
Brazil	Q1 96 1.0 / Q4 97 1.1	Q4 97 $51 billion / Q1 96 $53.8 billion	Q3 97 $219 billion / Q1 96 $190 billion	Q4 97 $13.3 billion / Q1 96 $10.3 billion	Q4 97 $16.8 billion / Q1 96 $11.5 billion	Q4 97 457.4 / Q1 96 343.0
Indonesia	Q1 96 2,337.0 / Q4 97 4,650.0	Q4 97 $16.7 billion / Q1 96 $15.1 billion	Q3 97 $323 trillion / Q1 96 $228 trillion	Q3 97 $14 billion / Q1 96 $11.2 billion	Q3 97 $10.4 billion / Q1 96 $9.6 billion	Q1 96 123.1 / Q4 97 51.3
Korea, Rep.	Q1 96 782.7 / Q4 97 1,695.8	Q1 96 $33.4 billion / Q4 97 $20.4 billion	Q3 97 $198 trillion / Q1 96 $152 trillion	Q4 97 $37.0 billion / Q1 96 $31.5 billion	Q4 97 $34.5 billion / Q1 96 $35.7 billion	Q1 96 122.2 / Q4 97 36.3
Malaysia	Q1 96 2.5 / Q4 97 3.9	Q3 97 $22.3 billion / Q1 96 $23.6 billion	Q3 96 $227 billion / Q1 96 $210 billion	Q3 97 $20.4 billion / Q1 96 $18.6 billion	Q3 97 $19.9 billion / Q1 96 $19.1 billion	Q4 97 728.2 / Q1 96 493.9
Mexico	Q1 96 7.5 / Q4 97 8.1	Q4 97 $28.1 billion / Q1 96 $16.9 billion	Q3 97 $802 billion / Q1 96 $595 billion	Q2 97 $16.5 billion / Q1 96 $13.9 billion	Q2 97 $18.8 billion / Q1 96 $13.9 billion	Q1 96 298.9 / Q4 97 108.1
Philippines	Q1 96 26.2 / Q4 97 35.4	Q3 97 $9.6 billion / Q1 96 $7.2 billion	Q3 97 $1.3 trillion / Q1 96 $990 billion	Q3 97 $6.5 billion / Q1 96 $4.6 billion	Q3 97 $10.2 billion / Q1 96 $7.9 billion	Q4 97 121.3 / Q1 96 280.6
Thailand	Q1 96 25.2 / Q4 97 47.2	Q1 96 $38.1 billion / Q4 97 $25.8 billion	Q3 97 $4.1 trillion / Q1 96 $3.4 trillion	Q4 97 $14.5 billion / Q1 96 $14.1 billion	Q1 96 $19.5 billion / Q3 97 $15.2 billion	Q1 96 390.2 / Q4 97 58.9

Source: International Monetary Fund, International Finance Statistics Database; International Finance Corporation, Emerging Markets Database.

The external environment for developing countries remains favorable in 1998, but growth was interrupted by the financial crisis in East Asia

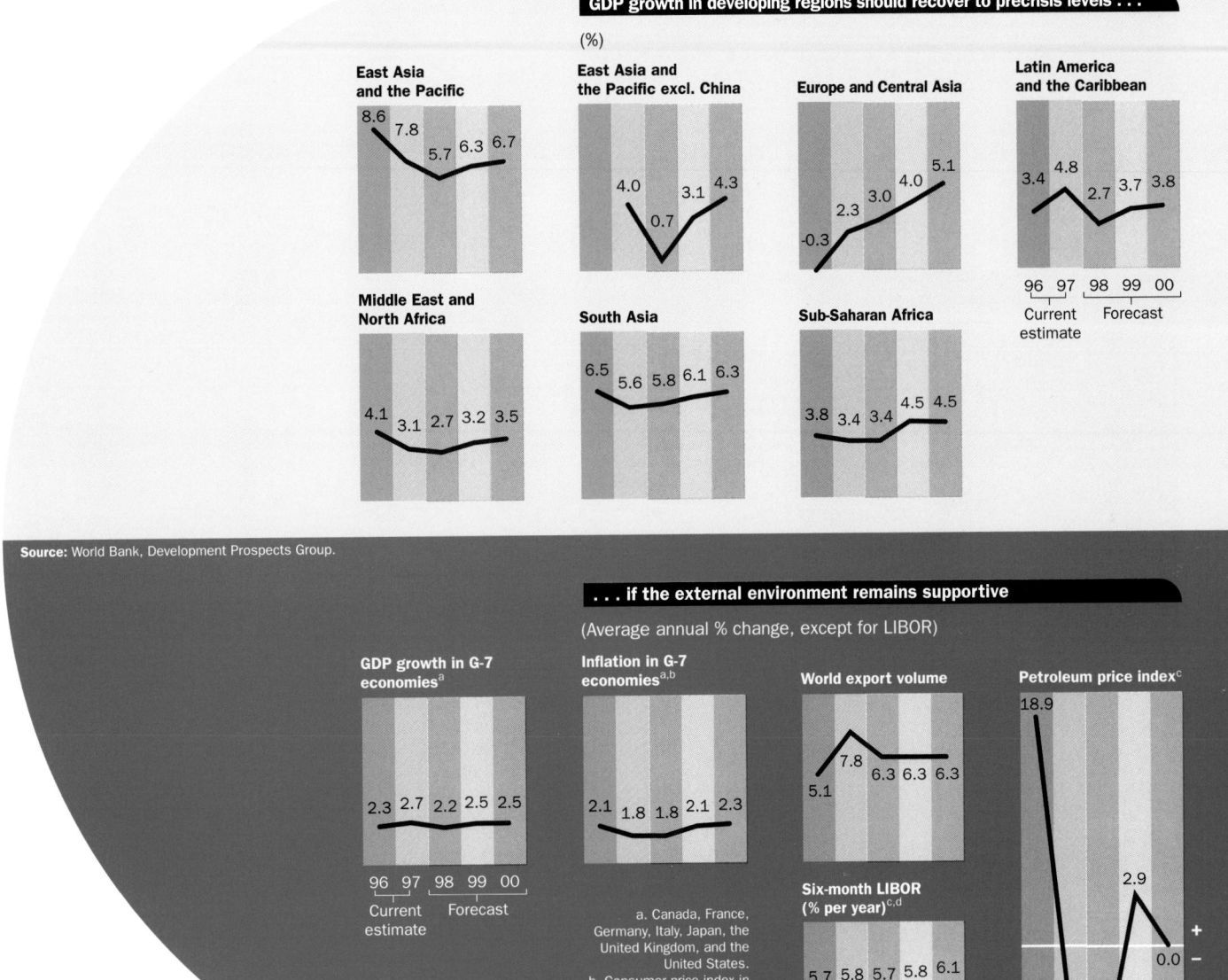

GDP growth in developing regions should recover to precrisis levels . . .

(%)

East Asia and the Pacific
8.6 7.8 5.7 6.3 6.7

East Asia and the Pacific excl. China
4.0 0.7 3.1 4.3

Europe and Central Asia
-0.3 2.3 3.0 4.0 5.1

Latin America and the Caribbean
3.4 4.8 2.7 3.7 3.8

96 97 98 99 00
Current estimate | Forecast

Middle East and North Africa
4.1 3.1 2.7 3.2 3.5

South Asia
6.5 5.6 5.8 6.1 6.3

Sub-Saharan Africa
3.8 3.4 3.4 4.5 4.5

Source: World Bank, Development Prospects Group.

. . . if the external environment remains supportive

(Average annual % change, except for LIBOR)

GDP growth in G-7 economies[a]
2.3 2.7 2.2 2.5 2.5

96 97 98 99 00
Current estimate | Forecast

Inflation in G-7 economies[a,b]
2.1 1.8 1.8 2.1 2.3

World export volume
5.1 7.8 6.3 6.3 6.3

Six-month LIBOR (% per year)[c,d]
5.7 5.8 5.7 5.8 6.1

Petroleum price index[c]
18.9 -6.1 -11.5 2.9 0.0
+
−

a. Canada, France, Germany, Italy, Japan, the United Kingdom, and the United States.
b. Consumer price index in local currency aggregated using 1988–90 GDP weights.
c. Measured in $.
d. London interbank offered rate.

World inflation and interest rates will remain low and stable

	Gross domestic product		Exports of goods and services		Imports of goods and services		GDP deflator		Current account balance		Gross international reserves	
	average annual % growth		average annual % growth		average annual % growth		% growth		% of GDP		$ millions	months of import coverage
	1996	1997	1996	1997	1996	1997	1996	1997	1996	1997	1997	1997
Argentina	4.4	7.8	8.6	7.3	10.0	29.6	0.5	0.7	–1.4	–3.7	25,668	7.0
Bangladesh	5.3	5.7	10.6	23.0	16.9	9.7	5.6	2.0	–5.1	–2.7	1,719	2.6
Brazil	2.9	3.9	6.1	14.7	5.9	24.3	15.8	7.2	–3.2	–4.2	55,110	6.6
Bulgaria	–9.0	–7.4	–8.7	8.3	–7.8	8.6	113.6	997.7	–0.2	1.8	2,425	4.5
Chile	7.2	5.6	3.9	7.6	8.2	5.0	7.0	6.5	–3.9	–3.7	18,891	9.9
China	9.9	9.0	8.3	24.1	7.8	7.2	5.9	2.5	0.9	2.2	145,988	10.2
Colombia	2.0	3.0	4.4	5.9	3.0	5.6	17.7	19.8	–5.6	–5.3	10,095	5.5
Côte d'Ivoire	5.9	6.3	24.1	0.3	21.2	6.7	3.5	3.1	–1.9	–9.4	724	1.6
Ecuador	1.9	3.3	3.6	7.2	2.2	12.7	29.5	31.2	1.5	–1.1	2,086	3.8
Egypt, Arab Rep.	5.0	5.0	8.4	0.5	6.4	7.9	9.1	6.2	0.7	–0.3	19,147	10.3
Ghana	5.0	3.0	19.8	–5.5	12.5	0.2	33.3	29.6	–5.1	–7.3	473	2.3
Hungary	1.3	4.0	7.4	21.4	5.7	19.9	20.4	17.1	–3.8	–3.3	9,249	4.5
India	7.5	6.1	7.5	9.1	5.7	10.1	7.0	7.0	–1.1	–1.5	28,497	5.3
Indonesia	7.6	3.5	6.3	7.7	9.6	–0.1	8.7	13.5	–3.7	–2.6	18,476	4.4
Jamaica	–1.7	–1.0	1.0	–1.9	5.7	1.5	24.4	12.0	–1.9	–4.4	810	2.5
Jordan	5.2	5.0	12.2	4.3	15.7	4.0	5.1	4.0	–3.1	–2.6	2,743	5.4
Kenya	4.3	2.0	13.3	–3.4	5.8	7.2	9.6	12.5	–0.8	–2.0	763	2.5
Lithuania	3.6	3.6	3.6	6.5	3.6	4.2	26.1	12.3	–9.3	–7.6	1,178	2.8
Malaysia	8.0	7.3	10.7	7.6	8.5	6.9	8.2	2.9	–5.2	–4.5	21,700	3.5
Mauritius	6.1	5.5	6.1	5.8	5.8	5.4	6.0	5.0	0.4	–2.3	1,026	4.2
Mexico	5.9	6.9	–18.9	13.0	–16.6	18.8	30.8	17.4	–0.6	–1.6	27,581	2.4
Morocco	11.5	–2.2	6.3	2.8	–3.9	6.2	2.3	3.0	–1.7	–2.6	4,367	4.2
Nigeria	3.5	3.3	15.9	7.4	14.4	7.8	31.0	8.8	9.7	3.2	7,040	6.4
Pakistan	4.6	3.5	2.0	–6.2	13.6	–3.1	11.3	11.4	–6.5	–6.0	1,964	1.4
Peru	2.8	6.5	10.1	12.7	0.2	9.6	9.4	9.0	–5.9	–5.5	7,093	6.9
Philippines	5.7	5.1	20.3	8.9	21.1	8.7	9.0	7.0	–4.5	–4.5	8,800	2.0
Poland	5.9	6.0	9.8	12.2	23.6	19.0	18.6	14.1	–2.4	–4.1	21,030	6.0
Russian Federation	–4.9	0.0	–2.0	2.2	–1.6	4.6	45.5	18.0	2.6	0.9	25,721	3.3
Slovak Republic	6.7	5.5	–1.6	6.8	18.5	1.3	5.8	5.5	–11.0	–8.6	3,719	3.4
South Africa	3.3	2.3	7.8	6.0	7.5	5.9	8.2	8.0	–1.6	–1.9	3,595	1.2
Sri Lanka	3.8	6.0	3.9	3.0	2.8	8.6	10.9	9.5	–4.7	–5.1	..	..
Thailand	6.4	0.4	2.4	3.4	2.8	–7.8	5.1	6.0	–7.9	–2.6	23,000	4.3
Trinidad and Tobago	3.2	2.9	–7.7	4.0	–2.4	2.9	4.1	5.7	1.7	–0.6	396	1.7
Tunisia	7.0	5.7	0.5	5.2	–2.9	6.9	4.4	3.6	–2.7	–2.6	2,612	3.2
Turkey	6.7	6.5	21.7	14.8	19.0	14.6	78.3	85.0	–0.8	–2.4	30,721	6.2
Venezuela	–1.6	5.7	4.3	7.0	–5.9	19.4	111.4	..	13.1	7.7	21,601	12.8
Zimbabwe	7.3	4.5	12.2	–4.4	8.1	1.3	23.0	23.5	–2.1	–3.7	399	1.4

Note: Data for 1997 are preliminary.
Source: World Bank staff estimates.

| Table 4d | Key macroeconomic indicators |

	Nominal exchange rate			Real effective exchange rate		Money and quasi money		Gross domestic credit		Real interest rate		Short-term debt
	local currency units per $	% change		1990 = 100		average annual % growth		average annual % growth		%	%	% of exports
	1997	**1996**	**1997**	**1996**	**1997**	**1996**	**1997**	**1996**	**1997**	**1996**	**1997**	**1996**
Argentina	1.0	0.0	0.0	..	..	18.7	26.9	6.9	10.8	10.0	8.5	38.4
Bangladesh	45.5	4.2	7.1	..	..	10.8	10.7	12.3	13.8	8.0	11.8	2.8
Brazil	1.1	6.8	6.8	..	..	6.8	16.8	11.0	7.4	..	..	58.0
Bulgaria	1,767.0	589.3	262.6	..	..	117.8	518.6	251.0	..	4.7	..	14.4
Chile	435.4	4.4	2.5	123.8	135.1	19.6	14.9	17.0	13.1	9.7	8.6	36.0
China	8.3	−0.2	−0.2	..	..	25.3	..	24.5	..	3.8	4.8	14.1
Colombia	1,303.1	1.8	29.6	143.9	162.5	24.4	52.5	25.3	28.3	20.7	12.0	37.7
Côte d'Ivoire	590.0	6.9	12.7	70.4	69.6	3.9	..	3.4	..	..	..	87.3
Ecuador	4,329.0	24.3	19.1	128.8	137.6	43.7	34.9	29.3	46.2	19.4	9.0	27.3
Egypt, Arab Rep.	3.4	−0.1	0.0	..	..	10.8	11.3	13.9	14.9	6.0	6.7	11.8
Ghana	..	21.1	..	..	..	32.6	..	20.2	..	..	..	39.6
Hungary	199.6	18.3	21.0	128.0	132.3	..	..	..	..	..	..	16.5
India	38.6	2.1	7.3	..	..	18.7	..	19.7	..	8.4	6.1	12.8
Indonesia	3,648.0	3.2	53.1	..	..	27.2	18.9	22.7	28.0	9.7	11.6	55.3
Jamaica	36.2	−12.0	3.7	..	..	10.9	17.8	28.0	..	15.8	..	15.1
Jordan	0.7	0.0	0.0	..	..	−0.9	7.6	0.8	−6.2	4.2	6.0	11.2
Kenya	63.5	−1.6	15.5	..	..	26.2	15.0	11.5	6.7	22.0	15.4	17.5
Lithuania	4.0	0.0	0.0	..	..	−3.0	33.8	−1.4	24.9	−3.6	1.9	3.8
Malaysia	3.5	−0.5	38.3	107.0	108.4	..	..	..	..	0.8	..	11.8
Mauritius	21.7	1.7	20.5	..	..	7.6	17.2	6.1	23.9	14.0	14.3	15.3
Mexico	8.2	2.7	4.4	..	..	26.2	24.3	3.0	−4.2	..	..	26.1
Morocco	9.6	3.9	9.0	113.2	113.4	6.6	8.1	7.1	−20.8	..	..	5.2
Nigeria	21.9	0.0	0.0	149.2	168.8	20.1	..	−26.4	..	..	..	36.2
Pakistan	44.1	17.1	9.8	..	..	20.1	14.4	21.1	5.9	..	..	23.6
Peru	2.7	12.6	4.6	..	..	37.2	28.0	21.7	58.6	15.2	19.3	78.0
Philippines	34.7	0.3	31.8	129.9	132.3	23.2	1.6	32.2	..	5.3	14.4	18.9
Poland	..	16.5	..	211.0	216.7	29.3	..	29.7	..	6.5	..	0.2
Russian Federation	5,919.0	19.8	6.5	..	..	33.0	34.7	53.7	26.1	69.6	13.3	11.1
Slovak Republic	..	7.9	..	..	..	16.2	..	28.9	..	8.0	14.4	26.5
South Africa	4.9	28.4	3.7	90.3	97.2	14.3	..	13.5	..	10.3	11.1	25.8
Sri Lanka	60.6	4.9	6.9	..	..	10.5	14.4	10.7	6.0	4.8	0.5	9.7
Thailand	40.1	1.7	56.7	..	..	12.6	18.1	−18.5	32.9	..	..	49.9
Trinidad and Tobago	6.3	3.3	1.4	85.2	84.2	5.8	..	4.4	..	11.2	8.8	8.9
Tunisia	1.1	5.0	13.0	..	..	13.3	17.6	2.4	1.4	..	..	8.6
Turkey	193,885.0	80.7	79.9	..	..	117.3	..	127.5	..	..	..	40.7
Venezuela	500.8	64.3	5.1	117.9	150.3	69.1	64.8	14.0	39.2	−37.7	..	10.8
Zimbabwe	14.5	16.4	33.5	..	..	33.3	36.9	30.2	49.5	9.9	7.5	25.6

Note: Data for 1997 are preliminary and may not cover the entire year.
Source: International Monetary Fund, *International Finance Statistics*; World Bank, Debtor Reporting System.

	Gross domestic product		Agriculture		Industry		Manufacturing		Services	
	average annual % growth		average annual % growth		average annual % growth		average annual % growth		average annual % growth	
	1980–90	1990–96	1980–90	1990–96	1980–90	1990–96	1980–90	1990–96	1980–90	1990–96
Albania	1.5	1.5	1.9	8.2	2.1	–11.0	..	..	–0.4	7.2
Algeria	2.8	0.6	4.6	3.0	2.3	–0.4	3.3	–8.9	2.9	0.8
Angola	3.7	–0.9	0.5	–9.5	6.4	4.3	–11.1	5.3	2.2	–4.7
Argentina	–0.3	4.9	0.9	0.6	–0.9	5.0	–0.5	..	0.0	5.5
Armenia	3.3	–21.2	–3.9	–0.6	5.1	–28.7	..	..	4.6	–19.7
Australia	3.4	3.7	3.3	–1.2	2.9	2.2	2.0	1.8	3.7	4.6
Austria	2.2	1.6	1.1	–1.1	1.9	1.3	2.7	0.7	2.4	2.0
Azerbaijan	..	–17.7	..	–6.0	..	–10.8	..	..	..	–11.1
Bangladesh	4.3	4.3	2.7	1.2	4.9	7.2	2.8	7.3	5.7	5.7
Belarus	..	–8.3	..	–9.8	..	–10.0	..	..	..	–5.6
Belgium	1.9	1.2	2.0	3.3	..	..	2.8	0.3	..	..
Benin	3.3	4.4	5.1	5.1	1.3	4.2	..	..	2.6	3.8
Bolivia	0.0	3.8	2.0	..	–2.9	..	–1.6	..	–0.1	..
Bosnia and Herzegovina	..	..	..	..	..	..	..	..	..	..
Botswana	10.3	4.1	2.2	–1.2	11.1	1.8	8.8	2.6	11	7.1
Brazil	2.7	2.9	2.8	3.7	2.0	2.1	1.6	2.2	3.5	3.7
Bulgaria	4.0	–3.5	–2.1	–3.3	5.2	–4.9	..	..	4.8	–0.6
Burkina Faso	3.7	2.8	3.1	3.8	3.7	1.4	2.0	9.8	4.7	2.1
Burundi	4.4	–3.8	3.1	–3.4	4.5	–8.3	5.7	–17.3	5.4	–2.7
Cambodia	..	6.5	..	2.1	..	11.3	..	7.8	..	8.4
Cameroon	3.3	–1.0	2.1	2.6	5.9	–5.2	11.8	–2.1	2.6	–0.9
Canada	3.4	1.9	1.5	0.7	2.9	1.8	3.2	2.7	3.6	1.8
Central African Republic	1.7	0.7	2.3	2.6	1.8	–1.9	4.9	–5.5	0.6	–0.4
Chad	6.1	1.5	2.3	5.2	8.1	–2.5	4.4	–2.6	7.7	0.4
Chile	4.1	7.2	5.6	5.5	3.7	6.2	3.4	6.3	4.2	8.2
China	10.2	12.3	5.9	4.4	11.1	17.3	10.7	17.2	13.6	9.6
Hong Kong, China	6.9	5.5	..	..	..	..	..	..	..	..
Colombia	3.7	4.5	2.9	1.2	5.0	2.9	3.5	1.4	3.1	6.8
Congo, Dem. Rep.	1.6	–6.6	2.5	3.0	0.9	–15.9	1.6	–13.4	1.2	–17.4
Congo, Rep.	3.6	0.1	3.4	0.7	5.2	2.0	6.9	–1.9	2.5	–1.8
Costa Rica	3.0	4.3	3.1	3.2	2.8	4.0	3.0	4.3	3.1	4.8
Côte d'Ivoire	0.8	2.4	0.3	2.3	4.4	3.4	3.0	1.6	–0.2	2.0
Croatia	..	–1.0	..	–4.4	..	–8.2	..	..	..	–3.9
Cuba	..	..	..	..	..	..	..	..	..	..
Czech Republic	1.7	–1.0	..	–8.2	..	–5.2	..	..	..	–3.6
Denmark	2.4	2.2	3.1	1.7	2.9	1.9	1.4	1.3	2.2	2.0
Dominican Republic	3.0	4.7	0.4	3.3	3.6	5.2	2.9	3.8	3.5	4.8
Ecuador	2.0	3.2	4.4	2.6	1.2	4.4	0.0	3.2	1.8	2.6
Egypt, Arab Rep.	5.3	3.7	2.7	2.8	5.2	3.9	..	4.3	6.6	3.4
El Salvador	0.2	5.8	–1.1	1.2	0.1	5.3	–0.2	5.3	0.7	7.5
Eritrea	..	..	..	..	..	..	..	..	..	..
Estonia	2.1	–6.5	..	–6.5	..	–11.6	..	..	..	–1.7
Ethiopia[a]	2.3	3.9	1.4	2.3	1.8	3.0	2.0	3.3	3.1	6.7
Finland	3.3	0.3	–0.2	0.9	3.3	1.0	3.4	4.3	3.7	–1.6
France	2.4	1.1	2.0	0.1	1.1	–0.3	0.8	0.0	3.0	1.6
Gabon	–0.5	2.3	1.4	–2.7	0.6	2.7	–3.8	0.2	–1.8	2.8
Gambia, The	3.0	2.6	0.7	1.1	5.4	1.4	6.7	1.6	4.0	3.5
Georgia	..	–26.1	..	..	..	..	..	..	..	..
Germany[b]	2.2	..	1.7	..	1.2	..	..	..	2.9	..
Ghana	3.0	4.4	1.0	2.6	3.3	4.3	3.9	2.6	6.4	6.4
Greece	1.8	1.6	–0.1	3.1	1.3	–0.8	0.5	–1.7	2.3	1.6
Guatemala	0.8	4.2	2.3	2.8	2.1	4.0	..	..	2.1	4.8
Guinea	..	3.9	..	4.4	..	2.7	..	..	..	4.4
Guinea-Bissau	4.0	3.5	4.7	4.8	2.2	2.2	..	..	3.7	2.0
Haiti	–0.2	–5.0	..	–0.8	..	–13.7	..	..	..	–5.2
Honduras	2.7	3.5	2.7	3.1	3.3	4.0	3.7	3.7	2.5	3.7

Growth of output | 4.1

	Gross domestic product		Agriculture		Industry		Manufacturing		Services	
	average annual % growth		average annual % growth		average annual % growth		average annual % growth		average annual % growth	
	1980–90	1990–96	1980–90	1990–96	1980–90	1990–96	1980–90	1990–96	1980–90	1990–96
Hungary	1.6	–0.4	0.6	–5.0	–2.6	1.1	..	..	3.6	–3.2
India	5.8	5.8	3.1	3.1	7.1	6.8	7.4	7.5	6.7	7.0
Indonesia	6.1	7.7	3.4	2.8	6.9	10.2	12.6	11.1	7.0	7.4
Iran, Islamic Rep.	1.5	4.2	4.5	4.8	3.3	3.8	4.5	4.6	–0.3	4.2
Iraq	–6.8	..	..	..	..	..	..	..	..	..
Ireland	3.2	6.1	..	..	..	..	..	..	..	..
Israel	3.5	6.4	..	..	..	..	..	..	..	..
Italy	2.4	1.0	0.1	1.4	..	..	2.9	1.0	..	..
Jamaica	2.0	0.8	0.6	6.7	2.4	–0.2	2.7	–1.4	1.9	0.8
Japan	4.0	1.4	1.3	–2.0	4.2	0.2	4.8	0.0	3.9	2.0
Jordan	2.6	7.6	6.8	–3.7	1.7	10.9	0.5	10.8	2.1	6.3
Kazakhstan	..	–10.5	..	–15.3	..	–15.7	..	..	..	4.1
Kenya	4.2	1.9	3.3	0.6	3.9	1.8	4.9	2.5	4.9	3.5
Korea, Dem. Rep.	..	..	..	..	..	..	..	..	..	..
Korea, Rep.	9.4	7.3	2.8	1.8	13.1	7.5	13.2	7.9	8.2	8.0
Kuwait	0.9	12.2	14.7	..	1.0	..	2.3	..	0.9	..
Kyrgyz Republic	..	–12.3	..	–4.6	..	–21.7	..	..	..	–13.4
Lao PDR	3.7	6.7	3.5	4.4	6.1	12.1	8.9	12.9	3.3	8.0
Latvia	3.4	–10.7	2.3	–13	4.3	–20.2	4.4	–20.1	3.1	–1.2
Lebanon	78.2	9.3	..	..	..	..	..	..	5.1	5.9
Lesotho	4.3	6.7	2.0	–0.8	7.1	12.5	13.6	9.4	5.1	5.9
Libya	–5.7	..	..	..	..	..	..	..	..	..
Lithuania	..	–6.0	..	8.7	..	–10.4	..	..	..	–4.6
Macedonia, FYR	..	–9.1	..	..	..	..	..	..	..	..
Madagascar	1.1	0.4	2.5	1.8	0.9	0.5	1.9	3.5	0.3	0.4
Malawi	2.3	2.6	2.0	5.1	2.9	1.1	3.6	0.7	3.5	–2.5
Malaysia	5.2	8.7	3.8	1.9	7.2	11.2	8.9	13.2	4.2	8.5
Mali	2.9	2.8	3.3	3.3	4.3	5.6	6.8	4.9	2.1	1.5
Mauritania	1.7	4.1	1.7	4.7	4.9	4.1	–2.1	2.1	0.3	3.8
Mauritius	6.2	5.0	2.9	0.2	10.3	5.7	11.1	5.7	5.4	6.3
Mexico	1.1	1.8	0.8	1.2	1.1	1.8	1.5	2.0	1.2	1.9
Moldova	..	–16.7	..	–14.7	..	–23.7	..	..	..	–12.3
Mongolia	..	..	..	..	..	..	..	..	..	..
Morocco	4.2	2.1	6.7	–0.7	3.0	2.1	4.1	2.5	4.2	2.9
Mozambique	1.7	7.1	5.5	4.2	–5.2	1.4	..	..	13.6	11.9
Myanmar	0.6	6.8	0.5	6.2	0.5	10.7	–0.2	8.5	0.7	6.4
Namibia	1.2	4.1	1.2	4.6	–0.1	2.9	7.8	3.8	0.1	4.2
Nepal	4.6	5.1	4.0	1.9	6.0	8.5	3.7	12.0	4.8	6.9
Netherlands	2.3	2.2	3.4	3.7	1.6	1.2	2.3	1.7	2.6	2.3
New Zealand	1.7	3.3	3.9	0.9	1.1	3.8	0.4	4.4	1.8	3.4
Nicaragua	–2.0	2.1	–2.2	4.4	–1.7	2.1	–3.1	0.6	–2.0	0.4
Niger	–1.1	1.0	1.8	2.8	–1.7	0.2	..	1.2	–2.9	–0.2
Nigeria	1.8	2.6	3.3	2.4	–1.1	0.5	0.7	–1.5	4.4	4.8
Norway	2.8	3.9	–0.2	4.4	3.3	5.2	0.1	2.2	2.7	2.8
Oman	8.3	6.0	7.9	..	10.3	..	20.6	..	6.0	..
Pakistan	6.3	4.6	4.3	3.8	7.3	5.5	7.7	5.5	6.8	5.0
Panama	0.5	4.8	2.5	2.2	–1.3	7.9	0.4	5.1	0.6	4.5
Papua New Guinea	1.9	7.6	1.8	4.8	1.9	13.6	0.1	5.2	2.0	4.0
Paraguay	2.5	3.1	3.6	2.7	–0.3	2.2	2.1	1.2	3.4	3.7
Peru	–0.3	6.0	..	5.6	..	6.5	..	..	..	5.8
Philippines	1.0	2.9	1.0	1.7	–0.9	3.1	0.2	2.6	2.8	3.3
Poland	1.9	3.2	–0.1	–1.6	–0.9	4.7	..	..	5.1	3.0
Portugal	2.9	1.4	..	–1.8	..	..	..	–0.3	..	..
Puerto Rico	4.1	3.0	1.8	..	3.6	..	1.5	..	4.6	..
Romania	0.5	0.0	..	–0.4	..	–2.1	..	..	..	–2.8
Russian Federation	2.8	–9.0	..	–8.2	..	–11.0	..	..	..	–8.4

4.1 | Growth of output

	Gross domestic product		Agriculture		Industry		Manufacturing		Services	
	average annual % growth		average annual % growth		average annual % growth		average annual % growth		average annual % growth	
	1980–90	1990–96	1980–90	1990–96	1980–90	1990–96	1980–90	1990–96	1980–90	1990–96
Rwanda	2.5	–9.7	0.5	–8.4	2.5	–14.9	2.6	–10.5	5.5	–9.0
Saudi Arabia	–1.2	1.7	13.4	..	–2.3	..	7.5	..	–1.2	..
Senegal	3.1	1.8	3.3	2.3	4.1	3.1	4.6	3.5	2.8	1.3
Sierra Leone	0.6	–3.3	3.1	–1.5	1.7	–6.4	..	..	–2.8	–3.9
Singapore	6.6	8.7	–6.2	1.8	5.4	9.1	6.6	7.9	7.6	8.5
Slovak Republic	2.0	–1.0	1.6	1.9	2.0	–7.2	..	..	0.8	6.4
Slovenia	..	4.3	..	–0.2	..	0.1	..	..	..	3.6
South Africa	1.2	1.2	2.9	1.4	0.0	0.5	–0.1	0.8	2.3	1.6
Spain	3.2	1.3	..	–4.8	..	–1.1	..	–0.7	..	1.7
Sri Lanka	4.2	4.8	2.2	1.7	4.6	6.6	6.3	8.8	4.7	6.1
Sudan	0.6	6.8	0.0	..	2.8	..	3.7	..	0.4	..
Sweden	2.3	0.6	1.5	–1.9	2.8	–0.7	2.6	0.8	2.1	–0.6
Switzerland	2.2	–0.1	..	..	..	..	..	..	..	..
Syrian Arab Republic	1.5	7.4	–0.6	..	6.6	..	..	..	0.4	..
Tajikistan	..	–16.4	..	..	..	..	..	..	..	..
Tanzania^c	..	..	..	..	..	..	..	..	..	..
Thailand	7.6	8.3	4.0	3.6	9.9	10.3	9.5	10.7	7.3	7.9
Togo	1.8	–0.6	5.6	0.1	1.1	0.6	1.7	–0.1	–0.3	–2.0
Trinidad and Tobago	–2.5	1.2	–5.8	1.3	–5.5	0.8	–10.1	1.1	–3.3	0.8
Tunisia	3.3	4.1	2.8	–0.1	3.1	4.3	3.7	5.2	3.6	5.2
Turkey	5.3	3.6	1.3	1.2	7.8	4.6	7.9	5.3	5.5	3.7
Turkmenistan	..	–9.6	..	..	..	..	..	..	..	..
Uganda	3.1	7.2	2.3	4.0	6.0	12.2	4.0	13.4	3.0	8.6
Ukraine	..	–13.6	..	–26.1	..	–20.0	..	..	..	–6.0
United Arab Emirates	–2.0	..	9.6	9.3	–4.2	–1.8	3.1	1.3	2.0	..
United Kingdom	3.2	1.6	..	..	..	..	..	..	..	..
United States	2.9	2.4	4.0	3.6	2.8	1.2	3.1	1.6	2.9	1.6
Uruguay	0.4	3.7	0.1	4.4	–0.2	0.4	0.4	–1.0	0.9	5.6
Uzbekistan	..	–3.5	..	–1.8	..	–6.0	..	–5.6	..	–2.3
Venezuela	1.1	1.9	3.0	1.1	1.6	3.1	4.3	1.5	0.5	1.0
Vietnam	4.6	8.5	4.3	5.2	..	13.3	..	..	..	8.5
West Bank and Gaza	..	..	..	..	..	..	..	..	..	..
Yemen, Rep.	..	2.8	..	2.9	..	2.9	..	..	..	2.7
Yugoslavia, FR (Serb./Mont.)	..	..	..	..	..	..	..	..	..	..
Zambia	0.8	–1.1	3.6	0.5	1.0	–3.2	4.0	–1.9	0.1	0.5
Zimbabwe	3.0	1.3	2.4	4.5	3.0	–2.1	2.9	–3.7	2.5	2.5
World	**3.1w**	**2.2w**	**2.8w**	**1.7w**	**3.3w**	**1.6w**	**3.6w**	**1.4w**	**3.3w**	**2.3w**
Low income	6.1	7.6	3.7	3.5	8.0	12.2	8.7	13.2	7.1	6.8
Excl. China & India	3.2	3.1	3.1	2.9	..	..	..	..	3.4	..
Middle income	2.1	0.8	..	..	..	..	..	..	2.5	3.8
Lower middle income	2.7	–0.7	..	..	..	..	..	..	..	..
Upper middle income	1.4	2.9	2.5	1.7	0.7	2.9	1.3	3.4	2.0	3.7
Low & middle income	3.1	2.9	3.2	2.8	4.3	..	4.5	..	3.5	4.6
East Asia & Pacific	7.7	10.2	4.8	4.0	8.9	14.5	9.7	15.0	8.9	8.3
Europe & Central Asia	2.9	–5.4	..	..	..	..	..	..	..	..
Latin America & Carib.	1.8	3.2	2.0	2.5	1.4	2.8	1.2	2.6	1.9	3.8
Middle East & N. Africa	0.4	2.6	4.6	3.2	1.3	..	..	2.7	1.2	..
South Asia	5.7	5.6	3.2	3.0	6.9	6.7	7.2	7.4	6.6	6.7
Sub-Saharan Africa	1.7	2.0	1.8	2.1	1.1	0.9	1.3	0.8	2.2	2.0
High income	3.2	2.0	2.2	0.8	3.2	0.7	3.5	0.4	3.3	1.9

a. Data prior to 1992 include Eritrea. b. Data prior to 1990 refer to the Federal Republic of Germany before unification. c. Data cover mainland Tanzania only.

Growth rates are calculated using constant price data in the local currency. Regional and income group growth rates are calculated after converting local currencies to U.S. dollars using the average official exchange rate reported by the International Monetary Fund for the year shown or, occasionally, an alternative conversion factor determined by the World Bank's Development Data Group. The growth rates in the table are annual average compound growth rates. Methods of computing growth rates and the alternative conversion factor are described in *Statistical methods*.

Measuring growth

An economy's growth is measured by the increase in value added produced by the individuals and enterprises operating in that economy. Thus measuring real growth requires estimates of GDP and its components valued in constant prices from one period to the next. In principle, real value added can be estimated by measuring the quantity of goods produced in a period, valuing them at an agreed set of base year prices, and subtracting the cost of inputs, also in constant prices. This double deflation method, recommended by the United Nations (UN) System of National Accounts, requires detailed information on the structure of prices of inputs and outputs. In some sectors, however, value added is extrapolated from the base year using volume indexes of inputs and outputs. In other sectors, particularly services, real output is imputed from labor inputs, such as real wages or the number of employees. The real output of governments and other unpriced services are calculated in the same way. In the absence of well-defined measures of output, measuring the real growth of services remains problematic.

Technical progress can lead to improvements in production and the quality of goods. If not properly accounted for, either effect can distort measures of value added and thus of growth. When inputs are used to estimate output, as in services, unmeasured technical progress leads to underestimates of the quantity and real value of output. Unmeasured changes in the quality of goods produced also lead to underestimates of real value. The result can be underestimates of real growth and productivity change and overestimates of inflation.

Nonmarket services pose a particular problem, especially in developing countries, where much economic activity may go unrecorded. Obtaining a complete picture of the economy requires estimating household outputs produced for local sale and for home use, barter exchanges, and illicit or deliberately unreported activity. How consistent and complete such estimates will be depends on the skill of the compiling statisticians and the resources available to them.

Rebasing national accounts

Countries occasionally rebase their national accounts by collecting a complete set of observations on the value and volume of production in a new base year. Using these data, they update price indexes to reflect the relative importance of inputs and outputs in total output, and generate volume indexes to reflect relative price levels. The new base year should represent normal operation of the economy—that is, a year without major shocks or distortions. But the choice of base year and the timing of economic surveys are also determined by administrative convenience, resource availability, and international agreement. Some developing countries have not rebased their national accounts for many years. Using an old base year can be misleading because implicit price and volume weights become progressively less relevant and useful.

The World Bank collects constant price national accounts series in national currencies and the country's original base year. To obtain comparable series of constant price data, GDP and its main sectoral components by industrial origin (agriculture, industry, and services) are rescaled to a common reference year, currently 1987. This process gives rise to a discrepancy between the rescaled GDP and the sum of the rescaled components. This discrepancy is allocated to the estimate of services value added on the output side and to private consumption expenditure on the expenditure side.

Changes in the System of National Accounts

Most countries use the definitions of the UN System of National Accounts (SNA), series F, no. 2, version 3, referred to as the 1968 SNA. Version 4 of the SNA was completed in 1993. Until new economic surveys can be implemented, most countries will continue to follow the 1968 SNA. A few low-income countries still use concepts from older SNA guidelines, including valuations such as factor cost and market prices, in describing major economic aggregates.

- **Gross domestic product** at purchasers' prices is the sum of the gross value added by all resident and nonresident producers in the economy plus any taxes and minus any subsidies not included in the value of the products. It is calculated without making deductions for depreciation of fabricated assets or for depletion and degradation of natural resources. Value added is the net output of a sector after adding up all outputs and subtracting intermediate inputs. The industrial origin of value added is determined by the International Standard Industrial Classification (ISIC), rev. 2. • **Agriculture** corresponds to ISIC divisions 1–5 and includes forestry and fishing. • **Industry** comprises value added in mining, manufacturing (also reported as a separate subgroup), construction, electricity, water, and gas. • **Manufacturing** refers to industries belonging to divisions 15–37. • **Services** correspond to ISIC divisions 50–99.

National accounts data for developing countries are collected from national statistical organizations and central banks by visiting and resident World Bank missions. Data for industrial countries come from OECD data files. The World Bank rescales constant price data to a common reference year. The complete national accounts time series is available on the *World Development Indicators* CD-ROM. For information on the OECD national accounts series see OECD, *National Accounts, 1960–1995*, volumes 1 and 2.

4.2 | Structure of output

	Gross domestic product ($ millions)		Agriculture value added (% of GDP)		Industry value added (% of GDP)		Manufacturing value added (% of GDP)		Services value added (% of GDP)	
	1980	1996	1980	1996	1980	1996	1980	1996	1980	1996
Albania	..	2,689	34	55	45	21	..	..	21	23
Algeria	42,345	45,699	10	13	54	48	9	8	36	38
Angola	..	6,721	..	7	..	69	..	7	..	24
Argentina	76,962	294,687	6	6	41	31	29	20	52	63
Armenia	..	1,454	18	44	58	35	..	25	25	20
Australia	160,109	392,507	5	4	36	28	19	15	58	68
Austria	78,539	226,100	4	2	36	31	25	20	60	68
Azerbaijan	..	3,650	..	23	..	19	..	18	..	58
Bangladesh	12,950	31,824	50	30	16	18	11	10	34	52
Belarus	..	19,346	..	16	..	41	..	35	..	43
Belgium	118,915	264,400	2	1	..	..	22	19	..	..
Benin	1,405	2,210	35	38	12	14	8	8	52	49
Bolivia	3,074	6,131	18	..	35	..	15	..	47	..
Bosnia and Herzegovina	..	..	..	..	..	..	..	..	..	..
Botswana	1,035	4,936	11	4	45	46	4	5	43	50
Brazil	235,025	748,916	11	14	44	36	33	23	45	50
Bulgaria	20,040	9,484	14	10	54	33	..	..	32	57
Burkina Faso	1,709	2,538	33	35	22	25	16	19	45	40
Burundi	920	1,137	62	57	13	17	7	17	25	26
Cambodia	..	3,125	..	51	..	15	..	5	..	35
Cameroon	6,741	9,252	29	40	23	22	9	10	48	39
Canada	263,193	579,300	..	..	..	..	..	..	..	..
Central African Republic	797	1,062	40	56	20	18	7	7	40	26
Chad	727	1,172	54	46	12	16	..	15	34	38
Chile	27,572	74,292	7	..	37	..	21	..	55	..
China	201,688	815,412	30	21	49	48	41	38	21	31
Hong Kong, China	28,495	154,767	1	0	32	16	24	9	67	84
Colombia	33,397	85,202	19	16	32	20	23	16	49	64
Congo, Dem. Rep.	14,922	6,904	25	64	33	13	14	5	42	23
Congo, Rep.	1,706	2,388	12	10	47	34	7	6	42	56
Costa Rica	4,815	9,015	18	16	27	24	19	18	55	60
Côte d'Ivoire	10,175	10,688	29	28	20	21	11	13	51	51
Croatia	..	19,081	..	12	..	25	..	20	..	62
Cuba	..	..	..	..	..	..	..	..	..	..
Czech Republic	29,123	54,890	7	6	63	39	..	..	30	55
Denmark	66,322	174,247	..	..	..	..	..	..	..	..
Dominican Republic	6,631	13,169	20	13	28	32	15	17	52	55
Ecuador	1,733	19,040	12	12	38	37	18	21	50	51
Egypt, Arab Rep.	22,913	67,691	18	17	37	32	12	24	45	51
El Salvador	3,574	10,469	38	13	22	27	16	21	40	60
Eritrea	..	..	..	10	..	27	..	15	..	63
Estonia	..	4,353	..	7	..	28	..	16	..	65
Ethiopia[a]	5,179	5,993	56	55	12	10	8	..	32	36
Finland	51,306	123,966	..	..	..	..	..	..	..	..
France	664,595	1,540,100	4	2	34	26	24	19	62	71
Gabon	4,279	5,704	7	7	60	52	5	6	33	41
Gambia, The	233	363	30	28	16	15	7	7	53	58
Georgia	..	4,308	24	35	36	35	28	20	40	29
Germany	..	2,353,200	..	1	..	..	..	24	..	..
Ghana	4,445	6,344	58	44	12	17	8	9	30	39
Greece	48,613	122,946	..	..	..	..	..	..	..	..
Guatemala	7,879	15,817	..	24	..	20	..	14	..	56
Guinea	..	3,934	..	26	..	36	..	5	..	39
Guinea-Bissau	105	271	44	54	20	11	..	0	36	35
Haiti	1,462	2,617	..	42	..	13	..	9	..	45
Honduras	2,566	4,011	24	22	24	31	15	18	52	47

	Gross domestic product		Agriculture value added		Industry value added		Manufacturing value added		Services value added	
	$ millions		% of GDP		% of GDP		% of GDP		% of GDP	
	1980	1996	1980	1996	1980	1996	1980	1996	1980	1996
Hungary	22,163	44,845	..	7	..	32	..	24	..	61
India	172,321	356,027	38	28	26	29	18	20	36	43
Indonesia	78,013	225,828	24	16	42	43	13	25	34	41
Iran, Islamic Rep.	92,664	..	18	25	32	34	9	14	50	40
Iraq	47,562	..	..	..	..	..	..	..	..	..
Ireland	20,080	69,600	..	..	..	..	..	..	..	..
Israel	22,579	91,965	..	..	..	..	..	..	..	..
Italy	449,913	1,207,700	6	3	..	..	28	21	..	..
Jamaica	2,679	4,426	8	8	38	36	17	17	54	55
Japan	1,059,254	4,599,700	4	2	42	38	29	25	54	60
Jordan	3,962	7,343	8	5	28	30	13	16	64	64
Kazakhstan	..	20,761	..	13	..	30	..	6	..	57
Kenya	7,265	9,222	33	29	21	16	13	10	47	55
Korea, Dem. Rep.	..	..	..	..	..	..	..	..	..	..
Korea, Rep.	63,661	484,777	15	6	40	43	29	26	45	51
Kuwait	28,639	26,650	0	0	75	53	6	11	25	46
Kyrgyz Republic	..	1,754	..	52	..	19	..	8	..	29
Lao PDR	..	1,857	..	52	..	21	..	15	..	28
Latvia	..	5,024	12	9	51	33	46	22	37	58
Lebanon	..	12,997	..	12	..	27	..	17	..	61
Lesotho	369	889	24	11	29	43	7	17	47	47
Libya	35,545	..	2	..	76	..	2	..	22	..
Lithuania	..	7,779	..	13	..	32	..	20	..	55
Macedonia, FYR	..	1,970	..	..	..	..	..	..	..	..
Madagascar	4,042	4,150	30	35	16	13	..	12	54	52
Malawi	1,238	2,204	44	40	23	21	14	14	34	39
Malaysia	24,488	99,213	22	13	38	46	21	34	40	41
Mali	1,686	2,660	48	48	13	17	7	7	38	35
Mauritania	709	1,094	30	25	26	32	..	12	44	44
Mauritius	1,132	4,292	12	10	26	32	15	23	62	58
Mexico	194,776	334,792	8	5	33	26	22	20	59	68
Moldova	..	1,805	..	50	..	23	..	8	..	27
Mongolia	..	972	15	31	33	35	..	..	52	34
Morocco	18,821	36,820	18	20	31	31	17	17	51	49
Mozambique	2,028	1,714	37	37	35	24	..	..	27	39
Myanmar	..	..	47	60	13	10	10	7	41	30
Namibia	2,172	3,230	12	14	53	34	5	12	35	52
Nepal	1,946	4,456	62	42	12	23	4	10	26	35
Netherlands	171,861	392,400	3	3	32	27	18	18	64	70
New Zealand	22,395	65,100	11	..	31	..	22	..	58	..
Nicaragua	2,144	1,971	23	34	31	22	26	16	45	44
Niger	2,538	1,987	43	39	23	18	4	6	35	43
Nigeria	64,202	31,995	21	43	41	25	8	8	39	31
Norway	63,419	157,802	4	2	35	30	15	12	61	68
Oman	5,982	12,102	3	..	69	..	1	..	28	..
Pakistan	23,690	64,846	30	26	25	25	16	17	46	50
Panama	3,810	8,244	10	8	21	18	12	9	69	73
Papua New Guinea	2,548	5,165	33	26	27	40	10	8	40	33
Paraguay	4,579	9,673	29	24	27	22	16	16	44	54
Peru	20,661	60,926	10	7	42	37	20	23	48	56
Philippines	32,500	83,840	25	21	39	32	26	23	36	47
Poland	57,068	134,477	..	6	..	34	..	22	..	59
Portugal	28,729	104,000	..	..	..	..	..	..	..	..
Puerto Rico	14,436	..	3	..	39	..	37	..	58	..
Romania	..	35,508	..	21	..	40	..	..	..	39
Russian Federation	..	440,562	9	7	54	39	..	..	37	54

	Gross domestic product ($ millions)		Agriculture value added (% of GDP)		Industry value added (% of GDP)		Manufacturing value added (% of GDP)		Services value added (% of GDP)	
	1980	1996	1980	1996	1980	1996	1980	1996	1980	1996
Rwanda	1,163	1,330	50	40	23	14	17	14	27	45
Saudi Arabia	156,487	126,266	1	..	81	..	5	..	18	..
Senegal	3,016	5,155	16	18	21	17	13	11	63	65
Sierra Leone	1,199	940	33	44	21	24	5	6	47	32
Singapore	11,718	94,063	1	0	38	36	29	26	61	64
Slovak Republic	..	18,963	..	5	..	31	..	..	..	64
Slovenia	..	18,558	..	5	..	38	..	28	..	57
South Africa	78,744	126,301	7	5	50	39	23	24	43	57
Spain	211,542	581,600	..	3	..	..	..	..	..	..
Sri Lanka	4,024	13,912	28	22	30	25	18	16	43	52
Sudan	6,760	..	34	..	14	..	7	..	52	..
Sweden	125,557	250,240	..	..	..	..	..	..	..	..
Switzerland	102,719	293,400	..	..	..	..	..	..	..	..
Syrian Arab Republic	13,062	17,587	20	..	23	..	..	..	56	..
Tajikistan	..	2,033	..	..	..	..	..	..	..	..
Tanzania[b]	..	5,838	..	48	..	21	..	7	..	31
Thailand	32,354	185,048	23	11	29	40	22	29	48	50
Togo	1,136	1,420	27	35	25	23	8	11	48	42
Trinidad and Tobago	6,236	5,464	2	2	60	45	9	9	38	53
Tunisia	8,742	19,516	14	14	31	28	12	18	55	58
Turkey	68,790	181,464	26	17	22	28	14	18	51	55
Turkmenistan	..	4,310	..	..	..	..	..	..	..	..
Uganda	1,245	6,115	72	46	4	16	4	8	23	39
Ukraine	..	44,007	..	13	..	39	..	..	..	48
United Arab Emirates	29,625	39,107	1	..	77	..	4	..	22	..
United Kingdom	537,382	1,145,801	..	..	..	..	..	..	..	..
United States	2,709,000	7,341,900	3	..	33	..	22	..	64	..
Uruguay	10,132	18,180	14	9	34	26	26	18	53	65
Uzbekistan	..	25,198	..	26	..	27	..	8	..	47
Venezuela	69,377	67,311	5	4	46	47	16	18	49	49
Vietnam	..	23,340	..	27	..	31	..	..	..	42
West Bank and Gaza	..	..	..	..	..	..	..	..	..	..
Yemen, Rep.	..	6,016	..	18	..	49	..	11	..	34
Yugoslavia, FR (Serb./Mont.)	..	..	..	..	..	..	..	..	..	..
Zambia	3,884	3,388	14	18	41	41	18	29	44	42
Zimbabwe	5,355	7,550	14	14	36	28	25	19	50	59
World	**10,704,631 t**	**28,583,721 t**	**7 w**	**.. w**	**38 w**	**.. w**	**24 w**	**.. w**	**55 w**	**.. w**
Low income	702,232	1,535,031	33	27	35	35	26	25	32	37
Excl. China & India	375,820	349,196	32	34	26	24	12	12	42	42
Middle income	2,373,585	4,374,039	12	11	45	36	..	..	43	53
Lower middle income	..	2,090,188	15	12	43	37	..	..	42	51
Upper middle income	1,021,067	2,258,327	8	9	48	34	24	23	44	57
Low & middle income	3,061,860	5,924,712	18	15	42	34	22	22	40	51
East Asia & Pacific	465,223	1,553,518	28	20	44	44	32	33	28	36
Europe & Central Asia	..	1,118,817	..	11	..	36	..	..	..	53
Latin America & Carib.	758,650	1,875,727	10	10	40	33	27	21	50	57
Middle East & N. Africa	459,114	..	12	..	48	..	9	..	40	..
South Asia	219,283	480,044	38	28	25	28	17	19	37	44
Sub-Saharan Africa	264,750	305,131	22	24	35	30	14	15	43	46
High income	7,810,607	22,756,455	3	..	36	..	24	..	61	..

a. Data prior to 1992 include Eritrea. b. Data cover mainland Tanzania only.

About the data

Output by industrial origin is the sum of the value of gross output of producers less the value of intermediate goods and services consumed in production. This concept is known as value added. A country's gross domestic product (GDP) represents the sum of value added by all producers in that country. Since 1968 the United Nations (UN) System of National Accounts (SNA) has called for estimates of GDP by industrial origin to be valued at either basic prices (excluding all indirect taxes on factors of production) or producer prices (including taxes on factors of production, but excluding indirect taxes on final output). Some countries, however, report such data at purchasers' prices—the prices at which final sales are made—which may affect estimates of the distribution of output. Total GDP as shown in the table and elsewhere in this book is measured at purchasers' prices. GDP components are measured at basic prices. When components are valued at purchasers' prices, this is noted in *Primary data documentation*.

While GDP by industrial origin is generally more reliable than estimates compiled from income or expenditure accounts, different countries use different definitions, methods, and reporting standards. World Bank staff review the quality of national accounts data and sometimes make adjustments to increase consistency with international guidelines. Nevertheless, significant discrepancies remain between international standards and actual practice. Many statistical offices, especially those in developing countries, face severe limits on the resources, time, training, and budgets required to produce reliable and comprehensive series of national accounts.

Data problems in measuring output

Among the difficulties faced by compilers of national accounts is the extent of unreported economic activity in the informal or secondary economy. In developing countries a large share of agricultural output is either not exchanged (because it is consumed within the household) or not exchanged for money. Financial transactions also may go unrecorded.

Agricultural production often must be estimated indirectly, using a combination of methods involving estimates of inputs, yields, and area under cultivation. This approach sometimes leads to crude approximations that can differ over time and across crops for reasons other than climatic conditions or farming techniques. Similarly, agricultural inputs, which cannot easily be allocated to specific outputs, are frequently "netted out" using equally crude and ad hoc approximations.

For further discussion of the measurement of agricultural production see the notes to table 3.3.

The output of industry ideally should be measured through regular censuses and surveys of firms. But in most developing countries such surveys are infrequent and quickly go out of date, so many results must be extrapolated. The sampling unit, which may be the enterprise (where responses may be based on financial records) or the establishment (where production units may be recorded separately), also affects the quality of the data. Moreover, much industrial production is organized not in firms but in unincorporated or owner-operated ventures that are not captured by surveys aimed at the formal sector. Even in large industries, where regular surveys are more likely, evasion of excise and other taxes lowers the estimates of value added. Such problems become more acute as countries move from state control of industry to private enterprise, because new firms enter business and growing numbers of established firms fail to report. In accordance with the SNA, output should include all such unreported activity as well as the value of illegal activities and other unrecorded, informal, or small-scale operations. Data on these areas need to be collected using techniques other than conventional surveys.

In sectors dominated by large organizations and enterprises, such as public utilities, data on output, employment, and wages are usually readily available and reasonably reliable. But in the service sector the many self-employed workers and one-person businesses are sometimes difficult to locate, and their owners have little incentive to respond to surveys, let alone report their full earnings. Compounding these problems are the many forms of economic activity that go unrecorded, including the work that women and children do for little or no pay. For further discussion of the problems of using national accounts data see Srinivasan (1994) and Heston (1994).

Dollar conversion

To produce national accounts aggregates that are internationally comparable, the value of output must be converted to a common currency. The World Bank conventionally uses the U.S. dollar and applies the average official exchange rate reported by the International Monetary Fund for the year shown. An alternative conversion factor is applied if the official exchange rate is judged to diverge by an exceptionally large margin from the rate effectively applied to domestic transactions in foreign currencies and traded products. Shares of output by industrial origin are calculated from data in local currencies and current prices.

Data sources

National accounts data for developing countries are collected from national statistical organizations and central banks by visiting and resident World Bank missions. Data for industrial countries come from OECD data files (see OECD, *National Accounts, 1960–1995*, volumes 1 and 2). The complete national accounts time series is available on the *World Development Indicators* CD-ROM.

4.3 | Structure of manufacturing

	Value added in manufacturing		Food, beverages, and tobacco		Textiles and clothing		Machinery and transport equipment		Chemicals		Other manufacturing[a]	
	$ millions		% of total		% of total		% of total		% of total		% of total	
	1980	1995	1980	1995	1980	1995	1980	1995	1980	1995	1980	1995
Albania	..	..	..	..	..	..	..	..	..	..	..	..
Algeria	3,257	3,154	27	13	18	14	10	15	3	5	43	54
Angola	..	383	..	..	..	..	..	..	..	..	..	..
Argentina	22,685	56,500	19	..	13	..	19	..	9	..	41	..
Armenia	..	350	..	..	..	..	..	..	..	..	..	..
Australia	30,722	51,706	17	..	7	..	21	..	7	..	46	..
Austria	19,263	46,451	16	15	10	5	25	29	7	8	42	42
Azerbaijan	..	725	..	..	..	..	..	..	..	..	..	..
Bangladesh	1,422	2,804	24	..	43	..	4	..	16	..	14	..
Belarus	..	6,152	..	..	..	..	..	..	..	..	..	..
Belgium	25,773	51,744	17	18	8	7	24	..	11	15	40	53
Benin	112	174	59	..	14	..	..	..	6	..	21	..
Bolivia	449	..	28	34	11	5	4	1	3	3	54	57
Bosnia and Herzegovina	..	..	..	..	..	..	..	..	..	..	..	..
Botswana	43	217	..	46	..	16	..	..	..	..	..	38
Brazil	71,098	140,179	14	..	11	..	25	..	11	..	40	..
Bulgaria	..	..	..	19	..	10	..	12	..	5	..	54
Burkina Faso	261	438	59	..	19	..	3	..	1	..	17	..
Burundi	63	212	..	..	..	..	..	..	..	..	..	..
Cambodia	..	149	..	..	..	..	..	..	..	..	..	..
Cameroon	593	794	56	31	9	8	4	1	3	3	29	56
Canada	47,077	..	14	14	7	5	23	30	8	9	48	42
Central African Republic	54	93	49	..	22	..	8	..	11	..	10	..
Chad	..	165	..	..	..	..	..	..	..	..	..	..
Chile	5,911	..	27	29	9	6	6	5	8	12	51	47
China	81,836	262,657	10	14	18	14	22	25	11	10	38	38
Hong Kong, China	6,392	11,598	5	10	42	32	18	23	2	2	34	33
Colombia	7,762	15,233	..	..	..	..	..	..	..	..	..	..
Congo, Dem. Rep.	2,144	337	..	..	..	..	..	..	..	..	..	..
Congo, Rep.	128	139	35	..	16	..	5	..	..	..	44	..
Costa Rica	895	1,725	46	46	10	7	8	9	7	12	28	26
Côte d'Ivoire	1,096	1,272	35	..	15	..	10	..	..	..	40	..
Croatia	..	2,847	..	..	..	..	..	..	..	..	..	..
Cuba	..	..	55	..	7	..	1	..	..	..	37	..
Czech Republic	..	..	..	..	..	..	..	..	..	..	..	..
Denmark	11,411	29,628	24	..	5	..	25	..	10	..	37	..
Dominican Republic	1,015	2,118	66	..	6	..	1	..	6	..	21	..
Ecuador	2,072	3,783	..	..	..	..	..	..	..	..	..	..
Egypt, Arab Rep.	2,678	13,740	19	..	30	..	11	..	9	..	31	..
El Salvador	589	2,002	37	29	22	18	4	6	11	16	27	32
Eritrea	..	..	..	..	..	..	..	..	..	..	..	..
Estonia	..	617	..	..	..	..	..	..	..	..	..	..
Ethiopia	381[b]	374	..	..	..	..	..	..	..	..	..	..
Finland	13,019	29,441	..	11	..	3	..	26	..	8	..	53
France	160,811	296,107	13	14	8	5	30	30	8	9	41	42
Gabon	195	297	24	..	4	..	9	..	4	..	58	..
Gambia, The	15	23	35	..	2	..	..	..	3	..	60	..
Georgia	..	854	..	..	..	..	..	..	..	..	..	..
Germany	..	581,335	..	..	..	..	..	..	..	..	..	..
Ghana	347	594	37	36	11	5	2	2	5	10	46	47
Greece	6,968	9,891	18	28	23	14	14	12	8	11	37	35
Guatemala	..	..	39	..	10	..	5	..	17	..	28	..
Guinea	..	76	..	..	..	..	..	..	..	..	..	..
Guinea-Bissau	..	..	..	..	..	..	..	..	..	..	..	..
Haiti	..	156	..	..	..	..	..	..	..	..	..	..
Honduras	344	614	51	51	9	16	2	2	5	4	34	27

Structure of manufacturing | 4.3

	Value added in manufacturing		Food, beverages, and tobacco		Textiles and clothing		Machinery and transport equipment		Chemicals		Other manufacturing[a]	
	$ millions		% of total		% of total		% of total		% of total		% of total	
	1980	1995	1980	1995	1980	1995	1980	1995	1980	1995	1980	1995
Hungary	..	8,841	11	21	11	7	28	23	11	8	38	40
India	27,422	57,942	9	11	21	13	25	25	14	19	30	32
Indonesia	10,133	48,650	32	23	14	19	13	15	11	9	30	34
Iran, Islamic Rep.	8,567	..	..	..	..	..	..	..	..	..	..	..
Iraq	..	..	..	..	..	..	..	..	..	..	..	..
Ireland	..	..	29	25	8	2	17	34	14	20	32	18
Israel	..	..	12	12	12	9	26	32	8	5	42	42
Italy	126,012	226,549	9	10	12	13	29	33	11	7	39	37
Jamaica	446	725	47	..	6	..	..	..	9	..	37	..
Japan	309,747	1,268,284	9	9	7	4	33	39	9	11	43	37
Jordan	447	883	23	29	7	6	1	6	7	14	62	44
Kazakhstan	..	1,196	..	..	..	..	..	..	..	..	..	..
Kenya	796	840	34	44	12	9	15	10	9	9	30	29
Korea, Dem. Rep.	..	..	..	..	..	..	..	..	..	..	..	..
Korea, Rep.	18,260	122,407	17	9	19	10	17	38	10	8	36	35
Kuwait	1,581	2,913	7	7	5	7	4	7	7	3	76	76
Kyrgyz Republic	..	124	..	..	..	..	..	..	..	..	..	..
Lao PDR	..	244	..	..	..	..	..	..	..	..	..	..
Latvia	..	953	..	39	..	11	..	20	..	5	..	26
Lebanon	..	1,788	..	..	..	..	..	..	..	..	..	..
Lesotho	21	116	73	..	7	..	..	..	4	..	16	..
Libya	682	..	31	..	10	..	..	..	16	..	43	..
Lithuania	..	1,879	..	..	..	..	..	..	..	..	..	..
Macedonia, FYR	..	..	..	24	..	22	..	16	..	10	..	29
Madagascar	..	372	34	..	45	..	3	..	6	..	13	..
Malawi	152	236	58	..	12	..	4	..	5	..	20	..
Malaysia	5,054	27,728	24	8	7	5	20	40	5	9	43	38
Mali	106	169	29	..	51	..	8	..	..	..	11	..
Mauritania	..	109	..	..	..	..	..	..	..	..	..	..
Mauritius	147	819	36	31	30	46	6	1	6	4	23	18
Mexico	43,089	54,630	..	29	..	3	..	22	..	17	..	28
Moldova	..	153	..	..	..	..	..	..	..	..	..	..
Mongolia	..	..	..	25	..	35	..	..	..	1	..	12
Morocco	3,167	6,046	..	25	..	19	..	8	..	15	..	32
Mozambique	..	..	..	..	..	..	..	..	..	..	..	..
Myanmar	..	..	..	..	..	..	..	..	..	..	..	..
Namibia	90	419	..	..	..	..	..	..	..	..	..	..
Nepal	78	392	..	35	..	34	..	3	..	4	..	23
Netherlands	30,866	70,407	23	24	4	3	27	24	14	14	32	35
New Zealand	4,950	7,836	26	..	11	..	17	..	6	..	40	..
Nicaragua	549	309	53	..	8	..	1	..	10	..	28	..
Niger	94	121	30	..	25	..	2	..	16	..	28	..
Nigeria	5,195	1,879	21	..	13	..	13	..	13	..	39	..
Norway	9,196	17,873	15	23	4	2	27	26	7	8	48	41
Oman	39	..	..	..	..	..	..	..	..	..	..	..
Pakistan	3,389	9,317	32	..	22	..	9	..	12	..	25	..
Panama	408	694	49	50	10	7	2	1	6	8	34	33
Papua New Guinea	242	389	40	..	1	..	16	..	3	..	41	..
Paraguay	733	1,230	..	..	..	..	..	..	..	..	..	..
Peru	4,176	13,625	25	..	13	..	13	..	10	..	40	..
Philippines	8,354	17,052	30	30	13	9	12	17	14	17	31	26
Poland	..	25,751	12	31	17	8	32	19	8	7	31	35
Portugal	..	19,607	13	15	22	21	16	13	7	5	42	46
Puerto Rico	5,306	14,132	..	17	..	5	..	13	..	52	..	14
Romania	..	..	..	27	..	12	..	10	..	7	..	44
Russian Federation	..	..	..	17	..	5	..	25	..	9	..	45

	Value added in manufacturing		Food, beverages, and tobacco		Textiles and clothing		Machinery and transport equipment		Chemicals		Other manufacturing[a]	
	$ millions		% of total		% of total		% of total		% of total		% of total	
	1980	1995	1980	1995	1980	1995	1980	1995	1980	1995	1980	1995
Rwanda	178	179	..	..	..	..	..	..	..	..	..	..
Saudi Arabia	7,740	..	..	..	..	..	..	..	..	..	..	..
Senegal	389	574	50	48	19	5	4	3	8	23	20	21
Sierra Leone	54	51	51	69	5	1	..	..	..	..	44	30
Singapore	3,415	22,428	5	4	5	1	44	59	5	8	41	28
Slovak Republic	..	..	..	13	..	10	..	20	..	10	..	47
Slovenia	..	4,589	..	15	..	11	..	18	..	13	..	43
South Africa	16,607	28,839	12	15	9	8	21	20	9	9	48	48
Spain	..	97,182	16	18	12	7	23	27	9	11	41	38
Sri Lanka	668	1,836	..	..	..	..	..	..	..	..	..	..
Sudan	424	..	..	..	..	..	..	..	..	..	..	..
Sweden	26,293	38,821	10	9	3	1	33	34	7	11	47	44
Switzerland	..	..	..	10	..	4	..	29	..	..	..	57
Syrian Arab Republic	..	..	..	..	..	..	..	..	..	..	..	..
Tajikistan	..	..	..	..	..	..	..	..	..	..	..	..
Tanzania[c]	..	334	23	..	33	..	8	..	6	..	30	..
Thailand	6,960	47,963	55	..	8	..	9	..	7	..	21	..
Togo	89	139	47	..	13	..	..	..	8	..	32	..
Trinidad and Tobago	557	441	22	33	4	2	9	3	4	24	61	38
Tunisia	1,030	3,390	18	20	19	24	7	6	15	6	42	45
Turkey	9,333	32,158	18	19	15	15	14	16	10	10	42	39
Turkmenistan	..	..	..	..	..	..	..	..	..	..	..	..
Uganda	53	359	..	..	..	..	..	..	..	..	..	..
Ukraine	..	..	..	..	..	..	..	..	..	..	..	..
United Arab Emirates	1,130	2,967	12	..	2	..	2	..	7	..	77	..
United Kingdom	125,830	185,594	13	14	6	5	33	31	10	13	38	37
United States	593,000	1,126,200	11	12	6	5	34	33	10	12	40	38
Uruguay	2,626	3,143	28	..	17	..	10	..	7	..	38	..
Uzbekistan	..	1,782	..	..	..	..	..	..	..	..	..	..
Venezuela	11,112	13,200	19	22	7	2	9	10	8	11	57	54
Vietnam	..	2,760	..	..	..	..	..	..	..	..	..	..
West Bank and Gaza	..	..	..	..	..	..	..	..	..	..	..	..
Yemen, Rep.	..	697	..	..	..	..	..	..	..	..	..	..
Yugoslavia, FR (Serb./Mont.)	..	..	..	..	..	..	..	..	..	..	..	..
Zambia	718	891	44	44	13	10	9	5	9	16	25	25
Zimbabwe	1,248	1,260	23	38	17	13	8	11	9	3	42	36
World	**2,447,474 t**	**4,983,629 t**	..	..	..	..	..	..	..	..	..	..
Low income	146,716	372,808	..	..	..	..	..	..	..	..	..	..
Excl. China & India	..	37,451	..	..	..	..	..	..	..	..	..	..
Middle income	..	..	..	..	..	..	..	..	..	..	..	..
Lower middle income	..	..	..	..	..	..	..	..	..	..	..	..
Upper middle income	204,869	393,534	..	..	..	..	..	..	..	..	..	..
Low & middle income	578,773	1,317,597	..	..	..	..	..	..	..	..	..	..
East Asia & Pacific	124,514	427,789	..	..	..	..	..	..	..	..	..	..
Europe & Central Asia	..	..	..	..	..	..	..	..	..	..	..	..
Latin America & Carib.	186,150	345,442	..	..	..	..	..	..	..	..	..	..
Middle East & N. Africa	32,562	..	..	..	..	..	..	..	..	..	..	..
South Asia	33,695	73,885	..	..	..	..	..	..	..	..	..	..
Sub-Saharan Africa	33,918	44,575	..	..	..	..	..	..	..	..	..	..
High income	1,891,432	4,003,267	..	..	..	..	..	..	..	..	..	..

a. Includes unallocated data. b. Includes Eritrea. c. Data cover mainland Tanzania only.

Structure of manufacturing | 4.3

About the data

Data on the distribution of manufacturing value added by industry are provided by the United Nations Industrial Development Organization (UNIDO). The classification of manufacturing industries is in accordance with the United Nations International Standard Industrial Classification (ISIC), rev. 2. Manufacturing comprises all of ISIC major division 3.

UNIDO obtains data on manufacturing value added from a variety of national and international sources, including the Statistical Division of the United Nations Secretariat, the World Bank, the Organisation for Economic Co-operation and Development, and the International Monetary Fund. To improve comparability over time and across countries, UNIDO supplements these data with information from industrial censuses, statistics supplied by national and international organizations, unpublished data that it collects in the field, and estimates by the UNIDO Secretariat. Nevertheless, coverage may be less than complete, particularly for the informal sector. To the extent that direct information on inputs and outputs is not available, estimates may be used that may result in errors in industry totals. Moreover, countries use different reference periods (calendar or fiscal year) and valuation methods (basic, producers', or purchasers' prices) to estimate value added. (See also the notes to table 4.2).

Data on manufacturing value added in U.S. dollars are from the World Bank's national accounts files (see *About the data* for table 4.2). These figures may differ from those used by UNIDO to calculate the shares of value added by industry. Thus estimates of value added in a particular industry group calculated by applying the shares to total value added will not match those from UNIDO sources.

Definitions

- **Value added in manufacturing** is the sum of gross output less the value of intermediate inputs used in production for industries classified in ISIC major division 3. • **Food, beverages, and tobacco** comprise ISIC division 31. • **Textiles and clothing** comprise ISIC division 32. • **Machinery and transport equipment** comprise ISIC groups 382–84. • **Chemicals** comprise ISIC groups 351 and 352. • **Other manufacturing** includes wood and related products (ISIC division 33), paper and related products (ISIC division 34), petroleum and related products (ISIC groups 353–56), basic metals and mineral products (ISIC divisions 36 and 37), fabricated metal products and professional goods (ISIC groups 381 and 385), and other industries (ISIC group 390). When data for textiles and clothing, machinery and transport equipment, or chemicals are shown as not available, they are included in other manufacturing.

Figure 4.3a

Manufacturing takes off in China

Value added to manufacturing, 1987 $ billions

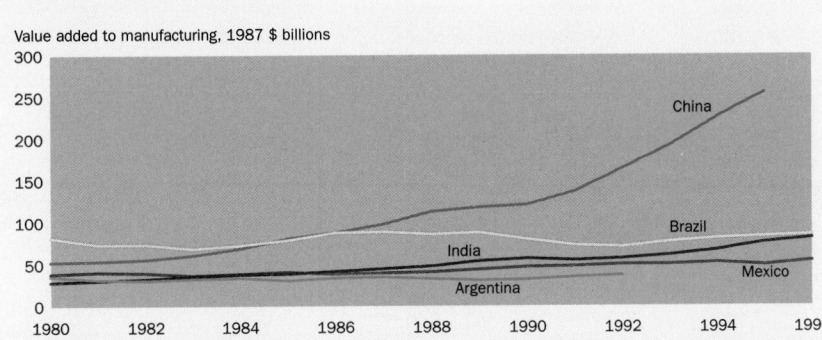

Source: World Bank staff estimates.

China's growth as a manufacturing power has paralleled its rise as an exporter of manufactured goods. In contrast, Brazil, which began the 1980s as a leading manufacturer among developing countries, has fallen far behind and is being overtaken by India.

Data sources

Data on value added in manufacturing in U.S. dollars are from the World Bank's national accounts files. The data used to calculate shares of value added by industry are provided to the World Bank in electronic files by UNIDO. The most recent published source is UNIDO's *International Yearbook of Industrial Statistics 1997*.

4.4 Structure of merchandise exports

	Merchandise exports		Food		Agricultural raw materials		Fuels		Ores and metals		Manufactures	
	$ millions		% of total		% of total		% of total		% of total		% of total	
	1980	1996	1980	1996	1980	1996	1980	1996	1980	1996	1980	1996
Albania	..	296 ª	..	..	..	..	..	..	..	..	..	..
Algeria	15,624	12,609 ª	1	1	0	0	98	95	0	1	0	4
Angola	1,902	4,472 ª	9	..	0	..	78	..	0	..	13	..
Argentina	8,019	23,810	65	52	6	4	3	13	2	1	23	30
Armenia	..	290 ª	..	..	..	..	..	..	..	..	..	..
Australia	21,279	53,252	34	25	11	7	11	19	17	16	22	30
Austria	17,478	57,822	4	4	8	3	2	1	4	3	83	88
Azerbaijan	..	618 ª	..	..	..	..	..	..	..	..	..	..
Bangladesh	740	3,297 ª	12	..	19	..	0	..	0	..	68	..
Belarus	..	5,122 ª	..	..	..	..	..	..	..	..	..	..
Belgiumᵇ	63,967	168,010	9	10	2	1	8	3	7	3	69	77
Benin	49	255 ª	62	..	25	..	4	..	1	..	3	..
Bolivia	1,036	1,087	8	29	3	10	24	13	62	32	3	16
Bosnia and Herzegovina	..	..	..	..	..	..	..	..	..	..	..	..
Botswana	..	..	..	..	..	..	..	..	..	..	..	..
Brazil	20,132	47,164	46	30	4	4	2	1	9	10	37	54
Bulgaria	10,372	4,543 ª	..	..	..	..	..	..	..	..	..	..
Burkina Faso	90	216 ª	41	..	48	..	0	..	0	..	11	..
Burundi	129	37 ª	70	..	6	..	..	..	6	..	4	..
Cambodia	15	300 ª	2	..	7	..	0	..	27	..	64	..
Cameroon	1,321	1,758	48	24	16	25	31	36	2	6	4	8
Canada	63,105	199,071	12	8	11	8	14	10	14	6	48	63
Central African Republic	111	115	31	1	43	24	0	0	0	26	26	43
Chad	72	125 ª	4	..	81	..	0	..	0	..	15	..
Chile	4,584	14,979	15	28	10	9	1	0	64	46	9	15
China†	18,136 ª	151,047	..	8	..	2	..	4	..	2	..	84
Hong Kong, Chinaᶜ	19,703	180,744	3	3	2	1	0	1	2	2	91	92
Colombia	3,945	10,976	72	26	5	5	3	34	0	1	20	34
Congo, Dem. Rep.	2,507	1,465 ª	11	..	3	..	8	..	47	..	6	..
Congo, Rep.	955	1,833 ª	1	1	2	12	90	83	0	1	7	2
Costa Rica	1,032	2,882	64	63	1	6	1	1	0	1	28	24
Côte d'Ivoire	2,979	4,996 ª	64	..	28	..	2	..	0	..	5	..
Croatia	..	4,512	..	11	..	5	..	9	..	2	..	72
Cuba	5,541	1,834 ª	89	..	0	..	3	..	5	..	0	..
Czech Republic	..	21,882	..	5	..	3	..	4	..	3	..	84
Denmark	16,407	48,868	33	23	5	3	3	4	2	1	55	59
Dominican Republic	704	3,893 ª	73	20	0	0	0	0	3	0	24	77
Ecuador	2,481	4,762	33	51	1	3	63	36	0	0	3	9
Egypt, Arab Rep.	3,046	3,534	7	10	16	4	64	48	2	6	11	32
El Salvador	720	1,023	47	52	12	1	3	3	3	2	35	41
Eritrea	..	..	..	..	..	..	..	..	..	..	..	..
Estonia	..	2,074	..	15	..	8	..	6	..	2	..	68
Ethiopiaᵈ	424	494 ª	74	..	18	..	7	..	0	..	0	..
Finland	14,140	40,520	3	3	19	7	4	3	4	3	70	83
France	110,865	283,318	16	14	2	1	4	3	4	2	73	79
Gabon	2,189	3,146	1	0	7	13	88	83	12	2	5	2
Gambia, The	36	22 ª	99	..	0	..	..	..	3	..	7	..
Georgia	..	261 ª	..	..	..	..	..	..	..	..	..	..
Germanyᵉ	191,647	511,728	5	5	1	1	4	1	3	2	85	87
Ghana	942	1,684 ª	78	..	4	..	0	..	17	..	1	..
Greece	5,142	9,558 ª	26	30	2	4	16	7	9	7	47	50
Guatemala	1,486	2,031	53	61	16	5	1	3	5	1	24	31
Guinea	374	774 ª	4	..	0	..	0	..	95	..	1	..
Guinea-Bissau	11	56 ª	85	..	2	..	0	..	0	..	8	..
Haiti	376	180 ª	31	..	1	..	0	..	4	..	63	..
Honduras	813	845	75	65	5	3	0	0	6	2	12	31
† Data for Taiwan, China	19,837	115,646	..	..	..	..	..	..	..	..	..	..

Structure of merchandise exports | 4.4

	Merchandise exports ($ millions)		Food (% of total)		Agricultural raw materials (% of total)		Fuels (% of total)		Ores and metals (% of total)		Manufactures (% of total)	
	1980	1996	1980	1996	1980	1996	1980	1996	1980	1996	1980	1996
Hungary	8,677	13,138	22	19	3	2	5	3	4	4	65	68
India	7,511	32,325 [a]	28	*19*	5	*1*	0	2	7	3	59	*74*
Indonesia	21,909	49,727	8	11	14	6	72	26	4	6	2	51
Iran, Islamic Rep.	13,804	22,102 [a]	1	..	1	..	93	..	0	..	5	..
Iraq	28,321	502 [a]	0	..	0	..	99	..	..	..	0	..
Ireland	8,473	45,565	37	16	2	1	1	0	3	1	54	82
Israel	5,540	20,504	12	5	4	2	0	1	2	1	82	91
Italy	77,640	250,718	7	7	1	1	6	1	2	1	84	89
Jamaica	942	1,347	14	24	0	0	2	0	21	6	63	69
Japan	129,542	410,481	1	0	1	1	0	1	2	1	95	95
Jordan	402	1,466 [a]	25	*25*	1	*2*	0	*0*	40	*24*	34	*49*
Kazakhstan	..	6,230 [a]	..	..	..	..	..	..	..	..	..	..
Kenya	1,313	2,203 [a]	44	..	8	..	33	..	2	..	12	..
Korea, Dem. Rep.	..	1,007 [a]	..	..	..	..	..	..	..	..	..	..
Korea, Rep.	17,446	124,404	7	2	1	1	0	3	1	1	90	92
Kuwait	20,435	13,420 [a]	1	*0*	0	*0*	89	*95*	0	*0*	10	5
Kyrgyz Republic	..	507	..	28	..	11	..	15	..	6	..	38
Lao PDR	9	334 [a]	*13*	..	*41*	..	..	..	4	..	*34*	..
Latvia	..	1,443	..	15	..	19	..	2	..	1	..	61
Lebanon	930	1,153 [a]	28	..	2	..	0	..	9	..	58	..
Lesotho	..	..	..	..	..	..	..	..	..	..	0	..
Libya	21,910	10,126 [a]	..	..	..	..	100	..	..	..	..	..
Lithuania	..	3,356	..	17	..	6	..	15	..	2	..	60
Macedonia, FYR	..	1,119 [a]	..	..	..	..	..	..	..	..	..	..
Madagascar	387	616 [a]	80	*69*	4	*6*	6	*1*	4	7	6	*14*
Malawi	269	501 [a]	91	*90*	2	*2*	0	*0*	0	0	6	7
Malaysia	12,939	78,151	15	9	31	5	25	8	10	1	19	76
Mali	235	288 [a]	30	..	69	..	0	..	..	..	0	..
Mauritania	255	574 [a]	16	..	1	..	0	..	83	..	0	..
Mauritius	420	1,699	72	31	0	1	0	0	0	0	27	68
Mexico	15,442	95,199	12	6	2	1	67	12	6	2	12	78
Moldova	..	1,104 [a]	..	72	..	2	..	1	..	3	..	23
Mongolia	..	424	..	2	..	28	..	0	..	60	..	10
Morocco	2,403	4,742	28	33	3	3	5	2	41	13	24	50
Mozambique	511	226	68	69	7	9	2	1	5	4	18	17
Myanmar	460	1,187 [a]	40	..	33	..	9	..	10	..	7	..
Namibia	..	..	..	..	..	..	..	..	..	..	..	..
Nepal	94	358 [a]	21	*1*	48	*0*	..	..	0	0	30	*99*
Netherlands	73,871	177,228	20	19	3	4	22	8	4	2	50	63
New Zealand	5,262	13,789	48	47	26	17	1	2	4	4	20	29
Nicaragua	414	653	75	60	8	5	2	1	1	1	14	34
Niger	580	79 [a]	11	..	1	..	1	..	85	..	2	..
Nigeria	25,057	15,610 [a]	2	..	0	..	97	..	0	..	0	..
Norway	18,481	48,922	7	8	3	1	48	55	10	7	32	23
Oman	3,748	6,395 [a]	1	*5*	0	*0*	96	*79*	0	*2*	3	*14*
Pakistan	2,588	9,266	24	9	20	6	7	1	0	0	48	84
Panama	353	558	67	73	0	0	23	5	1	2	9	20
Papua New Guinea	1,133 [a]	2,554 [a]	*33*	..	7	..	0	..	50	..	3	..
Paraguay	310	1,043	38	58	50	24	0	1	0	0	12	17
Peru	3,266	5,226	16	32	4	3	21	7	43	42	17	16
Philippines	5,751	20,328	36	10	6	1	1	2	21	3	21	84
Poland	16,997	24,387	6	11	3	2	13	7	7	6	61	74
Portugal	4,629	23,184	12	7	9	3	6	2	2	1	70	86
Puerto Rico	..	..	..	..	..	..	..	..	..	..	..	..
Romania	12,230	8,084	..	9	..	3	..	7	..	3	..	77
Russian Federation	..	81,438 [a]	..	..	..	..	..	..	..	..	..	..

4.4 Structure of merchandise exports

	Merchandise exports ($ millions)		Food (% of total)		Agricultural raw materials (% of total)		Fuels (% of total)		Ores and metals (% of total)		Manufactures (% of total)	
	1980	1996	1980	1996	1980	1996	1980	1996	1980	1996	1980	1996
Rwanda	138	168ᵃ	82	..	7	..	0	..	10	..	0	..
Saudi Arabia	109,113	58,177ᵃ	0	..	0	..	99	..	0	..	1	..
Senegal	477	655ᵃ	43	16	3	9	19	15	20	11	15	50
Sierra Leone	302	214ᵃ	24	..	1	..	0	..	34	..	40	..
Singapore	19,375	124,794	8	4	10	1	25	8	2	2	47	84
Slovak Republic	..	8,824	..	4	..	3	..	5	..	4	..	68
Slovenia	..	8,309	..	4	..	1	..	1	..	3	..	90
South Africaᶠ	25,539	18,132	9	14	2	5	4	9	7	10	18	49
Spain	20,827	101,417	18	16	2	1	4	2	5	2	72	78
Sri Lanka	1,043	4,097ᵃ	47	21	18	4	15	0	1	1	19	73
Sudan	584	468ᵃ	47	..	51	..	1	..	1	..	1	..
Sweden	30,788	82,704	2	2	10	5	4	2	5	3	78	80
Switzerland	29,471	80,756	3	3	1	1	0	0	5	2	90	94
Syrian Arab Republic	2,108	3,980ᵃ	4	..	9	..	79	..	1	..	7	..
Tajikistan	..	770ᵃ	..	..	..	..	..	..	..	..	..	..
Tanzania	528	828ᵃ	58	..	18	..	5	..	5	..	14	..
Thailand	6,369	55,789ᵃ	47	19	11	5	0	1	14	1	25	73
Togo	335	363ᵃ	21	..	2	..	26	..	40	..	11	..
Trinidad and Tobago	4,077	2,456	2	8	0	0	93	53	0	0	5	39
Tunisia	2,234	5,517	7	7	1	1	52	11	4	2	36	80
Turkey	2,910	23,045	51	20	14	2	1	1	7	2	27	74
Turkmenistan	..	1,693ᵃ	..	..	..	..	..	..	..	..	..	..
Uganda	465ᵃ	568ᵃ	96	..	2	..	1	..	1	..	1	..
Ukraine	..	16,040ᵃ	..	..	..	..	..	..	..	..	..	..
United Arab Emirates	21,618ᵃ	28,096ᵃ	..	..	..	..	..	..	..	..	..	..
United Kingdom	114,422	259,039	7	7	1	1	13	7	5	2	71	82
United States	212,887	575,477	18	11	5	3	4	2	5	2	66	78
Uruguay	1,059	2,391	39	47	22	15	0	1	1	1	38	36
Uzbekistan	..	2,671ᵃ	..	..	..	..	..	..	..	..	..	..
Venezuela	19,293	22,633	0	2	0	0	94	82	4	4	2	12
Vietnam	123	7,016ᵃ	30	..	23	..	32	..	1	..	14	..
West Bank and Gaza	..	..	..	..	..	..	..	..	..	..	..	..
Yemen, Rep.	23	4,538ᵃ	45	3	4	1	0	95	0	1	47	1
Yugoslavia, FR (Serb./Mont.)	8,977	1,842	12	28	6	4	3	2	7	15	73	49
Zambia	1,330	1,020ᵃ	1	..	0	..	0	..	82	..	16	..
Zimbabwe	433	2,094	40	51	3	8	3	2	17	10	36	30
World	1,875,309 t	5,398,224 t	12 w	9 w	4 w	2 w	12 w	5 w	5 w	3 w	65 w	78 w
Low income	..	302,497	..	14	..	2	..	4	..	3	..	77
Excl. China & India	..	..	..	..	..	..	..	..	..	..	..	..
Middle income	..	..	..	..	..	..	..	..	..	..	..	..
Lower middle income	..	..	..	..	..	..	..	..	..	..	..	..
Upper middle income	193,240	446,282	19	18	6	4	37	8	8	7	24	60
Low & middle income	..	..	..	..	..	..	..	..	..	..	..	..
East Asia & Pacific	..	371,815	..	9	..	3	..	9	..	3	..	75
Europe & Central Asia	..	..	..	..	..	..	..	..	..	..	..	..
Latin America & Carib.	102,403	261,905	36	24	4	3	32	18	10	8	19	45
Middle East & N. Africa	..	..	..	..	..	..	..	..	..	..	..	..
South Asia	12,464	50,819	28	17	11	2	3	1	5	2	53	76
Sub-Saharan Africa	..	..	..	..	..	..	..	..	..	..	..	..
High income	1,333,696	4,048,665	10	8	4	2	7	4	4	2	73	81

a. Data are from IMF, *Direction of Trade Statistics*. b. Includes Luxembourg. c. Includes reexports. d. Data prior to 1992 include Eritrea. e. Data prior to 1990 refer to the Federal Republic of Germany before unification. f. Data are for the South African Customs Union, which includes Botswana, Lesotho, Namibia, and South Africa.

Structure of merchandise exports | 4.4

About the data

Data on merchandise trade come from customs reports of goods entering an economy or from reports of the financial transactions related to merchandise trade recorded in the balance of payments. Because of differences in timing and definitions, estimates of trade flows from customs reports are likely to differ from those based on the balance of payments. Furthermore, several international agencies process trade data, each making estimates to correct for unreported or misreported data, and this leads to other differences in the available data.

The most detailed source of data on international trade in goods is the COMTRADE database maintained by the United Nations Statistical Division (UNSD). The International Monetary Fund (IMF) also collects customs-based data on exports and imports of goods.

The value of exports is recorded as the cost of the goods delivered to the frontier of the exporting country for shipment—the f.o.b. (free on board) value. Many countries collect and report trade data in U.S. dollars. When countries report in local currency, the UNSD applies the average official exchange rate for the period shown.

Countries may report trade according to the general or special system of trade (see *Primary data documentation*). Under the general system exports comprise outward-moving goods that are (a) goods wholly or partly produced in the country; (b) foreign goods, neither transformed nor declared for domestic consumption in the country, that move outward from customs storage; and (c) goods previously included as imports for domestic consumption but subsequently exported without transformation. Under the special system exports comprise categories a and c. In some compilations categories b and c are classified as reexports. Because of differences in reporting practices, data on exports may not be fully comparable across economies.

Total exports and the shares of exports by major commodity groups were estimated by World Bank staff from the COMTRADE database. Where necessary, data on total exports were supplemented from the IMF's *Direction of Trade Statistics*. The classification of commodity groups is based on the Standard International Trade Classification (SITC), revision 1. Shares may not sum to 100 percent because of unclassified trade. See table 6.2 for data on the growth of merchandise exports.

Figure 4.4a

Manufactured exports dominate trade in goods

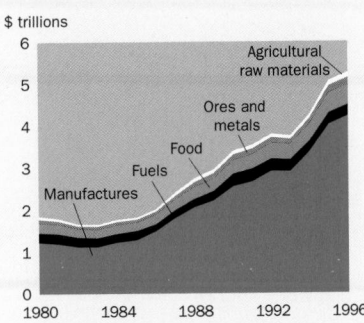

Source: World Bank staff estimates.

The value of merchandise exports has more than doubled since 1980, while manufactured exports have more than tripled. Traditional exports of primary commodities remain important for many developing countries, but world trade is increasingly dominated by manufactured goods.

Definitions

• **Merchandise exports** show the f.o.b. value of goods provided to the rest of the world valued in U.S. dollars. • **Food** comprises the commodities in SITC sections 0 (food and live animals), 1 (beverages and tobacco), and 4 (animal and vegetable oils and fats) and SITC division 22 (oil seeds, oil nuts, and oil kernels). • **Agricultural raw materials** comprise SITC section 2 (crude materials except fuels) excluding divisions 22, 27 (crude fertilizers and minerals excluding coal, petroleum, and precious stones), and 28 (metalliferous ores and scrap). • **Fuels** comprise SITC section 3 (mineral fuels). • **Ores and metals** comprise the commodities in SITC divisions 27, 28, and 68 (nonferrous metals). • **Manufactures** comprise the commodities in SITC sections 5 (chemicals), 6 (basic manufactures), 7 (machinery and transport equipment), and 8 (miscellaneous manufactured goods), excluding division 68.

Data sources

The United Nations Conference on Trade and Development publishes data on the structure of exports and imports in its *Handbook of International Trade and Development Statistics*. Estimates of total exports of goods are also published in the IMF's *International Financial Statistics* and *Direction of Trade Statistics* and in the United Nations *Monthly Bulletin of Statistics*.

4.5 | Structure of merchandise imports

	Merchandise imports		Food		Agricultural raw materials		Fuels		Ores and metals		Manufactures	
	$ millions		% of total		% of total		% of total		% of total		% of total	
	1980	1996	1980	1996	1980	1996	1980	1996	1980	1996	1980	1996
Albania	..	1,283 a	..	..	..	..	..	..	..	..	..	..
Algeria	10,524	8,372 a	21	29	3	3	2	1	2	2	72	65
Angola	873	2,039 a	18	..	1	..	7	..	1	..	73	..
Argentina	10,539	23,762	6	5	4	2	10	4	3	2	77	87
Armenia	..	862 a	..	..	..	..	..	..	..	..	..	..
Australia	19,870	60,897	5	5	3	1	14	6	2	1	75	86
Austria	24,415	67,142	6	6	4	2	15	5	5	3	69	82
Azerbaijan	..	1,255 a	..	..	..	..	..	..	..	..	..	..
Bangladesh	1,980	6,898 a	24	..	6	..	9	..	3	..	58	..
Belarus	..	6,778 a	..	..	..	..	..	..	..	..	..	..
Belgium b	71,192	157,860	11	11	3	2	17	7	8	4	58	73
Benin	302	869 a	26	..	1	..	8	..	1	..	62	..
Bolivia	655	1,601	19	11	1	1	1	3	2	2	78	83
Bosnia and Herzegovina	..	..	..	..	..	..	..	..	..	..	..	..
Botswana	..	..	..	..	..	..	..	..	..	..	..	..
Brazil	24,949	53,736	10	11	1	3	43	12	5	3	41	71
Bulgaria	9,650	4,313 a	..	..	..	..	..	..	..	..	..	..
Burkina Faso	358	783 a	20	..	2	..	13	..	1	..	64	..
Burundi	106	125 a	13	..	2	..	19	..	2	..	61	..
Cambodia	108	1,647 a	51	..	3	..	2	..	0	..	26	..
Cameroon	1,538	1,204	9	14	0	2	12	16	1	1	78	67
Canada	57,707	170,265	7	6	2	1	12	4	5	3	72	82
Central African Republic	80	180	21	12	1	14	2	8	2	1	75	61
Chad	37	217 a	23	24	2	1	2	18	1	1	72	56
Chile	5,123	16,810	15	7	2	1	18	11	2	1	60	78
China †	19,501 a	138,833 a	..	6	..	5	..	5	..	4	..	79
Hong Kong, China	22,027	198,543	12	6	4	2	6	2	2	2	75	88
Colombia	4,663	13,863	12	9	3	3	12	3	3	3	69	78
Congo, Dem. Rep.	1,117	1,331 a	9	..	2	..	12	..	1	..	75	..
Congo, Rep.	418	1,590 a	19	26	1	1	14	1	2	1	65	71
Costa Rica	1,596	3,871	9	12	2	1	15	9	2	2	68	77
Côte d'Ivoire	2,552	2,909 a	13	..	0	..	16	..	2	..	68	..
Croatia	..	7,788 a	..	11	..	2	..	11	..	3	..	69
Cuba	1,656	3,004 a	32	..	1	..	11	..	2	..	53	..
Czech Republic	..	27,709 a	..	7	..	2	..	9	..	3	..	79
Denmark	19,315	43,093	11	12	5	3	22	4	3	2	57	71
Dominican Republic	1,426	6,300 a	17	..	2	..	25	..	2	..	54	..
Ecuador	2,215	3,733	8	10	2	3	1	4	2	2	87	81
Egypt, Arab Rep.	4,860	13,020	32	29	6	6	1	1	1	3	59	60
El Salvador	976	2,670	18	17	2	3	18	12	2	1	61	66
Eritrea	..	..	..	..	..	..	..	..	..	..	..	..
Estonia	..	3,196 a	..	15	..	3	..	9	..	1	..	72
Ethiopia c	721	1,492 a	8	..	3	..	25	..	1	..	64	..
Finland	15,632	30,853	7	7	3	3	29	11	5	5	56	73
France	134,328	274,088	10	10	4	2	27	8	5	3	54	76
Gabon	674	898 a	19	19	0	1	1	4	1	1	78	75
Gambia, The	169	272 a	26	..	1	..	9	..	0	..	61	..
Georgia	..	884 a	..	..	..	..	..	..	..	..	..	..
Germany d	185,922	443,043	12	10	4	2	23	8	6	3	52	71
Ghana	1,129	3,219 a	10	..	1	..	27	..	2	..	59	..
Greece	10,531	26,881 a	9	16	5	2	23	7	2	3	60	71
Guatemala	1,559	3,146	8	14	2	1	24	15	1	1	65	68
Guinea	299	810 a	12	..	1	..	19	..	4	..	62	..
Guinea-Bissau	55	107 a	20	..	0	..	6	..	2	..	69	..
Haiti	536	865 a	22	..	1	..	13	..	1	..	62	..
Honduras	1,009	1,922	10	15	1	1	16	14	1	1	72	69
† Data for Taiwan, China	19,791	101,338	..	..	..	..	..	..	..	..	..	..

Structure of merchandise imports | 4.5

	Merchandise imports ($ millions)		Food (% of total)		Agricultural raw materials (% of total)		Fuels (% of total)		Ores and metals (% of total)		Manufactures (% of total)	
	1980	1996	1980	1996	1980	1996	1980	1996	1980	1996	1980	1996
Hungary	9,212	16,207	8	5	7	3	16	14	6	4	62	73
India	13,819	36,055 a	9	4	2	4	45	24	6	7	39	54
Indonesia	10,834	42,925	13	11	4	5	16	9	2	4	65	71
Iran, Islamic Rep.	9,330	13,926 a	21	..	4	..	1	..	2	..	72	..
Iraq	11,534	492 a	13	..	2	..	0	..	1	..	83	..
Ireland	11,133	35,750	12	8	3	1	15	4	2	2	66	77
Israel	8,023	29,796	11	7	3	1	26	6	4	2	57	82
Italy	98,119	202,908	13	12	7	5	28	9	6	4	45	68
Jamaica	1,178	2,916	20	15	1	2	38	15	2	1	39	65
Japan	139,892	347,496	12	16	9	5	50	17	10	6	19	55
Jordan	2,394	4,293 a	18	21	2	2	17	13	1	3	61	61
Kazakhstan	..	4,261 a	..	..	..	..	..	..	..	..	..	..
Kenya	2,590	3,480 a	8	..	1	..	34	..	1	..	56	..
Korea, Dem. Rep.	..	2,201 a	..	..	..	..	..	..	..	..	..	..
Korea, Rep.	22,228	144,724	10	6	11	5	30	17	6	5	43	67
Kuwait	6,554	8,113 a	15	16	1	1	1	1	1	2	81	81
Kyrgyz Republic	..	838 a	..	21	..	1	..	29	..	1	..	48
Lao PDR	85	642 a	21	..	0	..	19	..	1	..	56	..
Latvia	..	2,319 a	..	13	..	2	..	22	..	1	..	62
Lebanon	3,132	7,560 a	16	..	2	..	15	..	4	..	63	..
Lesotho	..	..	..	..	..	..	..	..	..	..	..	..
Libya	6,776	5,191 a	19	..	1	..	1	..	1	..	78	..
Lithuania	..	4,559 a	..	13	..	3	..	18	..	3	..	61
Macedonia, FYR	..	1,941 a	..	..	..	..	..	..	..	..	..	..
Madagascar	676	671 a	9	16	3	2	15	14	1	1	73	65
Malawi	440	687 a	8	14	1	1	15	11	1	1	75	73
Malaysia	10,735	76,082	12	5	2	1	15	3	4	3	67	85
Mali	491	1,159 a	19	..	1	..	35	..	0	..	45	..
Mauritania	287	616 a	30	..	1	..	14	..	0	..	52	..
Mauritius	619	2,255	26	16	4	3	14	8	1	1	54	71
Mexico	19,591	97,630 a	16	6	3	2	2	2	4	2	75	80
Moldova	..	1,522 a	..	8	..	3	..	46	..	2	..	42
Mongolia	..	451 a	..	14	..	1	..	19	..	1	..	65
Morocco	4,182	8,254	20	19	6	5	24	16	4	4	47	57
Mozambique	550	783 a	14	22	3	2	9	11	3	1	70	62
Myanmar	577	2,524 a	6	..	1	..	3	..	2	..	87	..
Namibia	..	..	..	..	..	..	..	..	..	..	..	..
Nepal	226	664 a	4	15	1	5	18	20	1	5	73	47
Netherlands	76,889	160,700	15	14	3	2	24	9	4	3	53	72
New Zealand	5,515	14,716	6	8	2	1	22	6	4	2	65	83
Nicaragua	882	1,076	15	19	1	1	20	9	1	1	63	71
Niger	608	567 a	14	..	0	..	26	..	3	..	55	..
Nigeria	13,408	6,433 a	15	..	0	..	7	..	2	..	76	..
Norway	16,952	34,290	8	7	3	2	17	5	5	5	67	80
Oman	1,732	4,610 a	15	20	1	1	11	2	0	2	66	70
Pakistan	5,350	11,812	13	15	3	4	27	21	3	1	58	71
Panama	1,447	2,778	10	11	1	1	31	16	1	..	61	..
Papua New Guinea	958	1,866 a	21	..	0	..	15	..	1	..	61	..
Paraguay	615	3,107	11	21	1	0	28	8	1	3	60	67
Peru	2,573	7,947	20	17	3	1	2	10	2	1	73	71
Philippines	8,295	34,663	8	8	2	2	28	9	3	3	48	78
Poland	19,089	37,092	14	10	5	3	18	9	6	3	51	75
Portugal	9,293	33,979	14	13	7	3	24	8	4	2	52	74
Puerto Rico	..	..	..	..	..	..	..	..	..	..	..	..
Romania	13,201	11,435	..	7	..	2	..	21	..	4	..	65
Russian Federation	..	43,318 a	..	..	..	..	..	..	..	..	..	..

	Merchandise imports		Food		Agricultural raw materials		Fuels		Ores and metals		Manufactures	
	$ millions		% of total		% of total		% of total		% of total		% of total	
	1980	1996	1980	1996	1980	1996	1980	1996	1980	1996	1980	1996
Rwanda	155	385 a	10	..	3	..	13	..	0	..	72	..
Saudi Arabia	29,957	27,764 a	14	13	1	1	1	0	1	4	82	79
Senegal	1,038	1,672 a	25	32	1	2	25	10	0	2	48	53
Sierra Leone	268	334 a	24	..	1	..	2	..	1	..	71	..
Singapore	24,003	131,083	8	4	6	1	29	9	2	2	54	83
Slovak Republic	..	10,924 a	..	7	..	2	..	12	..	4	..	61
Slovenia	..	9,412 a	..	8	..	3	..	8	..	4	..	77
South Africae	18,551	26,861	3	6	3	2	0	1	2	1	62	72
Spain	33,901	122,842	13	12	5	3	39	9	6	3	38	72
Sri Lanka	2,035	5,028 a	20	16	1	2	24	6	2	1	52	75
Sudan	1,499	1,439 a	26	..	1	..	13	..	1	..	60	..
Sweden	33,426	63,970	7	7	2	2	24	7	5	3	62	79
Switzerland	36,148	79,192	8	7	3	2	11	4	7	3	71	85
Syrian Arab Republic	4,124	6,399 a	14	..	3	..	26	..	2	..	55	..
Tajikistan	..	668 a	..	..	..	..	..	..	..	..	..	..
Tanzania	1,211	1,642 a	13	..	1	..	21	..	2	..	63	..
Thailand	9,450	73,289 a	5	4	3	4	30	7	4	3	51	81
Togo	550	1,032 a	17	..	1	..	23	..	0	..	59	..
Trinidad and Tobago	3,178	2,204	11	14	2	1	38	19	1	4	49	62
Tunisia	3,509	7,681	14	10	4	3	21	8	4	3	58	75
Turkey	7,573	42,733	4	7	2	5	48	14	3	5	43	69
Turkmenistan	..	1,313 a	..	..	..	..	..	..	..	..	..	..
Uganda	417	725 a	11	..	1	..	23	..	0	..	65	..
Ukraine	..	24,042 a	..	..	..	..	..	..	..	..	..	..
United Arab Emirates	8,098	30,374 a	11	..	1	..	11	..	2	..	74	..
United Kingdom	117,632	283,682	13	10	4	2	13	4	7	3	61	80
United States	250,280	814,888	8	5	3	2	33	9	5	2	50	78
Uruguay	1,652	3,322	8	11	4	3	29	11	3	1	56	74
Uzbekistan	..	4,761 a	..	..	..	..	..	..	..	..	..	..
Venezuela	10,669	8,902	14	16	3	3	2	1	2	3	79	77
Vietnam	618	13,910 a	37	..	2	..	5	..	0	..	55	..
West Bank and Gaza	..	..	..	..	..	..	..	..	..	..	..	..
Yemen, Rep.	1,853	3,443 a	28	29	0	2	7	8	1	1	63	59
Yugoslavia, FR (Serb./Mont.)	15,064	4,101	8	14	7	4	24	14	5	7	57	60
Zambia	1,100	1,106 a	5	..	1	..	22	..	1	..	71	..
Zimbabwe	193	2,808	6	10	2	2	12	10	1	1	73	73
World	**2,004,907 t**	**5,555,200 t**	**11 w**	**9 w**	**4 w**	**3 w**	**25 w**	**8 w**	**5 w**	**3 w**	**54 w**	**75 w**
Low income	..	294,697	..	9	..	5	..	11	..	4	..	69
Excl. China & India	..	..	..	..	..	..	..	..	..	..	..	..
Middle income	..	..	..	..	..	..	..	..	..	..	..	..
Lower middle income	..	..	..	..	..	..	..	..	..	..	..	..
Upper middle income	174,248	457,717	12	8	3	2	14	7	3	3	64	77
Low & middle income	..	..	..	..	..	..	..	..	..	..	..	..
East Asia & Pacific	..	395,405	..	7	..	4	..	6	..	4	..	78
Europe & Central Asia	..	..	..	..	..	..	..	..	..	..	..	..
Latin America & Carib.	110,273	315,627	13	10	2	2	18	7	3	2	63	76
Middle East & N. Africa	100,712	..	18	..	3	..	7	..	2	..	70	..
South Asia	..	62,294	..	8	..	4	..	22	..	6	..	55
Sub-Saharan Africa	..	..	..	..	..	..	..	..	..	..	..	..
High income	1,488,876	4,145,913	11	9	4	2	26	8	5	3	52	75

a. Data are from IMF, *Direction of Trade Statistics*. b. Includes Luxembourg. c. Data prior to 1992 include Eritrea. d. Data prior to 1990 refer to the Federal Republic of Germany before unification. e. Data are for the South African Customs Union, which includes Botswana, Lesotho, Namibia, and South Africa.

Structure of merchandise imports | 4.5

About the data

Data on imports of goods are derived from the same sources as data on exports. In principle, world exports and imports should be identical. Similarly, exports from an economy should equal the sum of imports by the rest of the world from that economy. But differences in timing and definitions result in discrepancies in reported values at all levels. For further discussion of indicators of merchandise trade see *About the data* for tables 4.4 and 6.2.

The value of imports is generally recorded as the cost of the goods when purchased by the importer plus the cost of transport and insurance to the frontier of the importing country—the c.i.f. (cost, insurance, and freight) value. A few countries, including Australia, Canada, and the United States, collect import data on an f.o.b. (free on board) basis and adjust them for freight and insurance costs. Many countries collect and report trade data in U.S. dollars. When countries report in local currency, the United Nations Statistical Division applies the average official exchange rate for the period shown.

Countries may report trade according to the general or special system of trade (see *Primary data documentation*). Under the general system imports include goods imported for domestic consumption and imports into bonded warehouses and free trade zones. Under the special system imports comprise goods imported for domestic consumption (including transformation and repair) and withdrawals for domestic consumption from bonded warehouses and free trade zones. Goods transported through a country en route to another country are excluded.

Total imports and the share of imports by major commodity groups were estimated by World Bank staff from the COMTRADE database. Where necessary, data on total imports were supplemented from the IMF's *Direction of Trade Statistics*. The classification of commodity groups is based on the Standard International Trade Classification (SITC), revision 1. Shares may not sum to 100 percent because of unclassified trade. See table 6.2 for data on the growth of merchandise imports.

Figure 4.5a

Regional imports follow similar patterns

East Asia and the Pacific, 1996

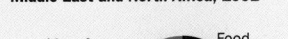

Latin America and the Caribbean, 1995

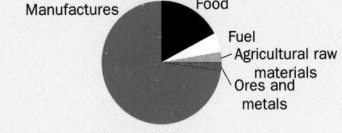

Middle East and North Africa, 1992

South Asia, 1995

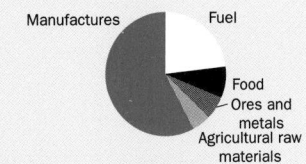

High-income economies, 1996

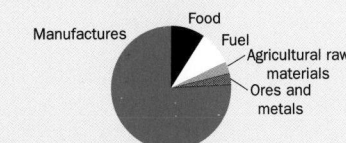

Source: World Bank staff estimates.

Because manufactured goods dominate world trade, the structure of regional imports and exports tend to look rather similar. A few exceptional cases include the large portion of fuel and the relatively smaller share of manufactured imports in South Asia and the larger share of food imports in the Middle East and North Africa.

Definitions

• **Merchandise imports** show the c.i.f. value of goods purchased from the rest of the world valued in U.S. dollars. • **Food** comprises the commodities in SITC sections 0 (food and live animals), 1 (beverages and tobacco), and 4 (animal and vegetable oils and fats) and SITC division 22 (oil seeds, oil nuts, and oil kernels). • **Agricultural raw materials** comprise SITC section 2 (crude materials except fuels) excluding divisions 22, 27 (crude fertilizers and minerals excluding coal, petroleum, and precious stones), and 28 (metalliferous ores and scrap). • **Fuels** comprise SITC section 3 (mineral fuels). • **Ores and metals** comprise the commodities in SITC divisions 27, 28, and 68 (nonferrous metals). • **Manufactures** comprise the commodities in SITC sections 5 (chemicals), 6 (basic manufactures), 7 (machinery and transport equipment), and 8 (miscellaneous manufactured goods), excluding division 68 (nonferrous metals).

Data sources

The United Nations Conference on Trade and Development publishes data on the structure of exports and imports in its *Handbook of International Trade and Development Statistics*. Estimates of total imports of goods are also published in the IMF's *International Financial Statistics* and *Direction of Trade Statistics* and in the United Nations *Monthly Bulletin of Statistics*.

4.6 | Structure of service exports

	Service exports		Transport		Travel		Communications, computer, information, and other services		Insurance and financial services	
	$ millions		% of total		% of total		% of total		% of total	
	1980	1996	1980	1996	1980	1996	1980	1996	1980	1996
Albania	11	129	42.1	23.4	6.4	59.4	46.8	13.5	4.7	3.6
Algeria	476	..	41.0	..	24.1	..	29.5	..	5.4	..
Angola	..	150	..	21.4	..	0.0	..	68.5	..	10.1
Argentina	1,876	3,221	42.9	..	18.3	..	38.4	..	0.3	..
Armenia	..	78	..	49.9	..	4.9	..	45.2	..	..
Australia	3,860	18,424	49.3	28.1	29.5	49.3	19.9	17.8	1.3	4.8
Austria	9,423	24,315	7.3	11.5	68.9	52.0	21.3	23.6	2.6	12.9
Azerbaijan	..	186	..	..	..	..	..	..	..	..
Bangladesh	163	626	18.5	10.1	7.5	3.6	74.0	86.3	0.0	0.0
Belarus	..	613	..	..	..	..	..	..	..	..
Belgium[a]	12,925	36,325	32.7	26.1	14.1	17.7	48.6	41.8	4.7	14.4
Benin	62	104	56.8	56.8	14.1	22.5	26.8	20.6	2.3	..
Bolivia	88	207	33.5	40.2	41.0	29.3	16.6	21.7	8.9	8.8
Bosnia and Herzegovina	..	..	..	..	..	..	..	..	..	..
Botswana	101	260	41.7	14.7	22.2	62.2	32.0	16.0	4.1	7.1
Brazil	1,737	6,135	46.8	42.4	7.3	15.8	38.1	25.3	7.9	16.5
Bulgaria	1,211	1,357	36.3	31.7	28.7	28.6	31.0	39.6	4.0	..
Burkina Faso	49	56	17.3	11.8	10.2	32.6	72.5	55.6	..	..
Burundi	..	17	..	12.0	..	8.4	..	78.5	..	1.0
Cambodia	..	163	..	30.5	..	50.2	..	19.3	..	..
Cameroon	374	438	48.4	38.5	15.5	11.8	30.9	44.0	5.2	5.7
Canada	7,441	28,512	34.1	20.3	34.2	31.1	31.7	48.6	..	..
Central African Republic	54	33	6.5	0.0	5.1	0.0	86.1	100.0	2.4	..
Chad	0	55	0.0	1.9	100.0	21.3	..	76.1	0.0	0.7
Chile	1,263	3,356	32.2	39.7	13.9	27.6	51.9	29.0	2.1	3.8
China	2,512	20,601	52.3	14.9	28.0	49.5	11.7	35.0	8.0	0.6
Hong Kong, China	..	..	..	..	..	..	..	..	..	..
Colombia	1,342	3,867	31.1	36.6	35.6	23.5	27.6	26.1	5.6	13.9
Congo, Dem. Rep.	57	..	40.4	..	15.8	..	42.1	..	1.8	..
Congo, Rep.	111	96	46.0	46.7	6.7	6.1	42.7	47.3	4.6	..
Costa Rica	194	1,310	24.9	13.7	43.7	51.2	30.7	35.1	0.7	..
Côte d'Ivoire	564	734	50.0	27.3	14.4	10.3	25.7	58.1	9.9	4.3
Croatia	..	3,496	..	19.5	..	62.5	..	18.0	..	..
Cuba	..	..	..	..	..	..	..	..	..	..
Czech Republic	..	8,181	..	16.3	..	49.9	..	31.6	..	2.2
Denmark	5,853	15,699	44.4	45.6	21.1	21.8	31.8	32.6	2.7	..
Dominican Republic	309	2,132	7.8	1.7	55.8	86.4	35.9	11.9	0.5	..
Ecuador	367	858	35.1	40.0	35.6	32.8	13.3	17.1	16.0	10.1
Egypt, Arab Rep.	2,393	10,636	52.4	29.0	24.8	34.6	22.4	35.4	0.4	1.0
El Salvador	139	388	18.3	24.9	9.6	22.0	50.5	46.3	21.6	6.9
Eritrea	..	105	..	..	..	..	..	..	..	..
Estonia	..	1,108	..	39.8	..	43.7	..	15.4	..	1.1
Ethiopia[b]	110	373	53.2	55.5	5.7	9.4	40.9	33.4	0.2	1.7
Finland	2,733	7,276	35.1	29.3	25.0	21.2	37.0	50.0	2.9	-0.5
France	43,506	88,891	24.2	22.8	19.0	31.9	53.4	36.2	3.4	9.1
Gabon	325	273	21.5	35.2	5.2	2.6	67.7	56.2	5.6	6.0
Gambia, The	18	101	0.0	15.5	100.0	52.4	0.0	31.9	0.0	0.2
Georgia	..	122	..	..	..	..	..	..	..	..
Germany[c]	33,062	84,639	26.6	23.3	15.1	20.8	57.4	48.9	0.8	7.0
Ghana	107	157	33.6	53.1	0.4	8.1	64.9	36.0	1.2	2.9
Greece	3,947	9,348	23.6	4.0	43.9	39.8	32.4	55.9	0.1	0.3
Guatemala	211	559	18.9	11.2	29.2	38.7	46.5	44.1	5.4	6.1
Guinea	..	124	..	12.2	..	5.2	..	80.2	..	2.3
Guinea-Bissau	6	..	8.9	..	12.5	..	78.6	..	..	..
Haiti	90	109	5.9	5.5	85.0	87.6	7.8	6.9	1.2	0.5
Honduras	82	258	36.9	21.9	30.1	31.1	18.5	45.3	14.5	1.7

Structure of service exports | 4.6

	Service exports		Transport		Travel		Communications, computer, information, and other services		Insurance and financial services	
	$ millions		% of total		% of total		% of total		% of total	
	1980	1996	1980	1996	1980	1996	1980	1996	1980	1996
Hungary	633	5,004	5.4	8.3	63.5	44.9	30.5	42.2	0.6	4.6
India	2,949	10,087	15.0	27.9	52.2	38.1	31.5	31.5	1.2	2.5
Indonesia	449	5,681	15.1	0.0	50.8	95.9	34.1	4.1	..	..
Iran, Islamic Rep.	731	593	4.5	23.3	4.0	11.3	91.5	57.5	..	7.9
Iraq	..	..	..	..	..	..	..	..	..	..
Ireland	1,381	5,563	36.6	20.4	42.0	44.4	21.4	35.2	..	..
Israel	2,722	8,004	38.1	24.3	36.0	36.3	24.9	39.3	1.0	0.2
Italy	19,192	69,910	23.9	21.6	46.7	42.9	22.9	28.1	6.5	7.3
Jamaica	401	1,388	28.0	18.1	61.2	73.6	6.7	7.4	4.2	0.9
Japan	20,240	67,724	62.9	31.9	3.2	6.0	32.4	55.4	1.6	4.9
Jordan	750	1,846	27.0	20.4	51.9	40.3	21.1	39.3	..	..
Kazakhstan	..	674	..	..	..	..	..	..	..	..
Kenya	577	956	38.0	33.5	41.4	46.9	19.8	18.6	0.8	1.0
Korea, Dem. Rep.	..	..	..	..	..	..	..	..	..	..
Korea, Rep.	4,710	26,806	33.5	36.8	7.8	18.2	53.1	41.8	5.6	3.2
Kuwait	1,225	1,613	57.7	73.3	30.8	8.9	11.5	17.8	..	..
Kyrgyz Republic	..	22	..	..	..	..	..	..	..	..
Lao PDR	..	97	..	15.5	..	52.8	..	31.2	..	0.4
Latvia	..	1,126	..	62.8	..	19.1	..	6.8	..	11.4
Lebanon	..	630	..	..	..	..	..	..	..	..
Lesotho	32	38	2.0	8.2	37.8	46.0	60.2	45.8	..	..
Libya	164	..	64.5	..	6.2	..	29.4	..	..	..
Lithuania	..	798	..	44.9	..	39.6	..	14.4	..	1.1
Macedonia, FYR	..	185	..	..	..	..	..	..	..	..
Madagascar	79	293	49.4	27.1	6.3	22.1	44.0	49.2	0.4	1.6
Malawi	32	22	49.8	58.5	29.5	20.6	19.8	20.6	0.9	0.3
Malaysia	1,135	11,269	41.6	21.3	28.0	35.2	29.8	43.4	0.6	0.1
Mali	58	67	30.9	38.2	25.8	27.0	42.2	34.0	1.0	0.9
Mauritania	56	28	26.3	6.3	11.9	40.1	61.8	53.6	0.0	..
Mauritius	140	908	38.4	22.9	30.2	55.5	31.2	21.6	0.2	0.0
Mexico	4,591	10,901	9.7	..	69.7	..	10.4	..	10.2	..
Moldova	..	104	..	49.3	..	31.6	..	14.5	..	2.8
Mongolia	37	57	26.5	26.2	8.6	36.0	64.9	33.5	..	4.4
Morocco	783	2,360	20.3	18.1	57.9	56.7	20.7	23.4	1.1	1.8
Mozambique	118	242	78.5	24.8	0.0	..	21.5	75.2	..	..
Myanmar	60	309	34.5	..	19.7	..	43.6	..	2.3	..
Namibia	..	242	..	..	..	86.0	..	12.6	..	1.4
Nepal	127	643	5.9	10.0	40.8	21.2	53.3	68.8	..	..
Netherlands	17,150	49,185	51.5	41.0	13.1	13.3	34.3	44.2	1.2	1.4
New Zealand	1,009	4,708	58.2	33.9	21.1	51.7	19.6	14.6	1.1	−0.1
Nicaragua	44	132	36.0	13.0	48.6	44.2	14.9	40.8	0.5	2.0
Niger	41	33	33.5	1.2	15.2	21.3	50.9	77.5	0.4	0.0
Nigeria	1,127	640	80.9	10.5	6.0	4.8	6.5	84.4	6.6	0.3
Norway	8,615	13,918	74.5	56.9	8.8	16.7	16.3	19.0	0.4	7.4
Oman	9	13	100.0	100.0	0.0	0.0	0.0	0.0	0.0	0.0
Pakistan	617	1,665	41.3	44.7	22.8	5.9	34.1	48.7	1.8	0.8
Panama	902	1,537	47.0	51.7	19.0	22.3	25.8	20.5	8.2	5.5
Papua New Guinea	43	436	33.8	7.7	28.3	3.1	36.9	86.0	1.0	3.1
Paraguay	118	1,229	2.0	2.9	55.3	56.5	42.6	40.5	0.2	..
Peru	715	1,371	30.9	24.9	40.9	46.1	24.9	21.4	3.2	7.6
Philippines	1,447	9,348	14.2	2.9	22.1	12.2	63.6	84.3	..	0.7
Poland	2,018	9,833	59.2	28.0	11.9	32.1	24.1	32.4	4.8	7.5
Portugal	2,006	8,141	23.5	17.9	57.3	57.7	18.1	19.5	1.2	4.9
Puerto Rico	..	..	..	..	..	..	..	..	..	..
Romania	1,063	1,563	37.6	36.6	30.5	33.8	27.8	24.8	4.2	4.7
Russian Federation	..	12,217	..	25.6	..	56.3	..	16.1	..	0.7

	Service exports		Transport		Travel		Communications, computer, information, and other services		Insurance and financial services	
	$ millions		% of total		% of total		% of total		% of total	
	1980	1996	1980	1996	1980	1996	1980	1996	1980	1996
Rwanda	32	21	42.3	33.1	10.8	16.3	46.3	50.7	0.6	..
Saudi Arabia	5,191	3,518	15.3	0.0	25.9	0.0	58.8	100.0	..	..
Senegal	337	556	19.1	10.1	29.3	30.2	51.3	59.3	0.3	0.4
Sierra Leone	49	100	31.4	12.3	25.5	69.0	43.1	18.5	..	0.2
Singapore	4,856	30,040	26.9	17.3	29.5	26.6	42.5	54.5	1.1	1.6
Slovak Republic	..	2,066	..	31.1	..	32.6	..	30.9	..	5.4
Slovenia	..	2,127	..	22.6	..	57.8	..	19.1	..	0.5
South Africa	2,929	4,253	41.8	27.0	47.1	52.3	2.5	11.0	8.6	9.6
Spain	11,593	44,364	25.9	15.2	60.0	62.3	11.6	18.8	2.4	3.6
Sri Lanka	231	765	18.8	44.3	42.9	21.7	37.4	30.4	1.0	3.6
Sudan	216	115	9.2	1.6	17.9	16.6	72.4	81.9	0.5	0.0
Sweden	7,489	16,930	40.5	30.2	12.9	21.8	44.0	45.7	2.6	2.4
Switzerland	6,888	26,225	18.8	9.4	46.0	34.2	30.5	26.9	4.7	29.5
Syrian Arab Republic	365	1,833	17.2	13.4	42.9	65.8	39.9	20.8	..	..
Tajikistan	..	2	..	..	..	..	..	..	..	..
Tanzania	165	667	39.5	11.5	12.6	77.8	46.4	7.7	1.5	0.4
Thailand	1,490	17,008	20.1	15.4	58.2	53.4	21.2	29.0	0.5	0.7
Togo	74	73	38.3	14.2	35.2	32.6	25.0	52.7	1.4	0.5
Trinidad and Tobago	411	343	27.7	56.6	37.3	22.6	35.0	12.0	..	8.9
Tunisia	1,067	2,632	19.4	24.4	64.1	60.3	14.8	13.4	1.6	1.9
Turkey	711	13,051	37.4	13.5	45.9	43.3	16.3	40.9	0.4	2.3
Turkmenistan	..	..	..	..	..	..	..	..	..	..
Uganda	10	135	0.0	13.7	40.4	81.1	59.6	5.2	..	..
Ukraine	..	4,799	..	75.6	..	6.7	..	15.0	..	2.7
United Arab Emirates	..	..	..	..	..	..	..	..	..	..
United Kingdom	36,452	79,389	38.9	22.6	19.0	25.2	42.1	41.0	..	11.2
United States	47,550	234,687	29.9	20.4	22.3	34.2	44.6	41.1	3.2	4.3
Uruguay	468	1,359	18.6	29.5	63.7	47.5	14.8	21.7	2.9	1.2
Uzbekistan	..	380	..	..	..	..	..	..	..	..
Venezuela	693	1,565	41.1	30.4	35.1	56.5	9.5	13.0	14.4	0.1
Vietnam	..	2,364	..	..	..	..	..	..	..	..
West Bank and Gaza	..	..	..	..	..	..	..	..	..	..
Yemen, Rep.	..	191	..	17.3	..	27.8	..	54.9	..	..
Yugoslavia, FR (Serb./Mont.)	..	..	..	..	..	..	..	..	..	..
Zambia	152	..	56.8	..	13.9	..	25.3	..	4.1	..
Zimbabwe	169	383	56.9	24.3	14.6	46.7	26.1	28.7	2.4	0.3
World	**413,965 t**	**1,355,168 t**	..	..	..	..	..	..	..	..
Low income	*11,623*	46,899	..	..	..	..	..	..	..	..
Excl. China & India	7,064	15,956	..	..	..	..	..	..	..	..
Middle income	*68,474*	242,524	..	..	..	..	..	..	..	..
Lower middle income	..	..	..	..	..	..	..	..	..	..
Upper middle income	24,317	67,872	..	..	..	..	..	..	..	..
Low & middle income	*79,442*	287,903	..	..	..	..	..	..	..	..
East Asia & Pacific	*8,864*	*64,989*	..	..	..	..	..	..	..	..
Europe & Central Asia	..	..	..	..	..	..	..	..	..	..
Latin America & Carib.	17,799	48,369	..	..	..	..	..	..	..	..
Middle East & N. Africa	13,985	30,113	..	..	..	..	..	..	..	..
South Asia	4,180	14,128	..	..	..	..	..	..	..	..
Sub-Saharan Africa	9,009	13,997	..	..	..	..	..	..	..	..
High income	336,480	1,066,134	..	..	..	..	..	..	..	..

a. Includes Luxembourg. b. Data prior to 1992 include Eritrea. c. Data prior to 1990 refer to the Federal Republic of Germany before unification.

Structure of service exports | 4.6

About the data

Balance of payments statistics, the main source of information for international trade in services, have many weaknesses. Until recently some large economies—such as the former Soviet Union—did not report data on trade in services. Disaggregation of important components may be limited and varies significantly across countries. There are inconsistencies in the methods used to report items. And the recording of major flows as net items is common (for example, insurance transactions are often recorded as premiums less claims). These factors contribute to a downward bias in the value of the service trade reported in the balance of payments.

Efforts are being made to improve the coverage, quality, and consistency of these data. Eurostat and the Organisation for Economic Co-operation and Development, for example, are working together to improve the collection of statistics on trade in services in member countries. In addition, the International Monetary Fund (IMF) is implementing the new classification of trade in services introduced in the fifth edition of its *Balance of Payments Manual* (1993).

Still, difficulties in capturing all the dimensions of international trade in services mean that the record is likely to remain incomplete. Cross-border intrafirm service transactions, which are usually not captured in the balance of payments, are increasing rapidly as foreign direct investment expands and electronic networks become pervasive. One example of such transactions is transnational corporations' use of mainframe computers around the clock for data processing, exploiting time zone differences between their home country and the host countries of their affiliates. Another important dimension of services trade not captured by conventional balance of payments statistics is establishment trade—sales in the host country by foreign affiliates. By contrast, cross-border intrafirm transactions in merchandise may be reported as exports or imports in the balance of payments.

Definitions

- **Service exports** refer to economic output of intangible commodities that may be produced, transferred, and consumed at the same time. International transactions in services are defined by the IMF's *Balance of Payments Manual* (1993), but definitions may nevertheless vary among reporting economies.
- **Transport** covers all transport services (sea, air, land, internal waterway, space, and pipeline) performed by residents of one economy for those of another and involving the carriage of passengers, movement of goods (freight), rental of carriers with crew, and related support and auxiliary services. Excluded are freight insurance, which is included in insurance services; goods procured in ports by nonresident carriers and repairs of transport equipment, which are included in goods; repairs of railway facilities, harbors, and airfield facilities, which are included in construction services; and rental of carriers without crew, which is included in other services. • **Travel** covers goods and services acquired from an economy by travelers in that economy for their own use during visits of less than one year for business or personal purposes. • **Communications, computer, information, and other services** cover international telecommunications and postal and courier services; computer data; news-related service transactions between residents and nonresidents; construction services; royalties and license fees; miscellaneous business, professional, and technical services; personal, cultural, and recreational services; and government services not included elsewhere.
- **Insurance and financial services** cover various types of insurance provided to nonresidents by resident insurance enterprises and vice versa, and financial intermediary and auxiliary services (except those of insurance enterprises and pension funds) exchanged between residents and nonresidents.

Figure 4.6a

Service exports are increasing

% of regional exports of goods and services

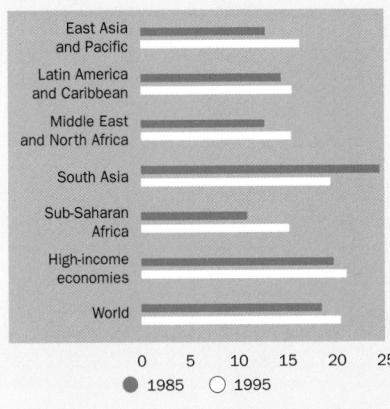

% of global service exports, 1996

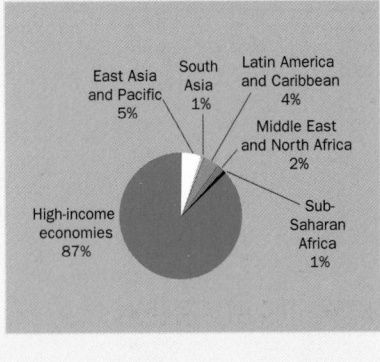

Source: International Monetary Fund, balance of payments data files.

The importance of service exports has increased in every region except South Asia, which already had the largest share of services in its export trade. East Asia and the Pacific was the fastest-growing service exporter during 1985–95, but service exports remain overwhelmingly a business of high-income countries.

Data sources

Data on exports and imports of services come from the IMF's balance of payments data files. The IMF publishes balance of payments data in its *International Financial Statistics* and *Balance of Payments Statistics Yearbook*.

4.7 | Structure of service imports

	Service imports $ millions		Transport % of total		Travel % of total		Communications, computer, information, and other services % of total		Insurance and financial services % of total	
	1980	1996	1980	1996	1980	1996	1980	1996	1980	1996
Albania	18	189	43.7	43.3	0.0	6.2	51.4	34.5	4.9	15.9
Algeria	2,697	..	39.9	..	12.4	..	40.9	..	6.8	..
Angola	..	1,563	..	20.1	..	5.6	..	72.0	..	2.3
Argentina	3,788	5,716	33.6	..	47.3	..	19.1	..	..	..
Armenia	..	129	..	71.7	..	17.0	..	3.6	..	7.7
Australia	6,532	18,495	47.4	36.4	28.1	31.3	23.2	27.0	1.3	5.3
Austria	6,204	22,455	12.7	8.6	50.6	48.9	32.4	26.7	4.2	15.8
Azerbaijan	..	399	..	..	..	..	..	..	..	..
Bangladesh	173	733	64.3	65.5	3.3	11.7	26.4	16.3	6.0	6.4
Belarus	..	183	..	..	..	..	..	..	..	..
Belgium[a]	12,827	33,811	29.7	22.3	25.7	28.3	39.4	38.4	5.2	11.0
Benin	109	111	56.9	65.8	7.1	5.3	25.9	23.0	10.1	5.8
Bolivia	259	377	53.3	61.7	21.3	13.4	16.2	15.8	9.1	9.1
Bosnia and Herzegovina	..	..	..	..	..	..	..	..	..	..
Botswana	216	444	42.4	42.2	26.0	32.7	27.8	17.1	3.7	8.0
Brazil	4,871	13,630	56.5	42.6	7.5	24.9	35.0	23.3	1.0	9.2
Bulgaria	549	1,234	51.3	38.5	8.6	16.1	34.4	45.4	5.7	..
Burkina Faso	209	138	58.0	46.9	15.3	16.4	21.9	32.6	4.8	4.0
Burundi	..	102	..	30.0	..	24.9	..	41.2	..	3.9
Cambodia	..	222	..	44.9	..	6.5	..	44.6	..	4.0
Cameroon	377	657	39.0	34.4	11.5	21.1	44.1	37.5	5.4	7.0
Canada	10,805	35,772	29.5	22.6	30.5	31.0	40.0	46.4	..	..
Central African Republic	142	114	47.3	43.7	24.5	38.0	23.5	10.4	4.7	7.9
Chad	24	199	6.4	48.1	57.5	13.0	35.4	37.6	0.7	1.3
Chile	1,583	3,587	52.4	52.9	12.6	22.4	32.3	26.8	2.7	–2.0
China	2,024	22,585	61.6	45.7	3.3	19.8	30.7	33.5	4.4	1.0
Hong Kong, China	..	..	..	..	..	..	..	..	..	..
Colombia	1,170	4,094	45.3	28.2	20.5	22.1	24.0	32.7	10.2	16.9
Congo, Dem. Rep.	608	..	41.9	..	8.1	..	46.1	..	3.9	..
Congo, Rep.	480	708	27.0	33.3	6.1	7.5	63.5	57.1	3.5	2.1
Costa Rica	286	947	58.2	40.1	21.1	34.6	13.9	18.7	6.7	6.6
Côte d'Ivoire	1,531	1,502	38.6	40.1	15.8	16.5	37.7	41.6	7.9	1.8
Croatia	..	3,185	..	42.4	..	32.4	..	25.2	..	..
Cuba	..	..	..	..	..	..	..	..	..	..
Czech Republic	..	6,264	..	11.2	..	47.3	..	37.1	..	4.5
Denmark	4,663	14,819	47.7	44.7	27.4	27.9	22.9	27.4	2.0	..
Dominican Republic	399	962	39.6	56.8	41.6	21.1	14.7	9.9	4.1	12.2
Ecuador	704	975	36.0	43.0	32.4	22.5	19.1	19.5	12.4	15.1
Egypt, Arab Rep.	2,343	6,253	40.3	32.6	7.2	25.9	49.1	37.7	3.4	3.8
El Salvador	273	490	29.3	52.7	38.8	14.8	20.7	21.5	11.2	11.0
Eritrea	..	..	..	..	..	..	..	..	..	..
Estonia	..	608	..	45.0	..	16.6	..	34.1	..	4.3
Ethiopia[b]	90	234	72.2	60.9	2.2	8.2	25.6	25.1	..	5.8
Finland	2,555	8,773	39.4	23.1	23.1	25.3	35.1	49.6	2.4	1.9
France	32,148	72,087	28.4	28.6	18.7	24.3	48.1	35.8	4.8	11.3
Gabon	789	949	22.0	29.6	12.2	18.3	60.1	48.3	5.7	3.8
Gambia, The	42	77	55.8	40.3	3.5	20.6	33.2	35.1	7.4	3.9
Georgia	..	105	..	..	..	..	..	..	..	..
Germany[c]	42,378	128,060	25.1	19.3	41.2	39.7	33.2	38.5	0.6	2.6
Ghana	270	456	39.7	47.0	12.1	4.9	45.8	41.1	2.3	7.0
Greece	1,428	4,238	41.5	29.8	21.6	28.6	31.1	36.9	5.8	4.7
Guatemala	487	660	37.0	43.5	33.6	20.5	26.0	28.4	3.3	7.6
Guinea	..	422	..	34.7	..	6.3	..	52.6	..	6.3
Guinea-Bissau	14	21	47.9	52.6	11.0	..	36.0	41.5	5.0	5.8
Haiti	162	283	49.3	66.8	25.1	13.1	23.0	20.1	2.7	..
Honduras	174	334	53.3	58.9	17.8	17.1	16.6	21.6	12.3	2.4

Structure of service imports | 4.7

	Service imports		Transport		Travel		Communications, computer, information, and other services		Insurance and financial services	
	$ millions		% of total		% of total		% of total		% of total	
	1980	1996	1980	1996	1980	1996	1980	1996	1980	1996
Hungary	524	3,506	60.3	9.0	26.9	27.3	6.1	56.9	6.7	6.7
India	1,516	8,287	60.0	55.5	3.8	9.7	31.0	29.3	5.3	5.4
Indonesia	4,998	13,475	40.1	35.2	11.9	16.1	44.2	45.4	3.8	3.2
Iran, Islamic Rep.	5,223	2,339	43.6	40.3	32.5	10.3	17.4	40.1	6.4	9.3
Iraq	..	..	..	..	..	..	..	..	..	..
Ireland	1,593	13,260	43.9	14.3	36.6	16.4	16.1	68.1	3.4	1.3
Israel	2,310	10,080	44.1	33.6	35.6	36.3	18.5	28.0	1.8	2.1
Italy	16,249	67,445	43.8	35.0	11.8	23.4	31.3	33.1	13.1	8.5
Jamaica	370	1,034	55.4	47.6	8.9	14.3	23.8	28.7	11.8	9.4
Japan	32,360	129,962	52.2	25.9	14.2	28.5	31.3	40.0	2.3	3.8
Jordan	819	1,603	32.8	45.0	33.1	23.9	28.1	25.2	6.0	6.0
Kazakhstan	..	928	..	..	..	..	..	..	..	..
Kenya	502	860	66.2	53.8	4.6	19.4	18.0	21.0	11.2	5.8
Korea, Dem. Rep.	..	..	..	..	..	..	..	..	..	..
Korea, Rep.	4,089	32,154	55.8	35.9	8.6	23.3	30.0	38.0	5.6	2.8
Kuwait	3,067	5,107	38.8	33.4	43.7	48.8	16.9	16.5	0.6	1.3
Kyrgyz Republic	..	49	..	..	..	..	..	..	..	..
Lao PDR	..	119	..	34.5	..	24.9	..	39.8	..	0.8
Latvia	..	742	..	23.4	..	50.3	..	19.4	..	7.0
Lebanon	..	604	..	..	..	..	..	..	..	..
Lesotho	50	64	31.6	50.8	15.8	10.6	49.7	33.6	2.8	5.0
Libya	2,303	..	51.4	..	20.4	..	23.2	..	5.0	..
Lithuania	..	677	..	44.2	..	39.3	..	15.6	..	0.9
Macedonia, FYR	..	379	..	..	..	..	..	..	..	..
Madagascar	311	373	57.3	46.7	9.9	19.3	28.0	33.2	4.8	0.8
Malawi	179	234	81.7	83.0	5.6	6.5	5.3	2.0	7.4	8.5
Malaysia	2,957	14,442	44.3	38.2	24.5	16.0	31.2	45.7	6.2	5.6
Mali	212	324	65.8	51.1	9.6	16.7	18.3	26.7	6.2	5.6
Mauritania	128	217	59.1	55.7	13.6	10.5	24.2	32.4	3.1	1.3
Mauritius	174	680	64.7	41.1	12.9	26.3	15.2	28.5	7.2	4.1
Mexico	6,514	10,819	28.2	38.6	47.0	31.3	16.3	19.0	8.5	8.9
Moldova	..	182	..	49.7	..	21.0	..	25.0	..	3.9
Mongolia	31	95	48.4	63.7	0.3	20.4	51.3	15.8	..	..
Morocco	1,436	1,984	34.4	36.5	6.8	15.1	55.5	42.4	3.3	5.9
Mozambique	124	350	79.0	32.7	0.0	..	14.5	65.1	6.5	2.2
Myanmar	85	122	56.8	..	4.6	..	31.7	..	6.9	..
Namibia	..	494	..	38.2	..	15.1	..	39.9	..	6.8
Nepal	81	261	30.1	22.9	29.2	51.4	38.2	25.7	2.5	..
Netherlands	18,148	45,736	43.9	30.2	26.6	25.2	27.1	41.7	2.4	2.9
New Zealand	1,843	5,037	39.4	41.2	28.3	29.4	31.8	25.9	0.6	3.5
Nicaragua	104	246	50.7	36.9	29.9	24.4	14.2	35.1	5.2	3.7
Niger	279	152	43.0	58.9	6.6	8.8	43.7	30.3	6.7	2.0
Nigeria	5,285	4,215	33.7	9.4	18.7	27.0	44.8	62.6	2.8	1.0
Norway	6,996	13,465	52.2	38.5	21.1	28.5	23.3	22.9	3.4	10.1
Oman	518	1,037	34.1	42.8	6.2	4.5	55.9	47.9	3.8	4.8
Pakistan	853	3,159	64.5	55.4	9.6	15.2	22.8	25.8	3.0	3.6
Panama	588	1,012	65.4	66.7	9.5	13.4	14.9	11.5	10.2	8.3
Papua New Guinea	302	747	60.4	24.1	5.9	9.6	28.9	58.3	4.8	7.9
Paraguay	260	960	58.0	53.8	21.0	23.6	13.2	12.5	7.9	10.1
Peru	880	2,050	55.4	41.2	12.2	17.1	25.2	30.9	7.2	10.7
Philippines	1,439	6,926	52.1	29.6	7.4	6.1	39.8	62.7	0.8	1.6
Poland	2,023	6,429	59.9	26.3	12.9	9.1	25.3	50.2	2.0	14.4
Portugal	1,525	6,943	48.8	25.1	19.1	33.9	27.2	33.8	4.9	7.2
Puerto Rico	..	..	..	..	..	..	..	..	..	..
Romania	1,045	1,948	76.8	35.5	7.0	34.1	7.7	25.8	8.5	4.5
Russian Federation	..	18,595	..	13.4	..	57.7	..	28.2	..	0.5

	Service imports		Transport		Travel		Communications, computer, information, and other services		Insurance and financial services	
	$ millions		% of total		% of total		% of total		% of total	
	1980	1996	1980	1996	1980	1996	1980	1996	1980	1996
Rwanda	123	150	63.5	30.0	9.3	8.3	27.3	61.7	..	..
Saudi Arabia	30,231	22,049	17.1	9.8	8.1	0.0	73.3	89.1	1.5	1.1
Senegal	340	578	46.9	39.9	17.6	12.4	29.0	42.8	6.5	4.9
Sierra Leone	85	108	54.8	14.5	9.8	57.5	23.4	24.8	11.9	3.2
Singapore	2,912	18,730	38.3	33.8	11.4	32.5	46.1	28.3	4.3	5.3
Slovak Republic	..	2,027	..	19.4	..	23.8	..	47.8	..	9.0
Slovenia	..	1,423	..	28.4	..	38.1	..	31.9	..	1.5
South Africa	3,805	5,689	48.4	46.7	20.3	27.5	20.0	17.0	11.3	8.8
Spain	5,732	24,352	38.6	28.4	21.5	20.2	34.6	44.4	5.4	7.1
Sri Lanka	351	1,202	60.4	57.3	9.5	14.6	23.5	22.1	6.5	6.0
Sudan	258	193	34.3	64.0	16.8	13.8	44.5	21.7	4.5	0.3
Sweden	7,018	18,755	35.9	26.2	31.6	33.3	28.1	39.1	4.4	1.4
Switzerland	4,885	15,387	30.4	24.9	48.8	49.0	19.3	25.1	1.6	1.0
Syrian Arab Republic	521	1,555	26.6	51.3	33.9	33.0	37.3	15.7	2.2	..
Tajikistan	..	..	..	..	..	..	..	..	..	..
Tanzania	132	1,018	62.1	23.1	6.7	43.2	25.6	30.1	5.5	3.4
Thailand	1,644	19,585	64.4	40.1	14.8	21.9	14.8	31.7	5.9	4.9
Togo	167	78	62.7	48.4	14.1	29.7	16.7	11.5	6.6	10.4
Trinidad and Tobago	645	242	45.7	38.9	21.6	28.7	23.5	25.1	9.2	7.3
Tunisia	600	1,259	51.1	40.1	17.7	19.9	25.5	33.6	5.7	6.4
Turkey	569	6,396	50.5	26.7	18.3	19.8	27.1	46.9	4.2	6.6
Turkmenistan	..	..	..	..	..	..	..	..	..	..
Uganda	123	383	58.2	34.7	14.6	20.4	22.8	41.0	4.4	3.9
Ukraine	..	1,625	..	34.0	..	15.7	..	42.9	..	7.3
United Arab Emirates	..	..	..	..	..	..	..	..	..	..
United Kingdom	27,933	68,153	47.5	28.0	22.9	38.2	29.6	32.7	..	1.2
United States	40,970	152,774	37.5	29.0	25.4	32.7	35.0	33.4	2.1	5.0
Uruguay	476	820	31.8	43.7	42.6	27.2	18.1	27.7	7.5	1.5
Uzbekistan	..	463	..	..	..	..	..	..	..	..
Venezuela	4,253	4,900	31.7	25.0	47.0	46.4	16.5	25.8	4.8	2.8
Vietnam	..	2,390	..	..	..	..	..	..	..	..
West Bank and Gaza	..	..	..	..	..	..	..	..	..	..
Yemen, Rep.	..	757	..	45.4	..	12.8	..	41.8	..	..
Yugoslavia, FR (Serb./Mont.)	..	..	..	..	..	..	..	..	..	..
Zambia	651	..	53.5	..	8.5	..	33.9	..	4.0	..
Zimbabwe	395	712	43.3	50.7	40.3	16.9	13.9	29.8	2.6	2.6
World	**456,685 t**	**1,340,766 t**	..	..	..	..	..	..	..	..
Low income	22,029	64,428	..	..	..	..	..	..	..	..
Excl. China & India	19,929	31,200	..	..	..	..	..	..	..	..
Middle income	161,500	279,679	..	..	..	..	..	..	..	..
Lower middle income	..	..	..	..	..	..	..	..	..	..
Upper middle income	62,647	100,655	..	..	..	..	..	..	..	..
Low & middle income	157,286	336,906	..	..	..	..	..	..	..	..
East Asia & Pacific	14,719	86,060	..	..	..	..	..	..	..	..
Europe & Central Asia	..	..	..	..	..	..	..	..	..	..
Latin America & Carib.	32,387	58,696	..	..	..	..	..	..	..	..
Middle East & N. Africa	55,014	48,471	..	..	..	..	..	..	..	..
South Asia	3,186	14,000	..	..	..	..	..	..	..	..
Sub-Saharan Africa	20,546	28,046	..	..	..	..	..	..	..	..
High income	313,872	1,019,680	..	..	..	..	..	..	..	..

a. Includes Luxembourg. b. Data prior to 1992 include Eritrea. c. Data prior to 1990 refer to the Federal Republic of Germany before unification.

Structure of service imports | 4.7

About the data

Although the data have many deficiencies, it is clear that trade in services has grown faster than trade in merchandise over the past 15 years. During 1980–95 service trade grew an average 8 percent a year, compared with 6 percent for merchandise trade (in nominal terms). This rapid growth boosted commercial services' share in global trade from 16 percent in 1980 to 18 percent in 1995. The most dynamic trade is in private services

such as financial, brokerage, and leasing services. Growing by an average 9.5 percent a year, trade in these services rose from 37 percent of commercial services trade in 1980 to 45 percent in 1993. Tourism is another rapidly growing sector (see table 6.15).

Data on service imports are taken from balance of payments statistics. For more information on trade in services see *About the data* for table 4.7.

Figure 4.7a

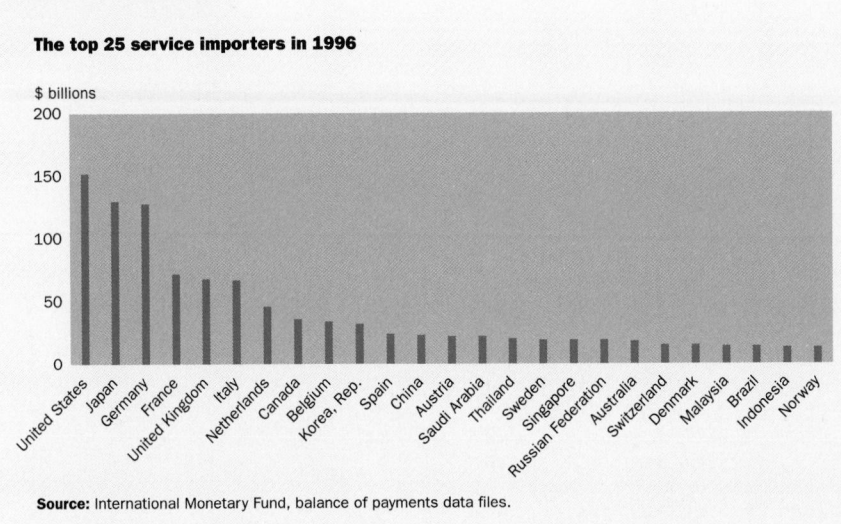

The top 25 service importers in 1996

$ billions

Source: International Monetary Fund, balance of payments data files.

The same countries that are the largest exporters of services also tend to be the largest importers. As trade in services continues to grow, more developing countries will begin to enter the ranks of major traders.

Definitions

• **Services imports** refer to economic output of intangible commodities that may be produced, transferred, and consumed at the same time. International transactions in services are defined by the International Monetary Fund's (IMF) *Balance of Payments Manual* (1993), but definitions may nevertheless vary among reporting economies. • **Transport** covers all transport services (sea, air, land, internal waterway, space, and pipeline) performed by residents of one economy for those of another and involving the carriage of passengers, movement of goods (freight), rental of carriers with crew, and related support and auxiliary services. Excluded are freight insurance, which is included in insurance services; goods procured in ports by nonresident carriers and repairs of transport equipment, which are included in goods; repairs of railway facilities, harbors, and airfield facilities, which are included in construction services; and rental of carriers without crew, which is included in other services. • **Travel** covers goods and services acquired from an economy by travelers in that economy for their own use during visits of less than one year for business or personal purposes. • **Communications, computer, information, and other services** cover international telecommunications and postal and courier services; computer data; news-related service transactions between residents and nonresidents; construction services; royalties and license fees; miscellaneous business, professional, and technical services; personal, cultural, and recreational services; and government services not included elsewhere. • **Insurance and financial services** cover various types of insurance provided to nonresidents by resident insurance enterprises and vice versa, and financial intermediary and auxiliary services (except those of insurance enterprises and pension funds) exchanged between residents and nonresidents.

Data sources

Data on exports and imports of services come from the IMF's balance of payments data files. The IMF publishes balance of payments data in its *International Financial Statistics* and *Balance of Payments Statistics Yearbook*.

4.8 | Structure of demand

	Private consumption		General government consumption		Gross domestic investment		Exports of goods and services		Imports of goods and services		Gross domestic savings	
	% of GDP		% of GDP		% of GDP		% of GDP		% of GDP		% of GDP	
	1980	1996	1980	1996	1980	1996	1980	1996	1980	1996	1980	1996
Albania	56	94	9	13	35	21	..	13	..	40	..	–7
Algeria	43	52	14	14	39	27	34	32	30	25	43	34
Angola	..	19	..	33	..	11	..	77	..	40	..	48
Argentina	76	82	.. [a]	.. [a]	25	19	5	9	6	9	24	18
Armenia	47	115	16	13	29	10	..	24	..	62	37	–28
Australia	59	61	18	18	25	21	16	21	18	21	24	21
Austria	55	56	18	20	29	25	36	39	38	39	27	24
Azerbaijan	..	87	..	9	..	24	..	21	..	42	..	4
Bangladesh	92	79	6	14	15	17	6	14	18	24	2	7
Belarus	..	59	..	23	..	25	..	44	..	52	..	17
Belgium	63	62	18	15	22	18	62	73	65	68	19	23
Benin	96	80	9	11	15	17	23	25	43	33	–5	9
Bolivia	67	79	14	13	15	15	21	20	17	27	19	8
Bosnia and Herzegovina	..	..	..	..	..	..	..	..	..	..	..	..
Botswana	47	28	18	29	35	24	50	51	49	33	36	43
Brazil	70	66	9	16	23	19	9	7	11	8	21	18
Bulgaria	55	71	6	12	34	14	36	65	31	62	39	17
Burkina Faso	95	79	10	13	17	25	10	12	33	29	–6	8
Burundi	91	91	9	10	14	9	9	4	23	14	–1	–1
Cambodia	..	87	..	8	..	21	..	27	..	43	..	5
Cameroon	70	71	10	8	21	16	27	19	27	13	20	21
Canada	55	60	19	20	24	18	28	38	27	35	25	21
Central African Republic	94	89	15	8	7	6	25	19	41	22	–9	3
Chad	99	85	8	13	4	19	24	28	41	44	–6	3
Chile	71	65	12	9	21	28	23	27	27	29	17	26
China	51	45	15	11	35	42	6	21	7	19	35	44
Hong Kong, China	60	60	6	9	35	32	90	142	91	143	34	31
Colombia	70	72	10	10	19	21	16	17	16	20	20	17
Congo, Dem. Rep.	82	88	8	4	10	6	16	35	16	33	10	8
Congo, Rep.	47	59	18	11	36	61	60	67	60	97	36	30
Costa Rica	66	60	18	17	27	23	26	45	37	46	16	22
Côte d'Ivoire	63	68	17	12	27	14	35	45	41	38	20	20
Croatia	..	66	..	30	..	15	..	42	..	53	..	3
Cuba	..	..	..	..	..	..	..	..	..	..	..	..
Czech Republic	..	51	..	22	..	35	..	55	..	62	..	27
Denmark	56	54	27	25	19	17	33	34	34	30	17	21
Dominican Republic	77	75	8	6	25	24	19	29	29	34	15	19
Ecuador	60	66	15	12	26	17	25	31	25	26	26	22
Egypt, Arab Rep.	69	78	16	10	28	17	31	21	43	25	15	12
El Salvador	72	87	14	9	13	16	34	21	33	33	14	3
Eritrea	..	127	..	.. [a]	..	26	..	32	..	85	..	–27
Estonia	..	61	..	25	..	27	..	73	..	86	..	14
Ethiopia[b]	83	82	14	12	9	21	11	13	17	27	3	7
Finland	54	53	18	22	29	16	33	38	34	30	28	25
France	59	60	18	19	24	18	22	23	23	21	23	21
Gabon	26	46	13	11	28	20	65	60	32	37	61	43
Gambia, The	79	77	20	18	26	21	47	58	72	74	1	5
Georgia	56	100	13	7	29	4	..	17	..	28	31	–7
Germany	..	57	..	20	..	23	..	24	..	23	..	23
Ghana	84	79	11	12	6	19	8	27	9	38	5	8
Greece	71	81	12	14	24	14	16	16	22	27	18	5
Guatemala	79	87	8	5	16	13	22	18	25	23	13	8
Guinea	..	82	..	8	..	13	..	19	..	22	..	10
Guinea-Bissau	77	93	29	7	30	22	8	10	44	32	–6	1
Haiti	82	101	10	9	17	2	22	7	31	28	8	–7
Honduras	70	63	13	9	25	32	36	48	44	52	17	27

Structure of demand | 4.8

	Private consumption		General government consumption		Gross domestic investment		Exports of goods and services		Imports of goods and services		Gross domestic savings	
	% of GDP		% of GDP		% of GDP		% of GDP		% of GDP		% of GDP	
	1980	1996	1980	1996	1980	1996	1980	1996	1980	1996	1980	1996
Hungary	61	64	10	10	31	27	39	39	41	40	29	26
India	73	66	10	10	21	27	7	12	10	15	17	24
Indonesia	51	59	11	8	24	32	34	26	20	25	38	33
Iran, Islamic Rep.	53	53	21	13	30	29	13	21	16	16	26	34
Iraq	..	..	..	..	..	..	..	..	..	..	..	..
Ireland	..	55	19	15	..	15	48	75	61	59	..	30
Israel	50	58	39	29	22	24	40	29	51	40	11	13
Italy	61	61	15	16	27	18	22	28	25	23	24	22
Jamaica	64	71	20	16	16	27	51	55	51	68	16	14
Japan	59	60	10	10	32	29	14	9	15	8	31	30
Jordan	79	66	29	23	37	35	40	50	84	75	−8	11
Kazakhstan	..	68	..	12	..	23	..	31	..	34	..	20
Kenya	62	69	20	15	29	20	28	33	39	37	18	17
Korea, Dem. Rep.	..	..	..	..	..	..	..	..	..	..	..	..
Korea, Rep.	64	55	12	11	32	38	34	32	41	36	25	34
Kuwait	31	49	11	33	14	12	78	55	34	49	58	18
Kyrgyz Republic	..	87	..	17	..	19	..	31	..	55	..	−4
Lao PDR	..	..	..	..	..	31	..	23	..	42	..	12
Latvia	59	70	8	20	26	19	..	46	..	55	33	10
Lebanon	..	102	..	15	..	30	..	11	..	58	..	−17
Lesotho	133	68	26	20	43	104	20	22	122	114	−59	12
Libya	21	..	22	..	22	..	66	..	31	..	57	..
Lithuania	..	70	..	18	..	21	..	52	..	62	..	11
Macedonia, FYR	..	82	..	14	..	15	..	37	..	49	..	4
Madagascar	89	90	12	6	15	10	13	18	30	24	−1	4
Malawi	70	72	19	17	25	17	25	21	39	27	11	11
Malaysia	51	47	17	11	30	41	58	92	55	91	33	42
Mali	92	78	10	11	16	27	16	21	34	36	−2	12
Mauritania	68	72	25	14	36	22	37	54	67	62	7	14
Mauritius	75	68	14	10	21	26	51	61	61	65	10	22
Mexico	65	66	10	10	27	21	11	22	13	20	25	23
Moldova	..	66	..	20	..	28	..	52	..	66	..	14
Mongolia	44	64	29	16	63	22	21	44	57	46	27	20
Morocco	68	68	18	16	24	21	17	25	28	30	14	16
Mozambique	103	68	21	12	0	48	21	28	45	56	−24	20
Myanmar	82	89	..[a]	..[a]	21	11	9	1	13	2	18	11
Namibia	44	59	17	30	29	20	76	49	67	58	39	11
Nepal	82	82	7	10	18	23	12	23	19	37	11	9
Netherlands	61	60	17	14	22	19	51	53	52	47	22	26
New Zealand	62	63	18	14	21	22	30	30	32	29	20	23
Nicaragua	83	84	20	13	17	28	24	41	43	66	−2	3
Niger	67	85	10	11	37	10	24	16	38	22	23	4
Nigeria	56	64	12	11	21	19	29	17	19	11	31	24
Norway	50	..	19	21	25	..	43	41	37	31	31	..
Oman	28	42	25	31	22	17	63	49	38	40	47	27
Pakistan	83	73	10	12	18	19	12	17	24	21	7	14
Panama	45	53	18	15	28	29	98	94	89	91	38	32
Papua New Guinea	61	36	24	24	25	27	43	57	53	44	15	40
Paraguay	76	73	6	10	32	23	15	21	29	26	18	17
Peru	57	73	11	8	29	24	22	12	19	16	32	19
Philippines	67	74	9	12	29	24	24	42	28	52	24	14
Poland	67	64	9	18	26	20	28	23	31	26	23	18
Portugal	67	65	13	18	33[c]	25[c]	25	33	38	41	20	17
Puerto Rico	75	..	16	..	17	..	65	..	73	..	10	..
Romania	60	70	5	11	40	25	35	27	40	33	35	19
Russian Federation	62	63	15	11	22	22	..	23	..	19	22	25

	Private consumption		General government consumption		Gross domestic investment		Exports of goods and services		Imports of goods and services		Gross domestic savings	
	% of GDP		% of GDP		% of GDP		% of GDP		% of GDP		% of GDP	
	1980	1996	1980	1996	1980	1996	1980	1996	1980	1996	1980	1996
Rwanda	83	92	12	10	16	14	14	6	26	22	4	−3
Saudi Arabia	22	42	16	26	22	20	71	42	30	30	62	32
Senegal	78	78	22	10	15	17	28	31	44	36	0	11
Sierra Leone	79	99	21	11	17	9	28	12	45	31	0	−10
Singapore	53	41	10	9	46	35	215	187	224	169	38	50
Slovak Republic	..	49	..	24	..	38	..	57	..	69	..	27
Slovenia	..	57	..	20	..	23	..	55	..	56	..	22
South Africa	50	61	13	21	28	18	36	26	28	26	36	18
Spain	66	62	13	17	23	21	16	24	18	23	21	21
Sri Lanka	80	73	9	10	34	25	32	35	55	44	11	17
Sudan	81	..	16	..	15	..	12	..	24	..	3	..
Sweden	51	52	29	26	21	15	29	40	31	33	19	22
Switzerland	63	..	14	15	27 c	..	36	36	40	32	23	..
Syrian Arab Republic	67	..	23	..	28	..	18	..	35	..	10	..
Tajikistan	..	71	..	11	..	17	..	114	..	114	..	18
Tanzania[d]	..	83	..	13	..	18	..	22	..	36	..	3
Thailand	65	55	12	10	29	41	24	39	30	44	23	35
Togo	54	80	22	13	28	14	51	31	56	38	23	6
Trinidad and Tobago	46	63	12	12	31	15	50	53	39	42	42	26
Tunisia	62	61	14	16	29	24	40	42	46	44	24	3
Turkey	77	71	12	12	18	24	5	22	12	27	11	18
Turkmenistan	..	..	..	..	..	..	..	..	..	..	..	..
Uganda	89	84	11	10	6	16	19	12	26	22	0	6
Ukraine	..	58	..	22	..	23	..	46	..	48	..	20
United Arab Emirates	17	54	11	18	28	27	78	70	34	69	72	27
United Kingdom	59	64	22	21	17	16 c	27	28	25	29	19	15
United States	64	68	17	16	20	18	10	11	11	13	19	16
Uruguay	76	76	12	13	17	12	15	18	21	20	12	11
Uzbekistan	..	66	..	25	..	16	..	31	..	38	..	9
Venezuela	55	66	12	5	26	17	29	37	22	24	33	30
Vietnam	..	86	..	.. a	..	28	..	42	..	55	..	14
West Bank and Gaza	..	..	..	..	..	..	..	..	..	..	..	..
Yemen, Rep.	..	70	..	15	..	25	..	40	..	50	..	16
Yugoslavia, FR (Serb./Mont.)	..	..	..	..	..	..	..	..	..	..	..	..
Zambia	55	74	26	18	23	15	41	38	45	45	19	8
Zimbabwe	46	63	27	19	23	18	30	41	26	41	27	18
World	61 w	63 w	15 w	16 w	24 w	22 w	19 w	21 w	20 w	21 w	24 w	22 w
Low income	64	59	13	11	27	31	10	20	13	23	24	29
Excl. China & India	74	73	13	13	20	20	21	26	28	31	13	14
Middle income	59	64	13	14	26	23	25	27	23	28	27	22
Lower middle income	60	65	14	13	26	23	..	30	..	30	25	23
Upper middle income	58	63	12	15	25	22	26	22	22	23	30	21
Low & middle income	61	63	13	13	26	25	20	25	20	27	26	24
East Asia & Pacific	53	50	14	11	32	39	16	29	15	29	33	38
Europe & Central Asia	64	64	13	15	25	23	..	31	..	33	23	21
Latin America & Carib.	67	67	10	12	24	20	16	17	17	16	23	20
Middle East & N. Africa	46	55	18	17	29	26	35	28	28	26	36	28
South Asia	75	68	9	11	21	25	8	13	13	17	15	22
Sub-Saharan Africa	62	67	14	15	23	18	30	28	29	28	23	18
High income	61	63	16	16	24	21	19	20	20	20	23	21

a. General government consumption figures are not available separately; they are included in private consumption. b. Data prior to 1992 include Eritrea. c. Includes statistical discrepancy. d. Data cover mainland Tanzania only.

Structure of demand | 4.8

Because policymakers have tended to focus on fostering the growth of output, and because data on production are easier to collect than data on spending, many countries generate their primary estimate of GDP using the production approach. Moreover, many countries do not estimate all the separate components of national expenditures or, if they do, derive some of the main aggregates indirectly using GDP (output) as the control total.

Expenditures from GDP include private consumption, general government consumption, gross domestic fixed capital formation (private and public investment), changes in inventories, and exports (minus imports) of goods and services. Such expenditures are generally recorded in purchasers' prices and so include net indirect taxes.

Private consumption is often estimated as a residual, by subtracting from GDP all other known expenditures. The resulting aggregate may incorporate fairly large discrepancies. When private consumption is calculated separately, the household surveys on which a large component of the estimates are based tend to be one-year studies with limited coverage. Thus the estimates quickly become outdated and must be supplemented by price- and quantity-based statistical estimating procedures. Complicating the issue, in many developing countries the distinction between cash outlays for personal business and those for household use may be blurred. General government consumption usually includes expenditures on national defense and security, some of which are now considered to be part of investment.

Gross domestic investment consists of outlays on additions to the economy's fixed assets plus net changes in the level of inventories. Under the revised (1993) guidelines for the United Nations System of National Accounts (SNA), gross domestic investment also includes capital outlays on defense establishments that may be used by the general public, such as schools and hospitals, and on certain types of private housing for family use. All other defense expenditures are treated as current spending. Investment data may be estimated from direct surveys of enterprises and administrative records or based on the commodity flow method using data from trade and construction activities. While the quality of public fixed investment data depends on the quality of government accounting systems (which tend to be weak in developing countries), measures of private fixed investment—particularly capital outlays by small, unincorporated enterprises—are usually very unreliable.

Estimates of changes in inventories are rarely complete but usually include the most important activities or commodities. In some countries these estimates are derived as a composite residual along with aggregate private consumption. According to national accounts conventions, adjustments should be made for appreciation of the value of inventories due to price changes, but this is not always done. In highly inflationary economies this element can be substantial.

Exports and imports are compiled from customs returns and from balance of payments data obtained from central banks. Although the data on exports and imports from the payments side provide reasonably reliable records of cross-border transactions, they may not adhere strictly to the appropriate valuation and timing definitions of the balance of payments or, more important, correspond with the change-of-ownership criterion. This issue has assumed greater significance with the increasing globalization of international business. Neither customs nor balance of payments data capture the illegal transactions that occur in many countries. Goods carried by travelers across borders in legal but unreported shuttle trade may further distort trade statistics.

For further discussion of the problems of building and maintaining national accounts see Srinivasan (1994), Heston (1994), and Ruggles (1994). For a classic analysis of the reliability of foreign trade and national income statistics see Morgenstern (1963).

• **Private consumption** is the market value of all goods and services, including durable products (such as cars, washing machines, and home computers) purchased or received as income in kind by households and nonprofit institutions. It excludes purchases of dwellings but includes imputed rent for owner-occupied dwellings. In practice it may include any statistical discrepancy in the use of resources relative to the supply of resources. • **General government consumption** includes all current spending for purchases of goods and services (including wages and salaries) by all levels of government, excluding most government enterprises. It also includes most expenditures on national defense and security. • **Gross domestic investment** consists of outlays on additions to the fixed assets of the economy plus net changes in the level of inventories. Fixed assets include land improvements (fences, ditches, drains, and so on); plant, machinery, and equipment purchases; and the construction of roads, railways, and the like, including commercial and industrial buildings, offices, schools, hospitals, and private residential dwellings. Inventories are stocks of goods held by firms to meet temporary or unexpected fluctuations in production or sales. • **Exports and imports of goods and services** represent the value of all goods and other market services provided to the world. Included is the value of merchandise, freight, insurance, travel, and other nonfactor services. Factor and property income (formerly called factor services), such as investment income, interest, and labor income, is excluded. Transfer payments are excluded from the calculation of GDP. • **Gross domestic savings** are calculated as the difference between GDP and total consumption.

National accounts data for developing countries are collected from national statistical organizations and central banks by visiting and resident World Bank missions. Data for industrial countries come from Organisation for Economic Co-operation and Development (OECD) data files. For information on the OECD national accounts series see OECD, *National Accounts, 1960–1995*, volumes 1 and 2. The complete national accounts time series is available on the *World Development Indicators* CD-ROM.

4.9 Growth of consumption and investment

	Private consumption				Private consumption per capita		General government consumption		Gross domestic investment	
	$ millions		average annual % growth		average annual % growth		average annual % growth		average annual % growth	
	1980	1996	1980–90	1990–96	1980–90	1990–96	1980–90	1990–96	1980–90	1990–96
Albania	..	2,532	..	..	..	..	..	..	−0.3	41.8
Algeria	18,293	24,081	1.9	0.8	−1.1	−1.5	4.7	4.2	−2.3	−4.8
Angola	..	1,269	0.3	−3.8	−2.5	−6.7	2.7	−3.9	−6.8	1.8
Argentina	..	..	..	..	..	..	..	..	−4.7	12.7
Armenia	..	1,692	3.5	−17.6	2.2	−18.6	5.9	−8.9	6.2	−17.7
Australia	94,360	224,020	3.0	3.5	1.5	2.3	3.6	2.6	2.7	5.2
Austria	43,264	129,065	2.4	2.0	2.2	1.3	1.4	2.0	2.2	2.3
Azerbaijan	..	3,187	..	8.2	..	..	..	..	..	−5.3
Bangladesh	11,857	25,178	4.1	0.8	1.5	−0.8	..a	..a	1.4	13.6
Belarus	..	11,462	..	−7.8	..	−7.9	..	−7.2	..	−17.1
Belgium	75,166	167,800	1.7	1.2	1.6	0.8	0.4	1.2	3.1	−0.7
Benin	1,356	1,775	2.3	..	−0.8	..	2.4	..	−4.0	6.6
Bolivia	2,064	4,353	0.6	3.1	−1.4	0.7	−3.1	5.8	−9.9	4.2
Bosnia and Herzegovina	..	..	..	..	..	..	..	..	..	..
Botswana	483	1,402	..	−3.3	9.8	−5.7	13.4	9.7	..	−1.7
Brazil	163,832	492,282	1.6	4.4	−0.4	2.9	7.3	−0.8	0.2	3.7
Bulgaria	11,089	6,777	2.5	0.1	2.6	0.8	9.1	−10.4	2.4	−15.4
Burkina Faso	1,631	2,014	2.6	2.7	0.0	−0.1	6.2	2.2	8.6	1.1
Burundi	840	1,033	3.7	−2.9	0.8	−5.4	6.4	−1.5	4.5	−4.7
Cambodia	..	2,720	..	..	..	..	..	..	..	..
Cameroon	4,710	6,570	3.9	−2.8	1.1	−5.6	5.3	−7.6	−2.7	−2.5
Canada	145,745	334,215	3.5	1.5	2.3	0.3	2.4	−0.1	5.2	1.8
Central African Republic	747	945	0.4	0.1	−2.0	−2.1	−2.0	−5.6	9.9	−12.8
Chad	579	991	3.7	0.2	1.2	−2.3	12.4	−7.2	25.1	−1.3
Chile	19,489	50,559	1.9	8.8	0.3	7.1	0.4	3.3	6.1	11.5
China	103,442	366,169	9.7	10.9	8.1	9.6	7.8	12.3	11.0	15.5
Hong Kong, China	17,013	93,004	6.7	5.9	5.3	4.2	5.0	5.8	4.0	11.3
Colombia	23,456	58,348	2.6	4.6	0.7	2.7	4.2	8.3	0.5	20.8
Congo, Dem. Rep.	12,167	6,049	3.4	−7.0	0.0	−9.9	0.0	−19.6	−5.1	−5.0
Congo, Rep.	797	1,403	2.7	5.5	−0.5	2.5	4.0	−9.9	−11.9	−3.2
Costa Rica	3,156	5,430	2.9	3.6	0.0	1.4	1.1	2.3	5.3	2.4
Côte d'Ivoire	6,388	7,232	1.6	0.4	−2.2	−2.6	−0.1	0.9	−9.8	13.6
Croatia	..	12,647	..	..	..	..	..	..	..	..
Cuba	..	..	..	..	..	..	..	..	..	..
Czech Republic	..	27,821	..	1.4	..	1.4	..	−2.6	2.3	0.9
Denmark	37,050	93,823	1.8	3.0	1.8	2.6	1.1	1.5	4.0	2.4
Dominican Republic	5,109	9,847	1.9	4.5	−0.3	2.5	10.1	−3.2	5.2	8.9
Ecuador	6,995	12,640	1.9	2.8	−0.7	0.5	−1.4	−0.9	−3.8	3.2
Egypt, Arab Rep.	15,848	51,192	4.1	4.2	1.5	2.1	2.7	1.9	0.0	0.4
El Salvador	2,567	9,160	0.8	7.2	−0.2	4.7	0.1	2.0	2.2	11.8
Eritrea	..	..	..	..	..	..	..	..	..	..
Estonia	..	2,637	..	−3.5	..	−2.3	..	4.5	..	−10.1
Ethiopia[b]	4,282	4,893	0.7	3.5	−2.5	1.6	4.4	−3.9	4.4	22.2
Finland	27,761	67,324	3.8	−0.5	3.4	−0.9	3.4	−0.9	3.0	−5.4
France	391,263	925,063	2.6	1.3	2.1	0.8	2.2	2.2	2.8	−2.1
Gabon	1,119	2,626	−0.7	−2.4	−3.9	−4.9	0.2	2.0	−7.1	0.1
Gambia, The	185	277	3.6	7.6	−0.1	3.4	2.5	−10.0	0.8	3.0
Georgia	..	4,302	..	..	..	..	..	..	..	..
Germany	..	1,377,876	..	..	..	..	..	..	..	..
Ghana	3,730	5,043	2.8	3.5	−0.6	0.7	2.4	7.3	3.3	3.0
Greece	32,706	85,184	2.4	1.4	1.9	0.8	2.7	1.2	−0.9	1.0
Guatemala	6,217	13,731	1.1	4.5	−1.7	1.5	2.7	3.9	−1.8	4.0
Guinea	..	3,222	..	4.4	..	1.7	..	1.0	..	−0.5
Guinea-Bissau	81	252	−0.8	7.1	−2.6	4.9	5.6	−1.1	..	−6.6
Haiti	1,197	1,684	0.9	−0.6	−1.0	−2.6	−4.4	−2.8	−0.6	−10.1
Honduras	1,806	2,534	2.7	3.3	−0.7	0.3	3.3	−2.3	2.9	9.2

Growth of consumption and investment | 4.9

	Private consumption				Private consumption per capita		General government consumption		Gross domestic investment	
	$ millions		average annual % growth		average annual % growth		average annual % growth		average annual % growth	
	1980	1996	1980–90	1990–96	1980–90	1990–96	1980–90	1990–96	1980–90	1990–96
Hungary	13,562	28,772	0.8	0.2	1.2	0.5	2.5	–6.3	–0.4	8.1
India	125,809	233,232	4.6	4.7	2.5	2.8	7.7	4.0	6.5	8.8
Indonesia	40,821	131,695	5.6	7.3	3.7	5.5	4.6	2.7	6.7	9.9
Iran, Islamic Rep.	48,854	..	2.8	2.2	–0.4	–0.4	–5.0	8.6	–2.5	–0.8
Iraq	..	..	..	..	..	..	..	..	..	..
Ireland	13,585	35,360	2.4	3.8	2.1	3.4	–0.3	2.5	..	–2.4
Israel	11,397	53,387	5.3	7.7	3.5	4.2	0.5	2.5	2.2	11.5
Italy	273,819	667,582	3.0	0.3	2.9	0.1	2.5	0.3	1.9	–2.2
Jamaica	1,710	3,128	4.5	5.6	3.3	4.6	6.3	–2.3	–0.1	4.9
Japan	623,286	3,080,624	3.7	2.0	3.2	1.7	2.4	2.5	5.3	0.2
Jordan	3,123	4,842	2.2	6.2	–1.5	1.2	2.3	7.3	–1.5	11.9
Kazakhstan	..	13,241	..	–5.0	..	..	..	–10.5	..	–13.3
Kenya	4,506	6,348	4.7	3.1	1.1	0.4	2.6	12.1	0.8	1.1
Korea, Dem. Rep.	..	..	..	..	..	..	..	..	..	..
Korea, Rep.	40,534	267,434	8.1	7.2	6.9	6.1	5.5	4.7	11.9	7.8
Kuwait	8,836	13,045	–1.4	..	–5.5	..	2.2	..	–4.5	..
Kyrgyz Republic	..	1,522	..	–1.5	..	..	..	–11.7	..	–7.8
Lao PDR	..	..	..	..	..	..	..	..	..	..
Latvia	..	3,358	..	–12.9	..	–11.8	5.0	2.3	3.4	–32.0
Lebanon	..	13,263	..	..	..	..	..	..	..	..
Lesotho	492	604	1.9	–3.0	–0.7	–5.0	2.9	8.4	6.3	10.7
Libya	7,171	..	..	..	..	..	..	..	..	..
Lithuania	..	5,461	..	0.0	..	–0.0	..	0.0	..	0.0
Macedonia, FYR	..	1620	..	–5.4	..	–6.0	..	–1.8	..	–24.3
Madagascar	3,611	3,719	–0.6	0.7	–3.5	–2.0	0.5	–2.1	4.9	–3.0
Malawi	866	1,576	1.7	1.8	–1.6	–1.0	6.3	–1.4	–2.8	–5.3
Malaysia	12,378	46,507	3.7	7.3	1.0	4.8	2.7	7.7	2.6	15.6
Mali	1,547	2,064	1.9	1.1	–0.6	–1.7	7.3	–2.0	7.1	6.3
Mauritania	481	790	3.0	1.1	0.4	–1.4	–6.9	8.8	–4.1	4.0
Mauritius	854	2,930	6.7	4.9	5.8	3.7	3.3	4.4	9.0	0.1
Mexico	126,745	222,549	1.0	0.3	–1.3	–1.5	2.0	9.8	–3.3	0.1
Moldova	..	1,190	..	–14.9	..	–14.8	..	–13.8	..	–21.3
Mongolia	..	621	..	..	..	..	..	..	..	..
Morocco	12,788	24,966	3.7	2.8	1.5	0.8	5.5	1.0	2.5	–0.1
Mozambique	2,081	1,168	0.3	1.8	–1.2	–2.6	–2.1	–0.2	2.7	4.0
Myanmar	..	..	0.6	5.2	..	..	–4.1	13.1	–2.7	12.5
Namibia	950	1,902	1.3	4.2	–1.5	1.5	3.7	2.7	–3.9	3.8
Nepal	1,600	3,643	6.9	9.0	4.2	6.1	4.7	6.1	1.8	5.2
Netherlands	104,571	236,978	1.7	2.1	1.2	1.4	2.1	1.1	3.1	–0.5
New Zealand	13,801	36,867	2.1	3.0	1.3	1.6	1.4	0.4	0.2	8.1
Nicaragua	1,770	1,661	–3.4	3.2	–6.2	0.1	3.0	–8.3	–4.7	11.2
Niger	1,704	1,688	–3.2	..	–6.3	..	9.8	..	–2.2	..
Nigeria	36,258	20,557	–2.6	3.2	–5.5	0.2	–3.5	–3.5	–8.5	0.1
Norway	29,694	75,083	2.2	2.6	1.9	2.0	2.3	2.8	0.6	..
Oman	1,657	4,732	..	..	..	..	..	..	..	..
Pakistan	19,688	47,628	4.7	6.1	1.6	3.1	10.3	1.3	5.9	4.2
Panama	1,709	4,201	4.2	2.8	2.1	1.0	1.2	1.1	–8.9	15.0
Papua New Guinea	1,568	1,876	0.4	7.1	–1.7	4.7	–0.1	–1.3	–0.9	3.7
Paraguay	3,467	7,045	2.4	6.1	–0.6	3.3	1.5	9.2	–0.8	3.1
Peru	12,006	44,236	1.0	4.8	–1.2	2.7	–1.4	4.9	–4.5	13.6
Philippines	20,910	66,998	2.6	4.2	0.0	1.8	0.6	3.7	–2.1	4.7
Poland	38,182	86,502	1.1	4.9	0.4	4.6	1.2	4.6	0.9	6.1
Portugal	19,166	65,262	2.3	2.2	2.2	2.1	4.9	2.3	..	0.5
Puerto Rico	10,756	..	3.5	2.3	2.5	1.5	5.1	0.2	6.9	3.7
Romania	..	24,917	..	1.0	..	1.4	..	0.8	..	–7.5
Russian Federation	..	279,314	..	5.6	..	5.7	..	–13.8	..	–13.2

	Private consumption				Private consumption per capita		General government consumption		Gross domestic investment	
	$ millions		average annual % growth		average annual % growth		average annual % growth		average annual % growth	
	1980	1996	1980–90	1990–96	1980–90	1990–96	1980–90	1990–96	1980–90	1990–96
Rwanda	969	1,227	1.4	−6.2	−1.6	−4.6	5.2	−12.9	4.3	−2.4
Saudi Arabia	34,538	52,897	..	..	..	..	..	..	..	..
Senegal	2,365	4,031	2.5	1.3	−0.3	−1.3	3.0	−4.3	3.6	4.8
Sierra Leone	951	928	0.1	1.2	−2.0	−1.3	0.0	1.6	−6.7	−12.8
Singapore	6,030	38,252	5.8	7.2	4.1	5.1	6.6	7.4	3.1	9.0
Slovak Republic	..	9,333	3.8	−3.7	3.5	−3.9	4.8	−0.7	1.1	−1.0
Slovenia	..	10,626	..	5.4	..	5.5	..	2.0	..	7.8
South Africa	39,543	76,867	2.3	1.6	0.1	−0.1	3.5	2.6	−4.7	5.4
Spain	139,348	346,651	2.7	0.9	2.4	0.7	5.3	2.0	5.7	−1.5
Sri Lanka	3,230	10,302	3.8	6.0	2.4	4.7	7.3	5.3	0.6	6.4
Sudan	5,447	..	0.2	..	−2.3	..	−0.3	..	−1.1	..
Sweden	64,624	132,017	1.8	−0.1	1.5	−0.7	1.5	−0.2	..	..
Switzerland	65,117	183,474	1.7	0.4	1.2	−0.4	3.0	0.9	..	..
Syrian Arab Republic	8,690	..	3.4	..	0.1	..	−2.9	..	−7.0	..
Tajikistan	..	1,300	..	..	..	..	..	..	..	..
Tanzania^c	..	4,870	..	2.7	..	−0.3	..	−4.1	..	−21.0
Thailand	21,175	100,112	5.9	7.9	4.1	6.6	4.2	4.3	9.4	10.3
Togo	619	1,142	2.8	1.1	−0.3	−1.9	−1.7	−3.2	5.9	−11.5
Trinidad and Tobago	2,860	3,418	−1.3	−0.3	−2.6	−1.1	−1.7	0.4	−10.1	10.1
Tunisia	5,380	11,931	2.9	3.7	0.3	1.8	3.8	3.9	−1.8	1.7
Turkey	42,067	122,590	−4.1	3.4	−6.3	1.5	2.7	3.0	4.9	4.0
Turkmenistan	..	..	..	..	..	..	..	..	..	..
Uganda	1,935	5,089	2.9	6.9	0.2	3.6	1.8	7.2	9.6	10.6
Ukraine	..	25,524	..	..	..	..	..	..	..	..
United Arab Emirates	5,116	19,423	4.6	..	−0.5	..	−3.9	..	−8.7	..
United Kingdom	320,290	701,563	4.1	1.5	3.8	1.1	1.1	1.0	6.4	..
United States	1,720,600	4,780,301	3.4	2.5	2.5	1.5	2.8	−0.3	..	..
Uruguay	7,681	14,597	0.5	10.0	−0.1	9.4	1.8	1.2	−7.8	6.0
Uzbekistan	..	16,698	..	..	..	..	..	−4.1	..	−7.6
Venezuela	38,066	44,308	1.3	1.4	−1.4	−0.8	2.0	−1.2	−5.3	2.8
Vietnam	..	15,537	..	8.1	..	..	..	a	..	..
West Bank and Gaza	..	..	..	..	..	..	..	..	..	..
Yemen, Rep.	..	4,209	..	−2.8	..	−6.8	..	−1.6	..	13.2
Yugoslavia, FR (Serb./Mont.)	..	..	..	..	..	..	..	..	..	..
Zambia	2,145	2,907	3.5	−4.5	0.4	−7.1	−3.4	−15.0	−4.4	2.5
Zimbabwe	2,488	4,757	3.2	2.5	−0.2	0.1	3.2	−1.9	2.7	−3.4
World	6,359,077 t	17,480,772 t	3.1 w	2.5 w	1.4 w	1.0 w	2.7 w	1.2 w	.. w	.. w
Low income	413,442	873,223	4.9	5.9	2.8	4.1	5.9	5.8	6.7	11.2
Excl. China & India	197,866	271,927	1.9	2.1	−0.8	−0.4	3.6	..	−1.1	3.7
Middle income	..	2,766,558	..	..	..	..	..	..	..	..
Lower middle income	..	1,410,874	..	..	..	..	..	..	..	..
Upper middle income	572,571	1,356,709	1.6	3.5	−0.3	2.0	4.3	2.9	−1.3	5.8
Low & middle income	1,735,865	3,633,586	3.1	4.2	1.1	2.5	..	3.5	1.8	6.8
East Asia & Pacific	222,019	793,405	7.1	8.9	5.4	7.5	5.8	9.5	8.6	13.3
Europe & Central Asia	..	759,734	..	..	..	..	..	..	..	..
Latin America & Carib.	504,693	1,178,615	1.5	3.9	−0.5	2.1	4.3	2.7	−1.6	5.9
Middle East & N. Africa	178,805	..	..	..	..	..	..	..	..	..
South Asia	165,188	325,891	4.6	4.7	2.3	2.7	8.0	3.7	6.1	8.5
Sub-Saharan Africa	158,841	203,785	1.4	1.6	−1.6	−1.0	2.0	0.3	−3.7	3.1
High income	4,723,048	14,195,083	3.4	2.1	2.7	1.4	2.5	0.8	..	..

a. General government consumption figures are not available separately; they are included in private consumption. b. Data prior to 1992 include Eritrea. c. Data cover mainland Tanzania only.

Growth of consumption and investment | 4.9

Measures of consumption and investment growth are subject to two kinds of inaccuracy. The first stems from the difficulty of measuring expenditures at current price levels, as described in *About the data* for table 4.8. The second arises in deflating current price data to measure growth in real terms, where results depend on the relevance and reliability of the price indexes used. Measuring price changes is more difficult for investment goods than for consumption goods because of the one-time nature of many investments and because the rate of technological progress in capital goods makes capturing quality change difficult. (A classic example is computers—prices have fallen as quality has improved.) Many countries estimate investment from the supply side, identifying capital goods entering an economy directly from detailed production and international trade statistics. This means that the price indexes used in deflating production and international trade, reflecting delivered or offered prices, will determine the deflator for investment expenditures on the demand side.

The data in the table on private consumption in current U.S. dollars are converted from national currencies using official exchange rates or an alternative conversion factor as noted in *Primary data documentation*. (Alternative conversion factors are discussed in *Statistical methods*.) These exchange rates and

conversion factors differ from the purchasing power parity conversion factors used to calculate private consumption per capita in table 4.10, which provide better estimates of comparative domestic purchasing power. Growth rates of private consumption per capita, general government consumption, and gross domestic investment are estimated using constant price data. Consumption and investment as shares of current GDP are shown in table 4.8.

To obtain government consumption in constant prices, countries may adjust current values by applying deflators that use a weighted index of government wages and salaries, or simply take a government employment index as a measure of output. Neither technique captures improvements in productivity or changes in the quality of government services. Deflators for private consumption are usually calculated from consumer price series. Many countries estimate private consumption as a residual that includes statistical discrepancies accumulated from other domestic sources; thus these estimates lack detailed breakdowns of expenditures.

Because the methods used to deflate consumption and investment can vary widely among countries, comparisons between countries in a given year, perhaps even more than those over time, should be treated with caution.

• **Private consumption** is the market value of all goods and services, including durable products (such as cars, washing machines, and home computers) purchased or received as income in kind by households and nonprofit institutions. It excludes purchases of dwellings but includes imputed rent for owner-occupied dwellings. In practice it may include any statistical discrepancy in the use of resources relative to the supply of resources. • **Private consumption per capita** is calculated using World Bank population estimates. • **General government consumption** includes all current expenditures for purchases of goods and services (including wages and salaries) by all levels of government, excluding most government enterprises. It also includes most expenditures on national defense and security. • **Gross domestic investment** consists of outlays on additions to the fixed assets of the economy plus net changes in the level of inventories. Fixed assets cover land improvements (fences, ditches, drains, and so on); plant, machinery, and equipment purchases; and the construction of roads, railways, and the like, including commercial and industrial buildings, offices, schools, hospitals, and private residential dwellings. Inventories are stocks of goods held by firms to meet temporary or unexpected fluctuations in production or sales.

 National accounts data for developing countries are collected from national statistical organizations and central banks by visiting and resident World Bank missions. Data for industrial countries come from Organisation for Economic Co-operation and Development (OECD) data files. For information on the OECD national accounts series see OECD, *National Accounts, 1960–1995*, volumes 1 and 2.

Figure 4.9a

Government consumption has risen steadily in most regions

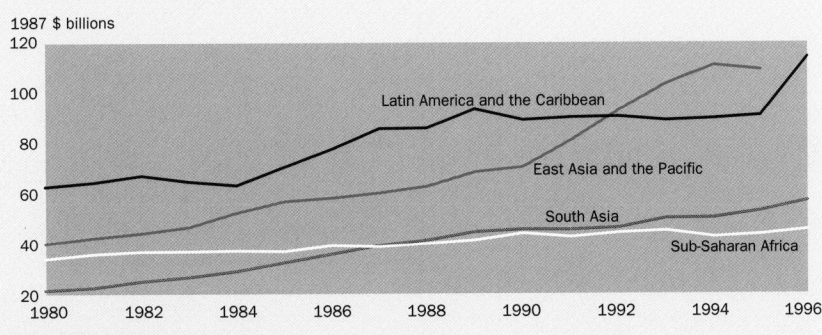

Source: World Bank staff estimates.

After rising dramatically since 1980, government consumption in East Asia and the Pacific began tapering off in 1995. Latin America and the Caribbean, by contrast, registered a substantial increase in 1995. Sub-Saharan Africa's government consumption has not increased much since 1980.

4.10 | Structure of consumption in PPP terms

	Private consumption per capita	Household consumption							
	PPP 1996	All food % 1996	Bread and cereals % 1996	Clothing and footwear % 1996	Fuel and power % 1996	Health care % 1996	Education % 1996	Transport and communications % 1996	Other consumption % 1996
Antigua and Barbuda	4,616	33	9	3	2	12	17	6	27
Australia	14,177	14	2	4	3	12	9	13	46
Austria	14,293	13	2	7	4	13	11	13	41
Bahamas	7,784	31	5	4	2	10	11	6	36
Bangladesh	722	41	21	4	9	7	23	3	14
Belarus	2,851	16	0	6	4	15	21	3	35
Belgium	15,107	15	2	6	3	14	11	11	40
Belize	2,789	28	4	10	2	5	10	11	34
Benin	1,074	45	13	8	3	3	8	14	17
Botswana	3,067	25	9	4	1	7	22	21	19
Bulgaria	3,462	15	0	5	5	9	24	6	36
Cameroon	1,270	38	7	14	2	6	9	8	24
Canada	15,176	9	2	5	4	11	9	11	51
Congo, Rep.	865	36	5	3	1	10	15	18	17
Côte d'Ivoire	1,302	35	8	9	..	8	26	11	11
Croatia	3,279	17	1	3	4	12	13	10	41
Czech Republic	8,163	15	1	4	6	10	15	5	46
Denmark	15,295	10	1	4	3	9	13	9	52
Dominica	2,777	32	5	6	2	9	16	11	23
Egypt, Arab Rep.	1,822	44	11	7	3	8	8	5	26
Fiji	2,814	30	5	4	2	5	12	7	40
Finland	12,402	11	2	3	5	12	11	10	49
France	15,260	12	2	4	3	21	8	12	40
Gabon	2,370	37	7	3	2	8	8	20	23
Germany	15,186	11	2	6	3	15	6	12	47
Greece	9,486	28	2	5	2	7	6	15	37
Grenada	2,955	26	8	4	1	24	14	4	26
Guinea	1,399	32	6	19	2	14	9	9	15
Hong Kong, China	16,435	10	1	18	2	11	4	10	45
Hungary	5,187	14	0	4	5	10	17	6	43
Iceland	15,656	13	2	5	5	13	10	10	44
Indonesia	1,766	45	19	3	3	7	4	15	22
Iran, Islamic Rep.	3,819	23	6	7	12	13	11	11	23
Ireland	10,943	14	3	6	3	11	13	8	45
Italy	14,301	14	2	7	3	14	7	9	46
Jamaica	2,125	26	7	8	1	9	8	24	24
Japan	14,929	11	3	5	2	17	8	9	47
Kenya	849	38	11	8	2	5	22	10	16
Korea, Rep.	6,418	21	4	3	5	13	16	17	24
Luxembourg	22,369	10	2	4	6	9	8	17	46
Malawi	598	45	17	18	2	8	9	8	10
Mali	646	48	13	13	..	2	7	14	15
Mauritius	6,371	24	4	8	3	10	5	19	30
Moldova	866	28	1	5	7	11	25	3	20
Morocco	2,441	45	11	9	2	5	10	10	18
Nepal	804	37	24	8	5	15	20	2	13
Netherlands	14,228	11	2	6	3	16	8	9	47
New Zealand	11,716	12	2	4	3	11	9	13	47
Nigeria	700	48	15	8	6	3	4	5	27
Norway	14,109	13	2	5	10	14	11	8	38
Pakistan	1,078	40	10	6	5	12	7	4	27
Philippines	2,735	33	9	3	1	3	3	4	52
Poland	4,689	20	1	3	5	12	19	6	34
Portugal	9,390	20	4	6	2	7	16	11	39
Romania	3,384	24	1	7	5	6	15	4	39
Russian Federation	2,752	18	1	7	11	13	22	7	22

Structure of consumption in PPP terms | 4.10

	Private consumption per capita	Household consumption							
		All food	Bread and cereals	Clothing and footwear	Fuel and power	Health care	Education	Transport and communications	Other consumption
	PPP 1996	% 1996	% 1996	% 1996	% 1996	% 1996	% 1996	% 1996	% 1996
Senegal	1,499	52	11	14	2	2	11	6	13
Sierra Leone	332	48	13	12	3	13	14	8	2
Singapore	15,043	14	2	7	3	11	7	18	42
Slovak Republic	5,594	17	1	5	7	15	20	4	32
Slovenia	8,864	13	0	4	4	11	12	11	45
Spain	10,395	17	2	7	2	11	8	12	43
Sri Lanka	1,304	38	11	0	4	8	7	28	14
St. Kitts and Nevis	4,171	30	8	4	4	24	9	5	24
St. Lucia	3,339	39	5	5	3	13	7	6	26
St. Vincent and Grenadines	2,903	24	7	4	2	28	14	5	23
Swaziland	2,321	27	6	6	4	10	17	18	19
Sweden	13,411	10	2	5	5	11	9	11	50
Switzerland	16,577	12	2	6	4	13	8	11	45
Thailand	3,265	23	7	8	3	22	10	17	17
Trinidad and Tobago	4,593	20	3	11	4	9	7	15	35
Tunisia	3,322	35	7	6	2	6	7	15	29
Turkey	4,100	23	6	7	4	5	9	7	44
Ukraine	1,470	21	1	5	14	13	26	4	17
United Kingdom	14,929	11	2	6	3	10	8	11	52
United States	20,890	8	1	6	3	12	7	14	49
Vietnam	1,140	40	17	5	4	17	10	..	20
Zambia	591	47	7	8	1	3	12	10	19
Zimbabwe	1,544	28	7	11	2	9	23	14	12

About the data

Cross-country comparisons of consumption expenditures must be made in a common currency. But when expenditures in different countries are converted to a single currency using official exchange rates, the comparisons do not account for the sometimes substantial differences in relative prices. Thus the results tend to undervalue real consumption in economies with relatively low prices and to overvalue consumption in countries with high prices. Differences in the structure of prices also distort the apparent structure of consumption—for example, services (such as health care or education) tend to be relatively cheaper than goods in low- and middle-income economies. Thus when domestic prices are used to calculate consumption patterns, services appear to be undervalued. The problem of making consistent comparisons of real consumption across countries has led to the use of purchasing power parities (PPPs) to convert reported values to a common unit of account.

PPPs measure the relative purchasing power of different currencies over equivalent goods and services. They are international price indexes that allow comparisons of the real value of consumption expenditures between countries in the same way that consumer price indexes allow comparisons of real values over time within countries. To calculate PPPs, data on prices and spending patterns are collected through surveys in each country. Then prices within a region, such as Africa, or a group, such as the Organisation for Economic Co-operation and Development (OECD), are compared. Finally, regions are linked by comparing regional prices, to create a globally consistent set of comparisons. The resulting PPP indexes measure the purchasing power of national currencies in "international dollars" that have the same purchasing power over GDP as the U.S. dollar has in the United States.

Because the goods and services that make up consumption are valued at uniform prices, PPP-based expenditure shares also provide a consistent view of differences in the real structure of consumption between countries. In other words, the shares shown in the table reflect the relative quantities of goods and services consumed rather than their nominal cost. Table 4.11 provides the corresponding data on the structure of prices within countries.

Private consumption refers to private (that is, household) and nonprofit (nongovernmental) consumption as defined in the United Nations (UN) System of National Accounts (SNA). Estimates of private consumption of education and health services include government as well as private outlays. The International Comparison Programme's (ICP) concept of enhanced consumption, or total consumption of the population, focuses on who consumes goods and services rather than on who pays for them. That is, it emphasizes consumption rather than expenditure. This approach, adopted in the 1993 SNA, improves international comparability because aggregate measures based on consumption are less sensitive to differences in national practices in financing health and education services.

Because national statistical offices tend to concentrate on the production side of national accounts, data on the detailed structure of consumption in low- and middle-income economies are generally weak. Consumption estimates are typically obtained through household budget surveys or other, similar surveys. These surveys are carried out irregularly and may be targeted at specific income groups or geographic areas. In some countries surveys are limited to urban areas or even to capital cities and so do not reflect national spending patterns. Urban surveys tend to show lower-than-average shares for food and higher-than-average shares for gross rent, fuel and power, transport and communications, and other consumption. Controlled food prices and incomplete accounting of subsistence activities may also contribute to low measured shares of food consumption.

The ICP collects price data from different outlets on several hundred consumption items that are carefully reviewed to ensure comparability. ICP surveys are conducted about every five years, but because not all countries have participated in all surveys, regression methods are used to extrapolate results from earlier surveys and to provide a complete set of estimates in a given year. See Ahmad (1994) for an extensive discussion of the ICP and its methods.

Although PPPs are more useful than official exchange rates in comparing consumption patterns, caution should be used in interpreting PPP results. PPP estimates are based on price comparisons of comparable items, but not all items can be matched perfectly in quality across countries and over time. Services are particularly difficult to compare, in part because of differences in productivity. Many services, such as government services, are not sold on the open market in all countries, so they are compared using input prices (mostly wages). Because this approach ignores productivity differences, it may inflate estimates of real quantities in lower-income countries.

Definitions

• **Private consumption** includes the consumption expenditures of individuals, households, and nongovernmental organizations. In the ICP goods and services accruing to households are included in private consumption whether they are financed by individuals, governments, or nonprofit institutions. Thus private consumption as defined by the ICP includes government expenditures on education, health, social security, and welfare services. • **Household consumption** shows the percentage shares of selected components of consumption computed from details of GDP converted using PPPs. • **All food** includes all food purchased for household consumption. • **Bread and cereals** comprise the main staple products—rice, flour, bread, all other cereals, and cereal preparations. • **Clothing and footwear** include purchases of new and used clothing and footwear and repair services. • **Fuel and power** exclude energy used for transport (rarely reported to be more than 1 percent of total consumption in low- and middle-income economies). • **Health care** and **education** include government as well as private expenditures. • **Transport and communications** cover all personal costs of transport, telephones, and the like. • **Other consumption** covers gross rent (including repair and maintenance charges); beverages and tobacco; nondurable household goods, household services, recreational services, services (including meals) supplied by hotels and restaurants, and purchases of carryout food; and consumer durables, such as household appliances, furniture, floor coverings, recreational equipment, and watches and jewelry.

Data sources

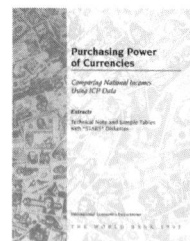

PPP data come from the ICP, which is coordinated by the regional economic commissions of the United Nations and other international organizations. The World Bank collects detailed ICP benchmark data from regional sources, establishes global consistency across the regional data sets, and computes regression-based estimates for nonbenchmark countries. For detailed information on the regional sources and compilation of benchmark data see the World Bank's *Purchasing Power of Currencies: Comparing National Incomes Using ICP Data* (1993b).

Relative prices in PPP terms | 4.11

| | International price level | Relative price level (price level of GDP = 100) | | | | | | | | | |
| | ratio of PPP rate to $ exchange rate 1993 | Private consumption | | | | | | | Transport and communications 1993 | Government consumption 1993 | Gross fixed capital formation 1993 |
		Private consumption 1993	All food 1993	Bread and cereals 1993	Clothing and footwear 1993	Fuel and power 1993	Health care 1993	Education 1993			
Antigua and Barbuda	86	106	105	107	178	187	116	43	154	63	112
Australia	92	100	77	97	113	75	88	83	102	93	105
Austria	119	100	102	97	108	101	92	82	115	90	102
Bahamas	110	100	90	93	103	112	111	94	166	112	93
Bangladesh	24	101	154	153	122	61	37	17	66	52	201
Belarus	..	101	149	151	135	13	67	36	114	78	119
Belgium	108	101	95	89	125	114	72	80	110	90	102
Belize	71	95	90	80	75	167	80	71	118	79	128
Botswana	37	125	103	113	157	256	105	98	65	113	108
Bulgaria	33	85	128	136	103	62	58	21	149	70	133
Cameroon	50	93	87	98	102	173	88	60	97	123	141
Canada	98	103	96	93	101	77	110	104	109	115	85
Congo, Rep.	64	98	115	105	109	288	58	55	67	78	193
Côte d'Ivoire	52	98	101	112	97	..	89	60	78	116	151
Croatia	..	83	116	106	118	65	85	38	119	68	183
Czech Republic	39	85	106	63	123	74	61	41	157	73	174
Denmark	136	105	108	107	104	130	111	85	123	90	87
Dominica	81	101	112	125	98	179	58	60	147	62	147
Egypt, Arab Rep.	35	90	88	97	124	80	54	49	92	92	106
Fiji	54	90	100	112	124	146	89	66	140	59	170
Finland	107	108	121	138	112	79	99	79	126	82	85
France	116	101	102	95	124	124	71	94	116	101	97
Gabon	80	126	147	95	131	284	90	119	79	152	91
Germany	127	99	90	91	107	106	84	96	110	104	103
Greece	80	105	104	126	161	149	80	68	98	74	98
Grenada	76	96	120	110	109	191	47	53	189	65	142
Guinea	33	97	106	118	75	173	37	68	94	80	127
Hong Kong, China	95	91	76	78	73	57	59	145	80	136	106
Hungary	69	66	74	78	87	50	50	29	122	84	114
Iceland	123	103	120	109	136	42	89	68	115	87	94
Indonesia	30	93	93	82	99	98	38	31	89	34	163
Iran, Islamic Rep.	24	79	105	115	113	17	35	47	38	68	182
Ireland	97	100	101	90	106	110	100	61	146	85	104
Italy	97	99	105	100	125	122	82	102	108	105	101
Jamaica	55	101	119	100	118	181	57	64	85	69	111
Japan	161	99	130	135	113	111	61	83	90	86	104
Kenya	21	90	91	123	85	140	45	41	71	72	196
Korea, Rep.	70	112	137	173	157	66	61	72	53	95	87
Luxembourg	115	95	93	84	128	86	80	109	97	119	109
Malawi	34	87	98	109	54	148	31	50	73	79	279
Mali	38	90	92	131	86	..	59	31	90	63	179
Mauritius	39	93	81	65	72	111	59	87	67	108	134
Moldova	32	87	131	127	102	60	62	41	122	88	137
Morocco	37	93	83	77	68	282	81	81	77	106	182
Nepal	22	93	129	129	116	94	25	30	61	61	..
Netherlands	115	97	92	79	101	100	71	86	119	92	112
New Zealand	82	98	91	93	102	65	96	74	100	78	119
Nigeria	36	101	150	171	58	56	46	40	47	61	115
Norway	126	106	117	114	108	50	97	87	139	98	88
Pakistan	28	101	115	103	133	113	35	60	108	42	167
Philippines	35	82	105	124	122	277	30	29	114	58	183
Poland	..	83	90	84	120	82	46	31	140	73	123
Portugal	73	107	125	102	162	175	117	54	140	56	116
Romania	33	82	132	74	106	59	59	22	116	78	132
Russian Federation	25	84	122	61	154	2	36	30	75	79	145

| | International price level | Relative price level (price level of GDP = 100) | | | | | | | | | |
| | | Private consumption | | | | | | | | | |
	ratio of PPP rate to $ exchange rate **1993**	Private consumption **1993**	All food **1993**	Bread and cereals **1993**	Clothing and footwear **1993**	Fuel and power **1993**	Health care **1993**	Education **1993**	Transport and communications **1993**	Government consumption **1993**	Gross fixed capital formation **1993**
Senegal	48	90	87	119	74	227	75	45	79	81	237
Sierra Leone	29	115	141	192	107	185	34	39	76	37	147
Singapore	85	95	67	72	88	46	46	86	100	88	107
Slovak Republic	40	80	87	55	111	68	53	33	123	74	124
Slovenia	76	93	108	104	146	71	70	55	110	76	109
Spain	92	102	100	112	120	111	83	76	125	85	102
Sri Lanka	34	86	123	109	113	58	26	33	53	31	170
St. Kitts and Nevis	80	95	105	118	108	121	47	81	145	69	129
St. Lucia	83	101	111	140	107	157	49	61	114	77	111
St. Vincent and Grenadines	69	97	132	125	113	236	43	55	146	56	173
Swaziland	35	92	84	100	138	187	53	66	54	91	..
Sweden	126	103	103	103	86	86	103	86	110	93	93
Switzerland	144	106	108	96	102	70	89	102	106	115	85
Thailand	43	122	106	79	158	74	40	71	103	68	93
Trinidad and Tobago	59	98	89	98	87	38	106	98	108	87	129
Tunisia	39	92	81	62	142	105	80	109	79	124	159
Turkey	55	106	123	100	140	141	75	53	116	62	106
Ukraine	..	86	145	65	187	21	56	29	64	60	134
United Kingdom	96	102	86	73	83	115	89	89	126	90	98
United States	100	100	88	82	84	90	145	125	89	113	94
Vietnam	..	98	125	116	90	114	30	59	..	39	266
Zambia	43	113	126	185	98	282	29	46	89	34	213
Zimbabwe	26	105	81	97	91	221	51	77	62	89	..

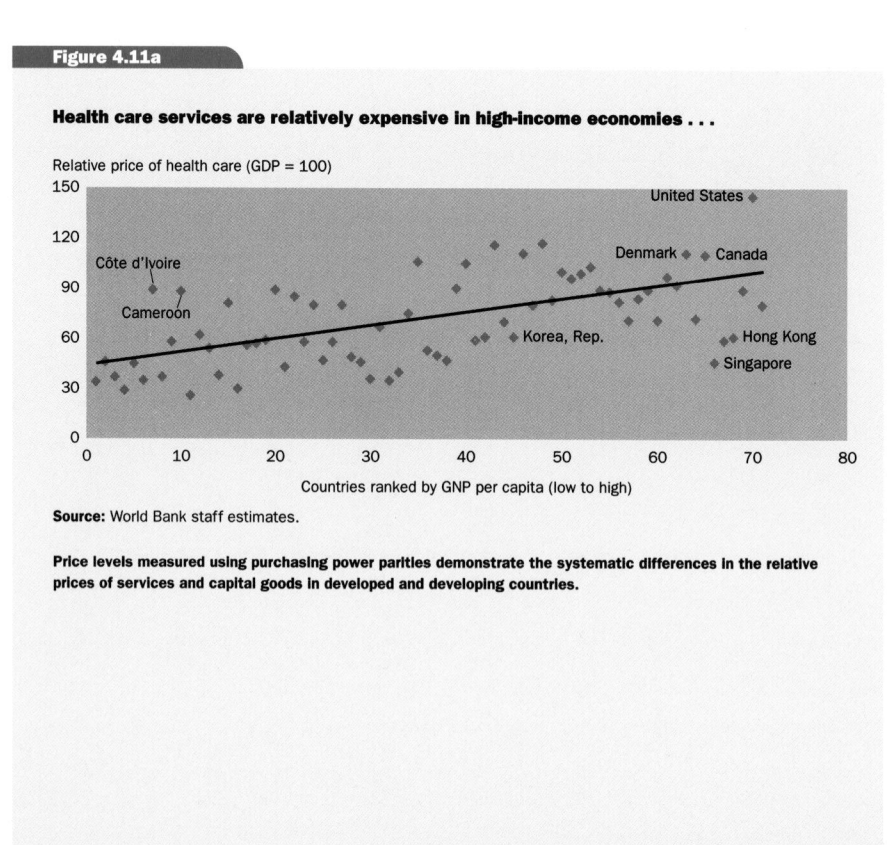

Figure 4.11a

Health care services are relatively expensive in high-income economies . . .

Source: World Bank staff estimates.

Price levels measured using purchasing power parities demonstrate the systematic differences in the relative prices of services and capital goods in developed and developing countries.

Relative prices in PPP terms 4.11

About the data

The International Comparison Programme (ICP) collects data on prices paid for a large set of comparable items in more than 100 countries. Purchasing power parities (PPPs) computed from these data allow comparisons of prices and real GNP expenditures across countries. PPPs are used in table 1.1 to measure GNP at internationally comparable prices and in table 4.10 to evaluate the structure of consumption. This table presents information on the relative prices of components of GDP based on the most recent ICP data.

A country's international price level is the ratio of its PPP rate to its official exchange rate for U.S. dollars. PPPs can be thought of as the exchange rate of dollars for goods in the local economy, while the U.S. dollar exchange rate measures the relative cost of domestic currency in dollars. Thus the international price level is an index measuring the cost of goods in one country relative to a numeraire country, in this case the United States. An international price level above 100 means that the general price level in the country is higher than that in the United States. For example, Japan's international price level of 161 implies that the price of goods and services in Japan is 61 percent higher than the price of comparable goods and services in the United States. By contrast, Kenya's price level of 21 means that a bundle of goods and services purchased for $100 in the United States costs only $21 in Kenya.

The relative prices of components of GDP shown in the table are calculated from their international prices measured relative to each country's price level of GDP. A figure above 100 indicates that the price of that component is higher than the average price level of GDP. This is not the same as saying that the component is more expensive in that country than in the United States. It indicates only that the price for that component is higher than the general price level prevailing in the country.

Relative prices for consumption items tend to be close to the overall price level of GDP. This is to be expected because consumption accounts for a large share of GDP. The data also indicate that the relative price of investment goods in developing countries tends to be higher than for other components of GDP. For example, Indonesia's relative price level of 163 indicates that the price level of investment goods is 63 percent higher than the overall price level. This reflects the fact that a large share of physical capital must be imported from high-income economies with higher price levels.

Figure 4.11b

. . . but capital investment is far more expensive in low- and middle-income economies

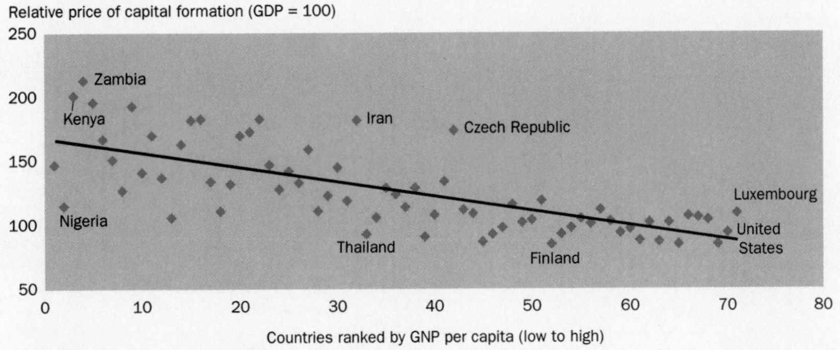
Relative price of capital formation (GDP = 100)
250
200
150
100
50
Zambia
Kenya
Nigeria
Iran
Czech Republic
Thailand
Finland
Luxembourg
United States
0
10
20
30
40
50
60
70
80
Countries ranked by GNP per capita (low to high)

Source: World Bank staff estimates.

Price levels measured using purchasing power parities demonstrate the systematic differences in the relative prices of services and capital goods in developed and developing countries.

Definitions

• **International price level** is the ratio of a country's PPP rate to its official exchange rate for U.S. dollars. • **Private consumption** includes the consumption expenditures of individuals, households, and nongovernmental organizations. • **All food** includes all food purchased for household consumption. • **Bread and cereals** comprise the main staple products—rice, flour, bread, all other cereals, and cereal preparations. • **Clothing and footwear** include purchases of new and used clothing and footwear and repair services. • **Fuel and power** exclude energy used for transport (rarely reported to be more than 1 percent of total consumption in low- and middle-income economies). • **Health care** and **education** include government as well as private expenditures. • **Transport and communications** cover all personal costs of transport, telephones, and the like. • **Government consumption** includes spending on goods and services for collective consumption less spending on recreational and other related cultural services, education, health, and housing. Expenditure on government final consumption consists of compensation of employees, consumption of intermediate goods and services, and consumption of fixed capital and indirect taxes paid less proceeds from sales of goods and services to other sectors (such as fees charged by municipalities and other government agencies, school fees, fees for medical and hospital treatment and drug sales, and sales of maps and charts). • **Gross fixed capital formation** comprises expenditures on construction, producer durables, and changes in stocks. Construction includes residential and nonresidential buildings and roads, bridges, and other civil engineering activities. Producer durables include machinery and nonelectrical equipment, electrical machinery and appliances, and transport equipment. Changes in stocks cover increases in the value of materials and supplies, works in progress, and livestock (including breeding stock and dairy cattle).

Data sources

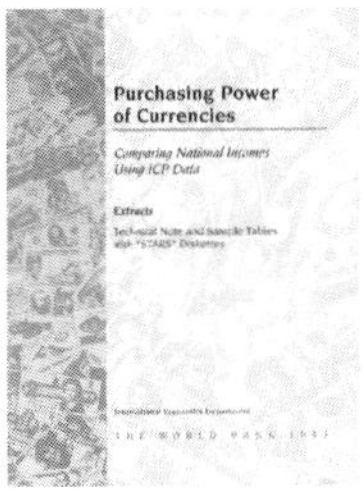
Purchasing Power of Currencies

PPP data come from the ICP, which is coordinated by the regional economic commissions of the United Nations. The World Bank collects detailed ICP benchmark data from regional sources, establishes global consistency across the regional data sets, and computes regression-based estimates for nonbenchmark countries.

4.12 Central government finances

	Current revenue[a]		Total expenditure		Overall budget deficit (including grants)		Financing from abroad		Domestic financing		Debt and interest payments	
	% of GDP		% of GDP		% of GDP		% of GDP		% of GDP		Total debt % of GDP	Interest % of current revenue
	1980	1995	1980	1995	1980	1995	1980	1995	1980	1995	1995	1995
Albania	..	21.2	..	31.0	..	−9.0	..	1.9	..	7.1	35.3	11.1
Algeria	..	..	..	..	..	..	..	..	..	..	..	..
Angola	..	..	..	..	..	..	..	..	..	..	..	..
Argentina	15.6	12.9	18.2	14.5	−2.6	−1.1	0.0	1.3	2.6	−0.3	..	11.3
Armenia	..	..	..	..	..	..	..	..	..	..	..	..
Australia	21.7	24.5	22.7	27.4	−1.5	−2.5	0.2	0.7	1.3	1.8	22.7	7.0
Austria	33.9	36.2	36.6	42.2	−3.3	−5.2	0.8	1.8	2.5	3.4	58.4	10.9
Azerbaijan	..	..	..	..	..	..	..	..	..	..	..	..
Bangladesh	11.3	..	10.0	..	2.5	..	2.5	..	0.7	..	..	..
Belarus	..	..	..	..	..	..	..	..	..	..	..	..
Belgium	43.4	44.6	50.6	49.4	−8.1	−3.9	2.4	−3.3	5.7	7.2	127.9	19.4
Benin	..	..	..	..	..	..	..	..	..	..	..	..
Bolivia	..	17.5	..	23.1	..	−2.5	..	5.4	..	−2.8	62.9	14.1
Bosnia and Herzegovina	..	..	..	..	..	..	..	..	..	..	..	..
Botswana	31.8	42.8	31.8	38.0	−0.2	2.8	1.3	−0.4	−1.2	−2.4	11.5	1.7
Brazil	22.6	25.6	20.2	37.4	−2.4	−9.4	0.0	..	2.4	..	..	75.6
Bulgaria	..	36.0	..	41.6	..	−5.3	..	−0.8	..	6.1	..	41.2
Burkina Faso	11.8	..	12.2	..	0.2	..	0.4	..	0.0	..	..	..
Burundi	13.9	15.7	21.5	24.9	−3.9	−3.7	2.0	1.7	1.9	2.0	105.0	7.7
Cambodia	..	..	..	..	..	..	..	..	..	..	..	..
Cameroon	16.4	13.0	15.7	12.7	0.5	0.2	0.7	0.3	−1.2	−0.3	139.6	23.0
Canada	18.7	20.8	21.3	24.6	−3.5	−3.7	0.6	..	2.9	..	..	21.3
Central African Republic	16.5	..	22.0	..	−3.5	..	2.1	..	1.5	..	..	..
Chad	..	..	..	..	..	..	..	..	..	..	..	..
Chile	32.0	21.5	28.0	19.2	5.4	2.5	−0.8	..	−4.7	..	19.0	3.4
China	..	5.7	..	8.3	..	−1.8	..	0.0	..	1.7	..	..
Hong Kong, China	..	..	..	..	..	..	..	..	..	..	..	..
Colombia	12.0	16.3	13.4	14.4	−1.8	−0.5	..	..	..	..	..	10.5
Congo, Dem. Rep.	9.4	4.9	12.4	7.6	−0.8	0.0	0.3	0.0	0.5	0.0	214.1	1.0
Congo, Rep.	35.3	..	49.4	..	−5.2	..	3.8	..	1.4	..	..	..
Costa Rica	17.8	26.3	25.0	29.1	−7.4	−2.9	1.1	−1.1	6.3	4.0	..	21.4
Côte d'Ivoire	22.9	..	31.7	..	−10.8	..	6.5	..	4.4	..	..	..
Croatia	..	44.9	..	46.5	..	−0.9	..	0.8	..	0.1	32.5	3.3
Cuba	..	..	..	..	..	..	..	..	..	..	..	..
Czech Republic	..	38.8	..	39.9	..	0.4	..	−0.5	..	0.1	15.5	3.2
Denmark	35.5	40.6	39.4	43.4	−2.7	−2.0	..	..	..	..	..	14.2
Dominican Republic	14.2	16.2	16.9	15.6	−2.6	0.8	1.4	−1.0	1.2	0.3	..	5.8
Ecuador	12.8	15.7	14.2	15.7	−1.4	0.0	0.5	..	0.9	..	..	21.8
Egypt, Arab Rep.	45.5	36.9	45.6	37.4	−6.3	0.3	2.1	−1.2	4.2	0.8	..	25.8
El Salvador	11.4	12.6	17.1	13.7	−5.7	−0.1	0.3	−0.5	5.5	0.6	27.7	11.3
Eritrea	..	..	..	..	..	..	..	..	..	..	..	..
Estonia	..	34.2	..	35.2	..	0.0	..	0.3	..	−0.4	..	0.6
Ethiopia[b]	16.3	11.8	19.5	18.1	−3.1	−5.9	1.2	2.7	1.9	3.2	..	16.8
Finland	27.2	32.7	28.1	42.7	−2.2	−9.8	0.8	0.0	1.4	9.9	66.1	14.5
France	39.6	40.5	39.5	46.4	−0.1	−6.5	0.0	0.7	0.1	5.8	..	7.9
Gabon	35.5	..	36.5	..	6.1	..	0.0	..	−6.1	..	..	..
Gambia, The	23.4	24.5	32.1	21.5	−4.5	3.7	1.2	3.0	3.3	−6.8	..	..
Georgia	..	..	..	..	..	..	..	..	..	..	..	..
Germany	..	32.1	..	33.9	..	−1.8	..	1.6	..	0.2	37.3	7.8
Ghana	6.9	17.9	10.9	22.1	−4.2	−2.6	0.7	1.4	3.5	1.2	..	20.5
Greece	25.3	21.9	29.3	33.6	−4.1	−9.6	1.6	1.1	2.6	8.5	116.9	57.8
Guatemala	9.4	8.5	12.1	8.9	−3.4	−0.7	1.4	..	3.0	..	..	11.1
Guinea	..	..	..	..	..	..	..	..	..	..	..	..
Guinea-Bissau	..	..	..	..	..	..	..	..	..	..	..	..
Haiti	10.6	..	17.4	..	−4.7	..	..	..	..	..	..	..
Honduras	14.6	..	..	..	..	..	..	..	..	..	..	..

Central government finances | 4.12

	Current revenue[a] (% of GDP)		Total expenditure (% of GDP)		Overall budget deficit (including grants) (% of GDP)		Financing from abroad (% of GDP)		Domestic financing (% of GDP)		Total debt % of GDP	Interest % of current revenue
	1980	1995	1980	1995	1980	1995	1980	1995	1980	1995	1995	1995
Hungary	53.4	..	56.2	..	–2.8	..	2.1	..	0.7	..	..	..
India	11.7	13.3	13.3	16.4	–6.5	–6.0	0.5	0.2	6.0	5.8	52.2	33.6
Indonesia	21.3	17.8	22.1	14.7	–2.3	2.2	2.1	–0.4	0.2	–1.9	30.9	8.9
Iran, Islamic Rep.	21.6	24.5	35.7	23.2	–13.8	1.4	–0.6	0.0	14.4	–1.4	..	0.0
Iraq	..	..	..	..	..	..	..	..	..	..	..	..
Ireland	34.7	36.7	45.1	40.3	–12.5	–2.0	..	..	..	..	..	15.3
Israel	50.4	38.5	70.2	44.7	–15.6	–4.7	7.9	0.2	7.8	4.6	113.9	15.9
Italy	31.4	41.2	41.3	48.6	–10.8	–7.6	0.2	..	10.6	..	..	27.7
Jamaica	29.0	..	41.5	..	–15.5	..	..	..	..	..	..	..
Japan	11.6	20.9	18.4	23.7	–7.0	–1.5	..	..	..	1.5	44.7	..
Jordan	17.9	28.6	41.3	31.6	–9.3	1.1	5.7	1.5	3.6	–2.6	90.2	10.0
Kazakhstan	..	..	..	..	..	..	..	..	..	..	..	..
Kenya	21.9	23.6	25.3	29.8	–4.5	–3.4	2.4	..	2.1	..	..	31.7
Korea, Dem. Rep.	..	..	..	..	..	..	..	..	..	..	..	..
Korea, Rep.	17.4	20.1	17.0	17.7	–2.2	0.3	0.8	–0.1	1.4	–0.2	9.0	3.0
Kuwait	89.3	..	27.7	51.4	58.7	..	..	..	..	..	..	..
Kyrgyz Republic	..	..	..	..	..	..	..	..	..	..	..	..
Lao PDR	..	..	..	..	..	..	..	..	..	..	..	..
Latvia	..	28.3	..	32.2	..	–4.2	..	1.7	..	2.5	16.0	3.6
Lebanon	..	16.8	..	32.5	..	–15.7	..	2.1	..	14.0	77.9	59.8
Lesotho	34.2	55.0	..	50.7	..	6.4	..	6.6	4.8	–12.9	..	..
Libya	..	..	..	..	..	..	..	..	..	..	..	..
Lithuania	..	23.7	..	25.5	..	–5.3	..	4.1	..	1.2	..	1.8
Macedonia, FYR	..	..	..	..	..	..	..	..	..	..	..	..
Madagascar	13.2	8.4	..	17.2	..	–1.6	..	1.8	..	–0.3	118.4	59.9
Malawi	19.1	..	34.6	..	–15.9	..	8.3	..	7.7	..	..	..
Malaysia	26.3	25.4	28.5	22.9	–6.0	2.3	0.6	–0.8	5.4	–1.4	42.8	12.1
Mali	10.5	..	20.6	..	–4.5	..	4.1	..	0.4	..	..	..
Mauritania	..	..	..	..	..	..	..	..	..	..	..	..
Mauritius	20.8	21.0	27.2	22.6	–10.3	–1.2	2.5	–1.2	7.8	2.4	33.7	11.5
Mexico	15.1	15.3	15.7	15.9	–3.0	–0.5	–0.4	5.4	3.4	–4.9	40.9	18.5
Moldova	..	..	..	..	..	..	..	..	..	..	..	..
Mongolia	..	24.3	..	21.5	..	–3.5	..	8.7	..	0.4	..	1.7
Morocco	23.3	..	33.1	..	–9.7	..	5.3	..	4.4	..	..	..
Mozambique	..	..	..	..	..	..	..	..	..	..	..	..
Myanmar	16.0	6.3	15.8	10.6	1.2	–4.1	1.2	0.0	–2.4	4.1	..	..
Namibia	..	33.4	..	38.5	..	–4.5	..	0.1	..	4.4	..	2.4
Nepal	7.8	10.7	14.3	17.5	–3.0	–4.6	1.9	4.1	1.2	0.5	66.2	..
Netherlands	49.4	46.0	52.9	50.8	–4.6	–4.9	0.0	3.2	4.6	1.7	63.0	10.4
New Zealand	34.2	35.8	38.3	32.9	–6.7	0.4	3.6	..	3.1	..	59.1	11.8
Nicaragua	23.3	25.4	30.4	33.2	–6.8	–0.6	3.6	0.2	3.2	0.5	..	15.8
Niger	14.4	..	18.4	..	–4.7	..	4.0	..	0.7	..	..	..
Nigeria	..	..	..	..	..	..	..	..	..	..	..	..
Norway	37.2	41.3	34.4	39.0	–1.7	1.6	–0.7	0.4	2.4	–2.0	28.1	5.5
Oman	38.2	31.7	38.5	42.4	0.4	–10.1	–3.6	9.1	3.1	1.0	31.3	7.8
Pakistan	16.2	19.4	17.5	23.2	–5.7	–4.8	2.3	2.4	3.4	2.4	..	28.8
Panama	25.3	26.1	30.5	24.7	–5.2	2.9	5.4	0.3	–0.2	–3.2	..	6.9
Papua New Guinea	23.0	22.0	34.4	29.4	–1.9	–4.1	2.5	–0.2	–0.5	4.3	43.0	12.2
Paraguay	10.7	14.1	9.9	13.0	0.3	1.2	2.2	–0.8	–2.5	–0.4	12.8	5.6
Peru	17.1	15.2	19.5	17.2	–2.4	–1.3	0.6	2.2	1.8	–0.8	45.9	19.9
Philippines	14.0	17.7	13.4	17.9	–1.4	0.6	0.9	–0.7	0.5	0.1	61.1	21.5
Poland	..	40.7	..	43.0	..	–2.0	..	0.3	..	1.7	57.9	11.8
Portugal	26.0	35.5	33.1	44.1	–8.4	–5.5	1.9	3.8	6.5	1.6	..	14.7
Puerto Rico	..	..	..	..	..	..	..	..	..	..	..	..
Romania	45.3	29.9	44.8	32.0	0.5	–2.5	..	0.0	..	2.5	..	4.4
Russian Federation	..	18.5	..	24.0	..	–4.4	..	1.5	..	2.9	..	17.0

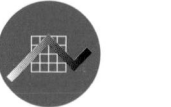

	Current revenue[a]		Total expenditure		Overall budget deficit (including grants)		Financing from abroad		Domestic financing		Debt and interest payments	
	% of GDP		% of GDP		% of GDP		% of GDP		% of GDP		Total debt % of GDP	Interest % of current revenue
	1980	1995	1980	1995	1980	1995	1980	1995	1980	1995	1995	1995
Rwanda	12.8	..	14.3	*25.8*	−1.7	−7.4	2.6	..	−0.9	..	..	..
Saudi Arabia	..	..	..	..	..	..	..	..	..	..	..	..
Senegal	24.1	..	23.1	..	0.9	..	−2.7	..	1.8	..	..	..
Sierra Leone	15.1	9.4	26.5	16.4	−11.8	−6.1	3.5	5.7	8.3	0.4	114.1	20.4
Singapore	25.4	25.9	20.0	15.9	2.1	14.3	−0.2	0.0	−2.0	−14.3	74.8	3.9
Slovak Republic	..	..	..	..	..	..	..	..	..	..	..	..
Slovenia	..	..	..	..	..	..	..	..	..	..	..	..
South Africa	23.5	27.6	22.1	33.7	−2.3	−5.9	−0.2	0.4	2.5	5.6	57.4	22.3
Spain	24.2	*31.2*	26.7	*38.2*	−4.2	−7.2	0.0	−4.3	4.2	*11.5*	*52.8*	*14.8*
Sri Lanka	20.2	20.4	41.4	29.3	−18.3	−8.3	4.5	3.2	13.8	5.1	94.6	28.1
Sudan	13.8	..	19.6	..	−3.3	..	2.8	..	0.5	..	..	..
Sweden	35.0	38.4	39.3	49.5	−8.1	−11.1	3.2	−1.3	4.9	12.4	71.1	16.6
Switzerland	19.5	23.2	20.1	26.6	−0.2	−1.0	..	0.0	..	1.0	22.5	4.0
Syrian Arab Republic	26.8	22.6	48.2	24.5	−9.7	−1.7	−0.2	..	9.8	..	..	..
Tajikistan	..	..	..	..	..	..	..	..	..	..	..	..
Tanzania	..	..	..	..	..	..	..	..	..	..	..	..
Thailand	14.3	18.6	18.8	15.8	−4.9	2.9	1.1	0.2	3.7	−3.1	4.6	1.8
Togo	30.3	..	30.8	..	−2.0	..	1.6	..	0.4	..	..	..
Trinidad and Tobago	43.2	28.2	30.9	29.2	7.4	0.2	..	2.7	..	−2.9	53.5	18.3
Tunisia	31.3	30.1	31.6	32.8	−2.8	−3.2	2.3	2.9	0.5	0.3	57.7	12.4
Turkey	18.1	17.9	21.3	22.2	−3.1	−4.1	0.4	−1.0	2.6	5.1	41.4	15.2
Turkmenistan	..	..	..	..	..	..	..	..	..	..	..	..
Uganda	3.2	..	6.2	..	−3.1	..	0.0	..	3.1	..	..	..
Ukraine	..	..	..	..	..	..	..	..	..	..	..	..
United Arab Emirates	0.2	*2.5*	12.1	*11.8*	2.1	*0.2*	0.0	*0.0*	−2.1	*−0.2*	..	..
United Kingdom	35.2	36.4	38.3	42.0	−4.6	−5.3	0.3	..	4.3	..	..	10.2
United States	20.2	20.5	22.0	22.7	−2.8	−2.2	0.0	2.7	2.8	−0.5	51.3	16.1
Uruguay	22.3	30.1	21.8	31.5	0.0	−1.3	0.9	*1.1*	−0.9	*1.7*	*26.3*	5.9
Uzbekistan	..	..	..	..	..	..	..	..	..	..	..	..
Venezuela	22.3	16.6	18.7	18.8	0.0	−3.7	1.8	0.1	−1.9	3.5	..	29.0
Vietnam	..	..	..	..	..	..	..	..	..	..	..	..
West Bank and Gaza	..	..	..	..	..	..	..	..	..	..	..	..
Yemen, Rep.	..	19.9	..	24.7	..	−5.5	..	−0.2	..	5.8	..	17.1
Yugoslavia, FR (Serb./Mont.)	..	..	..	..	..	..	..	..	..	..	..	..
Zambia	25.0	21.2	37.1	25.0	−18.5	−7.2	8.8	4.2	9.7	3.0	*161.8*	*8.7*
Zimbabwe	24.1	27.4	34.8	*34.1*	−10.9	*−10.7*	2.3	*7.2*	8.6	*3.5*	*69.5*	*23.4*
World	**22.6 w**	**25.9 w**	**25.7 w**	**29.1 w**	**−4.1 w**	**−3.3 w**	**1.3 m**	**.. m**	**2.6 m**	**.. m**	**.. m**	***13.6 m***
Low income	..	9.9	..	12.9	..	..	..	..	..	..	..	..
Excl. China & India	..	..	..	..	..	..	..	..	..	..	..	..
Middle income	..	..	..	..	..	..	..	*−0.1*	..	0.4	..	11.3
Lower middle income	..	20.0	..	22.6	..	−4.9	..	..	..	..	..	11.1
Upper middle income	21.1	*23.3*	20.3	29.0	−2.2	−6.6	*0.9*	0.6	2.4	0.1	37.3	11.4
Low & middle income	..	*18.9*	..	21.9	..	..	..	..	..	..	..	..
East Asia & Pacific	..	10.7	..	11.5	..	..	1.2	−0.2	0.3	−0.7	..	*11.5*
Europe & Central Asia	..	25.5	..	30.9	..	−7.7	..	..	..	..	..	..
Latin America & Carib.	19.3	*19.8*	18.8	24.5	−2.0	−5.8	1.0	−0.4	1.5	*0.0*	..	12.7
Middle East & N. Africa	..	..	..	..	..	..	2.6	0.4	3.1	0.8	..	*11.2*
South Asia	12.4	14.2	14.2	17.6	..	−6.2	2.1	2.8	4.7	3.7	66.2	28.8
Sub-Saharan Africa	20.0	..	22.2	..	..	..	2.4	..	1.9	..	..	..
High income	23.1	28.0	26.3	31.3	−4.1	−3.4	0.4	0.5	2.9	1.8	60.7	11.4

a. Excluding grants. b. Data prior to 1992 include Eritrea.

About the data

Tables 4.12–4.14 present an overview of the size and role of central governments relative to national economies. The International Monetary Fund's (IMF) *Manual on Government Finance Statistics* describes the government as the sector of the economy responsible for "implementation of public policy through the provision of primarily nonmarket services and the transfer of income, supported mainly by compulsory levies on other sectors" (1986, p. 3).

Data on government revenues and expenditures are collected by the IMF through questionnaires distributed to member governments and by the Organisation for Economic Co-operation and Development (OECD). Despite the IMF's efforts to systematize and standardize the collection of public finance data, statistics on public finance are often incomplete, untimely, and noncomparable.

In general, the definition of government excludes nonfinancial public enterprises and public financial institutions (such as the central bank). Units of government meeting this definition exist at many levels, from local administrative units to the highest level of national government. Inadequate statistical coverage precludes the presentation of subnational data, however, making cross-country comparisons potentially misleading.

Central government can refer to one of two accounting concepts: consolidated or budgetary. For most countries central government finance data have been consolidated into one account, but for others only budgetary central government accounts are available. Countries reporting budgetary data are noted in *Primary data documentation*. Because budgetary accounts do not necessarily include all central government units, the picture they provide of central government activities is usually incomplete. A key issue is the failure to include the quasi-fiscal operations of the central bank. Central bank losses arising from monetary operations and subsidized financing can result in sizable quasi-fiscal deficits. Such deficits may also result from the operations of other financial intermediaries, such as public development finance institutions. Also missing from the data are governments' contingent liabilities for unfunded pension and insurance plans.

Government finance statistics are reported in local currency. The indicators here are shown as percentages of GDP. Many countries report government finance data according to fiscal years; see *Primary data documentation* for the timing of these years. For further discussion of government finance statistics see the notes to tables 4.13 and 4.14.

Figure 4.12a

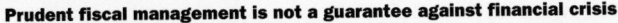

Prudent fiscal management is not a guarantee against financial crisis

Average overall budget deficit (including grants), 1991–96
% of GDP

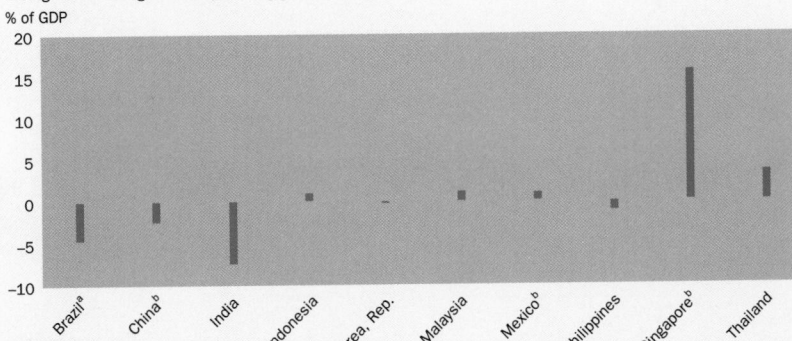

a. 1991–93. b. 1991–95.
Source: IMF, *Government Finance Statistics.*

The five Asian countries that experienced financial crisis in 1997—Indonesia, the Republic of Korea, Malaysia, the Philippines, and Thailand—ran small fiscal surpluses or modest deficits in the preceding period. The problems in Asia did not arise from fiscal management but from excessive short-term borrowing by the private sector that helped to fuel an investment bubble. Brazil, China, India, Mexico, and Singapore are shown for reference.

Data sources

Data on central government finances are from the IMF's *Government Finance Statistics Yearbook* (1996) and IMF data files. Each country's accounts are reported using the system of common definitions and classifications found in the IMF's *Manual on Government Finance Statistics* (1986). See these sources for complete and authoritative explanations of concepts, definitions, and data sources.

4.13 Central government expenditures

	Goods and services		Wages and salaries[a]		Interest payments		Subsidies and other current transfers		Capital expenditure	
	% of total expenditure		% of total expenditure		% of total expenditure		% of total expenditure		% of total expenditure	
	1980	1995	1980	1995	1980	1995	1980	1995	1980	1995
Albania	..	26	..	12	..	8	..	48	..	18
Algeria	..	..	..	..	..	..	..	..	..	..
Angola	..	..	..	..	..	..	..	..	..	..
Argentina	57	22	..	17	0	10	43	60	..	7
Armenia	..	..	..	..	..	..	..	..	..	..
Australia	22	23	..	..	7	6	65	67	7	4
Austria	26	25	11	9	5	9	60	59	9	7
Azerbaijan	..	..	..	..	..	..	..	..	..	..
Bangladesh	..	..	..	..	..	..	..	..	..	..
Belarus	..	..	..	..	..	..	..	..	..	..
Belgium	23	19	16	15	10	17	59	59	8	5
Benin	..	..	..	..	..	..	..	..	..	..
Bolivia	..	57	..	34	..	11	..	13	..	19
Bosnia and Herzegovina	..	..	..	..	..	..	..	..	..	..
Botswana	47	52	29	26	2	2	19	29	32	16
Brazil	20	*13*	16	7	8	*52*	64	*40*	8	*2*
Bulgaria	..	24	..	6	..	36	..	37	..	4
Burkina Faso	67	..	..	..	*3*	..	13	..	19	..
Burundi	*39*	39	*25*	22	*2*	5	7	12	*46*	42
Cambodia	..	..	..	..	..	..	..	..	..	..
Cameroon	55	53	32	37	1	23	11	13	33	8
Canada	22	10	..	..	12	*18*	65	..	1	..
Central African Republic	*67*	..	*54*	..	*1*	..	*16*	..	*6*	..
Chad	..	..	..	..	..	..	..	..	..	..
Chile	41	29	29	19	3	4	46	52	10	16
China	..	..	..	..	..	..	..	..	..	..
Hong Kong, China	..	..	..	..	..	..	..	..	..	..
Colombia	36	*29*	23	*19*	4	*11*	38	*43*	31	*18*
Congo, Dem. Rep.	65	94	42	58	8	1	8	2	20	3
Congo, Rep.	..	..	..	..	..	..	..	..	45	..
Costa Rica	53	48	44	36	9	19	24	25	21	8
Côte d'Ivoire	..	..	..	..	..	..	..	..	28	..
Croatia	..	59	..	25	..	3	..	30	..	8
Cuba	..	..	..	..	..	..	..	..	..	..
Czech Republic	..	15	..	8	..	3	..	69	..	12
Denmark	22	19	13	11	7	13	65	64	7	4
Dominican Republic	50	38	39	27	6	6	12	12	31	42
Ecuador	28	*47*	26	*42*	9	*22*	34	*9*	16	*21*
Egypt, Arab Rep.	*38*	*31*	*19*	*17*	6	*25*	*36*	*24*	20	*19*
El Salvador	50	48	40	38	3	10	16	25	16	12
Eritrea	..	..	..	..	..	..	..	..	..	..
Estonia	..	50	..	..	..	1	..	43	..	7
Ethiopia[b]	85	*63*	37	*39*	3	*11*	4	*12*	15	*19*
Finland	22	17	11	7	2	11	66	67	11	5
France	30	24	20	16	2	7	62	64	5	5
Gabon	..	..	..	..	..	..	..	..	..	..
Gambia, The	46	..	23	..	1	..	4	..	48	*23*
Georgia	..	..	..	..	..	..	..	..	..	..
Germany[c]	34	31	9	8	3	7	55	57	7	5
Ghana	48	*45*	27	*28*	16	*17*	26	*23*	10	*15*
Greece	45	29	29	23	8	38	35	20	16	13
Guatemala	50	52	35	35	5	11	6	12	42	26
Guinea	..	..	..	..	..	..	..	..	..	..
Guinea-Bissau	..	..	..	..	..	..	..	..	..	..
Haiti	*82*	..	..	..	2	..	5	..	20	..
Honduras	..	..	..	..	..	..	..	..	..	..

Central government expenditures | 4.13

	Goods and services		Wages and salaries[a]		Interest payments		Subsidies and other current transfers		Capital expenditure	
	% of total expenditure		% of total expenditure		% of total expenditure		% of total expenditure		% of total expenditure	
	1980	1995	1980	1995	1980	1995	1980	1995	1980	1995
Hungary	20	..	7	..	3	..	64	..	13	..
India	29	23	14	9	13	27	47	38	12	11
Indonesia	25	27	15	13	4	11	24	16	47	46
Iran, Islamic Rep.	57	52	45	36	1	0	19	15	22	33
Iraq	..	..	..	..	..	..	..	..	..	..
Ireland	19	18	13	13	14	14	57	59	10	9
Israel	50	33	12	15	11	14	35	45	4	9
Italy	18	15	13	11	11	24	63	57	5	5
Jamaica	..	..	..	..	..	..	..	..	..	..
Japan	13	..	..	..	13	..	54	..	19	..
Jordan	43	62	..	46	3	9	17	10	29	19
Kazakhstan	..	..	..	..	..	..	..	..	..	..
Kenya	57	51	27	32	7	25	13	5	23	19
Korea, Dem. Rep.	..	..	..	..	..	..	..	..	..	..
Korea, Rep.	45	28	16	14	7	3	34	49	14	20
Kuwait	45	59	22	27	0	4	23	22	32	14
Kyrgyz Republic	..	..	..	..	..	..	..	..	..	..
Lao PDR	..	..	..	..	..	..	..	..	..	..
Latvia	..	39	..	20	..	3	..	54	..	4
Lebanon	..	29	..	23	..	31	..	22	..	18
Lesotho	..	..	..	..	..	..	..	..	..	30
Libya	..	..	..	..	..	..	..	..	..	..
Lithuania	..	42	..	16	..	2	..	47	..	10
Macedonia, FYR	..	..	..	..	..	..	..	..	..	..
Madagascar	..	28	..	18	..	29	..	7	..	35
Malawi	37	..	15	..	9	..	6	..	48	..
Malaysia	38	44	28	26	10	13	19	21	35	23
Mali	46	..	33	..	1	..	11	..	9	..
Mauritania	..	..	..	..	..	..	..	..	..	..
Mauritius	42	49	32	38	14	11	28	24	17	17
Mexico	32	26	25	18	11	18	32	43	32	12
Moldova	..	..	..	..	..	..	..	..	..	..
Mongolia	..	34	..	10	..	2	..	40	..	23
Morocco	47	..	33	..	7	..	15	..	31	..
Mozambique	..	..	..	..	..	..	..	..	24	49
Myanmar	..	..	..	..	..	2	..	9	..	15
Namibia	..	74	..	..	..	..	..	..	..	..
Nepal	..	..	..	..	..	..	..	..	..	..
Netherlands	16	15	11	9	4	9	72	71	9	5
New Zealand	29	48	21	10	10	13	55	37	6	2
Nicaragua	60	30	..	19	8	12	13	25	19	33
Niger	30	..	17	..	6	..	14	..	49	..
Nigeria	..	..	..	..	..	..	..	..	..	..
Norway	20	20	9	8	7	6	67	69	6	5
Oman	71	74	13	24	3	6	5	5	21	15
Pakistan	47	43	..	..	12	24	23	15	17	18
Panama	50	54	33	40	18	7	14	27	18	11
Papua New Guinea	58	48	37	28	5	9	23	32	15	11
Paraguay	61	56	34	46	3	6	12	23	24	15
Peru	45	34	..	16	18	18	14	31	23	17
Philippines	61	46	27	32	7	21	7	18	26	15
Poland	..	26	..	14	..	11	..	59	..	3
Portugal	34	39	24	29	8	12	45	37	13	12
Puerto Rico	..	..	..	..	..	..	..	..	..	..
Romania	11	34	2	16	0	4	55	49	33	13
Russian Federation	..	40	..	14	..	13	..	49	..	5

	Goods and services		Wages and salaries[a]		Interest payments		Subsidies and other current transfers		Capital expenditure	
	% of total expenditure		% of total expenditure		% of total expenditure		% of total expenditure		% of total expenditure	
	1980	1995	1980	1995	1980	1995	1980	1995	1980	1995
Rwanda	58	..	30	..	2	..	5	..	35	..
Saudi Arabia	..	..	..	..	..	..	..	..	..	..
Senegal	72	..	45	..	6	..	18	..	8	..
Sierra Leone	..	37	..	21	..	12	..	28	20	24
Singapore	58	59	29	30	15	6	6	11	22	23
Slovak Republic	..	..	..	..	..	..	..	..	..	..
Slovenia	..	..	..	..	..	..	..	..	..	..
South Africa	47	27	20	26	8	18	31	46	14	9
Spain	40	16	32	11	1	12	48	66	11	5
Sri Lanka	31	39	13	18	8	20	20	21	40	21
Sudan	46	..	12	..	6	..	28	..	23	..
Sweden	17	15	8	5	7	13	71	70	5	3
Switzerland	27	30	6	5	3	3	63	63	7	4
Syrian Arab Republic	..	..	..	..	..	..	..	..	37	38
Tajikistan	..	..	..	..	..	..	..	..	..	..
Tanzania	52	..	19	..	7	..	4	..	40	..
Thailand	55	57	21	32	8	2	14	7	23	33
Togo	52	..	28	..	9	..	12	..	27	..
Trinidad and Tobago	34	51	28	33	3	18	24	21	39	10
Tunisia	42	38	29	32	5	11	24	31	30	20
Turkey	47	37	32	27	3	12	23	42	28	8
Turkmenistan	..	..	..	..	..	..	..	..	..	..
Uganda	..	..	..	..	..	..	..	..	13	..
Ukraine	..	..	..	..	..	..	..	..	..	..
United Arab Emirates	80	87	..	35	..	..	12	9	8	5
United Kingdom	32	30	14	9	11	9	53	56	5	5
United States	29	23	11	9	10	15	54	59	6	3
Uruguay	47	27	30	15	2	6	43	60	8	6
Uzbekistan	..	..	..	..	..	..	..	..	..	..
Venezuela	50	27	41	21	8	26	22	32	22	16
Vietnam	..	..	..	..	..	..	..	..	..	..
West Bank and Gaza	..	..	..	..	..	..	..	..	..	..
Yemen, Rep.	..	67	..	58	..	14	..	7	..	11
Yugoslavia, FR (Serb./Mont.)	..	..	..	..	..	..	..	..	..	..
Zambia	55	48	27	24	9	8	25	19	11	34
Zimbabwe	56	49	31	29	7	19	32	20	5	13
World	45 m	35 m	.. m	.. m	7 m	11 m	24 m	28 m	18 m	12 m
Low income	..	..	..	..	..	..	..	..	..	..
Excl. China & India	..	..	..	..	..	..	..	..	..	..
Middle income	..	40	..	26	..	11	..	30	24	15
Lower middle income	..	41	..	27	..	11	..	26	..	17
Upper middle income	41	28	22	19	3	10	32	45	14	11
Low & middle income	..	..	..	..	..	..	..	..	..	..
East Asia & Pacific	..	45	..	27	..	11	..	18	25	28
Europe & Central Asia	..	..	..	..	..	..	..	..	..	..
Latin America & Carib.	50	36	31	24	6	11	24	26	20	14
Middle East & N. Africa	..	55	..	32	..	11	..	13	32	19
South Asia	31	39	..	..	12	24	23	21	17	18
Sub-Saharan Africa	..	..	..	..	..	..	..	..	20	..
High income	28	25	13	11	7	9	56	59	7	5

Note: Includes expenditures financed by grants in kind and other cash adjustments.
a. Part of goods and services. b. Data prior to 1992 include Eritrea. c. Data prior to 1990 refer to the Federal Republic of Germany before unification.

Central government expenditures | 4.13

About the data

Government expenditures include all nonrepayable payments, whether current or capital, requited or unrequited. Total central government expenditure as presented in the International Monetary Fund's (IMF) *Government Finance Statistics* is a more limited measure of general government consumption than that shown in the national accounts (see table 4.9) because it excludes consumption expenditures by state and local governments. At the same time, the IMF's concept of central government expenditure is broader than the national accounts definition because it includes government gross domestic investment and transfer payments.

Expenditures can be measured either by function (education, health, defense) or by economic type (wages and salaries, interest payments, purchases of goods and services). Functional data are often incomplete, and coverage varies by country because functional responsibilities stretch across levels of government for which no data are available. Defense expenditures, which are usually the central government's responsibility, are shown in table 5.7. For more information on education expenditures see table 2.9; for more information on health expenditures see table 2.13.

The classification of expenditures by economic type can also be problematic. For example, the distinction between current and capital expenditure may be arbitrary, and subsidies to state-owned enterprises or banks may be disguised as capital financing subsidies may also be hidden in special contractual pricing for goods and services. For further discussion of government finance statistics see the notes to tables 4.12 and 4.14.

Definitions

• **Total expenditure** of the central government includes both current and capital (development) expenditures and excludes lending minus repayments. • **Goods and services** include all government payments in exchange for goods and services, whether in the form of wages and salaries to employees or other purchases of goods and services. • **Wages and salaries** consist of all payments in cash, but not in kind, to employees in return for services rendered, before deduction of withholding taxes and employee contributions to social security and pension funds. • **Interest payments** are payments made to domestic sectors and to nonresidents for the use of borrowed money. (Repayment of principal is shown as a financing item, and commission charges are shown as purchases of services.) Interest payments do not include payments by government as guarantor or surety of interest on the defaulted debts of others, which are classified as government lending. • **Subsidies and other current transfers** include all unrequited, nonrepayable transfers on current account to private and public enterprises, and the cost to the public of covering the cash operating deficits on sales to the public by departmental enterprises. • **Capital expenditure** is spending to acquire fixed capital assets, land, intangible assets, government stocks, and nonmilitary, nonfinancial assets. Also included are capital grants.

Figure 4.13a

A large portion of government spending goes to transfers and subsidies

Subsidies and transfers as % of
central government expenditure, 1995

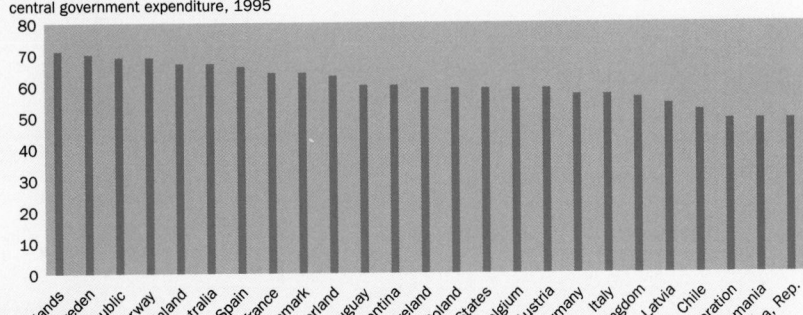

Source: International Monetary Fund, government finance statistics data files.

In high-income economies a large part of the central government's budget goes to subsidies and transfer payments. Subsidies and transfers are also quite high in some middle-income economies (Uruguay, Argentina, Chile) and transition economies (the Czech Republic, Poland, Latvia, Russia, Romania).

Data sources

Data on central government expenditures are from the IMF's *Government Finance Statistics Yearbook* (1996) and IMF data files. Each country's accounts are reported using the system of common definitions and classifications found in the IMF's *Manual on Government Finance Statistics* (1986). See these sources for complete and authoritative explanations of concepts, definitions, and data sources.

4.14 Central government revenues

	Taxes on income, profit, and capital gains		Social security taxes		Taxes on goods and services		Taxes on international trade		Other taxes		Nontax revenue	
	% of total current revenue		% of total current revenue		% of total current revenue		% of total current revenue		% of total current revenue		% of total current revenue	
	1980	1995	1980	1995	1980	1995	1980	1995	1980	1995	1980	1995
Albania	..	8	..	15	..	40	..	14	..	1	..	21
Algeria	..	..	..	..	..	..	..	..	..	..	..	..
Angola	..	..	..	..	..	..	..	..	..	..	..	..
Argentina	0	10	17	35	17	36	0	5	33	5	33	8
Armenia	..	..	..	..	..	..	..	..	..	..	..	..
Australia	61	64	0	0	23	21	5	3	0	2	10	10
Austria	21	21	35	40	26	23	2	0	9	7	8	9
Azerbaijan	..	..	..	..	..	..	..	..	..	..	..	..
Bangladesh	10	..	0	..	25	..	29	..	4	..	32	..
Belarus	..	..	..	..	..	..	..	..	..	..	..	..
Belgium	39	35	31	34	24	25	0	0	2	3	4	3
Benin	..	..	..	..	..	..	..	..	..	..	..	..
Bolivia	..	3	..	7	..	40	..	7	..	11	..	33
Bosnia and Herzegovina	..	..	..	..	..	..	..	..	..	..	..	..
Botswana	33	21	0	0	1	4	39	15	0	0	27	59
Brazil	11	15	25	31	32	19	7	2	4	5	21	28
Bulgaria	..	17	..	21	..	28	..	8	..	3	..	22
Burkina Faso	18	..	8	..	16	..	44	..	4	..	11	..
Burundi	19	19	1	7	25	41	40	26	8	1	6	7
Cambodia	..	..	..	..	..	..	..	..	..	..	..	..
Cameroon	22	17	8	0	18	25	38	28	5	3	8	27
Canada	53	48	10	19	17	19	7	2	0	0	14	11
Central African Republic	16	..	6	..	21	..	40	..	8	..	9	..
Chad	..	..	..	..	..	..	..	..	..	..	..	..
Chile	18	17	17	6	36	45	4	9	6	5	20	17
China	..	11	..	0	..	72	..	9	..	0	..	8
Hong Kong, China	..	..	..	..	..	..	..	..	..	..	..	..
Colombia	25	34	11	0	23	41	21	9	7	0	14	17
Congo, Dem. Rep.	30	33	2	0	12	19	38	33	5	8	12	8
Congo, Rep.	49	..	4	..	8	..	13	..	3	..	24	..
Costa Rica	14	11	29	26	30	33	19	15	2	1	6	14
Côte d'Ivoire	13	..	6	..	25	..	43	..	6	..	8	..
Croatia	..	11	..	33	..	42	..	9	..	1	..	4
Cuba	..	..	..	..	..	..	..	..	..	..	..	..
Czech Republic	..	15	..	40	..	32	..	4	..	1	..	8
Denmark	36	39	2	4	47	41	0	0	3	3	12	13
Dominican Republic	19	16	4	4	22	34	31	37	2	1	22	9
Ecuador	45	50	0	0	17	26	31	11	3	1	4	12
Egypt, Arab Rep.	19	19	9	10	9	13	20	10	7	10	37	39
El Salvador	23	27	0	0	30	51	37	17	8	1	4	5
Eritrea	..	..	..	..	..	..	..	..	..	..	..	..
Estonia	..	20	..	31	..	39	..	0	..	2	..	7
Ethiopia[a]	21	22	0	0	25	23	34	23	4	3	16	29
Finland	29	28	10	14	49	41	2	0	3	2	8	15
France	18	18	41	44	31	28	0	0	3	4	7	6
Gabon	40	..	0	..	5	..	20	..	2	..	34	..
Gambia, The	15	14	0	0	3	32	65	42	2	5	15	6
Georgia	..	..	..	..	..	..	..	..	..	..	..	..
Germany[b]	19	16	54	48	23	23	0	0	0	7	4	6
Ghana	20	17	0	0	28	34	44	27	0	0	7	23
Greece	17	33	26	2	32	59	5	0	10	–5	11	11
Guatemala	13	19	0	0	31	46	36	23	12	3	8	9
Guinea	..	..	..	..	..	..	..	..	..	..	..	..
Guinea-Bissau	..	..	..	..	..	..	..	..	..	..	..	..
Haiti	14	..	0	..	15	..	48	..	9	..	13	..
Honduras	31	..	0	..	24	..	37	..	2	..	7	..

Central government revenues | 4.14

	Taxes on income, profit, and capital gains		Social security taxes		Taxes on goods and services		Taxes on international trade		Other taxes		Nontax revenue	
	% of total current revenue		% of total current revenue		% of total current revenue		% of total current revenue		% of total current revenue		% of total current revenue	
	1980	1995	1980	1995	1980	1995	1980	1995	1980	1995	1980	1995
Hungary	19	..	15	..	38	..	7	..	5	..	16	..
India	18	22	0	0	42	29	22	24	1	0	17	25
Indonesia	78	46	0	6	9	33	7	4	1	1	5	9
Iran, Islamic Rep.	4	9	7	6	4	5	12	3	5	3	68	73
Iraq	..	..	..	..	..	..	..	..	..	..	..	..
Ireland	34	40	13	14	30	32	9	5	2	4	11	5
Israel	41	39	10	10	25	34	4	0	8	4	14	13
Italy	30	33	35	31	25	26	0	0	4	3	8	7
Jamaica	34	..	4	..	49	..	3	..	6	..	4	..
Japan	71	36	0	27	21	15	2	1	3	6	22	27
Jordan	13	11	0	0	7	26	48	25	10	10	22	27
Kazakhstan	..	..	..	..	..	..	..	..	..	..	..	..
Kenya	29	28	0	0	39	46	19	16	1	1	13	10
Korea, Dem. Rep.	..	..	..	..	..	..	..	..	..	..	..	..
Korea, Rep.	22	31	1	8	46	32	15	7	3	10	12	12
Kuwait	2	..	0	..	0	..	1	..	0	..	97	..
Kyrgyz Republic	..	..	..	..	..	..	..	..	..	..	..	..
Lao PDR	..	..	..	..	..	..	..	..	..	..	..	..
Latvia	..	7	..	35	..	41	..	3	..	0	..	13
Lebanon	..	6	..	0	..	5	..	44	..	14	..	31
Lesotho	13	13	0	0	10	15	61	59	2	0	14	13
Libya	..	..	..	..	..	..	..	..	..	..	..	..
Lithuania	..	13	..	30	..	50	..	3	..	0	..	4
Macedonia, FYR	..	..	..	..	..	..	..	..	..	..	..	..
Madagascar	17	15	11	0	39	26	28	55	3	2	2	2
Malawi	34	..	0	..	31	..	22	..	0	..	13	..
Malaysia	38	37	0	1	17	26	33	12	2	5	11	19
Mali	18	..	4	..	37	..	18	..	15	..	8	..
Mauritania	..	..	..	..	..	..	..	..	..	..	..	..
Mauritius	15	13	0	6	17	24	52	35	4	6	12	16
Mexico	34	27	12	14	50	54	7	4	−11	−15	7	16
Moldova	..	..	..	..	..	..	..	..	..	..	..	..
Mongolia	..	33	..	15	..	19	..	9	..	0	..	24
Morocco	19	..	5	..	35	..	21	..	7	..	12	..
Mozambique	..	..	..	..	..	..	..	..	..	..	..	..
Myanmar	3	20	0	0	42	26	15	12	0	0	40	41
Namibia	..	29	..	0	..	29	..	31	..	1	..	11
Nepal	6	12	0	0	37	40	33	30	8	4	16	15
Netherlands	30	25	36	42	21	23	0	0	3	4	11	7
New Zealand	67	59	0	0	18	25	3	2	1	1	10	12
Nicaragua	8	11	9	13	37	43	25	21	8	6	0	6
Niger	24	..	4	..	18	..	36	..	3	..	15	..
Nigeria	..	..	..	..	..	..	..	..	..	..	..	..
Norway	27	18	22	22	39	38	1	1	1	1	9	21
Oman	26	21	0	0	0	1	1	3	0	2	72	73
Pakistan	14	16	0	0	34	37	34	25	0	1	18	21
Panama	21	20	21	16	17	17	10	11	4	3	27	34
Papua New Guinea	60	50	0	0	12	10	16	24	1	2	10	14
Paraguay	15	10	13	0	18	36	25	12	19	6	9	35
Peru	26	17	0	9	37	49	27	10	2	3	8	11
Philippines	21	33	0	0	42	26	24	29	2	4	11	8
Poland	..	28	..	25	..	28	..	8	..	1	..	10
Portugal	19	25	26	25	34	37	5	0	9	3	7	10
Puerto Rico	..	..	..	..	..	..	..	..	..	..	..	..
Romania	0	30	13	29	0	23	0	4	9	2	78	12
Russian Federation	..	15	..	32	..	36	..	9	..	..	..	6

	Taxes on income, profit, and capital gains		Social security taxes		Taxes on goods and services		Taxes on international trade		Other taxes		Nontax revenue	
	% of total current revenue		% of total current revenue		% of total current revenue		% of total current revenue		% of total current revenue		% of total current revenue	
	1980	1995	1980	1995	1980	1995	1980	1995	1980	1995	1980	1995
Rwanda	18	..	4	..	19	..	42	..	2	..	14	..
Saudi Arabia	..	..	..	..	..	..	..	..	..	..	..	..
Senegal	18	..	4	..	26	..	34	..	11	..	6	..
Sierra Leone	22	16	0	0	16	38	50	42	2	0	10	4
Singapore	32	26	0	0	16	21	7	1	14	15	31	37
Slovak Republic	..	..	..	..	..	..	..	..	..	..	..	..
Slovenia	..	..	..	..	..	..	..	..	..	..	..	..
South Africa	56	51	1	2	24	36	3	2	3	3	13	6
Spain	23	30	48	40	13	23	4	0	4	0	8	7
Sri Lanka	16	13	0	0	27	53	50	18	2	4	5	13
Sudan	14	..	0	..	26	..	43	..	1	..	16	..
Sweden	18	12	33	37	29	31	1	1	4	7	14	12
Switzerland	14	12	48	54	19	18	9	6	2	2	7	7
Syrian Arab Republic	10	23	0	0	5	37	14	13	10	8	61	19
Tajikistan	..	..	..	..	..	..	..	..	..	..	..	..
Tanzania	32	..	0	..	41	..	17	..	2	..	8	..
Thailand	18	31	0	1	46	39	26	16	2	3	8	9
Togo	34	..	6	..	15	..	32	..	–2	..	16	14
Trinidad and Tobago	72	50	1	2	4	26	7	6	1	1	16	14
Tunisia	15	16	9	15	24	20	25	28	4	4	22	17
Turkey	49	31	0	0	20	40	6	4	5	3	21	22
Turkmenistan	..	..	..	..	..	..	..	..	..	..	..	..
Uganda	11	..	0	..	41	..	44	..	0	..	3	..
Ukraine	..	..	..	..	..	..	..	..	..	..	..	..
United Arab Emirates	0	0	0	2	0	23	0	0	0	0	100	75
United Kingdom	38	36	16	17	28	33	0	0	6	7	13	8
United States	57	52	28	33	4	4	1	1	1	1	8	9
Uruguay	11	10	23	31	43	32	14	4	3	16	6	8
Uzbekistan	..	..	..	..	..	..	..	..	..	..	..	..
Venezuela	67	38	5	4	4	33	7	9	2	0	15	19
Vietnam	..	..	..	..	..	..	..	..	..	..	..	..
West Bank and Gaza	..	..	..	..	..	..	..	..	..	..	..	..
Yemen, Rep.	..	18	..	..	..	..	..	10	..	19	..	50
Yugoslavia, FR (Serb./Mont.)	..	..	..	..	..	..	..	..	..	3	..	..
Zambia	38	33	0	0	43	48	8	12	3	0	7	7
Zimbabwe	46	49	0	0	28	20	4	19	1	1	20	11
World	21 m	22 m	3 m	5 m	25 m	29 m	15 m	8 m	3 m	3 m	11 m	11 m
Low income	19	..	0	..	26	..	34	..	3	..	11	..
Excl. China & India	19	..	0	..	26	..	34	..	3	..	11	..
Middle income	19	17	4	6	26	34	15	9	5	3	16	14
Lower middle income	19	17	4	6	20	36	21	13	6	3	14	15
Upper middle income	26	19	12	10	24	32	7	5	3	3	16	12
Low & middle income	19	..	0	..	25	..	25	..	3	..	12	..
East Asia & Pacific	29	33	0	1	29	26	20	12	1	1	10	9
Europe & Central Asia	..	..	..	..	..	..	..	..	..	..	..	..
Latin America & Carib.	20	17	7	8	27	40	20	9	6	3	9	13
Middle East & N. Africa	15	16	6	0	8	10	20	19	6	4	37	31
South Asia	14	14	0	0	34	39	33	25	2	2	17	18
Sub-Saharan Africa	22	..	1	..	25	..	34	..	3	..	10	..
High income	29	28	19	22	24	28	2	0	3	3	10	10

Note: Includes adjustments to tax revenue.
a. Data prior to 1992 include Eritrea. b. Data prior to 1990 refer to the Federal Republic of Germany before unification.

The International Monetary Fund (IMF) classifies government transactions as receipts or payments and according to whether they are repayable or nonrepayable. If nonrepayable, they are classified as capital (meant to be used in production for more than a year) or current, and as requited (involving payment in return for a benefit or service) or unrequited. Revenues include all nonrepayable receipts (other than grants), the most important of which are taxes. Grants are unrequited, nonrepayable, noncompulsory receipts from other governments or international organizations. Transactions are generally recorded on a cash rather than an accrual basis. Measuring the accumulation of arrears on revenues or payments on an accrual basis would normally result in a higher deficit. Transactions within the same level of government are not included, but transactions between levels are included. In some instances the government budget may include transfers used to finance the deficits of autonomous, extra-budgetary agencies.

The IMF's *Manual on Government Finance Statistics* (1986) describes taxes as compulsory, unrequited payments made to governments by individuals, businesses, or institutions. Taxes traditionally have been classified as either direct (those levied directly on the income or profits of individuals and corporations) or indirect (sales and excise taxes and duties) levied on goods and services. This distinction may be a useful simplification, but it has no particular analytical significance. For further discussion of taxes and tax policies see the notes to table 5.5. For further discussion of government revenues and expenditures see the notes to tables 4.12 and 4.13.

• **Taxes on income, profit, and capital gains** are levied on the actual or presumptive net income of individuals, on the profits of enterprises, and on capital gains, whether realized on land, securities, or other assets. Intragovernmental payments are eliminated in consolidation. • **Social security taxes** include employer and employee social security contributions and those of self-employed and unemployed people. • **Taxes on goods and services** include general sales and turnover or value added taxes, selective excises on goods, selective taxes on services, taxes on the use of goods or property, and profits of fiscal monopolies. • **Taxes on international trade** include import duties, export duties, profits of export or import monopolies, exchange profits, and exchange taxes. • **Other taxes** include employer payroll or labor taxes, taxes on property, and taxes not allocable to other categories. They may include negative values that are adjustments (for example, for taxes collected on behalf of state and local governments and not allocable to individual tax categories). • **Nontax revenue** includes requited, nonrepayable receipts for public purposes, such as fines, administrative fees, or entrepreneurial income from government ownership of property, and voluntary, unrequited, nonrepayable receipts other than from government sources. Proceeds of grants and borrowing, funds arising from the repayment of previous lending by governments, incurrence of liabilities, and proceeds from the sale of capital assets are not included.

Figure 4.14a

Direct taxes account for a larger share of government revenue in high-income economies

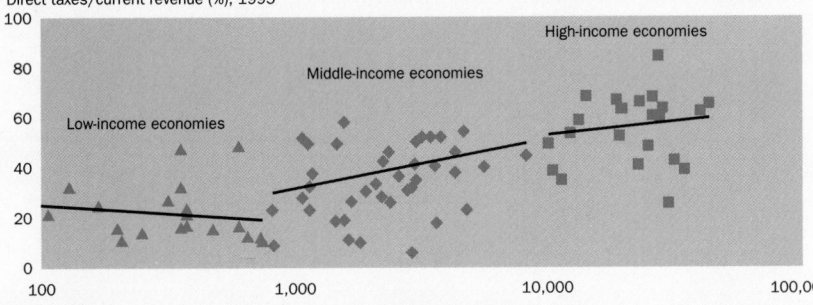

Direct taxes/current revenue (%), 1995

Source: International Monetary Fund, government finance statistics data files.

Governments of low-income economies typically raise a large share of their revenues through indirect taxes, such as tariffs and sales taxes. High-income economies usually rely more on direct taxes on income, wages, and profits—taxes that require a more sophisticated administrative apparatus but yield higher revenues. The figure shows the share of current revenues raised through direct taxes, including social security taxes, plotted against GNP per capita. Although the clustering of points and the fitted curves confirm the general observation, there are many exceptions in each group.

Data on central government revenues are from the IMF's *Government Finance Statistics Yearbook* (1996) and IMF data files. Each country's accounts are reported using the system of common definitions and classifications found in the IMF's *Manual on Government Finance Statistics* (1986). The IMF receives additional information from the Organisation for Economic Co-operation and Development (OECD) on the tax revenues of some of its members. See the IMF sources for complete and authoritative explanations of concepts, definitions, and data sources.

	Money and quasi money		Claims on private sector		Claims on governments and other public entities		GDP implicit deflator		Consumer price index		Food price index	
	annual % growth of M2		annual growth as % of M2		annual growth as % of M2		average annual % growth		average annual % growth		average annual % growth	
	1990	1996	1990	1996	1990	1996	1980–90	1990–96	1980–90	1990–96	1980–90	1990–96
Albania	..	43.8	..	2.4	..	35.4	–0.4	67.9	..	..	..	50.9
Algeria	11.4	–59.6	12.2	4.3	3.2	2.5	8.0	25.4	9.1	27.0	6.8	*28.6*
Angola	..	..	..	..	..	..	5.9	1,103.2	..	..	..	1,401.0
Argentina	1,113.3	18.7	1,444.7	6.3	1,573.2	2.5	389.1	15.8	390.6	20.1	206.9	19.5
Armenia	..	32.8	..	–1.9	..	33.8	0.3	*896.6*	..	..	..	..
Australia	12.8	10.6	15.3	10.9	–2.2	3.3	7.3	1.1	7.9	2.4	7.4	2.2
Austria	9.7	2.6	12.1	6.8	1.6	–0.4	3.3	3.0	3.2	3.1	2.6	2.2
Azerbaijan	..	17.1	..	2.5	..	31.1	..	589.9	..	..	1.5	585.5
Bangladesh	10.2	10.8	9.2	5.0	–0.2	4.4	9.5	4.9	10.5	3.6	10.4	4.0
Belarus	..	52.4	..	27.5	..	30.2	..	714.9	..	..	2.4	784.4
Belgium	4.1	6.8	3.5	5.1	4.8	1.1	4.4	2.8	4.2	2.4	4.0	0.6
Benin	28.6	13.0	–1.3	8.9	12.4	–6.6	0.9	10.8	..	..	..	..
Bolivia	52.8	24.2	40.7	15.3	17.5	–2.2	317.4	10.5	322.6	11.1	322.0	11.9
Bosnia and Herzegovina	..	..	..	..	..	..	..	..	..	..	..	..
Botswana	–14.0	30.4	12.6	2.1	–52.4	–25.3	13.1	9.7	10.0	12.4	10.7	12.9
Brazil	1,289.2	6.8	1,566.4	3.7	2,704.2	9.2	284.5	675.4	371.1	643.9	238.2	646.5
Bulgaria	*50.4*	117.8	*1.8*	76.4	*88.9*	238.1	1.2	80.3	..	..	..	114.7
Burkina Faso	–0.5	5.2	3.6	3.8	–1.5	3.2	3.1	7.1	3.4	7.0	–0.5	5.4
Burundi	10.4	*33.3*	16.3	*6.5*	–5.3	*11.1*	4.4	14.3	7.1	12.6	6.1	*6.7*
Cambodia	..	40.4	..	21.7	..	–3.1	..	44.7	..	..	..	..
Cameroon	–1.7	–10.1	0.9	2.2	–3.0	–6.7	5.6	6.4	8.7	8.9	3.9	..
Canada	7.8	5.0	9.2	11.9	0.5	–1.5	4.4	1.3	5.3	1.8	4.6	1.4
Central African Republic	–3.7	4.9	–1.6	0.1	2.3	–2.2	7.3	6.6	3.2	7.4	2.0	7.8
Chad	–2.4	27.9	–1.3	2.3	–17.3	18.9	1.3	8.6	0.6	8.7	..	*5.3*
Chile	23.5	19.6	21.4	26.5	16.4	–1.7	20.9	13.6	20.6	12.5	20.8	12.4
China	28.9	25.3	26.5	20.7	1.5	0.8	5.8	12.0	..	13.5	8.8	..
Hong Kong, China	*8.5*	12.5	*7.9*	14.1	*–1.0*	1.2	7.7	7.0	..	..	6.8	8.1
Colombia	32.7	24.4	*8.7*	21.2	–7.5	0.9	24.7	23.0	22.7	23.9	24.5	21.3
Congo, Dem. Rep.	195.4	147.9	18.0	92.8	429.7	21.2	62.9	2,746.5	57.1	2,824.8	..	..
Congo, Rep.	18.5	15.7	5.1	7.8	–12.6	1.7	0.3	8.3	6.1	15.3	4.1	10.2
Costa Rica	27.5	47.6	7.3	19.6	8.2	32.0	23.5	18.4	23.0	17.8	23.0	16.3
Côte d'Ivoire	–2.6	3.9	–3.9	1.6	–3.0	2.1	2.6	9.8	5.4	9.4	6.0	..
Croatia	..	49.4	..	3.8	..	–1.9	..	..	304.1	232.3	246.3	228.9
Cuba	..	..	..	..	..	..	..	..	..	..	..	..
Czech Republic	..	6.4	..	7.2	..	–1.5	1.5	17.7	..	..	..	11.2
Denmark	6.5	8.1	3.0	3.9	–3.1	–1.6	5.5	1.8	5.5	1.9	4.8	1.6
Dominican Republic	39.1	21.2	20.8	17.2	0.8	6.0	21.6	12.3	22.4	10.9	22.5	..
Ecuador	101.6	43.7	46.7	27.4	–22.4	–1.8	36.4	35.0	35.8	36.6	43.0	34.8
Egypt, Arab Rep.	28.7	10.8	6.3	10.5	25.3	2.5	13.7	11.3	17.4	12.4	19.0	9.5
El Salvador	32.4	15.5	8.8	14.8	9.6	–0.1	16.4	10.8	19.6	12.6	21.4	15.5
Eritrea	..	..	..	..	..	..	..	..	..	..	..	..
Estonia	*71.1*	36.4	*27.6*	33.0	*–13.5*	3.7	2.4	116.7	..	..	..	116.2
Ethiopia[a]	18.5	9.4	–1.0	19.9	21.7	–3.4	3.4	9.7	4.0	8.9	3.7	*12.4*
Finland	5.0	–2.9	17.1	–1.0	–0.1	–2.3	6.8	1.8	6.2	1.8	5.8	–1.1
France	3.3	4.0	15.7	–1.3	0.3	6.3	6.0	2.0	5.8	2.1	5.7	1.0
Gabon	3.3	17.2	0.7	–1.2	–20.6	–2.9	3.7	9.8	5.1	6.6	2.8	5.1
Gambia, The	8.4	5.8	7.8	0.1	–35.4	2.6	16.7	5.4	20.0	5.7	20.4	5.6
Georgia	..	..	..	..	..	..	1.9	*2,279.3*	..	..	..	..
Germany[b]	18.6	7.5	26.6	11.9	2.0	3.8	..	*2.9*	2.2	3.3	..	2.2
Ghana	13.3	32.6	4.9	21.2	–0.8	8.4	42.0	26.9	39.1	29.8	33.1	27.5
Greece	14.3	13.3	4.6	8.0	16.3	1.6	18.0	12.2	18.7	12.8	18.0	12.4
Guatemala	25.8	13.8	15.0	7.3	0.5	2.0	14.6	13.0	14.0	12.7	14.6	13.2
Guinea	*23.3*	3.6	*13.1*	2.0	*7.3*	10.4	..	8.8	..	..	..	9.1
Guinea-Bissau	65.3	48.4	57.4	8.8	109.9	11.9	56.9	47.8	..	44.5	..	..
Haiti	2.5	0.6	–0.6	6.1	0.4	1.5	7.5	25.0	5.2	26.8	4.1	*19.2*
Honduras	21.4	41.2	13.0	30.6	–10.9	–8.5	5.7	20.0	6.3	19.6	5.1	*19.1*

Monetary indicators and prices | 4.15

	Money and quasi money		Claims on private sector		Claims on governments and other public entities		GDP implicit deflator		Consumer price index		Food price index	
	annual % growth of M2		annual growth as % of M2		annual growth as % of M2		average annual % growth		average annual % growth		average annual % growth	
	1990	1996	1990	1996	1990	1996	1980–90	1990–96	1980–90	1990–96	1980–90	1990–96
Hungary	29.2	18.4	22.8	5.5	2.0	–7.3	8.6	22.5	9.6	24.1	9.5	24.4
India	15.1	18.7	5.9	10.5	10.5	9.3	8.0	9.2	8.6	9.9	8.4	10.8
Indonesia	44.6	27.2	66.9	23.9	–6.7	–0.4	8.5	8.1	8.3	8.8	8.6	9.5
Iran, Islamic Rep.	18.0	32.5	14.7	11.3	5.8	26.5	14.6	32.3	18.2	29.3	16.3	32.6
Iraq	..	..	..	..	..	..	10.3	..	..	..	14.3	..
Ireland	8.9	13.6	0.6	9.8	1.9	–2.6	6.6	1.9	6.8	2.3	10.5	2.0
Israel	19.4	15.1	18.5	17.0	4.9	2.2	101.5	12.2	101.7	12.1	102.4	9.6
Italy	10.2	2.2	11.8	0.8	1.8	3.7	10.0	4.7	9.1	4.8	8.2	4.3
Jamaica	21.5	10.9	12.5	8.8	–16.0	5.6	18.6	36.1	15.1	36.5	16.2	37.1
Japan	8.2	2.3	9.7	1.2	1.5	0.4	1.7	0.7	1.7	1.1	1.6	0.8
Jordan	8.3	–0.9	4.7	3.4	1.0	–2.8	4.2	4.0	5.7	4.2	4.7	4.1
Kazakhstan	..	..	..	..	..	..	..	604.9	..	..	..	777.8
Kenya	20.1	26.2	8.0	15.1	21.5	2.6	9.0	16.6	11.1	23.5	..	24.5
Korea, Dem. Rep.	..	..	..	..	..	..	..	..	..	..	..	..
Korea, Rep.	17.2	15.8	36.1	27.5	–1.5	–0.7	5.9	5.8	4.9	5.8	5.0	6.1
Kuwait	0.7	–0.6	3.3	10.0	–3.1	–11.3	–2.4	–2.0	2.9	2.3	1.6	..
Kyrgyz Republic	..	..	..	..	..	..	..	256.2	..	..	..	171.8
Lao PDR	7.8	26.7	3.6	13.9	7.0	–17.9	37.5	11.1	..	..	..	..
Latvia	..	18.6	..	3.4	..	–0.3	0.0	110.6	..	..	..	61.2
Lebanon	55.1	26.4	27.6	11.3	18.5	12.9	1.6	32.8	75.5	36.8	..	..
Lesotho	8.4	18.1	6.8	–6.3	–17.4	–16.9	13.9	8.8	13.6	12.1	13.2	13.0
Libya	20.3	1.8	0.9	–4.5	8.5	–2.5	0.2	..	..	..	..	..
Lithuania	..	–3.0	..	–4.8	..	3.9	..	179.3	..	..	..	154.7
Macedonia, FYR	..	..	..	..	..	..	..	286.4	..	..	242.1	197.0
Madagascar	4.5	16.2	23.8	0.9	–14.8	–2.7	17.0	25.4	16.6	23.9	15.7	24.3
Malawi	11.1	39.6	15.8	3.6	–12.8	6.8	14.9	33.2	16.9	34.2	16.3	39.7
Malaysia	10.6	20.0	20.8	26.3	–1.2	–0.6	1.7	4.4	2.6	4.2	1.3	4.8
Mali	–4.9	24.5	0.1	16.3	–13.4	–16.0	3.6	10.6	..	..	..	1.8
Mauritania	11.5	–5.1	20.2	15.3	1.5	–60.6	8.6	6.2	7.1	7.0	..	..
Mauritius	21.2	7.6	10.8	2.9	0.8	2.2	9.4	6.5	6.9	7.2	7.4	7.5
Mexico	75.8	26.2	64.0	–17.7	12.1	–0.6	71.5	18.5	73.8	18.2	73.1	17.3
Moldova	358.0	14.8	53.3	13.5	447.0	9.6	..	307.7	..	..	..	230.3
Mongolia	31.6	17.2	40.2	13.2	38.5	37.3	–1.8	106.2	..	..	..	..
Morocco	21.5	6.6	12.4	4.7	–4.9	1.2	7.2	4.0	7.0	5.5	6.7	8.4
Mozambique	37.2	19.0	22.0	19.2	–6.8	–25.6	35.9	47.2	..	48.5	24.4	..
Myanmar	37.7	..	12.8	..	23.4	..	12.2	21.9	11.5	25.5	11.9	27.1
Namibia	30.3	29.0	15.4	19.7	–4.7	8.4	13.3	9.6	12.6	11.2	14.9	11.1
Nepal	18.5	12.2	5.7	12.5	7.3	2.2	11.1	10.1	10.2	10.4	10.0	10.5
Netherlands	6.9	5.6	6.7	13.4	0.1	0.6	1.6	1.9	2.0	2.6	1.2	1.6
New Zealand	74.0	17.6	76.6	14.8	0.1	–5.2	10.8	1.8	10.9	2.0	9.9	0.7
Nicaragua	7,677.8	40.6	4,932.9	–8.6	12,679.3	–56.7	422.6	70.9	536.0	63.2	..	..
Niger	–4.1	–6.6	–5.1	1.0	1.4	5.0	3.0	7.4	0.7	7.2	–1.5	..
Nigeria	32.7	20.1	7.8	13.1	26.3	–52.9	16.5	37.6	21.5	48.8	21.6	36.2
Norway	5.6	4.5	5.0	14.2	–0.1	–4.9	5.6	1.7	7.4	2.1	7.8	0.9
Oman	10.0	8.1	9.6	13.7	–10.9	–1.2	–3.6	–2.9	..	..	..	0.2
Pakistan	11.6	20.1	5.9	9.1	7.7	15.4	6.7	11.3	6.3	11.1	6.6	11.9
Panama	36.6	6.1	0.8	10.5	–25.7	5.2	1.9	2.7	1.4	1.1	1.9	1.4
Papua New Guinea	4.3	30.7	1.3	–0.3	7.2	5.7	5.3	6.6	5.6	7.4	4.6	6.8
Paraguay	52.5	18.2	33.1	17.2	–9.5	–5.3	24.4	17.4	21.9	17.0	24.9	18.6
Peru	6,384.9	37.2	2,123.7	37.8	2,127.1	–25.1	231.3	49.1	246.3	54.3	..	50.2
Philippines	22.5	23.2	15.7	36.3	3.4	3.7	14.9	9.0	14.4	9.5	14.1	8.7
Poland	160.1	29.3	20.8	18.6	75.6	9.6	53.7	32.4	50.9	37.5	52.4	32.7
Portugal	9.4	6.1	7.4	10.9	3.2	–1.3	18.1	7.0	17.1	6.4	16.9	4.9
Puerto Rico	..	..	..	..	..	..	3.4	2.3	..	..	2.8	7.1
Romania	26.4	64.1	..	..	133.3	–68.6	2.5	132.7	..	..	1.8	138.3
Russian Federation	..	33.0	..	12.1	..	58.8	2.4	394.0	..	..	..	390.9

	Money and quasi money		Claims on private sector		Claims on governments and other public entities		GDP implicit deflator		Consumer price index		Food price index	
	annual % growth of M2		annual growth as % of M2		annual growth as % of M2		average annual % growth		average annual % growth		average annual % growth	
	1990	1996	1990	1996	1990	1996	1980–90	1990–96	1980–90	1990–96	1980–90	1990–96
Rwanda	5.6	10.9	–10.0	0.5	26.8	–1.6	3.6	19.5	3.9	22.6	6.1	..
Saudi Arabia	4.6	7.7	–4.5	1.0	4.2	–3.2	–3.7	1.1	–0.8	1.8	–0.4	1.7
Senegal	–4.8	11.7	–8.4	11.2	–5.3	–3.7	6.5	8.4	6.2	7.6	5.3	8.3
Sierra Leone	74.0	29.6	4.9	6.2	228.7	32.4	63.7	37.7	72.4	37.5	71.0	..
Singapore	20.0	9.8	13.7	17.1	–4.9	–4.6	2.2	3.4	1.6	2.4	0.9	2.1
Slovak Republic	..	16.2	..	11.6	..	9.5	1.8	14.2	..	..	1.6	18.1
Slovenia	123.6	21.3	96.1	15.2	–10.4	–2.7	..	..	..	..	252.3	50.6
South Africa	11.4	14.3	13.7	20.2	1.8	1.9	14.9	10.6	14.8	10.4	15.1	13.0
Spain	13.6	2.9	8.4	6.5	5.3	1.4	9.3	5.0	9.0	4.9	9.3	3.7
Sri Lanka	21.1	10.5	16.2	7.5	6.8	4.3	10.8	10.4	10.9	10.7	10.9	10.9
Sudan	48.8	65.3	12.6	27.1	29.4	48.4	37.1	86.2	37.6	114.3	38.0	..
Sweden	..	..	..	..	..	..	7.4	2.8	7.0	3.4	8.2	–0.5
Switzerland	0.8	9.6	11.7	–0.8	1.0	0.5	3.7	2.3	2.9	2.6	3.1	0.7
Syrian Arab Republic	26.1	–70.1	3.4	1.5	11.4	3.9	15.3	8.5	23.2	11.3	24.5	8.9
Tajikistan	..	..	..	..	..	..	..	394.3	..	..	..	..
Tanzania	41.9	8.4	22.6	–11.1	80.6	–2.4	..	..	31.0	26.8	30.2	26.0
Thailand	26.7	12.6	30.0	18.1	–4.0	–1.6	3.9	4.8	3.5	4.8	2.7	6.0
Togo	9.5	–6.3	1.8	4.8	6.9	3.8	4.7	9.4	2.5	9.8	1.2	..
Trinidad and Tobago	6.2	5.8	2.7	10.5	–1.9	–6.5	4.1	6.5	10.7	7.0	14.6	14.7
Tunisia	7.6	13.3	5.9	1.3	1.8	–1.3	7.4	5.1	7.4	5.3	8.3	5.0
Turkey	53.2	117.3	42.9	73.4	9.7	35.9	45.3	78.2	44.9	80.2	..	81.4
Turkmenistan	..	449.1	..	666.5	..	–520.5	..	1,074.2	..	..	..	..
Uganda	60.2	17.3	..	13.3	..	0.5	125.6	20.4	102.5	16.9	..	13.4
Ukraine	..	35.4	..	4.7	..	47.1	..	800.5	..	..	2.0	..
United Arab Emirates	–8.2	6.9	1.3	8.8	–4.8	1.7	0.7	..	..	..	..	..
United Kingdom	..	..	..	..	..	..	5.7	3.3	5.8	3.0	4.6	2.6
United States	4.9	6.1	1.1	7.1	0.7	0.7	4.3	2.5	4.2	3.0	3.8	4.0
Uruguay	115.8	33.6	58.5	25.8	28.0	2.3	61.3	49.8	61.1	53.7	62.0	48.7
Uzbekistan	..	..	..	..	..	..	..	546.5	..	..	..	..
Venezuela	71.2	69.1	17.0	33.9	42.8	–20.4	19.3	46.7	20.9	50.6	29.7	48.1
Vietnam	..	24.8	..	14.7	..	6.4	210.8	22.7	..	..	..	..
West Bank and Gaza	..	..	..	..	..	..	..	..	..	..	..	..
Yemen, Rep.	11.3	8.1	1.4	–0.6	10.2	–10.8	..	27.1	..	..	2.6	..
Yugoslavia, FR (Serb./Mont.)	..	..	..	..	..	..	..	..	..	..	..	..
Zambia	52.3	35.0	21.2	22.7	175.3	52.8	42.4	86.8	72.5	93.3	42.8	98.4
Zimbabwe	15.1	33.3	13.5	19.1	5.0	13.3	12.1	26.4	13.8	26.8	14.6	33.5

a. Data prior to 1992 include Eritrea. b. Data prior to 1990 refer to the Federal Republic of Germany before unification.

Monetary indicators and prices | 4.15

Money and the financial accounts that record the supply of money lie at the heart of a country's financial system. There are several commonly used definitions of the money supply. The narrowest, M1, encompasses currency held by the public and demand deposits with banks. M2 includes M1 plus time and savings deposits with banks that require a notice for withdrawal. M3 includes M2 as well as various money market instruments, such as certificates of deposit issued by banks, bank deposits denominated in foreign currency, and deposits with financial institutions other than banks. However defined, money is a liability of the banking system, distinguished from other bank liabilities by the special role it plays as a medium of exchange, a unit of account, and a store of value.

The banking system's assets include its net foreign assets and net domestic credit. Net domestic credit comprises credit to the private sector and credit extended to the nonfinancial public sector in the form of investments in short- and long-term government securities and loans to state enterprises; public and private sector deposits with the banking system are netted out. Domestic credit is the main vehicle through which changes in the money supply are regulated, with central bank lending to the government often playing the most important role. The central bank can regulate lending to the private sector in several ways—for example, by adjusting the cost of the refinancing facilities it provides to banks, by changing market interest rates through open market operations, or by controlling the availability of credit through changes in the reserve requirements imposed on banks and ceilings on the credit provided by banks to the private sector.

Monetary accounts are derived from the balance sheets of financial institutions—the central bank, commercial banks, and nonbank financial intermediaries. Although these balance sheets are usually reliable, they are subject to errors of classification and valuation and differences in accounting practices. For example, whether interest income is recorded on an accrual or a cash basis can make a substantial difference, as can the treatment of nonperforming assets. Valuation errors typically arise with respect to foreign exchange transactions, particularly in countries with flexible exchange rates or in those that have undergone a currency deval-

uation during the reporting period. The valuation of financial derivations can also be difficult.

The quality of commercial bank reporting also may be adversely affected by delays in reports from bank branches, especially in countries where branch accounts are not computerized. Thus the data in the balance sheets of commercial banks may be based on preliminary estimates subject to constant revision. This problem is likely to be even more serious for nonbank financial intermediaries.

Controlling inflation is one of the primary goals of monetary policy. Inflation is measured by the rate of change in a price index. Which index is used depends on which set of prices in the economy is being examined. The GDP deflator, the most general measure of the overall price level, takes into account changes in government costs, inventory appreciation, and investment expenditures. The GDP deflator reflects changes in prices for all the final demand categories, such as government consumption, capital formation, and international trade, as well as the main component, private final consumption. It is usually derived implicitly as the ratio of current to constant price GDP. It may also be calculated explicitly as a Laspeyres price index in which the weights are base period quantities of output.

Consumer price indexes are constructed explicitly, based on surveys of the cost of a defined basket of consumer goods and services. Indexes of consumer prices should be interpreted with caution. The definition of a household and the geographic (urban or rural) and income group coverage of consumer price surveys can vary widely across countries. Moreover, the weights are derived from household expenditure surveys, which for budgetary reasons tend to be conducted infrequently in developing countries, leading to poor comparability over time. Consumer price indexes should be distinguished from retail price indexes, which are used in a few countries. Retail price indexes are based on prices at retail outlets weighted by sales turnover, so the weights may differ by country and over time. In addition, the basket of goods chosen varies by country. Although a useful indicator for measuring consumer price inflation within a country, the consumer price index is of less value in making comparisons across countries. The food price index should be interpreted with similar caution.

• **Money and quasi money** comprise the sum of currency outside banks, demand deposits other than those of the central government, and the time, savings, and foreign currency deposits of resident sectors other than the central government. This definition of the money supply is frequently called M2; it corresponds to lines 34 and 35 in the International Monetary Fund's (IMF) *International Financial Statistics* (IFS). The change in money supply is measured as the difference in end-of-year totals relative to the level of M2 in the preceding year. • **Claims on private sector** (IFS line 32d) include gross credit from the financial system to individuals, enterprises, nonfinancial public entities not included under net domestic credit, and financial institutions not included elsewhere. • **Claims on governments and other public entities** (IFS line 32an + 32b + 32bx + 32c) usually comprise direct credit for specific purposes such as financing of the government budget deficit, loans to state enterprises, advances against future credit authorizations, and purchases of treasury bills and bonds. Public sector deposits with the banking system also include sinking funds for the service of debt and temporary deposits of government revenues. • **GDP implicit deflator** measures the average annual rate of price change in the economy as a whole for the periods shown. The least-squares method is used to calculate the growth rate of the GDP deflator. • **Consumer price index** reflects changes in the cost to the average consumer of acquiring a fixed basket of goods and services. The Laspeyres formula is generally used. • **Food price index** is a subindex of the consumer price index.

 The IMF collects data on the financial systems of its member countries. The data in the table are published in the monthly *International Financial Statistics* and the annual *International Financial Statistics Yearbook*. The World Bank receives data from the IMF in electronic files that may contain more recent revisions than the published sources. GDP data are from the World Bank's national accounts files. The discussion of monetary indicators draws from an IMF publication by Marcello Caiola, *A Manual for Country Economists* (1995).

	Goods and services				Net income		Net current transfers		Current account balance		Gross international reserves	
	Exports $ millions		Imports $ millions		$ millions		$ millions		$ millions		$ millions	
	1980	1996	1980	1996	1980	1996	1980	1996	1980	1996	1980	1996
Albania	378	373	371	1,111	4	72	6	559	16	–107	..	323
Algeria	14,128	13,960	12,311	..	–1,869	..	301	..	249	..	7,064	6,296
Angola	..	5,201	..	3,017	..	–735	..	245	..	–340	..	..
Argentina	9,897	27,032	13,182	27,905	–1,512	–3,591	23	334	–4,774	–4,130	9,297	19,719
Armenia	..	368	..	888	..	44	..	185	..	–291	..	168
Australia	25,752	78,488	27,053	79,450	–2,733	–15,015	–416	107	–4,450	–15,870	6,366	17,452
Austria	26,650	97,310	29,921	100,172	–528	–552	–66	–788	–3,865	–4,202	17,725	26,833
Azerbaijan	..	757	..	1,443	..	–60	..	80	..	–666	..	211
Bangladesh	885	4,508	2,545	7,614	14	–6	802	1,475	–844	–1,637	331	1,869
Belarus	..	6,017	..	6,922	..	–65	..	62	..	–909	..	469
Belgium[a]	70,498	190,732	74,259	179,072	61	6,944	–1,231	–4,217	–4,931	14,387	27,974	22,610
Benin	226	544	421	477	8	–41	151	149	–36	36	15	266
Bolivia	1,030	1,280	833	1,759	–263	–186	60	287	–6	–378	553	1,302
Bosnia and Herzegovina	..	..	..	..	..	..	..	..	..	..	..	..
Botswana	645	2,610	818	2,023	–33	–32	55	–27	–151	342	344	5,098
Brazil	21,869	53,950	27,826	63,293	–7,018	–11,105	144	3,621	–12,830	–24,300	6,875	59,685
Bulgaria	9,302	6,081	7,994	5,813	–412	–395	58	104	954	–23	..	864
Burkina Faso	210	360	577	483	–3	–29	322	255	–49	15	75	343
Burundi	..	50	..	277	..	–9	..	151	..	–6	105	146
Cambodia	..	806	..	1,294	..	–45	..	235	..	–298	..	266
Cameroon	1,792	2,159	1,829	1,857	–628	–595	102	74	–564	–220	206	3
Canada	74,973	234,311	70,399	211,509	–10,764	–20,311	95	318	–6,095	2,808	15,462	21,562
Central African Republic	201	196	327	244	3	–23	81	63	–43	–25	62	232
Chad	71	308	79	411	–4	–7	24	191	12	–38	12	164
Chile	5,968	18,709	7,052	20,086	–1,000	–2,016	113	472	–1,971	–2,921	4,128	15,520
China[†]	23,637	171,680	18,900	154,127	451	–12,437	486	2,129	5,674	7,243	10,091	111,728
Hong Kong, China	..	..	..	..	..	..	..	..	..	..	..	63,833
Colombia	5,328	14,518	5,454	16,878	–245	–2,925	165	532	–206	–4,754	6,474	9,690
Congo, Dem. Rep.	1,658	2,001	1,905	..	–496	..	150	..	–593	..	380	83
Congo, Rep.	1,021	1,584	1,025	2,133	–162	–455	–1	–30	–167	–1,034	93	91
Costa Rica	1,195	3,980	1,661	3,901	–212	–186	15	154	–664	–143	197	1,001
Côte d'Ivoire	3,577	5,110	4,145	4,017	–553	–915	–706	–381	–1,826	–204	46	622
Croatia	..	8,008	..	10,194	..	–45	..	779	..	–1,452	..	2,440
Cuba	..	..	..	..	..	..	..	..	..	..	..	..
Czech Republic	..	29,874	..	33,834	..	–722	..	384	..	–4,299	..	13,085
Denmark	21,989	64,359	21,727	56,166	–1,977	–4,447	–161	–1,826	–1,875	1,920	4,347	14,754
Dominican Republic	1,271	3,936	1,919	4,609	–277	–596	205	1,158	–720	–110	279	357
Ecuador	2,887	5,748	2,946	4,463	–613	–1,282	30	290	–642	293	1,257	2,011
Egypt, Arab Rep.	6,246	15,245	9,157	18,951	–318	539	2,791	3,666	–438	499	2,480	18,296
El Salvador	1,214	2,202	1,170	3,673	–62	–87	52	1,389	34	–322	382	1,110
Eritrea	..	200	..	567	..	–7	..	244	..	–131	..	..
Estonia	..	3,172	..	3,730	..	2	..	109	..	–447	..	640
Ethiopia[b]	569	783	782	1,646	7	–44	80	446	–126	–461	262	733
Finland	16,802	47,815	17,307	38,277	–783	–3,651	–114	–1,098	–1,403	4,790	2,451	7,507
France	153,197	362,954	155,915	331,050	2,680	–5,095	–4,170	–6,297	–4,208	20,511	75,592	57,020
Gabon	2,409	3,398	1,475	1,848	–426	–770	–124	–198	384	100	115	249
Gambia, The	66	220	179	294	–2	–3	28	30	–87	–48	6	102
Georgia	..	479	..	798	..	–87	..	190	..	–216	..	..
Germany[c]	224,224	604,077	225,599	576,283	914	–4,469	–12,858	–36,397	–13,319	–13,072	104,702	118,323
Ghana	1,210	1,728	1,178	2,393	–83	–140	81	482	30	–324	330	930
Greece	8,122	15,238	11,145	25,633	–273	–2,181	1,087	8,022	–2,209	–4,554	3,607	18,782
Guatemala	1,731	2,796	1,960	3,540	–44	–230	110	523	–163	–452	753	948
Guinea	..	761	..	948	..	–93	..	102	..	–177	..	87
Guinea-Bissau	17	23	75	80	–8	–15	–14	46	–80	–26	..	12
Haiti	306	192	481	782	–14	–10	89	463	–101	–138	27	115
Honduras	942	1,775	1,128	1,852	–152	–226	22	243	–317	–201	159	257
†Data for Taiwan, China	21,495	126,126	22,361	121,082	48	2,814	–95	–2,202	–913	5,656	4,055	93,047

Balance of payments current account | 4.16

	Goods and services				Net income		Net current transfers		Current account balance		Gross international reserves	
	Exports $ millions		Imports $ millions		$ millions		$ millions		$ millions		$ millions	
	1980	1996	1980	1996	1980	1996	1980	1996	1980	1996	1980	1996
Hungary	9,671	19,188	9,152	20,342	−1,113	−1,456	63	921	−531	−1,689	..	9,832
India	11,265	43,855	17,378	53,087	356	−4,429	2,860	9,780	−2,897	−3,881	12,010	24,889
Indonesia	23,797	56,130	21,540	53,244	−3,073	−5,778	250	839	−566	−7,023	6,803	19,396
Iran, Islamic Rep.	13,069	18,953	16,111	15,113	606	−478	−2	−4	−2,438	3,358	12,783	..
Iraq	..	..	..	..	..	..	..	..	..	..	..	..
Ireland	9,610	54,063	12,044	46,566	−902	−8,279	1,204	2,184	−2,132	1,402	3,071	8,338
Israel	8,668	28,421	11,511	38,566	−757	−2,491	2,729	6,338	−871	−6,298	4,055	11,418
Italy	97,298	320,752	110,265	257,467	1,278	−14,967	1,101	−7,280	−10,587	41,040	62,428	70,566
Jamaica	1,363	3,275	1,408	3,640	−212	−320	121	535	−136	−245	105	880
Japan	146,980	468,002	156,970	446,679	770	53,553	−1,530	−8,993	−10,750	65,884	38,919	225,594
Jordan	1,181	3,663	2,417	5,420	36	−282	1,481	1,834	281	−206	1,745	2,055
Kazakhstan	..	6,966	..	7,546	..	−222	..	50	..	−752	..	1,961
Kenya	2,007	3,027	2,846	3,441	−194	−221	157	561	−876	−74	539	776
Korea, Dem. Rep.	..	..	..	..	..	..	..	..	..	..	..	..
Korea, Rep.	21,924	155,110	25,687	175,763	−2,102	−2,526	592	119	−5,273	−23,060	3,101	34,158
Kuwait	21,857	16,309	9,823	12,769	4,847	4,916	−1,580	−1,683	15,302	6,773	5,425	4,452
Kyrgyz Republic	..	548	..	950	..	−80	..	78	..	−404	..	5,229
Lao PDR	..	457	..	660	..	−6	..	106	..	−106	..	176
Latvia	..	2,628	..	3,171	..	41	..	87	..	−415	..	746
Lebanon	..	1,413	..	7,596	..	290	..	2,550	..	−3,343	7,025	9,337
Lesotho	90	205	475	874	266	330	175	471	56	108	50	461
Libya	22,084	..	12,671	..	−65	..	−1,134	..	8,214	2,550	14,905	..
Lithuania	..	4,211	..	4,986	..	−91	..	144	..	−723	..	841
Macedonia, FYR	..	1,302	..	1,816	..	−84	..	236	..	−288	..	268
Madagascar	516	803	1,075	1,002	−44	−163	47	210	−556	−153	9	241
Malawi	313	469	487	873	−149	−86	63	124	−260	−450	76	230
Malaysia	14,098	91,387	13,526	86,595	−836	−4,236	−2	148	−266	−7,362	5,755	27,892
Mali	263	535	520	746	−17	−36	150	231	−124	−164	26	438
Mauritania	253	550	449	510	−27	−48	90	76	−133	22	146	145
Mauritius	574	2,701	690	2,767	−23	−40	22	123	−117	17	113	919
Mexico	22,622	106,900	27,601	100,288	−6,277	−13,067	834	4,531	−10,420	−1,923	4,175	19,527
Moldova	..	906	..	1,238	..	−27	..	59	..	−300	..	314
Mongolia	475	481	1,272	521	−11	−25	0	77	−808	39	..	161
Morocco	3,233	9,247	5,207	10,980	−562	−1,309	1,130	2,416	−1,407	−627	814	4,054
Mozambique	399	480	844	1,055	22	−140	56	339	−367	−445	..	344
Myanmar	539	1,120	806	1,669	−48	−101	92	478	−222	−173	409	315
Namibia	..	1,591	..	1,868	..	97	..	263	..	84	..	194
Nepal	224	1,003	365	1,653	13	−3	36	84	−93	−569	272	628
Netherlands	90,380	224,733	91,622	201,317	1,535	3,644	−1,148	−6,647	−855	20,414	37,549	39,607
New Zealand	6,403	18,876	6,934	18,712	−538	−4,665	96	553	−973	−3,948	365	5,953
Nicaragua	495	807	907	1,299	−124	−300	124	357	−411	−435	75	203
Niger	617	315	956	457	−33	−47	97	31	−276	−152	132	83
Nigeria	27,071	14,743	20,014	9,836	−1,304	−2,639	−576	824	5,178	3,092	10,640	4,329
Norway	27,264	63,870	23,749	49,500	−1,922	−1,638	−515	−1,488	1,079	11,246	6,746	26,954
Oman	3,757	7,352	2,298	5,423	−257	−536	−260	−1,659	942	−265	704	1,497
Pakistan	2,958	10,317	5,709	15,174	−281	−1,956	2,163	2,605	−869	−4,208	1,568	1,307
Panama	3,422	7,426	3,394	7,530	−397	−108	40	152	−329	−60	117	867
Papua New Guinea	1,029	2,966	1,322	2,260	−179	−465	184	72	−289	313	458	607
Paraguay	701	3,936	1,314	4,951	−4	306	0.3	42	−618	−668	783	882
Peru	4,631	7,268	3,970	9,947	−909	−1,575	147	647	−101	−3,607	2,804	10,990
Philippines	7,235	34,330	9,166	33,317	−420	3,662	447	880	−1,904	−1,980	3,978	11,747
Poland	16,061	37,390	17,842	41,273	−2,357	−1,075	721	1,694	−3,417	−3,264	574	18,019
Portugal	6,674	33,412	10,136	41,818	−608	−1,078	3,006	6,828	−1,064	−2,657	13,863	21,851
Puerto Rico	..	..	..	..	..	..	..	..	..	..	..	..
Romania	12,087	9,648	13,730	12,503	−777	−309	0	593	−2,420	−2,571	2,511	3,143
Russian Federation	..	102,450	..	86,001	..	−5,213	..	164	..	11,399	..	16,258

	Goods and services				Net income		Net current transfers		Current account balance		Gross international reserves	
	Exports $ millions		Imports $ millions		$ millions		$ millions		$ millions		$ millions	
	1980	1996	1980	1996	1980	1996	1980	1996	1980	1996	1980	1996
Rwanda	165	86	319	363	2	−13	104	291	−48	1	187	155
Saudi Arabia	106,765	60,221	55,793	47,407	526	3,214	−9,995	−15,813	41,503	215	26,129	8,491
Senegal	807	1,588	1,215	1,821	−98	−168	120	382	−386	−58	25	299
Sierra Leone	275	111	471	296	−22	−56	53	47	−165	−89	31	27
Singapore	24,285	156,052	25,312	142,461	−429	1,702	−106	−1,010	−1,563	14,283	6,567	76,847
Slovak Republic	..	10,889	..	13,134	..	−47	..	201	..	−2,090	..	3,895
Slovenia	..	10,497	..	10,674	..	155	..	62	..	39	..	2,297
South Africa	28,627	33,309	22,073	32,716	−3,285	−2,552	239	−74	3,508	−2,033	7,888	2,341
Spain	32,140	146,404	38,004	141,304	−1,362	−5,928	1,646	2,584	−5,580	1,756	20,474	63,699
Sri Lanka	1,293	4,861	2,197	6,074	−26	−203	274	764	−655	−653	283	1,985
Sudan	810	609	1,597	1,341	−70	−868	293	143	−564.1	−1,457	49	107
Sweden	38,151	101,620	39,878	84,809	−1,380	−8,303	−1,224	−2,616	−4,331	5,892	6,996	20,843
Switzerland	48,595	121,738	51,843	109,064	4,186	11,597	−1,140	−3,801	−201	20,470	64,748	69,183
Syrian Arab Republic	2,477	6,131	4,531	6,071	785	−399	1,520	624	251	285	828	..
Tajikistan	..	772	..	808	..	−68	..	20	..	−84	..	..
Tanzania	673	1,363	1,221	2,183	−14	−124	22	20	−540	−924	20	440
Thailand	7,939	71,416	9,996	83,482	−229	−3,385	210	760	−2,076	−14,690	3,026	38,645
Togo	550	490	691	444	−40	−45	86	30	−95	−57	85	93
Trinidad and Tobago	3,139	2,900	2,434	2,110	−306	−390	−42	−4	357	294	2,813	564
Tunisia	3,262	8,151	3,766	8,582	−259	−965	410	860	−353	−536	700	1,689
Turkey	3,621	45,354	8,082	48,331	−1,118	−2,920	2,171	4,447	−3,408	−1,450	3,298	17,819
Turkmenistan	..	1,691	..	1,532	..	..	..	4	..	43	..	..
Uganda	329	726	441	1,601	−7	−46	−2	421	−121	−500	3	528
Ukraine	..	20,346	..	21,468	..	−572	..	509	..	−1,185	..	1,972
United Arab Emirates	..	..	..	..	..	..	..	..	..	..	2,355	8,350
United Kingdom	146,072	339,301	134,200	347,532	−418	15,027	−4,592	−7,247	6,862	−451	31,755	46,700
United States	271,800	848,664	290,730	956,004	29,580	−897	−8,500	−40,489	2,150	−148,726	171,413	160,660
Uruguay	1,526	3,799	2,144	3,962	−100	−206	9	74	−709	−296	2,401	1,892
Uzbekistan	..	4,161	..	5,175	..	−69	..	8	..	−1,075	..	..
Venezuela	19,968	25,258	15,130	14,837	329	−1,735	−439	138	4,728	8,824	13,360	16,020
Vietnam	..	9,695	..	12,870	..	−505	..	1,045	..	−2,636	..	1,324
West Bank and Gaza	..	..	..	..	..	..	..	..	..	..	..	..
Yemen, Rep.	..	2,409	..	3,044	..	−617	..	1,182	..	−70	..	1,036
Yugoslavia, FR (Serb./Mont.)	..	..	..	..	..	..	..	..	..	..	..	..
Zambia	1,609	1,296	1,765	..	−205	..	−155	..	−516	..	206	163
Zimbabwe	1,610	3,092	1,730	2,515	−61	−294	31	40	−149	−425	419	834
World	2,400,597 t	6,689,040 t	2,405,428 t	6,522,540 t	..	..	..	..	..	..	..	..
Low income	80,994	308,987	110,030	340,761	..	..	..	..	..	..	..	..
Excl. China & India	68,382	93,703	83,939	130,133	..	..	..	..	..	..	..	..
Middle income	640,017	1,309,634	575,683	1,365,247	..	..	..	..	..	..	..	..
Lower middle income	..	..	..	..	..	..	..	..	..	..	..	..
Upper middle income	276,851	496,947	223,528	518,764	..	..	..	..	..	..	..	..
Low & middle income	633,124	1,623,100	671,734	1,690,263	..	..	..	..	..	..	..	..
East Asia & Pacific	77,284	447,383	85,129	422,216	..	..	..	..	..	..	..	..
Europe & Central Asia	..	..	..	..	..	..	..	..	..	..	..	..
Latin America & Carib.	121,191	320,894	142,086	316,469	..	..	..	..	..	..	..	..
Middle East & N. Africa	205,272	169,488	148,981	160,504	..	..	..	..	..	..	..	..
South Asia	17,450	65,583	29,271	85,500	..	..	..	..	..	..	..	..
Sub-Saharan Africa	89,966	98,101	83,985	100,832	..	..	..	..	..	..	..	..
High income	1,729,293	5,080,637	1,775,216	4,923,115	..	..	..	..	..	..	..	..

a. Includes Luxembourg. b. Data prior to 1992 include Eritrea. c. Data prior to 1990 refer to the Federal Republic of Germany before unification.

About the data

The balance of payments is divided into two groups of accounts. The current account refers to goods and services, income, and current transfers. The capital and financial account refers to capital transfers, the acquisition or disposal of nonproduced, nonfinancial assets, and financial assets and liabilities. This table presents data from the current account with the addition of gross international reserves from the capital and financial account.

The balance of payments is a double-entry accounting system that shows all flows of goods and services into and out of a country; all transfers that are the counterpart of real resources or financial claims provided to or by the rest of the world without a quid pro quo, such as donations and grants; and all changes in residents' claims on, and liabilities to, nonresidents that arise from economic transactions. All transactions are recorded twice—once as a credit and once as a debit. In principle the net balance should be zero, but in practice the accounts often do not balance. In these cases a balancing item, net errors and omissions, is included in the capital and financial account.

Discrepancies may arise in the balance of payments because there is no single source for balance of payments data and therefore no way to ensure that the data are fully consistent. Sources include customs data, monetary accounts of the banking system, external debt records, information provided by enterprises, surveys to estimate service transactions, and foreign exchange records. Differences in collection methods—such as in timing, definitions of residence and ownership, and the exchange rate used to value transactions—contribute to net errors and omissions. In addition, smuggling and other illegal or quasi-legal transactions may be unrecorded or misrecorded.

The concepts and definitions underlying the data here are based on the fifth edition of the International Monetary Fund's (IMF) *Balance of Payments Manual* (1993). The fifth edition redefined as capital transfers some transactions previously included in the current account, such as debt forgiveness, migrants' capital transfers, and foreign aid to acquire capital goods. Thus the current account balance now reflects more accurately net current transfer receipts in addition to transactions in goods, services (previously nonfactor services), and income (previously factor income). Many countries maintain their data collection systems according to the fourth edition. Where necessary, the IMF converts data reported in earlier systems to conform with the fifth edition (see *Primary data documentation*). Values are in U.S. dollars converted at market exchange rates.

Figure 4.16a

Current account balances for the three biggest traders

$ billions

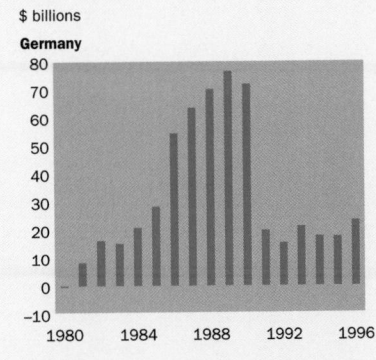

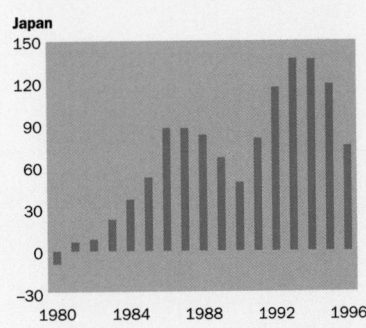

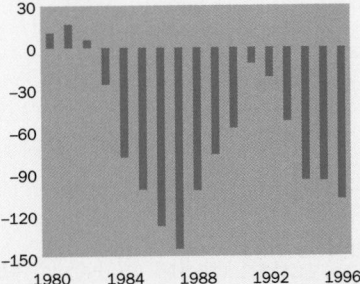

Source: International Monetary Fund, balance of payments data files.

The United States has run deficits and Japan and Germany have run surpluses for most of the past two decades. Most other countries exhibit greater variability in their current account balances, which respond to changes in demand and the terms of trade.

Definitions

• **Exports and imports of goods and services** comprise all transactions between residents of a country and the rest of the world involving a change in ownership of general merchandise, goods sent for processing and repairs, nonmonetary gold, and services. • **Net income** refers to employee compensation paid to nonresident workers and investment income (receipts and payments on direct investment, portfolio investment, other investments, and receipts on reserve assets). Income derived from the use of intangible assets is recorded under business services. • **Net current transfers** are recorded in the balance of payments whenever an economy provides or receives goods, services, income, or financial items without a quid pro quo. All transfers not considered to be capital are current. • **Current account balance** is the sum of net exports of goods and services, income, and current transfers. • **Gross international reserves** comprise holdings of monetary gold, special drawing rights, reserves of IMF members held by the IMF, and holdings of foreign exchange under the control of monetary authorities. The gold component of these reserves is valued at year-end (December 31) London prices ($589.50 an ounce in 1980 and $369.25 an ounce in 1996).

Data sources

More information about the design and compilation of the balance of payments can be found in the IMF's *Balance of Payments Manual*, fifth edition (1993), *Balance of Payments Textbook* (1996a), and *Balance of Payments Compilation Guide* (1995). The data come from the IMF's balance of payments database, *Balance of Payments Statistics,* and *International Financial Statistics* (IFS). The World Bank exchanges data with the IMF through electronic files that in most cases are more timely and cover a longer period than the published sources. The IFS is also available on CD-ROM.

4.17 External debt

	Total external debt ($ millions)		Long-term debt ($ millions)		Public and publicly guaranteed debt				Private nonguaranteed external debt ($ millions)		Use of IMF credit ($ millions)	
					Total ($ millions)		IBRD loans and IDA credits ($ millions)					
	1980	1996	1980	1996	1980	1996	1980	1996	1980	1996	1980	1996
Albania	..	781	..	673	..	673	..	137	..	0	..	54
Algeria	19,365	33,260	17,040	30,808	17,040	30,808	253	1,939	0	0	0	2,031
Angola	..	10,612	..	9,400	..	9,400	0	115	..	0	..	0
Argentina	27,157	93,841	16,774	75,348	10,181	62,392	404	5,372	6,593	12,956	0	6,293
Armenia	..	552	..	434	..	434	..	184	..	0	..	117
Australia	..	..	..	..	..	..	..	..	..	..	..	..
Austria	..	..	..	..	..	..	..	..	..	..	..	..
Azerbaijan	..	435	..	245	..	245	..	64	..	0	..	175
Bangladesh	4,230	16,083	3,594	15,403	3,594	15,403	981	5,759	0	0	424	517
Belarus	..	1,071	..	695	..	665	..	121	..	30	..	274
Belgium	..	..	..	..	..	..	..	..	..	..	..	..
Benin	424	1,594	334	1,449	334	1,449	52	520	0	0	16	99
Bolivia	2,702	5,174	2,274	4,523	2,182	4,238	239	904	92	285	126	276
Bosnia and Herzegovina	..	..	..	..	..	..	..	..	..	..	..	45
Botswana	147	613	143	607	143	607	66	80	0	0	0	0
Brazil	71,520	179,047	57,981	143,541	41,375	94,587	2,035	5,876	16,605	48,953	0	68
Bulgaria	..	9,819	..	8,334	..	8,138	..	453	..	196	..	586
Burkina Faso	330	1,294	281	1,160	281	1,160	77	636	0	0	15	81
Burundi	166	1,127	118	1,081	118	1,081	37	588	0	0	36	38
Cambodia	..	2,111	..	2,023	..	2,023	0	108	..	0	0	69
Cameroon	2,588	9,515	2,251	8,184	2,073	8,001	298	1,033	178	183	59	72
Canada	..	..	..	..	..	..	..	..	..	..	..	..
Central African Republic	195	928	147	844	147	844	29	422	0	0	24	28
Chad	284	997	259	914	259	914	36	433	0	0	14	65
Chile	12,081	27,411	9,399	20,421	4,705	4,890	184	1,103	4,693	15,531	123	0
China	4,504	128,817	4,504	103,410	4,504	102,260	0	15,195	0	1,150	0	0
Hong Kong, China	..	..	..	..	..	..	..	..	..	..	..	..
Colombia	6,941	28,859	4,604	22,975	4,089	14,814	1,012	2,187	515	8,162	0	0
Congo, Dem. Rep.	4,770	12,826	4,071	9,262	4,071	9,262	246	1,370	0	0	373	433
Congo, Rep.	1,526	5,240	1,257	4,665	1,257	4,665	61	253	0	0	22	38
Costa Rica	2,744	3,454	2,112	3,082	1,700	2,889	183	248	412	193	57	1
Côte d'Ivoire	7,462	19,713	6,339	14,720	4,327	11,367	314	2,323	2,012	3,353	65	503
Croatia	..	4,634	..	3,960	..	3,101	..	195	..	859	..	209
Cuba	..	..	..	..	..	..	..	..	..	..	..	..
Czech Republic	..	20,094	..	14,145	..	12,017	..	435	..	2,128	..	0
Denmark	..	..	..	..	..	..	..	..	..	..	..	..
Dominican Republic	2,002	4,310	1,473	3,520	1,220	3,515	83	261	254	5	49	96
Ecuador	5,997	14,491	4,422	12,755	3,300	12,435	146	1,005	1,122	320	0	145
Egypt, Arab Rep.	19,131	31,407	14,693	29,045	14,428	28,918	728	2,165	265	127	411	16
El Salvador	911	2,894	659	2,298	499	2,297	114	302	161	2	32	0
Eritrea	..	46	..	46	..	46	..	27	..	0	..	0
Estonia	..	405	..	220	..	217	..	62	..	4	..	78
Ethiopia[a]	824	10,077	688	9,483	688	9,483	304	1,555	0	0	79	92
Finland	..	..	..	..	..	..	..	..	..	..	..	..
France	..	..	..	..	..	..	..	..	..	..	..	..
Gabon	1,514	4,213	1,272	3,874	1,272	3,874	19	92	0	0	15	120
Gambia, The	137	452	97	412	97	412	16	166	0	0	16	18
Georgia	..	1,356	..	1,100	..	1,100	..	157	..	0	..	192
Germany	..	..	..	..	..	..	..	..	..	..	..	..
Ghana	1,398	6,202	1,162	4,955	1,152	4,684	213	2,574	10	271	105	543
Greece	..	..	..	..	..	..	..	..	..	..	..	..
Guatemala	1,166	3,785	831	2,887	549	2,766	144	200	282	121	0	0
Guinea	1,133	3,240	1,019	2,981	1,019	2,981	87	863	0	0	35	82
Guinea-Bissau	140	937	133	856	133	856	5	216	0	0	1	8
Haiti	303	897	242	836	242	836	66	442	0	0	46	25
Honduras	1,473	4,453	1,168	3,981	976	3,855	216	774	191	126	33	58

External debt 4.17

	Total external debt		Long-term debt		Public and publicly guaranteed debt				Private nonguaranteed external debt		Use of IMF credit	
					Total		IBRD loans and IDA credits					
	$ millions		$ millions		$ millions		$ millions		$ millions		$ millions	
	1980	1996	1980	1996	1980	1996	1980	1996	1980	1996	1980	1996
Hungary	9,764	26,958	6,416	23,428	6,416	18,423	0	1,650	0	5,005	0	171
India	20,581	89,827	18,333	81,788	17,997	74,406	5,969	26,384	336	7,382	977	1,313
Indonesia	20,938	129,033	18,163	96,803	15,021	60,108	1,606	11,874	3,142	36,694	0	0
Iran, Islamic Rep.	4,500	21,183	4,500	16,153	4,500	15,917	622	387	0	236	0	0
Iraq	..	..	..	..	..	..	..	..	..	..	..	..
Ireland	..	..	..	..	..	..	..	..	..	..	..	..
Israel	..	..	..	..	..	..	..	..	..	..	..	..
Italy	..	..	..	..	..	..	..	..	..	..	..	..
Jamaica	1,913	4,041	1,505	3,306	1,430	3,183	176	515	75	123	309	161
Japan	..	..	..	..	..	..	..	..	..	..	..	..
Jordan	1,971	8,118	1,486	7,182	1,486	7,137	102	844	0	45	0	340
Kazakhstan	..	2,920	..	2,147	..	1,932	..	490	..	215	..	552
Kenya	3,383	6,893	2,489	6,022	2,052	5,647	528	2,375	437	375	254	337
Korea, Dem. Rep.	..	..	..	..	..	..	..	..	..	..	..	..
Korea, Rep.	..	..	..	..	..	..	..	..	..	..	..	..
Kuwait	..	..	..	..	..	..	..	..	..	..	..	..
Kyrgyz Republic	..	789	..	640	..	640	..	197	..	0	..	140
Lao PDR	350	2,263	333	2,186	333	2,186	6	335	0	0	16	67
Latvia	..	472	..	298	..	298	..	75	..	0	..	130
Lebanon	510	3,996	216	2,343	216	1,933	27	132	0	410	0	0
Lesotho	72	654	58	612	58	612	24	216	0	0	6	34
Libya	..	..	..	..	..	..	..	..	..	..	..	..
Lithuania	..	1,286	..	856	..	792	..	101	..	64	..	273
Macedonia, FYR	..	1,659	..	1,387	..	863	..	203	..	524	..	68
Madagascar	1,249	4,175	919	3,589	919	3,589	152	1,153	0	0	87	73
Malawi	831	2,312	635	2,092	635	2,092	156	1,388	0	0	80	119
Malaysia	6,611	39,777	5,256	28,709	4,008	15,701	504	907	1,248	13,008	0	0
Mali	727	3,020	664	2,776	664	2,776	121	915	0	0	39	165
Mauritania	843	2,363	717	2,073	717	2,073	38	368	0	0	62	107
Mauritius	467	1,818	318	1,399	294	1,153	55	140	24	246	102	0
Mexico	57,378	157,125	41,215	113,778	33,915	93,438	2,063	12,568	7,300	20,340	0	13,279
Moldova	..	834	..	560	..	560	..	142	..	0	..	248
Mongolia	..	524	..	474	..	474	..	68	..	0	..	44
Morocco	9,247	21,767	8,013	21,165	7,863	20,774	578	3,764	150	392	457	3
Mozambique	..	5,842	..	5,475	..	5,433	0	1,076	..	43	0	181
Myanmar	1,500	5,184	1,390	4,804	1,390	4,804	146	742	0	0	106	0
Namibia	..	..	..	..	..	..	..	..	..	..	..	..
Nepal	205	2,414	156	2,349	156	2,349	76	1,049	0	0	42	39
Netherlands	..	..	..	..	..	..	..	..	..	..	..	..
New Zealand	..	..	..	..	..	..	..	..	..	..	..	..
Nicaragua	2,189	5,929	1,668	5,122	1,668	5,122	135	379	0	0	49	29
Niger	863	1,557	687	1,460	383	1,350	66	609	305	110	16	53
Nigeria	8,921	31,407	5,368	25,731	4,271	25,431	554	3,110	1,097	300	0	0
Norway	..	..	..	..	..	..	..	..	..	..	..	..
Oman	599	3,415	436	2,649	436	2,646	14	19	0	3	0	0
Pakistan	9,931	29,901	8,520	25,690	8,502	23,694	1,151	6,486	18	1,995	674	1,396
Panama	2,975	6,990	2,271	5,211	2,271	5,136	133	199	0	75	23	131
Papua New Guinea	719	2,359	624	2,275	486	1,522	110	375	139	752	31	51
Paraguay	955	2,141	780	1,398	630	1,377	124	172	151	21	0	0
Peru	9,386	29,176	6,828	21,793	6,218	20,415	359	1,633	610	1,378	474	924
Philippines	17,417	41,214	8,817	32,839	6,363	27,937	960	4,859	2,454	4,902	1,044	405
Poland	..	40,895	..	40,819	..	39,217	0	2,175	..	1,602	0	0
Portugal	..	..	..	..	..	..	..	..	..	..	..	..
Puerto Rico	..	..	..	..	..	..	..	..	..	..	..	..
Romania	9,762	8,291	7,131	6,825	7,131	6,456	807	1,009	0	369	328	651
Russian Federation	..	124,785	..	100,463	..	100,463	0	2,509	..	0	0	12,508

4.17 External debt

	Total external debt $ millions		Long-term debt $ millions		Public and publicly guaranteed debt				Private nonguaranteed external debt $ millions		Use of IMF credit $ millions	
					Total $ millions		IBRD loans and IDA credits $ millions					
	1980	1996	1980	1996	1980	1996	1980	1996	1980	1996	1980	1996
Rwanda	190	1,034	150	977	150	977	58	536	0	0	14	24
Saudi Arabia	..	..	..	..	..	..	..	..	..	..	..	..
Senegal	1,473	3,663	1,114	3,142	1,105	3,103	156	1,217	9	39	140	326
Sierra Leone	469	1,167	357	892	357	892	43	260	0	0	59	171
Singapore	..	..	..	..	..	..	..	..	..	..	..	..
Slovak Republic	..	7,704	..	4,437	..	3,891	0	250	..	546	0	319
Slovenia	..	4,031	..	3,972	..	2,038	0	155	..	1,935	0	1
South Africa	..	23,590	..	13,907	..	10,348	0	0	..	3,559	0	884
Spain	..	..	..	..	..	..	..	..	..	..	..	..
Sri Lanka	1,841	7,995	1,231	6,898	1,227	6,818	129	1,556	3	80	391	531
Sudan	5,177	16,972	4,147	9,865	3,822	9,369	236	1,250	325	496	431	893
Sweden	..	..	..	..	..	..	..	..	..	..	..	..
Switzerland	..	..	..	..	..	..	..	..	..	..	..	..
Syrian Arab Republic	3,552	21,420	2,921	16,698	2,921	16,698	257	426	0	0	0	0
Tajikistan	..	707	..	672	..	672	..	30	..	0	..	22
Tanzania	2,452	7,412	1,963	6,149	1,879	6,104	440	2,298	84	45	171	206
Thailand	8,297	90,824	5,646	53,210	3,943	17,039	703	1,707	1,703	36,171	348	0
Togo	1,049	1,463	896	1,285	896	1,285	47	576	0	0	33	90
Trinidad and Tobago	829	2,242	713	1,949	713	1,871	57	79	0	78	0	24
Tunisia	3,526	9,887	3,390	8,877	3,210	8,689	337	1,657	180	188	0	237
Turkey	19,131	79,789	15,575	58,591	15,040	48,172	1,347	4,385	535	10,419	1,054	662
Turkmenistan	..	825	..	538	..	538	..	3	..	0	..	0
Uganda	689	3,674	537	3,151	537	3,151	47	1,849	0	0	89	417
Ukraine	..	9,335	..	6,629	..	6,451	..	859	..	178	..	2,262
United Arab Emirates	..	..	..	..	..	..	..	..	..	..	..	..
United Kingdom	..	..	..	..	..	..	..	..	..	..	..	..
United States	..	..	..	..	..	..	..	..	..	..	..	..
Uruguay	1,660	5,899	1,338	4,232	1,127	4,097	72	446	211	135	0	9
Uzbekistan	..	2,319	..	1,990	..	1,990	..	155	..	0	..	238
Venezuela	29,345	35,344	13,795	30,266	10,614	28,452	133	1,408	3,181	1,814	0	2,196
Vietnam	..	26,764	..	22,344	..	22,344	2	412	..	0	..	539
West Bank and Gaza	..	..	..	..	..	..	..	..	..	..	..	..
Yemen, Rep.	1,684	6,356	1,453	5,622	1,453	5,622	137	893	0	0	48	121
Yugoslavia, FR (Serb./Mont.)[b]	18,486	13,439	15,586	11,239	4,581	8,480	1,359	1,178	11,005	2,759	760	81
Zambia	3,261	7,113	2,227	5,323	2,141	5,307	348	1,510	87	16	447	1,198
Zimbabwe	786	5,005	696	3,766	696	3,338	3	848	0	428	0	437

World	.. S	.. S	.. S	.. S	.. S	.. S	.. S	.. S	.. S	.. S	.. S	.. S
Low income	106,308	537,017	87,607	451,688	82,516	435,296	14,028	97,656	5,091	16,392	5,803	13,097
Excl. China & India	81,223	318,374	64,770	266,490	60,015	258,631	8,060	56,077	4,755	7,860	4,826	11,784
Middle income [c]	497,014	1,558,411	357,693	1,198,409	294,517	961,800	18,179	82,958	63,176	236,609	5,761	47,010
Lower middle income	251,877	881,547	191,656	685,473	165,166	578,201	11,413	50,277	26,491	107,273	5,516	25,626
Upper middle income [c]	245,137	676,863	166,037	512,935	129,351	383,599	6,766	32,682	36,686	129,336	245	21,384
Low & middle income	603,321	2,095,428	445,300	1,650,097	377,032	1,397,096	32,208	180,615	68,268	253,000	11,564	60,107
East Asia & Pacific	64,600	477,219	48,438	356,170	39,688	263,394	4,077	36,704	8,751	92,776	1,551	1,175
Europe & Central Asia [c]	75,503	370,172	56,283	297,042	44,743	270,109	3,513	18,070	11,540	26,933	2,143	20,053
Latin America & Carib.	257,263	656,388	187,253	517,632	144,795	406,990	8,134	36,400	42,458	110,642	1,413	23,892
Middle East & N. Africa	83,793	212,389	61,734	162,197	61,139	158,468	3,053	12,226	595	3,729	916	2,748
South Asia	38,015	152,098	33,053	137,971	32,696	128,513	8,306	41,295	357	9,458	2,508	3,795
Sub-Saharan Africa	84,148	227,163	58,539	179,085	53,973	169,621	5,125	35,921	4,567	9,463	3,033	8,445
High income	..	..	..	..	..	..	..	..	..	..	..	..

a. Data prior to 1992 include Eritrea. b. Data refer to the former Yugoslavia. c. Includes data for Gibraltar not included in other tables.

About the data

Data on the external debt of low- and middle-income economies are gathered by the World Bank through its Debtor Reporting System. World Bank staff calculate the indebtedness of developing countries using loan-by-loan reports submitted by these countries on long-term public and publicly guaranteed borrowing, along with information on short-term debt collected by the countries or from creditors through the reporting systems of the Bank for International Settlements and the Organisation for Economic Co-operation and Development (OECD). These data are supplemented by information on loans and credits from major multilateral banks and loan statements from official lending agencies in major creditor countries and by estimates from World Bank country economists and International Monetary Fund (IMF) desk officers. In addition, some countries provide data on private nonguaranteed debt. In 1996, 34 countries reported their private nonguaranteed debt to the World Bank; estimates were made for 28 additional countries known to have significant private debt. For estimates of total financial flows to developing countries see table 6.8.

Despite an ongoing effort to standardize the reporting of external debt (see, for example, International Working Group of External Debt Compilers 1987), the coverage, quality, and timeliness of debt data vary across countries. Coverage varies for both debt instruments and borrowers. With a widening spectrum of debt instruments and investors and the expansion of private nonguaranteed borrowing, comprehensive coverage of long-term external debt becomes more complex. Reporting countries differ in their ability to monitor debt, especially private nonguaranteed debt. Even public and publicly guaranteed debt is affected by coverage and accuracy in reporting—again, because of monitoring capacity and, sometimes, willingness to provide information. A key part that is often underreported is military debt.

Variations in reporting rescheduled debt also affect cross-country comparability. For example, rescheduling under the auspices of the Paris Club of official creditors may be subject to lags between the completion of the general rescheduling agreement and the completion of the specific, bilateral agreements that define the terms of the rescheduled debt. The World Bank estimates the effects of the general agreements and then revises the data when countries report their bilateral agreements. Other areas of inconsistency include country differences in treatment of arrears, reporting of debt owed to the Russian Federation, and treatment of nonresident national deposits denominated in foreign currency.

Definitions

• **Total external debt** is debt owed to nonresidents repayable in foreign currency, goods, or services. Total external debt is the sum of public, publicly guaranteed, and private nonguaranteed long-term debt, use of IMF credit, and short-term debt. Short-term debt includes all debt having an original maturity of one year or less and interest in arrears on long-term debt. • **Long-term debt** is debt that has an original or extended maturity of more than one year. It has three components: public, publicly guaranteed, and private nonguaranteed debt. • **Public and publicly guaranteed debt** comprises long-term external obligations of public debtors, including the national government, political subdivisions (or an agency of either), and autonomous public bodies, and external obligations of private debtors that are guaranteed for repayment by a public entity. • **IBRD loans and IDA credits** are extended by the World Bank Group. The International Bank for Reconstruction and Development (IBRD) lends at market rates. Credits from the International Development Association (IDA) are at concessional rates. • **Private nonguaranteed external debt** comprises long-term external obligations of private debtors that are not guaranteed for repayment by a public entity. • **Use of IMF credit** denotes repurchase obligations to the IMF for all uses of IMF resources (excluding those resulting from drawings on the reserve tranche). These obligations, shown for the end of the year specified, comprise purchases outstanding under the credit tranches, including enlarged access resources, and all special facilities (the buffer stock, compensatory financing, extended fund, and oil facilities), trust fund loans, and operations under the structural adjustment and enhanced structural adjustment facilities.

Data sources

The main sources of external debt information are reports to the World Bank through its Debtor Reporting System from member countries that have received IBRD loans or IDA credits. Additional information has been drawn from the files of the World Bank and the IMF. Summary tables of the external debt of developing countries are published annually in the World Bank's *Global Development Finance* and *Global Development Finance* CD-ROM.

4.18 | External debt management

	Present value of debt		Total debt service				Public and publicly guaranteed debt service		Short-term debt	
	% of GNP	% of exports of goods and services	% of GNP		% of exports of goods and services		% of central government current revenue		% of total debt	
	1996	1996	1980	1996	1980	1996	1980	1996	1980	1996
Albania	32	101	..	1.2	..	3.5	..	..	..	7.0
Algeria	71	228	9.9	9.7	27.4	27.7	..	..	12.0	1.3
Angola	310	219	..	20.1	..	13.3	..	..	0.0	11.4
Argentina	31	323	5.5	4.8	37.3	44.2	6.1	..	38.2	13.0
Armenia	27	114	..	3.0	..	10.7	..	..	..	0.3
Australia	..	..	..	..	..	..	..	..	..	..
Austria	..	..	..	..	..	..	..	..	..	..
Azerbaijan	10	45	..	0.3	..	1.3	..	..	..	3.6
Bangladesh	30	166	2.1	2.2	23.7	11.7	7.9	..	5.0	1.0
Belarus	4	21	..	0.6	..	2.0	..	..	..	9.5
Belgium	..	..	..	..	..	..	..	..	..	..
Benin	57[a]	215[a]	1.4	2.0	6.3	6.8	..	..	17.3	2.9
Bolivia	57[a]	270[a]	12.6	6.5	35.0	30.9	..	23.9	11.2	7.2
Bosnia and Herzegovina	53	408	..	..	..	..	..	..	..	..
Botswana	11	17	1.6	3.1	2.1	4.9	4.2	..	2.7	1.0
Brazil	26	292	6.5	3.4	63.3	41.1	15.3	..	18.9	19.8
Bulgaria	89	151	0.2	14.1	0.5	20.5	..	29.2	0.0	9.2
Burkina Faso	31[a]	241[a]	1.3	1.9	5.9	10.8	8.4	..	10.6	4.1
Burundi	47	538	0.9	2.7	..	54.6	4.8	13.8	7.2	0.7
Cambodia	54	191	..	0.3	..	1.2	..	..	14.3	0.9
Cameroon	106	399	4.6	6.3	15.3	23.6	17.0	..	10.7	13.2
Canada	..	..	..	..	..	..	..	..	..	..
Central African Republic	51	242	1.3	1.2	4.9	6.3	..	..	12.6	6.0
Chad	51	181	0.8	2.7	8.3	9.5	..	..	4.0	1.8
Chile	48	166	10.2	8.7	43.1	32.3	15.6	21.7	21.2	25.5
China	17	76	0.5	2.0	..	8.7	..	..	0.0	19.7
Hong Kong, China	..	..	..	..	..	..	..	..	..	..
Colombia	40	206	2.9	6.6	16.0	34.6	13.2	..	33.7	20.4
Congo, Dem. Rep.	127	693	3.8	0.8	..	2.4	29.2	..	6.8	24.4
Congo, Rep.	260	342	7.1	18.0	10.6	21.3	11.7	..	16.2	10.2
Costa Rica	37	83	7.7	6.5	29.1	14.1	23.9	..	21.0	10.8
Côte d'Ivoire	171[a]	299[a]	14.6	13.8	38.7	26.2	37.4	..	14.2	22.8
Croatia	24	56	..	2.4	..	5.5	..	2.9	..	10.0
Cuba	..	..	..	..	..	..	..	..	..	..
Czech Republic	42	70	1.6	4.8	..	8.3	..	10.0	100.0	29.6
Denmark	..	..	..	..	..	..	..	..	..	..
Dominican Republic	33	108	5.9	3.5	25.3	11.4	16.3	..	24.0	16.1
Ecuador	78	246	9.0	7.4	33.9	22.6	37.2	..	26.3	11.0
Egypt, Arab Rep.	35	117	5.8	3.4	13.4	11.6	..	..	21.1	7.5
El Salvador	26	78	2.7	3.0	7.5	9.5	10.3	23.6	24.1	20.6
Eritrea	3	6	..	..	..	0.0	..	..	..	0.0
Estonia	9	14	..	1.0	..	1.3	..	1.4	..	26.4
Ethiopia[b]	149	1,093	..	5.8	7.6	42.2	4.6	..	6.9	5.0
Finland	..	..	..	..	..	..	..	..	..	..
France	..	..	..	..	..	..	..	..	..	..
Gabon	86	123	11.2	7.8	17.7	11.1	26.2	..	15.1	5.2
Gambia, The	64	113	1.9	..	6.3	12.7	1.4	..	17.0	5.0
Georgia	26	209	..	0.3	..	..	..	..	..	4.7
Germany	..	..	..	..	..	..	..	..	..	..
Ghana	56[a]	208[a]	3.6	7.6	13.1	26.4	10.0	..	9.4	11.4
Greece	..	..	..	..	..	..	..	..	..	..
Guatemala	23	110	1.8	2.3	7.9	11.0	6.0	..	28.7	23.7
Guinea	61	298	..	3.0	..	14.7	..	..	7.0	5.5
Guinea-Bissau	248	2,312	4.5	4.2	..	48.7	..	..	3.7	7.8
Haiti	20	297	1.8	1.0	6.2	13.8	13.2	..	4.6	4.0
Honduras	92	200	8.5	14.1	21.4	28.8	26.0	..	18.5	9.3

External debt management | 4.18

	Present value of debt		Total debt service				Public and publicly guaranteed debt service		Short-term debt	
	% of GNP	% of exports of goods and services	% of GNP		% of exports of goods and services		% of central government current revenue		% of total debt	
	1996	1996	1980	1996	1980	1996	1980	1996	1980	1996
Hungary	62	158	8.8	19.3	..	41.0	..	..	34.3	12.5
India	22	152	0.8	3.6	9.3	24.1	5.6	21.6	6.2	7.5
Indonesia	64	236	4.1	9.9	..	36.8	10.6	30.3	13.3	25.0
Iran, Islamic Rep.	15	99	1.0	2.4	6.8	..	4.7	8.8	..	23.7
Iraq	..	..	..	..	..	..	..	..	..	..
Ireland	..	..	..	..	..	..	..	..	..	..
Israel	..	..	..	..	..	..	..	..	..	..
Italy	..	..	..	..	..	..	..	..	..	..
Jamaica	92	101	11.4	15.9	19.0	18.0	26.6	..	5.1	14.2
Japan	..	..	..	..	..	..	..	..	..	..
Jordan	110	148	5.2	9.2	11.2	12.3	25.7	..	24.6	7.4
Kazakhstan	14	48	..	3.4	..	9.9	..	..	..	7.6
Kenya	64	177	6.2	9.4	21.0	27.5	14.5	..	18.9	7.7
Korea, Dem. Rep.	..	..	..	..	..	..	..	..	..	..
Korea, Rep.	..	..	..	..	..	..	..	..	..	..
Kuwait	..	..	..	..	..	..	..	..	..	..
Kyrgyz Republic	37	130	..	3.1	..	9.2	..	..	..	1.1
Lao PDR	45	177	..	1.6	..	6.3	..	..	0.4	0.5
Latvia	9	20	..	1.3	..	2.3	..	2.0	..	9.4
Lebanon	33	90	..	2.3	..	6.4	..	..	57.6	41.4
Lesotho	33	69	0.9	2.9	1.5	6.1	..	..	11.1	1.3
Libya	..	..	..	..	..	..	..	..	..	..
Lithuania	16	35	..	1.6	..	2.9	..	4.2	..	12.2
Macedonia, FYR	74	106	..	2.6	..	3.9	..	..	..	12.3
Madagascar	97	426	2.6	1.9	20.3	9.4	13.8	16.0	19.5	12.3
Malawi	76 [a]	294 [a]	7.7	4.1	27.8	18.6	28.9	..	14.0	4.3
Malaysia	52	50	4.0	8.1	6.3	8.2	5.8	16.8	20.5	27.8
Mali	56 [a]	261 [a]	1.0	4.5	5.1	17.9	5.3	..	3.3	2.6
Mauritania	157	318	7.1	11.6	17.3	21.7	..	..	7.7	7.8
Mauritius	45	73	4.7	4.7	9.1	7.2	14.7	17.3	10.1	23.1
Mexico	44	154	5.8	12.7	44.4	35.4	26.9	..	28.2	19.1
Moldova	39	92	..	3.2	..	6.2	..	..	..	3.2
Mongolia	36	65	..	5.0	..	9.7	..	15.9	..	1.3
Morocco	61	185	7.9	8.9	33.4	27.7	27.5	..	8.4	2.7
Mozambique	411 [a]	1,344 [a]	..	11.3	..	32.3	..	..	0.0	3.2
Myanmar	34	296	..	..	25.4	..	11.9	..	0.3	7.3
Namibia	..	..	..	..	..	..	..	..	..	..
Nepal	26	102	0.4	1.9	3.2	7.7	2.6	14.3	3.4	1.1
Netherlands	..	..	..	..	..	..	..	..	..	..
New Zealand	..	..	..	..	..	..	..	..	..	..
Nicaragua	322 [a]	763 [a]	5.7	13.2	22.3	24.2	26.3	..	21.6	13.1
Niger	45 [a]	284 [a]	5.6	2.9	21.7	17.3	10.6	..	18.5	2.8
Nigeria	114	240	1.9	8.1	4.1	16.0	..	..	39.8	18.1
Norway	..	..	..	..	..	..	..	..	..	..
Oman	31	48	4.7	..	6.4	9.9	9.7	17.4	27.2	22.4
Pakistan	39	206	3.7	5.1	18.3	27.4	15.4	..	7.4	9.4
Panama	80	69	14.4	11.8	6.2	10.7	48.3	..	22.9	23.6
Papua New Guinea	37	61	6.0	7.9	13.8	12.6	10.5	..	8.9	1.4
Paraguay	22	47	3.1	2.5	18.6	5.5	16.0	..	18.3	34.7
Peru	43	318	10.9	4.9	44.5	35.4	42.7	24.3	22.2	22.1
Philippines	51	116	6.7	6.6	26.6	13.7	13.1	28.8	43.4	19.3
Poland	31	102	..	1.9	..	6.4	..	4.1	..	0.2
Portugal	..	..	..	..	..	..	..	..	..	..
Puerto Rico	..	..	..	..	..	..	..	..	..	..
Romania	23	89	..	3.5	12.6	12.6	7.5	..	23.6	9.8
Russian Federation	25	97	..	1.6	..	6.6	..	..	..	9.5

	Present value of debt		Total debt service				Public and publicly guaranteed debt service		Short-term debt	
	% of GNP	% of exports of goods and services	% of GNP		% of exports of goods and services		% of central government current revenue		% of total debt	
	1996	1996	1980	1996	1980	1996	1980	1996	1980	1996
Rwanda	47	682	0.7	1.4	4.2	20.3	2.9	..	13.7	3.2
Saudi Arabia	..	..	..	..	..	..	..	..	..	..
Senegal	53	150	8.9	5.4	28.7	15.9	30.1	..	14.9	5.3
Sierra Leone	78	515	5.6	6.4	23.8	52.6	22.7	70.8	11.3	9.0
Singapore	..	..	..	..	..	..	..	..	..	..
Slovak Republic	41	66	..	7.0	..	11.9	..	..	..	38.3
Slovenia	21	36	..	5.1	..	8.7	..	..	..	1.4
South Africa	18	67	..	3.1	..	11.1	..	..	..	37.3
Spain	..	..	..	..	..	..	..	..	..	..
Sri Lanka	41	97	4.5	3.1	12.0	7.3	10.3	13.0	11.9	7.1
Sudan	260	1,964	3.9	..	25.5	5.0	9.3	..	11.6	36.6
Sweden	..	..	..	..	..	..	..	..	..	..
Switzerland	..	..	..	..	..	..	..	..	..	..
Syrian Arab Republic	120	301	2.9	1.5	11.4	3.8	8.6	..	17.8	22.0
Tajikistan	24	69	..	0.0	..	0.1	..	..	..	1.9
Tanzania[c]	114	499	..	4.5	23.5	18.7	8.1	..	13.0	14.3
Thailand	56	131	5.0	4.8	18.9	11.5	9.5	6.3	27.8	41.4
Togo	80	191	4.8	4.0	9.0	10.8	11.0	..	11.5	6.1
Trinidad and Tobago	46	80	3.9	9.5	6.8	15.6	8.4	..	14.0	12.0
Tunisia	53	106	6.4	8.0	14.8	16.5	15.6	23.1	3.9	7.8
Turkey	47	184	2.3	5.9	28.0	21.7	8.5	25.2	13.1	25.7
Turkmenistan	18	39	..	4.1	..	10.6	..	..	..	34.8
Uganda	32[a]	294[a]	4.6	2.5	17.3	20.0	6.7	..	9.1	2.9
Ukraine	18	48	..	2.9	..	6.1	..	..	..	4.8
United Arab Emirates	..	..	..	..	..	..	..	..	..	..
United Kingdom	..	..	..	..	..	..	..	..	..	..
United States	..	..	..	..	..	..	..	..	..	..
Uruguay	33	143	3.1	3.7	18.8	15.6	8.8	9.7	19.4	28.1
Uzbekistan	9	56	..	1.2	..	8.1	..	..	..	3.9
Venezuela	51	147	8.7	6.9	27.2	16.8	19.1	22.5	53.0	8.2
Vietnam	123	322	..	1.5	..	3.5	..	..	..	14.5
West Bank and Gaza	..	..	..	..	..	..	..	..	..	..
Yemen, Rep.	88	160	..	1.6	..	2.4	..	..	10.8	9.7
Yugoslavia, FR (Serb./Mont.)[d]	..	..	5.2	..	..	..	10.9	..	11.6	15.8
Zambia	161	389	11.4	9.8	25.3	24.6	29.6	50.4	18.0	8.3
Zimbabwe	67	154	1.2	9.2	3.8	21.2	3.9	..	11.5	16.0
World			.. w	.. w	.. w	.. w			.. w	.. w
Low income			1.6	2.9	9.4	13.3			12.1	13.5
Excl. China & India			..	..	..	..			..	..
Middle income			3.7	5.1	13.5	18.3			26.9	20.1
Lower middle income			..	..	..	..			..	..
Upper middle income			..	..	..	..			..	..
Low & middle income			..	..	..	..			..	..
East Asia & Pacific			2.3	4.3	11.5	13.0			22.6	25.1
Europe & Central Asia			1.6	3.7	6.8	11.4			22.6	14.3
Latin America & Carib.			6.5	6.1	36.3	32.3			26.7	17.5
Middle East & N. Africa			2.5	3.2	5.6	11.4			25.2	22.3
South Asia			1.3	3.1	11.7	22.0			6.5	6.8
Sub-Saharan Africa			3.6	5.0	9.8	14.2			26.8	17.4
High income			..	..	..	..			..	..

a. Data are from debt sustainability analyses undertaken as part of the Heavily Indebted Poor Countries Debt Initiative. Present value estimates for these countries are for public and publicly guaranteed debt only, and export figures exclude worker remittances. b. Data prior to 1992 include Eritrea. c. Data refer to mainland Tanzania only. d. Data refer to the former Yugoslavia.

External debt management | 4.18

About the data

The indicators in the table measure the relative burden on developing countries of servicing external debt. The present value of external debt provides a measure of future debt service obligations that can be compared with the current value of such indicators as GNP and exports of goods and services. In this table the present value of total debt service in the most recent year (1996) is presented as a percentage of average GNP in 1994, 1995, and 1996 or the average of exports in the same three-year period. The ratios of total debt service and public and publicly guaranteed debt service compare current obligations with the size of the economy and its ability to obtain foreign exchange through exports. Because worker remittances are an important source of foreign exchange for many countries, they are included in the value of exports used to calculate debt indicators. The ratios shown here may differ from those published elsewhere because estimates of exports and GNP have been revised to incorporate data available as of February 1, 1998.

The present value of external debt is calculated by discounting the debt service (interest plus amortization) due on long-term external debt over the life of existing loans. Short-term debt is included at its face value. Data on debt are in U.S. dollars converted at official exchange rates. The discount rate applied to long-term debt is determined by the currency of repayment of the loan and is based on the Organisation for Economic Co-operation and Development's (OECD) commercial interest reference rates. Loans from the International Bank for Reconstruction and Development (IBRD) and credits from the International Development Association (IDA) are discounted using an SDR (special drawing rights) reference rate, as are obligations to the International Monetary Fund (IMF). When the discount rate is greater than the interest rate of the loan, the present value is less than the nominal sum of future debt service obligations.

The ratios in the table are used to assess the sustainability of a country's debt service obligations, but there are no absolute rules that determine what values are too high. Empirical analysis of the experience of developing countries and their debt service performance has shown that debt service difficulties become increasingly likely when the ratio of the present value of debt to exports reaches 200 percent and the ratio of debt service to GNP exceeds 20 percent. Still, what constitutes a sustainable debt burden varies from one country to another. Countries with fast-growing economies and exports are likely to be able to sustain higher debt levels.

Figure 4.18a

All regions except Asia have reduced their reliance on short-term debt

● 1980 ○ 1996

Source: World Bank 1998.

Definitions

• **Present value of debt** is the sum of short-term external debt plus the discounted sum of total debt service payments due on public, publicly guaranteed, and private nonguaranteed long–term external debt over the life of existing loans. • **Total debt service** is the sum of principal repayments and interest actually paid in foreign currency, goods, or services on long-term debt, interest paid on short-term debt, and repayments (repurchases and charges) to the IMF. • **Public and publicly guaranteed debt service** is the sum of principal repayments and interest actually paid on long-term obligations of public debtors and long-term private obligations guaranteed by a public entity. • **Short-term debt** includes all debt having an original maturity of one year or less and interest in arrears on long-term debt.

Data sources

The main sources of external debt information are reports to the World Bank through its Debtor Reporting System from member countries that have received IBRD loans or IDA credits. Additional information has been drawn from the files of the World Bank and the IMF. Data on GNP and exports of goods and services are from the World Bank's national accounts files. Summary tables of the external debt of developing countries are published annually in the World Bank's *Global Development Finance* and *Global Development Finance* CD-ROM.

The state grew everywhere in the 20th century—in very different ways, shaped by two world wars, the Russian revolution, the 1930s' depression, and decolonization. Today state involvement in economic activities ranges from 0 percent (Somalia) to 100 percent (Democratic People's Republic of Korea). Few would deny that "without an effective state, sustainable development, both economic and social, is impossible" (World Bank 1997g). But the diversity of experience makes it difficult to draw conclusions about what makes states and markets effective—and what size state is right for any set of social and economic circumstances and objectives.

Ratios of government spending to total GDP in 1996 range from less than 10 percent up to about 50 percent (table 5.1). The average for the OECD was 46 percent, having risen steadily since World War I. More than half of this goes for transfer payments and subsidies, the trappings of the modern welfare state. Investment spending by the state accounts for less than 5 percent.

Governments of different persuasions are looking for ways to freeze and eventually reduce this spending without losing votes. The market is often the most acceptable and efficient solution—with private health and unemployment insurance, private contributions to higher education costs and the care of the elderly, and even private prison services among reforms initiated in the 1990s. Industrial and agricultural subsidies are also being cut back, thus directing private investment where profitability does not depend on the profitability of handouts.

The challenge for the developing world is to provide as good an institutional framework for development as its capabilities will allow. Governments should not intervene where markets can operate more efficiently. But they do, everywhere.

State-owned enterprises, prominent in many low-income economies, rarely operate as efficiently as private firms. They also absorb a large share of the economy's resources. Their investments may be financed with government guaranteed external loans, adding to debt burdens. They are heavy users of domestic credit, often granted on preferential terms, crowding out private borrowers and giving the state-owned firms a competitive cost advantage. Their loan servicing is generally poor, and many are far in arrears on payments for goods and services purchased from other public sector entities. Many state-owned enterprises benefit from tax reductions or exemptions, distorting competition and further reducing the income to government. More often than not, they are a drain on the budget, and so constrain official allocations to more worthy social causes (table 5.8).

Containing corruption and making government more effective

Until recently a preoccupation with shrinking the size of the state led many development economists to neglect the vital task of understanding how to improve the state. One lesson from successful states is that they use market-like mechanisms to improve their efficiency.

The World Bank's *World Development Report 1997* showed that an effective state is vital for development. Using data from 94 countries over three decades, the study shows that it is not just economic policies and human capital but the quality of a country's institutions that determines economic outcomes. Institutions determine the environment in which markets operate. A weak institutional environment allows greater arbitrariness on the part of state agencies and public officials.

If laws are applied arbitrarily, or not applied at all, the market will consider them irrelevant. Similarly, if rules are changed frequently or unexpectedly, they lose credibility. The wider the discretionary controls of politicians and bureaucrats, the greater the opportunities for bribery. In these circumstances, noncompliant behavior may be the most efficient way to operate.

Governments with poor records in enforcing law and order can soon ruin even a thriving economy. If such countries attract investors, it is into areas that the state cannot afford to develop on its own. Infrastructure projects and those that exploit natural resources are usually set up under one-off, all-inclusive agreements that insulate the investor from the deficiencies of the regulatory system.

International investors can stay away from states perceived to be incompetent and corrupt. Local business communities cannot. They must operate within the system, with all its faults, or not operate at all. The credibility of the state in the eyes of local entrepreneurs has a direct bearing on economic performance.

In many developing countries local business people believe the state to be both incompetent and corrupt (box 5a). They invest less and operate less efficiently than they would in a more supportive policy environment. The main reason for this underperformance is that, to survive, they must divert time and money from their businesses into negotiations with bureaucrats and unreliable institutions.

The need for better governance, everywhere

Corruption, the abuse of public power for private gain, is a problem in every country. And the way it undermines the legitimacy of government has become a central issue for discussion in the 1990s. A series of widely publicized scandals helped to put the problem at the center of international concerns. Politicians and bureaucrats were prosecuted for taking bribes in several OECD countries, in the former Soviet Union, in Asia, and in Latin America. The crimes involved illegal payments for public contracts for infrastructure, defense, manufacturing, and services projects. They also involved campaign finance excesses—and protection money for the drug trade and other organized crime payoffs.

Transparency International—an anticorruption nonprofit group—has developed a corruption perception index that draws on surveys by Gallup International, the *World Competitiveness Yearbook*, Political & Economic Risk Consultancy in Hong Kong, DRI/McGraw Hill Global Risk Service, Political Risk Services in Syracuse, New York, and data gathered from Internet sources.

Low-income countries often have weaker markets and weaker institutions. As a result they may be more corrupt. And the connection runs both ways: countries are corrupt because they are poor, and they are poor because they are corrupt. Corruption damages economic performance by depressing investment in an unpredictable policy environment. On a corruption perception index with a range of zero to 10, a 2.4-point improvement led to a 4-point increase in a country's investment rate and a real rise in per capita income (Mauro 1997).

Reducing entrenched corruption requires far-reaching policy reforms, starting with strengthening institutions and liberalizing markets. Other measures include clearly separating executive and judiciary powers, revising public pay scales, and launching tax and tariff reforms.

Since the early 1990s the World Bank has been promoting good governance in its country assistance strategies and lending. Its research programs have improved understanding of the causes and effects of corruption in developing countries. And it has established a framework for addressing corruption as part of a broader promotion of good governance practices (World Bank 1997c).

A two-part strategy aims to help countries match the state's role to its capability and improve the quality of its institutions. This requires taking stock of a state's capabilities and ranking the economic and social fundamentals that can be supplied efficiently with existing resources. States can improve their capabilities by reinvigorating their institutions. This means not only building administrative or technical capacity but also instituting rules and norms that provide officials with incentives to act in the collective interest while restraining arbitrary action and corrup-

---

Box 5a

A credible state promotes private sector development

A survey of 3,600 firms in 60 countries prepared for the World Bank's *World Development Report 1997* concluded that the credibility accorded by the private sector to state institutions was linked to cross-country differences in economic growth and investment. Entrepreneurs were asked about their perceptions of the stability of laws and policies—and of the institutions set up to administer them. They were also asked about crime and corruption.

The predictability of rulemaking. The fewer surprises, the better. Worst off were entrepreneurs in the Commonwealth of Independent States (CIS), where 80 percent found their business seriously hampered by unpredictable changes in rules and policies.

The continuity of institutions. Business communities in the CIS, the Middle East and North Africa, and Sub-Saharan Africa were beset by institutional upheavals stemming from political changes. Entrepreneurs in both industrial and developing countries worried about crime against persons and property. And few respondents in developing countries trusted the system of law enforcement.

Freedom from corruption. Of the 3,600 firms, 40 percent had to resort to bribery in the normal course of business—30 percent in Asia, 60 percent in the CIS. More than half the respondents knew that the bribes would not bring forth the necessary permits, but would lead to yet further bribes to other officials.

tion. An independent judiciary, institutional checks and balances, and effective watchdogs can restrain arbitrary state action and corruption. Competitive wages for civil servants can attract more talented people and increase professionalism and integrity.

Privatization

Since the 1980s the privatization of public sector enterprises has been central in economic policy reforms—helping economies better match the state's role to its capabilities. The process was accelerated by far-reaching changes in political systems and by information technology that facilitated the globalization of private business operations. Governments of former socialist countries opened the way for greater private participation by lifting restrictions on local and foreign private investments in their economies. Developing countries elsewhere took steps to reduce the state's involvement in production and trade. Technological advances have fundamentally changed the nature of the telecommunications sector. What was once seen as a public good with a single dimension (the telephone call) is today a broad array of services (fixed, wireless, facsimile and other value-added) that can be efficiently and effectively provided by the private sector.

Some important lessons: First, the ownership change from states to markets must be seen as politically desirable. Second, it must be politically feasible—benefits to the reforming government must outweigh the foreseeable costs, including the potential loss of political support. Third, the reforms must be credible. The proposals must include adequate compensation for losers, particularly with forced layoffs and evictions. Equally, the reforms must protect investors' property rights. Enterprise reform is unlikely to succeed if these three conditions are not present (World Bank 1995b).

State enterprises have shrunk most visibly in the former socialist countries and in some middle- and high-income economies. In Sub-Saharan Africa, by contrast, reforms could have done much for macroeconomic stabilization, but the governments in power saw the risks as too great. Sell-offs met least resistance in the wake of economic crises, as in the former Soviet Union and Latin America. Although proceeds from privatization provide governments with valuable revenues at no immediate cost to taxpayers, governments (and therefore taxpayers) lose an asset that might have provided future cash flows, and governments often provide guarantees to the new owners that later cost taxpayers.

The privatization of infrastructure has proceeded worldwide at great speed since the 1980s. Today, more than 25 countries in Sub-Saharan Africa are transferring all or part of their telecommunications monopolies from the state to the private sector (table 5a). Most had looked at what was happening elsewhere and liked what they saw—faster growth, new technology, lower costs, and better services. Privatizations were most likely to succeed if they had support at the highest government levels—and from the workforce.

The state monopolies being transformed into private-led, competitive markets in Africa and Latin America show how important it is to establish lines of authority—and to set clear policies, rules, and procedures—before calling for tenders. For sales strategies, privatization's long-run benefits deserve more emphasis than cash up front. Allowing competition is the best solution. Even though the government gets less revenue from the sale of firms and licenses, it acquires growing future flows of tax revenues from the successful businesses. Phased privatizations that commit a government to sell only enough shares to give the private partner a

---

Table 5a

Privatization of state-owned telecommunications companies in 1997, selected countries

Country	Company name	% sold	Price paid ($ millions)	Type of sale
Australia	Telstra	33.3	10,882	International public offering
Côte d'Ivoire	CI-Telcom	51.0	210	Strategic equity partner
France	France Télécom	23.2	7,270	International public offering
Kazakhstan	Kazakhtelecom	40.0	370	International public offering
Panama	INTEL	49.0	652	Strategic equity partner
Senegal	Sonatel	33.3	107	International public offering
South Africa	Telkom SA	30.0	1,261	Strategic equity partner
Spain[a]	Telefónica de España	20.9	4,359	International public offering
Sri Lanka	Sri Lanka Telekom	35.0	225	Strategic equity partner
Yugoslavia FR (Serb./Mont.)	Telekom Srbija	49.0	1,254	Private sale

a. With this sale of the state's remaining stake, Telefónica de España is now 100 percent privately owned.
Source: International Telecommunication Union; World Bank.

controlling interest have also proved successful. If the partly privatized utility is making satisfactory profits, and if share prices increase, the remaining government shareholding may be sold at higher prices than the original shares, providing the government with a windfall. Of course, the downside is that share prices may fall, resulting in revenue losses for the government.

Private monopoly should not succeed public monopoly, but some infrastructure businesses, such as water distribution, may inevitably be monopolistic. Even so, privatization may still be desirable—regulated private monopoly may be better than public monopoly. Governments should ensure that there are no barriers to new market entrants; the use of limited-duration concessions is one way of limiting market power. The rules of competition must be clear, with the amount of discretion reflecting the government's capacity and credibility to regulate such matters as tariffs and service obligations. The climate for infrastructure projects can be greatly improved if the state agrees to abide by an international treaty, such as the 1997 World Trade Organization Agreement on Basic Telecommunications.

Many private infrastructure firms may need to be permanently regulated to facilitate an efficient market. As new technologies (wireless, satellite, Internet) enter the market, the agenda of regulatory agencies is likely to change. Few developing countries had fully operational regulatory systems in 1997, though several Latin American and Eastern European states were creating them. The problem is that if newly privatized utilities are poorly regulated, then the supply and quality of telecommunications, water, or energy may be as unsatisfactory as it was when provided by a state entity—and possibly worse. One unsuccessful privatization acts as a disincentive to investors. Governments that change the rules unpredictably—or fail to abide by them for one privatization—have difficulty finding buyers for the next privatization.

In telecommunications, wireless services create a competitive market where consumers have a real choice (table 5.11). The push of new technology has also improved competition in other infrastructure sectors so that regulatory requirements at the start of a privatization program run the risk of being overtaken by events unless adapted to the new technological developments. International cooperation in regulatory standards will help achieve this flexibility.

A parallel need has been created by the globalization of banking and financial services. Since the 1980s regulators have cooperated internationally to fight fraud and the laundering of proceeds from crime and corruption. Recent financial scandals and bank liquidity crises helped push governments normally protective of their authority closer to agreement on common standards and supervision systems—essential for an efficient international financial market and for the flow of investment funds. Here again, better-regulated markets—where investment decisions are not distorted by corruption—attract the greatest private capital flows.

A recent survey of local entrepreneurs in 69 countries found that many states are performing their core functions poorly; they are failing to ensure law and order, protect property, and apply rules and policies predictably. Because investors do not consider such states credible, growth and investment suffer.

Local entrepreneurs were asked to rank each of several indicators on a scale from 1 (no problem) to 6 (extreme problem); survey results for four indicators are shown below

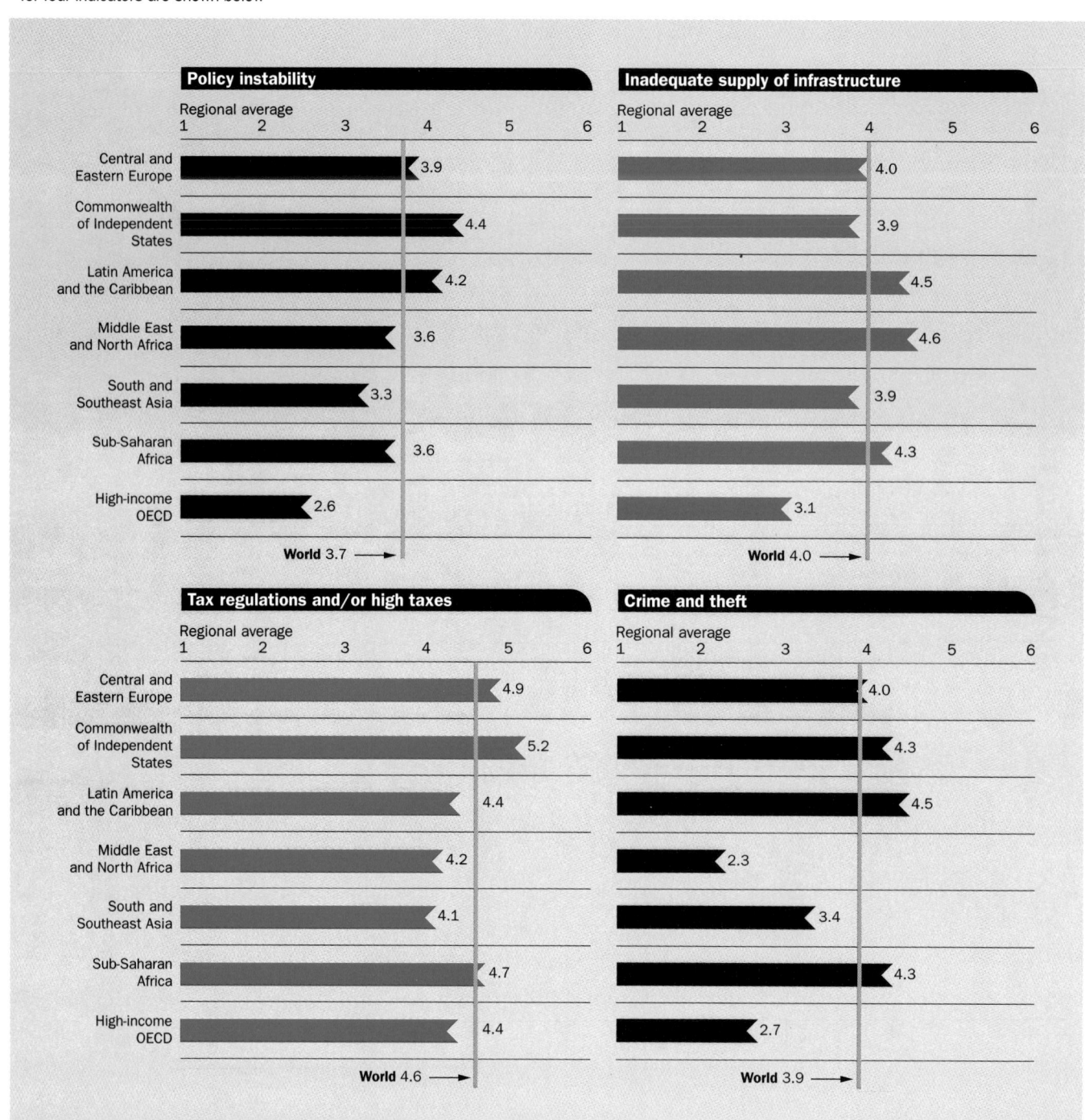

Policy instability

Regional average

Region	Value
Central and Eastern Europe	3.9
Commonwealth of Independent States	4.4
Latin America and the Caribbean	4.2
Middle East and North Africa	3.6
South and Southeast Asia	3.3
Sub-Saharan Africa	3.6
High-income OECD	2.6

World 3.7 →

Inadequate supply of infrastructure

Regional average

Region	Value
Central and Eastern Europe	4.0
Commonwealth of Independent States	3.9
Latin America and the Caribbean	4.5
Middle East and North Africa	4.6
South and Southeast Asia	3.9
Sub-Saharan Africa	4.3
High-income OECD	3.1

World 4.0 →

Tax regulations and/or high taxes

Regional average

Region	Value
Central and Eastern Europe	4.9
Commonwealth of Independent States	5.2
Latin America and the Caribbean	4.4
Middle East and North Africa	4.2
South and Southeast Asia	4.1
Sub-Saharan Africa	4.7
High-income OECD	4.4

World 4.6 →

Crime and theft

Regional average

Region	Value
Central and Eastern Europe	4.0
Commonwealth of Independent States	4.3
Latin America and the Caribbean	4.5
Middle East and North Africa	2.3
South and Southeast Asia	3.4
Sub-Saharan Africa	4.3
High-income OECD	2.7

World 3.9 →

Countries must establish mechanisms that give state agencies the flexibility and the incentive to act for the common good—while at the same time restraining arbitrary and corrupt behavior in dealings with businesses and citizens.

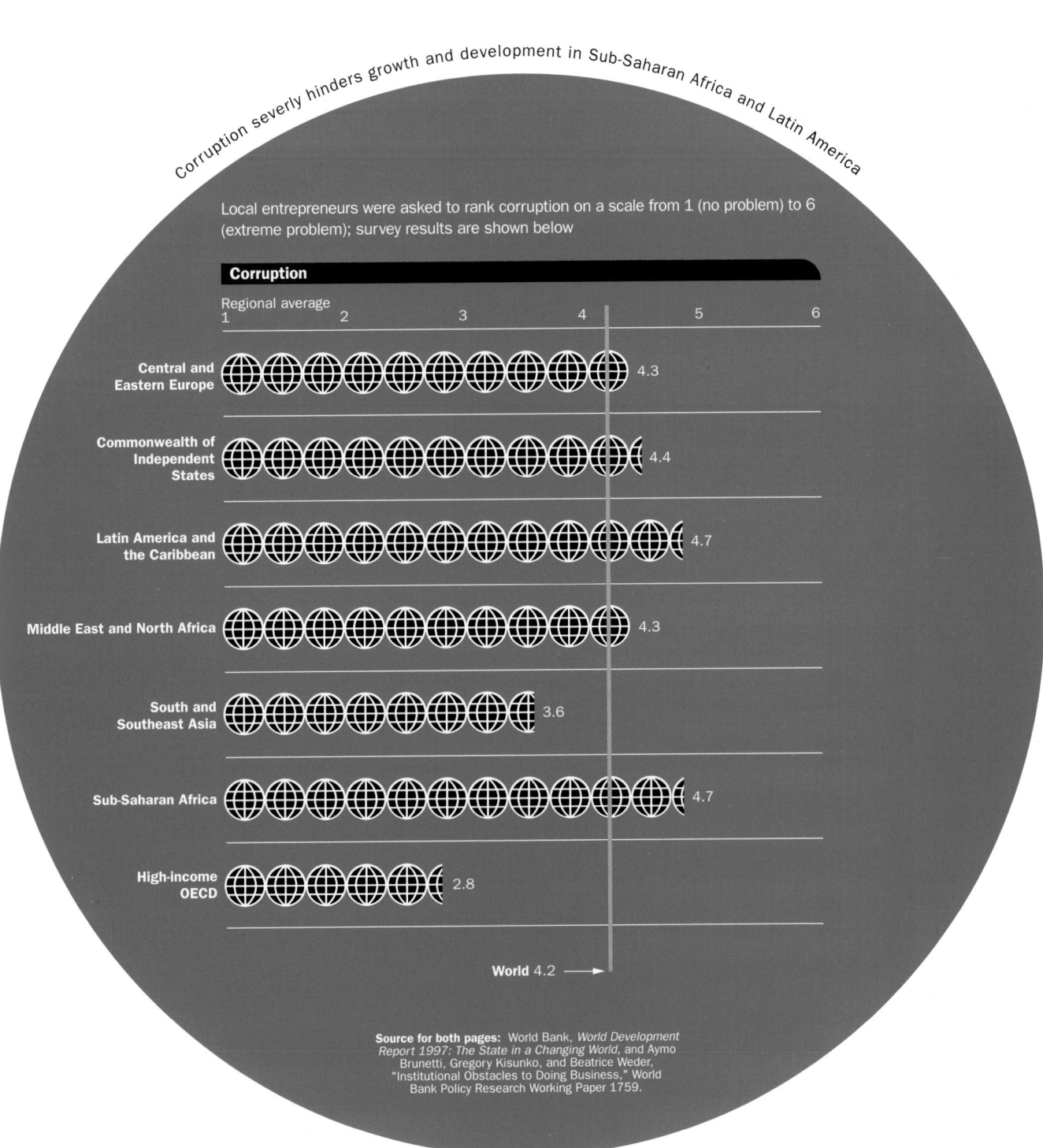

Corruption severely hinders growth and development in Sub-Saharan Africa and Latin America

Local entrepreneurs were asked to rank corruption on a scale from 1 (no problem) to 6 (extreme problem); survey results are shown below

Corruption

Regional average
1   2   3   4   5   6

Central and Eastern Europe — 4.3

Commonwealth of Independent States — 4.4

Latin America and the Caribbean — 4.7

Middle East and North Africa — 4.3

South and Southeast Asia — 3.6

Sub-Saharan Africa — 4.7

High-income OECD — 2.8

World 4.2 →

Source for both pages: World Bank, *World Development Report 1997: The State in a Changing World,* and Aymo Brunetti, Gregory Kisunko, and Beatrice Weder, "Institutional Obstacles to Doing Business," World Bank Policy Research Working Paper 1759.

	Private investment		Foreign direct investment				Credit to private sector		Private non-guaranteed debt		Central government expenditure	
	% of gross domestic fixed investment		% of gross domestic investment		% of GDP		% of GDP		% of external debt		% of GDP	
	1980	1996	1980	1996	1980	1996	1980	1996	1980	1996	1980	1995
Albania	..	..	..	16.3	..	3.4	..	3.9	..	0.0	..	31.0
Algeria	67.4	74.8	2.1	0.0	0.8	0.0	42.2	5.5	0.0	0.0	..	..
Angola	..	68.9	..	39.7	..	4.5	..	..	..	0.0	..	..
Argentina	..	85.8	3.5	7.9	0.9	1.5	25.4	18.6	24.3	13.8	18.2	14.5
Armenia	..	33.6	..	9.4	..	1.0	..	7.3	..	0.0	..	..
Australia	..	..	4.6	18.9	1.2	1.6	51.9	77.4	..	..	22.7	27.4
Austria	..	..	1.1	1.1	0.3	1.7	74.2	99.3	..	..	36.6	42.2
Azerbaijan	..	..	..	67.3	..	16.5	..	1.0	..	0.0	..	..
Bangladesh	58.9	62.5	0.0	0.3	0.0	0.0	8.1	20.6	0.0	0.0	10.0	..
Belarus	..	..	..	0.4	..	0.1	..	6.9	..	2.8	..	..
Belgium	..	..	..	..	..	..	29.1	67.5	..	..	50.6	49.4
Benin	..	61.7	2.0	0.5	0.3	0.1	28.6	9.1	0.0	0.0	..	..
Bolivia	51.3	41.9	10.3	17.9	1.5	6.4	15.5	52.8	3.4	5.5	..	23.1
Bosnia and Herzegovina	..	..	..	..	..	..	..	..	..	..	..	..
Botswana	62.1	..	30.5	6.3	10.8	1.5	11.3	11.1	0.0	0.0	31.8	38.0
Brazil	89.7	86.2	3.5	6.8	0.8	1.3	42.5	30.7	23.2	27.3	20.2	37.4
Bulgaria	85.9	85.0	0.0	8.6	0.0	1.2	..	36.9	..	2.0	..	41.6
Burkina Faso	..	57.9	0.0	0.0	0.0	0.0	16.7	7.0	0.0	0.0	12.2	..
Burundi	8.1	15.7	0.0	1.0	0.0	0.1	9.8	15.8	0.0	0.0	21.5	24.9
Cambodia	..	68.6	..	45.4	..	9.4	..	5.3	0.0	0.0	..	..
Cameroon	77.8	95.5	9.2	2.4	1.9	0.4	29.5	8.3	6.9	1.9	15.7	12.7
Canada	..	..	9.4	10.5	2.2	1.1	68.7	88.8	..	..	21.3	24.6
Central African Republic	46.5	41.8	9.5	8.4	0.7	0.5	13.9	4.3	0.0	0.0	22.0	..
Chad	4.8	35.8	0.0	8.0	0.0	1.5	24.4	5.0	0.0	0.0	..	..
Chile	72.2	80.0	3.7	19.9	0.8	5.5	46.8	55.2	38.8	56.7	28.0	19.2
China	43.4	47.0	0.0	11.6	0.0	4.9	53.4	94.7	0.0	0.9	..	8.3
Hong Kong, China	85.1	86.8	..	..	..	..	..	161.7	..	..	..	..
Colombia	58.3	47.8	2.5	18.7	0.5	3.9	30.5	41.4	7.4	28.3	13.4	14.4
Congo, Dem. Rep.	42.4	..	0.0	0.5	0.0	0.0	2.8	1.0	0.0	0.0	12.4	7.6
Congo, Rep.	..	91.4	6.6	0.6	2.3	0.3	15.5	8.0	0.0	0.0	49.4	..
Costa Rica	61.3	75.1	4.1	19.7	1.1	4.5	27.9	17.6	15.0	5.6	25.0	29.1
Côte d'Ivoire	53.2	69.1	3.5	1.4	0.9	0.2	40.8	19.7	27.0	17.0	31.7	..
Croatia	..	59.6	..	12.4	..	1.8	..	30.1	..	18.5	..	46.5
Cuba	..	..	..	..	..	..	..	..	..	..	..	..
Czech Republic	..	..	0.0	7.6	0.0	2.6	..	59.0	0.0	10.6	..	39.9
Denmark	..	..	1.1	14.1	0.2	0.4	42.1	33.4	..	..	39.4	43.4
Dominican Republic	68.4	66.5	5.6	12.3	1.4	3.0	30.8	26.6	12.7	0.1	16.9	15.6
Ecuador	59.7	78.3	2.3	13.4	0.6	2.3	22.8	33.5	18.7	2.2	14.2	15.7
Egypt, Arab Rep.	30.1	59.1	8.7	5.7	2.4	0.9	15.2	41.5	1.4	0.4	45.6	37.4
El Salvador	44.8	78.0	1.2	1.5	0.2	0.2	33.6	36.6	17.6	0.1	17.1	13.7
Eritrea	..	52.7	..	..	..	..	..	..	..	0.0	..	..
Estonia	..	80.2	..	12.9	..	3.5	..	18.1	..	0.9	..	35.2
Ethiopia	..	63.9	0.0	0.4	0.0	0.1	20.1	21.7	0.0	0.0	19.5	18.1
Finland	..	..	0.2	5.1	0.1	0.9	48.5	61.5	..	..	28.1	42.7
France	..	..	2.0	8.5	0.5	1.4	104.8	81.7	..	..	39.5	46.4
Gabon	80.1	72.0	2.7	-5.7	0.7	-1.1	15.8	6.6	0.0	0.0	36.5	..
Gambia, The	35.4	63.1	0.0	10.2	0.0	2.2	24.2	10.6	0.0	0.0	32.1	21.5
Georgia	..	73.7	..	4.6	..	0.2	..	..	..	0.0	..	..
Germany	..	..	..	2.2	..	-0.1	..	104.9	..	..	..	33.9
Ghana	..	26.3	6.2	10.1	0.4	1.9	2.2	6.6	0.7	4.4	10.9	22.1
Greece	..	..	5.9	7.0	1.4	0.7	43.9	35.7	..	..	29.3	33.6
Guatemala	63.8	81.3	8.9	3.8	1.4	0.5	16.2	18.9	24.2	3.2	12.1	8.9
Guinea	..	57.7	..	4.7	..	0.6	..	4.8	0.0	0.0	..	..
Guinea-Bissau	..	32.5	0.0	1.7	0.0	0.4	..	0.1	0.0	0.0	..	..
Haiti	..	27.6	5.3	0.0	0.9	0.2	15.7	14.3	0.0	0.0	17.4	..
Honduras	62.1	62.7	0.9	5.8	0.2	1.9	28.8	26.7	13.0	2.8	..	..

Credit, investment, and expenditure | 5.1

	Private investment		Foreign direct investment				Credit to private sector		Private non-guaranteed debt		Central government expenditure	
	% of gross domestic fixed investment		% of gross domestic investment		% of GDP		% of GDP		% of external debt		% of GDP	
	1980	1996	1980	1996	1980	1996	1980	1996	1980	1996	1980	1995
Hungary	..	..	0.0	16.5	0.0	4.4	48.3	22.3	0.0	18.6	56.2	..
India	55.5	66.1	0.2	2.7	0.0	0.7	25.4	23.7	1.6	8.2	13.3	16.4
Indonesia	..	60.5	1.0	11.1	0.2	3.5	8.8	55.8	15.0	28.4	22.1	14.7
Iran, Islamic Rep.	..	..	0.0	..	0.0	0.0	43.8	23.2	..	1.1	35.7	23.2
Iraq	..	..	..	..	0.0	..	..	..	0.0	0.0	..	..
Ireland	..	..	..	23.6	1.7	3.5	44.0	78.3	..	..	45.1	40.3
Israel	..	..	1.0	7.3	0.2	1.7	68.3	66.8	..	..	70.2	44.7
Italy	..	..	0.5	2.5	0.1	0.3	55.9	52.1	..	..	41.3	48.6
Jamaica	..	..	6.5	14.9	1.0	4.0	21.9	31.1	3.9	3.0	41.5	..
Japan	..	..	0.1	0.0	0.0	0.0	132.7	207.1	..	..	18.4	23.7
Jordan	51.4	77.1	2.3	0.6	0.9	0.2	51.0	71.5	0.0	0.5	41.3	31.6
Kazakhstan	..	98.8	..	6.4	..	1.5	..	7.1	..	7.4	..	..
Kenya	54.7	44.5	3.7	0.7	1.1	0.1	29.5	34.6	12.9	5.4	25.3	29.8
Korea, Dem. Rep.	..	..	..	..	..	..	..	..	..	..	..	..
Korea, Rep.	76.2	76.0	0.0	1.3	0.0	0.5	50.9	74.8	..	..	17.0	17.7
Kuwait	..	..	0.0	..	0.0	..	33.1	38.6	..	..	27.7	51.4
Kyrgyz Republic	..	87.5	..	13.8	..	2.6	..	..	..	0.0	..	..
Lao PDR	..	..	..	18.4	..	5.6	..	9.1	0.0	0.0	..	..
Latvia	..	89.3	..	34.9	..	6.5	..	7.3	..	0.0	..	32.2
Lebanon	..	71.8	..	2.0	..	0.6	116.4	63.8	0.0	10.3	..	32.5
Lesotho	..	36.8	2.9	3.0	1.2	3.2	9.8	17.5	0.0	0.0	..	50.7
Libya	8.3	..	-13.9	..	-3.1	..	11.2	..	..	..	..	..
Lithuania	..	86.3	..	9.3	..	2.0	..	11.9	..	5.0	..	25.5
Macedonia, FYR	..	..	..	4.6	..	0.7	..	..	..	31.6	..	..
Madagascar	..	42.5	-0.2	2.4	0.0	0.2	19.2	9.4	0.0	0.0	..	17.2
Malawi	21.4	84.3	3.1	0.3	0.8	0.0	20.7	4.2	0.0	0.0	34.6	..
Malaysia	62.6	69.8	12.5	11.0	3.8	4.5	49.9	129.5	18.9	32.7	28.5	22.9
Mali	..	54.4	0.9	3.3	0.1	0.9	23.0	13.0	0.0	0.0	20.6	..
Mauritania	..	68.3	10.6	2.1	3.8	0.5	31.0	23.6	0.0	0.0	..	..
Mauritius	64.0	64.8	0.5	3.3	0.1	0.9	21.6	44.7	5.1	13.5	27.2	22.6
Mexico	57.0	79.1	4.1	10.9	1.1	2.3	19.7	21.6	12.7	12.9	15.7	15.9
Moldova	..	78.5	..	8.0	..	2.2	..	7.3	..	0.0	..	..
Mongolia	..	..	..	2.3	..	0.5	..	12.3	..	0.0	..	21.5
Morocco	44.0	57.8	2.0	4.1	0.5	0.8	27.0	45.8	1.6	1.8	33.1	..
Mozambique	27.0	65.3	0.0	3.5	0.0	1.7	..	17.8	..	0.7	..	..
Myanmar	20.6	60.6	..	..	..	..	5.5	..	..	..	15.8	10.6
Namibia	42.0	62.2	0.0	21.4	0.0	4.2	..	50.5	0.0	0.0	..	38.5
Nepal	60.2	67.8	0.0	1.9	0.0	0.4	8.6	24.6	0.0	0.0	14.3	17.5
Netherlands	..	..	6.0	18.7	1.3	2.0	93.6	105.7	..	..	52.9	50.8
New Zealand	..	..	3.7	31.5	0.8	0.4	18.4	93.3	..	..	38.3	32.9
Nicaragua	..	38.6	0.0	8.2	0.0	2.3	48.3	28.6	0.0	0.0	30.4	33.2
Niger	..	50.6	5.3	0.0	1.9	0.0	16.9	3.1	35.3	7.0	18.4	..
Nigeria	..	62.5	-5.4	23.2	-1.2	4.3	12.2	10.6	12.3	1.0	..	..
Norway	..	..	0.4	..	0.1	2.5	51.4	72.3	..	..	34.4	39.0
Oman	34.1	..	7.4	4.0	1.6	0.4	13.7	29.2	0.0	0.1	38.5	42.4
Pakistan	36.1	52.5	1.4	5.7	0.3	1.1	24.0	26.7	0.2	6.7	17.5	23.2
Panama	..	83.8	-4.4	9.9	-1.2	2.9	58.1	80.6	0.0	1.1	30.5	24.7
Papua New Guinea	58.6	85.8	11.8	16.2	3.0	4.4	17.6	18.2	19.3	31.9	34.4	29.4
Paraguay	85.1	83.4	2.2	10.1	0.7	2.3	18.4	29.3	15.8	1.0	9.9	13.0
Peru	75.6	82.9	0.4	25.0	0.1	5.9	12.9	19.6	6.5	4.7	19.5	17.2
Philippines	69.0	81.1	-1.1	6.9	-0.3	1.7	42.2	54.2	14.1	11.9	13.4	17.9
Poland	..	81.9	0.1	16.4	0.0	3.3	6.4	15.5	..	3.9	..	43.0
Portugal	..	..	..	8.3	0.5	0.6	54.0	64.3	..	..	33.1	44.1
Puerto Rico	..	..	..	..	..	..	..	..	..	..	..	..
Romania	..	73.8	..	2.9	..	0.7	..	..	0.0	4.5	44.8	32.0
Russian Federation	..	91.1	..	2.5	..	0.6	..	7.4	..	0.0	..	24.0

5.1 Credit, investment, and expenditure

	Private investment		Foreign direct investment				Credit to private sector		Private non-guaranteed debt		Central government expenditure	
	% of gross domestic fixed investment		% of gross domestic investment		% of GDP		% of GDP		% of external debt		% of GDP	
	1980	1996	1980	1996	1980	1996	1980	1996	1980	1996	1980	1995
Rwanda	..	70.0	8.7	0.6	1.4	0.1	5.7	7.1	0.0	0.0	14.3	*25.8*
Saudi Arabia	..	..	−9.4	−7.7	−2.0	−1.5	22.8	*63.8*	..	..	..	..
Senegal	62.1	70.3	3.1	5.3	0.5	0.9	42.3	15.8	0.6	1.1	23.1	..
Sierra Leone	..	64.4	−9.0	5.7	−1.6	0.5	7.2	2.5	0.0	0.0	26.5	16.4
Singapore	75.6	..	22.8	28.6	10.5	10.0	81.0	108.8	..	..	20.0	15.9
Slovak Republic	..	..	..	3.9	..	1.5	..	31.7	0.0	7.1	..	..
Slovenia	..	26.7	..	4.3	..	1.0	..	28.5	..	48.0	..	..
South Africa	50.8	..	−0.1	0.6	0.0	0.1	60.3	137.1	..	15.1	22.1	33.7
Spain	..	..	3.0	5.2	0.7	1.1	78.2	74.9	..	..	26.7	*38.2*
Sri Lanka	77.4	..	3.2	3.4	1.1	0.9	17.2	25.2	0.2	1.0	41.4	29.3
Sudan	38.9	..	0.0	..	0.0	..	14.9	*5.3*	6.3	2.9	19.6	..
Sweden	..	..	0.9	41.2	0.2	2.2	78.0	37.1	..	..	39.3	49.5
Switzerland	..	..	..	..	..	1.2	113.7	167.7	..	..	20.1	26.6
Syrian Arab Republic	36.1	..	0.0	..	0.0	0.6	5.7	*10.9*	0.0	0.0	48.2	24.5
Tajikistan	..	..	..	4.7	..	0.8	..	..	..	0.0	..	..
Tanzania	..	..	..	14.2	..	2.6	..	3.4	3.4	0.6	..	..
Thailand	68.1	77.6	2.0	3.1	0.6	1.3	41.7	100.0	20.5	39.8	18.8	15.8
Togo	28.3	78.2	13.1	0.0	3.7	0.0	27.5	19.2	0.0	0.0	30.8	..
Trinidad and Tobago	..	88.0	9.7	38.1	3.0	5.9	28.7	46.3	0.0	3.5	30.9	29.2
Tunisia	46.9	51.0	9.1	6.8	2.7	1.6	46.4	63.5	5.1	1.9	31.6	32.8
Turkey	..	81.4	0.1	1.7	0.0	0.4	13.6	22.8	2.8	13.1	21.3	22.2
Turkmenistan	..	..	..	..	..	2.5	..	*17.5*	..	0.0	..	..
Uganda	..	63.9	0.0	12.1	0.0	2.0	3.9	4.7	0.0	0.0	6.2	..
Ukraine	..	..	..	3.5	..	0.8	..	1.4	..	1.9	..	..
United Arab Emirates	..	..	..	..	..	..	22.9	*50.0*	..	..	12.1	*11.8*
United Kingdom	..	..	11.2	..	1.9	2.8	27.6	124.1	..	..	38.3	42.0
United States	..	..	3.1	5.4	0.6	1.0	80.3	115.4	..	..	22.0	22.7
Uruguay	*67.9*	71.1	16.5	7.7	2.9	0.9	37.2	30.3	12.7	2.3	21.8	31.5
Uzbekistan	..	..	..	1.4	..	0.2	..	..	..	0.0	..	..
Venezuela	51.5	31.5	0.3	16.3	0.1	2.7	48.2	9.6	10.8	5.1	18.7	18.8
Vietnam	..	76.3	..	23.0	..	6.4	..	*8.2*	0.0	0.0	..	..
West Bank and Gaza	..	..	..	..	..	..	..	..	..	..	..	..
Yemen, Rep.	..	67.6	..	6.6	..	1.7	..	3.4	..	0.0	..	24.7
Yugoslavia, FR (Serb./Mont.)	..	..	0.0	..	0.0	..	..	..	59.5	20.5	..	..
Zambia	..	48.7	6.8	11.4	1.6	1.7	19.9	9.3	2.7	0.2	37.1	25.0
Zimbabwe	77.1	90.4	0.1	4.7	0.0	0.8	33.2	35.4	0.0	8.6	34.8	*34.1*
World	.. w	*68.1* w	2.3 w	8.1 w	0.6 w	1.1 w	69.8 w	107.7 w	.. w	.. w	25.7 w	29.1 w
Low income	48.5	*53.2*	0.0	9.8	0.0	3.3	32.3	62.3	4.8	3.1	..	12.9
Excl. China & India	..	*63.2*	−0.1	10.6	0.0	2.1	17.4	17.7	..	..	..	..
Middle income	..	*79.1*	1.1	7.8	0.3	1.8	30.4	35.0	12.7	15.2	..	..
Lower middle income	..	*75.0*	1.5	6.5	0.4	1.6	29.9	34.8	..	..	..	22.6
Upper middle income	66.8	*81.3*	0.8	9.2	0.2	2.0	30.7	35.1	..	..	20.3	*29.0*
Low & middle income	59.8	*66.6*	0.8	8.5	0.2	2.2	16.7	*18.7*	..	..	..	*21.9*
East Asia & Pacific	50.5	*56.9*	1.1	10.4	0.4	4.0	41.5	85.4	13.5	19.4	..	11.5
Europe & Central Asia	..	*84.5*	0.1	5.7	0.0	1.3	*17.0*	15.7	15.3	7.3	..	*30.9*
Latin America & Carib.	70.6	80.2	3.4	10.4	0.8	2.1	32.5	27.5	16.5	16.9	18.8	*24.5*
Middle East & N. Africa	..	..	3.1	3.3	0.7	0.7	28.6	37.5	0.7	1.8	..	..
South Asia	54.4	*64.4*	0.4	2.9	0.1	0.7	23.6	23.9	0.9	6.2	14.2	17.6
Sub-Saharan Africa	..	*64.8*	0.0	6.0	0.0	1.1	31.3	68.1	5.4	4.2	22.2	..
High income	..	..	2.9	5.4	0.7	0.9	82.7	82.9	..	..	26.3	31.3

Credit, investment, and expenditure | 5.1

About the data

The indicators in the table measure the relative size of states and markets in national economies. There is no ideal size for states, and size alone does not capture their full effect on markets. Large states may support prosperous and effective markets; small states may be predatory toward markets. The resources of a large state may be used to correct genuine market failures—or merely to subsidize state enterprises making goods or providing services that the private sector might have produced more efficiently. A large share of private domestic investment in total investment may reflect a highly competitive and efficient private sector—or one that is subsidized and protected. Thus, like other indicators in this book, the indicators here provide an important but incomplete picture of what they measure—in this case the roles of states and markets.

Because data on subnational units of government—state, provincial, and municipal—are not readily available, the size of the public sector is measured here by the size of the central government. While the central government is usually the largest economic agent in a country and typically accounts for most public sector revenues, expenditures, and deficits, in some countries—especially large ones—state, provincial, and local governments are important participants in the economy. In addition, "central government" activities can vary depending on the accounting practice followed. In most countries central government finance data are consolidated into one overall account, but in others only budgetary central government accounts are available, which often omit the operations of state-owned enterprises (see *Primary data documentation*).

When direct estimates of private gross domestic fixed investment are not available, such investment is estimated as the difference between total gross domestic investment and consolidated public investment. Total investment may be estimated directly from surveys of enterprises and administrative records or indirectly using the commodity flow method. Consolidated measures of public investment may omit important subnational units of government. In addition, public investment data may include financial as well as physical capital investment. As the difference between two estimated quantities, private investment may be undervalued or overvalued and subject to large errors over time. (See the notes to table 4.9 for further discussion on measuring domestic investment.)

Statistics on foreign direct investment are based on balance of payments data reported by the International

Monetary Fund (IMF), supplemented by data on net foreign direct investment reported by the Organisation for Economic Co-operation and Development and official national sources. The data suffer from deficiencies relating to definitions, coverage, and cross-country comparability. (See the notes to table 6.8 for a detailed discussion of data on foreign direct investment.)

Data on domestic credit to the private sector are taken from the banking survey of the IMF's *International Financial Statistics* or, when the broader aggregate is not available, from its monetary survey. The monetary survey includes monetary authorities (the central bank) and deposit money banks. In addition to these, the banking survey includes other banking institutions such as savings and loan institutions, finance companies, and development banks. In some cases credit to the private sector may include credit to state-owned or partially state-owned enterprises.

Figure 5.1a

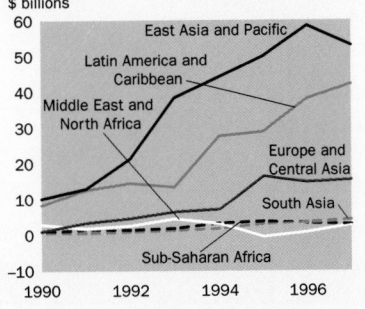

Foreign direct investment has retreated in East Asia

Source: World Bank 1998.

The surge in net foreign direct investment (FDI) flows during the 1990s has been spurred by three main factors: increasing liberalization of developing economies, strong growth in GDP and trade of the main developing economy recipients of FDI, and falling costs and improving quality of communication and transportation services.

FDI leveled off in 1997 to $120 billion after several years of rapid growth. The ratio of FDI to GDP in developing economies quadrupled from 0.6 percent in 1990 to 2.5 percent in 1997, while nominal levels quintupled.

The deceleration in 1997 reflected a reversal in East Asia and the Pacific, where flows fell 9 percent, to $53 billion. In contrast, flows to Latin America and the Caribbean increased 10 percent, to $42 billion.

Definitions

- **Private investment** covers gross outlays by the private sector (including private nonprofit agencies) on additions to its fixed domestic assets. Gross domestic fixed investment includes similar outlays by the public sector. No allowance is made for the depreciation of assets. • **Foreign direct investment** is net inflows of investment to acquire a lasting management interest (10 percent or more of voting stock) in an enterprise operating in an economy other than that of the investor. It is the sum of equity capital, reinvestment of earnings, other long-term capital, and short-term capital as shown in the balance of payments. Gross domestic investment (used in the denominator) is gross domestic fixed investment plus net changes in stocks inventories. • **Credit to private sector** refers to financial resources provided to the private sector—such as through loans, purchases of nonequity securities, and trade credits and other accounts receivable—that establish a claim for repayment. For some countries these claims include credit to public enterprises. • **Private nonguaranteed debt** consists of external obligations of private debtors that are not guaranteed for repayment by a public entity. Total external debt is the sum of public and publicly guaranteed long-term debt, private nonguaranteed long-term debt, IMF credit, and short-term debt. • **Central government expenditure** comprises the expenditures of all government offices, departments, establishments, and other bodies that are agencies or instruments of the central authority of a country. It includes both current and capital (development) expenditures.

Data sources

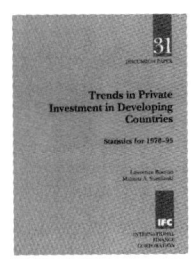

Private investment data are from the International Finance Corporation's *Trends in Private Investment in Developing Countries 1997* and World Bank estimates. Data on foreign direct investment are based on estimates compiled by the IMF in the *Balance of Payments Statistics Yearbook,* supplemented by World Bank staff estimates. Data on domestic credit are from the IMF's *International Financial Statistics,* and data on government expenditure are from the IMF's *Government Finance Statistics Yearbook.* External debt figures are from the World Bank's Debtor Reporting System as reported in *Global Development Finance 1998.*

5.2 | Stock markets

	Market capitalization				Value traded		Turnover ratio value of shares traded as % of capitalization		Listed domestic companies		IFC Investable index % change in price index	
	$ millions		% of GDP		% of GDP							
	1990	1997	1990	1996	1990	1996	1990	1997	1990	1996	1996	1997
Albania	..	..	..	..	..	..	..	..	..	..	..	..
Algeria	..	..	..	..	..	..	..	..	..	..	..	..
Angola	..	..	..	..	..	..	..	..	..	..	..	..
Argentina	3,268	59,252	2.3	15.2	0.6	1.5	33.6	49.5	179	147	18.7	17.4
Armenia	..	7	..	0.2	..	0.1	..	0.8	..	10	..	..
Australia	107,611	311,988	36.4	79.5	13.3	37.1	31.6	52.2	1,089	1,135		
Austria	11,476	33,953	7.2	15.0	11.7	9.1	110.3	61.8	97	106		
Azerbaijan	..	..	..	..	..	..	..	..	..	..	..	..
Bangladesh	321	4,551	1.4	14.3	0.0	2.3	1.5	24.2	134	186		
Belarus	..	..	..	..	..	..	..	..	..	..	..	..
Belgium	65,449	119,831	33.8	45.3	3.3	9.9	..	..	182	139		
Benin	..	..	..	..	..	..	..	..	..	..	..	..
Bolivia	..	114	..	1.6	..	0.0	..	0.6	..	10		
Bosnia and Herzegovina	..	..	..	..	..	..	..	..	..	..	..	..
Botswana	261	326	6.6	6.6	0.2	0.6	6.1	9.0	9	12		
Brazil	16,354	255,478	3.4	29.0	1.2	15.0	23.6	85.8	581	551	29.8	21.8
Bulgaria	..	7	..	0.1	..	0.0	..	0.1	..	15		
Burkina Faso	..	..	..	..	..	..	..	..	..	..	..	..
Burundi	..	..	..	..	..	..	..	..	..	..	..	..
Cambodia	..	..	..	..	..	..	..	..	..	..	..	..
Cameroon	..	..	..	..	..	..	..	..	..	..	..	..
Canada	241,920	486,268	42.6	83.9	12.5	45.8	26.7	62.2	1,144	1,265		
Central African Republic	..	..	..	..	..	..	..	..	..	..	..	..
Chad	..	..	..	..	..	..	..	..	..	..	..	..
Chile	13,645	72,046	44.9	88.8	2.6	11.4	6.3	10.5	215	291	–17.1	3.6
China	2,028	206,366	0.5	14.0	0.2	31.4	158.9	231.0	14	540	36.3	–25.0
Hong Kong, China	83,397	449,381	111.5	216.6	46.3	76.2	43.1	44.2	284	561		
Colombia	1,416	19,530	3.5	20.1	0.2	1.6	5.6	10.2	80	189	5.9	30.4
Congo, Dem. Rep.	..	..	..	..	..	..	..	..	..	..	..	..
Congo, Rep.	..	..	..	..	..	..	..	..	..	..	..	..
Costa Rica	311	782	5.5	8.7	0.1	0.2	5.8	3.5	82	114	..	..
Côte d'Ivoire	549	914	5.1	8.6	0.2	0.2	3.4	2.2	23	31	..	..
Croatia	..	581	..	3.2	..	0.3	..	0.0	..	61	..	..
Cuba	..	..	..	..	..	..	..	..	..	..	..	..
Czech Republic	..	12,786	..	32.9	..	15.4	..	47.9	..	1,588	16.9	–22.0
Denmark	39,063	71,688	30.3	41.1	8.6	19.9	28.0	54.2	258	237		
Dominican Republic	..	..	..	..	..	..	..	..	..	..	..	..
Ecuador	69	1,946	0.5	10.2	..	0.6	0.0	5.2	65	42	..	..
Egypt, Arab Rep.	1,765	20,830	4.1	20.9	0.3	3.6	..	33.5	573	646		
El Salvador	..	450	..	4.3	..	0.1	..	..	..	49		
Eritrea	..	..	..	..	..	..	..	..	..	..	..	..
Estonia	..	..	..	..	..	..	..	..	..	..	..	..
Ethiopia	..	..	..	..	..	..	..	..	..	..	..	..
Finland	22,721	63,078	16.9	50.9	2.9	18.1	..	..	73	71		
France	314,384	591,123	26.3	38.4	9.8	18.0	..	..	578	686		
Gabon	..	..	..	..	..	..	..	..	..	..	..	..
Gambia, The	..	..	..	..	..	..	..	..	..	..	..	..
Georgia	..	..	..	..	..	..	..	..	..	..	..	..
Germany	355,073	670,997	22.9	28.5	22.1	32.7	139.3	123.2	413	681		
Ghana	76	1,492	1.2	23.5	0.0	0.3	0.0	1.1	13	21		
Greece	15,228	34,164	18.4	19.7	4.7	6.7	36.3	73.8	145	224	0.3	33.7
Guatemala	..	168	..	1.1	..	0.0	..	2.5	..	9		
Guinea	..	..	..	..	..	..	..	..	..	..	..	..
Guinea-Bissau	..	..	..	..	..	..	..	..	..	..	..	..
Haiti	..	..	..	..	..	..	..	..	..	..	..	..
Honduras	40	338	1.3	8.5	..	0.0	0.0	0.0	26	111	..	..

Stock markets | 5.2

	Market capitalization				Value traded		Turnover ratio		Listed domestic companies		IFC Investable index	
	$ millions		% of GDP		% of GDP		value of shares traded as % of capitalization				% change in price index	
	1990	1997	1990	1996	1990	1996	1990	1997	1990	1996	1996	1997
Hungary	505	14,975	1.5	11.8	0.4	3.7	6.4	73.4	21	45	99.9	60.0
India	38,567	128,466	12.9	34.4	7.3	30.7	65.9	41.6	6,200	8,800	–2.0	5.8
Indonesia	8,081	29,105	7.1	40.3	3.5	14.2	75.8	64.2	125	253	16.4	–73.6
Iran, Islamic Rep.	34,282	17,008	28.0	..	4.3	..	30.4	22.2	97	220	..	..
Iraq	..	..	..	..	..	..	..	..	..	..	..	..
Ireland	..	12,243	..	17.6	..	6.7	..	24.5	..	76	..	..
Israel	3,324	45,268	6.0	39.6	10.1	10.0	95.8	26.7	216	655	..	21.7
Italy	148,766	258,160	13.6	21.4	3.9	8.5	26.8	43.8	220	244	..	..
Jamaica	911	1,887	21.4	42.6	0.8	0.9	3.4	2.5	44	46	..	..
Japan	2,917,679	3,088,850	98.4	67.2	54.0	27.2	43.8	37.1	..	53	13.4 [a]	..
Jordan	2,001	5,446	49.8	62.7	10.1	4.1	20.0	9.7	105	98	1.1	12.6
Kazakhstan	..	..	..	..	..	..	..	..	..	..	..	..
Kenya	453	1,846	5.3	20.0	0.1	0.7	2.2	3.7	54	56	..	..
Korea, Dem. Rep.	..	..	..	..	..	..	..	..	..	..	..	..
Korea, Rep.	110,594	41,881	43.6	28.6	29.9	36.6	61.3	172.3	669	760	–38.7	–68.9
Kuwait	..	18,817	..	51.1	..	24.0	..	..	2,071	2,334	..	..
Kyrgyz Republic	..	5	..	0.3	..	0.0	..	3.7	..	27	..	..
Lao PDR	..	..	..	..	..	..	..	..	..	..	..	..
Latvia	..	148	..	2.9	..	0.3	..	14.6	..	34	..	..
Lebanon	..	..	..	..	..	..	..	..	..	..	..	..
Lesotho	..	..	..	..	..	..	..	..	..	..	..	..
Libya	..	..	..	..	..	..	..	..	..	..	..	..
Lithuania	..	900	..	11.6	..	0.6	..	8.9	..	460	..	..
Macedonia, FYR	..	..	..	..	..	..	..	..	..	..	..	..
Madagascar	..	..	..	..	..	..	..	..	..	..	..	..
Malawi	..	..	..	..	..	..	..	..	..	..	..	..
Malaysia	48,611	93,608	113.6	309.6	25.4	174.9	24.6	72.6	282	621	24.1	–72.9
Mali	..	..	..	..	..	..	..	..	..	..	..	..
Mauritania	..	..	..	..	..	..	..	..	..	..	..	..
Mauritius	268	1,676	10.1	39.0	0.2	1.8	1.9	5.4	13	40	..	..
Mexico	32,725	156,595	12.5	31.8	4.6	12.9	44.1	39.7	199	193	16.9	48.8
Moldova	..	..	..	..	..	..	..	..	..	..	..	..
Mongolia	..	..	..	..	..	..	..	..	..	..	..	..
Morocco	966	12,177	3.7	23.6	0.2	1.2	..	10.2	71	47	..	..
Mozambique	..	..	..	..	..	..	..	..	..	..	..	..
Myanmar	..	..	..	..	..	..	..	..	..	..	..	..
Namibia	21	473	0.7	14.6	0.0	1.2	0.0	12.1	3	12	..	..
Nepal	..	208	..	4.7	..	0.1	..	2.3	..	90	..	..
Netherlands	119,825	378,721	46.4	96.5	15.6	86.5	29.0	92.4	260	217	..	..
New Zealand	8,835	38,288	24.5	58.8	5.4	15.2	17.3	28.1	171	158	..	..
Nicaragua	..	..	..	..	..	..	..	..	..	..	..	..
Niger	..	..	..	..	..	..	..	..	..	..	..	..
Nigeria	1,372	3,646	4.8	11.1	0.0	0.2	0.9	3.9	131	183	..	..
Norway	26,130	57,423	21.7	36.4	11.6	22.7	54.4	70.3	112	158	..	..
Oman	945	2,673	9.0	16.4	1.1	1.8	12.3	21.3	55	143	..	..
Pakistan	2,850	10,966	7.1	16.4	0.6	9.3	8.7	103.7	487	782	–19.3	26.9
Panama	226	831	3.4	10.5	0.0	0.1	0.9	1.1	13	16	..	..
Papua New Guinea	..	..	..	..	..	..	..	..	..	..	..	..
Paraguay	..	383	..	4.0	..	0.3	..	11.3	..	60	..	..
Peru	812	17,586	2.5	20.2	0.3	6.2	19.3	25.6	294	231	–0.2	12.5
Philippines	5,927	31,361	13.4	96.2	2.7	30.4	13.6	34.8	153	216	13.1	–61.6
Poland	144	12,135	0.2	6.2	0.0	4.1	89.7	78.4	9	83	71.8	–18.5
Portugal	9,201	38,954	13.7	23.7	2.5	6.9	16.9	66.8	181	158	26.2	44.4
Puerto Rico	..	..	..	..	..	..	..	..	..	..	..	..
Romania	..	61	..	0.2	..	0.0	..	7.2	..	17	..	..
Russian Federation	244	128,207	0.0	8.5	..	0.7	..	19.4	13	73	..	..

	Market capitalization				Value traded		Turnover ratio		Listed domestic companies		IFC Investable index	
	$ millions		% of GDP		% of GDP		value of shares traded as % of capitalization				% change in price index	
	1990	1997	1990	1996	1990	1996	1990	1997	1990	1996	1996	1997
Rwanda	..	..	..	..	..	..	..	..	..	..	..	..
Saudi Arabia	..	40,961	..	32.7	..	5.0	..	15.6	..	69	..	..
Senegal	..	..	..	..	..	..	..	..	..	..	..	..
Sierra Leone	..	..	..	..	..	..	..	..	..	..	..	..
Singapore	34,308	150,215	91.6	159.7	54.2	45.4	..	28.7	150	223	..	..
Slovak Republic		1,826	..	11.5	..	12.2	..	109.4	..	816	..	..
Slovenia	..	663	..	3.6	..	2.2	..	68.8	24	21	..	..
South Africa	137,540	232,069	128.9	191.3	7.6	21.5	..	18.3	732	626	–19.2	–13.8
Spain	111,404	242,779	22.6	41.7	8.3	42.8	..	..	427	357	..	..
Sri Lanka	917	2,096	11.4	13.3	0.5	1.0	5.8	15.6	175	235	–8.6	22.3
Sudan	..	..	..	..	..	..	..	..	..	..	..	..
Sweden	97,929	247,217	43.2	98.8	7.8	54.7	14.9	64.4	258	229	..	..
Switzerland	160,044	402,104	70.1	137.0	29.6	133.9	0.0	94.0	182	213	..	..
Syrian Arab Republic	..	..	..	..	..	..	..	..	..	..	..	..
Tajikistan	..	..	..	..	..	..	..	..	..	..	..	..
Tanzania	..	..	..	..	..	..	..	..	..	..	..	..
Thailand	23,896	23,538	27.9	53.9	26.7	24.0	92.6	39.2	214	454	–41.1	–78.8
Togo	..	..	..	..	..	..	..	..	..	..	..	..
Trinidad and Tobago	696	1,405	13.7	25.7	1.1	2.0	10.0	8.3	30	23	..	..
Tunisia	533	4,263	4.3	21.8	0.2	1.4	3.3	6.8	13	30	..	..
Turkey	19,065	61,090	12.7	16.5	3.9	20.3	42.5	113.5	110	229	42.3	109.9
Turkmenistan	..	..	..	..	..	..	..	..	..	..	..	..
Uganda	..	..	..	..	..	..	..	..	..	..	..	..
Ukraine	..	..	..	..	..	..	..	..	..	..	..	..
United Arab Emirates	..	..	..	..	..	..	..	..	..	..	..	..
United Kingdom	848,866	1,740,246	87.0	151.9	28.6	50.5	33.3	36.8	1,701	2,433	23.0 [b]	..
United States	3,059,434	8,484,433	55.1	115.6	31.5	97.0	53.4	92.8	6,599	8,479	20.3 [c]	..
Uruguay	38	266	0.5	1.5	0.0	0.0	0.0	1.7	36	18	..	..
Uzbekistan		128	..	0.5	..	0.3	..	..	..	4	..	..
Venezuela	8,361	14,581	17.2	14.9	4.6	1.9	43.0	31.0	76	88	117.7	25.7
Vietnam	..	..	..	..	..	..	..	..	..	..	..	..
West Bank and Gaza	..	..	..	..	..	..	..	..	..	..	..	..
Yemen, Rep.	..	..	..	..	..	..	..	..	..	..	..	..
Yugoslavia, FR (Serb./Mont.)	434		..	..	..	..	..	..	24	25	..	..
Zambia	..	229	..	6.8	..	0.1	..	1.0	..	5	..	..
Zimbabwe	2,395	1,969	35.2	48.1	0.7	3.4	2.9	17.1	57	64	72.4	–46.8

World	9,399,355 s	20,177,662 s	51.7 w	70.6 w	29.0 w	48.2 w	47.6 w	77.1 w	29,189 s	42,404 s	..	..
Low income	47,424	265,322	8.3	19.5	3.4	27.4	100.9	205.2	7,261	11,150		
Excl. China & India	8,857	28,962	..	..	..	..	..	..	1,061	1,810		
Middle income	328,104	1,461,083	21.6	40.0	10.1	15.0	..	39.5	4,195	8,895		
Lower middle income	73,734	436,341	9.8	25.8	..	9.4	..	34.5	1,859	3,664		
Upper middle income	254,370	1,024,742	23.3	51.6	3.5	19.6	29.6	43.5	2,336	5,231		
Low & middle income	375,528	1,726,405	18.5	34.4	5.5	18.4	..	85.1	11,456	20,045		
East Asia & Pacific	86,515	692,427	16.3	49.1	6.6	37.7	118.3	208.5	774	2,084		
Europe & Central Asia	19,065	103,563	2.1	10.5	..	6.0	..	45.5	110	3,428		
Latin America & Carib.	78,506	481,799	7.4	27.4	2.1	9.9	29.5	39.0	1,748	2,191		
Middle East & N. Africa	6,210	51,373	..	24.2	..	2.6	..	14.4	817	1,184		
South Asia	42,655	139,879	11.6	29.7	6.0	24.7	54.5	23.1	6,996	10,102		
Sub-Saharan Africa	142,577	257,364	56.3	123.1	4.1	13.3	..	8.1	1,011	1,056		
High income	9,023,827	18,451,257	56.4	78.5	32.5	54.6	48.8	75.1	17,733	22,359	..	..

a. Data refer to the Nikkei index. b. Data refer to the FT 100 index. c. Data refer to the S&P 500 index.

Stock markets | 5.2

About the data

Financial market development is closely related to an economy's overall development. At low levels of development, commercial banks tend to dominate the financial system. As economies grow, specialized financial intermediaries and equity markets develop.

A variety of measures are needed to gauge a country's level of stock market development Single measures suffer from conceptual and statistical weaknesses such as inaccurate reporting and different accounting standards. The stock market indicators presented in the table include measures of size (market capitalization and number of listed domestic companies) and liquidity (value traded as a percentage of GDP and turnover ratio). The percentage change in stock market prices in U.S. dollars comes from the International Finance Corporation's Investable (IFCI) index, an important measure of performance. Regulatory and institutional factors that can boost investor confidence, such as the existence of a securities and exchange commission and the quality of investor protection laws, influence the functioning of stock markets but are not included in this table.

Stock market size can be measured in a number of ways, each of which may produce a different ranking among countries. Market capitalization in U.S. dollars gives the overall size of the stock market and as a percentage of GDP. The number of listed domestic companies is an additional measure of market size. Market size is positively correlated with the ability to mobilize capital and diversify risk.

Market liquidity, the ability to easily buy and sell securities, is measured by dividing the total value traded by GDP. This indicator complements the market capitalization ratio by showing whether market size is matched by trading. The turnover ratio—shares traded as a percentage of market capitalization—is also a measure of liquidity as well as of transactions costs. (High turnover indicates low transactions costs.) The turnover ratio also complements the ratio of value traded to GDP, because turnover is related to the size of the market and the value traded ratio to the size of the economy. A small, liquid market will have a high turnover ratio but a small value traded ratio. Liquidity is an important attribute of stock market development because, in theory, liquid markets improve the allocation of capital and enhance prospects for long-term economic growth. A more comprehensive measure of liquidity would include trading costs and the time and uncertainty in finding a counterpart in settling trades.

The IFC has developed a series of indexes for investors interested in investing in stock markets in developing countries. The IFC Investable (IFCI) index series includes country indexes (shown in table 5.2), regional indexes, and the Composite index. The IFCI Composite index tracks 1,426 stocks with a market value of more than $769 billion in 31 emerging markets. The IFCI Composite index is the broadest index available, designed to measure returns on emerging market stocks that are legally and practically open to foreign portfolio investment. It is a widely used benchmark for international portfolio management purposes. See IFC's *The IFC Indexes: Methodology, Definitions, and Practices* booklet for further information on the IFCI indexes.

Figure 5.2a

Emerging stock markets have been volatile

% change

| | Composite emerging markets index | Latin America | Asia | Developing Europe, the Middle East, and Africa |

● 1996   ○ 1997

Source: IFC Emerging Markets Database.

The International Finance Corporation's Investable (IFCI) index covers 31 emerging markets. The figure shows the percentage change (in U.S. dollar terms) in emerging market stock prices for the composite index, Latin America, Asia, and developing Europe, the Middle East, and Africa.

It was a rollercoaster year for emerging markets in 1997. A stock market boom in Latin America and Eastern Europe was accompanied by substantial portfolio equity flows, but the onset of the East Asian financial crisis in July 1997—and its effects on other regions later in the year—led to a general retreat from new investments in emerging markets.

Definitions

• **Market capitalization** (also known as market value) is the share price times the number of shares outstanding. • **Value traded** refers to the total value of shares traded during the period. • **Turnover ratio** is the total value of shares traded during the period divided by the average market capitalization for the period. Average market capitalization is calculated as the average of the end-of-period values for the current period and the previous period. **Listed domestic companies** is the number of domestically incorporated companies listed on the country's stock exchanges at the end of the year. This indicator does not include investment companies, mutual funds, or other collective investment vehicles. • **IFC investable index price change** is the U.S. dollar price change in the stock markets covered by the IFCI index.

Data sources

Data are from the IFC's *Emerging Stock Markets Factbook 1997*, with supplemental data from IFC. IFC collects data through an annual survey of the world's stock exchanges, supplemented by information provided by Reuters and IFC's network of correspondents. GDP data are from the World Bank's national accounts data files. *About the data* is based on Demirgüç-Kunt and Levine (1996b).

	Entry and exit regulations			Composite ICRG risk rating	Institutional Investor credit rating[a]	Euromoney country credit-worthiness rating	Moody's sovereign long-term debt rating		Standard & Poor's sovereign long-term debt rating	
							Foreign currency	Domestic currency	Foreign currency	Domestic currency
	Entry 1996	Repatriation of income 1996	Repatriation of capital 1996	December 1997	September 1997	September 1997	January 1998	January 1998	January 1998	January 1998
Albania	..	..	..	52.8	11.6	21.7	..	..	..	..
Algeria	..	..	..	59.0	24.5	37.2	..	..	..	..
Angola	..	..	..	44.8	13.6	22.7	..	..	..	..
Argentina	Free	Free	Free	75.5	41.3	63.2	Ba3	Ba3	BB	BBB–
Armenia	..	..	..	..	..	25.6	..	..	..	..
Australia	..	..	..	83.5	73.3	91.5	Aa2	Aaa	AA	AAA
Austria	..	..	..	82.5	86.5	94.4	Aaa	Aaa	AAA	AAA
Azerbaijan	..	..	..	..	..	25.1	..	..	..	..
Bangladesh	Free	Free	Free	66.0	28.5	44.6	..	..	..	..
Belarus	..	..	..	..	14.2	29.1	..	..	..	..
Belgium	..	..	..	78.5	81.3	92.2	Aa1	Aa1	AA+	AAA
Benin	..	..	..	..	17.4	25.0	..	..	..	..
Bolivia	..	..	..	68.5	26.2	44.1	..	..	..	..
Bosnia and Herzegovina	..	..	..	..	..	..	..	..	..	..
Botswana	Free	Free	Free	81.0	51.2	53.4	..	..	..	..
Brazil	Free	Free	Free	71.3	39.5	56.7	B1	..	BB–	BB+
Bulgaria	Free	Free	Free	56.5	22.2	37.8	B3	..	..	..
Burkina Faso	..	..	..	60.3	19.7	33.8	..	..	..	..
Burundi	..	..	..	..	..	..	..	..	..	..
Cambodia	..	..	..	..	..	20.3	..	..	..	..
Cameroon	..	..	..	58.8	18.8	32.5	..	..	..	..
Canada	..	..	..	84.8	82.1	95.4	Aa2	Aa1	AA+	AAA
Central African Republic	..	..	..	..	..	13.7	..	..	..	..
Chad	..	..	..	..	..	23.1	..	..	..	..
Chile	Rel. free	Free	Delayed [b]	80.3	63.5	78.2	..	..	A–	AA
China	Special	Free	Free	75.0	57.8	71.3	A3	..	BBB+	..
Hong Kong, China	..	..	..	82.5	63.9	85.5	..	..	A+	AA–
Colombia	Auth. only	Free	Free	58.8	47.2	60.3	Baa3	..	BBB–	A+
Congo, Dem. Rep.	..	..	..	35.3	..	..	..	..	..	..
Congo, Rep.	..	..	..	52.0	7.0	27.7	..	..	..	..
Costa Rica	Free	Free	Free	74.3	36.0	50.3	Ba1	..	BB	BB+
Côte d'Ivoire	Rel. free	Free	Free	63.8	20.1	37.0	..	..	..	..
Croatia	Free	Free	Free	..	33.6	52.7	Baa3	..	BBB–	A–
Cuba	..	..	..	59.3	11.3	9.5	..	..	..	..
Czech Republic	Free	Free	Free	76.5	..	71.7	Baa1	..	A	..
Denmark	..	..	..	85.3	82.6	95.1	Aa1	Aaa	AA+	AAA
Dominican Republic	..	..	..	74.3	24.8	42.8	..	..	B+	BB
Ecuador	Free	Free	Free	62.0	26.3	37.1	B1	..	..	..
Egypt, Arab Rep.	Free	Free	Free	71.3	39.7	55.4	..	..	BBB–	A–
El Salvador	..	..	..	73.5	27.5	54.2	..	..	BB	BBB+
Eritrea	..	..	..	..	..	..	..	..	..	..
Estonia	..	..	..	..	36.9	56.5	..	..	BBB+	A–
Ethiopia	..	..	..	66.3	17.1	25.2	..	..	..	..
Finland	..	..	..	86.0	76.6	92.8	Aa1	Aaa	AA	AAA
France	..	..	..	81.3	88.4	93.0	Aaa	Aaa	AAA	AAA
Gabon	..	..	..	68.3	24.5	36.4	..	..	..	..
Gambia, The	..	..	..	71.5	..	24.7	..	..	..	..
Georgia	..	..	..	..	9.5	17.0	..	..	..	..
Germany	..	..	..	80.5	91.3	95.4	..	Aaa	AAA	AAA
Ghana	Free	Free	Free	63.5	31.5	45.6	..	..	..	..
Greece	Free	Free	Free	76.3	53.0	77.8	Baa1	A2	..	..
Guatemala	..	..	..	73.5	26.8	51.0	..	..	..	..
Guinea-Bissau	..	..	..	43.8	..	15.4	..	..	..	..
Guinea	..	..	..	55.5	14.9	28.7	..	..	..	..
Haiti	..	..	..	54.0	14.0	23.6	..	..	..	..
Honduras	..	..	..	59.5	18.9	41.2	..	..	..	..

Portfolio investment regulation and risk | 5.3

	Entry and exit regulations			Composite iCRG risk rating	Institutional Investor credit rating[a]	Euromoney country credit-worthiness rating	Moody's sovereign long-term debt rating		Standard & Poor's sovereign long-term debt rating	
	Entry 1996	Repatriation of income 1996	Repatriation of capital 1996	December 1997	September 1997	September 1997	Foreign currency January 1998	Domestic currency January 1998	Foreign currency January 1998	Domestic currency January 1998
Hungary	Free	Free	Free	75.8	49.7	70.0	Baa3	..	BBB–	A–
India	Auth. only	Free	Free	68.8	46.9	61.4	..	..	BB+	BBB+
Indonesia	Rel. free	Restricted	Restricted	66.3	51.8	68.9	Baa3	..	BBB–	A–
Iran, Islamic Rep.	..	..	..	70.8	27.5	33.1	..	..	..	..
Iraq	..	..	..	32.3	7.9	3.6	..	..	..	..
Ireland	..	..	..	88.0	76.7	93.8	Aa1	Aaa	AA	AAA
Israel	..	..	..	67.5	52.9	76.6	A3	..	A–	AA–
Italy	..	..	..	81.3	75.4	86.9	Aa3	Aa3	AA	AAA
Jamaica	Rel. free	Free	Free	73.5	29.7	39.3	..	..	..	..
Japan	..	..	..	87.3	91.5	93.0	..	Aaa	AAA	AAA
Jordan	Free	Free	Free	74.8	34.9	53.4	Ba3	..	BB–	BBB–
Kazakhstan	..	..	..	..	24.0	45.9	Ba3	..	BB–	BB+
Kenya	Rel. free	Free	Free	62.3	28.6	40.5	..	..	..	..
Korea, Dem. Rep.	..	..	..	38.3	4.7	5.8	..	..	..	..
Korea, Rep.	Rel. free	Free	Free	79.3	69.7	80.3	..	..	B+	BBB–
Kuwait	..	..	..	76.5	55.0	76.2	..	..	A	A+
Kyrgyz Republic	..	..	..	..	..	20.7	..	..	..	..
Lao PDR	..	..	..	..	..	24.8	..	..	..	..
Latvia	Free	Free	Free	..	32.6	57.7	..	..	BBB	A–
Lebanon	..	..	..	61.3	32.4	50.8	B1	..	BB–	BB
Lesotho	..	..	..	..	..	32.1	..	..	..	..
Libya	..	..	..	63.0	27.8	16.5	..	..	..	..
Lithuania	Rel. free	Free	Free	..	31.1	56.7	Ba2	..	BBB–	BBB+
Macedonia, FYR	..	..	..	..	..	23.3	..	..	..	..
Madagascar	..	..	..	61.8	..	28.4	..	..	..	..
Malawi	..	..	..	66.5	21.0	30.5	..	..	..	..
Malaysia	Free	Free	Free	76.3	66.7	79.4	A1	..	A	AA
Mali	..	..	..	61.5	17.4	34.2	..	..	..	..
Mauritania	..	..	..	..	..	19.7	..	..	..	..
Mauritius	Free	Free	Free	..	51.9	71.3	Baa2	..	..	..
Mexico	Free	Free	Free	70.8	43.5	63.5	Ba2	Baa3	BB	BBB+
Moldova	..	..	..	..	..	36.3	Ba2	..	..	..
Mongolia	..	..	..	65.8	..	38.4	..	..	..	..
Morocco	..	..	..	67.5	40.9	53.8	..	..	..	..
Mozambique	..	..	..	48.5	14.6	24.1	..	..	..	..
Myanmar	..	..	..	61.0	21.0	32.2	..	..	..	..
Namibia	Free	Free	Free	81.0	..	33.0	..	..	..	..
Nepal	..	..	..	..	25.9	36.6	..	..	..	..
Netherlands	..	..	..	86.5	90.6	98.0	..	..	AAA	AAA
New Zealand	..	..	..	81.8	73.1	92.2	Aa1	Aaa	AA+	AAA
Nicaragua	..	..	..	52.8	13.5	25.1	..	..	..	..
Niger	..	..	..	..	..	30.6	..	..	..	..
Nigeria	Free	Free	Free	57.0	15.3	30.6	..	..	..	..
Norway	..	..	..	91.0	85.8	96.6	Aaa	Aaa	AAA	AAA
Oman	Rel. free	Free	Free	74.5	53.0	66.7	..	..	BBB–	..
Pakistan	Free	Free	Free	60.5	27.2	44.5	B2	..	B+	..
Panama	Free	Free	Free	72.8	33.6	57.2	Ba1	..	BB+	BB+
Papua New Guinea	..	..	..	68.5	32.3	47.0	..	..	..	..
Paraguay	..	..	..	72.8	33.5	48.8	..	..	BB–	BBB–
Peru	Free	Free	Free	64.8	33.7	50.1	..	..	BB	BBB–
Philippines	Special	Free	Free	72.3	44.3	64.6	Ba1	..	BB+	A–
Poland	Free	Free	Free	78.3	50.2	66.7	Baa3	..	BBB–	A–
Portugal	Free	Free	Free	82.3	71.2	91.1	Aa3	Aa2	AA–	AAA
Puerto Rico	..	..	..	..	..	..	..	..	..	..
Romania	..	..	..	60.3	34.1	50.5	Ba3	..	BB–	BBB–
Russian Federation	Free	Free	Free	66.0	27.5	49.7	Ba2	..	BB–	..

5.3 Portfolio investment regulation and risk

	Entry and exit regulations			Composite ICRG risk rating	Institutional Investor credit rating[a]	Euromoney country credit-worthiness rating	Moody's sovereign long-term debt rating		Standard & Poor's sovereign long-term debt rating	
	Entry 1996	Repatriation of income 1996	Repatriation of capital 1996	December 1997	September 1997	September 1997	Foreign currency January 1998	Domestic currency January 1998	Foreign currency January 1998	Domestic currency January 1998
Rwanda	..	..	..	..	..	20.5	..	..	..	..
Saudi Arabia	..	..	..	76.8	54.8	74.3	..	..	..	..
Senegal	..	..	..	64.0	21.2	34.5	..	..	..	..
Sierra Leone	..	..	..	46.8	6.5	19.1	..	..	..	..
Singapore	..	..	..	92.5	84.2	93.7	..	..	AAA	AAA
Slovak Republic	Free	Free	Free	75.5	44.8	60.4	Baa3	..	BBB–	A
Slovenia	Closed	Restricted	Restricted	..	36.9	73.0	A3	..	A	AA
South Africa	Free	Free	Free	73.8	46.4	67.8	Baa3	Baa1	BB+	BBB+
Spain	..	..	..	79.5	75.5	91.3	Aa2	Aa2	AA	AAA
Sri Lanka	Rel. free	Restricted	Restricted	64.3	32.1	46.3	..	..	..	..
Sudan	..	..	..	37.5	9.1	13.7	..	..	..	..
Sweden	..	..	..	81.3	76.2	90.7	Aa3	Aa1	AA+	AAA
Switzerland	..	..	..	86.3	92.2	93.7	..	..	AAA	AAA
Syrian Arab Republic	..	..	..	67.5	24.3	36.8	..	..	..	..
Tajikistan	..	..	..	..	..	25.6	..	..	..	..
Tanzania	..	..	..	63.3	18.7	32.6	..	..	..	..
Thailand	Rel. free	Free	Free	71.3	59.9	66.8	Baa3	..	BBB	A
Togo	..	..	..	58.3	16.9	30.7	..	..	..	..
Trinidad and Tobago	Rel. free	Free	Free	77.5	42.9	51.0	Ba1	..	BB+	BBB+
Tunisia	..	..	..	70.5	47.9	59.9	Baa3	..	BBB–	A
Turkey	Free	Free	Free	52.3	38.6	52.8	B1	..	B	..
Turkmenistan	..	..	..	..	..	24.6	..	..	..	..
Uganda	..	..	..	63.5	20.1	36.9	..	..	..	..
Ukraine	..	..	..	..	19.8	29.7	..	..	..	..
United Arab Emirates	..	..	..	77.8	60.1	77.3	..	..	..	..
United Kingdom	..	..	..	83.3	88.4	97.6	Aaa	Aaa	AAA	AAA
United States	..	..	..	80.8	92.1	100.0	..	Aaa	AAA	AAA
Uruguay	..	..	..	68.3	43.4	58.5	Baa3	..	BBB–	BBB+
Uzbekistan	..	..	..	..	19.5	39.4	..	..	..	..
Venezuela	Rel. free	Free	Free	70.8	35.4	52.2	Ba2	..	B+	..
Vietnam	..	..	..	66.8	32.5	50.3	..	..	..	..
West Bank and Gaza	..	..	..	..	..	..	..	..	..	..
Yemen, Rep.	..	..	..	..	..	..	..	..	..	..
Yugoslavia, FR (Serb./Mont.)	..	..	..	54.8	..	17.5	..	..	..	..
Zambia	Free	Free	Free	64.3	16.0	24.0	..	..	..	..
Zimbabwe	Free	Free	Free	63.3	33.8	40.9	..	..	..	..
World	..	..	..	**70.8 m**	**38.6 m**	**53.8 m**				
Low income	..	..	..	61.5	18.8	32.3	..	..	..	..
Excl. China & India	..	..	..	61.3	18.7	32.2	..	..	..	..
Middle income	..	..	..	70.8	35.4	53.4	..	..	..	..
Lower middle income	..	..	..	68.0	32.4	50.3	..	..	..	..
Upper middle income	..	..	..	75.5	48.1	66.7	..	..	..	..
Low & middle income	..	..	..	66.0	28.6	45.6	..	..	..	..
East Asia & Pacific	..	..	..	66.6	32.5	50.3	..	..	..	..
Europe & Central Asia	..	..	..	66.0	38.6	52.8	..	..	..	..
Latin America & Carib.	..	..	..	70.8	33.6	50.6	..	..	..	..
Middle East & N. Africa	..	..	..	70.5	34.9	53.4	..	..	..	..
South Asia	..	..	..	65.2	30.3	45.5	..	..	..	..
Sub-Saharan Africa	..	..	..	62.3	18.1	31.6	..	..	..	..
High income	..	..	..	82.4	76.7	92.2	..	..	..	..

Note: For explanations of the terms used to describe entry and exit regulations see *Definitions*.

a. This copyrighted material is reprinted with permission from Institutional Investor, Inc., 488 Madison Avenue, New York, NY 10022.   b. After one year.

Portfolio investment regulation and risk | 5.3

As investment portfolios become increasingly global, investors and governments seeking to attract foreign investment must have a good understanding of country risk. Risk, by its nature, is perceived differently by different groups. This table presents information on country risk and creditworthiness from several major international rating services.

The information on the regulation of entry to and exit from stock markets is reported by the International Finance Corporation (IFC) for emerging markets only. In many economies certain industries are considered strategic and are not open to foreign or nonresident investors. Or foreign investment in a company or in certain classes of stocks may be limited by national law or corporate policy. The regulations summarized in the table refer to "new money" investment by foreign institutions; other regulations may apply to capital invested through debt conversion schemes or to capital from other sources. The regulations shown here are formal ones. But even formal regulations may have very different effects in different countries because of the prevailing bureaucratic culture, the speed with which applications are processed, and the extent of red tape. The effect of entry and exit regulations may also be influenced by graft and corruption, which are impossible to quantify.

Most risk ratings are numerical or alphabetical. For numerical ratings, a higher number means lower risk. For alphabetical ratings, a letter closer to the beginning of the alphabet means lower risk. Readers should refer to the data sources for more details on the rating processes of the rating agencies. Risk ratings may be highly subjective, reflecting external perceptions that do not always capture the actual situation in a country. But these subjective perceptions are the reality that policymakers face in the climate they create for foreign private inflows. Countries that are not rated by credit risk rating agencies typically do not attract registered flows of private capital. Note that the risk ratings presented here are not endorsed by the World Bank but are included for their analytic usefulness.

Political Risk Services' *International Country Risk Guide* (ICRG) collects information on 22 components of risk, groups it into three major categories (political, financial, and economic), and converts it into a single numerical risk assessment ranging from 0 to 100. Ratings below 50 are considered very high risk, and those above 80 very low risk. Ratings are updated every month.

Institutional Investor country credit ratings are based on information provided by leading international banks. Responses are weighted using a formula that gives more importance to responses from banks with greater worldwide exposure and more sophisticated country analysis systems. Countries are rated on a scale of 0 to 100, and ratings are updated every six months.

Euromoney country creditworthiness ratings are based on analytical, credit, and market indicators. The ratings, also on a scale of 0 to 100, are based on polls of economists and political analysts supplemented by quantitative data such as debt ratios and access to capital markets.

Ratings of sovereign foreign and domestic currency debt by Moody's Investors Service are presented for obligations that extend longer than one year. These long-term ratings measure total expected credit loss over the life of the security; they are not intended to measure other risks in fixed income investment, such as market risk.

Standard & Poor's ratings of sovereign long-term foreign and domestic currency debt are based on current information furnished by the issuer or obtained by Standard & Poor's from other sources it considers reliable. The ratings reflect several risk factors, such as the likelihood of default and the capacity and willingness of the debtor to make timely payments of interest and repayments of principal in accordance with the terms of the obligation. The ratings measure the creditworthiness of the debtor and do not take into account exchange-related uncertainties for foreign currency debt.

• **Regulations on entry into emerging stock markets** are evaluated using the following terms: *free* (no significant restrictions), *relatively free* (some registration procedures required to ensure repatriation rights), *special classes* (foreigners restricted to certain classes of stocks designated for foreign investors), *authorized investors* only (only approved foreign investors may buy stocks), and *closed* (closed or access severely restricted, as for nonresident nationals only). • **Regulations on repatriation of income** (dividends, interest, and realized capital gains) **and repatriation of capital** from emerging markets are evaluated as free (repatriation done routinely) or restricted (repatriation requires registration with or permission of a government agency that may restrict the timing of exchange release). • **Composite International Country Risk Guide (ICRG) risk rating** is an overall index, ranging from 0 to 100, based on 22 components of risk. • **Institutional Investor credit rating** ranks, from 0 to 100, the chances of a country's default. • **Euromoney country creditworthiness rating** ranks, from 0 to 100, the riskiness of investing in an economy. • **Moody's sovereign foreign and domestic currency long-term debt rating** assesses the risk of lending to governments. Aaa bonds are judged to be of the best quality and C bonds of the lowest quality. Numerical modifiers 1–3 are applied to classifications from Aa to B, with 1 indicating that the obligation ranks at the high end of its rating category. • **Standard & Poor's sovereign foreign and domestic currency long-term debt ratings** are categorized as investment grade (AAA through BBB) and speculative grade (BB through C). Ratings from AA to CCC may be modified by the addition of a plus (+) or minus (–) sign to show relative standing within the rating category.

Data on emerging stock markets' entry and exit regulations are from the IFC's *Emerging Stock Markets Factbook 1997*. Information on country risk and creditworthiness are from several sources: Political Risk Services' monthly *International Country Risk Guide;* the monthly *Institutional Investor;* the monthly *Euromoney;* Moody's Investors Service's *Sovereign, Subnational and Sovereign-Guaranteed Issuers;* and Standard & Poor's *Sovereign List in Credit Week.*

5.4 | Financial depth and efficiency

	Domestic credit provided by banking sector		Liquid liabilities		Quasi-liquid liabilities		Interest rate spread		Spread over LIBOR	
							Lending minus deposit rate percentage points		Lending rate minus LIBOR percentage points	
	% of GDP		% of GDP		% of GDP					
	1990	1996	1990	1996	1990	1996	1990	1996	1990	1996
Albania	..	44.9	..	52.5	..	23.6	2.1	7.2	16.7	18.4
Algeria	74.7	42.4	73.6	12.9	24.9	12.9	..	..	..	..
Angola	..	..	..	..	..	..	..	..	..	..
Argentina	32.4	26.0	11.5	20.8	7.1	14.4	..	3.2	..	5.0
Armenia	62.3	9.1	83.5	7.9	42.9	1.9	..	34.2	..	60.8
Australia	104.4	88.7	63.4	63.7	51.1	45.9	6.8	..	12.2	..
Austria	123.0	130.9	89.4	91.7	75.7	75.2	..	..	..	..
Azerbaijan	57.2	11.3	33.5	9.8	11.6	2.7	..	..	..	156.5
Bangladesh	32.5	30.7	31.8	37.5	22.8	26.6	3.9	6.7	7.7	8.5
Belarus	..	16.1	..	15.2	..	6.5	..	31.9	..	58.8
Belgium	77.2	154.1	48.6	82.2	29.6	62.2	6.9	4.4	4.7	1.7
Benin	22.4	10.8	26.7	24.8	5.9	8.0	9.0	..	7.7	..
Bolivia	33.3	57.5	26.5	46.5	19.5	33.2	18.0	36.8	33.5	50.5
Bosnia and Herzegovina	..	..	..	..	..	..	..	..	..	..
Botswana	−49.7	−37.7	23.6	28.4	14.7	21.9	1.8	4.1	−0.4	9.0
Brazil	87.2	36.8	25.6	28.0	17.9	22.7	..	..	..	..
Bulgaria	126.8	150.6	73.4	73.8	53.6	59.7	9.9	48.8	43.4	118.0
Burkina Faso	13.7	6.7	21.3	23.1	7.5	5.4	9.0	..	7.7	..
Burundi	24.5	21.1	18.2	17.8	6.5	5.1	..	10.0	4.0	9.2
Cambodia	..	6.9	..	11.1	..	7.1	..	10.0	..	13.3
Cameroon	31.2	16.4	22.6	12.6	10.1	5.8	11.0	10.5	10.2	10.0
Canada	86.6	102.1	74.9	78.7	60.3	61.1	1.3	1.7	5.7	0.5
Central African Republic	13.1	11.2	15.6	23.2	1.9	1.7	11.0	10.5	10.2	10.0
Chad	14.4	15.3	20.7	19.9	0.8	0.9	11.0	10.5	10.2	10.0
Chile	72.9	59.6	40.6	40.8	32.7	32.1	8.6	3.9	40.5	11.9
China	90.0	98.0	79.2	112.2	41.4	67.0	0.7	2.6	1.0	4.6
Hong Kong, China	132.1	156.7	181.7	175.3	166.8	160.7	3.3	3.9	1.7	3.0
Colombia	36.2	45.5	29.8	38.6	19.3	27.3	8.8	10.8	36.9	36.5
Congo, Dem. Rep.	25.3	1.6	12.9	2.2	2.1	2.2	..	..	..	..
Congo, Rep.	29.1	16.2	22.0	15.1	6.1	2.5	11.0	10.5	10.2	10.0
Costa Rica	29.9	33.1	42.7	32.9	30.0	23.8	11.4	9.0	24.2	20.8
Côte d'Ivoire	44.5	29.1	28.8	27.2	10.9	9.5	9.0	..	7.7	..
Croatia	..	46.4	..	35.2	..	24.2	499.3	16.9	1,153.9	17.0
Cuba	..	..	..	..	..	..	..	..	..	..
Czech Republic	..	78.6	..	91.4	..	55.1	..	5.8	..	7.0
Denmark	65.1	58.7	60.9	59.6	30.3	29.3	6.2	5.9	5.8	3.2
Dominican Republic	31.3	29.5	25.7	27.0	13.1	16.9	..	..	..	..
Ecuador	17.2	31.7	23.3	33.3	12.9	24.3	−6.0	13.0	29.2	49.0
Egypt, Arab Rep.	106.8	82.8	87.9	81.7	60.7	62.3	7.0	5.0	10.7	10.1
El Salvador	32.0	41.4	30.6	44.2	19.6	33.4	3.2	4.6	12.9	13.1
Eritrea	..	..	..	..	..	..	..	..	..	..
Estonia	65.0	20.1	136.2	27.0	93.5	6.4	..	7.6	26.6	8.2
Ethiopia	67.5	45.1	42.2	42.1	12.6	17.7	3.6	4.5	−2.3	8.4
Finland	84.3	63.8	55.2	58.6	46.6	26.5	4.1	3.8	3.3	0.6
France	106.3	102.1	64.6	67.7	38.7	44.0	6.0	3.1	2.2	1.3
Gabon	20.0	15.0	17.8	14.4	6.6	5.0	11.0	10.5	10.2	10.0
Gambia, The	3.6	8.2	21.8	26.6	9.3	12.9	15.2	13.0	18.2	20.0
Georgia	..	..	..	..	..	..	..	..	..	..
Germany	110.0	136.7	64.4	68.1	44.2	45.4	4.5	7.2	3.3	4.5
Ghana	13.2	21.4	14.1	17.3	3.4	5.6	..	..	..	..
Greece	103.8	84.2	90.1	82.5	72.5	64.2	8.1	7.5	19.3	15.4
Guatemala	17.4	19.4	21.2	24.7	11.8	15.7	5.1	15.1	15.0	17.2
Guinea	5.5	8.2	9.3	8.9	1.2	2.0	0.2	4.0	12.9	15.5
Guinea-Bissau	0.7	0.1	0.3	0.2	0.1	0.1	13.1	4.5	37.4	46.2
Haiti	32.9	27.4	31.4	48.0	15.9	26.2	..	..	..	..
Honduras	40.9	26.0	33.6	34.0	18.8	21.1	8.3	13.0	8.7	24.2

Financial depth and efficiency | 5.4

	Domestic credit provided by banking sector		Liquid liabilities		Quasi-liquid liabilities		Interest rate spread		Spread over LIBOR	
	% of GDP		% of GDP		% of GDP		Lending minus deposit rate percentage points		Lending rate minus LIBOR percentage points	
	1990	1996	1990	1996	1990	1996	1990	1996	1990	1996
Hungary	82.6	49.1	43.8	42.9	19.0	24.1	4.1	6.5	20.5	26.6
India	54.7	49.8	45.8	49.4	29.9	32.3	..	..	8.2	10.4
Indonesia	45.5	54.6	40.4	52.5	29.1	42.7	3.3	2.0	12.5	13.7
Iran, Islamic Rep.	70.8	45.7	57.6	40.8	31.1	22.2	..	..	..	..
Iraq	..	..	..	..	..	..	..	..	..	..
Ireland	58.0	84.5	45.1	75.6	32.8	61.5	5.0	5.6	3.0	0.3
Israel	101.3	79.4	67.0	73.9	60.6	67.8	12.0	6.2	18.1	15.2
Italy	90.6	95.0	71.0	62.0	35.7	29.4	7.3	5.6	5.8	6.5
Jamaica	34.8	33.5	51.0	53.4	37.8	36.4	6.6	18.8	22.2	38.5
Japan	266.8	295.1	187.5	203.3	159.6	167.6	3.4	2.4	−1.4	−2.9
Jordan	117.9	86.1	131.2	91.6	77.8	61.9	3.3	3.5	1.7	4.0
Kazakhstan	..	9.5	..	..	..	..	..	..	..	..
Kenya	52.9	52.8	43.3	50.6	29.3	35.6	5.1	16.2	10.4	28.3
Korea, Dem. Rep.	..	..	..	..	..	..	..	..	..	..
Korea, Rep.	65.3	74.5	54.3	84.0	45.5	73.8	0.0	1.3	1.7	3.3
Kuwait	243.0	103.2	196.8	92.7	158.4	77.8	0.4	2.7	4.1	3.3
Kyrgyz Republic	..	..	..	..	..	..	..	27.9	..	53.1
Lao PDR	5.1	8.7	7.2	14.3	3.1	9.9	2.5	11.0	20.0	21.5
Latvia	..	13.0	..	23.5	..	8.3	..	14.1	..	20.3
Lebanon	132.6	104.2	193.7	141.6	170.9	133.0	23.1	9.7	31.6	19.7
Lesotho	30.1	−15.2	38.8	34.7	22.4	18.1	7.4	5.0	12.1	12.2
Libya	..	..	..	..	..	..	1.5	..	−1.3	..
Lithuania	..	11.5	..	17.4	..	5.8	..	18.7	..	16.1
Macedonia, FYR	..	..	..	..	..	..	..	..	..	..
Madagascar	26.8	14.6	18.2	21.4	5.4	8.5	5.3	13.8	17.5	27.2
Malawi	19.9	9.1	21.3	17.2	11.8	9.0	8.9	19.0	12.7	39.8
Malaysia	77.9	131.9	66.3	121.5	44.3	91.8	1.3	1.9	−1.1	3.5
Mali	13.4	10.1	20.0	23.3	5.4	5.6	9.0	..	7.7	..
Mauritania	54.7	14.0	28.5	16.5	7.0	5.6	5.0	..	1.7	..
Mauritius	45.1	63.9	63.3	76.2	49.1	63.4	5.4	10.0	9.7	15.3
Mexico	42.5	40.6	24.9	37.7	18.1	28.3	..	..	..	..
Moldova	62.8	22.0	70.3	16.3	35.4	4.7	..	11.2	..	31.2
Mongolia	68.5	17.5	52.4	22.4	13.8	11.0	..	55.5	..	86.4
Morocco	60.1	74.8	61.0	72.1	18.4	27.3	0.5	..	0.7	5.3
Mozambique	29.5	2.8	46.1	34.6	9.1	5.2	..	..	..	..
Myanmar	32.7	..	27.9	..	7.8	..	2.1	4.0	−0.3	11.0
Namibia	19.2	60.1	23.1	48.6	13.5	28.4	10.6	6.6	17.4	13.6
Nepal	28.9	36.0	32.2	38.3	18.5	23.9	..	..	6.1	7.4
Netherlands	107.5	124.6	84.1	84.3	60.1	57.1	8.4	2.4	3.4	0.4
New Zealand	74.4	89.7	65.4	79.5	32.2	41.3	4.4	3.8	7.7	6.8
Nicaragua	206.6	144.8	5.7	40.6	2.3	31.2	12.5	8.4	13.7	15.2
Niger	16.2	8.8	19.8	12.3	8.3	3.4	9.0	..	7.7	..
Nigeria	23.7	16.3	23.6	18.6	10.3	6.7	5.5	6.7	17.0	14.2
Norway	89.5	74.6	59.9	56.6	27.0	17.8	4.6	2.9	5.9	1.6
Oman	16.6	29.2	28.9	32.5	19.3	22.4	1.4	2.4	1.4	3.7
Pakistan	50.9	52.5	39.8	45.6	10.0	21.3	..	..	..	..
Panama	52.7	74.4	43.6	70.5	35.0	59.5	3.6	3.4	3.7	5.1
Papua New Guinea	35.8	28.0	35.2	36.4	24.0	20.7	6.9	1.1	7.2	7.8
Paraguay	14.9	28.2	21.4	31.2	12.8	22.6	8.1	11.7	22.7	23.4
Peru	16.2	12.1	19.9	18.8	9.5	13.2	2,335.0	11.2	4,766.2	20.6
Philippines	26.9	72.2	36.8	58.7	28.2	48.1	4.6	5.2	15.8	9.3
Poland	19.5	35.4	34.0	37.2	17.2	23.6	462.5	6.1	495.9	20.6
Portugal	73.6	99.7	65.1	81.7	39.7	53.4	7.8	5.4	13.5	6.2
Puerto Rico	..	..	..	..	..	..	..	..	..	..
Romania	79.7	4.2	60.4	27.1	32.7	17.3	..	..	..	..
Russian Federation	..	24.8	..	16.3	..	7.5	..	91.8	..	141.3

	Domestic credit provided by banking sector		Liquid liabilities		Quasi-liquid liabilities		Interest rate spread		Spread over LIBOR	
	% of GDP		% of GDP		% of GDP		Lending minus deposit rate percentage points		Lending rate minus LIBOR percentage points	
	1990	1996	1990	1996	1990	1996	1990	1996	1990	1996
Rwanda	17.0	10.6	14.8	17.5	7.0	6.3	6.3	..	4.9	..
Saudi Arabia	58.7	38.0	47.9	51.2	21.9	24.6	..	..	..	..
Senegal	33.7	21.8	22.9	21.2	9.7	8.2	9.0	..	7.7	..
Sierra Leone	26.3	52.3	14.5	9.9	4.0	3.8	12.0	18.2	44.2	26.6
Singapore	74.1	78.0	120.9	112.5	98.4	92.1	2.7	2.9	−1.0	0.7
Slovak Republic	..	60.1	..	70.6	..	40.7	..	4.6	..	8.4
Slovenia	36.8	36.0	34.2	36.7	25.8	28.6	179.9	8.7	847.5	18.2
South Africa	102.7	160.2	47.1	47.2	28.9	20.0	2.1	4.6	12.7	14.0
Spain	109.0	105.9	76.6	80.5	45.3	53.0	5.4	2.4	7.7	3.0
Sri Lanka	43.0	35.0	35.2	41.3	22.9	31.2	−6.4	0.2	4.7	10.8
Sudan	29.9	18.9	29.4	24.0	4.2	9.4	..	..	..	..
Sweden	145.5	68.2	46.6	45.5	..	..	6.8	4.9	8.4	1.9
Switzerland	179.0	183.4	146.7	147.3	119.9	119.0	−0.9	3.6	−0.9	−0.5
Syrian Arab Republic	56.6	45.7	54.7	53.4	10.5	13.9	..	..	..	..
Tajikistan	..	..	..	..	..	..	..	..	..	..
Tanzania	39.2	17.4	22.6	24.3	7.2	11.0	..	23.6	..	31.7
Thailand	90.7	98.8	74.7	79.5	65.8	70.4	4.3	5.9	8.2	9.6
Togo	21.3	25.8	36.1	25.8	19.1	9.1	9.0	..	7.7	..
Trinidad and Tobago	58.5	54.2	54.6	49.7	42.7	37.2	6.9	9.1	4.6	10.3
Tunisia	62.5	65.4	51.5	48.4	26.7	27.1	..	..	..	..
Turkey	25.9	34.4	24.1	32.3	16.4	27.2	..	..	..	..
Turkmenistan	..	1.7	..	9.7	..	0.8	..	..	..	..
Uganda	17.7	4.6	7.6	10.7	1.4	3.3	7.4	9.7	30.4	14.8
Ukraine	83.2	15.0	0.0	0.0	0.0	0.0	..	46.3	..	74.4
United Arab Emirates	35.2	48.6	47.0	56.7	38.2	42.2	..	..	..	..
United Kingdom	123.0	131.0	..	..	..	..	2.2	2.9	6.4	0.4
United States	114.4	137.5	68.6	61.1	51.8	43.5	..	..	1.7	2.8
Uruguay	60.7	39.8	61.2	38.2	53.3	32.1	76.6	63.4	166.1	86.0
Uzbekistan	..	..	..	..	..	..	..	..	..	..
Venezuela	37.4	19.9	41.1	22.3	29.4	10.4	0.5	4.1	19.9	26.2
Vietnam	15.9	20.8	22.7	20.3	9.3	8.3	..	10.4	..	22.3
West Bank and Gaza	..	..	..	..	..	..	..	..	..	..
Yemen, Rep.	62.0	29.0	56.3	40.5	10.7	16.6	..	..	..	..
Yugoslavia, FR (Serb./Mont.)	..	..	..	..	..	..	..	..	..	..
Zambia	64.5	56.2	23.8	17.9	12.6	11.2	9.5	11.7	26.8	48.3
Zimbabwe	53.8	55.3	54.0	50.7	39.2	32.3	2.9	12.7	3.4	28.7
World	**126.0 w**	**139.1 w**	**71.1 w**	**72.4 w**	**64.6 w**	**66.9 w**	..	..	..	..
Low income	64.6	73.6	54.4	80.8	30.0	48.6	..	..	..	..
Excl. China & India	37.9	32.6	28.3	30.0	12.6	16.3	..	..	..	..
Middle income	60.6	46.0	36.6	35.4	24.0	26.3	..	..	..	..
Lower middle income	52.0	45.3	44.4	38.9	30.1	30.0	..	..	..	..
Upper middle income	65.8	46.7	30.6	32.0	19.6	22.8	..	..	..	..
Low & middle income	61.7	53.9	41.7	48.4	25.7	32.7	..	..	..	..
East Asia & Pacific	76.5	88.2	66.7	92.9	41.6	61.6	..	..	..	..
Europe & Central Asia	..	31.9	..	28.9	..	18.0	..	..	..	..
Latin America & Carib.	59.7	35.7	23.5	26.9	17.6	21.7	..	..	..	..
Middle East & N. Africa	69.4	70.0	58.6	60.5	30.4	44.0	..	..	..	..
South Asia	52.4	48.3	43.4	47.3	27.0	30.3	..	..	..	..
Sub-Saharan Africa	59.6	84.5	37.0	38.5	18.5	15.8	..	..	..	..
High income	138.9	157.9	77.5	78.1	73.2	74.9	..	..	..	..

Financial depth and efficiency | 5.4

About the data

Households and institutions save and invest independently. The financial system's role is to intermediate between them and cycle funds. Savers accumulate claims on financial institutions, which pass these funds to their final users. As an economy develops, this indirect lending by savers to investors becomes more efficient and gradually increases financial assets relative to GDP. This wealth allows increased saving and investment, facilitating and enhancing economic growth. As more specialized savings and financial institutions emerge, more financing instruments become available, reducing risks and costs to liability holders. As securities markets mature, savers can invest their resources directly in financial assets issued by firms.

The ratio of domestic credit provided by the banking sector to GDP is used to measure the growth of the banking system because it reflects the extent to which savings are financial. Liquid liabilities include bank deposits of generally less than one year plus currency. Their ratio to GDP indicates the ease with which their owners can use them to buy goods and services without incurring any cost. Quasi-liquid liabilities are long-term deposits and assets—such as certificates of deposit, commercial paper, and bonds—that can be converted into currency or demand deposits, but at a cost.

No less important than the size and structure of the financial sector is its efficiency, as indicated by the margin between the cost of mobilizing liabilities and the earnings on assets. Small margins are crucial for economic growth because they lower interest rates and thus the overall cost of investment. Interest rates reflect the responsiveness of financial institutions to competition and price incentives. The interest rate spread, also known as the intermediation margin, is a summary measure of a banking system's efficiency. It may not be a reliable measure of efficiency to the extent that information about interest rates is inaccurate, banks do not monitor all bank managers, or the government sets deposit and lending rates. The spread over LIBOR reflects the interest rate differential between a country's lending rate and the London interbank offered rate (ignoring expected changes in the exchange rate). Interest rates are expressed as annual averages.

In some countries financial markets are distorted by restrictions on foreign investment, selective credit controls, and controls on deposit and lending rates. Interest rates may reflect the diversion of resources to finance the public sector deficit through statutory reserve requirements and direct borrowing from the banking system. And where state-owned banks dominate the financial sector, noncommercial considerations may unduly influence credit allocation.

The indicators in the table provide quantitative assessments of each country's financial sector, but qualitative assessments of policies, laws, and regulations are needed to analyze overall financial conditions. In addition, the accuracy of financial data depends on the quality of accounting systems, which are weak in some developing economies. Some of these indicators are highly correlated, particularly the ratios of domestic credit, liquid liabilities, and quasi-liquid liabilities to GDP, because changes in liquid and quasi-liquid liabilities flow directly from changes in domestic credit. Moreover, the precise definition of the financial aggregates presented varies by country. Data on domestic credit and liquid and quasi-liquid liabilities are cited on an end-of-year basis.

The indicators reported here do not capture the activities of the informal sector—which remains an important source of finance in developing economies. Personal credit or credit extended through community-based pooling of assets may be the only source of credit available to small farmers, small businesses, or home-based producers. And in financially repressed economies the rationing of formal credit forces many borrowers and lenders to turn to the informal market and self-financing.

Definitions

• **Domestic credit provided by banking sector** includes all credit to various sectors on a gross basis, with the exception of credit to the central government, which is net. The banking sector includes monetary authorities, deposit money banks, and other banking institutions for which data are available (including institutions that do not accept transferable deposits but do incur such liabilities as time and savings deposits). Examples of other banking institutions include savings and mortgage loan institutions and building and loan associations. • **Liquid liabilities** are also known as broad money, or M3. They are the sum of currency and deposits in the central bank (M0), plus transferable deposits and electronic currency (M1), plus time and savings deposits, foreign currency transferable deposits, certificates of deposit, and securities repurchase agreements (M2), plus travelers checks, foreign currency time deposits, commercial paper, and shares of mutual funds or market funds held by residents. • **Quasi-liquid liabilities** are the M3 money supply less M1. • **Interest rate** spread is the interest rate charged by banks on loans to prime customers minus the interest rate paid by commercial or similar banks for demand, time, or savings deposits. • **Spread over LIBOR** (London interbank offered rate) is the interest rate charged by banks on loans to prime customers minus LIBOR. LIBOR is the most commonly recognized international interest rate and is quoted in several currencies. The average three-month LIBOR on U.S. dollar deposits is used here.

Data sources

Data on credit, liabilities, and interest rates are collected from central banks and finance ministries and are reported in the print and electronic versions of the International Monetary Fund's *International Financial Statistics*.

Figure 5.4a

Interest rate spreads vary dramatically

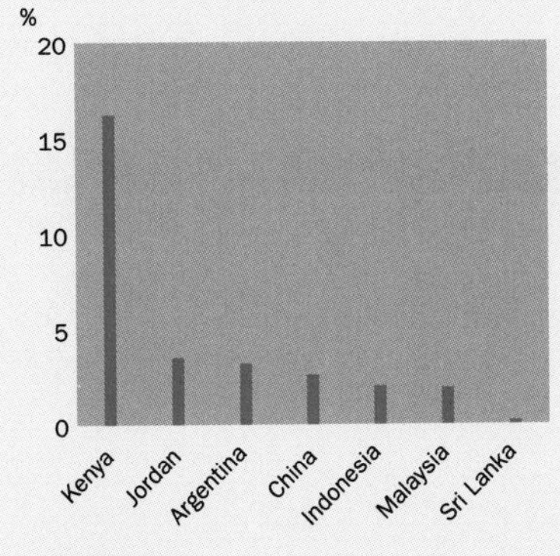

[bar chart: %, y-axis 0 to 20; Kenya ~16, Jordan ~4, Argentina ~3, China ~3, Indonesia ~2, Malaysia ~2, Sri Lanka ~2]

Source: Table 5.4.

The spread between banks' deposit and lending rates provides information about the efficiency of the financial sector, competition between institutions, the adequacy of margins for prudential purposes, the costs of reserve requirements and other banking regulations, and the effectiveness of monetary policy.

	Tax revenue	Taxes on income, profits, and capital gains		Domestic taxes on goods and services		Export duties		Import duties		Highest marginal tax rate		
	% of GDP	% of total taxes		% of value added of industry and services		% of exports		% of imports		Individual rate %	on income exceeding $	Corporate rate %
	1996	1980	1996	1980	1996	1980	1996	1980	1996	1997	1997	1997
Albania	16.6	..	10.7	..	19.0	..	..	..	10.0	..	..	..
Algeria	..	..	..	..	..	..	..	..	..	..	..	..
Angola	..	..	..	..	..	..	..	..	..	..	..	..
Argentina	11.9	0.0	11.2	2.8	5.0	0.0	0.2	0.0	8.5	33	120,000	33
Armenia	..	..	..	..	..	..	..	..	..	..	..	..
Australia	23.0	67.6	71.5	5.4	5.4	0.5	0.0	8.5	4.5	47	39,582	36
Austria	33.1	22.8	22.7	9.1	8.6	0.2	0.0	1.6	0.4	50	63,903	34
Azerbaijan	..	..	..	..	..	..	..	..	..	40	1,757	32
Bangladesh	..	14.8	..	5.7	..	3.9	..	16.4	..	..	..	..
Belarus	..	..	..	..	..	..	..	..	..	..	..	..
Belgium	43.3	40.2	35.9	..	..	..	..	..	..	55	75,507	39
Benin	..	..	..	..	..	1.6	..	..	..	..	..	..
Bolivia	11.8	..	7.6	..	..	..	0.0	..	5.1	13	..	25
Bosnia and Herzegovina	..	..	..	..	..	..	..	..	..	..	..	..
Botswana	17.5	45.5	51.3	0.3	2.0	0.1	0.0	21.8	17.1	30	16,680	15
Brazil	..	13.6	..	9.0	..	0.0	..	16.5	..	25	20,789	15
Bulgaria	25.2	..	25.4	..	9.6	..	..	..	5.5	40	2,630	36
Burkina Faso	..	20.1	..	2.9	..	3.6	..	20.7	..	..	..	..
Burundi	11.2	20.4	25.1	10.1	12.0	24.4	7.1	21.5	13.4	..	..	..
Cambodia	..	..	..	..	..	..	..	..	..	..	..	..
Cameroon	9.4	23.7	23.4	4.1	5.4	7.5	5.1	21.3	19.7	60	14,313	39
Canada	18.5	60.8	54.7	..	..	1.1	0.0	4.6	1.7	29	43,178	38
Central African Republic	..	17.7	..	6.0	..	9.2	..	23.9	..	..	..	..
Chad	..	..	..	..	..	..	..	..	..	..	..	..
Chile	18.3	22.0	21.9	12.4	..	..	..	..	..	45	6,588	15
China	5.2	..	12.0	..	5.1	..	..	..	3.2	45	12,051	30
Hong Kong, China	..	..	..	..	..	..	..	..	..	..	..	17
Colombia	13.6	28.9	40.8	3.4	7.6	6.7	0.0	12.3	8.7	35	49,934	35
Congo, Dem. Rep.	4.5	34.5	35.8	1.6	2.6	10.7	0.4	18.3	9.6	..	..	..
Congo, Rep.	..	63.8	..	3.1	..	0.1	..	14.0	..	..	..	45
Costa Rica	22.5	14.6	12.9	6.6	10.1	6.7	2.5	6.8	9.2	25	24,559	30
Côte d'Ivoire	..	14.0	..	7.9	..	8.0	..	28.9	..	10	4,489	35
Croatia	42.8	..	12.4	..	27.1	..	..	..	10.0	35	4,675	..
Cuba	..	..	..	..	..	..	..	..	..	..	..	..
Czech Republic	34.1	..	15.2	..	13.9	..	..	..	4.0	40	27,660	39
Denmark	35.3	40.6	45.4	..	..	..	..	0.1	0.1	60	..	34
Dominican Republic	14.7	24.8	17.2	3.8	6.3	6.2	0.0	15.0	27.1	..	..	..
Ecuador	13.9	46.6	56.5	2.5	4.6	3.0	0.0	16.3	8.3	25	61,861	20
Egypt, Arab Rep.	22.6	29.6	30.3	5.2	6.1	5.3	0.0	25.9	20.3	32	14,749	40
El Salvador	11.6	23.8	28.8	5.5	7.6	10.3	0.0	4.1	6.5	30	22,857	25
Eritrea	..	..	..	..	..	..	..	..	..	..	..	..
Estonia	31.0	..	17.5	..	17.7	..	..	..	0.2	26	..	26
Ethiopia	..	25.4	..	9.2	..	31.3	..	16.4	..	..	..	..
Finland	27.8	30.9	32.5	..	..	..	..	1.9	0.2	38	65,352	28
France	38.8	19.1	19.5	12.8	11.7	..	..	0.1	0.0	..	..	33
Gabon	..	60.2	..	1.8	..	1.4	..	31.7	..	55	..	40
Gambia, The	..	18.1	..	1.2	..	0.5	..	18.5	..	..	..	..
Georgia	..	..	..	..	..	..	..	..	..	..	..	..
Germany	29.3	19.4	17.1	..	..	..	..	0.0	0.0	53	77,406	30
Ghana	..	22.0	..	4.6	..	30.5	..	14.5	..	35	9,173	35
Greece	19.7	19.5	36.8	..	..	..	..	..	..	45	68,820	40
Guatemala	7.7	14.4	20.7	..	5.2	9.8	0.0	7.6	9.2	30	30,002	30
Guinea	..	..	..	..	..	..	..	..	..	..	..	..
Guinea-Bissau	..	..	..	..	..	..	..	..	..	..	..	..
Haiti	..	15.9	..	..	..	..	..	..	..	..	..	..
Honduras	..	32.9	..	5.1	..	7.5	..	7.8	..	40	196,382	15

Tax policies | 5.5

	Tax revenue	Taxes on income, profits, and capital gains		Domestic taxes on goods and services		Export duties		Import duties		Highest marginal tax rate		
	% of GDP	% of total taxes		% of value added of industry and services		% of exports		% of imports		Individual rate %	on income exceeding $	Corporate rate %
	1996	1980	1996	1980	1996	1980	1996	1980	1996	1997	1997	1997
Hungary	..	22.1	..	..	..	0.1	..	6.6	..	42	6,614	18
India	10.5	21.9	29.3	8.9	6.0	1.8	0.1	26.4	27.8	40	3,359	40
Indonesia	14.8	82.0	58.0	2.4	6.0	2.2	0.1	5.7	2.6	30	20,982	30
Iran, Islamic Rep.	6.7	12.2	29.0	1.0	1.7	..	..	20.9	5.4	54	172,063	10
Iraq	..	..	..	..	..	..	..	..	..	..	..	..
Ireland	34.7	38.6	42.1	..	..	..	..	6.2	3.9	48	15,732	36
Israel	33.4	47.3	42.7	17.4	..	..	..	4.6	..	50	57,730	36
Italy	40.7	32.1	34.8	..	..	..	..	0.1	0.0	51	196,005	37
Jamaica	..	35.1	..	15.6	..	..	..	2.1	..	25	1,449	33
Japan	..	74.7	..	2.5	..	..	..	2.3	..	50	258,398	38
Jordan	21.0	17.0	15.6	1.6	9.6	..	..	21.2	13.8	..	..	..
Kazakhstan	..	..	..	..	..	..	..	..	..	40	..	30
Kenya	21.3	33.3	30.7	14.8	18.7	1.4	0.0	10.7	14.2	35	374	35
Korea, Dem. Rep.	..	..	..	..	..	..	..	..	..	..	..	..
Korea, Rep.	18.6	25.5	33.3	9.4	7.5	..	..	7.7	4.7	40	94,764	28
Kuwait	1.2	63.6	15.4	0.2	0.0	..	..	3.0	3.4	0	..	55
Kyrgyz Republic	..	..	..	..	..	..	..	..	..	..	..	..
Lao PDR	..	..	..	..	..	..	..	..	..	..	..	..
Latvia	25.5	..	8.1	..	16.3	..	..	..	1.8	25	..	25
Lebanon	11.6	..	9.1	..	1.1	..	..	..	11.2	..	..	..
Lesotho	..	15.6	..	5.4	..	3.5	..	15.8	..	..	..	..
Libya	..	..	..	..	..	..	..	..	..	..	..	..
Lithuania	22.0	..	13.8	..	14.2	..	0.0	..	1.8	33	..	29
Macedonia, FYR	..	..	..	..	..	..	..	..	..	..	..	..
Madagascar	8.2	17.1	18.8	8.4	3.3	3.0	0.9	17.6	28.6	..	..	..
Malawi	..	38.9	..	11.7	..	0.0	..	16.8	..	38	2,763	38
Malaysia	20.1	41.9	45.3	5.7	7.3	9.1	0.6	9.1	3.1	30	58,893	30
Mali	..	20.5	..	7.9	..	3.0	..	8.0	..	..	..	..
Mauritania	..	..	..	..	..	..	..	..	..	..	..	..
Mauritius	16.3	17.3	15.7	4.8	6.1	8.6	0.0	16.2	14.1	30	2,764	35
Mexico	12.8	36.9	32.4	8.3	8.7	0.6	0.0	10.1	3.7	35	21,173	34
Moldova	..	..	..	..	..	..	..	..	..	..	..	..
Mongolia	18.7	..	29.2	..	7.3	..	..	..	5.0	..	..	..
Morocco	..	22.0	..	9.9	..	2.3	..	22.3	..	44	6,814	35
Mozambique	..	..	..	..	..	..	..	..	..	..	..	..
Myanmar	3.7	4.9	34.4	12.7	4.2	..	..	19.3	45.7	..	..	..
Namibia	..	..	..	..	..	..	..	..	..	35	17,152	35
Nepal	10.4	6.6	15.3	8.0	9.0	5.4	1.2	16.0	10.4	..	..	..
Netherlands	42.6	33.1	27.7	10.6	10.8	..	..	..	..	60	55,730	36
New Zealand	32.8	75.0	67.7	6.9	..	0.1	0.0	4.4	4.0	33	21,848	33
Nicaragua	23.9	8.9	11.8	11.3	16.4	3.9	0.0	8.0	11.7	30	20,202	30
Niger	..	28.1	..	4.5	..	2.6	..	17.0	..	..	..	..
Nigeria	..	..	..	..	..	..	..	..	..	25	754	30
Norway	32.5	30.3	22.3	15.2	15.9	0.1	0.0	0.8	1.0	..	..	28
Oman	8.5	92.4	77.4	0.2	..	..	..	1.4	..	0	..	50
Pakistan	15.3	16.8	20.3	8.6	10.8	1.9	0.0	25.2	28.4	35	7,485	46
Panama	17.2	29.0	29.7	5.2	5.2	..	1.4	..	8.7	30	200,000	30
Papua New Guinea	18.9	67.5	58.3	4.2	3.2	1.4	2.7	8.0	16.0	35	14,900	25
Paraguay	..	16.6	..	2.7	..	0.5	..	8.4	..	0	..	30
Peru	14.0	28.1	23.9	7.1	8.1	10.9	0.0	17.1	13.0	30	49,923	30
Philippines	16.7	23.6	37.1	7.8	6.4	1.0	0.0	13.4	12.2	35	19,016	35
Poland	36.0	..	28.3	..	12.3	..	0.0	..	10.7	44	14,542	40
Portugal	32.1	20.9	27.3	..	..	0.0	0.0	4.4	0.0	40	39,247	40
Puerto Rico	..	..	..	..	..	..	..	..	..	33	50,000	20
Romania	26.4	0.0	34.0	..	9.0	..	..	0.0	6.0	60	3,600	38
Russian Federation	17.4	..	15.8	..	7.8	..	4.3	..	3.1	35	8,587	35

	Tax revenue	Taxes on income, profits, and capital gains		Domestic taxes on goods and services		Export duties		Import duties		Highest marginal tax rate		
	% of GDP	% of total taxes		% of value added of industry and services		% of exports		% of imports		Individual rate %	on income exceeding $	Corporate rate %
	1996	1980	1996	1980	1996	1980	1996	1980	1996	1997	1997	1997
Rwanda	..	20.7	..	5.3	..	..	..	..	..	..	..	..
Saudi Arabia	..	..	..	..	..	..	..	..	..	0	..	45
Senegal	..	21.4	..	7.5	..	3.1	..	26.9	..	50	24,141	..
Sierra Leone	7.7	25.0	16.3	4.1	4.7	10.8	0.0	19.8	17.8	..	..	..
Singapore	16.2	47.0	41.5	4.1	5.3	..	..	0.9	..	28	285,836	26
Slovak Republic	..	..	..	..	..	..	..	..	..	42	33,861	..
Slovenia	..	..	..	..	..	..	..	..	..	..	..	..
South Africa	25.9	64.0	54.3	6.4	11.8	0.1	0.0	3.0	2.5	45	21,440	35
Spain	28.9	25.2	31.9	..	..	..	..	6.0	0.0	56	79,896	35
Sri Lanka	16.9	16.4	15.9	8.0	14.6	22.0	0.0	9.6	9.2	35	5,293	35
Sudan	..	17.2	..	6.0	..	3.3	..	33.5	..	..	..	..
Sweden	37.2	21.1	12.7	..	..	..	..	1.5	1.0	30	30,326	28
Switzerland	21.5	15.1	12.7	..	..	..	..	4.0	4.9	13	460,382	46
Syrian Arab Republic	18.4	24.7	28.9	1.8	..	1.7	..	11.6	..	..	..	..
Tajikistan	..	..	..	..	..	..	..	..	..	..	..	..
Tanzania	..	35.2	..	..	..	11.2	..	9.2	..	35	14,075	35
Thailand	16.9	19.3	35.0	8.6	8.5	4.4	0.1	11.1	6.6	37	158,479	30
Togo	..	38.6	..	6.4	..	3.3	..	15.2	..	..	..	..
Trinidad and Tobago	24.2	85.7	58.7	1.6	7.9	..	..	9.8	5.3	35	8,103	35
Tunisia	24.9	19.2	18.8	8.7	7.2	1.0	0.2	18.8	19.6	..	..	..
Turkey	15.2	61.8	38.6	5.1	11.4	..	..	8.9	3.2	55	14,877	25
Turkmenistan	..	..	..	..	..	..	..	..	..	..	..	..
Uganda	..	11.8	..	4.6	..	55.8	..	15.8	..	30	4,800	30
Ukraine	..	..	..	..	..	..	..	..	..	..	..	..
United Arab Emirates	0.6	..	0.0	0.0	..	..	..	..	..	..	..	..
United Kingdom	33.6	43.4	38.9	..	..	0.0	0.0	0.1	0.1	40	44,692	33
United States	19.4	61.6	58.2	0.9	..	..	..	3.0	2.6	40	271,050	35
Uruguay	29.2	11.5	14.2	11.2	11.3	0.0	0.1	19.2	6.1	0	..	30
Uzbekistan	..	..	..	..	..	..	..	..	..	..	..	..
Venezuela	14.5	79.4	54.1	1.0	5.9	..	..	9.6	10.2	34	..	34
Vietnam	..	..	..	..	..	..	..	..	..	50	6,278	25
West Bank and Gaza	..	..	..	..	..	..	..	..	..	..	..	..
Yemen, Rep.	9.9	..	35.2	..	2.9	..	..	..	20.3	..	..	..
Yugoslavia, FR (Serb./Mont.)	..	..	..	..	..	..	..	..	..	..	..	..
Zambia	16.7	41.1	39.6	12.5	9.6	0.0	0.0	7.2	6.0	30	1,376	35
Zimbabwe	..	57.9	..	8.4	..	..	..	18.4	..	40	5,597	38

Tax policies | 5.5

Taxes are compulsory, unrequited payments made to governments by individuals, businesses, or institutions. They are considered unrequited because governments provide nothing specifically in return for them, although they typically are used to provide goods or services to individuals or communities. The sources of the revenue received by governments and the relative contributions of these sources are determined by policy choices about where and how to impose taxes and by changes in the structure of the economy. Tax policy may reflect concerns about distributional effects, economic efficiency (including corrections for externalities), and the practical problems of administering a tax system. There is no ideal level of taxation. But taxes influence incentives, and hence the behavior of economic actors and country competitiveness.

The level of taxation is typically measured by tax revenue as a share of GDP. Comparing levels of taxation across countries provides a quick overview of the fiscal obligations and incentives facing the private sector. In this table tax data measured in local currencies are normalized by scaling variables in the same units to ease cross-country comparisons. The table refers only to central government data, which may considerably understate the total tax burden, particularly in countries where provincial and municipal governments are large or have considerable tax authority.

Low ratios of tax collections to GDP may reflect weak administration and large-scale tax avoidance or evasion. They also may reflect the presence of a sizable parallel economy with unrecorded and undisclosed incomes. Tax collection ratios tend to rise with income, with more developed countries relying on taxes to finance a much broader range of social services and social security than less developed countries are able to provide.

As countries develop, they typically expand their capacity to tax residents directly, and indirect taxes become less important as a source of revenue. Thus the share of taxes on income, profits, and capital gains is one measure of a tax system's level of development. In the early stages of development governments tend to rely on indirect taxes because the administrative costs of collecting them are relatively low. The two main indirect taxes are international trade taxes (including customs revenues) and domestic taxes on goods and services. The table shows these domestic taxes as a percentage of value added in industry and services. Agriculture and mining are excluded from the denominator because indirect taxation of these sectors is usually negligible. What is

missing here is a measure of the uniformity of these taxes across industries and along the value added chain of production. Without such data no clear inferences can be drawn about how neutral a tax system is between subsectors. "Surplus" revenues raised by some governments by charging higher prices for goods produced under monopoly by state-owned enterprises are not counted as tax revenues.

Export and import duties are shown separately because their burdens on the economy (and hence growth) are likely to be high. Export duties, typically levied on primary (particularly agricultural) products, often take the place of direct taxes on income and profits, but they reduce the incentive to export and encourage a shift to other products. High import duties penalize consumers, create protective barriers—which promote higher-priced output and inefficient production—and implicitly tax exports. By contrast, lower trade taxes enhance openness—to foreign competition, knowledge, technologies, and resources—energizing development in many ways. The economies growing fastest over the past 15 years have not relied on tax revenues from imports. Seeing this pattern, many developing countries have lowered tariffs over the past decade and, given the successful completion of the Uruguay Round of the General Agreement on Tariffs and Trade (GATT), this trend is expected to continue. In some countries, such as members of the European Union, most customs duties are collected by a supranational authority; these revenues are not reported in the individual countries' accounts.

The tax revenues collected by governments are the outcomes of systems that are often complex, containing many exceptions, exemptions, penalties, and other inducements that affect tax incidence and thus influence the decisions of workers, managers, and entrepreneurs. A potentially important influence on both domestic and international investors is a tax system's progressivity, as reflected in the highest marginal tax rate on individual and corporate income. Figures for individual marginal tax rates generally refer to employment income. For some countries, the highest marginal tax rate is also the basic or flat rate, and other surtaxes, deductions, and the like may apply.

• **Tax revenue** comprises compulsory, unrequited, non-repayable receipts collected by central governments for public purposes. It includes interest collected on tax arrears and penalties collected on nonpayment or late payment of taxes and is shown net of refunds and other corrective transactions. • **Taxes on income, profits, and capital gains** include taxes levied by central governments on the actual or presumptive net income of individuals and profits of enterprises. Also included are taxes on capital gains, whether realized or not, on the sale of land, securities, and other assets. Social security contributions based on gross pay, payroll, or number of employees are not included, but social security contributions based on personal income after deductions and personal exemptions are included. • **Domestic taxes on goods and services** include all taxes and duties levied by central governments on the production, extraction, sale, transfer, leasing, or delivery of goods and rendering of services, or on the use of goods or permission to use goods or perform activities. These include general sales taxes, turnover or value added taxes, excise taxes, and motor vehicle taxes. • **Export duties** include all levies collected on goods at the point of export. Rebates on exported goods—that is, repayments of previously paid general consumption taxes, excise taxes, or import duties—should be deducted from the gross receipts of the appropriate taxes, not from export duty receipts. • **Import duties** comprise all levies collected on goods at the point of entry into the country. They include levies for revenue purposes or import protection, whether on a specific or ad valorem basis, as long as they are restricted to imported products. • **Highest marginal tax rate** is the highest rate shown on the schedule of tax rates applied to the taxable income of individuals and corporations. Also presented are the income levels above which the highest marginal tax rates apply for individuals.

The definitions used here are from the International Monetary Fund (IMF) *Manual on Government Finance Statistics*. Data on tax revenues are from print and electronic editions of the IMF's *Government Finance Statistics Yearbook*. Data on individual and corporate tax rates are from Price Waterhouse's *Individual Taxes: A Worldwide Summary* (1997) and *Corporate Taxes: A Worldwide Summary (1997)*.

	Exchange rate arrangements		Official exchange rate	Ratio of official to parallel exchange rate	Real effective exchange rate	Purchasing power parity conversion factor		Interest rate			Key agricultural producer prices	
			local currency units to $		1990 = 100	local currency units to international $		Deposit %	Lending %	Real %	Wheat $ per metric ton	Maize $ per metric ton
	Classification 1996	Structure 1996	1996	1996	1996	1990	1996	1996	1996	1996	1995	1995
Albania	IF	U	104.5	1.0	..	..	..	16.8	24.0	8.2	347	178
Algeria	MF	U	54.7	0.4	..	4.9	17.9	..	..	..	399	336
Angola	P	U	59,515.0	..	..	22.3	38,483,348.5	..	..	..	..	..
Argentina	P	U	1.0	1.0	..	0.3	0.9	7.4	10.5	10.0	93	66
Armenia	IF	U	414.0	..	..	0.0	82.4	32.2	66.4	-18.9	..	..
Australia	IF	U	1.3	1.0	93.4	1.4	1.3	..	..	..	158	176
Austria	FL	U	10.6	1.0	86.2	13.4	13.7	1.7	..	..	270	159
Azerbaijan	IF	U	4,301.3	..	..	0.1	1,367.0	..	162.5	-63.0	..	..
Bangladesh	P	U	41.8	0.8	..	9.0	10.6	7.3	14.0	8.0	186	..
Belarus	MF	U	..	..	..	0.1	3,974.9	32.3	64.3	12.3	..	..
Belgium	FL	U	31.0	1.0	108.0	36.2	36.3	4.0	7.2	5.3	..	..
Benin	P	U	511.6	..	..	109.3	160.1	..	..	..	..	187
Bolivia	IF	U	5.1	1.0	95.4	0.9	1.5	19.2	56.0	35.4	187	137
Bosnia and Herzegovina	P	U	..	..	..	..	..	..	..	..	..	..
Botswana	P	D	3.3	1.0	..	0.8	1.3	10.4	14.5	4.2	..	151
Brazil	MF	D	1.0	1.0	..	0.0	0.7	26.4	..	..	155	122
Bulgaria	IF	U	177.9	0.8	..	1.1	45.3	74.7	123.5	4.7	55	61
Burkina Faso	P	U	511.6	..	..	106.8	127.6	..	..	..	..	..
Burundi	P	U	302.7	0.8	100.5	47.8	89.7	..	15.3	-5.9	..	..
Cambodia	MF	D	2,624.1	1.0	..	..	..	8.8	18.8	11.0	..	..
Cameroon	P	U	511.6	..	67.3	146.5	175.9	5.5	16.0	-0.9	..	..
Canada	IF	U	1.4	1.0	82.7	1.3	1.2	4.3	6.1	4.8	94	85
Central African Republic	P	U	511.6	1.0	64.5	96.2	111.0	5.5	16.0	4.2	..	543
Chad	P	U	511.6	..	..	74.1	99.7	5.5	16.0	5.4	318	127
Chile	MF	D	412.3	0.9	123.8	97.5	176.8	13.5	17.4	9.7	215	168
China	MF	U	8.3	1.0	..	1.1	1.7	7.5	10.1	3.8	126	98
Hong Kong, China	MF	U	7.7	1.0	..	6.1	7.8	4.6	8.5	3.0	..	..
Colombia	MF	U	1,036.7	0.9	143.9	115.0	336.9	31.2	42.0	20.7	218	185
Congo, Dem. Rep.	IF	U	7,024.4	..	82.6	0.0	7,299.4	..	..	..	..	..
Congo, Rep.	P	U	511.6	..	..	194.4	251.9	5.5	16.0	11.2	..	200
Costa Rica	MF	U	207.7	1.0	102.2	34.1	84.0	17.3	26.3	8.6	..	176
Côte d'Ivoire	P	U	511.6	..	70.4	158.7	221.0	..	..	..	..	..
Croatia	MF	U	5.4	0.6	..	..	5.1	5.6	22.5	16.6	..	..
Cuba	..	..	..	..	..	..	..	..	..	..	..	..
Czech Republic	P	U	27.1	1.0	..	5.8	13.1	6.8	12.5	-1.5	104	118
Denmark	FL	U	5.8	1.0	105.6	8.8	8.5	2.8	8.7	6.7	182	..
Dominican Republic	MF	D	13.8	1.0	117.7	2.6	5.0	..	..	..	..	311
Ecuador	MF	D	3,189.5	1.0	128.8	195.2	1,018.1	41.5	54.5	19.4	186	182
Egypt, Arab Rep.	MF	M	3.4	1.0	..	0.8	1.4	10.5	15.6	6.0	157	142
El Salvador	MF	U	8.8	0.9	..	3.6	5.6	14.0	18.6	10.3	..	164
Eritrea	MF	D	..	..	..	..	..	..	..	..	..	..
Estonia	P	U	12.0	0.8	..	0.1	7.7	6.1	13.7	-6.8	..	..
Ethiopia	IF	U	6.4	0.6	..	0.9	1.3	9.4	13.9	12.5	..	..
Finland	FL	U	4.6	1.0	68.3	6.2	5.9	2.4	6.2	5.1	199	..
France	FL	U	5.1	1.0	95.7	6.5	6.3	3.7	6.8	5.4	170	165
Gabon	P	U	511.6	..	63.9	265.5	348.1	5.5	16.0	9.7	..	..
Gambia, The	IF	U	9.8	0.9	94.0	2.1	2.4	12.5	25.5	18.1	..	..
Georgia	MF	U	..	..	..	0.0	576.6	..	..	..	..	..
Germany	FL	U	1.5	1.0	122.2	..	2.0	2.8	10.0	8.9	167	168
Ghana	IF	U	1,637.2	1.0	..	91.2	323.7	34.5	..	..	..	144
Greece	MF	U	240.7	1.0	105.4	130.4	226.4	13.5	21.0	11.1	273	213
Guatemala	IF	U	6.0	1.0	..	1.2	2.3	7.7	22.7	12.3	203	176
Guinea	IF	U	1,004.0	1.0	..	226.5	326.9	17.5	21.5	15.5	..	..
Guinea-Bissau	P	D	511.6	..	..	654.9	6,233.1	47.3	51.8	2.3	..	..
Haiti	IF	U	15.7	0.7	..	1.6	5.0	..	..	..	..	328
Honduras	MF	D	11.7	1.0	..	1.3	3.6	16.7	29.7	7.0	..	211

Relative prices and exchange rates | 5.6

	Exchange rate arrangement		Official exchange rate	Ratio of official to parallel exchange rate	Real effective exchange rate	Purchasing power parity conversion factor		Interest rate			Key agricultural producer prices	
			local currency units to $		1990 = 100	local currency units to international $		Deposit %	Lending %	Real %	Wheat $ per metric ton	Maize $ per metric ton
	Classification 1996	Structure 1996	1996	1996	1996	1990	1996	1996	1996	1996	1995	1995
Hungary	MF	U	152.6	1.0	128.0	31.1	96.6	*26.1*	*32.6*	*4.6*	89	95
India	IF	U	35.4	0.9	..	5.7	8.3	..	16.0	8.4	134	104
Indonesia	MF	U	2,342.3	1.0	..	564.6	776.7	17.3	19.2	9.7	..	168
Iran, Islamic Rep.	MF	D	1,750.8	0.3	..	161.3	745.0	..	..	..	153	138
Iraq	P	D	0.3	0.0	..	..	..	..	..	..	..	..
Ireland	FL	U	0.6	1.0	72.1	0.7	0.6	0.3	5.8	4.8	141	..
Israel	MF	U	3.2	1.0	..	1.8	*2.8*	14.5	20.7	*9.7*	..	..
Italy	FL	U	1,543.0	1.0	79.0	1,412.1	1,612.6	6.5	12.1	7.3	196	189
Jamaica	IF	U	37.1	0.9	..	4.0	22.3	25.2	44.0	15.8	..	383
Japan	IF	U	108.8	1.0	124.4	191.5	171.8	0.3	2.7	2.7	..	..
Jordan	P	U	0.7	1.0	..	0.3	0.3	6.0	9.5	4.2	212	137
Kazakhstan	IF	U	67.3	..	..	0.0	26.5	..	..	..	..	..
Kenya	IF	U	57.1	1.0	..	8.1	16.6	17.6	33.8	22.0	97	140
Korea, Dem. Rep.	..	..	..	..	..	..	..	..	..	..	..	..
Korea, Rep.	MF	U	804.5	1.0	..	534.5	649.0	7.5	8.8	5.2	..	569
Kuwait	P	U	0.3	..	..	0.2	*0.2*	6.0	8.8	*3.1*	..	..
Kyrgyz Republic	MF	U	12.8	..	..	0.0	2.4	30.7	58.6	20.3	..	..
Lao PDR	MF	U	921.1	1.0	..	176.7	289.5	16.0	27.0	12.5	..	..
Latvia	MF	U	0.6	..	..	0.0	0.3	11.7	25.8	9.7	73	..
Lebanon	IF	U	1,571.4	1.0	..	..	844.5	15.5	25.2	15.0	..	..
Lesotho	P	U	4.3	1.0	99.6	0.8	1.1	12.7	17.7	5.2	..	177
Libya	P	U	0.4	0.2	..	..	..	..	..	..	..	..
Lithuania	P	U	4.0	..	..	0.0	1.9	*8.4*	21.6	*–3.6*	..	..
Macedonia, FYR	MF	U	40.0	..	..	..	..	..	..	..	..	..
Madagascar	IF	U	4,061.3	0.9	..	417.2	1,312.6	19.0	32.8	9.6	..	..
Malawi	IF	U	15.3	0.5	82.3	1.1	4.8	26.3	45.3	1.3	85	47
Malaysia	MF	U	2.5	1.0	107.0	1.0	1.1	7.1	9.0	0.8	..	..
Mali	MF	U	511.6	..	..	128.5	187.2	..	..	..	..	..
Mauritania	IF	U	137.2	..	..	27.2	33.7	..	..	..	..	..
Mauritius	MF	U	17.9	0.9	..	6.0	7.5	10.8	20.8	14.0	..	283
Mexico	IF	U	7.6	1.0	..	1.3	3.4	25.1	..	..	181	194
Moldova	IF	U	4.6	..	..	0.0	1.3	25.4	36.7	13.1	..	..
Mongolia	IF	U	548.4	1.0	..	2.3	114.2	36.4	91.9	58.5	..	..
Morocco	P	U	8.7	1.0	113.2	3.2	3.5	..	*10.0*	*9.6*	274	246
Mozambique	IF	U	11,293.7	0.9	..	230.2	1,933.1	..	..	..	..	..
Myanmar	P	D	5.9	0.0	..	..	..	12.5	16.5	16.5	..	..
Namibia	P	U	4.3	..	..	1.2	1.7	12.6	19.2	8.4	199	197
Nepal	P	U	56.7	0.8	..	6.8	10.4	..	12.9	5.9	..	..
Netherlands	FL	U	1.7	0.9	97.2	2.2	2.1	3.5	5.9	4.4	..	..
New Zealand	IF	U	1.5	1.0	107.2	1.5	1.5	8.5	12.3	10.2	..	..
Nicaragua	MF	U	8.4	1.0	80.6	0.0	1.8	12.3	20.7	8.1	..	216
Niger	P	U	511.6	..	..	96.7	116.7	..	..	..	..	..
Nigeria	P	D	21.9	0.3	149.2	3.6	19.7	*13.5*	*20.2*	*–23.6*	580	254
Norway	MF	U	6.4	1.0	106.1	10.3	9.9	4.2	7.1	2.4	394	..
Oman	P	U	0.4	1.0	..	0.3	*0.2*	6.9	9.2	5.5	..	..
Pakistan	MF	U	36.1	0.9	..	6.2	10.1	..	..	..	126	..
Panama	P	U	1.0	..	..	0.4	0.4	7.2	10.6	8.8	..	257
Papua New Guinea	IF	U	1.3	1.0	93.0	0.4	0.5	12.2	13.3	9.8	..	..
Paraguay	IF	U	2,062.8	0.9	118.8	462.7	1,144.1	17.2	28.9	15.8	119	145
Peru	IF	U	2.5	1.0	..	0.1	1.4	14.9	26.1	15.2	280	204
Philippines	IF	U	26.2	0.9	129.9	6.0	8.9	9.7	14.8	5.3	..	194
Poland	MF	U	2.7	1.0	211.0	0.3	1.6	20.0	26.1	6.5	153	..
Portugal	FL	U	154.2	1.0	115.4	92.4	119.4	6.3	11.7	8.1	187	179
Puerto Rico	..	..	..	..	..	..	..	..	..	..	..	..
Romania	IF	U	3,084.2	0.8	..	9.1	1,045.7	..	..	..	127	86
Russian Federation	MF	U	5,120.8	1.0	..	0.7	3,577.0	55.1	146.8	69.6	..	..

	Exchange rate arrangement		Official exchange rate	Ratio of official to parallel exchange rate	Real effective exchange rate	Purchasing power parity conversion factor		Interest rate			Key agricultural producer prices	
			local currency units to $			local currency units to international $		Deposit %	Lending %	Real %	Wheat $ per metric ton	Maize $ per metric ton
	Classification 1996	Structure 1996	1996	1996	1990 = 100 1996	1990	1996	1996	1996	1996	1995	1995
Rwanda	IF	U	306.8	0.9	..	39.8	95.3	10.9	..	..	267	172
Saudi Arabia	FL	U	3.7	1.0	93.1	2.6	2.5	..	..	..	..	..
Senegal	P	U	511.6	..	..	142.5	182.6	..	..	..	..	140
Sierra Leone	IF	U	920.7	1.0	106.3	54.1	359.5	14.0	32.1	4.8	..	..
Singapore	MF	U	1.4	1.0	..	1.6	1.6	3.4	6.3	3.7	..	..
Slovak Republic	P	U	30.7	0.9	..	7.4	14.5	9.3	13.9	8.0	109	125
Slovenia	MF	U	135.4	0.9	..	..	105.1	15.0	23.7	12.8	194	144
South Africa	IF	U	4.3	1.0	90.3	1.2	1.9	14.9	19.5	10.3	217	164
Spain	FL	U	126.7	1.0	91.5	105.1	121.1	6.1	8.5	5.0	227	222
Sri Lanka	MF	U	55.3	1.0	..	11.5	18.1	16.0	16.3	4.8	..	180
Sudan	MF	U	1,250.8	..	..	..	..	..	..	..	209	..
Sweden	IF	U	6.7	1.0	79.1	9.5	9.7	2.5	7.4	6.4	142	..
Switzerland	IF	U	1.2	1.0	111.7	2.1	2.1	1.3	5.0	4.8	880	520
Syrian Arab Republic	P	M	11.2	0.2	..	10.4	14.0	..	..	..	..	..
Tajikistan	IF	U	2,204.3	..	..	0.0	57.7	..	..	..	..	..
Tanzania	IF	U	580.0	1.0	..	..	..	13.6	37.2	11.9	..	..
Thailand	P	U	25.3	1.0	..	9.9	11.4	10.3	14.4	9.4	..	142
Togo	P	U	511.6	..	73.6	72.6	101.8	..	..	..	..	112
Trinidad and Tobago	IF	U	6.0	1.0	85.2	3.2	3.8	6.9	15.8	11.2	..	371
Tunisia	MF	U	1.0	1.0	..	0.4	0.4	..	..	..	291	..
Turkey	MF	U	81,404.9	1.0	..	1,525.1	39,421.8	80.7	..	..	169	196
Turkmenistan	MF	D	404.4	..	..	0.0	907.1	..	..	..	..	..
Uganda	IF	U	1,046.1	0.9	87.7	118.2	302.3	10.6	20.3	14.0	..	..
Ukraine	MF	U	1.8	1.0	..	0.0	0.7	33.6	79.9	9.6	..	..
United Arab Emirates	FL	U	3.7	1.0	..	3.6	3.7	..	..	..	..	..
United Kingdom	IF	U	0.6	1.0	91.5	0.6	0.6	3.0	6.0	2.8	186	..
United States	IF	U	1.0	..	96.6	1.0	1.0	..	8.3	6.2	127	89
Uruguay	MF	U	8.0	1.0	160.4	0.6	5.8	28.1	91.5	55.7	153	154
Uzbekistan	MF	M	..	..	..	0.0	9.3	..	..	..	..	..
Venezuela	MF	U	417.3	0.9	117.9	16.5	151.3	27.6	31.7	−37.7	..	222
Vietnam	MF	U	10,962.1	..	..	479.4	2,185.9	17.9	28.3	7.4	..	..
West Bank and Gaza	..	..	..	..	..	..	..	..	..	..	..	..
Yemen, Rep.	IF	U	94.2	..	..	..	46.1	..	..	..	367	416
Yugoslavia, FR (Serb./Mont.)	..	..	..	..	..	..	..	..	..	..	..	..
Zambia	IF	U	1,203.7	0.9	97.7	15.6	500.3	42.1	53.8	13.5	..	..
Zimbabwe	IF	U	10.0	0.9	..	0.8	2.9	21.6	34.2	9.9	224	139

Note: Exchange rate arrangements are given for the end of the year in 1996. Exchange rate classifications: FL = flexibility limited, IF = independent floating, MF = managed floating, P = pegged. Exchange rate structures: D = dual exchange rates, M = multiple exchange rates, U = unitary rate.

Relative prices and exchange rates | 5.6

In a market-based economy the choices households, producers, and governments make about the allocation of resources are influenced by relative prices—the real exchange rate, real wages, real interest rates, and commodity prices. Relative prices also reflect, to a large extent, the choices of these agents. Thus relative prices convey vital information about the interaction of economic agents in an economy and in relation to the rest of the world.

The exchange rate is the price of one currency in terms of another. Official exchange rates and exchange rate arrangements are established by governments. Parallel, or "black market," exchange rates reflect unofficial rates negotiated by traders and are by their nature difficult to measure. Parallel exchange rate markets often account for only a small share of transactions and so may be both thin and volatile. But in countries with weak policies and financial systems, they often represent the "going" rate. The parallel rates reported here are collected by Currency Data & Intelligence, Inc., from a variety of sources, some within the country and some outside but doing business with entities based in the country.

Real effective exchange rates are derived by deflating a trade-weighted average of the nominal exchange rates that apply between trading partners. For most industrial countries the weights are based on trade in manufactured goods with other industrial countries during 1989–91, and an index of relative, normalized unit labor costs is used as the deflator. (Normalization smooths a time series by removing short-term fluctuations while retaining changes of a large amplitude over the longer economic cycle.) For other countries, prior to 1990, the weights take into account trade in manufactured and primary products during 1980–82; from January 1990 onward weights are based on trade in manufactured and primary products during 1988–90, and an index of relative changes in consumer prices is used as the deflator. An increase in the real effective exchange rate represents an appreciation of the local currency. Because of conceptual and data limitations, changes in real effective exchange rates should be interpreted with caution.

The exchange rate is often used as a basis for converting prices in different currencies because it is observable in the market and available universally. But because market imperfections are extensive and exchange rates reflect at most the relative prices of tradables, the volume of goods and services that a U.S. dollar buys in the United States may not correspond to what a U.S. dollar converted to another country's currency at

the official exchange rate would buy in that country. The alternative approach is to convert national currency estimates of GNP to a common currency by using conversion factors that reflect equivalent purchasing power. Purchasing power parity (PPP) conversion factors are based on price and expenditure surveys conducted by the International Comparison Programme (ICP) and represent the conversion factors applied to equalize price levels across countries. See the notes to tables 4.10 and 4.11 for further discussion of the PPP conversion factor.

Many interest rates coexist in an economy, reflecting competitive conditions, the terms governing loans and deposits, and differences in the position and status of creditors and debtors. In some economies interest rates are set by regulation or administrative fiat. In economies with imperfect markets or where reported nominal rates are not indicative of effective rates, it may be difficult to obtain data on interest rates that reflect actual market transactions. Deposit and lending rates are collected by the International Monetary Fund (IMF) as representative interest rates offered by banks to resident customers. The terms and conditions attached to these rates differ by country, however, limiting their comparability. Real interest rates are calculated by adjusting nominal rates by an estimate of the inflation rate in the economy. A negative real interest rate indicates a loss in the purchasing power of the principal. The real interest rates in the table are calculated as $(i - P)/(1 + P)$, where i is the nominal interest rate and P is the inflation rate (as measured by the GDP deflator).

The table also shows prices for two key agricultural commodities, wheat and maize. Prices received by farmers, as used here, are important determinants of the type and volume of agricultural production. In theory these prices should refer to national average farmgate, or first-point-of-sale, transactions. But depending on the country's institutional arrangements—that is, whether it relies on market wholesale prices, government-fixed prices, or support prices—the data might not always refer to the same selling points. These data come from the Food and Agriculture Organization (FAO), and most originated from official country publications or from FAO questionnaires. As the data show, prices received by farmers often are not equalized across international markets (even after adjusting for freight, transport, and insurance costs and differences in quality). Market imperfections such as taxes, subsidies, and trade barriers drive a wedge between domestic and international prices.

• **Exchange rate arrangement** describes the arrangement that an IMF member country has furnished to the IMF under Article IV, Section 2(a). *Exchange rate classification* indicates how the exchange rate is determined in the main market when there is more than one market: pegged (to a single currency or composite of currencies), flexibility limited, or floating (managed or independent). *Exchange rate structure* shows whether countries have unitary, dual, or multiple exchange rates. • **Official exchange rate** refers to the actual, principal exchange rate and is an annual average based on monthly averages (local currency units relative to U.S. dollars) determined by country authorities or on rates determined largely by market forces in the legally sanctioned exchange market. • **Ratio of official to parallel exchange rate** measures the premium people must pay, relative to the official exchange rate, to exchange the domestic currency for dollars in the black market. • **Real effective exchange rate** is the nominal effective exchange rate (a measure of the value of a currency against a weighted average of several foreign currencies) divided by a price deflator or index of costs. • **Purchasing power parity conversion factor** is the number of units of a country's currency required to buy the same amounts of goods and services in the domestic market as $1 would buy in the United States. • **Deposit interest rate** is the rate paid by commercial or similar banks for demand, time, or savings deposits. • **Lending interest rate** is the rate charged by banks on loans to prime customers. • **Real interest rate** is the lending interest rate adjusted for inflation as measured by the GDP deflator. • **Key agricultural producer prices** are the domestic producer prices per metric ton for wheat and maize converted to U.S. dollars using the official exchange rate.

Information on exchange rate arrangements is drawn from the IMF's *Exchange Arrangements and Exchange Restrictions Annual Report 1997*. Official and real effective exchange rates and deposit and lending interest rates are from the IMF's *International Financial Statistics*. Estimates of parallel market exchange rates are from Currency Data & Intelligence, Inc.'s *Global Currency Report*. PPP conversion factors are from the ICP and World Bank staff estimates. Real interest rates are calculated using World Bank data on the GDP deflator. Agricultural price data are from the FAO's *Production Yearbook*.

5.7 Defense expenditures and trade in arms

	Military expenditures				Armed forces personnel				Arms trade			
	% of GNP		% of central government expenditure		Total thousands		% of labor force		Exports % of total exports		Imports % of total imports	
	1985	1995	1985	1995	1985	1995	1985	1995	1985	1995	1985	1995
Albania	5.3	1.1	10.9	3.2	..	52	..	3.2	0.0	0.0	0.0	0.0
Algeria	2.5	3.2	6.3	6.9	170	120	2.9	1.4	0.0	0.0	5.1	2.2
Angola	19.9	3.0	..	..	66	82	1.7	1.7	0.0	0.0	118.0	13.7
Argentina	3.8	1.7	12.4	27.0	129	65	1.1	0.5	1.0	0.3	5.5	0.2
Armenia	..	0.9	..	..	..	60	..	3.3	..	0.0	..	4.5
Australia	2.7	2.5	9.4	8.8	70	58	0.9	0.6	0.6	0.1	3.6	1.5
Austria	1.3	0.9	3.3	2.2	40	45	1.1	1.2	0.9	0.1	0.1	0.1
Azerbaijan	..	2.8	..	..	..	87	..	2.7	..	0.0	..	0.0
Bangladesh	1.7	1.7	13.0	9.9	91	115	0.2	0.2	0.0	0.0	2.2	0.9
Belarus	..	0.8	..	..	..	115	..	2.1	..	4.0	..	0.0
Belgium	3.1	1.7	5.3	3.5	107	47	2.7	1.1	0.7	0.1	0.7	0.2
Benin	2.2	1.2	..	8.6	6	6	0.3	0.2	0.0	0.0	3.0	0.0
Bolivia	3.3	2.3	22.6	9.5	28	28	1.2	0.9	0.0	0.0	0.7	0.7
Bosnia and Herzegovina	..	..	..	..	..	50	..	2.4	..	0.0	..	20.0
Botswana	2.5	5.3	5.8	12.7	3	8	0.6	1.3	0.0	0.0	1.7	0.0
Brazil	0.8	1.7	2.1	3.9	496	285	0.9	0.4	1.3	0.0	0.4	0.3
Bulgaria	14.1	2.8	32.5	6.3	189	86	4.2	2.0	4.7	2.8	7.3	0.0
Burkina Faso	1.9	2.9	18.7	12.0	9	9	0.2	0.2	0.0	0.0	6.0	0.0
Burundi	3.0	4.4	20.8	24.8	9	22	0.3	0.7	0.0	0.0	2.6	0.0
Cambodia	..	3.1	..	..	35	90	0.9	1.8	0.0	0.0	..	3.2
Cameroon	1.9	1.9	8.3	10.2	15	22	0.4	0.4	0.0	0.0	1.7	0.8
Canada	2.2	1.7	8.6	7.1	83	70	0.6	0.5	0.6	0.1	0.1	0.1
Central African Republic	1.8	2.5	6.4	..	5	5	0.4	0.3	0.0	0.0	0.0	0.0
Chad	2.0	3.1	6.1	..	16	30	0.6	1.0	0.0	0.0	12.0	4.5
Chile	4.0	3.8	11.4	17.5	124	102	2.8	1.8	2.1	0.0	0.7	2.4
China	4.9	2.3	23.8	18.5	4,100	2,930	0.7	0.4	..	..	..	..
Hong Kong, China	..	..	..	..	..	..	..	..	..	..	..	..
Colombia	1.6	2.6	10.3	16.2	66	146	0.6	0.9	0.0	0.0	0.5	0.4
Congo, Dem. Rep.	1.2	0.3	9.8	3.7	62	49	0.4	0.3	0.0	0.0	5.0	0.0
Congo, Rep.	4.0	2.9	9.2	..	15	10	1.8	0.9	0.0	0.0	6.7	1.5
Costa Rica	0.7	0.6	2.8	2.7	8	8	0.8	0.6	0.0	0.0	1.8	0.0
Côte d'Ivoire	1.2	1.1	..	4.2	8	15	0.2	0.3	0.0	0.0	1.1	0.0
Croatia	..	10.5	..	32.0	..	60	..	2.7	..	0.0	..	1.5
Cuba	4.5	1.6	..	..	297	70	7.0	1.4	0.1	0.0	27.4	0.0
Czech Republic	..	2.3	..	6.6	..	68	..	1.2	..	0.6	..	0.0
Denmark	2.3	1.8	5.2	4.1	29	27	1.0	0.9	0.1	0.0	0.6	0.2
Dominican Republic	1.2	1.4	8.0	9.1	22	22	0.9	0.7	0.0	0.0	0.7	0.3
Ecuador	2.8	3.7	16.9	18.3	43	58	1.4	1.4	0.0	0.0	4.0	6.2
Egypt, Arab Rep.	12.8	5.7	22.1	13.7	466	430	2.9	2.1	4.9	0.0	30.9	16.2
El Salvador	5.7	1.1	29.1	7.4	48	22	2.9	1.0	0.0	0.0	12.5	0.7
Eritrea	..	..	..	..	..	..	..	..	..	0.0	..	0.0
Estonia	..	1.1	..	2.9	..	6	..	0.7	..	0.0	..	0.2
Ethiopia	6.7	2.2	28.9	9.2	240	120	1.2	0.5	0.0	0.0	80.6	0.0
Finland	1.7	2.0	5.4	5.1	40	32	1.6	1.2	0.0	0.0	0.8	0.1
France	4.0	3.1	8.8	6.6	563	504	2.3	2.0	7.0	0.8	0.1	0.1
Gabon	2.8	2.6	6.6	9.6	7	10	1.7	2.0	0.0	0.0	11.7	0.0
Gambia, The	..	4.6	..	16.2	1	1	0.3	0.2	0.0	0.0	10.8	0.0
Georgia	..	2.4	..	..	..	5	..	0.2	..	0.0	..	4.0
Germany[a]	3.2	1.9	10.3	5.0	495	352	1.3	0.9	..	0.2	..	0.1
Ghana	1.0	1.4	7.2	5.8	15	7	0.3	0.1	0.0	0.0	0.0	0.0
Greece	7.0	5.5	13.8	10.8	201	213	5.0	4.8	0.7	0.1	2.9	2.2
Guatemala	1.6	1.3	17.0	14.2	43	36	1.6	1.0	0.0	0.0	2.6	0.2
Guinea	..	1.5	..	..	28	12	1.1	0.4	0.0	0.0	20.8	0.0
Guinea-Bissau	2.9	2.8	4.7	..	11	7	2.5	1.4	0.0	0.0	41.0	0.0
Haiti	1.5	2.9	7.5	21.6	6	0	0.2	0.0	0.0	0.0	2.3	0.0
Honduras	3.5	1.4	14.0	8.7	21	18	1.4	0.8	0.0	0.0	3.4	0.8

Defense expenditures and trade in arms | 5.7

	Military expenditures				Armed forces personnel				Arms trade			
	% of GNP		% of central government expenditure		Total thousands		% of labor force		Exports % of total exports		Imports % of total imports	
	1985	1995	1985	1995	1985	1995	1985	1995	1985	1995	1985	1995
Hungary	7.2	1.5	15.3	4.6	117	71	1.8	1.1	2.6	0.2	1.0	0.2
India	3.5	2.4	15.7	12.7	1,260	1,265	0.4	0.3	0.1	0.0	16.3	1.2
Indonesia	2.4	1.8	10.3	8.9	278	280	0.4	0.3	0.0	0.0	1.6	0.4
Iran, Islamic Rep.	7.7	2.6	34.1	13.6	345	440	2.5	2.3	0.0	1.6	17.2	2.2
Iraq	41.2	..	..	..	788	390	19.5	7.2	0.2	0.0	46.4	0.0
Ireland	1.7	1.3	2.9	3.4	14	13	1.1	0.9	0.0	0.0	0.1	0.0
Israel	20.3	9.6	27.2	21.1	195	185	11.9	8.5	10.8	4.1	10.9	1.1
Italy	2.2	1.8	4.6	3.9	504	435	2.1	1.7	1.7	0.1	0.3	0.1
Jamaica	0.9	0.8	1.8	1.4	2	3	0.2	0.2	0.0	0.0	0.9	0.0
Japan	1.0	1.0	5.6	4.2	241	240	0.4	0.4	0.1	0.0	0.8	0.2
Jordan	15.5	7.7	39.7	21.7	81	112	12.3	10.2	0.0	0.0	22.9	1.9
Kazakhstan	..	0.9	..	..	..	28	..	0.3	..	0.4	..	7.2
Kenya	2.3	2.3	8.4	6.2	19	22	0.2	0.2	0.0	0.0	0.7	0.3
Korea, Dem. Rep.	20.0	28.6	..	..	784	1,040	8.5	8.8	25.4	3.6	22.1	5.8
Korea, Rep.	5.0	3.4	26.6	13.6	600	655	3.4	3.0	0.6	0.0	1.5	0.8
Kuwait	5.7	11.6	13.6	25.5	16	20	2.4	2.4	0.0	0.0	6.2	11.6
Kyrgyz Republic	..	0.7	..	..	..	13	..	0.7	..	2.5	..	0.0
Lao PDR	7.4	4.2	..	22.3	54	50	2.9	2.1	0.0	0.0	48.8	0.0
Latvia	..	0.9	..	..	..	7	..	0.5	..	0.0	..	0.3
Lebanon	..	3.7	..	9.7	21	55	2.2	4.2	0.0	0.0	2.3	0.5
Lesotho	5.3	1.9	..	2.5	2	2	0.3	0.2	0.0	0.0	0.0	0.0
Libya	12.0	6.0	..	..	91	76	8.2	4.9	0.8	0.0	39.0	0.0
Lithuania	..	0.5	..	2.1	..	12	..	0.6	..	0.0	..	0.2
Macedonia, FYR	..	3.3	..	..	..	16	..	1.6	..	0.0	..	0.0
Madagascar	1.9	0.9	8.0	5.0	27	21	0.6	0.3	0.0	0.0	7.5	1.0
Malawi	2.0	1.6	5.8	3.5	6	10	0.2	0.2	0.0	0.0	1.8	0.0
Malaysia	3.8	3.0	10.7	12.4	110	122	1.8	1.5	0.0	0.1	3.8	1.0
Mali	2.9	1.8	8.1	..	8	8	0.2	0.2	0.0	0.0	3.3	0.0
Mauritania	6.9	3.2	25.0	9.3	16	10	1.9	1.0	0.0	0.0	0.0	0.0
Mauritius	0.2	0.4	0.8	1.6	1	1	0.3	0.2	0.0	0.0	0.0	0.0
Mexico	0.7	1.0	2.6	5.1	140	175	0.5	0.5	0.0	0.0	0.2	0.0
Moldova	..	2.1	..	..	..	12	..	0.6	..	5.6	..	0.0
Mongolia	8.3	2.4	13.1	7.0	38	21	4.2	1.7	0.0	0.0	0.5	0.0
Morocco	6.0	4.3	20.0	13.8	165	195	2.1	1.9	0.0	0.0	3.1	0.6
Mozambique	9.9	5.4	38.0	..	35	12	0.5	0.1	0.0	0.0	63.7	0.0
Myanmar	1.9	2.9	18.7	12.0	9	9	0.0	0.0	0.0	0.0	17.7	10.5
Namibia	..	2.1	..	5.5	..	8	..	1.2	..	0.0	..	0.8
Nepal	1.1	0.9	6.2	5.8	25	40	0.3	0.4	0.0	0.0	1.1	0.0
Netherlands	3.0	2.1	5.4	4.4	103	67	1.7	0.9	0.2	0.1	0.8	0.1
New Zealand	2.0	1.3	4.5	3.3	13	10	0.9	0.6	0.0	0.0	1.3	0.3
Nicaragua	17.4	2.2	26.2	5.3	74	14	6.7	0.9	0.0	7.7	29.0	0.0
Niger	0.8	1.2	5.0	7.9	5	9	0.2	0.2	0.0	0.0	0.0	0.0
Nigeria	1.5	0.8	9.4	3.5	134	89	0.4	0.2	0.0	0.0	3.8	0.0
Norway	3.1	2.7	7.5	6.5	36	38	1.8	1.7	0.2	0.0	1.2	0.4
Oman	24.4	16.7	42.3	33.9	25	36	6.9	6.5	0.0	..	4.4	10.8
Pakistan	6.2	6.1	28.1	25.3	483	587	1.4	1.3	1.5	0.3	8.0	4.2
Panama	2.0	1.4	6.4	5.3	12	12	1.5	1.1	0.0	0.0	0.7	0.0
Papua New Guinea	1.5	1.4	4.5	5.6	3	5	0.2	0.2	0.0	0.0	1.0	0.0
Paraguay	1.1	1.4	11.9	7.3	14	12	1.0	0.7	0.0	0.0	4.0	0.0
Peru	6.7	1.7	36.3	9.3	128	115	2.0	1.3	0.0	0.0	3.8	3.0
Philippines	1.4	1.5	9.5	8.5	115	110	0.5	0.4	0.2	0.0	0.7	0.3
Poland	10.2	2.3	40.7	5.4	439	278	..	2.0	11.3	0.2	4.2	0.3
Portugal	2.9	2.6	6.5	5.9	102	78	2.2	1.6	3.7	0.0	3.1	0.3
Puerto Rico	..	..	..	..	..	..	..	..	..	..	..	..
Romania	6.9	2.5	20.0	11.2	237	209	2.2	2.0	3.6	0.3	0.4	0.0
Russian Federation	..	11.4	..	38.1	..	1,400	..	1.8	..	4.3	..	0.0

	Military expenditures				Armed forces personnel				Arms trade			
	% of GNP		% of central government expenditure		Total thousands		% of labor force		Exports % of total exports		Imports % of total imports	
	1985	1995	1985	1995	1985	1995	1985	1995	1985	1995	1985	1995
Rwanda	1.7	5.2	9.4	23.3	5	33	0.2	0.8	0.0	0.0	0.0	0.0
Saudi Arabia	22.7	13.5	27.0	41.0	80	175	2.1	2.8	0.0	0.0	30.1	31.3
Senegal	2.8	1.6	8.8	..	18	14	0.6	0.4	0.0	0.0	0.6	0.4
Sierra Leone	0.8	6.1	5.0	28.9	4	14	0.3	0.8	0.0	0.0	0.0	0.0
Singapore	5.9	4.7	17.0	24.0	56	60	4.7	4.1	0.2	0.0	0.5	0.2
Slovak Republic	..	3.0	..	6.8	..	52	..	1.9	..	0.8	..	3.1
Slovenia	..	1.5	..	3.5	..	10	..	1.0	..	0.1	..	0.3
South Africa	3.8	2.2	11.6	6.7	95	100	0.8	0.6	0.5	0.4	0.2	0.8
Spain	2.4	1.6	6.9	5.6	314	210	2.1	1.3	2.4	0.1	0.6	0.6
Sri Lanka	2.9	4.6	8.4	15.7	22	110	0.4	1.5	0.0	0.0	1.6	3.1
Sudan	3.2	6.6	..	37.6	65	89	0.8	0.9	0.0	0.0	7.8	8.4
Sweden	3.0	2.8	6.1	5.8	69	51	1.6	1.1	0.6	0.4	0.3	0.0
Switzerland	2.4	1.6	..	6.0	23	29	0.7	0.8	1.2	0.1	2.4	0.0
Syrian Arab Republic	21.8	7.2	42.0	..	402	320	13.9	8.0	0.0	0.0	37.8	1.5
Tajikistan	..	3.7	..	..	..	8	..	0.4	..	1.4	..	0.0
Tanzania	3.8	1.8	12.8	8.4	43	35	0.4	0.2	0.0	0.0	3.8	0.0
Thailand	4.2	2.5	19.7	15.2	235	288	0.8	0.9	0.0	0.0	2.1	1.6
Togo	2.6	2.3	6.9	10.2	7	12	0.5	0.7	0.0	0.0	0.0	0.0
Trinidad and Tobago	..	1.7	..	4.0	2	3	0.4	0.6	0.0	0.0	0.0	0.0
Tunisia	3.6	2.0	8.8	6.3	38	35	1.5	1.0	0.0	0.0	11.2	0.5
Turkey	4.6	4.0	17.9	17.6	814	805	3.8	2.9	1.5	0.3	4.4	2.0
Turkmenistan	..	1.7	..	..	..	21	..	1.1	..	0.0	..	4.2
Uganda	2.0	2.3	15.6	13.3	15	52	0.2	0.5	0.0	0.0	3.1	0.0
Ukraine	..	2.9	..	7.8	..	476	..	1.8	..	1.4	..	0.0
United Arab Emirates	6.7	4.8	43.5	38.4	44	60	6.1	5.3	0.0	0.0	3.4	1.8
United Kingdom	5.1	3.0	12.6	7.2	334	233	1.2	0.8	1.6	2.1	0.7	0.1
United States	6.1	3.8	25.7	17.4	2,244	1,620	1.9	1.2	6.3	2.7	0.5	0.1
Uruguay	2.9	2.4	10.6	7.3	30	25	2.4	1.7	0.0	0.0	0.0	0.2
Uzbekistan	..	3.8	..	..	..	21	..	0.2	..	0.3	..	0.0
Venezuela	2.1	1.1	9.2	6.3	71	75	1.2	0.9	0.0	0.0	6.5	0.8
Vietnam	19.4	2.6	..	10.9	1,027	550	3.5	1.5	2.7	0.0	94.3	2.7
West Bank and Gaza	..	..	..	..	..	..	..	..	..	..	..	..
Yemen, Rep.	..	15.7	..	29.4	..	68	..	1.5	..	0.0	..	13.9
Yugoslavia, FR (Serb./Mont.)	3.7	..	54.8	..	258	..	5.7	..	4.2	..	0.2	..
Zambia	..	2.8	..	12.6	16	22	0.6	0.6	0.0	0.0	4.6	0.6
Zimbabwe	5.7	4.0	14.4	10.5	46	40	1.2	0.8	0.0	0.0	0.0	0.0
World[b]	**5.2 w**	**2.8 w**	**18.1 w**	**9.9 w**	**28,070 t**	**22,790 t**	**..**	**..**	**2.7 w**	**0.7 w**	**2.7 w**	**0.7 w**

Note: Data for some countries are based on partial or uncertain data or rough estimates; see ACDA 1997.

a. Data prior to 1990 refer to the Federal Republic of Germany before unification. b. U.S. Arms Control and Disarmament Agency aggregate measures.

Defense expenditures and trade in arms | 5.7

Although national defense is an important function of government, and security from external threats contributes to economic development, high levels of defense spending place a burden on the economy and may impede growth. Determining the appropriate level of defense spending is not easy, but concerns about perceived vulnerability and risk play an important role. Comparisons between countries should take into account the many factors that influence such perceptions including historical and cultural traditions, the length of borders that need defending, the quality of relations with neighbors, and the role of the armed forces in the body politic.

Although the traditional secrecy shrouding data on defense spending and trade in arms is gradually lifting, data from governments are often incomplete and unreliable. Even in countries where parliaments vigilantly review government budgets and spending, defense spending and trade in arms often do not receive close scrutiny. Finance ministries also may not exercise due oversight, particularly in countries where the armed forces have a strong political voice. For a detailed critique of the quality of such data, see Ball (1984) and Happe and Wakeman-Linn (1994).

The International Monetary Fund's (IMF) *Government Finance Statistics* is the primary source for data on defense spending. It uses a consistent definition of defense spending based on the United Nations' classification of the functions of government and the North Atlantic Treaty Organization (NATO) definition. The IMF checks data on defense spending for broad consistency with other macroeconomic data reported to it but is not always able to verify the accuracy and completeness of such data. Moreover, country coverage is affected by delays or failure to report data. Thus most researchers supplement the IMF's data with independent assessments of military outlays by organizations such as the U.S. Arms Control and Disarmament Agency (ACDA), the Stockholm International Peace Research Institute (SIPRI), and the International Institute for Strategic Studies (IISS). However, these agencies rely heavily on reporting by governments, on confidential intelligence estimates of varying quality, on sources that they do not or cannot reveal, and on one another's publications. Data presented in this table are from the ACDA.

Definitions of military spending differ depending on whether they include civil defense, reserves and auxiliary forces, police and paramilitary forces, dual-purpose forces such as military and civilian police,

military grants in kind, pensions for military personnel, and social security contributions paid by one part of government to another. Official government data may omit parts of military spending, disguise financing through extrabudgetary accounts or unrecorded use of foreign exchange receipts, or fail to include military assistance or secret military equipment imports. Current spending is more likely to be reported than capital spending. In some cases a more accurate estimate of military spending can be obtained by adding the value of estimated arms imports and nominal military expenditures. This method may understate or overstate spending in a particular year, however, because payments for arms may not coincide with deliveries.

Data on armed forces refer to active-duty military personnel, including paramilitary forces. These data exclude payments to civilians from the defense budget and so are not consistent with the data on military spending. Moreover, since they exclude payments to personnel not on active duty, they underestimate the share of the labor force that works for the defense establishment. Because governments rarely report the size of their armed forces, such data typically come from intelligence sources. The ACDA attributes its data to unspecified U.S. government sources.

The Standard International Trade Classification (SITC) does not clearly distinguish trade in military goods. For this and other reasons, customs-based data on trade in arms are of little use, so most compilers rely on trade publications, confidential government information on third-country trade, and other sources. The construction of defense production facilities and licensing fees paid for the production of arms are included in trade data when they are specified in military transfer agreements. Grants in kind are usually included as well. Definitional issues include treatment of dual-use equipment such as aircraft, use of military establishments like hospitals and schools by civilians, and purchases by nongovernment buyers. ACDA data do not include arms supplied to subnational groups. Valuation problems arise when data are reported in volume terms and the purchase price must be estimated. Differences between sources may reflect reporting lags or differences in the period covered. Most compilers revise their time series data regularly, so estimates for the same year may not be consistent between publication dates.

- **Military expenditures** for NATO countries are based on the NATO definition, which covers military-related expenditures of the defense ministry (including recruiting, training, construction, and the purchase of military supplies and equipment) and other ministries. Civilian-type expenditures of the defense ministry are excluded. Military assistance is included in the expenditures of the donor country, and purchases of military equipment on credit are included at the time the debt is incurred, not at the time of payment. Data for other countries generally cover expenditures of the ministry of defense (excluded are expenditures on public order and safety, which are classified separately). • **Armed forces personnel** refers to active-duty military personnel, including paramilitary forces if those forces resemble regular units in their organization, equipment, training, or mission. • **Arms trade** is exports and imports of military equipment usually referred to as "conventional," including weapons of war, parts thereof, ammunition, support equipment, and other commodities designed for military use. See *About the data* for more details.

Data on military expenditures, armed forces, and arms trade are from the ACDA's *World Military Expenditures and Arms Transfers 1996* (1997).

	Economic activity		Investment		Credit		Net financial flows from government		Overall balance before transfers		Employment		Proceeds from privatization
	% of GDP		% of gross domestic investment		% of gross domestic credit		% of GDP		% of GDP		% of total		$ millions
	1985–90	1990–95	1985–90	1990–95	1985–90	1990–95	1985–90	1990–95	1985–90	1990–95	1985–90	1990–95	1990–96
Albania	..	..	..	..	..	..	..	..	..	..	..	..	..
Algeria	..	..	..	..	..	38.5	..	..	..	..	7.2ª	..	18.5
Angola	..	..	..	..	..	..	..	..	..	..	..	..	7.6
Argentina	2.7	1.3	9.4	3.0	..	..	..	..	–0.6	–0.1	2.6	..	16,323.7
Armenia	..	..	..	..	..	..	..	..	..	..	..	..	79.2
Australia	..	..	15.0ᵇ	..	..	..	..	..	..	..	..	..	..
Austria	..	..	..	..	..	..	..	..	..	..	..	..	..
Azerbaijan	..	..	..	..	..	65.4	..	..	..	..	..	..	..
Bangladeshᶜ	3.1	3.4	33.5	23.5	17.4	12.0	..	–0.7	–3.9	–3.6	..	..	66.1
Belarus	..	..	..	..	..	..	..	..	..	..	..	..	..
Belgium	2.8	..	7.8	..	..	0.5	..	..	..	..	..	..	..
Benin	..	..	21.8	..	..	..	..	..	..	..	..	..	31.6
Bolivia	13.9	13.8	25.9	26.0	12.3	6.1	–7.6ª	–8.2	7.9ª	7.2	2.3ª	..	89.7
Bosnia and Herzegovina	..	..	..	..	..	..	..	..	..	..	..	..	..
Botswana	5.6ª	5.6ª	16.5ª	23.2ª	9.4	6.0	–0.3ª	–0.3ª	–2.6ª	–6.2ª	6.3ª	5.8ª	..
Brazil	7.6	8.0	13.0	8.6	14.0	4.2	–0.6ᵇ	–2.2	0.9ᵇ	3.3	..	..	22,402.1
Bulgaria	..	..	..	..	..	38.9	..	..	..	..	..	..	434.3
Burkina Faso	..	..	..	..	..	..	..	..	..	..	..	..	12.0
Burundi	7.3	..	41.1	..	19.3	12.0	2.7	..	..	..	32.1ᵇ,ᵈ	..	4.2
Cambodia	..	..	..	..	..	1.2	..	..	..	..	..	..	..
Cameroon	18.0	..	..	..	13.0	7.5	1.3	..	..	..	..	..	82.1
Canada	..	..	..	..	..	..	..	..	..	..	..	..	..
Central African Republic	4.1	..	..	..	12.5	11.1	..	..	–3.1	..	..	..	..
Chad	..	..	..	..	32.5	15.9	..	..	..	..	..	..	..
Chile	14.4	8.1	15.3	6.1	2.0	1.7	–8.6ᵉ	–4.8ᵉ	8.7ᵉ	4.8ᵉ	..	..	1,107.3
Chinaᶠ	..	..	29.2	24.9	..	..	..	..	..	..	7.7	7.4	10,411.4
Hong Kong, China	..	..	..	..	..	..	..	..	..	..	..	..	..
Colombia	7.0	..	13.5	..	5.2	1.5	–0.6	..	0.8	..	..	..	4,606.3
Congo, Dem. Rep.	..	..	18.8	..	1.0	5.0	..	..	..	..	..	..	..
Congo, Rep.	15.1ª	..	..	..	12.8	8.1	..	..	..	..	..	..	..
Costa Rica	8.1	..	8.5	11.0	..	..	–1.9	..	1.9	..	..	..	77.7
Côte d'Ivoire	..	..	21.4ᵇ	..	..	..	..	..	..	..	..	..	261.1
Croatia	..	..	..	..	..	4.2	..	..	..	..	..	..	417.6
Cuba	..	..	..	..	..	..	..	..	..	..	..	..	1,412.0
Czech Republic	..	..	..	..	..	22.4	..	..	..	..	..	..	659.1
Denmark	..	..	..	..	..	..	..	..	..	..	..	..	..
Dominican Republic	..	..	11.6ᵇ	..	15.0	11.1	..	..	..	..	..	..	..
Ecuador	10.2	..	12.7	..	0.3	0.5	0.3	..	–1.2	..	..	..	191.9
Egypt, Arab Rep.	..	..	65.5	..	21.5	18.5	–0.7	..	–2.6	..	13.8	..	..
El Salvador	1.8ª	..	7.1	..	3.2	0.0	0.1ª	..	–0.3ª	..	..	..	..
Eritrea	..	..	..	..	..	..	..	..	..	..	..	..	..
Estonia	..	..	..	..	..	23.5	..	..	..	..	..	..	359.5
Ethiopia	..	..	..	..	..	..	..	..	..	..	..	..	..
Finland	..	..	..	..	..	..	..	..	..	..	..	..	..
France	11.2	..	..	..	..	..	..	..	..	..	..	..	..
Gabon	..	..	..	..	5.2	3.3	..	..	..	..	25.1	..	..
Gambia, The	3.8	..	..	..	..	..	2.9ª	..	..	..	..	..	..
Georgia	..	..	..	..	..	..	..	..	..	..	..	..	..
Germany	..	..	..	..	..	..	..	..	..	..	..	..	..
Ghana	8.5	..	18.5	..	12.4	10.0	0.4ª	..	0.1	..	34.3ᵈ	..	1,368.9
Greece	11.5	..	20.2	..	..	..	..	..	..	..	..	..	..
Guatemala	1.9	..	8.0	..	..	..	0.2ª	..	0.0	..	..	..	..
Guinea	..	..	..	..	..	0.8	..	..	..	..	..	..	..
Guinea-Bissau	..	..	..	..	13.2	8.9	..	..	..	..	..	..	0.7
Haiti	..	..	11.3	..	7.7	3.9	..	..	..	..	..	..	..
Honduras	5.5	..	15.1	..	..	..	0.0	..	–0.9	..	..	..	79.4

State-owned enterprises | 5.8

	Economic activity % of GDP		Investment % of gross domestic investment		Credit % of gross domestic credit		Net financial flows from government % of GDP		Overall balance before transfers % of GDP		Employment % of total		Proceeds from privatization $ millions
	1985–90	1990–95	1985–90	1990–95	1985–90	1990–95	1985–90	1990–95	1985–90	1990–95	1985–90	1990–95	1990–96
Hungary	..	..	..	..	..	..	..	..	..	..	..	..	8,135.4
India	13.4	13.4	35.4	32.4	..	..	–0.3[g]	–0.9	–2.5[g]	–1.2[g]	8.5[g]	8.1[g]	6,890.0
Indonesia	14.5	..	8.9	15.7	..	3.4	1.3	..	–0.5	–2.6	0.9	1.1	5,745.7
Iran, Islamic Rep.	..	..	..	..	..	..	..	..	..	..	..	..	..
Iraq	..	..	..	..	..	..	..	..	..	..	..	..	..
Ireland	..	..	..	..	..	..	..	..	..	..	..	..	..
Israel	..	..	..	..	..	..	..	..	..	..	..	..	..
Italy	..	..	12.9	..	..	..	..	..	..	..	..	..	..
Jamaica	..	..	23.6	..	3.3	1.8	..	..	–1.4	..	..	..	478.7
Japan	..	..	5.8	..	2.8	2.0	..	..	..	..	..	..	22.6
Jordan	..	..	..	..	9.5	7.7	..	..	..	..	..	..	..
Kazakhstan	..	..	..	..	..	..	..	..	..	..	..	..	1,659.1
Kenya	11.6	..	20.4	..	5.2	3.2	0.2	..	..	..	7.9	7.8	334.1
Korea, Dem. Rep.	..	..	..	..	..	..	..	..	..	..	..	..	..
Korea, Rep.	10.3	..	14.3	..	..	..	–0.2	..	0.7	..	1.9	..	..
Kuwait	..	..	..	..	..	..	..	..	..	..	..	..	..
Kyrgyz Republic	..	..	..	..	..	..	..	..	..	..	..	..	279.0
Lao PDR	..	..	..	..	..	29.8	..	..	..	..	..	..	..
Latvia	..	..	..	..	..	5.3	..	..	..	..	..	..	392.0
Lebanon	..	..	..	..	..	..	..	..	..	..	..	..	..
Lesotho	..	..	..	..	..	..	..	..	..	..	..	..	..
Libya	..	..	..	..	..	..	..	..	..	..	..	..	..
Lithuania	..	..	..	..	..	9.9	..	..	..	..	..	..	1,672.1
Macedonia, FYR	..	..	..	..	..	..	..	..	..	..	..	..	..
Madagascar	..	..	..	..	..	..	..	..	..	..	..	..	..
Malawi	4.3[a]	..	9.2	..	12.8	10.2	0.8[a]	..	–1.2[a]	..	..	..	17.9
Malaysia[h]	..	..	..	25.9	..	..	..	..	..	–1.9	..	..	7,443.8
Mali	..	..	..	..	..	..	..	..	..	..	..	..	43.7
Mauritania	..	..	19.3	..	..	0.1	..	..	..	..	..	..	..
Mauritius	1.9	..	..	..	..	..	–0.3	..	..	..	..	..	..
Mexico[i]	6.7	4.9	14.6	*10.5*	6.5	0.9	–2.4	–3.1	2.4	3.0	3.5	2.1	24,929.3
Moldova	..	..	..	..	..	38.9	..	..	..	..	..	..	3.1
Mongolia	..	..	..	..	..	32.8	..	..	..	..	..	..	..
Morocco	16.8	..	19.3	..	..	..	0.0[a]	..	..	..	..	..	1,507.8
Mozambique	..	..	..	..	8.9	5.9	..	..	..	..	..	..	98.5
Myanmar	..	..	37.1[a]	..	..	..	..	..	..	..	..	..	..
Namibia	..	..	11.9	..	..	1.3	2.5[a]	..	–1.0	..	..	..	..
Nepal	..	..	50.0	..	7.4	3.0	..	..	..	..	..	..	13.9
Netherlands	..	..	..	..	..	..	..	..	..	..	..	..	..
New Zealand	..	..	..	..	..	..	..	..	..	..	..	..	..
Nicaragua	..	..	..	..	43.2	13.1	..	..	..	..	..	..	141.5
Niger	5.1	..	..	..	..	..	..	..	–0.1	..	..	..	..
Nigeria	..	..	..	..	..	0.5	..	..	..	..	..	..	738.2
Norway	..	..	25.8[b]	..	2.0	1.2	..	..	..	..	..	..	..
Oman	..	..	..	..	1.3	0.7	..	..	..	..	..	..	86.5
Pakistan	..	..	28.8	*25.9*	..	..	..	..	..	..	..	..	3,268.1
Panama	8.2	..	10.2	..	..	..	–0.8	..	1.6	..	..	..	302.9
Papua New Guinea	..	..	7.8	..	..	..	..	..	..	..	..	..	447.2
Paraguay	4.8	4.5	11.2	5.1	16.6	7.9	0.2	–0.4	–2.0	1.2	..	..	44.0
Peru	6.4	5.7	9.7	4.4	..	..	–3.1	–3.1	1.5	3.0	2.5	..	9,524.6
Philippines	2.3	2.2	8.4	9.9	8.5	3.9	..	..	–2.2	–3.7	0.8	..	3,666.9
Poland	..	..	..	..	76.0	29.6	..	..	..	..	..	..	3,803.8
Portugal	15.1	..	16.6	..	12.2	6.4	..	..	..	..	..	..	..
Puerto Rico	..	..	..	..	..	..	..	..	..	..	..	..	..
Romania	..	..	..	..	..	75.2	..	..	..	..	..	..	410.6
Russian Federation	..	..	..	..	..	20.3	..	..	..	..	..	..	..

5.8 | State-owned enterprises

| | Economic activity | | Investment | | Credit | | Net financial flows from government | | Overall balance before transfers | | Employment | | Proceeds from privati- zation |
| | % of GDP | | % of gross domestic investment | | % of gross domestic credit | | % of GDP | | % of GDP | | % of total | | $ millions |
	1985–90	1990–95	1985–90	1990–95	1985–90	1990–95	1985–90	1990–95	1985–90	1990–95	1985–90	1990–95	1990–96
Rwanda	..	..	..	..	..	..	0.4	..	1.0	..	..	..	..
Saudi Arabia	..	..	..	..	..	12.9	..	..	..	..	..	..	..
Senegal	6.9	..	28.2	..	..	..	4.5	..	–1.0	..	20.4d	..	..
Sierra Leone	..	..	..	..	0.4	0.1	0.1b	..	..	..	..	..	..
Singapore	..	..	..	..	..	..	..	..	..	..	..	..	..
Slovak Republic	..	..	..	..	..	27.9	..	..	..	..	..	..	1,865.0
Slovenia	..	..	..	..	..	..	..	..	..	..	..	..	377.2
South Africa	14.9	..	16.5	..	..	..	..	..	..	..	..	..	1,317.3
Spain	..	..	..	..	0.1	..	..	..	..	..	..	..	..
Sri Lanka	..	..	26.8	..	..	..	2.8	1.4	..	..	..	12.4	398.2
Sudan	..	..	..	..	16.5	5.0	..	..	..	..	..	..	..
Sweden	..	..	10.3	..	..	..	..	..	..	..	..	..	..
Switzerland	..	..	..	..	..	..	..	..	..	..	..	..	..
Syrian Arab Republic	..	..	..	..	..	..	..	..	..	..	..	..	..
Tajikistan	..	..	..	..	..	..	..	..	..	..	..	..	..
Tanzania	12.9	..	30.0	..	..	..	..	..	–7.4	..	21.4	..	66.1
Thailand	..	..	11.5	10.4	1.5	2.3	–0.3	–0.4	–0.3	–0.8	1.0b	..	1,773.7
Togo	..	..	10.7	..	..	..	1.6	..	..	..	..	..	24.7
Trinidad and Tobago	9.1a	..	16.4	..	10.8	6.5	0.0	..	0.7a	..	..	..	..
Tunisia	..	..	31.0	..	..	..	7.6	..	..	..	..	..	149.0
Turkey	6.5	5.1	27.1	15.1	6.0	4.7	1.6	1.0	–3.2	–6.6	3.7	2.9	2,777.4
Turkmenistan	..	..	..	..	..	..	..	..	..	..	..	..	..
Uganda	..	..	..	..	..	..	..	..	..	..	..	..	138.7
Ukraine	..	..	..	..	..	49.2	..	..	..	..	..	..	31.5
United Arab Emirates	..	..	..	..	..	..	..	..	..	..	..	..	..
United Kingdom	3.4	..	6.5b	..	..	..	..	..	..	..	..	..	..
United States	1.1	..	3.6	..	..	..	..	..	..	..	..	..	..
Uruguay	5.0	..	14.2	..	11.1	11.1	–3.2	..	3.3	..	..	..	4.0
Uzbekistan	..	..	..	..	..	..	..	..	..	..	..	..	424.0
Venezuela	22.3	..	50.9	..	1.1	4.2	–11.1	..	9.3	..	..	..	6,503.1
Vietnam	..	..	..	..	..	44.0	..	..	..	..	..	..	2.6
West Bank and Gaza	..	..	..	..	..	..	..	..	..	..	..	..	..
Yemen, Rep.	..	..	..	..	..	3.0	..	..	..	..	..	..	..
Yugoslavia, FR (Serb./Mont.)	..	..	..	..	..	..	..	..	..	..	..	..	..
Zambia	32.2	..	..	..	..	..	..	..	..	..	..	..	89.7
Zimbabwe	10.8	11.3	..	..	29.7	7.5	..	..	..	..	..	..	25.0

Note: Averages for a period have been calculated only when three or more years' data are available. Figures in italics refer to an average for the period 1990–96.
a. Selected major state–owned enterprises only. b. Includes financial state–owned enterprises. c. Data for 1985–90 refer to the 10 largest state–owned enterprises. For 1990–95, data refer to 210 enterprises. d. As a percentage of formal sector employment. e. Nonoperating revenue before 1989 is split between current transfers and grants. Since 1989, all nonoperating revenue is classified as current transfers. f. Data refer to industrial state–owned enterprises. g. Data refer to central public enterprises only. h. Data prior to 1991 have not been shown due to lack of consistency. i. Data for economic activity and employment refer to nonfinancial enterprises in both the controlled and the noncontrolled sectors. Data on investment up to 1986 refer to nonfinancial enterprises in both the controlled and the noncontrolled sectors. Since 1987 financial enterprises in the noncontrolled sector are included. Data on overall balances before transfers and net financial flows from government refer only to nonfinancial enterprises in the controlled sector.

State-owned enterprises | 5.8

State-owned enterprises are government-owned or -controlled economic entities that generate most of their revenue by selling goods and services. This definition encompasses commercial enterprises directly operated by a government department and those in which the government holds a majority of shares directly or indirectly through other state enterprises. It also includes enterprises in which the state holds a minority of shares, if the distribution of the remaining shares leaves the government with effective control. It excludes public sector activity—such as education, health services, and road construction and maintenance—that is financed in other ways, usually from the government's general revenue. Because financial enterprises are of a different nature, they have generally been excluded from the data.

The definition of a state enterprise varies among countries and within countries over time. In exceptional cases governments include noncommercial activities, such as agricultural research institutes, in their data on state enterprises. But more often they omit activities that clearly are state enterprises. The most common omissions occur when governments use a narrow definition of state enterprises—for example, by excluding those with a particular legal form (such as departmental enterprises), those owned by local governments (typically utilities), or those considered unimportant in terms of size or need for fiscal resources. Accordingly, data on state enterprises tend to underestimate their relative importance in the economy.

Although attempts have been made to correct for differences in definitions and coverage across countries, inconsistencies remain. These cases are detailed in the country notes in *Bureaucrats in Business* (World Bank 1995b). The state enterprises covered in the table are limited to central or federal government enterprises because data on local government-owned enterprises are extremely limited. Another weakness in the data is that many state enterprises do not follow generally accepted accounting principals, so accounting rules can vary by country and enterprise. In many case small state enterprises are not audited by internationally accredited accounting firms, so there may be no independent check on their record keeping and reporting.

To assess the importance of government ownership, the table includes three measures of the economic size of state enterprises: share in economic activity, investment, and employment. Indicators that measure the performance of state enterprises and their effect on the macroeconomy and growth include credit, net financial flows from government, and overall balance before transfers. These indicators do not, however, allow for analysis of the relative efficiency of state enterprises and private firms because not enough data are available on ownership by economic sector. Data in the table are period averages for 1985–90 and 1990–95. The updating of data has necessitated revisions to earlier years in order to ensure consistency over the time series.

Data on proceeds from privatization are included in the table because privatization—that is, the transfer of productive assets from the public to the private sector—has been one of the defining economic changes of the 1980s and 1990s. Direct sales are the most common privatization method, accounting for 60 percent of privatization transactions in 1996. Direct sales enable governments to attract strategic investors who can transfer capital, technology, and managerial know-how to newly privatized enterprises. Share issues in domestic and international capital markets are the second most common method, accounting for about 33 percent of privatization transactions in 1996.

Large sales proceeds do not necessarily imply major changes in the control or stock of state-owned enterprises, however. For example, selling equity may not change effective control. It may only generate revenue, with no gains in efficiency. A preliminary analysis suggests that the increase in proceeds is due to a larger number of countries privatizing a few firms rather than to a radical restructuring of ownership in many countries (Haggarty and Shirley 1997).

State enterprises still play a sizable role in many developing economies

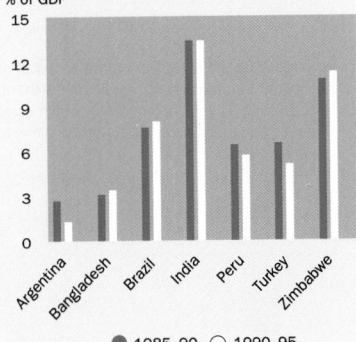

● 1985–90 ○ 1990–95

Source: Table 5.8.

Despite more than a decade of divestitures and the growing consensus that the private sector should play a larger role, the share of state-owned enterprises in GDP continues to rise in some countries.

• **Economic activity** is the value added of state enterprises, estimated as their sales revenue minus the cost of their intermediate inputs, or as the sum of their operating surplus (balance) and wage payments. • **Investment** refers to fixed capital formation by state enterprises. • **Credit** is credit extended to state enterprises by domestic financial institutions. • **Net financial flows from government** is the difference between total financial flows from the government to state enterprises (including government loans, equity, and subsidies) and total flows from state enterprises to the government (including dividends and taxes). Taxes paid by state enterprises are treated as a transfer of financial resources to the government. • **Overall balance before transfers** is the sum of net operating and net nonoperating revenues minus net capital expenditure. Net operating revenues (or operating surplus or balance) refer to gross operating profits, or operating revenues, minus the costs of intermediate inputs, wages, factor rentals, and depreciation. • **Employment** for many countries refers to the share of full-time state enterprise employees in total employment, but for some it refers to employment only in selected state enterprises, including financial ones, and for others it refers to employment as a share of total formal sector employment. Thus the data on state enterprise employment are not directly comparable. • **Proceeds from privatization** include all sales of public assets to private entities through public offers, direct sales, management and employee buyouts, concessions or licensing agreements, and joint ventures.

Data on state enterprises were collected from World Bank member country central banks, finance ministries, enterprises, and World Bank and International Monetary Fund reports. These data were then collated into a database for the World Bank Policy Research Report *Bureaucrats in Business: The Economics and Politics of Government Ownership* (1995b). Updates to this database have been made for several economies. Data on privatization are from the World Bank's *Global Development Finance 1998*. Data on credit are from the International Monetary Fund's *International Financial Statistics*.

5.9 Transport infrastructure

	Roads			Railways		Air		
	Paved roads % 1996	Normalized road index 1996	Goods transported million ton-km 1996	Passenger-km per PPP $ million of GDP 1996	Goods transported ton-km per PPP $ million of GDP 1996	Aircraft departures thousands 1996	Passengers carried thousands 1996	Air freight millions ton-km 1996
Albania	30.0	110	3	..	..	1	29	0
Algeria	68.9	165	20,000	12,987	17,681	45	3,494	16
Angola	25.0	..	2,187	..	..	8	207	62
Argentina	29.1	115	..	..	..	132	7,779	177
Armenia	100.0	100	18	..	..	2	358	12
Australia	38.7	128	128,000	..	..	371	30,075	1,834
Austria	100.0	136	64,400	51,084	79,531	118	4,719	192
Azerbaijan	..	..	11,459	..	..	19	1,233	28
Bangladesh	7.2	64	..	..	..	13	1,252	136
Belarus	70.1	85	350	262,101	619,342	26	596	2
Belgium	..	..	428	30,112	32,214	165	5,174	591
Benin	20.0	52	..	..	..	1	75	16
Bolivia	5.5	59	..	..	..	32	1,783	47
Bosnia and Herzegovina	52.3	..	..	..	..	..	..	..
Botswana	23.5	314	..	..	..	4	104	0
Brazil	9.3	145	384,000	1,265	50,730	483	22,004	1,645
Bulgaria	91.9	99	39	136,587	202,772	14	718	24
Burkina Faso	16.0	96	..	..	..	3	138	16
Burundi	7.1	108	..	..	..	1	9	0
Cambodia	7.5	..	..	..	..	..	..	..
Cameroon	12.5	59	..	13,165	33,723	4	362	42
Canada	..	..	182,000	2,076	266,190	306	22,856	1,781
Central African Republic	..	..	..	..	..	1	75	16
Chad	0.8	15	..	..	..	2	93	16
Chile	13.8	56	..	4,344	6,096	94	3,622	806
China	..	..	463,000	88,980	360,383	493	51,770	1,689
Hong Kong, China	100.0	..	14	..	..	..	..	..
Colombia	11.9	41	..	..	..	195	8,342	311
Congo, Dem. Rep.	..	..	..	705	4,387	..	..	..
Congo, Rep.	9.7	62	..	69,853	54,139	5	253	17
Costa Rica	17.0	211	..	..	..	21	918	45
Côte d'Ivoire	9.7	86	..	7,822	13,484	5	179	16
Croatia	81.5	..	4	49,544	103,711	15	727	2
Cuba	55.9	..	..	..	..	14	929	44
Czech Republic	100.0	..	686	71,434	196,511	27	1,394	22
Denmark	100.0	103	..	39,515	14,713	108	5,892	172
Dominican Republic	49.4	136	..	..	..	1	30	0
Ecuador	13.3	76	54,300	..	..	22	1,873	24
Egypt, Arab Rep.	78.1	149	..	327,998	27,908	41	4,282	198
El Salvador	19.9	80	4,273	..	..	21	1,800	16
Eritrea	21.8	..	..	..	..	..	..	..
Estonia	53.2	65	11	45,250	540,949	5	149	1
Ethiopia	15.0	55	..	..	..	25	743	118
Finland	64.0	79	374	33,706	70,489	103	5,597	237
France	100.0	143	1,275	47,243	39,290	542	40,300	4,811
Gabon	8.2	38	..	9,441	61,672	7	431	35
Gambia, The	35.4	251	..	..	..	..	..	..
Georgia	93.5	..	7	..	..	2	205	2
Germany	99.1	..	294,160	..	39,068	567	40,118	6,036
Ghana	24.1	106	..	..	..	3	197	30
Greece	91.8	155	201	13,406	1,913	92	6,396	119
Guatemala	27.6	62	..	..	..	5	300	26
Guinea	16.5	146	..	..	..	1	36	1
Guinea-Bissau	10.3	..	..	..	..	1	21	0
Haiti	24.3	..	..	..	..	..	..	..
Honduras	20.3	135	..	..	..	13	498	2

Transport infrastructure | 5.9

	Roads			Railways		Air		
	Paved roads % 1996	Normalized road index 1996	Goods transported million ton-km 1996	Passenger-km per PPP $ million of GDP 1996	Goods transported ton-km per PPP $ million of GDP 1996	Aircraft departures thousands 1996	Passengers carried thousands 1996	Air freight millions ton-km 1996
Hungary	43.1	147	39	88,789	103,268	25	1,563	29
India	50.2	566	..	231,700	177,267	151	13,255	565
Indonesia	45.5	211	..	23,842	6,843	311	16,173	749
Iran, Islamic Rep.	50.0	..	..	22,281	34,136	63	7,610	110
Iraq	86.0	..	..	..	..	..	..	..
Ireland	94.1	276	..	19,499	9,314	95	7,677	102
Israel	100.0	100	..	2,685	11,827	50	3,695	1,113
Italy	100.0	60	..	43,529	18,432	307	25,838	1,459
Jamaica	70.7	797	..	..	..	18	1,388	25
Japan	74.1	77	..	89,507	8,896	564	95,914	6,801
Jordan	100.0	98	..	..	47,815	17	1,299	297
Kazakhstan	80.5	171	803	..	..	10	568	17
Kenya	13.9	115	..	13,587	46,448	14	779	48
Korea, Dem. Rep.	6.4	..	..	..	..	6	254	3
Korea, Rep.	76.1	151	410	52,691	24,665	207	33,003	6,551
Kuwait	80.6	..	..	..	..	19	2,133	334
Kyrgyz Republic	91.1	..	110	..	..	13	488	2
Lao PDR	13.8	273	..	..	..	4	125	1
Latvia	38.3	161	30	157,014	1,115,793	17	407	2
Lebanon	95.0	..	..	..	..	10	775	80
Lesotho	17.9	137	..	..	..	1	17	0
Libya	57.2	..	..	..	..	6	639	0
Lithuania	87.6	436	89	53,960	491,829	7	214	2
Macedonia, FYR	63.8	79	3	..	..	5	287	1
Madagascar	11.6	158	..	..	..	17	542	25
Malawi	18.5	327	..	10,172	10,172	4	153	3
Malaysia	75.1	..	..	6,159	6,867	188	15,118	1,415
Mali	12.1	90	..	..	..	1	75	16
Mauritania	11.3	54	..	..	..	5	235	17
Mauritius	93.1	110	..	..	..	9	718	137
Mexico	37.4	106	..	2,674	52,983	223	14,678	169
Moldova	87.3	91	41	..	..	4	190	1
Mongolia	7.8	..	2	..	..	10	662	3
Morocco	50.4	144	54,671	18,995	55,334	32	2,301	57
Mozambique	18.7	141	230	..	..	4	163	5
Myanmar	12.2	..	..	..	..	15	334	1
Namibia	12.1	362	..	4,234	131,387	7	237	29
Nepal	41.5	76	..	..	..	28	755	18
Netherlands	90.1	73	..	44,295	9,816	..	..	..
New Zealand	58.1	100	..	..	..	187	9,597	745
Nicaragua	10.1	65	..	..	..	1	51	9
Niger	7.9	27	..	..	..	1	75	16
Nigeria	18.8	106	..	580	1	6	221	5
Norway	72.0	106	244	..	..	279	12,727	177
Oman	30.0	212	..	..	..	18	1,620	106
Pakistan	57.0	283	..	93,400	25,084	70	5,375	427
Panama	33.6	128	..	..	..	17	689	9
Papua New Guinea	3.5	27	..	..	..	27	970	18
Paraguay	9.5	..	..	..	..	4	213	0
Peru	10.1	39	..	2,007	5,176	35	2,328	14
Philippines	..	..	..	..	..	65	7,263	384
Poland	65.4	156	1,640	85,254	290,148	36	1,806	70
Portugal	86.0	94	369	33,497	13,832	80	4,806	211
Puerto Rico	100.0	..	..	..	..	..	..	..
Romania	51.0	108	616,044	175,264	230,933	17	913	15
Russian Federation	78.8	..	18,000	340,048	1,790,023	465	22,117	854

	Roads			Railways		Air		
	Paved roads % 1996	Normalized road index 1996	Goods transported million ton-km 1996	Passenger-km per PPP $ million of GDP 1996	Goods transported ton-km per PPP $ million of GDP 1996	Aircraft departures thousands 1996	Passengers carried thousands 1996	Air freight millions ton-km 1996
Rwanda	9.1	119	..	..	..	..	..	..
Saudi Arabia	42.7	142	..	875	4,384	101	11,706	863
Senegal	29.3	106	..	14,302	30,617	5	155	16
Sierra Leone	11.0	100	..	..	..	0	15	0
Singapore	97.4	..	..	..	..	57	11,841	4,115
Slovak Republic	98.5	..	34,745	93,911	298,678	3	63	0
Slovenia	82.0	100	5	27,278	115,975	8	393	23
South Africa	41.5	300	..	34,034	336,265	90	7,183	335
Spain	99.0	104	589	25,645	15,998	333	27,759	740
Sri Lanka	40.0	469	3,020	83,699	4,027	9	1,171	159
Sudan	36.3	..	..	..	..	11	491	46
Sweden	76.1	85	..	35,735	103,765	183	9,879	250
Switzerland	..	..	410,000	..	..	218	10,468	1,511
Syrian Arab Republic	23.0	..	..	11,108	29,013	9	599	16
Tajikistan	82.7	..	..	..	..	3	594	3
Tanzania	4.2	69	..	..	..	6	224	3
Thailand	97.5	173	..	..	..	92	14,078	1,348
Togo	31.6	180	..	..	..	1	75	16
Trinidad and Tobago	51.1	217	..	..	..	14	897	22
Tunisia	78.9	172	..	23,325	53,910	14	1,371	18
Turkey	25.0	53	135,781	13,978	17,619	85	8,464	207
Turkmenistan	81.2	..	..	..	..	9	523	2
Uganda	..	..	..	2,168	12,829	1	100	1
Ukraine	95.0	96	1,254,540	329,444	910,955	28	1,151	17
United Arab Emirates	100.0	32	..	..	..	38	4,063	651
United Kingdom	100.0	64	1,689	26,728	11,465	..	..	..
United States	60.8	112	..	1,387	365,655	8,032	571,072	21,676
Uruguay	90.0	167	..	22,273	18,789	7	504	4
Uzbekistan	87.3	..	..	..	..	12	1,566	8
Venezuela	39.4	108	..	..	..	83	4,487	117
Vietnam	25.1	..	..	24,649	20,223	29	2,505	2
West Bank and Gaza	..	..	..	..	..	..	..	..
Yemen, Rep.	8.1	..	..	..	..	..	..	..
Yugoslavia, FR (Serb./Mont.)	58.3	..	100	..	..	..	..	..
Zambia	18.3	189	..	34,856	60,312	..	..	..
Zimbabwe	47.4	144	..	22,994	200,217	10	654	153
World	**47.4 m**	**.. m**				**18,746 s**	**1,388,670 s**	
Low income	19	106.0				1,040	87,460	
Excl. China & India	18	105.8				396	22,435	
Middle income	55	..				3,604	222,117	
Lower middle income	53	109.1				1,995	125,563	
Upper middle income	57	..				1,609	96,554	
Low & middle income	32	110.1				4,644	309,577	
East Asia & Pacific	..	..				1,305	110,432	
Europe & Central Asia	82	99.8				869	47,754	
Latin America & Carib.	26	107.9				1,519	76,532	
Middle East & N. Africa	54	..				371	36,896	
South Asia	42	282.8				284	22,305	
Sub-Saharan Africa	17	107.1				296	15,658	
High income	..	..				14,102	1,079,094	

Transport infrastructure | 5.9

Transport infrastructure—highways, railways, ports and waterways, and airports and air traffic control systems—and the services that flow from it are crucial to the activities of households, producers, and governments. Because performance indicators vary significantly by transport mode and by measurement focus (whether on physical infrastructure or the services flowing from that infrastructure), highly specialized and carefully specified indicators are required. The table provides selected indicators of the size and extent of roads, railways, and air transport systems and the volume of freight and passengers carried.

Data for most transport sectors are not internationally comparable. Unlike demographic statistics, national income accounts, and international trade data, the collection of infrastructure data has not been "internationalized." Data on roads are collected by the International Road Federation (IRF) and data on air transport are collected by the International Civil Aviation Organization (ICAO). National road associations are the primary source of IRF data; in countries where such an association is absent or does not respond, other agencies are contacted, such as road directorates, ministries of transport or public works, or central statistical offices. As a result the compiled data are of uneven quality.

Even when data are available, they are often of limited value because of incompatible definitions, inappropriate geographical units of observation, or lack of timeliness. Data on passengers carried, for example, may be distorted because of "ticketless" travel or breaks in journeys; in such cases the statistics may report the number of passenger-kilometers for two passengers instead of one. Measurement problems are compounded because over time the mix of transported commodities changes, and in some cases shorter-haul traffic has been excluded from intercity traffic. Finally, the quality of transport service (reliability, transit time, and condition of goods delivered) is rarely measured but may be as important as quantity in assessing an economy's transport system. Serious efforts are needed to create internationally databases whose comparability and accuracy can be gradually improved.

Some form of normalization is required to measure the relative size of an indicator over time or across countries. The table presents normalized indicators for railway passenger-kilometers per million dollars of GDP measured in purchasing power parity (PPP) terms, goods transported by railway ton-kilo-meters per million dollars of GDP PPP (see tables 4.10 and 4.11 for a discussion of PPP) and the normalized road index. While the rail traffic indicators are normalized by a single indicator—the size of the economy—the normalized road index uses a multidimensional regression function to estimate a country's "normal," or expected, stock of roads (Armington and Dikhanov 1996). Normalizing variables include population, population density, per capita income, urbanization, and regional differences. The value of the normalized road index shows whether a country's stock of roads exceeds or falls short of the average for countries with similar characteristics.

• **Paved roads** are roads that have been sealed with asphalt or similar road-building materials. • **Normalized road index** is the total length of roads in a country compared with the expected length of roads, where the expectation is conditioned on population, population density, per capita income, urbanization, and region-specific dummy variables. A value of 100 is "normal." If the index is more than 100, the country's stock of roads exceeds the average. • **Goods transported by road** are the volume of goods transported by road vehicles, measured in millions of metric tons times kilometers traveled. • **Railway passengers** measures the total passenger-kilometers per million dollars of GDP measured in PPP terms (see tables 4.10 and 4.11 for a discussion of PPP). • **Goods transported by rail** measures the tonnage of goods transported times kilometers traveled per million dollars of GDP measured in PPP terms. • **Aircraft departures** are the number of domestic and international takeoffs of aircraft. • **Passengers carried** include both domestic and international aircraft passengers. • **Air freight** is the sum of the tons of freight, express, and diplomatic bags carried on each flight stage (the operation of an aircraft from takeoff to its next landing) multiplied by the stage distance.

Data on roads are from the International Road Federation's *World Road Statistics*. The normalized road index is based on World Bank staff estimates. Railway data are from a database maintained by the World Bank's Transportation, Water, and Urban Development Department, Transport Division. Air transport data are from the International Civil Aviation Organization's *Civil Aviation Statistics of the World*.

5.10 | Power and communications

	Electric power			Telephone mainlines							International telecommunications	
	Consumption per capita kwh 1995	Production average annual % growth 1980–95	Transmission and distribution losses % of output 1995	per 1,000 people 1996	In largest city per 1,000 people 1996	Waiting list thousands 1996	Waiting time years 1996	per employee 1996	Revenue per line $ 1996	Cost of local call $ per 3 minutes 1996	Outgoing traffic minutes per subscriber 1996	Cost of call to U.S. $ per 3 minutes 1996
Albania	623	–2.3	51	19	*51*	21.3	3.1	14	571	0.04	323	9.51
Algeria	513	7.3	17	44	*64*	702.2	>10	58	162	0.02	73	5.04
Angola	60	2.1	28	5	*23*	..	..	26	*3,698*	..	302	..
Argentina	1,519	2.3	18	174	*224*	110.6	0.2	*186*	982	0.10	30	7.37
Armenia	811	*–6.0*	39	154	*212*	54.7	..	*63*	*50*	..	*90*	..
Australia	8,033	4.5	7	519	*485*	..	0.0	118	1,411	0.19	101	3.00
Austria	5,800	2.7	6	466	*632*	3.7	0.1	219	709	0.20	251	3.77
Azerbaijan	1,806	*–1.5*	23	85	170	178.7	5.7	44	68	0.19	42	3.85
Bangladesh	57	10.4	32	3	*25*	155.2	6.6	*15*	634	0.04	81	..
Belarus	2,451	*1.6*	15	208	302	573.1	5.5	86	106	0.01	62	6.00
Belgium	6,752	3.4	5	465	*635*	1.7	0.0	180	948	0.20	248	3.05
Benin	43	2.2	50	6	33	5.2	1.3	25	1,107	0.13	212	8.26
Bolivia	356	4.0	12	*47*	*67*	50.1	1.0	*199*	*307*	..	65	..
Bosnia and Herzegovina	..	..	23	90	*429*	70.0	..	80	*2*	..	19	..
Botswana	..	..	..	48	180	9.3	1.0	43	972	0.03	427	6.06
Brazil	1,610	5.5	17	96	273	..	..	169	821	0.04	24	4.68
Bulgaria	3,415	–1.5	13	313	374	480.0	7.7	100	30	0.01	32	..
Burkina Faso	..	..	..	3	29	..	..	28	1,229	0.12	200	..
Burundi	..	..	..	2	45	0.0	..	26	1,079	0.04	169	13.67
Cambodia	..	..	..	1	*5*	13.4	>10	12	25,550	..	857	..
Cameroon	196	3.1	4	5	*30*	42.0	9.4	36	1,088	0.08	352	12.02
Canada	15,147	2.8	5	602	..	..	0.0	245	731	..	*168*	1.16
Central African Republic	..	..	..	3	*12*	1.1	1.5	23	1,933	0.20	389	24.04
Chad	..	..	..	1	6	1.4	3.0	16	1,264	0.20	314	14.07
Chile	1,698	6.1	10	156	*192*	72.5	0.3	*184*	743	0.09	72	2.79
China	637	8.2	7	45	*140*	812.0	0.1	114	318	..	26	..
Hong Kong, China	4,850	7.6	15	547	537	..	0.0	91	1,866	..	504	2.64
Colombia	948	5.4	21	118	*227*	728.0	1.5	228	439	0.01	31	4.12
Congo, Dem. Rep.	132	3.4	3	*1*	6	6.0	..	19	..	..	36	..
Congo, Rep.	207	9.4	0	*8*	..	0.8	0.6	23	*2,182*	0.13	*190*	..
Costa Rica	1,348	5.5	8	155	..	77.9	0.8	*106*	441	0.04	113	..
Côte d'Ivoire	159	2.7	4	9	36	74.2	5.6	38	1,393	0.20	285	..
Croatia	2,074	*8.5*	19	309	405	71.7	0.6	146	426	0.02	175	5.66
Cuba	818	0.7	13	32	74	..	..	25	1,217	..	35	..
Czech Republic	4,654	1.3	8	273	*525*	577.0	2.0	104	412	0.07	103	4.64
Denmark	5,975	2.9	6	618	..	..	0.0	199	1,236	0.17	175	3.35
Dominican Republic	588	4.5	25	83	*132*	..	..	158	..	1.52	124	..
Ecuador	600	5.4	21	73	*189*	60.2	0.7	160	296	0.01	56	8.21
Egypt, Arab Rep.	896	8.6	..	50	*95*	1,310.2	5.0	58	256	0.01	37	6.19
El Salvador	507	5.4	13	56	144	200.0	4.0	60	617	0.02	187	..
Eritrea	..	..	..	5	32	40.0	>10	35	922	0.03	92	8.25
Estonia	3,022	–0.9	20	299	245	96.1	3.2	116	337	0.04	133	4.99
Ethiopia	22	4.6	3	3	45	195.5	>10	27	526	0.03	70	..
Finland	12,785	4.2	2	549	*1,349*	0.0	0.0	178	956	0.14	118	3.22
France	5,892	3.9	6	564	..	..	0.0	198	830	0.15	90	3.03
Gabon	737	3.1	10	32	*90*	3.5	2.0	44	1,971	0.18	508	..
Gambia, The	..	..	..	19	*37*	19.2	>10	28	781	0.09	274	..
Georgia	1,057	*–3.8*	25	105	*183*	230.0	..	*72*	*13*	..	0	..
Germany	5,527	1.2	5	538	*558*	..	..	205	1,011	0.16	118	2.87
Ghana	318	4.6	4	4	*17*	28.3	2.9	38	1,283	0.08	267	..
Greece	3,259	3.8	7	509	*620*	78.6	0.4	224	587	..	97	3.05
Guatemala	264	5.5	13	31	*117*	100.0	2.7	56	615	0.03	101	..
Guinea	..	..	..	2	*5*	1.6	1.0	19	1,555	0.12	152	8.76
Guinea-Bissau	..	..	..	7	..	0.7	1.8	32	1,447	0.09	311	..
Haiti	32	–1.6	53	8	*23*	40.0	8.0	*23*	*1,301*	..	..	7.07
Honduras	333	6.5	28	31	*102*	268.9	>10	42	887	0.06	219	7.11

Power and communications | 5.10

	Electric power			Telephone mainlines							International telecommunications	
	Consumption per capita kwh 1995	Production average annual % growth 1980–95	Transmission and distribution losses % of output 1995	per 1,000 people 1996	In largest city per 1,000 people 1996	Waiting list thousands 1996	Waiting time years 1996	per employee 1996	Revenue per line $ 1996	Cost of local call $ per 3 minutes 1996	Outgoing traffic minutes per subscriber 1996	Cost of call to U.S. $ per 3 minutes 1996
Hungary	2,682	0.7	14	261	386	250.9	0.6	164	485	0.15	100	4.77
India	339	9.2	18	15	80	2,277.0	1.0	28	238	0.02	29	6.35
Indonesia	263	15.0	12	21	77	117.5	0.2	105	643	0.05	63	6.07
Iran, Islamic Rep.	1,059	8.7	20	95	222	1,379.0	1.9	121	148	0.01	29	6.02
Iraq	1,396	6.5	..	33	..	..	..	..	..	..	..	..
Ireland	4,139	4.0	9	395	..	..	..	120	1,273	0.16	311	3.32
Israel	4,836	5.9	4	446	414	12.0	0.1	287	843	0.08	108	3.43
Italy	4,163	3.1	7	440	529	32.0	0.1	253	911	0.20	84	3.36
Jamaica	2,049	8.4	11	142	168	183.1	3.5	88	995	0.06	183	..
Japan	6,937	4.0	4	489	..	..	0.0	276	1,532	0.09	28	5.10
Jordan	1,139	10.9	8	60	160	129.0	9.9	81	702	0.03	226	..
Kazakhstan	3,106	−2.8	15	118	275	668.4	4.2	39	126	0.00	8	..
Kenya	123	5.5	16	8	77	70.6	4.5	22	1,139	0.06	80	11.17
Korea, Dem. Rep.	261	−10.2	84	49	85	16.4	0.2	..	..	..	3	..
Korea, Rep.	3,606	11.7	5	430	466	..	0.0	294	469	0.04	35	4.88
Kuwait	13,185	5.5	..	232	..	2.6	0.2	52	572	..	364	5.51
Kyrgyz Republic	1,666	2.8	28	75	230	67.4	..	51	63	..	79	11.82
Lao PDR	..	..	..	6	13	2.6	0.4	29	709	..	240	..
Latvia	1,789	−1.6	32	298	371	102.2	6.8	116	191	0.08	59	6.43
Lebanon	1,224	1.4	13	149	..	..	..	115	580	0.04	85	7.14
Lesotho	..	..	..	9	64	5.4	2.5	22	742	0.04	..	..
Libya	3,569	10.2	..	59	36	..	..	..	..	0.03	147	..
Lithuania	1,711	−2.7	15	268	362	141.6	3.2	100	134	0.02	54	7.88
Macedonia, FYR	2,443	..	12	170	235	21.0	1.5	108	219	0.01	139	..
Madagascar	..	..	..	3	5	10.0	6.5	12	1,012	0.06	148	..
Malawi	..	..	..	4	13	29.9	>10	8	900	0.04	214	12.45
Malaysia	1,953	10.1	10	183	143	160.0	0.4	137	679	0.04	66	5.99
Mali	..	..	..	2	16	..	..	16	2,646	0.17	429	17.59
Mauritania	..	..	..	4	13	1.2	1.6	23	2,689	0.13	479	..
Mauritius	..	..	..	162	206	35.9	1.4	106	622	0.06	118	5.85
Mexico	1,305	4.9	14	95	..	196.9	0.5	180	786	0.08	107	3.01
Moldova	1,517	1.3	18	140	269	201.7	8.7	78	64	0.10	105	6.39
Mongolia	..	..	..	39	80	40.0	4.6	18	206	0.02	26	14.77
Morocco	407	6.5	4	45	121	48.0	0.3	85	556	0.09	104	6.88
Mozambique	67	−5.8	5	3	24	22.7	>10	25	948	0.04	217	..
Myanmar	52	5.3	34	4	24	55.0	2.7	25	1,792	0.17	82	26.86
Namibia	..	..	..	54	253	4.5	0.7	51	899	..	596	..
Nepal	39	10.9	26	5	69	136.2	>10	30	337	0.02	184	..
Netherlands	5,374	2.9	4	543	..	20.0	0.1	284	1,005	0.19	181	2.69
New Zealand	8,504	3.0	9	499	..	..	0.0	189	1,201	..	179	4.78
Nicaragua	272	1.7	28	26	55	..	..	30	473	0.04	304	..
Niger	..	..	..	2	16	1.4	0.9	13	1,138	0.15	273	..
Nigeria	85	4.6	32	4	16	98.1	3.5	28	1,904	0.26	233	..
Norway	23,892	2.2	7	555	732	6.0	0.0	132	1,471	0.11	189	2.36
Oman	2,891	15.7	..	86	150	3.3	0.2	92	1,260	0.07	317	..
Pakistan	304	10.1	23	18	61	209.5	0.7	44	442	0.05	32	..
Panama	1,089	2.9	19	122	214	28.8	1.4	90	764	..	127	5.20
Papua New Guinea	..	..	..	11	200	0.3	0.1	23	2,418	0.13	572	..
Paraguay	683	10.4	1	36	125	..	..	25	1,034	0.06	123	..
Peru	525	2.4	21	60	119	44.8	0.2	228	920	0.10	49	5.76
Philippines	337	2.8	16	25	96	900.2	2.9	98	610	..	108	6.22
Poland	2,324	0.2	13	169	..	2,327.4	3.3	89	389	0.06	67	4.12
Portugal	2,857	5.1	11	375	645	7.6	0.0	171	1,198	0.08	91	3.74
Puerto Rico	..	..	..	336	437	57.0	1.1	157	917	0.13	723	..
Romania	1,603	−3.5	11	140	..	1,299.0	7.0	59	177	0.01	29	11.57
Russian Federation	4,172	0.0	10	175	467	8,796.8	>10	58	203	0.27	9	8.61

	Electric power			Telephone mainlines							International telecommunications	
	Consumption per capita kwh 1995	Production average annual % growth 1980–95	Transmission and distribution losses % of output 1995	per 1,000 people 1996	In largest city per 1,000 people 1996	Waiting list thousands 1996	Waiting time years 1996	per employee 1996	Revenue per line $ 1996	Cost of local call $ per 3 minutes 1996	Outgoing traffic minutes per subscriber 1996	Cost of call to U.S. $ per 3 minutes 1996
Rwanda	..	..	..	3	..	..	..	20	1,016	..	89	..
Saudi Arabia	3,906	11.5	9	106	189	1,262.5	9.7	104	1,053	0.02	292	6.41
Senegal	91	3.1	13	11	29	17.8	1.7	65	1,275	0.10	260	9.36
Sierra Leone	..	..	..	4	20	14.0	>10	18	896	0.07	190	..
Singapore	6,018	8.6	4	513	513	0.2	0.0	224	1,785	0.03	541	4.02
Slovak Republic	4,075	0.8	8	232	515	144.9	1.2	79	382	0.06	109	5.45
Slovenia	4,710	1.2	5	333	610	48.7	1.1	215	492	0.02	160	5.56
South Africa	3,874	3.3	6	100	495	126.8	0.6	74	892	0.09	83	..
Spain	3,594	3.4	10	392	440	1.5	0.0	229	753	0.10	77	3.49
Sri Lanka	208	6.2	18	14	219	237.8	7.4	25	1,410	0.05	122	3.25
Sudan	37	0.8	19	4	3	75.0	6.4	38	221	0.03	130	10.55
Sweden	14,096	2.8	6	682	778	..	0.0	304	852	0.11	156	2.65
Switzerland	6,916	2.5	6	640	956	0.3	0.0	221	1,863	0.24	416	2.43
Syrian Arab Republic	698	7.2	..	82	176	2,945.4	>10	68	1,001	0.05	66	33.41
Tajikistan	2,367	2.6	12	42	222	72.0	..	50	41	..	1	26.87
Tanzania	52	5.5	13	3	21	107.9	>10	19	775	0.08	63	..
Thailand	1,199	12.4	8	70	339	821.6	1.2	120	545	0.12	61	7.39
Togo	..	..	..	6	22	7.6	3.4	28	1,618	0.12	359	13.20
Trinidad and Tobago	2,817	4.0	10	168	137	6.4	1.1	80	825	0.04	277	3.48
Tunisia	661	6.6	10	64	73	81.8	1.5	98	505	0.07	146	6.47
Turkey	1,057	8.6	16	224	438	752.7	0.7	194	180	0.06	33	4.37
Turkmenistan	1,109	0.5	10	74	198	79.6	3.3	41	..	..	17	..
Uganda	..	..	..	2	34	6.3	0.7	36	905	0.19	113	9.29
Ukraine	2,785	−1.1	10	181	395	3,103.2	7.2	71	119	..	9	..
United Arab Emirates	7,752	8.0	..	302	425	1.0	0.0	128	1,396	..	798	3.78
United Kingdom	5,081	2.0	7	528	..	..	0.0	218	932	0.19	148	1.86
United States	11,571	3.0	7	640	..	..	0.0	190	1,073	0.09	113	..
Uruguay	1,574	3.8	19	209	309	48.0	1.0	117	938	0.19	82	5.90
Uzbekistan	1,731	0.8	10	76	234	329.4	3.3	68	68,478	..	..	5.25
Venezuela	2,518	4.3	21	117	285	476.3	2.4	161	622	0.06	53	5.25
Vietnam	146	9.2	22	16	70	..	..	16	530	0.11	44	..
West Bank and Gaza	..	..	..	..	..	..	..	..	..	..	..	..
Yemen, Rep.	99	9.2	26	13	71	75.3	5.2	57	732	0.02	117	13.44
Yugoslavia, FR (Serb./Mont.)	2,921	−2.5	9	197	424	212.8	4.0	178	199	0.01	109	7.04
Zambia	574	−2.2	11	9	22	24.7	..	24	1,393	0.25	162	4.40
Zimbabwe	738	1.8	7	15	60	113.2	7.2	30	789	0.03	279	..
World	**1,978 w**	**4.5 w**	**8 w**	**133 w**	**271 w**	**28,093.8 s**	**1.5 m**	**179 w**	**791 w**	**0.07 m**	**161 m**	**5.81 m**
Low income	414	8.4	15	26	56	2,383.2	4.9	104	333	0.10	195	10.86
Excl. China & India	187	8.3	22	11	56	1,571.2	5.4	40	592	0.10	200	11.17
Middle income	1,619	7.8	14	105	305	25,612.6	1.5	116	462	0.06	121	5.99
Lower middle income	1,429	11.2	15	94	312	21,735.4	2.9	110	293	0.04	108	6.22
Upper middle income	2,072	4.2	13	130	283	3,877.1	1.0	131	738	0.06	175	5.45
Low & middle income	836	7.9	15	52	225	27,995.8	2.7	113	420	0.06	142	6.45
East Asia & Pacific	575	7.7	13	41	52	1,899.7	0.4	113	391	0.11	162	..
Europe & Central Asia	2,798	11.4	12	185	397	19,569.9	3.3	102	218	0.04	73	6.00
Latin America & Carib.	1,298	4.7	17	102	196	1,658.1	1.1	155	780	0.06	124	5.25
Middle East & N. Africa	1,122	8.8	15	64	121	3,780.7	5.1	93	427	0.05	104	6.47
South Asia	300	9.2	19	14	69	155.5	6.6	30	291	0.05	99	..
Sub-Saharan Africa	437	3.1	13	15	30	931.9	2.9	63	935	0.09	253	10.55
High income	7,755	3.2	6	540	569	98.0	0.0	210	1,002	0.16	199	3.27

Power and communications | 5.10

A country's production of electricity is a basic indicator of its size and level of development. Although a few countries export electrical power, most production is for domestic consumption. Expanding the supply of electricity to meet the growing demand of increasingly urbanized and industrialized economies without incurring unacceptable social, economic, and environmental costs is one of the great challenges facing developing countries.

Data on electric power production and consumption are collected from national energy agencies by the International Energy Agency (IEA) and adjusted by the IEA to meet international definitions. Adjustments are made, for example, to account for self-production by establishments that, in addition to their main activities, generate electricity wholly or partly for their own use. In some countries self-production by households and small entrepreneurs is substantial because of remoteness or unreliable public power sources, and in these cases may not be adequately reflected in these adjustments. Electricity consumption is equivalent to production less power plants' own use and transmission, distribution, and transformation losses. It includes consumption by auxiliary stations, losses in transformers that are considered integral parts of those stations, and electricity produced by pumping installations. It covers electricity generated by primary sources of energy—coal, oil, gas, nuclear, hydro, geothermal, wind, tide and wave, and combustible renewables—where data are available. Neither production nor consumption data capture the reliability of supplies, including frequency of outages, breakdowns, and load factors.

Over the past decade privatization and deregulation have spurred dramatic growth in telecommunications in many countries. The table presents some common performance indicators for telecommunications, including measures of supply and demand, service quality, productivity, economic and financial performance, capital investment, and tariffs.

Demand for telecommunications is often measured by the sum of telephone mainlines and the number of registered applicants for new connections. (A mainline is normally identified by a unique number that is the one billed). In some countries the list of registered applicants does not reflect real current pending demand, which is often hidden or suppressed, reflecting an extremely short supply that has discouraged potential applicants from applying for telephone service. And in some cases waiting lists may overstate demand because applicants have placed their names on the list several times to improve their chances.

Waiting time is calculated by dividing the number of applicants on the waiting list by the average number of mainlines added each year over the past three years. The number of mainlines no longer reflects a telephone system's full capacity because mobile telephones provide an alternative point of access. (See table 5.11 for data on mobile phones.)

The table includes two measures of efficiency in telecommunications: mainlines per employee and revenue per mainline. Caution should be used in interpreting the estimates of mainlines per employee because some firms may subcontract part of their work. The cross-country comparability of revenue per mainline may also be limited because, for example, some countries do not require telecommunications providers to submit financial information; the data usually do not include revenues from cellular and mobile phones or radio, paging, and data services; and there are definitional and accounting differences between countries.

Figure 5.10a

Where are the telephones?

% of country's total telephone mainlines in largest city, 1995

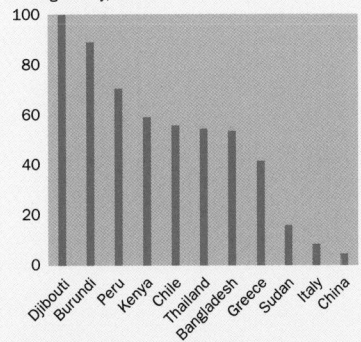

Source: International Telecommunication Union.

In some countries a single large city may dominate the economic and social life of the country, and access to telecommunications services is high. In other countries with widely dispersed populations and policies aimed at wider diffusion of telecommunications, the share of telephone mainlines in the largest city may be well under 5 percent of the country's total mainlines.

• **Electricity consumption** measures the production of power plants and combined heat and power plants less distribution losses and own use by heat and power plants. • **Electric power production** refers to gross production in kilowatt-hours by private companies, cooperative organizations, local and regional authorities, government organizations, and self-producers. Electric power production growth is average annual growth in power production. • **Electric power transmission and distribution losses** are losses in transmission between sources of supply and points of distribution and in distribution to consumers, including pilferage. • **Telephone mainlines** are telephone lines connecting a customer's equipment to the public switched telephone network. Data are presented for the entire country and the largest city. • **Waiting list** shows the number of applications for a connection to a mainline that have been held up by a lack of technical capacity. • **Waiting time** is the approximate number of years applicants must wait for a telephone line. • **Mainlines per employee** is calculated by dividing the number of mainlines by the number of telecommunications staff (with part-time staff converted to full-time equivalents) employed by telecommunications enterprises providing public telecommunications services. • **Revenue per line** is the revenues received by firms for providing telecommunication services. • **Cost of local call** is the cost of a three-minute call within the same exchange area using the subscriber's equipment (that is, not from a public phone). • **Outgoing traffic** is the telephone traffic, measured in minutes per subscriber, that originated in the country that has a destination outside the country. • **Cost of international call to U.S.** is the cost of a three-minute peak rate call from the country to the United States.

Data on electricity consumption, power growth, and losses are from the IEA's *Energy Statistics and Balances of Non-OECD Countries 1994–95*, the IEA's *Energy Statistics of OECD Countries 1994–95*, and the United Nations' *Energy Statistics Yearbook*. Telecommunications data are from the International Telecommunication Union's (ITU) *World Telecommunication Development Report* except for data on telephone traffic data, which are from *Direction of Traffic*, published by the ITU and TeleGeography, Inc.

5.11 The information age

	Daily newspapers	Radios	Television		Mobile phones	Fax machines	Personal computers	Internet hosts
			Sets	Cable subscribers				
	per 1,000 people 1994	per 1,000 people 1995	per 1,000 people 1996	per 1,000 people 1996	per 1,000 people 1996	per 1,000 people 1995	per 1,000 people 1996	per 10,000 people July 1997
Albania	54	..	173	..	1	..	..	0.32
Algeria	46	..	68	..	0	0.2	3.4	0.01
Angola	11	58	51	..	0	..	..	0.02
Argentina	138	..	347	116.4	16	1.4	24.6	5.32
Armenia	23	..	216	..	0	0.1	..	0.88
Australia	258	..	666	..	208	26.3	311.3	382.44
Austria	472	348	493	110.7	74	35.4	148.0	108.25
Azerbaijan	28	..	212	0.1	2	0.3	..	0.11
Bangladesh	6	48	7	..	0	0.0	..	0.00
Belarus	187	322	292	..	1	0.9	..	0.44
Belgium	321	..	464	360.0	47	17.8	167.3	84.64
Benin	2	1,461	73	..	0	0.1	..	0.02
Bolivia	69	..	202	3.4	4	..	..	0.69
Bosnia and Herzegovina	131	..	55	..	0	..	..	0.13
Botswana	24	816	27	..	..	2.2	6.7	1.58
Brazil	45	222	289	11.8	16	1.7	18.4	4.20
Bulgaria	..	354	361	..	3	1.8	295.2	6.65
Burkina Faso	0	31	6	..	0	..	..	0.04
Burundi	3	71	2	..	0	0.5	..	0.01
Cambodia	..	121	9	..	2	0.1	..	0.01
Cameroon	4	326	75	..	0	..	..	0.05
Canada	189	..	709	262.5	114	23.6	192.5	228.05
Central African Republic	1	93	5	..	0	0.0	..	0.02
Chad	0	620	2	..	..	0.0	..	0.00
Chile	100	..	280	39.9	23	1.8	45.1	13.12
China	23	161	252	28.9	6	0.2	3.0	0.21
Hong Kong, China	719	..	388	50.2	216	46.3	150.5	74.84
Colombia	64	..	188	3.1	13	2.6	23.3	1.81
Congo, Dem. Rep.	3	102	41	..	0	0.1	..	0.00
Congo, Rep.	8	318	8	..	..	..	..	0.02
Costa Rica	99	..	220	16.2	14	0.7	..	12.14
Côte d'Ivoire	7	..	60	..	1	..	1.4	0.17
Croatia	575	..	251	10.5	14	8.2	20.9	14.08
Cuba	120	241	200	0.0	0	..	..	0.06
Czech Republic	219	..	406	45.9	19	7.1	53.2	47.66
Denmark	365	..	533	263.4	250	47.6	304.1	259.73
Dominican Republic	34	..	84	15.5	8	0.3	..	0.03
Ecuador	72	..	148	3.9	5	2.7	3.9	0.90
Egypt, Arab Rep.	64	..	126	..	0	0.4	5.8	0.31
El Salvador	50	..	250	4.6	3	..	..	0.34
Eritrea	..	372	7	..	..	0.2	..	0.00
Estonia	242	..	449	12.9	47	8.8	6.7	45.35
Ethiopia	2	206	4	..	..	0.0	..	0.00
Finland	473	1,331	605	164.4	292	31.5	182.1	653.61
France	237	..	598	31.6	42	32.7	150.7	49.86
Gabon	16	..	76	..	6	0.3	6.3	0.00
Gambia, The	2	157	..	..	3	0.9	..	0.00
Georgia	..	62	474	2.4	0	0.1	..	0.55
Germany	317	..	493	203.9	71	19.5	233.2	106.68
Ghana	18	..	41	..	1	0.3	1.2	0.15
Greece	156	..	442	..	53	2.9	33.4	18.76
Guatemala	23	..	122	16.9	4	1.0	2.8	0.79
Guinea	..	76	8	..	0	..	0.3	0.00
Guinea-Bissau	6	40	..	..	..	0.5	..	0.09
Haiti	6	60	5	..	..	..	..	0.00
Honduras	44	108	80	8.0	0	..	..	0.94

The information age | 5.11

	Daily newspapers	Radios	Television		Mobile phones	Fax machines	Personal computers	Internet hosts
			Sets	Cable subscribers				
	per 1,000 people 1994	per 1,000 people 1995	per 1,000 people 1996	per 1,000 people 1996	per 1,000 people 1996	per 1,000 people 1995	per 1,000 people 1996	per 10,000 people July 1997
Hungary	228	..	*444*	*135.2*	46	*4.4*	44.1	33.29
India	..	119	64	*17.2*	0	*0.1*	1.5	0.05
Indonesia	20	..	232	..	3	*0.4*	4.8	0.54
Iran, Islamic Rep.	17	238	164	..	1	*0.5*	32.7	0.00
Iraq	27	..	*78*	..	..	..	..	0.00
Ireland	170	..	*469*	151.9	82	22.4	*145.0*	90.89
Israel	281	..	*303*	160.5	184	*25.0*	117.6	104.79
Italy	105	*102*	*436*	..	112	*31.4*	92.3	36.91
Jamaica	66	764	326	73.9	22	..	4.6	1.36
Japan	576	..	700	*87.9*	214	102.2	128.0	75.80
Jordan	48	*325*	*175*	..	3	7.3	7.2	0.38
Kazakhstan	..	..	*275*	..	0	*0.2*	..	0.70
Kenya	13	..	*19*	..	0	*0.1*	1.6	0.16
Korea, Dem. Rep.	213	*163*	*115*	..	..	0.1	..	*0.00*
Korea, Rep.	404	..	326	84.8	70	*8.9*	131.7	28.77
Kuwait	401	..	373	..	89	20.7	74.1	21.72
Kyrgyz Republic	11	..	238	..	..	..	..	0.23
Lao PDR	3	134	10	..	1	*0.1*	1.1	0.00
Latvia	228	..	598	*147.1*	11	*0.3*	7.9	21.03
Lebanon	172	..	355	0.4	65	..	24.3	2.72
Lesotho	7	*77*	*13*	..	1	*0.3*	..	0.08
Libya	13	..	143	..	..	..	..	0.01
Lithuania	136	485	376	80.9	14	1.0	*6.5*	7.46
Macedonia, FYR	21	..	*170*	..	0	0.8	..	2.15
Madagascar	4	*214*	*24*	..	0	..	..	0.03
Malawi	2	902	..	..	0	0.1	..	0.00
Malaysia	124	*473*	228	..	74	*5.0*	42.8	19.30
Mali	4	*168*	11	..	0	..	..	0.03
Mauritania	0	*188*	82	..	..	0.1	5.3	0.00
Mauritius	68	..	*219*	..	18	17.7	*31.9*	1.84
Mexico	113	..	*193*	13.3	11	*2.4*	29.0	3.72
Moldova	24	209	307	8.4	0	0.1	2.6	0.39
Mongolia	88	79	63	8.8	0	0.9	..	0.07
Morocco	13	..	*145*	..	2	0.3	*1.7*	0.32
Mozambique	5	*46*	*3*	..	..	0.4	0.8	0.02
Myanmar	23	71	7	..	0	0.0	..	0.00
Namibia	102	..	*29*	..	4	..	12.7	2.16
Nepal	8	57	4	0.2	..	*0.0*	..	0.07
Netherlands	334	..	*495*	*377.9*	52	*32.3*	232.0	219.01
New Zealand	297	..	517	0.3	138	*18.1*	266.1	424.34
Nicaragua	30	..	*170*	4.9	1	..	..	1.60
Niger	1	*61*	*23*	..	..	*0.0*	..	0.04
Nigeria	18	..	*55*	..	0	..	*4.1*	0.00
Norway	607	..	569	151.4	287	30.1	273.0	474.63
Oman	30	370	591	..	6	1.3	10.9	0.00
Pakistan	21	..	24	..	0	*1.2*	*1.2*	0.07
Panama	62	..	*229*	11.3	..	..	..	1.44
Papua New Guinea	15	..	4	..	1	*0.2*	..	0.18
Paraguay	42	..	*144*	7.1	7	..	..	0.47
Peru	86	..	142	6.3	8	*0.6*	*5.9*	2.63
Philippines	65	*168*	125	*5.9*	13	*0.7*	9.3	0.59
Poland	141	535	418	*70.4*	6	*1.4*	36.2	11.22
Portugal	41	..	367	17.3	67	*5.0*	*60.5*	18.26
Puerto Rico	184	*796*	*322*	69.1	45	150.2	..	0.30
Romania	297	..	226	110.6	1	*0.9*	*5.3*	2.66
Russian Federation	267	*341*	386	*69.4*	2	0.2	23.7	5.51

5.11 The information age

	Daily newspapers	Radios	Television		Mobile phones	Fax machines	Personal computers	Internet hosts
			Sets	Cable subscribers				per 10,000 people
	per 1,000 people 1994	per 1,000 people 1995	per 1,000 people 1996	per 1,000 people 1996	per 1,000 people 1996	per 1,000 people 1995	per 1,000 people 1996	July 1997
Rwanda	0	76	..	..	..	..	..	0.01
Saudi Arabia	54	..	263		10	8.4	37.2	0.15
Senegal	6	..	38		0	..	7.2	0.31
Sierra Leone	2	67	17	..	..	0.2		0.00
Singapore	364	..	361	13.2	141	25.1	216.8	196.30
Slovak Republic	256	917	384	83.7	5	8.3	186.1	20.47
Slovenia	185	..	375	130.8	20	7.8	47.8	85.66
South Africa	33	182	123	..	22	2.4	37.7	30.67
Spain	104	1,020	509	10.8	33	16.6	94.2	31.00
Sri Lanka	25	193	82	..	4	0.6	3.3	0.33
Sudan	23	311	80	..	0	0.2	0.7	0.00
Sweden	483	..	476	212.4	282	45.3	214.9	321.48
Switzerland	409	..	493	345.1	93	27.8	408.5	207.98
Syrian Arab Republic	18	..	91	..	..	0.3	1.4	0.00
Tajikistan	13	172	279	..	0	0.2	..	0.00
Tanzania	8	398	16	0.0	0	..	..	0.02
Thailand	48	204	167	3.5	28	1.7	16.7	2.11
Togo	2	362	14	..	..	2.4	..	0.01
Trinidad and Tobago	135	..	318	..	11	1.6	19.2	3.24
Tunisia	46	176	156	..	1	2.8	6.7	0.02
Turkey	44	126	309	7.6	13	1.6	13.8	3.60
Turkmenistan	..	..	163	..	..	..	..	0.00
Uganda	2	126	26	..	0	0.1	0.5	0.01
Ukraine	118	..	341	..	1	0.0	5.6	2.09
United Arab Emirates	161	..	276	..	79	16.8	65.5	7.66
United Kingdom	351	..	612	35.6	122	30.8	192.6	149.06
United States	228	..	806	239.5	165	64.6	362.4	442.11
Uruguay	237	..	305	22.0	25	3.5	22.0	3.18
Uzbekistan	7	..	190	..	0	0.1	..	0.06
Venezuela	215	..	180	10.1	35	1.1	21.1	2.06
Vietnam	8	..	180	..	1	0.2	3.3	0.00
West Bank and Gaza	..	..	..	..	..	..	..	..
Yemen, Rep.	17	45	278	..	1	0.2	..	0.00
Yugoslavia, FR (Serb./Mont.)	90	..	185	..	..	1.4	..	2.72
Zambia	8	112	80	..	0	0.1	..	0.27
Zimbabwe	18	..	29	..	..	0.4	6.7	0.24
World	**99 w**	**.. w**	**211 w**	**52.9 w**	**28 w**	**8.9 w**	**49.9 w**	**35.18 w**
Low income	19	143	147	23.7	3	0.2	2.3	0.12
Excl. China & India	13	..	47	..	0	..	..	0.07
Middle income	94	..	224	..	10	1.5	21.6	4.21
Lower middle income	92	..	246	..	7	0.8	17.1	1.85
Upper middle income	97	..	259	33.7	18	2.7	30.5	9.73
Low & middle income	50	163	177	..	5	0.6	8.7	1.53
East Asia & Pacific	29	160	228	26.6	7	0.4	4.5	0.57
Europe & Central Asia	171	..	350	..	6	1.2	17.4	6.53
Latin America & Carib.	83	..	216	18.4	14	1.9	23.2	3.48
Middle East & N. Africa	37	..	144	..	3	1.0	17.1	0.20
South Asia	..	119	53	17.2	0	0.2	1.5	0.06
Sub-Saharan Africa	11	..	43	..	..	..	..	2.03
High income	303	..	611	160.1	131	47.5	224.2	203.46

The information age | 5.11

The table includes indicators that measure the penetration of the information economy—newspapers, radios, television sets, mobile phones, fax machines, personal computers, and Internet hosts. Other important indicators of information and communications technology—such as the use of teleconferencing or the use of the Internet in organizing and mobilizing conferences, distance education, and commercial transactions—are not collected systematically and so are not reported here.

Data on the number of daily newspapers in circulation and radio receivers in use are obtained from statistical surveys carried out by the United Nations Educational, Scientific, and Cultural Organization (UNESCO). In some countries definitions, classifications, and methods of enumeration do not entirely conform to UNESCO standards. For example, newspaper circulation data should refer to the number of copies distributed, but in some cases the figures reported are the number of copies printed. In addition, many countries impose radio license fees to help pay for public broadcasting, discouraging radio owners from declaring ownership. Because of these and other data collection problems, estimates of the number of newspapers and radios vary widely in reliability and should be interpreted with caution.

Data presented for other electronic communications and information technology are from the International Telecommunication Union (ITU) and Network Wizards. Data on television sets and cable television subscribers are supplied to the ITU through annual questionnaires sent to national broadcasting authorities and industry associations. Some countries require that television sets

be registered. To the extent that households do not register their televisions or do not register all of their televisions, the number of licensed sets may understate the true number.

Because of different regulatory requirements for the provision of data, complete measurement of the telecommunications sector is not possible. Telecommunications data are compiled through annual questionnaires sent to telecommunications authorities and operating companies. The data are supplemented by annual reports and statistical yearbooks of telecommunications ministries, regulators, operators, and industry associations. In some cases estimates are derived from ITU documents or other references.

Data on fax machines exclude fax modems attached to computers. Some operators report only the equipment they sell, lease, or register, so the actual number is almost certainly much higher.

Estimates of the number of personal computers (PCs) are derived from an annual questionnaire, supplemented by other sources. In many countries mainframe computers are used extensively, and thousands of users can be connected to a single mainframe computer; thus the number of PCs understates the total use of computers.

Internet hosts are assigned to countries based on the host's country code, though this does not necessarily indicate that the host is physically located in the country. In addition, all hosts lacking a country code identification are assigned to the United States. Thus the number of Internet hosts shown for each country should be considered an approximation.

• **Daily newspapers** are the number of newspapers published at least four times a week, per 1,000 people. • **Radios** are the estimated number of radio receivers in use for broadcasts to the general public, per 1,000 people. • **Television sets** are the estimated number of television sets in use, per 1,000 people. • **Mobile phones** refer to users of portable telephones subscribing to an automatic public mobile telephone service using cellular technology that provides access to the public switched telephone network, per 1,000 people. • **Fax machines** are the estimated number of facsimile machines connected to the public switched telephone network, per 1,000 people. • **Personal computers** are the estimated number of self-contained computers designed to be used by a single individual, per 1,000 people. • **Internet hosts** are the number of computers directly connected to the worldwide network of interconnected computer systems, per 10,000 people. All hosts without a country code identification are assumed to be located in the United States.

Data on newspapers and radios are from UNESCO, which compiles data mainly from official replies by member states to UNESCO questionnaires and special surveys, but also from official reports and publications, supplemented by information from national and international sources. Data on television sets, mobile phones, fax machines, and personal computers are from the annual questionnaire sent to member countries by the ITU. These data are reported in the ITU's *World Telecommunication Development Report* or the Telecommunications Indicators database. The text also draws on ITU sources. Data on Internet hosts are from Network Wizards (http://www.nw.com).

Figure 5.11a

Mobile telephones play a big role in countries with limited traditional telephone services

Cellular subscribers as % of all telephone subscribers, 1995

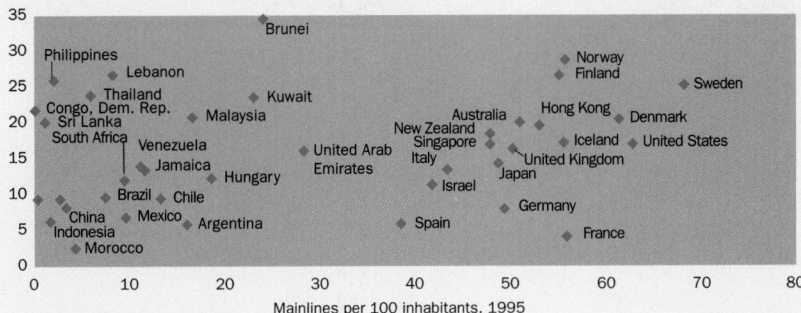

Source: International Telecommunication Union.

In some developing countries—mostly in Southeast Asia and South Asia—with low fixed-line penetration, cellular mobile services are developing as substitutes to traditional fixed-line services. Growth in cellular mobile services is reinforced by strong demand (as expressed, for example, through long waiting lists and times for fixed-line service) and competitive service provision. In industrial countries cellular mobile services mainly complement fixed-line services.

5.12 | Science and technology

	Scientists and engineers in R&D per million people 1981–95[b]	Technicians in R&D per million people 1981–95[b]	Expenditures for R&D % of GNP 1981–95[b]	High-technology exports $ millions 1996	High-technology exports % of manufactured exports 1996	Royalty and license fees Receipts $ millions 1990	Royalty and license fees Receipts $ millions 1996	Royalty and license fees Payments $ millions 1990	Royalty and license fees Payments $ millions 1996	Patent applications filed[a] Residents 1995	Patent applications filed[a] Non-residents 1995
Albania	..	..	..	..	..	0	0	0	0	..	1,564
Algeria	..	..	..	46	15	0	..	1	..	28	114
Angola	..	..	..	..	..	0	18	0	0	..	..
Argentina	..	..	0.3	1,195	17	4	6	409	221	..	..
Armenia	..	..	..	..	..	..	..	..	..	..	15,570
Australia	2,477	943	1.4	6,226	39	162	251	827	1,089	9,325	28,156
Austria	1,604	801	1.5	11,975	24	91	181	287	691	2,419	63,707
Azerbaijan	..	..	..	..	..	..	..	..	..	221	31
Bangladesh	..	..	..	..	..	0	0	0	5	70	156
Belarus	3,300	515	0.9	..	..	..	..	..	..	626	16,625
Belgium	1,814	2,200	1.7	..	..	682	683	1,328	1,197	1,464	52,187
Benin	..	..	..	..	..	0	0	0	0	..	..
Bolivia	250	154	1.7	71	41	0	0	3	5	17	106
Bosnia and Herzegovina	..	..	..	..	..	..	..	..	..	..	..
Botswana	..	..	..	..	..	0	0	8	6	1	50
Brazil	165	58	0.4	4,448	18	12	32	54	529	2,757	23,040
Bulgaria	4,240	1,205	1.7	..	..	0	..	0	..	370	16,953
Burkina Faso	..	..	..	..	..	0	0	0	0	..	..
Burundi	..	..	..	..	..	0	0	0	0	..	1
Cambodia	..	..	..	..	..	..	..	..	..	..	..
Cameroon	..	..	..	4	3	1	0	0	1	..	..
Canada	2,322	978	1.6	30,715	24	..	..	..	..	3,039	40,565
Central African Republic	55	31	..	0	0	0	..	0	..	..	..
Chad	..	..	..	..	..	0	0	0	0	..	..
Chile	..	..	0.8	388	18	1	63	37	51	181	1,535
China	537	187	0.6	26,938	21	0	..	0	..	10,066	31,707
Hong Kong, China	..	..	..	7,032	27	..	..	..	..	23	1,938
Colombia	..	..	..	815	21	21	59	13	49	141	1,093
Congo, Dem. Rep.	..	..	..	..	..	..	..	..	..	3	15
Congo, Rep.	..	..	..	2	12	0	..	0	..	..	..
Costa Rica	539	..	..	95	14	1	3	9	12	..	..
Côte d'Ivoire	..	..	..	..	..	0	0	0	0	..	..
Croatia	1,977	845	..	561	17	..	..	..	..	265	335
Cuba	1,369	878	0.9	..	..	..	..	..	..	104	33
Czech Republic	1,285	949	1.3	2,485	14	..	43	..	98	628	19,382
Denmark	2,647	2,656	1.9	7,386	25	0	..	0	..	2,257	59,810
Dominican Republic	..	..	..	295	19	0	0	0	11	..	..
Ecuador	169	215	0.1	45	11	0	0	37	68	8	270
Egypt, Arab Rep.	458	340	1.0	101	9	0	55	0	40	..	..
El Salvador	19	299	0.0	73	17	0	0	1	3	3	64
Eritrea	..	..	..	..	..	..	..	..	..	..	..
Estonia	3,296	550	0.6	272	19	0	1	0	3	16	14,751
Ethiopia	..	..	..	..	..	0	0	0	0	..	..
Finland	3,675	2,360	2.3	7,663	23	50	66	317	465	2,533	20,192
France	2,537	2,926	2.5	68,655	31	1,295	1,860	1,629	2,627	16,140	73,626
Gabon	..	..	..	20	32	0	0	0	0	..	..
Gambia, The	..	..	..	..	..	0	0	0	0	..	..
Georgia	..	..	..	..	..	..	..	..	..	288	15,660
Germany	3,016	1,607	2.6	110,000	25	1,987	3,320	3,797	5,866	51,948	84,667
Ghana	..	..	..	..	..	..	..	0	1	..	42
Greece	774	314	0.5	728	13	0	0	15	57	452	44,697
Guatemala	..	..	..	95	15	0	..	0	..	5	57
Guinea	..	..	..	..	..	..	..	..	..	..	..
Guinea-Bissau	..	..	..	..	..	0	..	0	..	..	..
Haiti	..	..	..	..	..	0	0	0	0	..	..
Honduras	..	..	..	8	3	0	0	3	9	7	40

Science and technology | 5.12

	Scientists and engineers in R&D	Technicians in R&D	Expenditures for R&D	High-technology exports		Royalty and license fees				Patent applications filed[a]	
						Receipts $ millions		Payments $ millions			
	per million people 1981–95[b]	per million people 1981–95[b]	% of GNP 1981–95[b]	$ millions 1996	% of manufactured exports 1996	1990	1996	1990	1996	Residents 1995	Non-residents 1995
Hungary	1,157	588	1.0	1,690	19	49	45	36	132	1,117	19,770
India	151	114	0.8	2,350	10	1	1	72	90	1,545	5,021
Indonesia	..	..	..	4,676	18	0	0	0	0	..	..
Iran, Islamic Rep.	..	..	..	..	..	0	0	0	0	278	129
Iraq	..	..	..	..	..	..	..	..	..	76	24
Ireland	1,871	510	1.4	23,192	62	38	94	591	3,434	927	44,660
Israel	..	..	2.2	5,654	30	63	142	73	172	1,266	3,159
Italy	1,303	796	1.3	34,442	15	1,040	381	1,959	1,027	1,625	63,330
Jamaica	..	..	..	619	67	3	4	7	19	7	54
Japan	5,677	869	3.0	151,000	39	2,866	6,683	6,051	9,834	335,061	53,896
Jordan	..	..	..	183	26	0	0	0	0	..	..
Kazakhstan	..	..	..	..	..	..	..	..	..	1,031	16,368
Kenya	..	..	..	..	..	9	5	6	0	..	28,728
Korea, Dem. Rep.	..	..	..	..	..	..	..	..	..	..	15,693
Korea, Rep.	2,636	317	2.8	44,433	39	37	185	136	2,431	59,249	37,308
Kuwait	..	..	..	77	13	..	..	..	..	..	..
Kyrgyz Republic	..	..	..	47	24	..	..	..	..	119	15,599
Lao PDR	..	..	..	..	..	0	0	0	0	..	..
Latvia	1,165	3	..	143	16	0	0	0	1	210	16,140
Lebanon	67	2	..	..	..	..	..	..	..	..	..
Lesotho	..	..	..	..	..	0	0	1	0	8	2,608
Libya	361	493	0.2	..	..	0	..	0	..	6	37
Lithuania	1,278	..	..	365	23	..	0	..	4	106	15,882
Macedonia, FYR	1,258	334	..	..	..	..	..	..	..	100	3,084
Madagascar	..	..	..	2	3	0	1	0	3	21	15,802
Malawi	..	..	..	1	3	0	0	0	0	5	28,868
Malaysia	87	88	0.4	39,448	67	0	0	0	0	141	3,911
Mali	..	..	..	..	..	0	0	0	0	..	..
Mauritania	..	..	..	..	..	0	0	0	0	..	..
Mauritius	361	158	0.4	14	1	0	0	0	0	3	4
Mexico	95	27	0.3	24,179	32	73	122	380	360	436	23,233
Moldova	..	..	..	15	9	..	..	..	..	271	15,606
Mongolia	..	..	..	1	2	0	..	0	..	130	15,847
Morocco	..	..	..	568	24	4	5	60	133	89	292
Mozambique	..	..	..	1	5	0	..	0	..	..	..
Myanmar	..	..	..	..	..	0	..	0	..	..	..
Namibia	..	..	..	..	..	1	0	3	3	..	..
Nepal	22	5	..	0	0	0	0	0	0	3	5
Netherlands	2,656	1,774	1.9	46,651	42	1,086	2,361	1,751	2,852	4,460	59,279
New Zealand	1,778	822	1.1	428	11	0	0	0	0	1,418	19,230
Nicaragua	..	..	..	88	40	0	0	0	0	..	35
Niger	..	..	..	..	..	..	..	0	1	..	..
Nigeria	..	..	..	..	..	0	0	0	0	..	..
Norway	3,434	1,705	1.9	2,703	24	133	729	148	942	1,278	20,398
Oman	..	..	..	65	8	0	..	0	..	..	..
Pakistan	54	76	..	269	3	0	6	0	21	21	678
Panama	..	..	..	14	..	..	..	9	11	16	62
Papua New Guinea	..	..	..	..	..	0	0	0	0	..	..
Paraguay	..	..	..	8	4	0	..	0	..	..	..
Peru	..	..	..	94	11	..	2	5	63	..	..
Philippines	..	..	..	10,561	62	1	2	38	99	..	..
Poland	1,083	1,380	0.9	1,926	11	0	24	0	144	2,598	19,491
Portugal	599	381	0.6	2,295	12	14	26	117	262	96	58,605
Puerto Rico	..	..	..	..	..	..	..	..	..	..	..
Romania	1,382	613	0.7	442	7	0	101	0	12	1,811	16,856
Russian Federation	4,358	905	0.8	..	..	..	..	..	..	17,611	23,746

	Scientists and engineers in R&D	Technicians in R&D	Expenditures for R&D	High-technology exports		Royalty and license fees				Patent applications filed[a]	
						Receipts $ millions		Payments $ millions			
	per million people 1981–95[b]	per million people 1981–95[b]	% of GNP 1981–95[b]	$ millions 1996	% of manufactured exports 1996	1990	1996	1990	1996	Residents 1995	Non-residents 1995
Rwanda	..	..	..	..	..	0	0	0	1	..	..
Saudi Arabia	..	..	..	..	..	0	0	0	0	28	718
Senegal	..	..	..	145	55	1	1	0	0	..	..
Sierra Leone	..	..	..	..	..	0	..	0	..	..	5
Singapore	2,512	1,524	1.1	73,701	71	..	..	..	..	10	11,871
Slovak Republic	1,922	796	1.1	982	16	..	18	..	83	273	17,659
Slovenia	2,998	2,390	1.5	1,180	16	4	6	5	27	318	16,267
South Africa	..	..	..	..	..	54	67	130	250	5,549	5,501
Spain	1,098	342	0.9	13,179	17	90	238	1,022	1,424	2,329	68,922
Sri Lanka	..	..	..	65	3	0	0	0	0	76	15,944
Sudan	..	..	..	..	..	0	..	0	..	..	28,951
Sweden	3,714	3,173	3.5	20,905	31	563	997	743	1,006	6,396	64,165
Switzerland	..	..	2.6	..	..	..	..	..	..	5,116	64,626
Syrian Arab Republic	..	..	..	..	..	0	..	0	..	43	12
Tajikistan	..	..	..	..	..	..	..	..	..	33	15,598
Tanzania	..	..	..	..	..	..	0	..	0	..	..
Thailand	173	51	0.2	14,746	36	0	25	170	717	..	..
Togo	..	..	..	..	..	0	0	0	0	..	..
Trinidad and Tobago	..	..	..	312	33	0	0	7	0	24	15,515
Tunisia	388	71	0.3	450	10	1	1	1	2	31	115
Turkey	209	23	0.8	1,326	8	..	..	..	..	206	1,506
Turkmenistan	..	..	..	..	..	..	..	..	..	..	8,420
Uganda	..	..	..	..	..	0	0	0	0	..	20,840
Ukraine	..	..	..	..	..	..	..	..	..	4,806	17,548
United Arab Emirates	..	..	..	..	..	..	..	..	..	..	..
United Kingdom	2,417	1,019	2.2	85,035	40	2,540	4,725	2,992	3,625	25,355	90,399
United States	3,732	..	..	198,000	44	16,635	29,973	3,138	7,322	127,476	107,964
Uruguay	..	..	..	91	10	0	0	0	8	..	..
Uzbekistan	1,760	313	..	..	..	..	..	..	..	1,039	15,873
Venezuela	208	32	0.5	377	14	..	..	..	2	..	..
Vietnam	..	..	..	..	..	..	..	..	..	23	16,959
West Bank and Gaza	..	..	..	..	..	..	..	..	..	..	..
Yemen, Rep.	..	..	..	0	0	..	..	..	..	..	..
Yugoslavia, FR (Serb./Mont.)	1,476	400	..	141	16	..	..	..	..	592	230
Zambia	..	..	..	..	..	0	..	0	..	4	90
Zimbabwe	..	..	..	33	5	1	1	8	6	56	177

a. Other patent applications filed in 1995 include those filed under the auspices of the African Intellectual Property Organization (27 by residents, 15,819 by nonresidents), African Regional Industrial Property Organization (4 by residents, 15,032 by nonresidents), and European Patent Office (35,390 by residents, 42,869 by nonresidents). Information was originally provided by the WIPO. The International Bureau of WIPO assumes no liability or responsibility with regard to the transformation of this data. b. See *Primary data documentation* for survey year.

Science and technology | 5.12

About the data

Rapid progress in science and technology is changing the global economy and increasing the importance of knowledge as a factor of production. It is also driving rapid shifts in comparative advantage between countries. The table shows several key indicators that provide a partial picture of the "technological base": the availability of skilled human resources (scientists, engineers, and technicians employed in research and development), the competitive edge countries enjoy in high-technology exports, sales and purchases of technology through royalties and licenses, and the number of patent applications filed.

The United Nations Educational, Scientific, and Cultural Organization (UNESCO) collects data on scientific and technical workers and research and development expenditures from member states, mainly from official replies to UNESCO questionnaires and special surveys, as well as from official reports and publications, supplemented by information from other national and international sources. UNESCO reports either the stock of scientists, engineers, and technicians (all qualified persons in those fields on a given reference date) or the number of economically active persons qualified to be scientists, engineers, or technicians (people engaged in or actively seeking work in any branch of the economy on a given date). Stock data generally come from censuses and are less timely than measures of the economically active population. UNESCO supplements these data with estimates of the number of qualified scientists and engineers by counting the number of people who have completed education at ISCED (International Standard Classification of Education) levels 6 and 7; qualified technicians are estimated using the number of people who have completed education at ISCED level 5. The data on scientists, engineers, and technicians, normally calculated in terms of full-time equivalent staff, cannot take into account the considerable variations in quality of training and education.

Data on R&D expenditures may reflect the different tax treatment of such expenditures. In some countries they may also reflect large and possibly unproductive outlays by governments or state-owned research establishments.

High-technology exports are those produced by a country's 10 most R&D-intensive industries. Industry rankings are based on a methodology developed by Davis (1982). Using input-output techniques, Davis estimated the technology intensity for U.S. industries in terms of the R&D expenditures required to produce a certain manufactured good. This methodology takes into account direct R&D expenditures made by final producers as well as indirect R&D expenditures made by suppliers of intermediate goods used in producing the final good. Industries classified on the basis of the U.S. Standard Industrial Classification (SIC) were ranked according to their R&D intensity, and the top 10 SIC groups (three-digit classification) were designated as high-technology industries. The industry ranked tenth had an R&D intensity index 30 percent greater than the industry ranked eleventh and was more than 100 percent greater than the average for manufacturing.

To translate Davis's industry classification into a definition of high-technology trade, Braga and Yeats (1992) used the concordance between the SIC grouping and the Standard International Trade Classification (SITC) revision 1 classification proposed by Hatter (1985). Given the imperfect match between SIC and SITC codes, Hatter estimated high-technology weights (the share of U.S. high-technology imports and exports in each SITC group, based on 1975–77 U.S. trade data) to highlight the relative importance of high-technology products in SITC groups. In preparing the data on high-technology trade, Braga and Yeats considered only SITC groups (at a four-digit level) that had a high-technology weight above 50 percent. Examples of high-technology exports include aircraft, office machinery, pharmaceuticals, and scientific instruments. It is worth noting that this methodology rests on the somewhat unrealistic assumption that using U.S. input-output relations and trade patterns for high-technology production does not introduce a bias in the classification.

Most countries have adopted systems that protect patentable inventions. Under most legislation concerning inventions, to be protected by law ("patentable"), an idea must be new in the sense that it has not already been published or publicly used; it must be nonobvious ("involve an inventive step") in the sense that it would not have occurred to any specialist in the particular industrial field, had such a specialist been asked to find a solution to the particular problem; and it must be capable of industrial application in the sense that it can be industrially manufactured or used. Data on patent applications filed by residents and nonresidents are shown in the table. The World Intellectual Property Organization (WIPO) estimates that at the end of 1995 about 3.7 million patents were in force in the world.

Definitions

- **Scientists and engineers in R&D** are people trained to work in any field of science who are engaged in professional R&D activity (including administrators). Most such jobs require completion of tertiary education. • **Technicians in R&D** are people engaged in professional R&D activity who have received vocational or technical training in any branch of knowledge or technology of a specified standard. Most of these jobs require three years beyond the first stage of secondary education. • **Expenditures for R&D** are current and capital expenditures (including overhead) on creative, systematic activity intended to increase the stock of knowledge and on the use of this knowledge to devise new applications. This includes fundamental and applied research and experimental development work leading to new devices, products, or processes. • **High-technology exports** are goods produced by industries (based on U.S. industry classifications) that rank among a country's top 10 in terms of R&D expenditures. Manufactured exports are commodities in the SITC, revision 1, sections 5–9 (chemicals and related products, basic manufactures, manufactured articles, machinery and transport equipment, and other manufactured articles and goods not elsewhere classified), excluding division 68 (nonferrous metals). • **Royalty and license fees** are payments and receipts between residents and nonresidents for the authorized use of intangible, nonproduced, nonfinancial assets and proprietary rights (such as patents, copyrights, trademarks, industrial processes, and franchises) and for the use, through licensing agreements, of produced originals of prototypes (such as manuscripts and films). • **Patents** are documents, issued by a government office, that describe the invention and create a legal situation in which the patented invention can normally only be exploited (made, used, sold, imported) by, or with the authorization of, the patentee. The protection of inventions is limited in time (generally 20 years from the filing date of the application for the grant of a patent). Information on patent applications filed is shown separately for residents and nonresidents of the country.

Data sources

Data on technical personnel and R&D expenditures are collected by UNESCO and published in its *Statistical Yearbook*. Information on high-technology exports are from the United Nations COMTRADE database. Data on royalty and license fees are from the IMF's *Balance of Payments Statistics Yearbook*. Data on patents are from WIPO's *Industrial Property Statistics*.

Global economic integration increases the ability of individuals and firms to undertake economic transactions with residents of other countries. Critics and proponents of globalization generally agree that the world is more integrated now than 50 years ago. But they disagree on whether integration is an opportunity or a danger and whether increasing integration is a strategic choice or an inevitable consequence—for better or worse—of economic and technological change. How much more integrated is the world? Which countries have been included, and which left out? Have new, market-based links (such as investment) replaced old, official ones (such as aid)? The answers to these questions are important for shaping future development strategies, and they depend in part on how integration is measured.

There are two broad approaches to measuring integration: evaluation of the barriers to integration and evaluation of the outcomes of integration. In a fully integrated world there would be no official barriers to negotiating and executing economic transactions—anywhere. And residents of one economy would face no higher transactions costs in another economy than in their own. Barriers to integration begin at the border with tariffs and nontariff barriers but are buttressed by a wide range of domestic policies and practices. Their outcomes can be seen in the volume of trade and capital flows or in the pattern of product and asset prices across countries.

Average tariffs, nontariff barrier coverage ratios, and indicators of capital controls are all useful indicators, and they are frequently cited as evidence of significant reductions in barriers since World War II—especially in the past two decades. But the story they tell may be misleading. Posted tariffs are not always collected. Capital controls can be evaded. Behind-the-border barriers such as domestic regulation, private collusive behavior, and information asymmetries—for which we lack even the simplest quantitative indicators—may be more restrictive. And obstacles to integration extend beyond official actions to include market structures, technology, geography, and access to information.

These difficulties in measuring the barriers to integration lead many to instead measure the outcomes of integration. Such studies focus on the effect integration has on trade or capital flows or product or asset prices. Such indicators suggest that global integration has increased in recent decades, but that considerable segmentation remains between national markets. One difficulty for outcome studies is disentangling the separate influences of the many forces affecting market outcomes. In what follows we review some of the techniques for measuring global integration. Such efforts may send contradictory signals, however. Globalization is far from complete. No measure—or even group of measures—suffices as an unambiguous indicator of what is occurring. This is especially the case when comparing countries that are only beginning to open their doors to the global market.

Border barriers

Indicators of average tariffs and nontariff barriers help to identify countries that have policies conducive to global integration. But such indicators may be incomplete or misleading. Widespread exemptions or rebates, sometimes granted in response to lobbying, lower effective barriers to well below official rates (figure 6a). Other problems arise in aggregating tariffs on individual products into a summary measure. Simple averages ignore the differing economic importance of product lines, and import-weighted averages understate the significance of the tariffs that have been most successful in reducing imports. Coverage ratios show the share of imports covered by nontariff barriers such as import quotas, but not the restrictiveness of the barriers. And measures based on tariff rates and nontariff barrier coverage ignore the effects of domestic taxes and subsidies, which are often used to replicate trade barriers.

Official controls on international capital movements are even less amenable to direct quantitative measurement or cross-country comparison. Without detailed qualitative analysis of the rules and regulations controlling capital account transactions in each country, often the most that can be said is whether a particular control is used.

Behind-the-border barriers

National standards and regulations can both help and hinder integration. Help because they allow products to be compared on a common basis, lowering the cost of collecting information about the product for consumers and producers and facilitating economies of scale and diffusion of new technologies embodied in standards. Many developing countries lack national standards that are compatible with the international norms developed by such bodies as the International Standards Organization. Moreover, the national institutions responsible for developing standards and assessing conformity are often weak.

But standards and regulations can also frustrate competition—if, say, they apply exclusively to foreign suppliers and require foreign products to undergo more costly health and safety tests. The fact that different countries pursue regulatory objectives in different ways can also handicap multinational firms operating in several countries, relative to those operating in one, by forcing them to comply with different regulations and thus lose economies of scale.

Measuring outcomes

Outcome measures of integration look either at quantities, such as volumes of international trade and capital flows, or at prices, such as product prices or assist prices and yields. Starting with

prices, the economist's "law of one price" suggests that in the absence of official barriers, and given a number of other assumptions, arbitrage should lead to equalization of the prices of products or financial assets, when stated in a common currency, wherever traded. But numerous studies have documented large and persistent deviations from the law of one price in product markets, even among narrowly defined and highly traded products. Several reasons have been suggested for the apparent lack of arbitrage in product markets:

- The goods compared are not exactly equivalent.
- Transportation costs drive a wedge between prices in different markets.
- Prices tend to be sticky in the currency in which the product is sold, remaining stable in local currency terms despite swings in nominal exchange rates.
- Tariff and nontariff trade barriers.
- Differences in national product standards—such as differences in electricity voltage or the side of the road on which automobiles are driven—make arbitrage more difficult.
- Noncompetitive market structures.
- The cost of local marketing and retailing. Inefficiencies in retail distribution are often cited as a reason for Japan's high retail prices.

Engel and Rogers (1995) reviewed several of these factors as explanations for price dispersion in a sample of 24 countries. They confirm the importance of distance and exchange rate movements (price stickiness). But they find formal trade barriers to be insignificant. After allowing for these factors, they find price dispersion to be significantly lower between countries in the same

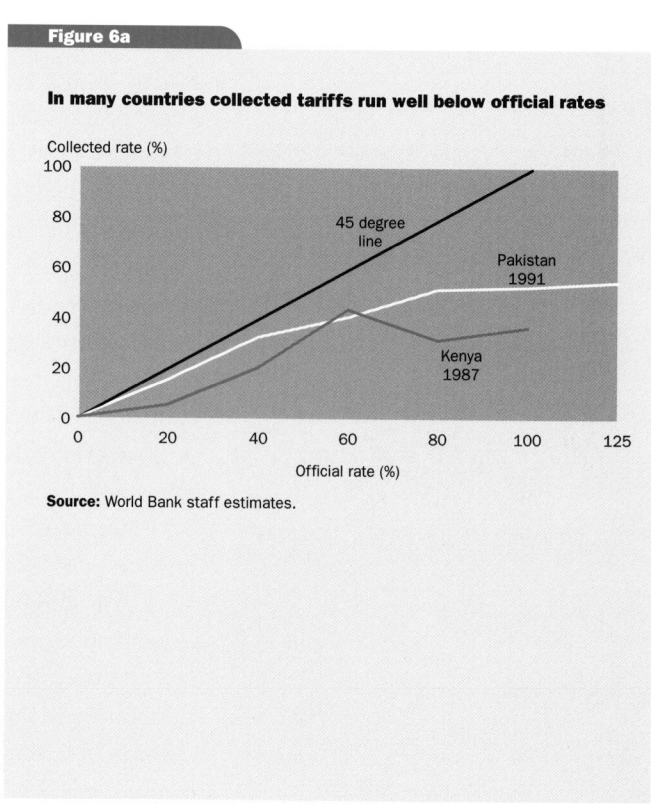

Figure 6a

In many countries collected tariffs run well below official rates

Source: World Bank staff estimates.

region, such as Canada and the United States or members of the European Union (though not Mexico and the United States or countries in Asia). They suggest that greater price uniformity within regions could reflect more integrated marketing and distribution systems.

Ratios of total trade (exports plus imports) to GDP are the most widely used quantitative measure of product market integration. Before these ratios can be used for cross-country comparisons of integration, they must be adjusted for the influence of structural factors—such as country size, factor endowment, geographic isolation, and stage of development:

- Large countries tend to trade less than small countries because they contain more diversified resources.
- Countries that are well endowed with natural resources, such as oil, export and import more.
- Countries with an abundance of labor, such as those in East Asia, may undertake more processing and assembly trade, with a high content of imported intermediate imports and less value added per dollar of gross output.
- Rich countries appear to be less integrated than they really are because they devote more of their output and consumption to services, which are harder to trade. They also tend to have higher prices for services, which again makes them seem less integrated by their trade-GDP ratios.
- Trade data in gross terms compared with GDP in value added terms can inflate trade-GDP ratios. This problem can be corrected by stating trade in value added terms or domestic product in gross output terms, but such data are available for only a few high-income countries.

An alternative indicator of trade integration is the "home bias" measure, which aims to provide an all-inclusive summary of barriers to trade. This measure is defined as purchases from domestic suppliers relative to purchases from other countries, after adjusting for such factors as the size of exporter and importer economies, bilateral distance, location of the importing country, and whether the countries share a common language or border. Shang-jin Wei (1996) found that in 1982–94 OECD countries purchased about 2.5 times more from themselves than from otherwise identical foreign countries. The United States has the lowest home bias—statistically indistinguishable from one. Mexico and Portugal have the highest, with domestic purchases running five to six times those from similar foreign countries (figure 6b). The average home bias for OECD countries fell slowly during 1982–94, but the drop was especially marked among EU members.

Shang-jin Wei's study also found that sharing a common language is a big determinant of trade—countries with language ties have 80 percent higher trade than otherwise. A common language greatly reduces the transactions costs associated with gathering information, making contacts, and conducting negotiations. Immigration also may foster trade between industrial and developing countries by helping to overcome obstacles created by weak international trade institutions in developing countries (Gould 1994). Immigrants know the language of their home countries and have detailed knowledge of home country tastes and products. And they often have access to networks of contacts with high levels of mutual trust, lowering the transactions costs of negotiating and enforcing contracts.

Integrating financial markets

When applied to financial markets, the law of one price implies that, with full integration, identical financial assets (except for their currency and political jurisdiction) should have identical prices or yields once exchange rate risk has been hedged or covered in the forward market. Covered interest rate differentials among most industrial countries are now quite small, reflecting extensive capital market liberalization during the 1970s and 1980s. In Europe the Single European Act, passed in 1987, appears to have turned the corner for such countries as France, where covered spreads on three-month interbank deposits fell from more than 200 basis points in 1982–86 to near zero in the early 1990s.

High explicit or implicit barriers to capital movements remain common in most developing countries, however. Among the small number of emerging markets with data on forward exchange rates, covered interest differentials averaged more than 600 basis points in 1982–88.

If all countries can borrow and lend in integrated global capital markets at the same expected real interest rate, there should be no connection between domestic investment and national savings. In other words, the regression coefficient of investment on savings rates—the savings retention ratio—should be zero under complete financial integration. But this ratio is typically much closer to one than to zero, leading some to argue that capital markets are much less integrated than is commonly supposed.

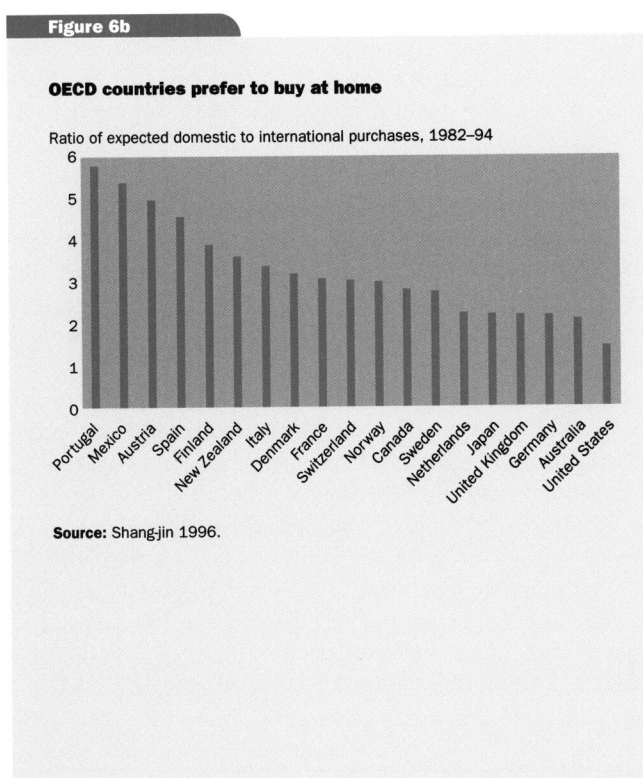

Figure 6b

OECD countries prefer to buy at home

Ratio of expected domestic to international purchases, 1982–94

Portugal, Mexico, Austria, Spain, Finland, New Zealand, Italy, Denmark, France, Switzerland, Norway, Canada, Sweden, Netherlands, Japan, United Kingdom, Germany, Australia, United States

Source: Shang-jin 1996.

Others deny such a conclusion. They say that savings is an endogenous variable and that both savings and investment reflect common factors—such as the economic cycle or demographic and productivity trends. Budget constraints may place bounds on how far savings and investment can diverge over long periods. In particular, developed economies may be much closer to their desired long-run capital stock than developing countries, which may have many unused investment opportunities and thus require large capital inflows—a feature that also helps explain why developing countries have lower savings retention ratios than developed ones.

Savings retention ratios for industrial countries rose sharply in the 1930s and remained high through the 1950s, a period characterized by extensive capital controls (figure 6c). But by the 1980s these ratios had fallen close to the levels at the end of the 19th century, a period of high capital mobility.

As in product markets, there is no reason for a high degree of financial integration or capital mobility to necessarily result in high gross capital flows. But there are reasons to think that it should (Montiel 1993). For example, with financial integration the geographical location of traders does not matter, so the volume of transactions crossing borders should be high. Moreover, if financial assets in different countries have different risk and return characteristics, individuals can insure themselves against risks to consumption by diversifying their asset portfolios internationally. Thus many countries with low or negative net capital flows with the rest of the world continue to have high two-way gross capital flows.

Although international capital flows have grown rapidly in recent years, they remain well below what financial models suggest should prevail under full international capital mobility. With perfect capital mobility, the proportion of loans by a country's residents that go to domestic borrowers should be about the same as the country's share in global lending (Golub 1990). For a small country whose share in global lending is close to zero, for example, very few loans by domestic residents will go to domestic borrowers. (Conversely, nearly all of the country's borrowing should come from foreign lenders.)

In 1980–86, however, the share of loans by domestic residents that went to domestic borrowers in OECD countries ranged from a low of 60 percent in Belgium to a high of 94 percent in the United States, with an average of 86 percent. In all cases this ratio was much higher than the countries' shares in overall OECD lending, suggesting a strong home bias in international portfolio allocation, similar to that in product markets, though the size of this bias appear to have fallen since the 1970s.

Does the strong home bias in financial portfolios mean that international capital market restrictions have resulted in significant segmentation of national financial markets? This may be a plausible explanation for low holdings of developing country assets in international portfolios, given that financial liberalization in these countries gathered pace only in the 1990s, and then in only a small group of countries. But it is less plausible for home bias among industrial countries, where capital market restrictions would have to be much higher to explain the observed facts. The same can be

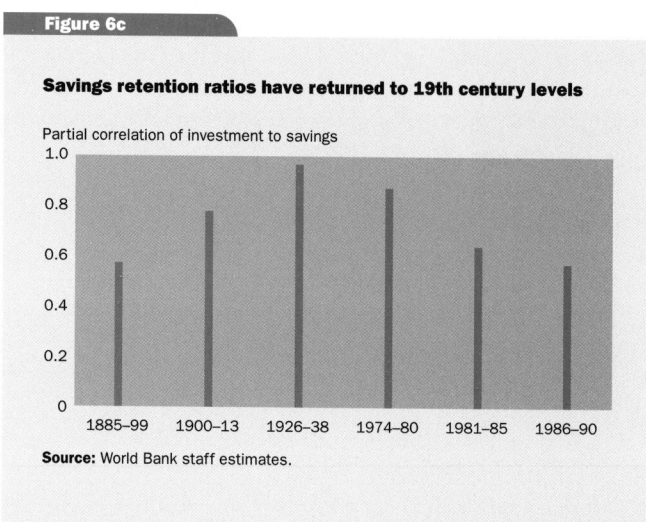

Figure 6c

Savings retention ratios have returned to 19th century levels

Partial correlation of investment to savings

Source: World Bank staff estimates.

said about transactions costs, exchange rate risks, and political risks. The fact that foreign assets in investor portfolios are turned over at a significantly higher rate than domestic assets also casts doubt on the idea of high transactions costs as a cause of home bias.

Thus there is an international diversification puzzle. Explanations include the possibility that investor expectations are less than fully rational—that is, investors systematically overestimate returns on domestic assets. Another interesting area of investigation concerns the role of information asymmetries. Gordon and Bovenberg (1996) argue that foreign investors may be handicapped relative to domestic investors by their poorer knowledge of the domestic market. Because they are poorly informed, they are vulnerable to being overcharged when they acquire shares in a firm or purchase inputs and services. They also risk misjudging markets and, therefore, investing real resources less efficiently. For example, foreign investors tend to pay much more than domestic investors to acquire publicly traded U.S. firms, and foreign subsidiaries earn much lower rates of return than domestic firms.

Links and chains

A more integrated world is not without risks. The recent financial crisis in East Asia demonstrates some of the risks, just as the region's spectacular earlier growth demonstrates some of the benefits. As economies become more closely linked, they become more dependent on one another's performance. Failures of management and governance in one economy may be transmitted to another as swiftly as electronic signals. But an integrated world is better able to diversify risk and to provide insurance against disasters, both natural and humanmade.

Ultimately, the value of integration must be assessed by its effect on people's lives. An integrated global economy may be more efficient, but it also may be less comfortable for many people. The continuing debates over tariff reductions and capital account liberalization reflect a deep suspicion that the benefits of globalization have been oversold. Concerns about environmental and social protection will also have to be resolved as globalization proceeds. Better measures of policies and their outcomes can inform this debate.

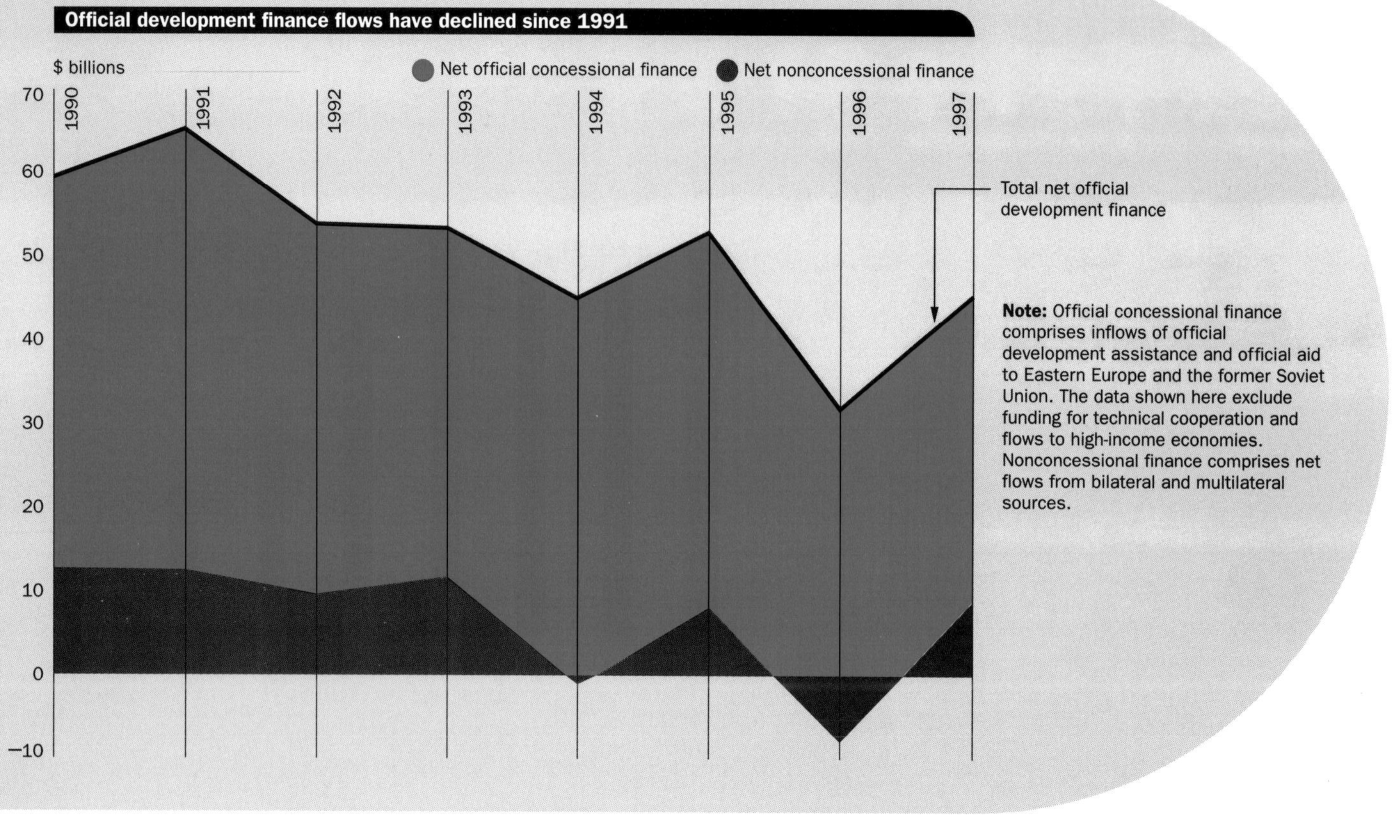
Official development finance flows have declined since 1991
$ billions
Net official concessional finance
Net nonconcessional finance
70
60
50
40
30
20
10
0
–10
1990
1991
1992
1993
1994
1995
1996
1997
Total net official development finance
Note: Official concessional finance comprises inflows of official development assistance and official aid to Eastern Europe and the former Soviet Union. The data shown here exclude funding for technical cooperation and flows to high-income economies. Nonconcessional finance comprises net flows from bilateral and multilateral sources.

Foreign direct investment was essentially flat in 1997 after jumping from $24 billion in 1990 to almost $120 billion in 1996

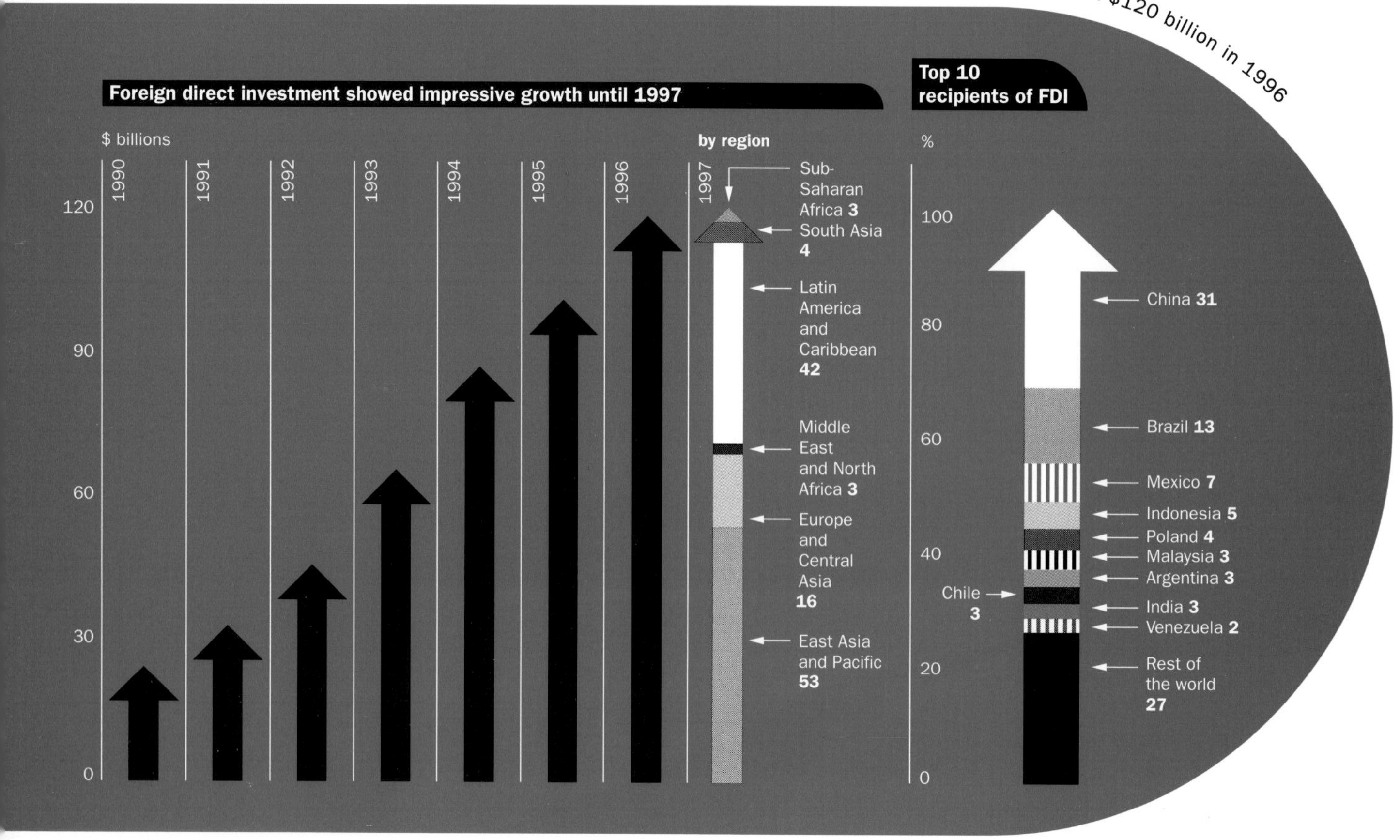
Foreign direct investment showed impressive growth until 1997
$ billions
120
90
60
30
0
1990
1991
1992
1993
1994
1995
1996
1997
by region
Sub-Saharan Africa 3
South Asia 4
Latin America and Caribbean 42
Middle East and North Africa 3
Europe and Central Asia 16
East Asia and Pacific 53
Top 10 recipients of FDI
%
100
80
60
40
20
0
China 31
Brazil 13
Mexico 7
Indonesia 5
Poland 4
Malaysia 3
Argentina 3
Chile 3
India 3
Venezuela 2
Rest of the world 27

Despite a leveling of foreign direct investment and a sharp drop in portfolio equities, private capital flows remained the largest source of finance to the developing world in 1997. But most of these flows went to a handful of countries. The rest continue to depend on a declining flow of aid.

Portfolio equities took a big hit in 1997, falling by nearly a third

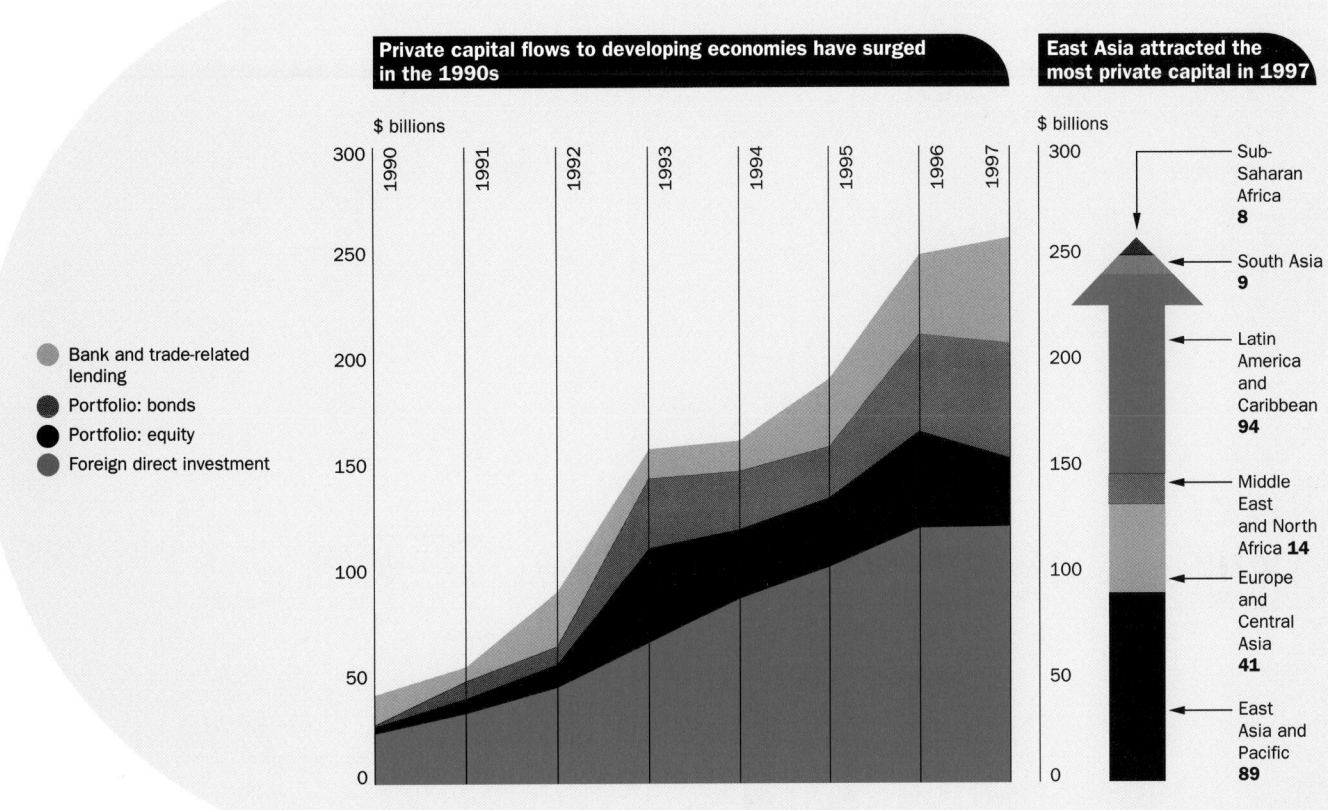

Private capital flows to developing economies have surged in the 1990s

East Asia attracted the most private capital in 1997

$ billions

Bank and trade-related lending
Portfolio: bonds
Portfolio: equity
Foreign direct investment

Sub-Saharan Africa **8**

South Asia **9**

Latin America and Caribbean **94**

Middle East and North Africa **14**

Europe and Central Asia **41**

East Asia and Pacific **89**

Source for both pages: World Bank, *Global Development Finance 1998*

6.1 | Integration with the global economy

	Trade		Trade in goods		Growth in real trade less growth in real GDP	Mean tariff	Gross private capital flows		Gross foreign direct investment	
	% of PPP GDP		% of goods GDP		percentage points	All products %	% of PPP GDP		% of PPP GDP	
	1986	1996	1986	1996	1986–96	1990–96[a]	1986	1996	1986	1996
Albania	..	..	30.0	76.7	..	15.9	..	..	..	..
Algeria	17.3	15.0	41.1	66.7	–0.8	..	0.8	..	0.0	..
Angola	23.3	29.3	79.8	113.6	5.8	..	4.2	12.8	2.1	0.9
Argentina	5.9	14.0	23.1	44.0	8.0	11.2	3.4	7.9	0.3	1.3
Armenia	..	14.0	..	..	..	..	..	2.9	..	0.0
Australia	24.2	34.0	78.5	127.7	4.4	6.0	12.0	14.3	4.5	4.2
Austria	48.7	71.6	120.8	142.0	3.1	6.8[b]	11.7	22.0	0.5	2.9
Azerbaijan	..	16.3	..	109.6	..	..	..	..	..	..
Bangladesh	5.7	8.3	39.4	67.3	5.7	..	1.5	0.9	0.0	0.0
Belarus	..	26.3	..	100.2	..	12.6	..	..	..	..
Belgium	..	..	..	..	..	3.0	6.8[b]	..	..	..
Benin	12.4	15.9	78.8	98.9	–1.9	..	2.5	2.9	0.0	0.0
Bolivia	11.7	12.0	50.0	..	3.0	9.7	9.5	3.3	0.1	1.8
Bosnia and Herzegovina	..	..	..	..	..	..	..	..	..	..
Botswana	..	..	..	..	–4.0	..	5.6	3.5	1.7	1.0
Brazil	5.8	10.2	26.0	24.9	6.6	12.2	3.2	4.6	0.1	0.7
Bulgaria	18.3	23.8	40.8	216.3	–17.6	..	4.9	6.4	0.0	0.4
Burkina Faso	8.5	9.8	45.7	63.1	–3.0	..	2.5	3.5	0.1	0.0
Burundi	12.5	4.2	41.6	18.7	0.3	..	4.1	1.2	0.1	0.1
Cambodia	..	..	4.3	95.2	..	..	..	..	..	..
Cameroon	9.6	13.0	40.3	60.8	3.2	..	7.0	10.2	0.5	0.1
Canada	45.6	58.5	112.4	..	5.1	8.5	12.6	15.1	1.9	2.3
Central African Republic	8.7	8.6	41.8	52.9	–3.5	18.6	3.7	2.6	0.3	0.2
Chad	4.7	5.7	34.1	45.5	–4.1	..	3.4	2.4	1.1	0.5
Chile	11.6	18.9	85.9	..	4.5	11.0	5.0	8.8	0.5	3.0
China	6.6	7.1	35.5	58.4	2.1	23.9	1.4	1.5	0.2	1.0
Hong Kong, China	111.8	247.6	513.0	1,227.0	8.9	..	..	..	..	..
Colombia	7.1	9.5	47.5	55.8	7.1	11.7	3.3	4.0	0.6	1.3
Congo, Dem. Rep.	4.2	6.9	43.0	60.6	–5.5	..	..	..	..	..
Congo, Rep.	51.4	70.6	162.4	323.1	2.1	..	41.1	93.8	0.9	0.0
Costa Rica	20.2	34.4	104.7	214.5	5.3	..	6.1	3.5	0.6	1.8
Côte d'Ivoire	36.0	32.0	118.5	151.6	0.7	4.8	4.9	3.4	0.5	0.1
Croatia	..	59.9	..	122.8	..	..	..	12.3	..	1.7
Cuba	..	..	..	..	..	10.7	..	..	..	..
Czech Republic	..	46.3	..	187.2	..	7.0	..	10.9	..	1.3
Denmark	58.8	73.7	125.0	128.1	2.2	6.8[b]	16.9	27.8	1.1	2.7
Dominican Republic	12.3	28.3	88.8	173.7	1.8	..	3.2	1.9	0.3	1.2
Ecuador	11.8	16.3	64.4	95.0	2.4	11.4	3.4	2.6	0.2	0.7
Egypt, Arab Rep.	13.6	14.8	60.9	70.6	–0.5	..	4.6	2.5	1.5	0.4
El Salvador	20.7	22.5	87.1	88.0	7.0	10.2	3.6	3.5	0.3	0.2
Eritrea	..	..	..	..	..	..	..	..	..	..
Estonia	..	77.4	..	280.4	..	0.1	..	13.7	..	3.2
Ethiopia[c]	10.5	6.8	32.7	41.2	–3.2	..	1.7	3.5	0.0	0.0
Finland	51.5	70.1	90.6	115.1	3.7	6.8[b]	15.8	29.7	2.0	5.6
France	33.7	45.4	91.2	111.6	2.5	6.8[b]	7.3	17.2	1.1	3.7
Gabon	40.4	45.5	112.1	113.3	–0.1	..	21.1	8.7	2.2	4.6
Gambia, The	22.4	11.9	266.8	92.8	2.1	..	6.6	2.9	0.0	0.5
Georgia	..	9.6	..	27.8	..	..	..	..	..	..
Germany	..	55.1	..	..	..	6.8[b]	..	16.9	..	2.0
Ghana	11.0	15.3	44.6	126.6	2.4	..	1.9	2.4	0.0	0.4
Greece	21.2	27.9	58.6	50.4	4.0	6.8[b]	4.3	10.9	0.6	0.8
Guatemala	9.7	12.2	75.1	74.9	4.3	..	3.1	4.1	0.3	0.2
Guinea	13.2	13.1	67.5	62.8	–1.7	..	5.3	2.1	0.1	0.2
Guinea-Bissau	11.4	14.2	74.2	85.0	–3.2	..	5.7	27.8	0.0	..
Haiti	6.9	12.6	..	48.9	9.0	..	1.6	1.0	0.1	0.0
Honduras	21.7	42.5	80.1	234.7	–0.7	..	6.2	3.9	0.4	0.4

Integration with the global economy | 6.1

	Trade		Trade in goods		Growth in real trade less growth in real GDP	Mean tariff	Gross private capital flows		Gross foreign direct investment	
	% of PPP GDP		% of goods GDP		percentage points	All products %	% of PPP GDP		% of PPP GDP	
	1986	1996	1986	1996	1986–96	1990–96[a]	1986	1996	1986	1996
Hungary	34.5	41.4	126.9	137.1	0.5	14.4	5.4	14.0	0.0	2.8
India	3.9	4.5	16.4	31.1	3.0	30.0	0.9	0.6	0.0	0.2
Indonesia	10.7	13.6	55.0	69.7	1.3	13.2	2.0	2.1	0.1	0.8
Iran, Islamic Rep.	9.7	9.6	17.1	..	1.2	..	2.0	1.5	0.0	0.0
Iraq	..	..	..	..	..	..	..	..	..	..
Ireland	86.7	121.6	375.0	773.1	3.1	6.8[b]	14.4	66.5	0.1	4.7
Israel	39.5	47.5	102.6	..	1.3	..	4.7	8.6	0.4	2.2
Italy	28.0	39.6	..	..	4.3	6.8[b]	3.3	19.1	0.4	0.8
Jamaica	28.8	53.7	146.3	299.3	1.5	..	10.4	9.2	0.5	1.8
Japan	21.5	26.1	41.2	39.9	2.8	6.0	12.7	15.8	0.9	0.9
Jordan	36.8	36.6	123.8	172.4	7.8	..	3.3	4.7	0.4	0.4
Kazakhstan	..	19.6	..	110.7	..	..	..	..	..	..
Kenya	16.8	17.9	67.1	115.2	5.5	..	2.9	2.5	0.2	0.0
Korea, Dem. Rep.	..	..	..	..	..	..	..	..	..	..
Korea, Rep.	33.6	46.7	115.0	118.0	4.5	11.3	3.5	11.1	0.8	1.1
Kuwait	54.3	45.8	158.2	132.4	..	..	41.1	16.8	1.0	1.7
Kyrgyz Republic	..	13.6	..	99.2	..	..	..	5.4	..	0.5
Lao PDR	2.9	16.5	..	72.6	..	..	4.3	4.1	0.0	1.6
Latvia	..	41.1	..	..	..	6.0	..	15.4	..	3.8
Lebanon	..	36.0	..	151.2	..	..	..	..	..	..
Lesotho	..	..	..	..	–4.1	..	3.9	2.3	0.2	0.7
Libya	..	..	91.9	..	..	..	..	..	..	..
Lithuania	..	46.6	..	195.2	..	4.6	..	6.5	..	0.9
Macedonia, FYR	..	..	..	..	..	..	..	..	..	..
Madagascar	7.8	10.0	36.6	60.6	3.3	..	3.7	3.0	0.0	0.1
Malawi	13.0	16.8	58.3	83.0	0.5	25.3	2.4	2.8	0.0	0.0
Malaysia	33.6	70.2	163.5	269.0	7.8	9.1	2.8	4.6	0.7	2.0
Mali	16.4	19.9	63.0	82.0	0.3	..	4.6	5.7	0.2	0.7
Mauritania	29.6	26.7	143.5	178.2	–4.7	..	10.5	11.8	0.2	0.2
Mauritius	30.9	35.7	164.5	176.2	0.7	29.1	1.9	1.8	0.2	0.4
Mexico	6.8	26.1	51.2	143.8	7.3	13.1	5.9	6.6	0.6	1.0
Moldova	..	41.4	..	191.5	..	..	..	7.6	..	0.6
Mongolia	3.8	19.5	..	141.6	..	..	31.2	2.0	0.0	0.2
Morocco	12.6	14.0	66.8	69.3	4.1	..	2.8	1.7	0.0	0.4
Mozambique	17.8	14.4	36.0	127.9	–4.6	15.6	2.5	4.1	0.0	0.5
Myanmar	..	..	..	..	..	..	..	..	..	..
Namibia	..	..	..	..	0.1	..	..	7.6	..	1.7
Nepal	4.2	4.3	22.8	34.3	13.3	..	1.0	1.6	0.0	0.1
Netherlands	86.7	106.4	139.9	541.5	1.9	6.8[b]	20.0	35.5	4.0	9.0
New Zealand	29.7	45.0	121.8	..	3.6	6.2	14.9	6.7	4.7	2.0
Nicaragua	15.5	19.4	64.3	164.1	6.9	..	18.6	61.0	0.0	0.9
Niger	9.6	7.4	52.4	57.0	..	..	3.4	1.7	0.8	0.2
Nigeria	17.2	21.5	65.0	98.6	–2.0	..	11.2	12.4	0.4	1.4
Norway	67.8	80.3	103.0	103.5	1.3	6.1	23.6	43.3	4.6	9.6
Oman	52.9	45.4	136.7	..	..	..	10.2	2.5	1.4	0.2
Pakistan	9.3	10.0	49.4	59.9	0.5	..	1.9	1.6	0.1	0.4
Panama	14.3	111.0	119.1	1,069.3	0.7	..	68.8	57.5	3.0	1.5
Papua New Guinea	31.2	33.0	123.2	128.3	–1.1	20.7	2.7	21.2	1.6	0.9
Paraguay	8.2	29.3	45.5	115.3	12.3	9.4	4.0	3.9	0.0	1.1
Peru	6.6	13.0	..	..	4.2	13.3	5.3	5.2	0.1	3.3
Philippines	8.0	21.3	57.4	98.8	7.3	21.6	2.3	4.8	0.1	0.8
Poland	15.9	26.5	50.6	107.2	8.6	18.4	3.8	9.3	0.0	2.0
Portugal	23.2	43.1	106.1	..	5.2	6.8[b]	4.5	19.0	0.4	1.0
Puerto Rico	..	..	..	..	..	..	..	..	..	..
Romania	20.3	16.8	76.1	83.1	..	..	1.6	3.9	0.0	0.3
Russian Federation	..	19.8	..	52.5	..	12.7	..	11.6	..	0.4

	Trade		Trade in goods		Growth in real trade less growth in real GDP	Mean tariff	Gross private capital flows		Gross foreign direct investment	
	% of PPP GDP		% of goods GDP		percentage points	All products %	% of PPP GDP		% of PPP GDP	
	1986	1996	1986	1996	1986–96	1990–96[a]	1986	1996	1986	1996
Rwanda	11.7	12.9	38.5	72.9	6.6	..	2.7	1.3	0.4	0.1
Saudi Arabia	36.2	41.2	110.2	..	..	..	14.2	5.5	0.9	1.0
Senegal	22.6	16.1	108.6	98.9	–0.6	..	7.7	4.5	0.3	0.4
Sierra Leone	14.1	22.8	41.9	82.8	–0.3	..	29.6	9.5	7.3	0.2
Singapore	191.0	316.0	697.4	763.6	4.8	..	31.9	61.0	7.5	17.5
Slovak Republic	..	52.2	..	243.6	10.0	..	..	10.7	..	0.8
Slovenia	..	74.0	..	184.5	..	..	..	9.5	..	0.8
South Africa	17.4	20.7	93.4	105.4	3.8	8.8[d]	2.2	3.5	0.1	0.1
Spain	18.4	36.8	64.8	..	5.6	6.8[b]	4.6	10.3	1.1	1.9
Sri Lanka	14.0	21.5	80.9	124.8	3.3	20.0	5.3	4.9	0.1	0.3
Sudan	..	..	16.7	..	–8.2	..	..	..	..	..
Sweden	61.5	87.2	117.0	158.3	3.2	6.8[b]	12.8	62.6	4.5	6.1
Switzerland	67.4	89.8	..	..	1.4	0.0	32.7	90.2	3.9	9.3
Syrian Arab Republic	20.0	19.6	64.5	..	–5.6	..	6.1	5.0	0.0	0.2
Tajikistan	..	26.9	..	..	..	..	..	..	..	..
Tanzania	..	..	28.8	59.4	..	21.6	..	..	..	..
Thailand	14.7	31.3	85.8	138.2	6.9	..	1.6	5.0	0.2	0.8
Togo	11.0	19.5	89.0	195.0	–5.0	..	2.2	2.1	0.2	0.0
Trinidad and Tobago	42.7	53.7	142.8	171.0	–4.7	..	7.3	11.5	1.4	3.6
Tunisia	20.6	30.2	84.6	..	1.6	..	3.8	5.8	0.3	0.6
Turkey	10.3	17.5	44.7	71.3	5.7	..	3.0	5.1	0.1	0.2
Turkmenistan	..	32.8	..	..	..	..	..	7.9	..	1.2
Uganda	10.1	6.3	28.9	32.6	–0.2	..	6.0	1.8	0.0	0.6
Ukraine	..	35.0	..	149.6	..	10.1	..	7.2	..	0.2
United Arab Emirates	83.6	135.7	160.8	..	..	..	..	..	..	..
United Kingdom	33.3	46.3	89.6	106.9	2.6	6.8[b]	38.7	59.9	3.6	6.6
United States	14.0	19.4	46.1	..	4.5	6.0	8.0	12.5	1.4	2.6
Uruguay	14.7	22.8	68.4	89.1	6.8	9.7	3.2	11.9	0.3	0.7
Uzbekistan	..	12.4	..	49.5	..	..	..	..	..	..
Venezuela	15.3	19.0	59.3	103.8	2.3	12.0	3.3	5.2	0.4	1.1
Vietnam	..	17.7	16.9	..	..	..	..	..	..	..
West Bank and Gaza	..	..	..	..	..	..	..	..	..	..
Yemen, Rep.	..	56.3	..	210.7	..	..	..	15.9	..	1.8
Yugoslavia, FR (Serb./Mont.)	..	..	..	..	..	..	..	..	..	..
Zambia	22.4	26.1	137.1	107.8	–0.7	13.6	17.1	..	0.5	..
Zimbabwe	13.8	19.8	76.1	139.1	3.6	24.3	2.0	3.8	0.1	0.2
World	**20.7 w**	**29.1 w**	**63.8 w**	**93.8 w**			**8.4 w**	**14.5 w**	**1.1 w**	**2.2 w**
Low income	7.1	7.9	33.8	56.9			2.0	2.1	0.2	1.0
Excl. China & India	12.0	15.7	50.8	92.4			4.9	4.2	0.3	0.5
Middle income	12.5	21.8	53.3	81.1			4.0	5.8	0.3	0.9
Lower middle income	12.5	20.0	47.1	84.5			3.3	4.8	0.3	0.6
Upper middle income	12.5	24.1	59.0	77.6			4.6	7.1	0.3	1.3
Low & middle income	10.4	15.2	46.1	76.8			3.2	4.0	0.2	0.8
East Asia & Pacific	9.1	13.0	48.1	127.3			1.7	1.9	0.2	1.0
Europe & Central Asia	..	25.5	57.2	79.7			..	9.2	..	0.8
Latin America & Carib.	7.9	17.3	40.6	61.7			4.6	6.6	0.3	1.1
Middle East & N. Africa	19.4	18.9	52.1	78.4			5.0	3.2	0.4	0.4
South Asia	4.9	5.8	22.1	39.2			1.2	0.9	0.0	0.2
Sub-Saharan Africa	15.8	18.9	70.3	102.5			4.8	5.7	0.3	0.4
High income	26.5	38.9	70.4	178.8			11.4	19.3	1.6	2.7

a. Estimates are for most recent year available (see table 6.7). b. Average tariff for European Union. c. Data prior to 1992 include Eritrea. d. Data are for the South African Customs Union, which includes Botswana, Lesotho, Namibia, and South Africa.

About the data

The growing importance of trade in the world's economies is one indication of increasing global economic integration. Another is the increased size and importance of private capital flows to developing countries that have liberalized their financial markets. The indicators in the table highlight key features of the ongoing expansion of global markets in goods and capital. For three of the indicators GDP measured in purchasing power parity (PPP) terms has been used in the denominator to adjust for differences in domestic prices. (No adjustment has been made to the numerators because goods and capital exchanged on international markets are assumed to be valued at international prices.) This is a conservative measure: because the GDP of many developing countries is larger in PPP terms than when converted at official exchange rates, the resulting ratios tend to be lower. Still, there is ample evidence of the increasing importance of trade and international capital flows.

The growth of services has also affected the historical record. Compared with the levels achieved at the end of the last century, trade in goods appears to have declined in importance relative to GDP, especially in economies with growing service sectors. Deducting value added by services from GDP thus provides a better measure of the relative size of merchandise trade than physical output, although it neglects the growing services component of most goods output.

Trade in services, traditionally called invisibles, is becoming an important element of global integration. The difference between the growth of real trade in goods and services and the growth of GDP helps to identify economies with dynamic trade regimes.

Tariffs provide one indication of an economy's openness, but they are not definitive. Countries typically have an array of tariffs that are applied to different partners. The mean tariffs shown in the table are based on applied most-favored-nation, ad valorem rates, but lower rates may apply to regional trading partners and others. Many countries also use an array of specific tariffs (based on physical units), nontariff barriers, and export taxes and subsidies to regulate trade.

In the financial account of the balance of payments inward investment is recorded as a credit and outward investment as a debit. Thus net flows, the sum of credits and debits, represent a balance in which many transactions are canceled out. Gross flows are a better measure of integration because they measure the total value of financial transactions during a given period. The investment indicators in the table were constructed from data recorded at the most detailed level available. Higher-level aggregates tend to be affected by the netting out of credits and debits and so produce a smaller total. The comparability of these indicators between countries and over time is affected by the accuracy and completeness of balance of payments records and by their level of detail.

Definitions

• **Trade as a share of PPP GDP** is the sum of merchandise exports and imports measured in current U.S. dollars divided by the value of GDP converted to international dollars using purchasing power parity conversion factors. • **Trade in goods as a share of goods GDP** is the sum of merchandise exports and imports divided by the current value of GDP in U.S. dollars after subtracting value added in services. • **Growth in real trade less growth in real GDP** is the difference between annual growth in trade of goods and services and growth in GDP. Growth rates are calculated using constant price series taken from national accounts, expressed in percentages. • **Mean tariff** is the simple (unweighted) average of applied most-favored-nation tariffs imposed by the country. • **Gross private capital flows** are the sum of the absolute values of direct, portfolio, and other investment inflows and outflows recorded in the balance of payments financial account, excluding changes in the assets and liabilities of monetary authorities and general government. The indicator is calculated as a ratio to GDP converted to international dollars using purchasing power parities. • **Gross foreign direct investment** is the sum of the absolute values of inflows and outflows of foreign direct investment recorded in the balance of payments financial account. It includes equity capital, reinvestment of earnings, other long-term capital, and short-term capital. Note that this indicator differs from the standard measure of foreign direct investment (see table 6.8), which captures only inward investment. The indicator is calculated as a ratio to GDP converted to international dollars using purchasing power parities.

Data sources

Data on merchandise trade are from the International Monetary Fund's (IMF) *Direction of Trade Statistics*. Data on GDP in PPP terms comes from the World Bank's International Comparison Programme database. Data on real trade and GDP growth come from the World Bank's national accounts files. Mean tariffs were calculated using the SMART (Software for Market Analysis and Restrictions on Trade) system developed jointly by the World Bank and the United Nations Conference on Trade and Development. Gross private capital flows and foreign direct investment were calculated from the IMF's Balance of Payments Statistics database.

Figure 6.1a

Gross foreign direct investment is one indicator of global integration

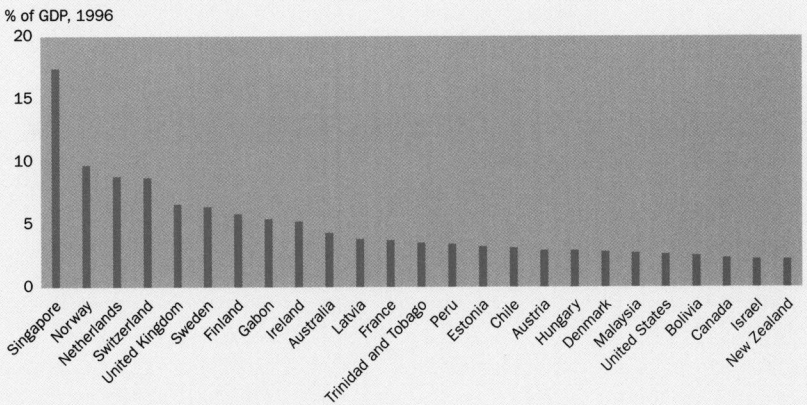

% of GDP, 1996

Note: GDP has been adjusted for purchasing power parity.
Source: World Bank staff estimates.

Gross foreign direct investment measures the two-way flow of investment assets and liabilities. Relative to its size, Singapore is the world's most integrated economy. Although high-income economies, which are active investors and sources of investment opportunities, dominate the list here, a number of low- and middle-income economies also appear on it.

	Export volume		Import volume		Export value		Import value		Net barter terms of trade	
	average annual % growth		average annual % growth		average annual % growth		average annual % growth		1987 = 100	
	1980–90	1990–96	1980–90	1990–96	1980–90	1990–96	1980–90	1990–96	1980	1996
Albania	−5.9	..	7.2	..	..	28.2	..	40.3	..	..
Algeria	4.9	−2.9	−6.9	−4.2	−0.8	−3.7	−2.3	0.0	..	..
Angola	13.8	3.7	−1.7	−2.5	10.0	1.2	3.1	−0.7	..	..
Argentina	2.8	0.9	−4.4	40.6	5.1	10.0	0.2	24.7	..	..
Armenia	..	..	..	..	..	−22.8	..	−14.4	..	..
Australia	6.2	7.2	5.5	9.2	6.6	6.7	6.4	8.5	122.6	102.5
Austria	6.6	2.8	5.7	2.2	10.2	6.0	8.7	5.5	..	..
Azerbaijan	..	..	..	..	..	−30.4	..	−9.6	..	..
Bangladesh	10.0	11.7	6.0	6.6	12.0	10.0	10.4	11.2	148.4	..
Belarus	..	..	..	..	..	18.9	..	24.1	..	..
Belgium	..	..	..	..	..	..	..	..	..	..
Benin	−6.2	−22.2	−6.6	22.2	−2.4	−13.9	−2.5	30.6	..	..
Bolivia	2.7	−3.9	−2.2	10.9	1.0	0.2	2.8	10.8	..	..
Bosnia and Herzegovina	..	..	..	..	..	..	..	..	..	..
Botswana	..	..	..	..	17.9	7.9	9.4	−1.8	..	..
Brazil	5.5	5.3	3.3	7.4	5.8	6.8	4.9	14.1	96.0	..
Bulgaria	−3.8	..	−3.3	..	−12.5	−15.7	..	2.4	..	..
Burkina Faso	7.7	−17.2	4.1	−4.1	13.4	−14.5	8.5	1.5	119.9	..
Burundi	..	..	..	..	2.5	−4.2	2.2	−6.5	..	..
Cambodia	..	..	..	..	..	..	..	..	..	..
Cameroon	10.5	−0.9	−1.1	−16.2	9.9	4.0	3.3	−13.2	..	..
Canada	6.4	9.7	7.4	9.3	6.8	9.0	7.9	6.8	113.3	101.7
Central African Republic	−3.9	24.0	0.9	2.6	−1.1	19.1	6.0	2.7	..	..
Chad	2.5	−4.0	2.0	−3.7	8.8	4.4	6.4	−2.2	..	..
Chile	6.8	9.4	10.4	10.3	15.0	13.6	14.3	13.1	595.3	..
China[†]	..	..	..	..	12.8	17.3	13.5	18.0	..	..
Hong Kong, China	15.4	13.1	13.7	14.6	16.8	14.3	15.0	16.2	100.7	100.2
Colombia	13.6	2.0	−1.7	4.9	11.3	7.0	2.4	4.1	123.8	89.7
Congo, Dem. Rep.	−3.7	−12.3	1.5	−13.7	0.6	−4.8	6.3	−5.5	..	..
Congo, Rep.	−0.6	2.4	−6.1	0.4	−4.0	−0.1	−2.2	1.8	..	..
Costa Rica	5.8	5.8	6.3	9.6	7.0	11.2	10.5	7.5	101.9	..
Côte d'Ivoire	3.1	−6.9	−1.8	5.7	3.5	0.9	2.4	14.0	133.2	..
Croatia	..	..	..	..	..	3.5	..	10.2	..	..
Cuba	−7.0	−31.4	−5.5	−15.1	−3.4	−30.4	−1.0	−14.9	..	..
Czech Republic	..	..	..	..	..	21.4	..	32.7	..	..
Denmark	4.1	4.0	3.1	3.7	8.4	6.2	6.3	5.6	90.8	101.0
Dominican Republic	0.1	−2.4	..	..	−2.1	2.6	3.3	9.7	..	..
Ecuador	2.2	11.2	−2.4	9.8	−0.1	10.7	2.6	10.7	..	..
Egypt, Arab Rep.	−1.2	13.2	−6.2	3.7	−2.6	8.5	−1.7	9.3	..	..
El Salvador	−5.5	5.8	−0.2	14.5	−4.7	6.1	3.4	10.5	..	..
Eritrea	..	..	..	..	..	..	..	..	..	..
Estonia	..	..	..	..	..	36.8	..	52.7	..	..
Ethiopia[a]	1.0	0.7	−1.0	1.3	−1.6	8.2	2.1	1.5	..	..
Finland	2.3	8.5	4.4	−2.0	7.4	8.9	6.9	3.0	86.2	..
France	3.6	5.0	3.7	4.3	7.5	5.2	6.5	2.9	90.0	105.2
Gabon	0.7	3.5	−5.8	−1.0	−1.2	−2.8	−0.9	2.0	..	..
Gambia, The	−2.9	9.6	4.4	−3.1	−3.3	−8.1	9.2	−4.6	..	..
Georgia	..	..	..	..	..	..	..	..	..	..
Germany[b]	4.5	3.0	4.9	1.8	9.2	4.6	7.1	4.1	85.9	..
Ghana	11.8	7.3	7.8	13.2	10.9	3.0	11.6	13.8	..	..
Greece	5.0	10.6	6.4	11.1	5.8	2.7	6.6	1.9	97.8	..
Guatemala	−1.1	2.7	2.7	11.5	−0.1	8.5	6.3	9.6	..	..
Guinea	..	..	..	..	4.1	..	10.0	..	..	..
Guinea-Bissau	..	..	..	..	−4.6	−2.8	−5.2	−15.2	..	..
Haiti	..	..	..	..	−1.2	−8.3	−2.9	11.2	..	..
Honduras	0.7	9.3	−2.9	8.9	3.0	10.0	0.8	12.1	..	..
†Data for Taiwan, China	11.6	5.6	9.9	6.5	14.8	9.5	12.4	11.3	78.0	98.7

Growth of merchandise trade | 6.2

	Export volume		Import volume		Export value		Import value		Net barter terms of trade	
	average annual % growth		average annual % growth		average annual % growth		average annual % growth		1987 = 100	
	1980–90	1990–96	1980–90	1990–96	1980–90	1990–96	1980–90	1990–96	1980	1996
Hungary	3.4	0.9	1.3	5.6	1.4	4.6	0.1	9.9	112.2	..
India	8.0	6.4	5.7	3.6	11.0	10.5	7.3	8.9	71.5	..
Indonesia	7.1	14.2	0.7	5.7	2.0	9.7	4.6	9.4	..	..
Iran, Islamic Rep.	1.1	12.0	–7.9	6.8	–4.2	6.1	–3.1	6.1	..	..
Iraq	2.1	–12.9	–12.3	–17.7	–3.2	–48.7	–7.7	–20.0	..	..
Ireland	9.3	13.1	4.8	8.8	12.7	13.4	7.0	10.1	93.0	90.6
Israel	6.9	9.6	5.8	11.8	8.3	10.9	5.9	11.5	95.0	109.3
Italy	4.4	6.9	5.4	2.3	8.7	6.9	6.9	1.9	85.1	108.7
Jamaica	4.5	2.8	4.1	3.4	8.2	4.4	6.0	7.5	..	..
Japan	5.1	0.9	6.6	6.0	8.9	7.1	5.1	7.6	65.5	..
Jordan	6.8	4.6	–6.2	5.4	8.9	9.0	–2.6	5.2	98.4	120.5
Kazakhstan	..	..	..	..	..	26.6	..	3.4	..	..
Kenya	3.2	8.1	5.0	2.5	0.1	8.5	7.5	7.5	144.3	109.6
Korea, Dem. Rep.	..	..	..	..	..	..	..	..	..	..
Korea, Rep.	13.0	7.3	12.6	8.5	16.7	11.2	15.6	13.0	84.7	89.5
Kuwait	–1.1	30.8	–10.4	13.6	–5.4	22.4	–5.8	8.8	..	..
Kyrgyz Republic	..	..	..	..	..	..	..	..	..	..
Lao PDR	..	..	..	..	11.2	30.3	7.4	27.1	..	..
Latvia	..	–4.1	..	..	..	15.7	..	39.0	..	..
Lebanon	1.2	2.9	–8.2	6.0	4.5	13.5	–3.8	9.4	..	..
Lesotho	..	..	..	..	3.0	21.7	3.4	5.6	..	..
Libya	3.1	–5.5	–7.0	4.9	–2.8	–6.1	–2.1	8.1	..	..
Lithuania	..	..	..	..	..	19.5	..	27.9	..	..
Macedonia, FYR	..	..	..	..	..	0.7	..	3.0	..	..
Madagascar	2.4	1.0	–1.4	–0.5	0.0	1.2	2.7	3.7	..	..
Malawi	1.8	–4.6	6.2	–0.8	5.0	–1.5	10.7	0.8	118.0	..
Malaysia	10.8	11.7	7.0	12.6	10.6	13.5	11.3	14.9	131.9	..
Mali	7.3	–1.2	2.9	9.9	13.4	6.2	7.2	13.1	..	..
Mauritania	..	..	..	..	8.0	..	–2.1	..	..	..
Mauritius	9.8	–6.9	17.9	3.6	20.9	–2.1	23.2	4.5	69.7	..
Mexico	3.2	8.9	11.3	16.6	1.2	11.5	16.7	8.7	..	..
Moldova	..	..	..	..	..	16.1	..	14.3	..	..
Mongolia	..	..	..	..	–0.5	–3.2	–3.8	–8.8	..	..
Morocco	5.4	2.0	5.5	2.0	11.2	5.3	8.7	10.8	103.4	77.2
Mozambique	–2.8	–0.7	3.9	–0.6	1.5	1.1	7.5	–3.9	..	..
Myanmar	–9.8	24.7	–7.0	21.0	–7.3	14.9	–2.2	20.3	..	..
Namibia	..	..	..	..	..	4.6	..	0.9	..	..
Nepal	..	..	..	..	8.1	9.2	6.9	14.9	..	..
Netherlands	4.5	7.1	4.5	6.5	4.6	7.7	4.4	6.6	96.9	102.3
New Zealand	3.5	5.6	4.3	6.6	6.2	8.1	5.4	9.7	95.9	105.6
Nicaragua	–5.6	–6.0	–7.7	14.5	–4.1	7.4	–4.5	14.0	87.9	..
Niger	–5.1	–1.0	–0.1	5.2	0.0	1.6	3.1	6.3	..	..
Nigeria	1.8	8.7	–14.8	8.2	–2.6	8.9	–10.3	9.4	..	..
Norway	4.1	8.3	3.4	6.9	5.3	5.5	6.2	5.0	122.8	103.1
Oman	..	..	..	..	2.9	1.6	0.7	8.2	..	..
Pakistan	10.0	12.5	1.4	7.3	11.1	9.6	4.4	7.5	95.2	88.1
Panama	1.5	21.2	–6.2	10.4	1.3	15.9	–3.7	9.6	..	..
Papua New Guinea	6.7	4.5	..	..	4.9	15.8	1.3	3.4	..	..
Paraguay	15.9	–4.0	5.7	–11.2	20.6	–12.1	9.3	–9.3	..	..
Peru	–3.5	8.4	0.5	12.7	1.1	11.1	5.6	18.4	..	..
Philippines	3.8	7.8	7.2	10.9	8.3	13.9	9.9	13.6	103.9	..
Poland	4.8	2.0	1.5	22.6	1.4	10.7	–3.2	23.8	95.5	..
Portugal	11.9	..	15.1	..	15.1	6.4	10.3	4.5	..	..
Puerto Rico	..	..	..	..	..	..	..	..	..	..
Romania	..	..	..	..	–4.0	9.7	–3.8	6.3	..	..
Russian Federation	..	..	..	..	..	23.4	..	17.6	..	..

	Export volume		Import volume		Export value		Import value		Net barter terms of trade	
	average annual % growth		average annual % growth		average annual % growth		average annual % growth		1987 = 100	
	1980–90	1990–96	1980–90	1990–96	1980–90	1990–96	1980–90	1990–96	1980	1996
Rwanda	4.3	−23.4	−1.8	14.5	2.4	−19.5	2.6	37.4	..	..
Saudi Arabia	2.8	4.7	−11.4	3.1	−2.0	4.4	−6.8	6.3	..	..
Senegal	0.3	6.4	1.2	4.5	4.4	8.4	4.1	6.4	81.7	..
Sierra Leone	1.6	−7.8	−3.2	−4.1	2.5	−23.4	0.7	1.6	..	..
Singapore	12.1	16.2	8.6	12.2	9.9	17.0	8.0	15.0	109.0	89.4
Slovak Republic	..	..	..	..	..	18.7	..	21.3	..	..
Slovenia	..	..	..	..	..	7.8	..	13.1	..	..
South Africa	3.3	7.4	−0.8	7.9	0.8	4.0	−1.3	9.8	108.8	117.0
Spain	3.0	12.1	8.4	6.0	10.9	10.5	10.6	4.9	92.2	114.7
Sri Lanka	5.6	14.5	2.8	14.8	6.0	12.9	5.9	12.7	93.8	109.4
Sudan	−3.7	−4.7	−6.6	4.6	1.0	−6.4	−2.8	4.0	..	..
Sweden	4.4	2.1	5.0	1.9	8.0	7.4	6.7	4.3	91.4	103.5
Switzerland	3.7	..	4.3	..	9.5	4.0	8.8	1.9	79.3	..
Syrian Arab Republic	11.1	4.5	−13.2	4.5	8.8	5.3	−10.5	6.4	214.9	97.0
Tajikistan	..	..	..	..	..	..	..	..	..	..
Tanzania	1.3	−10.1	−1.4	−0.4	−0.5	−4.1	2.1	−1.9	..	..
Thailand	18.3	17.4	17.3	7.9	22.0	13.2	20.2	10.3	116.5	..
Togo	8.8	10.7	8.6	−5.1	7.0	9.6	12.3	9.9	..	..
Trinidad and Tobago	0.2	3.9	−11.5	4.4	−4.3	6.3	−9.8	10.7	195.6	..
Tunisia	11.0	5.3	4.7	2.7	10.5	8.9	8.3	7.3	104.3	..
Turkey	..	6.6	..	8.2	14.0	10.8	9.3	11.5	..	..
Turkmenistan	..	..	..	..	..	..	..	..	..	..
Uganda	−6.2	24.5	−3.4	28.1	−11.3	29.1	1.5	26.5	..	..
Ukraine	..	..	..	..	..	14.6	..	16.6	..	..
United Arab Emirates	10.6	3.9	1.4	11.1	3.4	2.2	5.9	13.2	..	..
United Kingdom	4.5	5.5	6.7	3.8	5.8	6.0	8.4	4.5	105.3	102.9
United States	3.6	6.4	7.2	7.8	5.7	8.1	8.2	9.1	88.8	101.2
Uruguay	3.6	−2.4	6.5	15.2	9.6	5.2	9.7	11.8	..	..
Uzbekistan	..	..	..	..	..	..	..	..	..	..
Venezuela	4.0	2.1	−4.9	6.7	−0.9	6.7	0.0	4.3	215.2	148.9
Vietnam	..	..	..	..	18.9	12.4	8.7	18.7	..	..
West Bank and Gaza	..	..	..	..	..	..	..	..	..	..
Yemen, Rep.	..	..	..	..	..	..	..	..	..	..
Yugoslavia, FR (Serb./Mont.)[c]	1.9	−20.9	1.1	−24.7	5.5	−20.6	5.5	−25.2	..	..
Zambia	−0.7	21.6	−0.5	−2.5	8.2	8.7	3.3	4.9	..	..
Zimbabwe	2.0	..	−0.6	..	2.5	5.1	−0.4	2.7	..	..

a. Data prior to 1992 include Eritrea. b. Data prior to 1990 refer to the Federal Republic of Germany before unification. c. Data refer to the former Yugoslavia.

Growth of merchandise trade | 6.2

About the data

Data on international trade in goods are recorded in each country's balance of payments and by customs services. While the balance of payments focuses on the financial transactions that accompany trade, customs data record the direction of trade and the physical quantities and value of goods entering or leaving the customs area. Customs data may differ from those recorded in the balance of payments because of differences in valuation and the time of recording.

Trade in goods, or merchandise trade, includes all goods that add to or subtract from an economy's material resources. Currency in circulation, titles of ownership, and securities are excluded, but monetary gold is included. Trade data are collected on the basis of a country's customs area, which in most cases is the same as its geographic area. Goods provided as part of foreign aid are included, but goods destined for extraterritorial agencies (such as embassies) are not.

Collecting and tabulating trade statistics is difficult. Some developing countries lack the capacity to report timely data. As a result it is necessary to estimate their trade from the data reported by their partners. (See *About the data* for table 6.3 for further discussion of the use of partner country reports.) In some cases economic or political concerns may lead national authorities to suppress or misrepresent data on certain trade flows, such as military equipment, oil, or the exports of a dominant producer. In other cases reported trade data may be distorted by deliberate underinvoicing or overinvoicing to effect capital transfers or avoid taxes.

And in some regions smuggling and black market trading result in unreported trade flows.

By international agreement customs data are reported to the United Nations Statistical Division, which maintains the Commodity Trade, or COMTRADE, database. The International Monetary Fund (IMF) also maintains a database on the direction of trade. The United Nations Conference on Trade and Development (UNCTAD) compiles a variety of international trade statistics, including price and volume indexes, based on the COMTRADE data. The World Bank supplements data from UNCTAD with data from the IMF for high-income economies and, in some cases, with data taken directly from the COMTRADE database.

The growth rates and terms of trade for low- and middle-income economies were calculated from index numbers compiled by UNCTAD. Volume measures for high-income economies were derived by deflating the value of trade using deflators from the IMF's *International Financial Statistics*. Terms of trade were computed from the same indicators.

The terms of trade measure the relative prices of a country's exports and imports. There are a number of ways to calculate terms of trade. The most common is the net barter, or commodity, terms of trade, constructed as the ratio of the export price index to the import price index. When the net barter terms of trade increase, a country's exports are becoming more valuable or its imports cheaper.

- **Growth rates of export and import volumes** are average annual growth rates calculated from UNCTAD's quantum index series for low- and middle-income economies and from export and import data deflated by the IMF's trade price deflators for high-income economies. • **Growth rates of export and import values** are average annual growth rates calculated from UNCTAD's value indexes for low- and middle-income economies and from current values of exports and imports for high-income economies. • **Net barter terms of trade** are the ratio of the export price index to the corresponding import price index measured relative to the base year 1987.

Data sources

 The main source of trade data for developing countries is UNCTAD's annual *Handbook of International Trade and Development Statistics*. The IMF's *International Financial Statistics* includes data on the export and import values and deflators for high-income and selected developing economies. The United Nations publishes trade data in its *International Trade Statistics Yearbook*.

Figure 6.2a

Terms of trade have shifted since the 1970s

% change

[Bar chart showing % change from -10 to 25 across regions: High-income economies, East Asia and the Pacific, Europe and Central Asia, Latin America and the Caribbean, Middle East and North Africa, South Asia, Sub-Saharan Africa]

● 1974–80   ○ 1981–90   ● 1991–97

Source: World Bank staff estimates.

Changes in terms of trade reflect changes in the relative prices of exports and imports and the mix of goods traded by countries. Positive changes mean that imports cost less relative to exports. Although regional averages blur the many differences among economies, they are broadly representative of the economic forces that have shaped trade patterns over time. The effects of the oil shocks in the 1970s are unmistakable. Since then terms of trade have tended to shift against producers of primary commodities.

6.3 Direction and growth of merchandise trade

High-income importers

Direction of trade % of world trade, 1996	United States	European Union	Japan	Other industrial	All industrial	Other high income	All high income
Source of exports							
High-income economies	10.8	30.7	4.0	6.4	51.9	8.0	59.9
Industrial economies	8.5	29.2	2.7	6.0	46.3	6.1	52.4
United States		2.5	1.3	3.0	6.8	1.7	8.5
European Union	2.8	23.5	0.9	2.3	29.4	1.8	31.3
Japan	2.2	1.2		0.3	3.7	2.0	5.8
Other industrial economies	3.5	2.1	0.5	0.3	6.4	0.5	6.9
Other high-income economies	2.3	1.6	1.3	0.4	5.5	1.9	7.5
Low- & middle-income economies	4.5	5.5	2.2	0.7	12.9	2.7	15.7
Sub-Saharan Africa	0.3	0.6	0.1	0.0	1.0	0.1	1.1
East Asia & Pacific	1.3	1.1	1.4	0.3	4.1	1.7	5.8
South Asia	0.2	0.3	0.1	0.0	0.6	0.1	0.7
Europe & Central Asia	0.2	2.0	0.1	0.1	2.5	0.1	2.6
Middle East & North Africa	0.2	0.9	0.3	0.1	1.5	0.5	2.0
Latin America & Caribbean	2.2	0.7	0.2	0.2	3.3	0.1	3.4
World	15.3	36.3	6.1	7.1	64.8	10.7	75.5

Low- and middle-income importers

Direction of trade % of world trade, 1996	Sub-Saharan Africa	East Asia & Pacific	South Asia	Europe & Central Asia	Middle East & N. Africa	Latin America & Caribbean	All low & middle income	World
Source of exports								
High-income economies	1.0	6.8	0.8	3.9	1.7	4.0	18.2	78.1
Industrial economies	0.9	3.7	0.6	3.6	1.6	3.6	13.9	66.3
United States	0.1	0.8	0.1	0.2	0.3	2.1	3.6	12.0
European Union	0.6	1.1	0.3	3.2	1.1	1.0	7.3	38.6
Japan	0.1	1.5	0.1	0.1	0.1	0.3	2.2	8.0
Other industrial economies	0.0	0.4	0.1	0.1	0.1	0.1	0.9	7.8
Other high-income economies	0.1	3.1	0.2	0.3	0.2	0.3	4.3	11.8
Low- & middle-income economies	0.4	1.5	0.4	2.2	0.6	1.2	6.2	21.9
Sub-Saharan Africa	0.2	0.1	0.0	0.0	0.0	0.0	0.4	1.4
East Asia & Pacific	0.1	0.8	0.1	0.1	0.1	0.1	1.4	8.1
South Asia	0.0	0.1	0.0	0.0	0.0	0.0	0.2	1.0
Europe & Central Asia	0.0	0.2	0.1	1.9	0.2	0.1	2.3	4.9
Middle East & North Africa	0.1	0.2	0.1	0.1	0.2	0.0	0.7	2.7
Latin America & Caribbean	0.0	0.1	0.0	0.1	0.1	0.9	1.3	4.7
World	1.4	8.3	1.2	6.1	2.3	5.1	24.5	100.0

Direction and growth of merchandise trade | 6.3

High-income importers

Nominal growth of trade annual % growth, 1986–96	United States	European Union	Japan	Other industrial	All industrial	Other high income	All high income
Source of exports							
High-income economies	6.8	8.6	11.0	8.1	8.3	14.8	8.9
Industrial economies	6.4	8.4	10.4	8.0	8.0	13.9	8.6
United States		8.6	9.6	10.9	9.8	14.8	10.6
European Union	6.1	8.5	13.8	6.0	8.2	15.0	8.5
Japan	3.3	6.4		1.4	4.0	12.4	6.2
Other industrial economies	9.2	7.9	7.7	11.5	8.7	14.1	9.0
Other high-income economies	8.6	13.8	12.4	9.2	10.7	18.4	12.2
Low- & middle-income economies	14.1	9.4	11.6	13.0	11.4	17.0	12.2
Sub-Saharan Africa	9.3	5.5	6.5	3.3	6.4	17.3	7.1
East Asia & Pacific	19.4	16.8	14.5	18.2	16.8	17.7	17.0
South Asia	14.0	15.4	8.9	12.0	13.7	18.0	14.5
Europe & Central Asia	13.5	10.7	6.5	12.3	10.8	23.5	11.2
Middle East & North Africa	7.3	5.4	8.3	11.1	6.5	13.0	7.6
Latin America & Caribbean	13.4	6.4	7.6	12.8	11.1	18.5	11.3
World	8.4	8.7	11.2	8.5	8.8	15.3	9.5

Low- and middle-income importers

Nominal growth of trade annual % growth, 1986–96	Sub-Saharan Africa	East Asia & Pacific	South Asia	Europe & Central Asia	Middle East & N. Africa	Latin America & Caribbean	All low & middle income	World
Source of exports								
High-income economies	5.3	18.4	7.6	12.4	3.6	12.1	11.9	9.6
Industrial economies	4.8	15.1	5.9	12.0	3.5	11.7	10.3	8.9
United States	8.0	16.4	6.8	10.8	5.4	13.4	12.4	11.1
European Union	4.6	15.1	7.1	13.5	3.7	11.4	10.0	8.7
Japan	3.7	14.9	1.4	−1.9	−0.9	7.1	9.2	6.9
Other industrial economies	2.8	13.8	6.8	4.0	3.6	5.6	7.6	8.9
Other high-income economies	10.2	24.4	12.9	20.4	4.7	19.3	20.1	14.5
Low- & middle-income economies	12.6	17.2	12.5	7.0	3.6	14.0	10.0	11.5
Sub-Saharan Africa	13.6	12.4	18.5	2.2	10.7	11.4	11.9	8.1
East Asia & Pacific	15.3	20.4	14.9	6.6	6.9	21.8	15.7	16.8
South Asia	17.1	21.9	14.4	−0.7	6.5	23.0	9.8	13.1
Europe & Central Asia	2.6	8.9	7.1	8.3	−0.5	9.4	7.4	9.2
Middle East & North Africa	13.8	21.1	12.1	−6.3	4.0	0.4	6.5	7.3
Latin America & Caribbean	9.6	13.6	10.1	5.5	6.3	15.3	13.2	11.8
World	7.0	18.2	8.9	10.1	3.6	12.5	11.4	10.0

6.3 Direction and growth of merchandise trade

The data in table 6.3 were compiled from the International Monetary Fund's (IMF) *Direction of Trade Statistics,* which reports the value of exports and imports between its member countries.

Most countries report their trade data to the IMF in national currencies, which are converted using the IMF's published exchange rate series rf (official rate, period average) or rh (market rate, period average). Most industrial countries and about 22 developing countries report their trade data to the IMF each month. Together these countries account for about 80 percent of world exports. Trade from less timely reporters and from countries that do not report at all is estimated using reports of partner countries. Because the largest exporting and importing countries are reliable reporters, a large portion of the missing trade flows can be estimated from partner reports. Even so, a small amount of trade between developing countries, particularly in Africa, is not captured in partner data. Inter-European trade estimates have been significantly affected by changes in reporting methods following the creation of a common customs union.

Because imports are reported at c.i.f. (cost, insurance, and freight) valuations and exports are reported at f.o.b. (free on board) valuations, the IMF divides partner country reports of import values by 1.10 to estimate equivalent export values. This approximation is more or less accurate, depending on the set of partners and the items traded. Other factors affecting the accuracy of trade data include lags in reporting, recording differences across countries, and whether the country follows the general or special system of trade. (See *About the data* for table 4.5 for further discussion of systems of trade.)

The regional trade flows shown in this table were calculated from current price values. Growth rates therefore include the effects of changes in both volumes and prices.

• **Merchandise trade** includes all trade in goods. Trade in services is not included. • **Regional groupings** are based on World Bank definitions and may differ from those used by other organizations. Within the high-income group, • **European Union** refers to the 15 current members of the European Union. • **Other industrial economies** include Australia, Canada, Iceland, New Zealand, Norway, and Switzerland. • **Other high-income economies** include Cyprus, Hong Kong (China), Israel, the Republic of Korea, Kuwait, Qatar, Singapore, Taiwan (China), and the United Arab Emirates. Some small high-income economies such as Aruba, the Bahamas, and Bermuda have been included in the Latin America and Caribbean group.

Regional trade flows were calculated from intercountry trade data reported in the IMF's *Direction of Trade Statistics Yearbook* (1997). These were then updated from tables that appeared in the statistical appendix to *Global Economic Prospects and the Developing Countries 1997* (World Bank 1997b).

OECD trade with low- and middle-income economies | 6.4

Exports to low- and middle-income economies	High-income OECD countries		European Union		United States		Japan	
	1985	1996a	1985	1996	1985	1996	1985	1996
$ billions								
Food	24.7	58.9	10.6	28.2	8.7	21.1	0.3	0.4
Cereals	10.4	17.9	2.8	5.8	4.6	9.4	0.0	0.0
Agricultural raw materials	5.9	14.7	2.2	4.4	2.0	5.3	0.5	1.2
Ores and nonferrous metals	5.0	13.0	1.9	5.4	1.1	3.0	0.4	1.4
Fuels	5.4	14.7	2.3	6.0	2.0	4.2	0.2	0.8
Crude petroleum	0.1	0.8	0.0	0.5	..	0.0	..	..
Petroleum products	3.4	10.7	2.1	5.2	0.9	3.0	0.1	0.7
Manufactured goods	184.5	595.3	90.2	291.2	38.0	133.3	42.6	104.6
Chemical products	26.9	76.9	15.7	41.3	6.6	18.4	2.3	7.0
Mach. and transport equip.	103.4	350.1	45.8	160.5	24.5	80.9	26.6	74.2
Other	54.1	168.2	28.6	89.5	6.9	33.9	13.7	23.4
Miscellaneous goods	4.1	18.4	10.0	4.1	1.3	7.4	0.3	1.5
Total	**229.5**	**715.0**	**117.1**	**339.3**	**53.1**	**174.3**	**44.4**	**109.9**
% of total exports								
Food	10.8	8.2	9.0	8.3	16.5	12.1	0.8	0.3
Cereals	4.5	2.5	2.4	1.7	8.7	5.4	0.1	0.0
Agricultural raw materials	2.6	2.1	1.8	1.3	3.8	3.1	1.2	1.1
Ores and nonferrous metals	2.2	1.8	1.6	1.6	2.1	1.7	0.9	1.3
Fuels	2.3	2.1	2.0	1.8	3.8	2.4	0.5	0.7
Crude petroleum	0.0	0.1	0.0	0.2	..	0.0	..	..
Petroleum products	1.5	1.5	1.8	1.5	1.8	1.7	0.2	0.6
Manufactured goods	80.4	83.3	77.0	85.8	71.5	76.4	95.9	95.2
Chemical products	11.7	10.8	13.4	12.2	12.3	10.6	5.1	6.4
Mach. and transport equip.	45.1	49.0	39.1	47.3	46.2	46.4	59.9	67.5
Other	23.6	23.5	24.5	26.4	13.0	19.5	30.9	21.3
Miscellaneous goods	1.0	2.6	8.5	1.2	2.4	4.3	0.8	1.4
Total	**100.0**	**100.0**	**100.0**	**100.0**	**100.0**	**100.0**	**100.0**	**100.0**

a. Excludes Greece.

Imports from low- and middle-income economies	High-income OECD countries		European Union		United States		Japan	
	1985	1996a	1985	1996	1985	1996	1985	1996
$ billions								
Food	45.3	93.9	24.1	46.2	13.4	21.0	5.1	18.7
Cereals	1.4	2.1	0.6	0.7	0.1	0.5	0.6	0.5
Agricultural raw materials	12.6	27.2	6.5	13.0	1.7	4.7	3.4	6.2
Ores and nonferrous metals	19.3	43.0	9.1	19.3	3.8	7.5	5.5	10.7
Fuels	140.2	169.5	63.6	67.0	35.9	49.0	33.5	31.4
Crude petroleum	104.0	120.2	49.5	45.8	25.1	39.4	23.2	18.9
Petroleum products	22.9	23.7	8.6	9.3	10.4	8.7	3.0	3.2
Manufactured goods	60.7	459.1	22.7	171.7	28.4	189.2	4.1	58.6
Chemical products	7.4	28.5	3.5	13.7	2.3	7.1	0.9	3.4
Mach. and transport equip.	17.5	159.1	4.6	48.1	10.2	82.0	0.4	16.3
Other	35.8	271.6	14.6	109.9	16.0	100.1	2.9	38.9
Miscellaneous goods	2.3	11.5	0.5	4.6	1.3	5.2	0.4	1.3
Total	**280.4**	**804.3**	**126.5**	**321.9**	**84.4**	**276.7**	**52.0**	**126.8**
% of total exports								
Food	16.2	11.7	19.0	14.4	15.9	7.6	9.7	14.7
Cereals	0.5	0.3	0.5	0.2	0.1	0.2	1.2	0.4
Agricultural raw materials	4.5	3.4	5.1	4.0	2.0	1.7	6.6	4.9
Ores and nonferrous metals	6.9	5.3	7.2	6.0	4.4	2.7	10.5	8.5
Fuels	50.0	21.1	50.3	20.8	42.6	17.7	64.4	24.7
Crude petroleum	37.1	14.9	39.1	14.2	29.8	14.2	44.6	14.9
Petroleum products	8.2	2.9	6.8	2.9	12.3	3.2	5.7	2.6
Manufactured goods	21.7	57.1	17.9	53.3	33.7	68.4	8.0	46.2
Chemical products	2.6	3.5	2.8	4.2	2.7	2.6	1.8	2.7
Mach. and transport equip.	6.3	19.8	3.6	15.0	12.1	29.6	0.7	12.9
Other	12.8	33.8	11.5	34.1	18.9	36.2	5.5	30.6
Miscellaneous goods	0.8	1.4	0.4	1.4	1.5	1.9	0.8	1.0
Total	**100.0**	**100.0**	**100.0**	**100.0**	**100.0**	**100.0**	**100.0**	**100.0**

a. Excludes Greece.

OECD trade with low- and middle-income economies | 6.4

About the data

Trade flows between high-income members of the Organisation for Economic Co-operation and Development (OECD) and low- and middle-income economies reflect the changing mix of exports to and imports from developing economies. While food and primary commodities have continued to fall as a share of OECD imports, the share of manufactured goods supplied by developing countries has grown. At the same time, developing countries have increased their imports of manufactured goods from high-income countries—particularly capital-intensive goods such as machinery and transport equipment. Although trade between developing countries has grown substantially over the past decade (see table 6.6), high-income OECD countries remain the developing world's most important partners.

The aggregate flows in the table were compiled from intercountry flows recorded in the United Nations Statistical Division's Commodity Trade (COMTRADE) database. Partner country reports by high-income OECD countries were used for both exports and imports. Exports are recorded free on board (f.o.b.); imports include insurance and freight charges (c.i.f.).

For further discussion of merchandise trade statistics see *About the data* for tables 4.4, 6.2, and 6.3.

Figure 6.4a

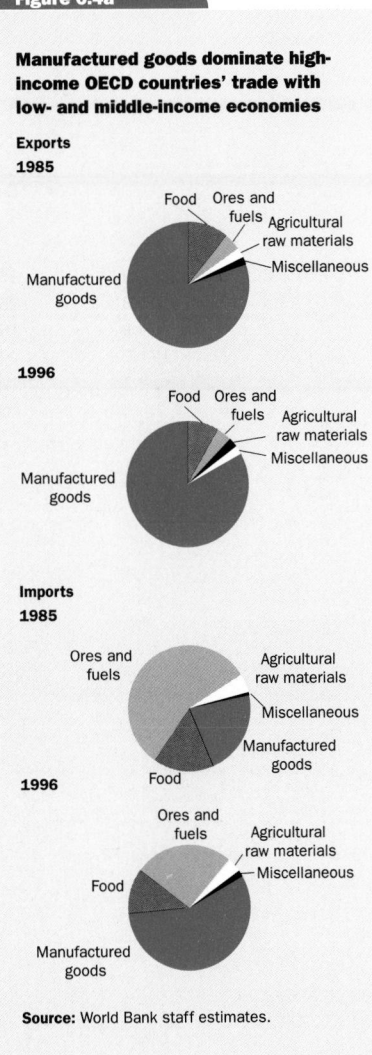

Manufactured goods dominate high-income OECD countries' trade with low- and middle-income economies

Source: World Bank staff estimates.

Definitions

The product groups in the table are defined in accordance with the Standard International Trade Classification (SITC), revision 1: *food* (0, 1, 22, and 4), cereals (04); *agricultural raw materials* (2 excluding 22, 27, and 28); *ores and nonferrous metals* (27, 28, and 68); *fuels* (3), *crude petroleum* (331) and *petroleum products* (332); *manufactured goods* (5–8 excluding 68), *chemical products* (5), *machinery and transportation equipment* (7), and *other manufactured goods* (6 and 8 excluding 68); and *miscellaneous goods* (9). • **Exports** are all merchandise exports by high-income OECD countries to low- and middle-income economies as recorded in the United Nations COMTRADE database. • **Imports** are all merchandise imports by high-income OECD countries from low- and middle-income economies as recorded in the United Nations COMTRADE database. • **High-income OECD countries** in 1996 were Australia, Austria, Belgium, Canada, Denmark, Finland, France, Germany, Iceland, Ireland, Italy, Japan, the Republic of Korea, Luxembourg, the Netherlands, New Zealand, Norway, Portugal, Spain, Sweden, Switzerland, the United Kingdom, and the United States. • **European Union** comprises Austria, Belgium, Denmark, Finland, France, Germany, Greece, Italy, Ireland, Luxembourg, the Netherlands, Portugal, Spain, Sweden, and the United Kingdom.

Data sources

COMTRADE data are available in machine-readable form from the United Nations Statistical Division. Although not as comprehensive as the underlying COMTRADE records, detailed statistics on international trade are published annually in the United Nations' Conference on Trade and Development's *Handbook of International Trade and Development Statistics* and the United Nations *International Trade Statistics Yearbook*.

6.5 | Primary commodity prices

	1980	1985	1990	1991	1992	1993	1994	1995	1996	1997
World Bank commodity price index										
(1990 = 100)										
Nonfuel commodities	174	133	100	93	86	86	101	103	101	107
Agriculture	191	145	100	96	88	93	112	110	110	117
Beverages	253	239	100	91	73	79	135	127	111	156
Food	191	124	100	97	94	93	97	98	108	106
Raw materials	145	103	100	97	92	104	114	113	112	104
Fertilizers	179	130	100	100	90	79	85	87	105	109
Metals and minerals	132	102	100	87	81	70	77	85	78	82
Petroleum	224	173	100	83	78	69	63	63	78	76
Steel products[a]	110	88	100	96	83	86	84	90	84	81
MUV G-5 index	72	69	100	102	107	106	110	119	114	108
Commodity prices										
(1990 $)										
Agricultural raw materials										
Cotton (cents/kg)	284.3	192.1	181.9	164.1	119.9	120.4	160.0	178.5	155.3	162.2
Logs, Cameroon ($/cu. m)[a]	349.4	253.5	343.5	309.2	310.8	291.9	299.7	284.8	237.8	238.9
Logs, Malaysian ($/cu. m)	271.5	177.5	177.2	187.4	196.5	366.7	279.1	214.4	220.8	221.2
Rubber (cents/kg)	197.9	110.6	86.5	80.8	80.8	78.2	102.2	132.6	122.1	94.5
Sawnwood, Malaysian ($/cu. m)	550.3	447.7	533.1	540.6	569.6	713.3	745.0	620.8	649.2	616.2
Tobacco ($/mt)	3,160.9	3,807.3	3,392.2	3,424.7	3,226.6	2,535.6	2,399.0	2,214.3	2,671.0	3,277.1
Beverages (cents/kg)										
Cocoa	361.6	328.6	126.7	116.9	103.2	105.1	126.7	120.2	127.5	150.3
Coffee, robustas	450.4	386.1	118.2	104.9	88.2	108.9	237.8	232.4	158.1	161.2
Coffee, other milds	481.4	471.0	197.2	183.3	132.4	146.8	300.2	279.6	235.9	386.9
Tea, auctions, avg.	250.3	263.6	205.1	167.9	159.8	157.8	143.1	128.1	147.9	195.0
Tea, London, all	309.9	289.1	203.2	180.3	187.6	175.3	166.2	137.8	155.3	206.9
Energy										
Coal, Australian ($/mt)	55.9	49.2	39.7	38.8	36.2	29.5	29.3	33.0	33.4	32.6
Coal, U.S. ($/mt)	59.9	67.9	41.7	40.6	38.1	35.7	33.1	32.9	32.6	33.8
Natural gas, Europe ($/mmbtu)	4.7	5.4	2.6	3.0	2.4	2.5	2.2	2.3	2.5	2.5
Natural gas, U.S. ($/mmbtu)	2.2	3.6	1.7	1.5	1.7	2.0	1.7	1.5	2.4	2.3
Petroleum ($/bbl)	51.2	39.6	22.9	19.0	17.8	15.8	14.4	14.4	17.9	17.8

About the data

Primary commodities are raw or partially processed materials that will be transformed into finished goods. They are often the most significant exports of developing countries, and revenues obtained from them have an important effect on living standards. Price data for primary commodities are collected from a variety of sources, including international study groups, trade journals, newspaper and wire service reports, government market surveys, and commodity exchange spot and near-term forward prices. This table uses the most reliable and frequently updated price reports. When possible, the prices received by exporters are used; if export prices are unavailable, the prices paid by importers are used. Annual price series are generally simple averages based on higher-frequency data. The constant price series in the table are deflated using the manufactures unit value (MUV) index for the G-5 countries (see below).

Commodity price indexes are calculated as Laspeyres index numbers in which the fixed weights are the 1987–89 export values for low- and middle-income economies, rebased to 1990. Each index represents a fixed basket of primary commodity exports. The nonfuel commodity price index contains 37 price series for 31 nonfuel commodities. Separate indexes are compiled for petroleum and steel products, which are not included in the nonfuel commodity price index.

The MUV index is a composite index of prices for manufactured exports from the five major (G-5) industrial countries (France, Germany, Japan, the United Kingdom, and the United States) to low- and middle-income economies, valued in U.S. dollars. The index covers products in Standard International Trade Classification (SITC) groups 5–8. To construct the MUV G-5 index, unit value indexes for each country are combined using weights determined by each country's export share.

Primary commodity prices | 6.5

	1980	1985	1990	1991	1992	1993	1994	1995	1996	1997
Fertilizers ($/mt)										
Phosphate rock	64.9	49.4	40.5	41.6	39.2	31.0	29.9	29.4	34.2	38.1
TSP	250.3	176.9	131.8	130.3	113.3	105.3	119.9	125.5	154.0	159.6
Food										
Fats and oils ($/mt)										
Coconut oil	935.8	860.1	336.5	423.7	541.8	423.6	551.3	562.1	658.1	609.8
Groundnut oil	1,193.1	1,319.2	963.7	875.5	572.1	695.3	928.1	831.4	785.7	938.0
Palm oil	810.4	730.3	289.8	331.7	369.1	355.4	479.5	526.8	464.9	506.8
Soybeans	411.4	326.5	246.8	234.4	220.9	240.0	228.5	217.3	266.9	274.2
Soybean meal	363.9	228.9	200.2	192.9	191.7	195.8	174.6	165.3	234.2	256.0
Soybean oil	830.0	833.8	447.3	444.4	402.4	451.9	558.6	524.3	482.9	524.3
Grains ($/mt)										
Grain sorghum	179.0	150.1	103.9	102.8	96.4	93.2	94.3	99.9	131.4	101.8
Maize	174.0	163.6	109.3	105.1	97.8	96.0	97.6	103.6	145.2	108.7
Rice	570.5	287.1	270.9	287.0	251.6	221.5	242.8	269.2	296.7	281.8
Wheat	239.9	198.0	135.5	125.9	141.8	131.9	135.9	148.5	181.8	148.1
Other food										
Bananas ($/mt)	526.4	554.4	540.9	547.5	443.8	416.8	398.8	373.4	411.2	466.7
Beef (cents/kg)	383.3	314.0	256.3	260.6	230.3	246.3	211.5	160.0	156.3	172.3
Oranges ($/mt)	542.5	580.8	531.1	509.8	458.9	406.9	373.1	445.9	430.5	426.1
Sugar, EU domestic										
(cents/kg)	67.6	51.1	58.3	59.9	58.9	58.3	56.4	57.7	59.9	58.2
Sugar, U.S. domestic										
(cents/kg)	92.0	65.4	51.3	46.5	44.1	44.8	44.1	42.6	43.2	44.9
Sugar, world (cents/kg)	87.7	13.1	27.7	19.3	18.7	20.7	24.2	24.6	23.1	23.3
Metals and minerals										
Aluminum ($/mt)	2,022.2	1,517.5	1,639.0	1,274.2	1,176.6	1,071.5	1,340.1	1,514.8	1,318.0	1,484.7
Copper ($/mt)	3,030.6	2,066.2	2,661.5	2,288.5	2,139.9	1,799.7	2,093.8	2,462.8	2,010.0	2,113.6
Iron ore (cents/DMTU)	39.0	38.7	30.8	32.5	29.6	26.5	23.1	22.6	25.0	26.8
Lead (cents/kg)	125.8	57.0	81.1	54.6	50.8	38.2	49.7	52.9	67.8	57.9
Nickel ($/mt)	9,053.8	7,141.5	8,864.1	7,980.0	6,567.8	4,979.7	5,753.0	6,902.7	6,568.0	6,430.9
Tin (cents/kg)	2,329.8	1,682.1	608.5	547.5	572.3	485.5	495.8	521.3	539.9	524.2
Zinc (cents/kg)	105.7	114.1	151.3	109.3	116.3	90.5	90.5	86.5	89.8	122.2

a. Series not included in the nonfuel index.

• **Nonfuel commodities price index** covers the 31 nonfuel primary commodities that make up the agriculture, fertilizer, and metals and minerals indexes. • **Agriculture,** in addition to food, beverages, and agricultural raw materials, includes sugar, bananas, beef, and oranges. • **Beverages** include cocoa, coffee, and tea. • **Food** includes rice, wheat, maize, sorghum, soybeans, soybean oil, soybean meal, palm oil, coconut oil, and groundnut oil. • **Agricultural raw materials** include timber (logs and sawnwood), cotton, natural rubber, and tobacco. • **Fertilizers** include phosphate rock and triple superphosphate (TSP). • **Metals and minerals** include aluminum, copper, iron ore, lead, nickel, tin, and zinc.

• **Petroleum price index** refers to the average spot price of Brent, Dubai, and West Texas Intermediate crude oil, equally weighted. • **Steel products price index** is the composite price index for eight steel products based on quotations f.o.b. (free on board) Japan excluding shipments to China and the United States, weighted by product shares of apparent combined consumption (volume of deliveries) for Germany, Japan, and the United States. • **MUV G-5 index** is the manufactures unit value index for G-5 country exports to developing countries. • **Commodity prices**—for definitions and sources see the World Bank's quarterly *Commodity Markets and the Developing Countries.*

Commodity price data are compiled by the World Bank's Development Prospects Group. More information can be obtained from its quarterly *Commodity Markets and the Developing Countries.* The MUV G-5 index is constructed by the Development Prospects Group. Monthly updates of commodity prices are available on the World Wide Web at www.worldbank.org/html/ieccp/ieccp.html.

Exports within bloc
$ millions

	1970	1980	1985	1990	1993	1994	1995	1996
High-income and low- and middle-income economies								
APEC	56,020	353,778	491,623	897,427	1,200,684	1,407,314	1,644,931	1,706,692
European Union	76,451	459,469	421,641	985,128	890,933	1,027,540	1,259,688	1,275,696
NAFTA	22,078	102,218	143,191	226,273	301,531	352,335	394,472	436,805
Latin America and the Caribbean								
Andean Group	97	1,161	768	1,312	2,892	3,752	4,751	4,806
CACM	287	1,174	544	671	1,088	1,175	1,365	1,566
CARICOM	51	431	358	395	441	193	228	841
LAIA	1,263	10,981	7,139	12,331	23,694	28,300	34,408	38,617
MERCOSUR	451	3,424	1,953	4,127	10,067	12,049	14,180	17,151
OECS	..	4	10	30	36	40	45	53
Africa								
CEMAC	22	75	84	139	102	115	129	142
CEPGL	3	2	9	7	10	10	8	9
COMESA	239	592	400	847	808	1,025	1,270	1,479
ECCAS	29	88	118	158	122	136	153	169
ECOWAS	86	692	1,026	1,533	1,686	1,628	1,949	2,345
MRU	1	7	4	0	1	1	1	1
SADC	76	96	294	942	2,245	2,671	3,872	4,231
UEMOA	52	444	395	603	401	402	515	638
Middle East and Asia								
ASEAN	1,201	12,016	13,130	26,367	41,749	56,199	71,094	77,221
Bangkok Agreement	47	612	984	1,511	2,790	2,399	3,901	3,968
ECO	24	369	2,431	1,239	2,770	2,662	3,782	4,496
GCC	117	4,632	3,101	6,906	5,023	5,296	5,782	5,723
SAARC	99	613	601	862	1,191	1,434	2,024	2,123
UMA	60	109	274	958	795	969	1,067	1,143

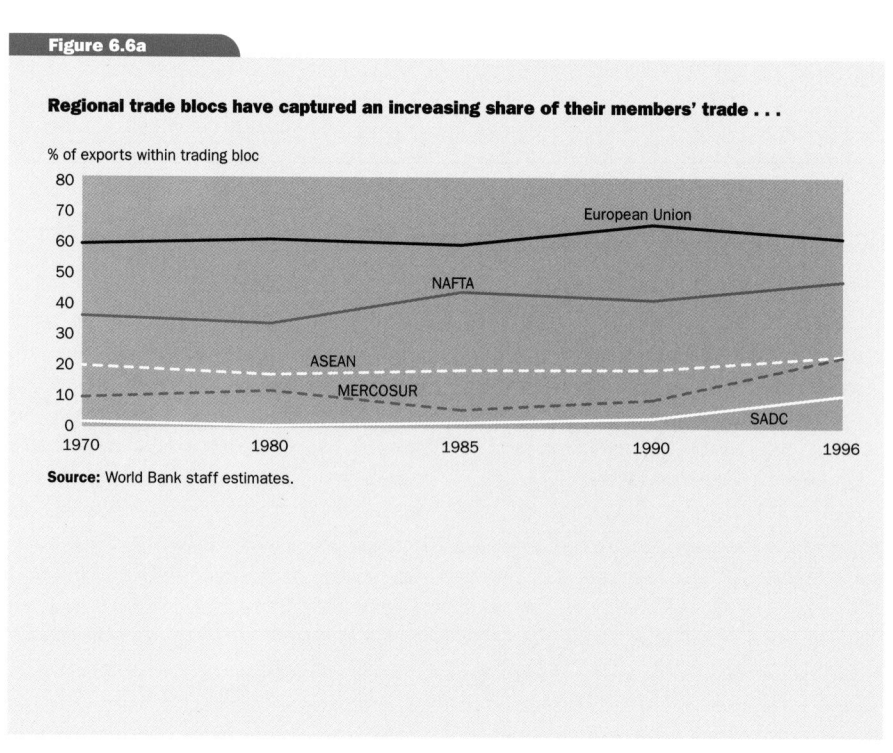

Figure 6.6a

Regional trade blocs have captured an increasing share of their members' trade . . .

% of exports within trading bloc

Source: World Bank staff estimates.

Regional trade blocs | 6.6

Exports within bloc
% of total exports

	1970	1980	1985	1990	1993	1994	1995	1996
High-income and low- and								
middle-income economies								
APEC	56.9	57.6	67.7	68.5	71.2	73.2	73.0	73.1
European Union	59.5	61.0	59.3	66.0	61.7	62.1	62.4	61.5
NAFTA	36.0	33.6	43.9	41.4	45.8	47.9	46.2	47.5
Latin America and the Caribbean								
Andean Group	1.8	3.8	3.2	3.8	9.8	10.5	11.8	10.4
CACM	26.0	24.4	14.4	15.3	16.9	16.7	14.1	15.7
CARICOM	4.6	4.2	5.8	7.8	8.8	3.7	3.8	12.9
LAIA	9.9	13.7	8.3	10.6	16.3	16.4	16.6	16.5
MERCOSUR	9.4	11.6	5.5	8.9	18.5	19.2	20.2	22.8
OECS	..	9.2	6.5	8.2	9.5	12.0	12.3	11.6
Africa								
CEMAC	4.9	1.6	1.9	2.3	2.0	2.1	2.2	1.9
CEPGL	0.4	0.1	0.8	0.5	0.8	0.7	0.5	0.5
ECCAS	2.4	1.4	2.1	2.0	1.9	2.0	2.0	1.9
ECOWAS	2.9	10.1	5.2	7.8	9.0	8.5	9.1	8.7
MRU	0.2	0.8	0.4	0.0	0.1	0.0	0.0	0.0
SADC	1.4	0.3	1.4	2.9	7.1	8.0	10.3	10.4
COMESA	7.5	10.3	5.4	7.6	7.9	8.7	9.3	9.3
UEMOA	6.6	9.3	8.7	12.7	9.4	9.5	9.5	9.3
Middle East and Asia								
ASEAN	19.7	16.9	18.4	18.7	20.0	22.0	22.8	23.2
Bangkok Agreement	1.5	2.2	2.4	1.7	2.6	1.9	2.4	2.3
ECO	1.7	5.9	9.9	3.2	6.3	5.4	6.4	6.7
GCC	4.6	3.0	4.9	8.0	5.6	5.8	5.4	4.6
SAARC	3.2	4.8	4.5	3.2	3.7	3.8	4.4	4.3
UMA	1.4	0.3	1.0	2.9	3.1	3.8	3.6	3.4

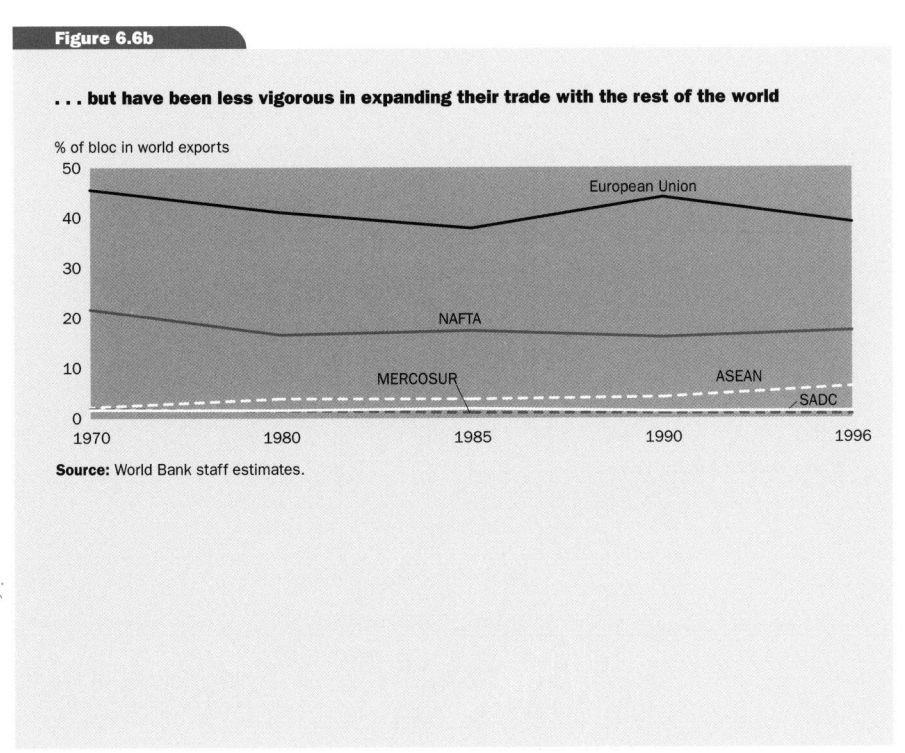

Figure 6.6b

. . . but have been less vigorous in expanding their trade with the rest of the world

Source: World Bank staff estimates.

Total exports by bloc
% of world exports

	1970	1980	1985	1990	1993	1994	1995	1996
High-income and low- and middle-income economies								
APEC	34.9	33.5	38.7	38.7	45.3	45.2	44.4	44.1
European Union	45.6	41.1	37.9	44.1	38.8	38.9	39.8	39.1
NAFTA	21.7	16.6	17.4	16.1	17.7	17.3	16.8	17.4
Latin America and the Caribbean								
Andean Group	1.9	1.7	1.3	1.0	0.8	0.8	0.8	0.9
CACM	0.4	0.3	0.2	0.1	0.2	0.2	0.2	0.2
CARICOM	0.4	0.6	0.3	0.1	0.1	0.1	0.1	0.1
LAIA	4.5	4.4	4.6	3.4	3.9	4.1	4.1	4.4
MERCOSUR	1.7	1.6	1.9	1.4	1.5	1.5	1.4	1.4
OECS	..	0.0	0.0	0.0	0.0	0.0	0.0	0.0
Africa								
CEMAC	0.2	0.3	0.2	0.2	0.1	0.1	0.1	0.1
CEPGL	0.3	0.1	0.1	0.0	0.0	0.0	0.0	0.0
COMESA	1.1	0.3	0.4	0.3	0.3	0.3	0.3	0.3
ECCAS	0.4	0.3	0.3	0.2	0.2	0.2	0.1	0.2
ECOWAS	1.1	0.4	1.0	0.6	0.5	0.5	0.4	0.5
MRU	0.1	0.0	0.1	0.1	0.0	0.0	0.0	0.0
SADC	1.9	1.5	1.1	1.0	0.9	0.8	0.7	0.8
UEMOA	0.3	0.3	0.2	0.1	0.1	0.1	0.1	0.1
Middle East and Asia								
ASEAN	2.2	3.9	3.8	4.2	5.6	6.0	6.1	6.3
Bangkok Agreement	1.1	1.5	2.2	2.6	2.9	3.0	3.2	3.2
ECO	0.5	0.3	1.3	1.1	1.2	1.2	1.2	1.3
GCC	0.9	8.5	3.4	2.5	2.4	2.2	2.1	2.4
SAARC	1.1	0.7	0.7	0.8	0.9	0.9	0.9	0.9
UMA	1.5	2.3	1.5	1.0	0.7	0.6	0.6	0.6

About the data

The table shows the value of exports by bloc members to one another—sometimes called intratrade—and the size of intratrade relative to the bloc's total exports of goods. Service exports are not included. Also shown is the share of the bloc's total exports in world exports.

Data on country exports are drawn from the International Monetary Fund's (IMF) *Direction of Trade Statistics* database and should be broadly consistent with other sources such as the United Nations Commodity Trade database. However, trade flows between many developing countries, particularly in Africa, are not well recorded. Thus the value of intra-trade within certain groups may be understated. Data on trade between developing countries and high-income countries are generally complete. Although bloc exports have been calculated back to 1970 based on current membership, most of the blocs came into existence at later dates and their membership may have changed over time.

Definitions

• **Exports within bloc** are the sum of exports by members of a trading bloc to other members of the bloc. Both the value in U.S. dollars and the share of exports within the bloc as a percentage of total exports by the bloc are shown. • **Total exports by bloc as a share of world exports** are the ratio of the bloc's total exports (exports within the bloc and to the rest of the world) to total exports by all economies in the world. • **Regional bloc memberships: Asia Pacific Economic Cooperation (APEC),** Australia, Brunei, Darussalam, Canada, Chile, China, Hong Kong (China), Indonesia, Japan, the Republic of Korea, Malaysia, Mexico, New Zealand, Papua New Guinea, the Philippines, Singapore, Taiwan (China), Thailand, and the United States; **European Union,** Austria, Belgium, Denmark, France, Finland, Germany, Greece, Ireland, Italy, the Netherlands, Luxembourg, Portugal, Spain, Sweden, and the United Kingdom; **North American Free Trade Association (NAFTA),** Canada, Mexico, and the United States; **Andean Group,** Colombia, Ecuador, Peru, and Venezuela; **Central American Common Market (CACM),** Costa Rica, El Salvador, Guatemala, Honduras, and Nicaragua; **Caribbean Community (CARICOM),** Antigua and Barbuda, Bahamas, Barbados, Belize, Dominica, Grenada, Guyana, Jamaica, Montserrat, St. Kitts and Nevis, St. Lucia, St. Vincent and the Grenadines, and Trinidad and Tobago; **Latin American Integration Association (LAIA),** Argentina, Bolivia, Brazil, Chile, Colombia, Ecuador, Mexico, Paraguay, Peru, Uruguay, and Venezuela; **Southern Common Market (MERCOSUR),** Argentina, Brazil, Paraguay, and Uruguay; **Organization of Eastern Caribbean States (OECS),** Antigua and Barbuda, Dominica, Grenada, Montserrat, St. Kitts and Nevis, St. Lucia, and St. Vincent and the Grenadines; **Economic and Monetary Community of Central Africa (CEMAC),** Cameroon, the Central African Republic, Chad, Republic of Congo, Equatorial Guinea, and Gabon; **Economic Community of the Great Lakes Countries (CEPGL),** Burundi, the Democratic Republic of Congo, and Rwanda; **Common Market for Eastern and Southern Africa (COMESA),** Angola, Burundi, Comoros, Djibouti, Eritrea, Ethiopia, Kenya, Lesotho, Madagascar, Malawi, Mauritius, Mozambique, Namibia, Rwanda, Seychelles, Somalia, Sudan, Swaziland, Uganda, Tanzania, Zambia, and Zimbabwe; **Economic Community of Central African States (ECCAS),** Burundi, Cameroon, the Central African Republic, Chad, the Democratic Republic of Congo, Republic of Congo, Equatorial Guinea, Gabon, Rwanda, and São Tomé and Principe; **Economic

Community of West African States (ECOWAS), Benin, Burkina Faso, Cape Verde, Côte d'Ivoire, Gambia, Ghana, Guinea, Guinea-Bissau, Liberia, Mali, Mauritania, Niger, Nigeria, Senegal, Sierra Leone, and Togo; **Mano River Union (MRU),** Guinea, Liberia, and Sierra Leone; **Southern African Development Community (SADC),** Angola, Botswana, Lesotho, Malawi, Mauritius, Mozambique, Namibia, Swaziland, South Africa, Tanzania, Zambia, and Zimbabwe; **West African Economic and Monetary Union (UEMOA),** Benin, Burkina Faso, Côte d'Ivoire, Somalia, Niger, Senegal, and Togo; **Association of South-East Asian Nations (ASEAN),** Brunei, Darussalam, Indonesia, Malaysia, the Philippines, Singapore, and Thailand; **Bangkok Agreement (First Agreement on Trade Negotiation Developing Member Countries of the Economic and Social Commission for Asia and the Pacific),** Bangladesh, India, the Republic of Korea, the Lao People's Democratic Republic, and Sri Lanka; **Economic Cooperation Organization (ECO),** Afghanistan, Azerbaijan, the Islamic Republic of Iran, Kazakhstan, the Kyrgyz Republic, Pakistan, Tajikistan, Turkmenistan, Turkey, and Uzbekistan; **Gulf Cooperation Council (GCC),** Bahrain, Kuwait, Oman, Qatar, Saudi Arabia, and the United Arab Emirates; **South Asian Association for Regional Cooperation (SAARC),** Bangladesh, Bhutan, India, Maldives, Nepal, Pakistan, and Sri Lanka; and **Arab Maghreb Union (UMA),** Algeria, Libya, Mauritania, Morocco, and Tunisia.

Data sources

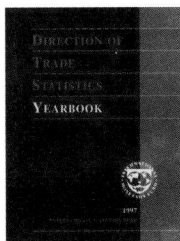

Data on merchandise trade flows are published in the IMF's *Direction of Trade Statistics Yearbook*. Definitions of the regional blocs are from UNCTAD's *Handbook of International Trade Statistics 1995* (1997). UNCTAD also publishes data on intratrade in the *Handbook.*

6.7 | Tariff barriers

		All products			Primary products			Manufactured products		
	Year	Mean tariff %	Standard deviation of tariff rates %	Weighted mean tariff %	Mean tariff %	Standard deviation of tariff rates %	Weighted mean tariff %	Mean tariff %	Standard deviation of tariff rates %	Weighted mean tariff %
Albania	1997	15.9	8.3	13.6	14.5	6.9	13.2	16.3	8.6	13.8
Argentina	1992	11.8	7.4	11.2	8.2	4.8	5.0	12.7	7.7	13.2
	1993	10.9	5.0	11.3	6.0	2.9	3.7	12.1	4.7	13.7
	1995	10.5	7.6	10.5	8.5	5.3	5.6	10.9	7.9	12.1
	1996	11.2	7.0	11.3	8.5	5.3	5.7	11.8	7.1	12.6
	1997	11.3	6.8	11.3	8.4	5.3	5.6	11.9	6.9	12.7
Australia	1991	12.9	15.1	10.4	2.6	5.3	1.4	15.5	15.7	13.1
	1993	9.8	11.9	7.7	2.5	4.8	1.3	11.7	12.4	9.7
	1996	6.0	9.0	4.0	1.2	2.3	0.7	7.2	9.6	4.8
	1997	5.7	8.2	3.9	1.2	2.2	0.7	6.8	8.7	4.6
Austria	1990	8.7	7.5	7.9	7.9	9.9	4.8	9.0	6.5	8.8
Bangladesh	1989	114.0	84.9	114.2	85.1	58.7	76.1	123.2	89.8	125.5
	1993	4.1	10.3	2.6	0.5	3.7	0.0	11.6	14.6	8.3
Belarus	1997	12.6	8.4	13.7	11.0	6.7	8.4	13.3	8.9	15.0
Bolivia	1993	9.8	1.0	9.8	10.0	0.2	10.0	9.7	1.1	9.7
	1994	10.0	0.2	10.0	10.0	0.1	10.0	10.0	0.2	10.0
	1995	9.7	1.1	9.7	10.0	0.1	10.0	9.7	1.3	9.7
	1996	9.7	1.3	9.7	10.0	0.2	10.0	9.6	1.4	9.6
	1997	9.7	1.3	9.7	10.0	0.2	10.0	9.6	1.4	9.6
Brazil	1991	25.1	17.3	26.7	19.8	19.8	9.4	26.3	16.3	32.2
	1994	11.9	8.2	14.6	8.2	7.1	7.2	12.8	8.2	17.0
	1995	12.0	6.9	12.9	8.7	5.4	8.2	12.7	7.0	14.4
	1996	12.2	8.5	15.8	8.7	6.0	7.4	13.0	8.8	18.8
	1997	11.9	7.5	14.8	8.7	5.7	6.9	12.6	7.7	17.6
Canada	1989	8.8	7.2	7.0	5.0	5.9	2.9	9.9	7.1	8.2
	1993	8.7	7.0	6.8	4.7	5.8	2.7	9.7	6.9	8.0
	1995	10.1	24.2	7.2	14.2	49.3	5.5	8.9	6.6	7.7
	1996	8.5	24.4	6.3	13.5	49.3	5.8	7.0	6.7	6.4
	1997	6.0	8.1	5.2	4.0	12.4	3.0	6.5	6.4	5.9
Central African Republic	1995	18.6	9.6	17.1	20.6	9.7	16.2	17.9	9.5	17.4
Chile	1992	11.0	0.7	10.9	11.0	0.0	11.0	10.9	0.8	10.8
	1993	11.0	0.7	10.9	11.0	0.0	11.0	10.9	0.8	10.8
	1994	11.0	0.7	10.9	11.0	0.0	11.0	10.9	0.8	10.8
	1995	11.0	0.7	10.9	11.0	0.0	11.0	10.9	0.8	10.8
	1997	11.0	0.6	10.9	11.0	0.0	11.0	10.9	0.7	10.8
China	1992	42.9	32.1	40.6	36.2	26.2	22.3	44.9	33.4	46.5
	1993	39.9	29.9	38.4	33.3	24.7	20.9	41.8	31.0	44.0
	1994	36.3	27.9	35.5	32.1	24.3	19.6	37.6	28.8	40.6
	1996	23.9	17.6	25.4	25.1	22.1	19.4	23.6	16.0	27.6
	1997	17.8	13.2	20.9	17.8	18.2	19.9	17.8	11.2	21.2
Colombia	1991	6.7	8.3	8.2	8.9	10.0	5.8	6.1	7.7	8.9
	1992	11.7	6.3	11.9	12.0	5.9	10.5	11.7	6.4	12.4
	1994	11.8	6.3	12.0	12.1	6.0	10.6	11.7	6.4	12.4
	1995	13.3	4.9	12.7	12.7	5.9	10.7	13.5	4.6	13.4
	1996	11.7	6.3	12.1	12.2	6.2	10.5	11.6	6.4	12.7
	1997	11.7	6.3	12.1	12.2	6.2	10.5	11.6	6.3	12.7
Côte d'Ivoire	1997	4.8	1.1	4.6	4.8	1.1	4.5	4.7	1.1	4.7
Cuba	1997	10.7	6.9	9.7	7.6	7.3	4.9	11.6	6.5	11.5
Czech Republic	1996	7.0	7.2	5.9	8.4	11.6	4.8	6.4	4.2	6.2
Ecuador	1993	9.3	6.0	8.5	9.4	5.8	7.6	9.2	6.1	8.8
	1994	11.8	6.3	11.2	12.2	6.1	10.3	11.7	6.4	11.5
	1995	12.3	5.6	11.9	12.2	6.2	10.4	12.4	5.4	12.4
	1996	11.4	6.4	11.8	11.6	6.5	9.9	11.3	6.4	12.5
	1997	11.4	6.4	11.8	11.6	6.5	9.9	11.3	6.4	12.5
El Salvador	1995	10.2	7.6	8.5	11.5	6.4	8.8	9.8	8.0	8.4
Estonia	1995	0.1	1.2	0.4	0.1	1.1	0.0	0.1	1.2	0.5
European Union	1989	8.7	7.2	7.0	10.1	10.4	5.0	8.2	5.9	7.5

Tariff barriers | 6.7

	Year	All products			Primary products			Manufactured products		
		Mean tariff %	Standard deviation of tariff rates %	Weighted mean tariff %	Mean tariff %	Standard deviation of tariff rates %	Weighted mean tariff %	Mean tariff %	Standard deviation of tariff rates %	Weighted mean tariff %
	1994	7.7	6.3	6.6	10.3	10.6	4.9	6.9	3.8	7.0
	1995	6.7	5.9	6.3	8.3	10.0	5.1	6.2	4.1	6.6
	1996	6.8	6.1	4.9	10.4	8.8	3.7	5.5	3.8	5.3
	1997	6.9	6.0	5.0	10.4	8.8	3.8	5.6	3.8	5.3
Finland	1988	7.7	10.1	5.6	6.6	11.2	2.7	8.0	9.7	6.6
	1990	7.8	10.2	5.6	6.8	11.4	2.6	8.1	9.8	6.6
Hungary	1991	12.6	11.0	12.3	14.7	16.2	10.2	12.1	8.9	13.0
	1997	14.4	16.9	10.9	28.3	25.3	12.6	8.9	6.1	10.3
Iceland	1996	4.6	13.0	3.1	8.8	22.2	4.1	2.8	4.8	2.8
India	1990	81.8	39.4	83.0	74.1	38.4	49.5	84.1	39.4	93.6
	1997	30.0	14.0	27.7	25.7	22.6	22.6	31.3	9.8	29.5
Indonesia	1993	19.4	16.1	21.7	16.7	12.3	10.0	20.3	17.0	25.4
	1996	13.2	17.0	17.6	12.3	19.4	8.5	13.5	16.1	20.9
Japan	1989	6.7	7.8	3.4	8.9	10.2	4.8	6.0	6.7	3.1
	1991	6.3	8.7	3.0	9.4	11.9	5.3	5.3	6.9	2.5
	1992	6.3	8.4	3.0	9.2	11.2	5.1	5.3	7.0	2.5
	1993	6.3	8.3	3.0	9.2	10.9	4.9	5.3	7.0	2.5
	1994	6.3	8.3	2.9	9.2	10.9	4.9	5.3	6.9	2.5
	1995	6.3	8.4	2.6	9.7	11.2	3.6	5.1	6.7	2.3
	1996	6.0	7.7	2.8	9.2	10.6	4.4	4.8	5.8	2.4
	1997	6.0	8.0	2.7	9.1	10.7	4.0	4.8	6.4	2.3
Korea, Rep.	1990	13.3	6.7	12.3	15.7	13.0	11.0	12.7	2.9	12.6
	1992	11.6	6.5	10.7	14.6	12.6	10.3	10.8	2.2	10.8
	1996	11.3	27.0	10.1	21.2	48.4	16.7	8.2	14.0	7.8
Latvia	1997	6.0	10.7	3.2	11.0	16.1	3.9	3.7	5.8	2.9
Lithuania	1995	4.5	9.0	2.5	8.7	13.0	3.6	2.8	5.9	2.1
	1997	4.6	9.2	2.6	8.6	13.1	3.6	2.8	5.9	2.2
Malawi	1994	30.8	15.5	27.1	24.6	15.6	14.8	32.8	15.0	32.2
	1997	25.3	11.6	23.5	21.2	12.6	13.6	26.5	11.0	26.7
Malaysia	1988	17.0	14.2	12.6	15.8	10.7	6.3	17.6	15.6	14.4
	1991	16.9	14.7	12.5	15.3	10.6	6.0	17.8	16.3	14.4
	1993	14.3	14.1	11.1	10.9	12.7	6.0	15.3	14.3	12.6
	1997	9.1	19.4	11.9	4.1	21.2	9.4	12.2	17.5	12.8
Mauritius	1997	29.1	26.2	31.9	19.7	19.1	19.1	31.7	27.3	36.0
Mexico	1991	13.0	4.4	13.1	12.1	5.7	10.8	13.2	4.2	13.8
	1995	12.6	5.4	11.8	12.3	6.0	10.8	12.6	5.3	12.2
	1996	13.1	10.6	13.2	14.5	19.6	12.9	12.8	7.8	13.3
	1997	13.1	10.6	13.2	14.5	19.6	13.0	12.8	7.8	13.3
Mozambique	1997	15.6	14.3	14.1	16.9	15.1	12.0	15.3	14.0	14.8
New Zealand	1992	8.7	10.6	8.1	4.4	6.3	2.3	9.9	11.3	9.9
	1993	8.5	10.3	7.7	4.3	6.0	2.1	9.7	11.0	9.4
	1996	6.2	8.0	6.0	3.0	4.2	1.4	7.2	8.6	7.6
	1997	5.4	7.1	5.2	2.5	3.7	1.2	6.3	7.6	6.7
Nigeria	1988	33.7	26.1	29.9	31.0	22.5	24.2	34.5	27.0	31.5
	1989	35.6	31.0	34.2	33.1	30.1	26.0	36.3	31.1	36.6
	1990	35.7	30.8	34.3	33.2	30.1	26.1	36.5	31.0	36.8
	1992	34.4	25.1	31.7	31.4	22.7	24.5	35.3	25.7	33.9
Norway	1988	6.0	7.1	5.0	1.6	4.9	0.8	7.1	7.1	6.4
	1993	6.1	7.0	5.1	1.7	5.0	0.8	7.1	7.1	6.5
	1995	5.9	10.7	4.9	2.4	20.6	0.9	6.7	6.6	6.0
	1996	6.1	15.5	4.3	4.9	30.9	1.6	6.3	8.5	5.2
Papua New Guinea	1997	20.7	19.1	21.3	30.4	23.0	23.1	17.8	16.7	20.6
Paraguay	1991	15.9	12.9	14.4	15.6	13.8	10.2	15.9	12.6	15.8
	1994	8.0	7.7	8.1	7.9	7.4	5.3	8.1	7.8	8.9
	1995	9.3	6.9	9.0	8.2	5.2	5.2	9.5	7.2	10.3
	1996	9.4	7.1	9.2	8.9	6.2	6.0	9.5	7.3	10.4
	1997	9.6	6.7	9.3	8.7	5.8	5.7	9.7	6.9	10.6

		All products			Primary products			Manufactured products		
	Year	Mean tariff %	Standard deviation of tariff rates %	Weighted mean tariff %	Mean tariff %	Standard deviation of tariff rates %	Weighted mean tariff %	Mean tariff %	Standard deviation of tariff rates %	Weighted mean tariff %
Peru	1993	17.6	4.4	17.1	17.3	4.2	16.3	17.7	4.4	17.3
	1997	13.3	2.9	12.8	13.8	3.4	13.0	13.1	2.8	12.7
Philippines	1988	28.2	15.1	27.2	28.9	16.2	22.0	28.0	14.7	28.8
	1989	28.2	15.1	27.2	29.0	16.2	22.0	28.0	14.7	28.9
	1990	19.7	9.2	18.3	20.4	9.8	15.7	19.4	9.0	19.2
	1992	19.6	9.2	18.3	20.1	9.8	15.5	19.4	9.0	19.2
	1993	22.5	14.1	20.2	23.9	15.3	17.9	22.1	13.7	21.0
	1994	21.6	13.3	19.5	22.3	14.0	16.8	21.5	13.1	20.4
	1995	27.6	4.9	27.0	28.9	4.5	22.7	27.2	5.0	27.9
Poland	1991	11.7	8.9	10.4	10.4	10.3	7.4	12.1	8.5	11.4
	1992	11.7	8.9	10.4	10.4	10.3	7.4	12.1	8.5	11.4
	1996	18.4	27.5	15.2	28.0	46.1	16.5	14.1	9.0	14.8
Russian Federation	1993	7.3	9.8	9.7	3.8	12.2	5.7	8.7	8.2	10.9
	1994	11.5	12.4	14.0	8.0	8.4	5.1	12.9	13.4	16.8
	1997	12.7	8.3	13.4	10.8	6.4	8.2	13.5	8.9	15.3
Singapore	1989	0.5	2.7	2.3	0.2	2.4	1.0	0.6	2.8	2.7
South Africa[a]	1988	12.7	11.8	11.3	7.2	9.8	5.0	13.6	11.9	13.1
	1990	11.0	11.3	10.9	6.9	9.6	4.7	11.7	11.4	12.7
	1991	10.5	11.8	11.3	6.8	10.6	4.8	11.2	11.9	13.3
	1993	19.7	21.9	14.3	9.4	11.7	5.2	21.2	22.6	16.9
	1997	8.8	11.0	8.4	8.0	11.4	4.2	9.0	10.9	9.9
Sri Lanka	1990	28.3	25.5	24.1	31.4	28.7	30.2	27.5	24.5	22.2
	1993	24.2	18.1	23.0	26.8	21.9	25.3	23.5	16.8	22.3
	1997	20.0	15.4	20.7	23.8	23.0	23.6	19.1	12.6	19.8
Sweden	1988	4.8	4.8	3.8	1.7	4.2	0.4	5.6	4.7	4.8
	1989	4.8	4.8	3.8	1.7	4.2	0.4	5.6	4.7	4.8
Switzerland	1990	0.0	0.0	0.0	0.0	0.0	0.0	0.0	0.0	0.0
	1993	0.0	0.0	0.0	0.0	0.0	0.0	0.0	0.0	0.0
	1995	0.0	0.0	0.0	0.0	0.0	0.0	0.0	0.0	0.0
	1996	0.0	0.0	0.0	0.0	0.0	0.0	0.0	0.0	0.0
	1997	0.0	0.0	0.0	0.0	0.0	0.0	0.0	0.0	0.0
Tanzania	1997	21.6	14.0	22.0	30.6	10.9	22.1	19.6	13.8	22.0
Thailand	1989	39.8	23.0	38.7	32.7	19.1	25.5	41.7	23.6	42.4
	1991	39.9	23.0	38.8	32.8	19.1	25.5	41.8	23.6	42.4
	1993	45.6	25.0	41.5	40.3	19.4	33.9	47.2	26.2	43.7
Trinidad and Tobago	1991	18.6	15.3	16.5	22.2	18.2	13.2	17.4	14.1	17.6
	1992	18.7	15.3	16.7	22.9	18.0	14.1	17.4	14.1	17.5
Ukraine	1997	10.1	11.0	7.2	15.4	12.1	5.9	7.7	9.5	7.7
United States	1989	6.3	6.1	4.4	4.5	6.4	2.4	6.7	5.9	4.8
	1990	6.3	6.1	4.4	4.6	6.5	2.4	6.7	5.9	4.8
	1991	6.3	6.1	4.4	4.6	6.5	2.4	6.7	5.9	4.8
	1992	6.3	6.1	4.4	4.6	6.5	2.4	6.7	5.9	4.8
	1993	6.4	6.1	4.4	4.6	6.5	2.4	6.7	5.9	4.8
	1995	5.9	7.8	4.1	5.5	10.9	2.7	6.0	6.9	4.4
	1996	6.0	12.4	4.2	6.9	25.7	3.4	5.8	5.8	4.4
Uruguay	1992	6.5	5.9	5.5	7.2	6.2	4.3	6.2	5.8	5.9
	1995	9.3	7.1	8.9	8.4	5.4	5.3	9.5	7.4	10.1
	1996	9.7	7.3	9.6	8.6	5.6	5.3	9.9	7.6	11.2
	1997	10.2	6.9	9.8	8.8	5.3	5.3	10.5	7.3	11.4
Venezuela	1992	15.7	11.3	16.1	15.8	10.3	12.7	15.7	11.5	16.9
	1995	13.4	4.8	12.8	12.8	5.8	10.9	13.5	4.5	13.4
	1997	12.0	6.1	12.4	12.3	6.1	10.7	11.9	6.1	13.0
Zambia	1997	13.6	9.3	14.0	15.7	8.7	12.1	13.1	9.3	14.7
Zimbabwe	1997	24.3	23.4	23.4	21.7	20.2	16.6	25.1	24.3	25.8

a. Data are for the South African Customs Union, which includes Botswana, Lesotho, Namibia, and South Africa.

Economies regulate their imports through a combination of tariff and nontariff measures. The most common form of tariff is an ad valorem duty, but tariffs may also be levied on a specific, or per unit, basis. Tariffs may be used to raise fiscal revenues or to protect domestic industries from foreign competition—or both. Nontariff barriers, which limit the quantity of imports of a particular good, take many forms. Some common ones are licensing schemes, quotas, prohibitions, export restraint arrangements, and health and quarantine measures.

Mean tariffs are calculated as the average ad valorem duty across all tariff lines. Specific duties—duties not expressed as a proportion of the declared value—are not included. Countries typically maintain a hierarchy of trade preferences applicable to specific trading partners. The rates used in calculating the indicators here are the applied most-favored-nation duties. Applied rates are less than or equal to the bound rates that countries have agreed to in World Trade Organization negotiations, but they may exceed the rates applied to partners in preferential trade agreements such as the North American Free Trade Agreement. (See table 6.6 for the membership of regional trade blocs and data on their exports.)

The table shows both simple average tariffs and average tariffs weighted by world imports. Simple averages are a better indicator of tariff protection than averages weighted by import values, which are biased downward, especially when tariffs are set so high as to discourage trade. Weights based on world imports provide an alternative measure of a country's tariff barriers that reflects average world trading patterns.

Some countries set fairly uniform tariff rates across all imports. Others are more selective, setting high tariffs to protect favored domestic industries and low tariffs on goods that have few domestic suppliers or that are necessary inputs for domestic industry. The standard deviation of tariffs is a measure of the dispersion of tariff rates around their mean value. Highly dispersed rates are evidence of discriminatory tariffs that may distort production and consumption decisions. But this tells only part of the story. The effective rate of protection—the degree to which the value added in an industry is protected—may exceed the nominal rate if the tariff system systematically differentiates among imports of raw materials, intermediate products, and finished goods.

Nontariff barriers are not shown in this table. (But see table 5.6 in *World Development Indicators 1997*

for estimates of the shares of tariff lines covered by them.) Nontarriff barriers are generally considered more detrimental to economic efficiency than tariffs because efficient foreign producers cannot undercut the barriers by reducing their costs and thus their prices. A high percentage of products subject to nontariff barriers indicates a protectionist trade regime, but the frequency of nontariff barriers does not measure their restrictiveness. Moreover, a wide range of domestic policies and regulations (such as health regulations) that are not measured by this indicator may act as nontariff barriers. A full evaluation would require careful analysis of the individual measures.

The indicators shown in this table were calculated using the new Software for Market Analysis and Restrictions on Trade (SMART) system. SMART contains tariff line data on bound, applied, and preferential duties for 66 countries. Data are classified using the Harmonized System of trade codes at the six- or eight-digit level. Tariff line data were matched to Standard International Trade Classification (SITC) revision 2 codes to define the commodity groups and global import weights. The SMART database is still under development. Data are shown only for those countries and years for which complete data are available.

• **Primary products** are commodities classified in SITC sections 0, 1, 2, 3, and 4 plus division 68 (nonferrous metals). • **Manufactured products** are commodities classified in SITC sections 5, 6, 7, 8, and 9, excluding division 68. • **Mean tariff** is the unweighted average of the applied rates for all products subject to tariffs. • **Standard deviation of tariff rates** measures the average dispersion of tariff rates around the simple mean. • **Weighted mean tariff** is the average of applied rates weighted by product shares in world imports.

Mean tariff rates and their standard deviations were calculated by World Bank staff using the SMART system. Data on tariffs come from the UNCTAD Trade Analysis Information System. Data on global imports come from the United Nations Commodity Trade database.

Figure 6.7a

Average tariffs are declining

Source: World Bank staff estimates.

Tariff rates have fallen in most countries since the start of the Uruguay Round of the General Agreement on Tariffs and Trade. Average tariffs are expected to fall still further as reductions negotiated during the Round are implemented.

6.8 Global financial flows

	Net private capital flows $ millions		Foreign direct investment $ millions		Portfolio investment flows				Bank and trade-related lending $ millions	
					Bonds $ millions		Equity $ millions			
	1990	1996	1990	1996	1990	1996	1990	1996	1990	1996
Albania	31	92	0	90	0	0	0	0	31	2
Algeria	−442	−72	0	4	−15	0	0	5	−427	−81
Angola	237	753	−335	300	0	0	0	0	572	453
Argentina	−203	14,417	1,836	4,285	−857	8,945	13	864	−1,196	323
Armenia	..	18	..	18	..	0	..	0	..	0
Australia	..	..	6,517	6,321	..	..	..	..	..	..
Austria	..	..	653	3,826	..	..	..	..	..	..
Azerbaijan	..	601	..	601	..	0	..	0	..	0
Bangladesh	70	92	3	15	0	0	0	30	67	47
Belarus	..	7	..	18	..	0	..	0	..	−11
Belgium	..	..	..	..	..	..	..	..	..	..
Benin	1	2	1	2	0	0	0	0	0	0
Bolivia	3	571	27	527	0	0	0	0	−24	44
Bosnia and Herzegovina	..	..	..	..	..	..	..	..	..	..
Botswana	77	66	95	75	0	0	0	0	−19	−9
Brazil	562	28,384	989	9,889	129	4,634	0	3,981	−556	9,880
Bulgaria	−42	300	4	115	65	−205	0	500	−111	−109
Burkina Faso	0	0	0	0	0	0	0	0	0	0
Burundi	−5	0	1	1	0	0	0	0	−6	−1
Cambodia	0	290	0	294	0	0	0	0	0	−3
Cameroon	−125	−28	−113	35	0	0	0	0	−12	−63
Canada	..	..	7,581	6,398	..	..	..	..	..	..
Central African Republic	0	5	1	5	0	0	0	0	−1	0
Chad	−1	18	0	18	0	0	0	0	−1	0
Chile	2,098	6,803	590	4,091	−7	1,859	320	103	1,194	750
China	8,107	50,100	3,487	40,180	−48	1,190	0	3,466	4,668	5,264
Hong Kong, China	..	..	..	..	..	..	..	..	..	..
Colombia	345	7,739	500	3,322	−4	1,844	0	290	−151	2,283
Congo, Dem. Rep.	−24	2	−12	2	0	0	0	0	−12	0
Congo, Rep.	−100	−7	0	8	0	0	0	0	−100	−15
Costa Rica	23	387	163	410	−42	−7	0	1	−99	−16
Côte d'Ivoire	57	160	48	21	−1	0	0	30	10	109
Croatia	..	915	..	349	..	22	..	111	..	433
Cuba	..	..	..	..	..	..	..	..	..	..
Czech Republic	876	4,894	207	1,435	0	171	0	164	669	3,124
Denmark	..	..	1,132	773	..	..	..	..	..	..
Dominican Republic	130	366	133	394	0	0	0	0	−3	−28
Ecuador	183	816	126	447	0	−10	0	1	57	377
Egypt, Arab Rep.	698	1,434	734	636	−1	0	0	1,233	−35	−435
El Salvador	8	48	2	25	0	0	0	0	6	23
Eritrea	..	0	..	0	..	0	..	0	..	0
Estonia	..	191	..	150	..	40	..	5	..	−4
Ethiopia[a]	−45	−205	12	5	0	0	0	0	−57	−210
Finland	..	..	812	1,118	..	..	..	..	..	..
France	..	..	13,183	21,972	..	..	..	..	..	..
Gabon	103	−114	74	−65	0	0	0	0	29	−49
Gambia, The	−7	11	0	11	0	0	0	0	−7	0
Georgia	..	40	..	40	..	0	..	0	..	0
Germany	..	..	2,532	−3,183	..	..	..	..	..	..
Ghana	−5	477	15	120	0	250	0	124	−20	−18
Greece	..	..	..	..	..	..	..	..	..	..
Guatemala	44	5	48	77	−11	−33	0	0	7	−39
Guinea	−1	41	18	24	0	0	0	0	−19	17
Guinea-Bissau	2	1	2	1	0	0	0	0	0	0
Haiti	8	4	8	4	0	0	0	0	0	0
Honduras	77	65	44	75	0	−13	0	0	33	3

Global financial flows | 6.8

	Net private capital flows $ millions		Foreign direct investment $ millions		Portfolio investment flows				Bank and trade-related lending $ millions	
					Bonds $ millions		Equity $ millions			
	1990	1996	1990	1996	1990	1996	1990	1996	1990	1996
Hungary	–308	1,618	0	1,982	921	–940	150	1,004	–1,379	–429
India	1,873	6,404	162	2,587	147	–457	105	4,398	1,459	–124
Indonesia	3,219	18,030	1,093	7,960	26	3,744	312	3,099	1,788	3,228
Iran, Islamic Rep.	–392	–352	–362	10	0	0	0	0	–30	–362
Iraq	..	..	..	..	..	..	..	..	..	..
Ireland	..	..	627	2,456	..	..	..	..	..	..
Israel	..	..	101	2,110	..	..	..	..	..	..
Italy	..	..	6,411	3,523	..	..	..	..	..	..
Jamaica	92	191	138	175	0	53	0	0	–46	–36
Japan	..	..	1,777	200	..	..	..	..	..	..
Jordan	254	–119	38	16	0	–5	0	25	216	–154
Kazakhstan	..	615	..	310	..	200	..	0	..	105
Kenya	124	–104	57	13	0	0	0	43	67	–160
Korea, Dem. Rep.	..	..	..	..	..	..	..	..	..	..
Korea, Rep.	..	..	788	2,325	..	..	..	..	..	..
Kuwait	..	..	..	..	..	..	..	..	..	..
Kyrgyz Republic	..	46	..	46	..	0	..	0	..	0
Lao PDR	6	104	6	104	0	0	0	0	0	0
Latvia	..	331	..	328	..	0	..	0	..	3
Lebanon	12	740	6	80	0	460	0	122	6	78
Lesotho	17	38	17	28	0	0	0	0	0	10
Libya	..	..	..	..	..	..	..	..	..	..
Lithuania	..	469	..	152	..	160	..	21	..	136
Macedonia, FYR	..	8	..	8	..	0	..	0	..	0
Madagascar	7	5	22	10	0	0	0	0	–15	–5
Malawi	2	–3	0	1	0	0	0	0	2	–4
Malaysia	769	12,096	2,333	4,500	–1,239	2,062	293	4,353	–617	1,180
Mali	–8	23	–7	23	0	0	0	0	–1	0
Mauritania	6	25	7	5	0	0	0	0	–1	20
Mauritius	85	112	41	37	0	0	0	34	44	41
Mexico	8,240	23,647	2,634	7,619	661	11,344	563	3,922	4,382	763
Moldova	..	115	..	41	..	0	..	0	..	74
Mongolia	16	–15	0	5	0	0	0	0	16	–20
Morocco	337	388	165	311	0	293	0	222	172	–438
Mozambique	35	23	9	29	0	0	0	0	26	–6
Myanmar	153	129	161	100	0	0	0	10	–8	19
Namibia	..	..	..	..	..	..	..	..	..	..
Nepal	–9	9	6	19	0	0	0	0	–15	–10
Netherlands	..	..	12,343	7,824	..	..	..	..	..	..
New Zealand	..	..	1,735	280	..	..	..	..	..	..
Nicaragua	21	41	0	45	0	–8	0	0	21	4
Niger	9	–24	–1	0	0	0	0	0	10	–24
Nigeria	467	706	588	1,391	0	0	0	5	–121	–690
Norway	..	..	1,003	3,960	..	..	..	..	..	..
Oman	–259	69	141	67	0	0	0	25	–400	–24
Pakistan	182	1,936	244	690	0	150	0	700	–63	396
Panama	127	301	132	238	–2	75	0	5	–4	–17
Papua New Guinea	204	414	155	225	0	0	0	187	49	2
Paraguay	67	202	76	220	0	0	0	0	–9	–18
Peru	59	5,854	41	3,581	0	0	0	2,740	18	–467
Philippines	639	4,600	530	1,408	395	2,319	0	1,333	–286	–460
Poland	71	5,333	89	4,498	0	216	0	722	–18	–103
Portugal	..	..	2,610	618	..	..	..	..	..	..
Puerto Rico	..	..	..	..	..	..	..	..	..	..
Romania	4	1,814	0	263	0	1,029	0	11	4	510
Russian Federation	5,604	7,454	0	2,479	310	21	0	5,008	5,294	–54

6.8 Global financial flows

	Net private capital flows		Foreign direct investment		Portfolio investment flows				Bank and trade-related lending	
					Bonds $ millions		Equity $ millions			
	$ millions		$ millions						$ millions	
	1990	1996	1990	1996	1990	1996	1990	1996	1990	1996
Rwanda	6	1	8	1	0	0	0	0	−2	0
Saudi Arabia	..	..	..	..	..	..	..	..	..	..
Senegal	42	34	57	45	0	0	0	0	−15	−11
Sierra Leone	36	5	32	5	0	0	0	0	4	0
Singapore	..	..	5,575	9,440	..	..	..	..	..	..
Slovak Republic	278	1,265	0	281	0	380	0	0	278	604
Slovenia	..	1,219	..	186	..	163	..	360	..	510
South Africa	..	1,417	..	136	..	367	..	1,759	..	−845
Spain	..	..	13,984	6,396	..	..	..	..	..	..
Sri Lanka	54	123	43	120	0	0	0	70	11	−67
Sudan	0	0	0	0	0	0	0	0	0	0
Sweden	..	..	1,982	5,492	..	..	..	..	..	..
Switzerland	..	..	4,961	3,512	..	..	..	..	..	..
Syrian Arab Republic	18	77	71	89	0	0	0	0	−53	−12
Tajikistan	..	16	..	16	..	0	..	0	..	0
Tanzania	5	143	0	150	0	0	0	0	5	−7
Thailand	4,498	13,517	2,444	2,336	−87	3,774	449	1,551	1,692	5,856
Togo	0	0	0	0	0	0	0	0	0	0
Trinidad and Tobago	−69	343	109	320	−52	125	0	0	−126	−102
Tunisia	−122	697	76	320	−60	0	0	0	−138	377
Turkey	1,782	5,635	684	722	597	1,578	35	799	466	2,536
Turkmenistan	..	355	..	108	..	0	..	0	..	247
Uganda	16	114	0	121	0	0	0	0	16	−7
Ukraine	..	395	..	350	..	−80	..	0	..	125
United Arab Emirates	..	..	..	..	..	..	..	..	..	..
United Kingdom	..	..	32,427	32,347	..	..	..	..	..	..
United States	..	..	47,918	76,955	..	..	..	..	..	..
Uruguay	−192	499	0	169	−16	59	0	5	−176	266
Uzbekistan	..	431	..	55	..	0	..	0	..	376
Venezuela	−126	4,244	451	1,833	345	−51	0	1,740	−922	721
Vietnam	16	2,061	16	1,500	0	0	0	390	0	171
West Bank and Gaza	..	..	..	..	..	..	..	..	..	..
Yemen, Rep.	30	100	−131	100	0	0	0	0	161	0
Yugoslavia, FR (Serb./Mont.)[b]	1,836	0	0	0	−2	0	0	0	1,838	0
Zambia	194	33	203	58	0	0	0	0	−9	−25
Zimbabwe	85	42	−12	63	−30	−30	0	17	128	−8
World	.. s	.. s	238,969 s	552,616 s	.. s	.. s	.. s	.. s	.. s	.. s
Low income	11,625	65,176	4,683	49,531	67	1,082	105	9,283	6,770	5,280
Excl. China & India	..	..	..	..	..	..	..	..	..	..
Middle income	29,271	181,769	19,004	69,429	32	44,602	2,134	36,547	8,100	31,191
Lower middle income	..	..	..	..	..	..	..	..	..	..
Upper middle income	..	..	..	..	..	..	..	..	..	..
Low & middle income	41,881	246,944	23,687	118,960	100	45,684	3,225	45,830	14,870	36,471
East Asia & Pacific	18,443	101,272	10,347	58,681	−952	13,089	1,750	14,389	7,299	15,113
Europe & Central Asia	7,787	35,005	1,097	14,941	1,893	2,755	235	8,705	4,561	8,604
Latin America & Carib.	12,601	95,569	8,188	38,015	101	28,812	1,099	13,893	3,213	14,850
Middle East & N. Africa	646	1,979	2,757	614	−148	748	0	1,632	−1,963	−1,015
South Asia	2,173	8,743	464	3,439	147	−307	105	5,198	1,457	413
Sub-Saharan Africa	195	4,376	834	3,271	−941	586	0	2,012	302	−1,494
High income	..	..	167,908	195,736	..	..	..	..	..	..

Note: Totals for low- and middle-income economies may not sum to regional totals because of unallocated amounts.

a. Includes Eritrea. b. Data for 1990 refer to the former Yugoslavia.

Global financial flows |6.8

The data on foreign direct investment are based on balance of payments data reported by the International Monetary Fund (IMF), supplemented by data on net foreign direct investment reported by the Organisation for Economic Co-operation and Development (OECD) and official national sources. The internationally accepted definition of foreign direct investment is that provided in the fifth edition of the IMF's *Balance of Payments Manual* (1993). The OECD has also published a definition, in consultation with the IMF, Eurostat, and the United Nations. Foreign direct investment has three components: equity investment, reinvested earnings, and short- and long-term intercompany loans between parent firms and foreign affiliates. However, many countries fail to report reinvested earnings, and the definition of long-term loans differs among countries. Foreign direct investment, as distinguished from other kinds of international investment, is made to establish a lasting interest in or effective management control over an enterprise in another country. As a guideline, the IMF suggests that investments should account for at least 10 percent of the voting stock to be counted as foreign direct investment. In practice many countries set a higher threshold. Because of the multiplicity of sources and differences in definitions and reporting methods, there may be more than one estimate of foreign direct investment for a country and data may not be comparable across countries.

Foreign direct investment data do not give a complete picture of international investment in an economy. Balance of payments data on foreign direct investment do not include capital raised in the host economies, which has become an important source of financing for investment projects in some developing countries. There is also increasing awareness that foreign direct investment data are limited because they capture only cross-border investment flows involving equity participation and omit nonequity cross-border transactions such as intrafirm flows of goods and services. For a detailed discussion of the data issues see the World Bank's *World Debt Tables 1993–94* (volume 1, chapter 3).

Portfolio flow data are compiled from several official and market sources, including Euromoney databases and publications, Micropal Inc., Lipper Analytical Services, published reports of private investment houses, central banks, national securities and exchange commissions, national stock exchanges, and the World Bank's Debtor Reporting System (DRS).

Gross statistics on international bond and equity issues are produced by aggregating individual transactions reported by market sources. Transactions of public and publicly guaranteed bonds are reported through the DRS by member economies that have received either International Bank for Reconstruction and Development (IBRD) loans or International Development Association (IDA) credits. Information on private nonguaranteed bonds is collected from market sources, because official national sources reporting to the DRS are not asked to report the breakdown between private nonguaranteed bonds and private nonguaranteed loans. Information on transactions by nonresidents in local equity markets is gathered from national authorities, investment positions of mutual funds, and market sources.

The volume of portfolio investment reported by the World Bank generally differs from that reported by other sources because of differences in the classification of economies, in the sources, and in the method used to adjust and disaggregate reported information. Differences in reporting arise particularly for foreign investments in local equity markets, where there is a lack of clarity, adequate disaggregation, and lack of comprehensive and periodic reporting in many developing economies. By contrast, capital flows through international debt and equity instruments are well recorded, and the differences in reporting lie primarily in differences in the classification of economies, in the exchange rates used, in whether particular tranches of the transactions are included, or in the treatment of certain offshore issuances.

• **Net private capital flows** consist of private debt and nondebt flows. Private debt flows include commercial bank lending, bonds, and other private credits; nondebt private flows are foreign direct investment and portfolio equity investment. • **Foreign direct investment** is net inflows of investment to acquire a lasting management interest (10 percent or more of voting stock) in an enterprise operating in an economy other than that of the investor. It is the sum of equity capital, reinvestment of earnings, other long-term capital, and short-term capital as shown in the balance of payments. • **Portfolio investment flows** are net and include non-debt-creating portfolio equity flows (the sum of country funds, depository receipts, and direct purchases of shares by foreign investors) and portfolio debt flows (bond issues purchased by foreign investors). • **Bank and trade-related lending** covers commercial bank lending and other private credits.

Global Development Finance

The principal source of information for the table is reports to the World Bank's DRS from member economies that have received IBRD loans or IDA credits. These data are compiled and published in the World Bank's annual *Global Development Finance*. Additional information has been drawn from the data files of the World Bank and the IMF.

Net flows to part I countries	Official development assistance				Other official flows	Private flows						Total net flows
	Total	Bilateral grants	Bilateral loans	Contributions to multilateral institutions		Total	Foreign direct investment	Bilateral portfolio investment	Multilateral portfolio investment	Private export credits	Net grants by NGOs	
$ millions, 1996												
Australia	1,121	899	0	223	220	0	0	0	0	0	76	1,417
Austria	557	353	59	145	335	938	247	0	0	691	47	1,878
Belgium	913	528	2	384	94	4,528	461	4,194	0	−127	60	5,595
Canada	1,795	1,392	−35	439	489	1,859	2,024	−154	0	−11	302	4,446
Denmark	1,772	1,074	−16	715	−48	188	199	0	0	−11	36	1,949
Finland	408	218	−4	194	244	472	257	162	0	53	0	1,124
France	7,451	5,634	120	1,697	−284	11,115	4,657	5,352	0	1,106	0	18,283
Germany	7,601	4,507	29	3,066	194	12,336	3,456	6,980	187	1,712	1,044	21,175
Ireland	179	114	0	65	0	125	0	125	0	0	68	371
Italy	2,416	530	281	1,604	1,978	289	457	1,642	0	−1,810	31	4,713
Japan	9,439	5,438	2,770	1,232	947	27,469	8,573	19,981	−599	−485	232	38,088
Luxembourg	82	57	0	26	0	0	0	0	0	0	16	99
Netherlands	3,246	2,509	−234	971	57	5,858	6,225	−912	1,044	−499	353	9,514
New Zealand	122	102	0	20	0	9	9	0	0	0	16	147
Norway	1,311	935	9	367	−1	294	202	0	0	92	80	1,685
Portugal	218	126	31	61	135	593	482	0	0	111	−1	944
Spain	1,251	563	325	364	0	2,865	2,865	0	0	0	122	4,239
Sweden	1,999	1,395	0	604	0	−17	339	0	0	−357	22	2,004
Switzerland	1,026	726	−4	304	0	395	1,316	0	−583	−338	182	1,604
United Kingdom	3,199	1,782	8	1,409	81	18,196	5,852	12,120	0	224	382	21,859
United States	9,377	7,672	−755	2,460	1,119	42,848	23,430	19,472	−997	943	2,509	55,853
Total	**55,485**	**36,553**	**2,585**	**16,347**	**5,562**	**130,360**	**61,051**	**68,963**	**−948**	**1,295**	**5,577**	**196,984**

Net flows to part II countries	Official aid				Other official flows	Private flows						Total net flows
	Total	Bilateral grants	Bilateral loans	Contributions to multilateral institutions		Total	Foreign direct investment	Bilateral portfolio investment	Multilateral portfolio investment	Private export credits	Net grants by NGOs	
$ millions, 1996												
Australia	10	7	0	2	0	0	0	0	0	0	0	10
Austria	226	185	0	40	4	355	355	0	0	0	5	590
Belgium	70	14	0	56	−4	4,109	169	4,007	0	−67	0	4,175
Canada	181	180	0	0	−132	3	0	0	0	3	0	52
Denmark	120	109	9	11	26	248	248	0	0	0	5	398
Finland	57	37	3	7	−7	146	194	−64	0	16	0	195
France	709	0	0	293	0	4,860	1,192	3,886	0	−218	0	709
Germany	1,329	832	11	443	908	4,671	3,648	171	0	852	61	6,969
Ireland	1	1	0	0	0	0	0	0	0	0	0	1
Italy	294	12	0	283	64	218	153	706	0	−641	0	577
Japan	184	141	10	27	898	1,928	1,315	1,652	0	−1,039	0	3,010
Luxembourg	2	2	0	0	0	0	0	0	0	0	0	2
Netherlands	13	13	0	0	−6	−36	45	−78	0	−2	0	−29
New Zealand	0	0	0	0	0	0	0	0	0	0	0	0
Norway	50	50	0	0	0	−193	−201	0	0	8	0	−143
Portugal	18	0	0	18	3	−4	3	0	0	−7	0	17
Spain	2	2	0	0	0	−102	−102	0	0	0	0	−100
Sweden	178	127	0	51	23	−107	−84	0	0	−23	0	94
Switzerland	97	76	0	21	4	705	705	0	0	0	0	807
United Kingdom	362	133	0	228	0	3,952	390	3,500	0	62	13	4,327
United States	1,694	1,612	31	82	−24	2,652	2,226	578	0	−152	295	4,617
Total	**5,596**	**3,535**	**63**	**1,561**	**1,758**	**23,406**	**10,255**	**14,358**	**0**	**−1,207**	**379**	**26,279**

Net financial flows from Development Assistance Committee members 6.9

About the data

The high-income members of the Organisation for Economic Co-operation and Development (OECD) are the main (though not only) source of external finance for developing countries. This table provides an overview of the flow of financial resources from members of the OECD's Development Assistance Committee (DAC) to official and private recipients in developing countries. DAC exists to help its member countries coordinate their development assistance and to encourage the expansion and improve the effectiveness of the aggregate resources made available to developing and transition economies. In this capacity DAC monitors the flow of all financial resources, but its main concern is official development assistance (ODA). DAC has three criteria for ODA: It is undertaken by the official sector. It promotes economic development or welfare as a main objective. It is provided on concessional terms, with a grant element of at least 25 percent on loans.

This definition excludes military aid and nonconcessional flows from official creditors, which are considered other official flows. It includes capital projects, food aid, emergency relief, peacekeeping efforts, and technical cooperation. Also included are contributions to multilateral institutions, such as the United Nations and its specialized agencies, and concessional funding to the multilateral development banks, including the World Bank's International Development Association.

DAC maintains a list of countries and territories that are aid recipients. Part I of the list comprises countries and territories considered by DAC members to be eligible for ODA. Part II of the list, created after the collapse of the Soviet Union, monitors the flow of concessional assistance to transition economies that are not considered eligible for ODA but that nevertheless receive ODA-like flows. Under a procedure agreed to in 1993, countries with relatively higher incomes are moving from part I to part II status. To differentiate assistance to the two groups of recipients, ODA-like flows to part II countries are termed official aid.

The data in the table were compiled from replies by DAC member countries to questionnaires issued by the DAC Secretariat. Net flows of resources are defined as gross disbursements of grants and loans minus repayments on earlier loans. Because the data are based on donor country reports, they do not provide a complete picture of the resources received by developing countries, for three reasons. First, flows from DAC members are only part of the aggregate resource flows to developing countries. Second, the data that record contributions to multilateral institutions measure the flow of resources made available to those institutions by DAC members, not the flow of resources from those institutions to developing countries. Third, because some of the countries and territories on the DAC recipient list are normally classified as high income, the reported flows may overstate the resources available to low- and middle-income economies. High-income countries receive only a small fraction of all development assistance, however.

Net disbursements of ODA by some important donor countries that are not DAC members are shown in table 6.9a.

Definitions

• **Official development assistance** (ODA) comprises grants and loans net of repayments that meet the DAC definition of ODA and are made to countries and territories in part I of the DAC list of aid recipients. • **Official aid** comprises grants and ODA-like loans net of repayments to countries and territories in part II of the DAC list of aid recipients. • **Bilateral grants** are transfers in money or in kind for which no repayment is required. • **Bilateral loans** are loans extended by governments or official agencies that have a grant element of at least 25 percent and for which repayment is required in convertible currencies or in kind. • **Contributions to multilateral agencies** are concessional funding received by multilateral institutions from DAC members in the form of grants or capital subscriptions. • **Other official flows** are transactions by the official sector whose main objective is other than development or whose grant element is less than 25 percent. • **Private flows** consist of flows at market terms financed from private sector resources. They include changes in holdings of private long-term assets by residents of the reporting country and private grants by nongovernmental organizations, net of subsidies from the official sector. • **Foreign direct investment** is investment by residents of DAC member countries to acquire a lasting management interest (at least 10 percent of voting stock) in an enterprise operating in the recipient country. The data in the table reflect changes in the net worth of subsidiaries in recipient countries whose parent company is in the DAC source country. • **Bilateral portfolio investment** covers bank lending and the purchase of bonds, shares, and real estate by residents of DAC member countries in recipient countries. • **Multilateral portfolio investment** records the transactions of private banks and nonbanks in DAC member countries in the securities issued by multilateral institutions. • **Private export credits** are loans that are extended to recipient countries by the private sector in DAC member countries for the purpose of promoting trade and are supported by an official guarantee. • **Net grants by NGOs** are grants by nongovernmental organizations, net of subsidies from the official sector. • **Total net flows** comprise ODA or official aid flows, other official flows, private flows, and net grants by NGOs.

Data sources

Data on financial flows are compiled by DAC and published in its annual statistical report, *Geographical Distribution of Financial Flows to Aid Recipients*, and the DAC chairman's annual report, *Development Co-operation*.

Table 6.9a

Official development assistance by non-DAC donors

	1992	1993	1994	1995	1996
OECD members (non-DAC)					
Czech Republic	..	20	25	..	..
Greece	55	90[a]	122[a]	152[a]	184
Iceland	5	7	6	..	..
Korea, Rep. of	77	112	140	116	159
Turkey	87	73	58	107	..
Arab countries					
Kuwait	203	395	555	371	412
Saudi Arabia	783	549	317	192	306
United Arab Emirates	172	239	100	65	31
Other donors					
India	86	25	28	..	..

($ millions)

Note: China also provides aid, but does not disclose the amount.

a. Comprises total aid disbursements to both part I countries (official development assistance) and part II countries (official aid).

Source: OECD.

Net flows to part I countries	Net official development assistance							Aid appropriations		Untied aid	
	$ millions		% of GNP		annual average % change in volume[a] 1991–92 to 1995–96	Per capita of donor country[a] $	$	% of central government budget		% of total ODA commitments	
	1991	1996	1991	1996		1991	1996	1991	1996	1991	1995
$ millions, 1996											
Australia	1,050	1,121	0.38	0.30	1.7	61	62	1.3	1.2	45.5	0.0
Austria	547	557	0.34	0.24	1.5	70	69	0.6	0.0	0.4	12.1
Belgium	831	913	0.41	0.34	–2.6	84	91	0.0	0.0	14.7	0.0
Canada	2,604	1,795	0.45	0.32	–3.3	96	60	1.9	1.4	6.0	14.9
Denmark	1,200	1,772	0.96	1.04	3.5	233	336	3.2	2.7	29.5	47.8
Finland	930	408	0.80	0.34	–14.2	186	80	2.0	1.0	13.2	25.9
France	7,386	7,451	0.62	0.48	–2.2	129	128	3.2	0.0	50.8	25.6
Germany	6,890	7,601	0.40	0.33	–2.2	86	93	0.0	0.0	45.5	23.0
Ireland	72	179	0.19	0.31	18.8	21	50	0.7	0.0	0.0	0.0
Italy	3,347	2,416	0.30	0.20	–9.4	58	42	0.6	0.0	2.3	22.0
Japan	10,952	9,439	0.32	0.20	–3.6	88	75	1.3	0.0	13.2	59.2
Luxembourg	42	82	0.33	0.44	9.2	108	211	0.0	0.0	0.0	0.0
Netherlands	2,517	3,246	0.88	0.81	0.5	167	209	0.0	3.6	34.1	42.6
New Zealand	100	122	0.25	0.21	0.5	29	34	0.5	0.6	31.7	0.0
Norway	1,178	1,311	1.13	0.85	–0.2	276	300	2.0	1.8	41.6	42.1
Portugal	205	218	0.30	0.21	–0.2	19	22	0.0	0.0	0.0	54.7
Spain	1,262	1,251	0.24	0.22	2.3	32	32	0.6	1.0	0.0	0.0
Sweden	2,116	1,999	0.90	0.84	–2.2	246	226	2.7	0.0	65.8	57.9
Switzerland	863	1,026	0.36	0.34	0.3	126	145	3.1	2.8	55.4	47.7
United Kingdom	3,201	3,199	0.32	0.27	1.1	56	54	1.2	1.1	15.3	19.2
United States	11,262	9,377	0.20	0.12	–8.0	44	35	0.0	0.0	20.8	0.0
Total	**56,670**	**55,485**	**0.34**	**0.25**	**–3.2**	**74**	**68**	**0.8**	**1.6**	**29.3**	**32.3**

Net official aid

Net flows to part II countries	$ millions		% of GNP		annual average % change in volume[a] 1991–92 to 1995–96	Per capita of donor country[a] $	$
	1991	1996	1991	1996		1991	1996
$ millions, 1996							
Australia	9	10	0.00	0.00	–2.0	0	1
Austria	290	226	0.18	0.10	–9.0	37	28
Belgium	274	70	0.14	0.03	–25.3	28	7
Canada	145	181	0.03	0.03	4.1	5	6
Denmark	65	120	0.05	0.07	12.2	13	23
Finland	114	57	0.10	0.05	–3.7	23	11
France	457	709	0.04	0.05	11.3	8	12
Germany	2,637	1,329	0.15	0.06	–5.3	33	16
Ireland	15	1	0.04	0.00	–5.3	4	0
Italy	382	294	0.03	0.02	–3.4	7	5
Japan	110	184	0.00	0.00	–1.1	1	1
Luxembourg	5	2	0.04	0.01	–36.7	12	5
Netherlands	152	13	0.05	0.00	–3.7	10	1
New Zealand	1	0	0.00	0.00	–73.8	0	0
Norway	25	50	0.02	0.03	10.6	6	11
Portugal	22	18	0.03	0.02	–4.3	2	2
Spain	162	2	0.03	0.00	–69.3	4	0
Sweden	50	178	0.02	0.07	–6.2	6	20
Switzerland	55	97	0.02	0.03	2.5	8	14
United Kingdom	327	362	0.03	0.03	3.8	6	6
United States	1,832	1,694	0.03	0.02	1.9	7	6
Total	**7,128**	**5,596**	**0.04**	**0.03**	**–2.4**	**9**	**7**

a. At 1996 exchange rates and prices.

Aid flows from Development Assistance Committee members | 6.10

As part of its work the Development Assistance Committee (DAC) of the Organisation for Economic Co-operation and Development (OECD) assesses the aid performance of member countries relative to the size of their economies. As measured here, aid comprises bilateral disbursements of concessional financing to recipient countries plus the provision by donor governments of concessional financing to multilateral institutions. Volume measures, in constant prices and exchange rates, are used to measure the change in real resources provided over time. Aid flows to part I recipients—official development assistance (ODA)—are tabulated separate from official aid to part II recipients (see *About the data* for table 6.9 for more information on the distinction between ODA and official aid).

Measures of aid flows from the perspective of donors differ from aid receipts by recipient countries. This is because the concessional funding received by multilateral institutions from donor countries is recorded as an aid disbursement by the donor when the funds are deposited with the multilateral institution and recorded as a resource receipt by the recipi

ent country when the multilateral institution makes a disbursement.

Aid-to-GNP ratios, aid per capita, and aid appropriations as a percentage of donor government budgets are calculated by the OECD. The denominators used in calculating these ratios may differ from corresponding values elsewhere in this book because of differences in timing or definition.

The proportion of untied aid is reported here because tying arrangements require recipients to purchase goods and services from the donor country or from a specified group of countries. Tying arrangements may be justified on the grounds that they prevent a recipient from misappropriating or mismanaging aid receipts, but they may also be motivated by a desire to benefit suppliers in the donor country. The same volume of aid may have different purchasing power, depending on the relative costs of suppliers in countries to which the aid is tied and the degree to which each recipient's aid basket is untied. Thus tying arrangements may prevent recipients from obtaining the best value for their money and so reduce the value of the aid received.

Definitions

• **Net official development assistance** and **net official aid** record the actual international transfer by the donor of financial resources or of goods or services valued at the cost to the donor, less any repayments of loan principal during the same period. Data are shown at current prices and dollar exchange rates. • **Aid as a percentage of GNP** shows the donor's contributions of ODA or official aid as a share of its GNP. • **Annual average percentage change in volume** and **aid per capita of donor country** are calculated using 1996 exchange rates and prices. • **Aid appropriations** are the share of ODA or official aid appropriations in the donor's national budget. • **Untied aid** is the share of aid that is not subject to restrictions by donors on procurement sources.

Data sources

The data in this table appear in the DAC chairman's report, *Development Co-operation*. The OECD also makes its data available on diskette, magnetic tape, and the Internet.

Figure 6.10a

Aid flows are falling

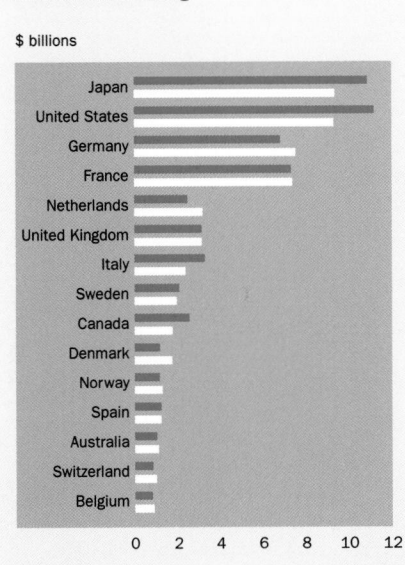
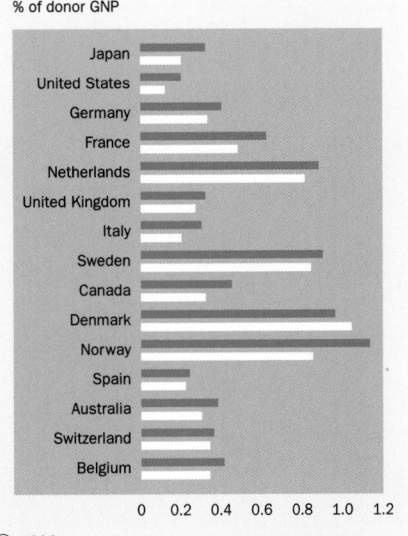

● 1991 ○ 1996

Source: OECD and World Bank staff estimates.

Official development assistance to developing countries has declined over the past five years. Not only has the dollar value fallen, but many DAC members now provide less aid as a portion of their gross domestic output.

6.11 Aid dependency

| | Net official development assistance and official aid ($ millions) | | Aid per capita ($) | | Aid dependency ratios | | | | | | | |
| | | | | | Aid as % of GNP | | Aid as % of gross domestic investment | | Aid as % of imports of goods and services | | Aid as % of central government expenditures | |
	1991	1996	1991	1996	1991	1996	1991	1996	1991	1996	1991	1996
Albania	324.2	222.0	99	68	29.2	8.1	470.1	40.2	95.3	19.8	..	24.1
Algeria	340.0	309.0	13	11	0.8	0.7	2.4	2.5	3.2	..	..	..
Angola	279.7	544.2	29	49	9.6	15.8	51.5	72.0	6.6	12.0	..	..
Argentina	299.5	277.4	9	8	0.2	0.1	1.1	0.5	1.7	0.8	1.4	0.6
Armenia	2.7	294.9	1	78	0.1	18.2	0.2	146.8	37.0	32.0	..	..
Australia												
Austria												
Azerbaijan	0.0	106.3	0	14	0.0	3.0	2.3	11.9	..	8.5	..	..
Bangladesh	1,889.1	1,254.5	17	10	8.1	3.9	70.2	23.2	47.3	16.1	..	..
Belarus	187.0	73.0	18	7	0.5	0.4	1.8	1.5	7.7	1.1	..	..
Belgium												
Benin	268.4	292.8	55	52	14.5	13.5	98.6	77.5	41.2	49.7	..	..
Bolivia	512.6	849.9	76	112	10.8	13.3	74.0	71.6	36.4	42.6	57.0	49.5
Bosnia and Herzegovina	0.0	811.6	0	184	..	..	..	..	..	..	..	..
Botswana	136.0	80.8	103	55	3.4	1.7	10.9	6.8	5.7	3.6	10.2	5.4
Brazil	182.6	408.2	1	3	0.0	0.1	0.2	0.3	0.5	0.5	0.2	..
Bulgaria	316.0	170.0	37	20	3.2	1.9	12.8	12.7	7.3	2.7	10.3	3.6
Burkina Faso	423.7	418.2	46	39	15.2	16.5	73.7	64.8	55.8	83.7	86.1	..
Burundi	259.1	203.8	46	32	22.4	18.1	154.0	203.2	72.4	96.9	77.2	81.8
Cambodia	91.0	452.9	10	44	5.6	14.5	59.3	70.0	39.2	33.5	..	..
Cameroon	518.5	413.3	44	30	4.5	4.9	25.0	28.4	18.7	16.8	20.3	42.2
Canada												
Central African Republic	174.7	166.9	58	50	12.8	16.1	99.8	280.7	52.3	62.3	..	..
Chad	265.8	305.2	46	46	20.2	26.9	274.0	134.9	56.7	50.7	65.9	..
Chile	125.5	203.4	9	14	0.4	0.3	1.5	1.0	1.1	0.9	1.7	1.4
China	1,998.7	2,617.3	2	2	0.5	0.3	1.5	0.8	3.5	1.5	5.3	6.1
Hong Kong, China	36.1	13.2	6	2	0.0	0.0	0.2	0.0	..	..	..	..
Colombia	122.5	250.8	4	7	0.3	0.3	1.9	1.4	1.4	1.2	2.6	..
Congo, Dem. Rep.	476.2	167.4	12	4	5.7	2.8	94.0	38.5	..	..	25.6	41.7
Congo, Rep.	133.7	429.7	57	159	5.9	22.9	24.9	29.6	7.9	16.6	..	..
Costa Rica	174.1	−6.8	56	−2	3.2	−0.1	12.3	−0.3	6.9	0.6	12.5	1.0
Côte d'Ivoire	632.7	967.6	51	67	6.9	9.9	82.0	66.1	14.6	19.5	..	..
Croatia	0.0	133.4	0	28	0.0	0.7	0.0	4.7	0.0	1.3	0.0	1.5
Cuba	37.6	67.8	4	6	..	..	..	..	..	..	..	..
Czech Republic	231.0	122.0	22	12	0.9	0.2	3.2	0.6	0.5	0.3	0.8	0.6
Denmark												
Dominican Republic	66.5	105.8	9	13	0.9	0.8	3.8	3.3	2.7	2.0	7.9	6.7
Ecuador	238.0	260.9	23	22	2.2	1.5	9.1	7.8	5.6	4.5	14.3	8.3
Egypt, Arab Rep.	5,024.7	2,211.8	94	37	14.3	3.3	64.2	19.7	26.4	10.9	44.4	14.0
El Salvador	294.2	317.2	57	55	5.6	3.1	36.0	19.3	16.7	8.0	48.1	22.6
Eritrea	0.0	157.2	0	43	..	..	..	..	0.0	27.1	..	..
Estonia	15.0	62.0	10	42	0.3	1.4	1.0	5.4	14.4	1.6	9.2	4.3
Ethiopia	1,097.3	849.4	21	15	20.6	14.3	287.8	67.6	84.2	49.1	51.4	..
Finland												
France												
Gabon	143.4	126.5	145	112	3.0	2.6	10.0	11.0	6.0	5.5	8.7	..
Gambia, The	102.7	38.5	107	34	31.6	13.4	157.2	62.6	38.3	12.7	132.1	..
Georgia	0.2	318.4	0	59	0.0	7.1	0.0	121.1	..	23.6	..	..
Germany												
Ghana	882.1	653.6	58	37	13.6	10.5	84.2	55.2	49.5	25.6	95.4	..
Greece												
Guatemala	198.6	216.1	21	20	2.1	1.4	14.8	10.7	9.0	5.7	23.4	16.5
Guinea	382.0	295.5	64	44	13.6	7.8	77.5	57.6	31.2	28.1	58.9	..
Guinea-Bissau	115.5	179.9	118	164	49.4	67.5	144.9	304.3	100.6	122.0	..	..
Haiti	181.9	375.2	28	51	5.6	14.4	49.4	2,117.2	32.9	47.4	..	..
Honduras	302.5	367.3	58	60	10.5	9.2	40.0	28.6	21.2	19.4	..	..

Aid dependency | 6.11

| | Net official development assistance and official aid | | Aid per capita | | Aid dependency ratios | | | | | | | |
| | $ millions | | $ | | Aid as % of GNP | | Aid as % of gross domestic investment | | Aid as % of imports of goods and services | | Aid as % of central government expenditures | |
	1991	1996	1991	1996	1991	1996	1991	1996	1991	1996	1991	1996
Hungary	626.0	185.0	61	18	2.0	0.4	9.2	1.5	4.8	0.8	..	..
India	2,745.0	1,936.2	3	2	1.1	0.6	4.8	2.1	9.3	3.3	5.9	3.3
Indonesia	1,874.4	1,120.6	10	6	1.5	0.5	5.0	1.6	4.9	2.3	8.8	3.4
Iran, Islamic Rep.	194.4	171.0	4	3	0.2	0.1	0.5	..	0.6	1.2	0.1	0.5
Iraq	552.4	387.4	30	18	..	..	..	..	..	..	..	..
Ireland												
Israel	1,749.5	2,216.7	353	389	2.8	0.4	11.6	1.6	6.8	5.2	7.8	4.8
Italy												
Jamaica	162.2	59.9	67	24	4.9	1.4	16.7	5.1	5.9	2.6	..	..
Japan												
Jordan	920.7	513.7	260	119	23.8	7.2	84.9	20.2	23.6	8.8	59.4	25.5
Kazakhstan	111.5	124.0	7	8	0.4	0.6	0.1	2.6	0.2	1.6	..	..
Kenya	921.2	606.1	38	22	12.1	6.8	53.9	32.4	33.3	16.5	39.6	33.3
Korea, Dem. Rep.	9.0	42.8	0	2	..	..	..	..	..	..		
Korea, Rep.	54.8	–146.9	1	–3	0.0	0.0	0.0	–0.1	0.1	–0.1	0.1	–0.2
Kuwait	4.5	3.1	3	2	0.0	0.0	0.0	0.2	0.0	0.0	0.0	0.0
Kyrgyz Republic	0.0	231.9	0	51	0.0	13.9	0.0	69.2	0.9	22.4	..	..
Lao PDR	143.3	338.6	35	72	13.9	18.2		59.8	57.3	46.4	..	..
Latvia	3.0	79.0	1	32	0.0	1.6	0.1	8.4	8.4	2.4	..	4.9
Lebanon	132.2	232.8	36	57	2.7	1.8	15.4	6.0	3.3	2.9	8.2	5.2
Lesotho	126.2	107.2	69	53	12.1	8.7	27.6	11.6	13.9	12.8	35.9	..
Libya	26.4	9.9	6	2								
Lithuania	4.0	89.0	1	24	0.0	1.2	0.1	5.5	2.6	1.7	10.9	4.5
Macedonia, FYR	0.0	105.5	0	53	0.0	5.3	0.0	26.0	0.0	5.7		
Madagascar	455.9	364.5	38	27	17.9	9.1	161.3	87.8	54.0	31.1	112.5	52.9
Malawi	524.6	500.8	60	50	24.6	23.2	119.6	132.4	60.6	48.9	..	..
Malaysia	289.5	–451.6	16	–22	0.6	–0.5	1.7	–1.1	0.7	0.1	2.1	–2.1
Mali	457.7	505.1	53	51	19.2	19.4	83.0	71.7	53.3	56.2	..	..
Mauritania	219.9	273.6	107	117	20.6	26.4	108.4	113.8	37.6	41.3	..	..
Mauritius	67.5	19.6	63	17	2.4	0.5	8.3	1.8	3.4	0.7	10.6	2.0
Mexico	278.3	289.1	3	3	0.1	0.1	0.4	0.4	0.4	0.2	0.6	0.9
Moldova	0	37.0	0	9	0.0	2.1	0.0	7.3	1.8	2.9	..	..
Mongolia	69.5	202.6	31	81	24.2	21.3	80.8	93.0	13.4	37.8	51.3	96.4
Morocco	1,232.4	650.8	50	24	4.6	1.8	19.5	8.6	12.8	5.2	15.9	..
Mozambique	1,070.3	922.9	74	51	83.8	59.8	163.4	111.5	88.4	87.8	..	..
Myanmar	179.4	56.2	4	1	..	..	..	..	17.3	9.1	4.1	1.3
Namibia	184.4	188.6	133	119	6.9	5.7	38.8	29.4	10.5	9.1	18.3	..
Nepal	453.4	401.4	24	18	12.0	8.9	58.8	38.8	52.2	23.8	74.3	46.0
Netherlands												
New Zealand												
Nicaragua	841.1	954.0	217	212	64.1	57.1	239.8	174.5	70.3	59.3	176.5	105.6
Niger	377.0	258.7	48	28	16.5	13.2	176.4	134.8	57.6	53.0	..	..
Nigeria	262.6	191.8	3	2	1.1	0.6	4.1	3.2	2.0	1.5	..	..
Norway												
Oman	15.2	61.6	9	28	0.2	0.6	0.9	5.0	0.3	1.0	0.4	1.3
Pakistan	1,370.9	876.8	12	7	2.9	1.4	15.9	7.3	11.6	5.1	13.7	6.0
Panama	101.9	89.5	42	33	1.9	1.1	9.1	3.7	1.5	1.0	7.3	2.5
Papua New Guinea	396.8	385.0	101	87	10.8	8.0	38.2	27.7	17.5	14.0	29.6	20.2
Paraguay	146.0	97.1	34	20	2.4	1.0	10.1	4.5	5.8	2.6	19.5	..
Peru	614.0	409.8	28	17	2.2	0.7	12.6	2.9	9.6	3.4	11.2	4.1
Philippines	1,053.0	883.2	16	12	2.3	1.0	11.5	4.4	6.5	2.5	12.1	5.8
Poland	2,508	830	0	22	0.0	0.6	0.0	3.0	0.0	1.9	..	1.5
Portugal												
Puerto Rico												
Romania	321.0	218.0	14	10	1.1	0.6	4.0	2.4	5.1	1.7	3.1	1.5
Russian Federation	564.0	1,225.0	4	0	0.1	0.0	0.3	0.0	0.0	0.0	..	0.0

	Net official development assistance and official aid $ millions		Aid per capita $		Aid dependency ratios							
					Aid as % of GNP		Aid as % of gross domestic investment		Aid as % of imports of goods and services		Aid as % of central government expenditures	
	1991	1996	1991	1996	1991	1996	1991	1996	1991	1996	1991	1996
Rwanda	363.6	674.3	51	100	19.9	51.2	166.8	371.9	101.9	176.5	95.5	..
Saudi Arabia	44.7	28.5	3	1	0.0	0.0	0.2	0.1	0.1	0.1	..	..
Senegal	639.0	581.5	85	68	12.0	11.6	94.8	68.3	33.4	32.9	..	..
Sierra Leone	104.9	195.5	26	42	14.8	21.2	123.3	223.2	47.8	78.3	100.3	140.4
Singapore	7.8	0.0	3	0	0.0	0.0	0.1	0.0	0.0	0.0	0.1	0.1
Slovak Republic	114.0	141.0	22	26	1.1	0.7	3.4	2.0	0.7	1.1	..	..
Slovenia	0.0	82.2	0	41	0.0	0.4	0.0	1.9	0.0	0.8	..	..
South Africa	0.0	361.1	0	10	0.0	0.3	0.0	1.6	0.0	1.0	0.0	0.9
Spain												
Sri Lanka	890.5	494.5	52	27	10.1	3.6	43.3	13.9	23.4	7.7	33.8	13.0
Sudan	880.9	230.3	36	8	12.3	..	94.9	..	70.7	15.8	..	..
Sweden												
Switzerland												
Syrian Arab Republic	381.4	225.3	30	16	3.0	1.4	18.6	..	8.6	3.2	5.6	2.8
Tajikistan	0.0	113.0	0	19	0.0	5.6	0.0	20.3	3.7	12.9	..	..
Tanzania	1,080.7	893.7	41	29	24.9	15.6	86.7	84.8	61.4	38.4	..	..
Thailand	721.5	832.0	13	14	0.7	0.5	1.7	1.1	1.6	0.9	5.1	2.8
Togo	202.2	166.0	56	39	12.9	12.0	73.7	85.4	22.2	25.5	..	..
Trinidad and Tobago	−1.5	16.9	−1	13	0.0	0.3	−0.2	2.0	−0.1	1.0	0.2	1.7
Tunisia	357.2	126.4	43	14	2.8	0.7	10.5	2.7	5.6	1.3	8.1	2.0
Turkey	1,622.5	232.5	28	4	1.1	0.1	4.7	0.5	5.8	0.4	5.1	0.5
Turkmenistan	0.0	23.8	0	5	0.0	0.5	0.0	..	0.0	1.6	..	..
Uganda	666.8	683.6	39	35	20.4	11.3	132.3	68.3	91.1	41.0	..	..
Ukraine	368.0	379.0	7	7	0.5	0.9	1.7	3.8	..	1.7	..	..
United Arab Emirates	−5.9	0.0	−3	0	0.0	0.0	−0.1	−0.1	..	..	−0.1	−0.2
United Kingdom												
United States												
Uruguay	51.5	51.5	17	16	0.5	0.3	4.3	2.3	2.1	1.1	1.9	0.9
Uzbekistan	0.0	87.2	0	4	0.0	0.4	0.0	2.1	0.2	1.7	..	..
Venezuela	30.7	44.2	2	2	0.1	0.1	0.3	0.4	0.2	0.2	0.3	0.4
Vietnam	237.5	927.2	4	12	2.5	4.0	16.5	14.2	8.6	6.8	..	..
West Bank and Gaza	0.0	593.0	..	..	..	..	..	..	..	..	..	..
Yemen, Rep.	300.1	260.4	22	17	6.2	4.9	37.4	17.1	9.1	7.0	8.2	6.4
Yugoslavia, FR (Serb./Mont.)[a]	159.0	681.0	15	64	..	..	..	..	..	..	..	..
Zambia	883.3	613.9	110	67	27.7	18.6	237.4	120.2	48.8	..	44.7	87.1
Zimbabwe	393.3	374.2	39	33	6.3	5.2	23.9	27.6	15.4	19.8	..	..
World[b]	68,110.3 s	63,773.6 s	15 w	13 w	1.4 w	1.0 w	.. w	.. w	.. w	4.1 w	..	..
Low income	28,568.2	28,186.8	10	8	3.1	1.8	11.6	5.1	15.7	7.8	..	..
Excl. China & India	23,824.5	23,633.2	25	21	8.5	7.1	48.5	33.6	26.1	18.8	..	..
Middle income	26,049.7	19,403.2	17	12	0.8	0.5	3.2	2.3	3.7	1.8	..	..
Lower middle income	20,832.9	16,145.9	20	14	1.2	0.8	4.0	3.2	6.3	2.6	..	..
Upper middle income	5,216.8	3,257.3	12	7	0.4	0.3	1.8	1.4	1.4	1.1	..	..
Low & middle income	61,160.0	54,035.8	14	11	1.5	0.9	5.7	3.4	7.0	3.6	..	..
East Asia & Pacific	7,541.2	8,359.5	5	5	1.0	0.6	..	..	3.6	2.6	..	..
Europe & Central Asia	8,890.3	8,938.2	19	17	0.8	0.6	2.6	2.3	2.8	1.7	..	..
Latin America & Carib.	5,850.2	8,025.1	13	17	0.5	0.5	2.6	1.7	2.7	1.9	..	..
Middle East & N. Africa	10,311.9	5,342.5	43	19	2.4	1.3	9.2	..	5.7	3.1	..	..
South Asia	8,114.1	5,499.9	7	4	2.3	1.1	10.6	4.8	15.1	5.6	..	..
Sub-Saharan Africa	18,206.9	17,299.5	33	26	6.3	5.3	35.3	27.8	18.0	12.7	..	..
High income	2,653.6	3,091.5	..	..	..	..	..	..	..	..	..	..

a. Data for 1991 include net flows to the states of the former Yugoslavia: Bosnia and Herzegovina, Croatia, Macedonia FYR, and Slovenia. b. Includes aid not allocated by country or region.

Aid dependency 6.11

About the data

Ratios of aid to GNP, investment, imports, or public spending provide a measure of the recipient country's dependency on aid. But care must be taken in drawing policy conclusions. For foreign policy reasons some countries have traditionally received large amounts of assistance. Thus aid dependency ratios may reveal as much about the interests of donors as they do about the need of recipients. In general, aid dependency ratios in Sub-Saharan Africa are much higher than those in other regions, and they increased during the 1980s. These high ratios are due only in part to the volume of aid flows. Many African countries experienced severe erosion in their terms of trade during the 1980s, which, along with weak policies, contributed to falling incomes, imports, and investment. Thus the increase in aid dependency ratios reflects events affecting both the numerator and the denominator.

As defined here, aid includes official development assistance (ODA) and official aid. The data cover bilateral loans and grants from Development Assistance Committee (DAC) countries, multilateral organizations, and certain Arab countries. They do not reflect aid given by recipient countries to other developing countries. As a result some countries that are net donors (such as Saudi Arabia) are shown in the table as aid recipients. (See table 6.9a for aid disbursement by some non-DAC countries.)

The data in the table do not distinguish among different types of aid (program, project, or food aid, emergency assistance, peacekeeping assistance, or technical cooperation), each of which may have a very different effect on the economy. Technical cooperation expenditures do not always directly benefit the economy to the extent that they defray costs incurred outside the country on the salaries and benefits of technical experts and the overhead costs of firms supplying technical services.

Because the table relies on information from donors, it is not consistent with information recorded by recipients in the balance of payments, which often excludes all or some technical assistance—particularly payments to expatriates made directly by the donor. Similarly, grant commodity aid may not always be recorded in trade data or in the balance of payments. Although ODA estimates in balance of payments statistics are meant to exclude purely military aid, the distinction is sometimes blurred. The definition used by the country of origin usually prevails.

The nominal values used here tend to overstate the amount of resources transferred. Changes in international prices and in exchange rates reduce the purchasing power of aid. The practice of tying aid, still prevalent though declining in importance, also reduces the purchasing power of aid (see *About the data* for table 6.10).

Aid not allocated by country or region—including administrative costs, research into development issues, and aid to nongovernmental organizations—is included in the world total. Thus regional and income group totals do not add up to the world total.

Figure 6.11a

Twenty countries received more than half the official development assistance and official aid provided in 1996

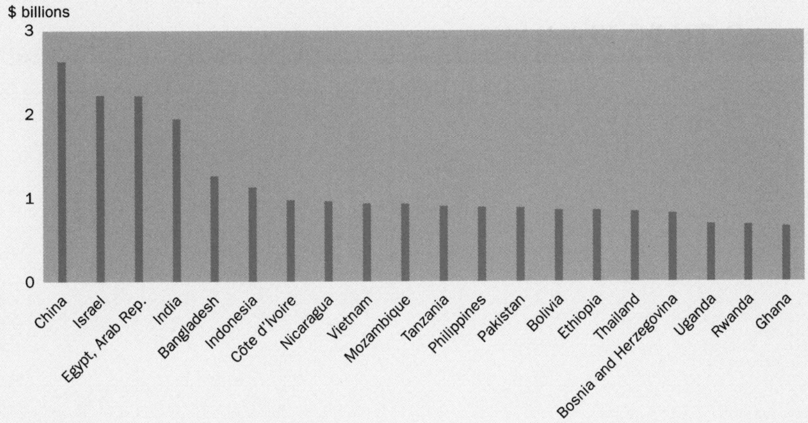
$ billions
3
2
1
0
China
Israel
Egypt, Arab Rep.
India
Bangladesh
Indonesia
Côte d'Ivoire
Nicaragua
Vietnam
Mozambique
Tanzania
Philippines
Pakistan
Bolivia
Ethiopia
Thailand
Bosnia and Herzegovina
Uganda
Rwanda
Ghana

Source: OECD.

Definitions

• **Net official development assistance** consists of disbursements of loans (net of repayments and principal) and grants made on concessional terms by official agencies of the members of DAC and certain Arab countries to promote economic development and welfare in recipient economies listed as developing by DAC. Loans with a grant element of more than 25 percent are included in ODA. ODA also includes technical cooperation and assistance. • **Official aid** refers to aid flows net of repayments from official donors to the transition economies of Eastern Europe and the former Soviet Union and to certain advanced developing countries and territories as determined by DAC. Official aid is provided under terms and conditions similar to those for ODA. • **Aid per capita** includes both ODA and official aid. • **Aid dependency ratios** are calculated using values in U.S. dollars converted at official exchange rates. See the notes to tables 1.1, 4.8, and 4.13 for definitions of GNP, gross domestic investment, imports of goods and services, and central government expenditures.

Data sources

Development Co-operation
OECD

Data on aid are compiled by DAC and published in its annual statistical report, *Geographical Distribution of Financial Flows to Aid Recipients*, and in the DAC chairman's report, *Development Co-operation*. The OECD also makes its data available on diskette, magnetic tape, and the Internet.

$ millions, 1996	Total	Ten major DAC donors										Others
		United States	Japan	France	Germany	Netherlands	United Kingdom	Canada	Sweden	Denmark	Norway	
Albania	114.3	17.0	2.7	3.1	33.4	8.0	1.9	0.0	0.4	2.8	0.6	44.4
Algeria	263.0	0.0	0.9	241.1	−23.8	1.1	0.1	1.2	3.0	0.0	0.2	39.1
Angola	294.4	25.0	5.2	11.9	25.3	30.1	16.2	2.2	36.2	1.4	25.2	115.9
Argentina	85.0	0.0	19.0	7.8	10.3	1.3	0.3	1.1	1.6	0.7	0.0	42.9
Armenia	115.3	88.0	0.0	5.8	6.7	7.5	1.4	0.0	2.7	0.0	1.8	1.3
Australia												
Austria												
Azerbaijan	25.3	9.0	0.3	0.4	4.1	4.4	2.5	0.0	1.6	0.0	1.3	1.7
Bangladesh	644.5	41.0	174.0	27.2	84.0	67.2	71.4	37.5	28.3	36.9	39.6	37.3
Belarus	58.6	30.0	0.2	0.0	17.9	0.0	0.0	0.0	1.2	1.3	0.1	7.9
Belgium												
Benin	164.9	7.0	44.7	44.3	22.4	8.4	0.0	8.3	0.0	14.6	−0.2	15.4
Bolivia	590.9	94.0	98.0	44.6	104.1	57.1	9.3	12.5	20.0	15.2	3.0	133.1
Bosnia and Herzegovina	593.8	135.0	25.0	7.2	39.9	88.4	0.6	0.0	30.1	0.0	46.8	220.9
Botswana	67.9	7.0	18.0	0.7	7.7	4.2	4.7	0.3	15.7	0.9	6.6	2.0
Brazil	190.8	−7.0	65.5	12.8	39.3	21.3	6.3	3.1	2.6	5.6	2.1	39.2
Bulgaria	57.4	9.0	13.0	0.0	25.0	1.9	1.9	0.0	0.0	0.5	0.2	6.1
Burkina Faso	269.2	9.0	14.9	100.0	41.0	37.6	0.5	7.2	0.8	22.8	2.4	33.1
Burundi	67.8	2.0	1.0	13.0	13.8	3.3	2.6	4.3	4.9	0.4	5.8	16.6
Cambodia	252.6	28.0	71.3	52.1	14.2	8.4	12.3	2.1	16.0	0.9	5.8	41.4
Cameroon	279.6	1.0	7.1	176.0	56.4	12.0	4.0	17.0	0.0	0.0	0.5	5.7
Canada												
Central African Republic	121.0	2.0	30.6	66.6	10.6	0.9	0.1	0.6	0.6	7.3	0.3	1.5
Chad	121.8	2.0	0.3	73.5	25.6	2.2	0.2	0.6	0.3	0.2	0.2	16.8
Chile	182.7	−3.0	52.9	43.5	43.0	8.3	3.0	1.5	5.3	0.4	1.3	26.6
China	1,671.1	0.0	861.7	97.2	461.1	2.7	57.1	38.4	17.7	20.9	9.1	105.1
Hong Kong, China	10.1	0.0	5.9	2.1	1.6	0.1	0.1	0.0	0.3	0.0	0.0	0.2
Colombia	159.8	17.0	36.6	16.5	32.2	8.5	3.3	3.3	3.5	2.0	2.4	34.5
Congo, Dem. Rep.	106.3	0.0	4.5	14.9	34.4	4.7	1.9	3.0	4.5	0.0	6.3	32.3
Congo, Rep.	394.6	10.0	0.2	211.4	6.7	2.3	1.5	6.8	0.6	0.0	0.0	155.0
Costa Rica	−12.5	−44.0	−17.5	5.1	2.2	16.0	3.4	2.3	3.1	1.0	0.7	15.0
Côte d'Ivoire	449.2	14.0	58.1	300.3	34.0	0.4	10.1	20.2	0.4	0.0	2.1	9.6
Croatia	126.5	10.0	0.1	2.3	92.5	0.9	0.4	0.0	3.4	0.0	5.5	11.4
Cuba	26.9	0.0	1.0	2.1	2.1	0.8	0.4	1.6	1.1	0.0	1.7	16.0
Czech Republic	64.0	7.0	1.9	0.0	32.0	0.0	5.6	1.3	0.2	2.0	0.4	13.7
Denmark												
Dominican Republic	57.3	3.0	20.0	3.9	7.0	2.5	0.0	0.3	0.7	0.2	0.5	19.3
Ecuador	207.1	12.0	47.5	11.4	27.1	11.9	2.7	1.8	2.5	0.6	1.6	87.9
Egypt, Arab Rep.	1,933.3	725.0	201.3	301.2	442.4	17.1	8.1	114.0	11.8	32.7	4.7	74.9
El Salvador	229.3	74.0	70.4	4.1	37.3	8.0	0.3	4.3	7.3	1.8	3.3	18.6
Eritrea	124.8	15.0	2.0	4.1	22.3	11.9	2.8	3.6	2.1	4.4	13.4	43.3
Estonia	37.8	1.0	0.2	0.0	9.4	0.0	4.4	1.2	9.3	2.3	2.8	7.3
Ethiopia	445.4	56.0	50.2	10.8	81.4	60.2	19.7	13.8	39.3	4.5	21.4	88.1
Finland												
France												
Gabon	113.4	2.0	0.2	102.4	2.4	0.1	0.0	3.6	0.0	0.0	0.0	2.6
Gambia, The	17.2	4.0	0.1	0.9	4.0	2.1	3.3	0.3	1.2	0.3	0.3	0.8
Georgia	112.3	55.0	0.2	2.6	33.6	9.0	2.8	0.0	1.4	0.0	1.1	6.6
Germany												
Ghana	348.9	30.0	110.0	15.9	37.1	22.7	33.6	21.0	3.9	42.3	1.6	30.9
Greece												
Guatemala	141.2	−1.0	44.6	2.1	29.3	12.3	0.7	2.5	8.2	2.0	15.1	25.4
Guinea	134.7	24.0	18.3	52.8	17.8	0.3	0.7	9.1	0.3	3.8	3.0	4.7
Guinea-Bissau	124.8	6.0	10.9	11.3	2.6	6.7	0.0	0.4	6.3	1.5	0.0	79.1
Haiti	150.1	67.0	7.2	29.6	4.8	3.9	0.0	24.4	0.2	0.0	1.1	11.9
Honduras	155.2	27.0	63.8	1.4	15.5	10.6	1.6	4.5	1.7	1.0	1.2	27.0

Distribution of net aid by Development
Assistance Committee members | 6.12

$ millions, 1996	Total	Ten major DAC donors										Others
		United States	Japan	France	Germany	Netherlands	United Kingdom	Canada	Sweden	Denmark	Norway	
Hungary	88.4	10.0	17.5	0.0	39.8	0.0	1.3	2.6	0.1	0.8	0.8	15.7
India	1,025.1	6.0	579.3	14.8	51.2	58.5	154.3	29.4	51.4	37.6	13.4	29.2
Indonesia	1,062.6	−57.0	965.5	28.4	−106.0	−62.5	46.1	18.0	1.1	0.8	13.7	214.5
Iran, Islamic Rep.	141.3	−21.0	58.1	12.5	70.1	0.1	0.5	0.0	8.3	0.0	1.0	11.8
Iraq	284.2	108.0	0.0	2.6	91.5	19.3	11.5	1.5	34.4	0.0	4.0	11.4
Ireland												
Israel	2,216.6	2,253.0	1.6	10.7	−57.3	6.6	0.0	0.0	0.6	0.0	0.0	1.4
Italy												
Jamaica	4.0	−7.0	12.4	−0.2	−9.6	5.6	−6.7	3.7	0.2	0.0	0.3	5.3
Japan												
Jordan	324.3	45.0	123.7	19.4	67.5	6.3	7.7	6.1	1.7	0.0	0.1	46.9
Kazakhstan	93.5	63.0	9.0	1.7	13.3	0.2	3.9	1.2	0.6	0.2	0.0	0.5
Kenya	345.7	11.0	92.8	17.0	53.5	39.9	43.8	5.6	23.3	20.0	2.7	36.2
Korea, Dem. Rep.	9.1	0.0	0.0	0.0	0.9	0.5	0.5	0.1	1.3	0.0	2.0	3.8
Korea, Rep.	−149.2	−54.0	−127.9	10.1	16.0	0.2	0.0	0.7	0.2	0.0	0.0	5.7
Kuwait	0.4	0.0	0.1	0.0	0.1	0.0	0.0	0.0	0.0	0.0	0.0	0.1
Kyrgyz Republic	99.4	28.0	44.3	0.3	10.5	8.7	0.9	0.0	0.0	3.0	0.2	3.6
Lao PDR	147.5	3.0	57.4	16.4	22.9	3.2	0.9	1.0	17.7	0.0	4.2	20.8
Latvia	47.9	4.0	0.2	0.0	12.0	0.0	7.9	1.2	11.4	4.0	2.8	4.5
Lebanon	87.1	6.0	0.6	41.3	7.7	1.0	0.8	3.2	2.8	0.0	3.7	20.1
Lesotho	49.3	3.0	7.9	2.9	12.3	0.5	8.0	0.3	2.6	0.7	0.5	10.7
Libya	2.1	0.0	0.1	0.4	1.1	0.0	0.0	0.0	0.2	0.0	0.0	0.3
Lithuania	55.7	14.0	0.3	0.0	11.5	0.0	1.3	1.2	17.4	6.4	2.2	1.7
Macedonia, FYR	26.3	5.0	6.1	1.4	8.2	0.5	1.2	0.0	0.5	0.0	0.0	3.4
Madagascar	229.8	33.0	50.1	101.8	16.3	4.0	1.0	0.7	0.1	0.0	7.8	15.1
Malawi	263.9	32.0	64.0	2.5	31.7	10.2	83.6	8.0	8.1	15.2	4.5	4.2
Malaysia	−453.1	0.0	−482.5	3.5	7.5	2.3	−4.0	3.4	0.7	8.4	0.1	7.4
Mali	297.5	5.0	38.1	82.3	65.7	42.8	13.9	14.3	0.8	4.6	9.9	20.2
Mauritania	98.8	2.0	29.8	45.2	12.3	0.2	0.5	1.3	0.3	1.0	0.4	5.8
Mauritius	−1.1	−1.0	4.8	7.3	−18.1	0.2	0.8	0.5	0.0	0.0	0.0	4.3
Mexico	274.3	26.0	212.8	6.5	12.4	4.0	5.7	3.5	0.3	0.8	0.4	2.0
Moldova	24.3	13.0	0.0	0.0	3.2	4.0	1.3	0.0	0.0	0.5	0.0	2.4
Mongolia	136.2	6.0	103.8	0.4	11.8	1.2	0.5	0.0	0.3	3.5	3.0	5.8
Morocco	391.4	−3.0	46.4	290.7	−0.9	0.2	0.5	5.0	0.7	−1.2	0.1	53.0
Mozambique	551.9	45.0	30.1	20.7	41.3	45.7	35.4	13.4	61.3	46.8	51.8	160.6
Myanmar	45.3	0.0	35.2	2.1	1.5	1.8	0.4	0.2	0.3	0.2	1.2	2.3
Namibia	136.4	9.0	4.7	3.8	43.1	5.0	9.6	0.4	16.8	6.6	16.6	20.8
Nepal	236.3	15.0	88.8	2.0	25.7	11.4	23.3	5.4	1.5	23.0	11.2	29.0
Netherlands												
New Zealand												
Nicaragua	764.0	30.0	70.5	10.2	403.0	38.1	1.5	12.9	49.4	33.6	24.0	90.8
Niger	163.2	12.0	4.8	86.8	18.3	9.3	0.7	2.8	0.0	1.8	0.6	26.1
Nigeria	47.3	5.0	−2.1	6.1	14.5	1.8	11.5	1.3	1.0	1.9	1.5	4.9
Norway												
Oman	15.7	4.0	9.9	0.6	0.9	0.0	0.1	0.0	0.0	0.1	0.0	0.0
Pakistan	338.6	−101.0	282.2	5.4	15.8	15.9	61.4	9.7	7.9	3.2	6.8	31.4
Panama	47.9	−5.0	37.7	0.4	4.6	0.5	0.8	0.5	0.0	0.0	0.0	8.4
Papua New Guinea	352.9	1.0	96.2	0.5	1.5	1.4	0.0	0.0	0.3	0.0	0.0	252.0
Paraguay	62.6	2.0	41.2	0.2	9.8	0.8	0.0	0.3	0.8	−0.5	0.7	7.4
Peru	277.7	51.0	56.4	11.6	42.4	31.5	4.2	18.7	3.9	1.9	2.3	53.9
Philippines	748.3	46.0	414.5	27.4	106.6	22.8	10.0	16.4	15.9	5.2	2.2	81.4
Poland	542.6	33.0	89.2	0.0	96.9	0.4	0.8	133.7	5.4	13.1	3.0	167.3
Portugal												
Puerto Rico												
Romania	94.9	20.0	6.2	0.0	28.7	5.9	20.6	2.4	0.2	0.5	0.6	9.9
Russian Federation	1,031.3	416.0	5.4	0.0	474.9	1.0	9.2	13.5	19.0	13.8	29.2	49.2

$ millions, 1996	Total	United States	Japan	France	Germany	Netherlands	United Kingdom	Canada	Sweden	Denmark	Norway	Others
Rwanda	251.9	10.0	0.6	10.3	45.6	41.1	19.3	20.4	5.4	0.4	24.5	74.3
Saudi Arabia	12.6	0.0	9.9	2.1	0.6	0.0	0.0	0.0	0.0	0.0	0.0	0.1
Senegal	392.0	43.0	58.0	177.6	35.8	9.7	1.2	16.0	1.8	2.4	14.0	32.6
Sierra Leone	67.0	11.0	−0.1	3.4	10.6	4.9	17.5	0.5	2.2	0.3	2.6	14.1
Singapore	11.9	0.0	8.5	0.0	2.5	0.1	0.0	0.6	0.0	0.0	0.0	0.1
Slovak Republic	83.8	9.0	1.2	0.0	12.9	0.0	51.9	0.9	0.1	1.1	0.4	6.4
Slovenia	32.3	2.0	1.1	0.7	21.8	0.2	1.5	0.0	0.1	0.0	0.0	4.9
South Africa	311.9	73.0	7.3	13.4	29.3	37.6	30.9	11.0	33.2	30.0	13.7	32.5
Spain												
Sri Lanka	279.3	4.0	173.9	−1.6	15.8	13.1	12.1	4.0	12.5	−0.8	31.7	14.4
Sudan	118.1	10.0	18.6	5.0	19.4	22.1	9.6	3.2	2.2	1.5	10.7	15.8
Sweden												
Switzerland												
Syrian Arab Republic	70.2	0.0	34.9	13.1	19.0	0.2	0.1	0.0	2.0	0.0	0.1	0.9
Tajikistan	44.5	21.0	0.3	0.0	5.4	3.8	7.9	0.0	0.8	0.0	1.0	4.4
Tanzania	605.4	13.0	105.7	3.5	58.7	74.9	67.3	9.2	65.2	91.2	54.4	62.4
Thailand	803.1	3.0	664.0	10.4	23.2	6.4	1.9	10.7	19.0	19.0	3.7	41.6
Togo	97.2	2.0	26.7	36.0	20.1	0.7	1.0	0.5	1.5	2.8	0.0	6.0
Trinidad and Tobago	−1.2	0.0	1.8	0.9	−4.9	0.2	0.5	0.3	0.0	0.0	0.0	0.2
Tunisia	41.5	−21.0	−3.4	58.8	−19.2	3.1	0.2	2.3	1.2	−0.2	0.0	19.8
Turkey	50.6	−75.0	2.7	31.5	82.8	−1.4	2.2	4.6	4.2	−0.4	0.2	−0.9
Turkmenistan	13.6	12.0	0.7	0.3	0.3	0.0	0.3	0.0	0.0	0.0	0.0	0.1
Uganda	369.9	29.0	26.9	12.8	40.4	32.6	69.4	1.7	32.7	68.0	21.2	35.3
Ukraine	353.2	256.0	0.5	0.0	67.6	0.0	4.9	13.6	2.7	1.5	0.3	6.0
United Arab Emirates	6.0	0.0	3.3	0.0	2.2	0.0	0.4	0.0	0.0	0.0	0.0	0.0
United Kingdom												
United States												
Uruguay	29.4	1.0	4.4	4.3	8.8	1.8	0.3	1.2	1.6	0.0	0.1	6.0
Uzbekistan	64.2	6.0	25.3	1.8	29.8	0.1	1.0	0.0	0.0	0.0	0.0	0.2
Venezuela	26.0	0.0	6.4	4.6	4.5	0.3	0.3	0.9	0.0	0.0	0.0	9.0
Vietnam	469.7	0.0	120.9	67.3	52.8	29.7	8.4	9.1	46.2	34.6	4.2	96.6
West Bank and Gaza	262.2	27.0	7.5	9.6	24.6	58.8	5.4	3.7	27.6	3.6	50.4	44.1
Yemen, Rep.	133.3	3.0	25.8	11.8	43.1	42.4	3.4	0.4	0.7	−0.5	0.1	3.0
Yugoslavia, FR (Serb./Mont.)[a]	69.4	0.0	0.1	3.2	41.6	0.0	0.0	0.0	13.0	0.0	2.3	9.3
Zambia	354.1	18.0	42.3	1.7	79.7	26.6	60.7	9.5	31.1	25.8	30.6	28.2
Zimbabwe	280.8	17.0	46.7	6.3	30.5	32.4	25.2	9.4	35.9	20.5	19.5	37.5
World	43,467 s	8,531 s	8,379 s	6,464 s	5,426 s	2,288 s	1,924 s	1,538 s	1,523 t	1,166 s	995 s	5,233 s
Low income	18,865	1,025	4,449	2,558	2,741	1,440	1,343	693	915	816	679	2,432
Excl. China & India	13,097	957	2,369	2,010	1,921	1,001	808	359	624	562	542	1,945
Middle income	16,100	2,433	3,885	1,950	2,661	540	407	718	433	243	194	2,410
Lower middle income	11,497	2,063	3,297	1,271	1,935	319	201	273	273	122	133	1,611
Upper middle income	1,982	223	31	347	428	79	109	169	55	63	28	451
Low & middle income	29,272	3,249	7,138	3,740	4,796	1,460	1,328	869	1,021	805	726	4,141
East Asia & Pacific	569	396	36	43	40	5	16	18	8	16	8	118
Europe & Central Asia	5,257	1,986	294	63	1,116	164	232	213	214	114	121	736
Latin America & Carib.	5,605	1,917	986	242	860	360	89	135	139	78	69	731
Middle East & N. Africa	3,762	846	526	1,016	704	92	33	139	67	31	15	293
South Asia	3,733	253	1,342	402	269	244	344	134	145	139	154	244
Sub-Saharan Africa	9,204	635	1,070	2,078	1,203	680	615	255	461	444	385	1,378
High income	3,065	2,205	−91	822	−33	144	1	1	1	0	0	13

Note: World and regional totals include aid to economies not specified elsewhere. World totals include aid not allocated by country or region.

a. Includes net flows to states of the former Yugoslavia: Bosnia and Herzegovina, Croatia, Macedonia FYR, and Slovenia.

Distribution of net aid by Development Assistance Committee members | 6.12

The data in the table show net bilateral aid to low- and middle-income economies from members of the Development Assistance Committee (DAC) of the Organisation for Economic Co-operation and Development (OECD). The DAC compilation includes aid to some countries and territories not shown in the table and small quantities to unspecified economies that are recorded only at the regional or global level. Aid to countries and territories not shown in the table has been assigned to regional totals based on the World Bank's regional classification system. Aid to unspecified economies has been included in regional totals, but not in totals for income groups. Aid not allocated by country or region—including administrative costs, research into development issues, and aid to nongovernmental organizations—is included in the world total; thus regional and income group totals do not add up to the world total.

Because these data are based on donor country reports of bilateral programs, they cannot be reconciled with recipient country reports. Nor do they reflect the full extent of aid flows from the reporting donor countries or to recipient countries. A full accounting would include donor country contributions to multilateral institutions and the flow of resources from multilateral institutions to recipient countries as well as flows from countries that are not members of DAC. In addition, the expenditures countries report as official development assistance (ODA) have changed. For example, some DAC members providing aid to refugees within their own borders have reported these expenditures as ODA.

Some of the aid recipients shown in the table are themselves significant donors. See table 6.9a for a summary of ODA from non-DAC countries.

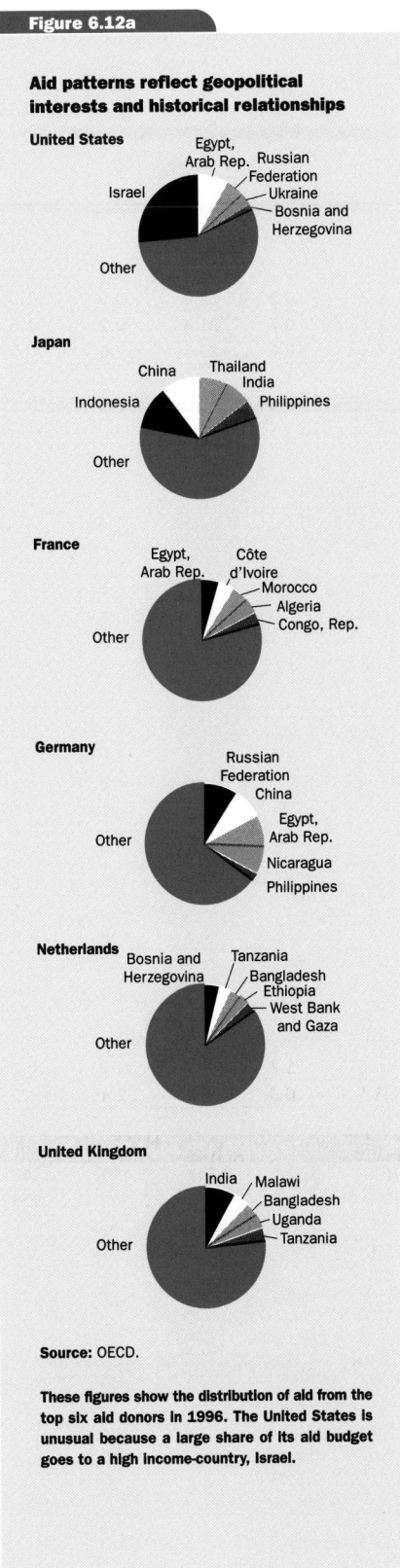

Figure 6.12a

Aid patterns reflect geopolitical interests and historical relationships

Source: OECD.

These figures show the distribution of aid from the top six aid donors in 1996. The United States is unusual because a large share of its aid budget goes to a high income-country, Israel.

Definitions

• **Net aid** comprises net bilateral ODA to part I recipients and net bilateral official aid to part II recipients (see *About the data* for table 6.9). • **Other DAC donors** are Australia, Austria, Belgium, Denmark, Finland, Ireland, Luxembourg, New Zealand, Portugal, Spain, and Switzerland.

Data sources

Data on aid are compiled by DAC and published in its annual statistical report, *Geographical Distribution of Financial Flows to Aid Recipients*, and in the DAC chairman's report, *Development Co-operation*. The OECD also makes its data available on diskette, magnetic tape, and the Internet.

$ millions, 1996	World Bank IDA	IBRD	IMF Concessional	Non-concessional	Regional development banks Concessional	Non-concessional	Others	WFP	UNDP	UNFPA	UNICEF	Others	Total
Albania	32.3	0.0	0.0	−8.3	0.0	4.7	3.6	0.0	2.2	0.4	2.0	2.4	39.2
Algeria	0.0	34.3	0.0	607.5	0.0	72.2	279.9	6.4	0.5	1.4	1.1	8.0	1,011.4
Angola	37.8	0.0	0.0	0.0	0.0	0.0	0.0	66.6	4.8	1.2	15.9	28.8	155.1
Argentina	0.0	794.8	0.0	365.3	−1.9	312.1	0.0	0.0	144.6	0.2	2.5	7.7	1,625.4
Armenia	87.0	5.4	49.0	0.0	0.0	29.2	−8.8	3.4	0.5	0.5	1.9	7.4	175.5
Australia													
Austria													
Azerbaijan	35.8	0.0	0.0	78.1	0.0	5.7	2.0	4.3	1.0	0.4	2.7	6.0	136.0
Bangladesh	2,29.1	−4.5	−85.3	0.0	259.4	2.9	66.2	27.6	10.8	8.0	24.5	15.2	553.9
Belarus	0.0	13.9	0.0	0.0	0.0	28.7	0.0	0.0	0.7	0.0	0.2	0.8	44.3
Belgium													
Benin	37.4	0.0	17.9	0.0	23.2	−0.1	4.9	3.6	6.7	0.9	2.4	4.3	101.2
Bolivia	96.9	−25.4	17.3	0.0	64.8	−9.1	18.5	4.0	16.2	1.9	9.9	5.3	200.3
Bosnia and Herzegovina	109.6	−25.1	0.0	−2.1	0.0	0.0	0.0	0.0	5.7	0.4	11.5	8.8	108.8
Botswana	−0.5	−20.9	0.0	0.0	−0.8	−4.4	3.6	3.0	3.7	0.7	0.9	1.4	−13.2
Brazil	0.0	278.2	0.0	−70.0	−0.5	490.5	0.0	0.1	123.3	1.8	21.9	21.4	866.8
Bulgaria	0.0	39.9	0.0	−108.7	0.0	0.0	40.7	0.0	1.0	0.0	0.0	3.2	−23.9
Burkina Faso	46.0	0.0	8.7	0.0	14.6	−1.7	6.4	6.6	8.0	2.4	4.7	3.1	98.7
Burundi	13.9	0.0	−8.7	0.0	14.5	−4.3	−1.4	10.3	4.9	1.3	7.2	83.5	121.3
Cambodia	45.6	0.0	0.0	0.0	31.1	0.0	4.8	13.7	37.5	2.3	8.8	7.8	151.5
Cameroon	79.6	−75.8	0.0	22.8	−0.2	−25.2	−18.2	1.6	1.0	1.3	1.9	3.4	−7.9
Canada													
Central African Republic	21.5	0.0	−6.2	0.0	0.0	0.0	0.8	0.1	3.3	1.0	1.4	5.9	27.8
Chad	66.6	0.0	17.8	0.0	23.5	0.0	0.8	13.8	5.7	1.3	3.6	1.2	134.3
Chile	−0.7	−187.8	0.0	0.0	−1.3	−403.0	0.0	0.0	7.8	0.4	0.8	2.9	−580.9
China	790.7	943.0	0.0	0.0	0.0	612.0	−4.1	22.4	28.7	0.0	18.3	14.1	2,425.1
Hong Kong, China													
Colombia	−0.7	−198.2	0.0	0.0	−12.0	14.2	11.0	1.9	63.4	0.6	5.0	3.1	−111.7
Congo, Dem. Rep.	0.0	0.0	−3.5	−32.9	0.0	0.0	0.0	3.6	9.4	0.6	10.2	6.9	−5.7
Congo, Rep.	0.8	−14.2	20.2	0.0	−0.3	−7.3	−6.9	0.0	0.4	1.1	1.2	1.7	−3.4
Costa Rica	−0.2	−31.7	0.0	−23.0	−11.2	23.5	37.7	0.9	2.8	0.3	1.1	5.4	5.6
Côte d'Ivoire	234.8	−159.5	138.3	−47.6	16.0	−22.9	−51.4	6.0	1.2	1.5	2.4	13.0	131.9
Croatia	0.0	88.9	0.0	−4.5	0.0	10.6	−46.6	0.0	1.2	0.1	3.5	0.4	53.6
Cuba								10.3	1.5	1.8	3.0	2.3	18.9
Czech Republic	0.0	32.2	0.0	0.0	0.0	−21.0	17.2	0.0	0.5	0.0	0.0	1.8	30.8
Denmark													
Dominican Republic	−0.7	−18.8	0.0	−59.5	16.8	26.5	7.2	3.2	5.8	1.2	1.1	1.8	−15.5
Ecuador	−1.1	−29.8	0.0	−23.0	13.8	47.3	42.8	3.0	18.3	1.0	4.7	4.2	81.3
Egypt, Arab Rep.	67.4	−151.8	0.0	−85.1	1.6	25.8	95.1	1.8	12.6	2.6	5.3	16.1	−8.8
El Salvador	−0.7	−1.1	0.0	0.0	24.3	199.3	19.2	4.6	18.0	0.6	3.5	2.8	270.6
Eritrea	1.5	0.0	0.0	0.0	0.0	0.0	2.3	0.0	8.7	1.1	8.7	3.1	25.3
Estonia	0.0	16.5	0.0	−11.1	0.0	26.6	5.7	0.0	1.1	0.0	0.0	0.3	39.0
Ethiopia	127.5	0.0	21.4	0.0	75.8	31.0	−4.6	46.4	39.6	3.1	18.0	16.8	374.9
Finland													
France													
Gabon	0.0	−10.5	0.0	26.6	0.3	26.9	0.7	0.0	0.8	0.5	0.8	1.1	47.2
Gambia, The	9.1	0.0	−7.4	0.0	8.8	−1.2	−0.6	0.9	3.1	0.4	1.3	2.5	16.8
Georgia	76.3	0.0	80.6	0.0	0.0	2.0	0.0	3.0	1.0	0.0	2.1	7.0	171.9
Germany													
Ghana	233.7	−10.0	−61.0	−24.5	17.5	−9.5	1.4	0.0	4.8	0.7	6.9	7.7	167.6
Greece													
Guatemala	0.0	55.9	0.0	0.0	8.7	−18.3	22.9	8.3	12.3	0.4	2.9	9.6	102.6
Guinea-Bissau	13.3	0.0	2.0	0.0	11.5	−1.5	0.8	4.5	3.0	0.5	1.8	1.9	37.7
Guinea	43.0	0.0	−8.4	0.0	7.1	16.0	9.8	0.8	3.4	2.4	3.1	24.2	101.5
Haiti	62.9	0.0	−2.6	0.0	28.6	0.0	−0.6	4.4	20.1	2.2	9.0	3.1	127.1
Honduras	50.0	−59.2	0.0	−37.9	111.8	−90.3	7.4	2.6	12.4	1.4	2.3	3.0	3.5

Net financial flows from multilateral institutions | 6.13

$ millions, 1996	International financial institutions — World Bank — IDA	IBRD	IMF Concessional	IMF Non-concessional	Regional development banks Concessional	Regional development banks Non-concessional	Others	United Nations WFP	UNDP	UNFPA	UNICEF	Others	Total
Hungary	0.0	−419.0	0.0	−203.3	0.0	17.3	−167.5	0.0	0.3	0.0	0.0	4.0	−768.3
India	671.9	−154.4	0.0	−972.5	0.0	502.4	−4.1	29.0	22.2	13.3	62.7	12.2	182.7
Indonesia	−20.4	−503.0	0.0	0.0	22.6	−839.7	50.8	0.0	12.2	2.9	13.9	8.3	−1,252.5
Iran, Islamic Rep.	0.0	110.0	0.0	0.0	0.0	−9.5		2.2	2.3	2.3	1.2	20.2	128.6
Iraq								46.7	0.6	0.0	16.0	10.1	73.4
Ireland													
Israel								0.0	0.0	0.0	0.0	0.1	0.1
Italy													
Jamaica	0.0	−39.2	0.0	−71.8	−4.8	7.6	−0.1	0.7	0.9	0.5	1.4	1.3	−103.6
Japan													
Jordan	−2.3	90.9	0.0	97.2	0.0	0.0	146.8	5.5	2.4	1.1	1.5	67.9	411.0
Kazakhstan	0.0	225.1	0.0	134.7	6.0	26.3	−31.6	0.0	1.1	1.4	2.1	0.9	366.0
Kenya	145.5	−88.5	−24.6	0.0	24.3	−7.5	−11.4	17.7	1.7	3.1	6.7	20.7	87.6
Korea, Dem. Rep.								22.4	3.8	0.8	3.1	1.8	31.8
Korea, Rep.								0.0	2.0	0.0	0.0	1.5	3.5
Kuwait								0.0	1.7	0.0	0.0	1.0	2.7
Kyrgyz Republic	61.2	0.0	23.4	−3.9	26.7	20.8	1.9	0.0	2.5	1.1	1.2	0.4	135.3
Lao PDR	59.0	0.0	5.5	0.0	82.5	0.0	44.5	4.5	11.5	1.1	3.8	2.4	214.8
Latvia	0.0	24.3	0.0	−25.5	0.0	7.8	0.0	0.0	1.3	0.0	0.0	0.2	8.1
Lebanon	0.0	27.1	0.0	0.0	0.0	0.0	111.5	1.2	6.3	1.0	2.5	38.4	187.9
Lesotho	10.2	7.2	−3.4	0.0	4.2	−2.4	6.1	6.8	3.7	0.7	1.3	1.8	36.2
Libya								0.0	1.9	0.0	0.0	6.0	7.9
Lithuania	0.0	43.8	0.0	20.6	0.0	44.2	2.2	0.0	1.6	0.0	0.0	0.7	113.0
Macedonia, FYR	44.1	−12.7	0.0	13.5	0.0	48.3	−7.6	0.0	0.3	0.0	1.4	2.9	90.1
Madagascar	68.5	−4.0	2.8	0.0	−0.7	−0.5	13.5	1.6	7.9	1.8	4.6	3.5	99.0
Malawi	132.8	−9.3	7.3	0.0	13.7	−3.0	−2.7	8.0	11.6	3.2	7.8	3.5	172.8
Malaysia	0.0	−76.4	0.0	0.0	0.0	−21.7	−3.8	0.0	3.9	0.4	0.8	2.3	−94.6
Mali	77.2	0.0	22.6	0.0	16.7	−0.9	3.5	3.9	14.4	1.8	5.6	9.6	154.5
Mauritania	34.1	−2.1	10.8	0.0	10.4	−4.5	−0.9	5.1	5.0	0.5	2.9	6.9	68.3
Mauritius	−0.6	−6.7	0.0	0.0	0.1	−3.2	13.4	0.1	0.7	0.3	0.6	0.8	5.5
Mexico	0.0	−358.8	0.0	−2,052.3	−5.9	764.9	0.0	0.1	3.3	1.7	4.6	10.2	−1,632.3
Moldova	0.0	0.0	0.0	25.3	0.0	35.1	19.0	0.0	1.4	0.1	0.7	0.3	82.0
Mongolia	11.0	0.0	8.1	−10.0	34.1	0.0	0.0	0.0	3.2	0.9	1.4	2.2	50.9
Morocco	−1.4	39.1	0.0	−47.2	3.9	66.7	40.3	0.7	3.8	3.6	2.3	4.8	116.5
Mozambique	220.2	0.0	−14.4	0.0	30.4	1.7	5.3	14.6	19.5	1.6	14.4	14.5	307.9
Myanmar	−10.8	0.0	0.0	0.0	−11.4	−0.9	−4.0	0.0	5.8	1.0	8.0	16.4	4.2
Namibia								0.7	1.8	1.9	3.6	4.2	12.2
Nepal	53.8	0.0	−7.6	0.0	57.8	0.0	6.0	12.7	8.3	6.2	8.5	10.6	156.2
Netherlands													
New Zealand													
Nicaragua	67.4	−16.4	0.0	−9.3	28.7	12.5	6.4	5.5	15.0	2.0	3.5	3.7	118.9
Niger	28.7	0.0	2.3	0.0	0.3	0.0	4.0	6.7	5.4	1.7	5.0	6.4	60.3
Nigeria	89.2	−230.7	0.0	0.0	4.0	−1.8	0.0	0.0	25.0	2.7	14.5	4.3	−92.8
Norway													
Oman	0.0	−4.6	0.0	0.0	0.0	0.0	31.3	0.0	0.0	0.0	0.7	0.8	28.1
Pakistan	241.3	144.4	−79.1	−86.6	350.3	−37.2	90.6	5.7	7.4	6.4	10.2	16.4	669.7
Panama	0.0	37.2	0.0	24.1	−9.5	61.2	6.9	0.1	43.5	0.3	0.9	0.9	165.6
Papua New Guinea	−1.8	−9.1	0.0	2.9	3.2	−3.8	−6.8	0.0	3.0	1.0	3.6	2.1	−5.6
Paraguay	−1.4	−6.3	0.0	0.0	18.8	36.7	11.4	1.5	12.6	0.6	1.2	1.2	76.3
Peru	0.0	29.3	0.0	0.0	−7.2	100.0	−20.5	2.0	80.7	2.4	8.2	3.5	198.4
Philippines	13.1	17.7	0.0	−301.3	45.8	27.2	14.6	0.0	6.2	7.6	7.6	6.7	−154.7
Poland	0.0	266.2	0.0	0.0	0.0	0.0	0.0	0.0	1.6	0.1	0.0	3.5	271.4
Portugal													
Puerto Rico													
Romania	0.0	227.5	0.0	−356.2	0.0	173.9	48.3	0.0	1.0	0.4	2.4	4.6	101.8
Russian Federation	0.0	1,097.1	0.0	3,235.1	0.0	85.9	−287.3	2.8	1.6	0.4	1.6	12.9	4,150.1

| $ millions, 1996 | International financial institutions | | | | | | | United Nations | | | | | Total |
| | World Bank | | IMF | | Regional development banks | | Others | | | | | | |
	IDA	IBRD	Concess-ional	Non-concessional	Concess-ional	Non-concessional		WFP	UNDP	UNFPA	UNICEF	Others	
Rwanda	38.1	0.0	−1.3	0.0	7.6	−0.5	9.4	183.8	22.8	0.7	22.4	89.5	372.6
Saudi Arabia								0.0	4.8	0.0	0.0	11.1	15.9
Senegal	102.9	−9.8	−10.0	0.0	10.5	−5.5	−32.3	1.8	3.5	3.5	5.0	10.3	79.9
Sierra Leone	33.7	−0.5	11.4	0.0	21.1	0.0	2.8	13.8	7.6	0.5	3.2	4.3	97.9
Singapore								0.0	0.0	0.0	0.0	0.4	0.4
Slovak Republic	0.0	6.0	0.0	−124.2	0.0	14.7	21.4	0.0	0.3	0.0	0.0	2.0	−79.8
Slovenia	0.0	−19.2	0.0	−2.6	0.0	38.3	29.3	0.0	0.2	0.0	0.0	2.7	48.7
South Africa	0.0	0.0	0.0	0.0	0.0	0.0	0.0	0.0	3.5	1.1	2.3	5.3	12.1
Spain													
Sri Lanka	94.9	−6.4	−45.3	0.0	137.4	0.0	0.3	2.5	5.6	1.7	4.4	7.4	202.5
Sudan	0.0	0.0	0.0	−35.6	2.4	14.9	0.0	18.7	7.7	3.1	28.5	13.1	52.7
Sweden													
Switzerland													
Syrian Arab Republic	0.0	0.0	0.0	0.0	0.0	0.0	50.8	6.6	2.4	1.5	0.8	26.2	88.4
Tajikistan	30.4	0.0	0.0	21.8	0.0	0.0	0.0	9.6	2.4	0.3	2.6	0.9	67.9
Tanzania	120.5	−26.6	15.6	0.0	50.1	−4.1	−14.4	7.0	10.4	4.0	10.8	18.0	191.4
Thailand	−1.8	−58.4	0.0	0.0	−1.4	35.1	39.1	0.0	5.1	1.8	3.0	9.5	32.0
Togo	51.8	0.0	−11.2	−0.3	5.9	−2.5	−5.5	0.3	5.4	1.0	1.6	5.7	52.2
Trinidad and Tobago	0.0	12.1	0.0	−25.1	−0.1	68.5	3.4	0.0	0.2	0.0	0.0	0.9	59.9
Tunisia	−2.1	14.6	0.0	−46.6	0.0	82.4	55.5	5.4	1.5	2.4	1.0	2.8	116.8
Turkey	−5.9	−325.9	0.0	0.0	0.0	0.0	−293.0	0.4	3.3	0.7	2.9	7.1	−610.4
Turkmenistan	0.0	2.5	0.0	0.0	0.0	0.5	−55.3	0.0	1.2	0.1	1.4	0.3	−49.3
Uganda	115.7	0.0	13.7	0.0	27.0	0.1	6.5	14.3	19.0	2.9	13.0	25.4	237.5
Ukraine	0.0	406.0	0.0	778.2	0.0	18.6	126.7	0.0	1.5	0.0	1.8	2.4	1,335.1
United Arab Emirates								0.0	2.4	0.0	0.0	0.1	2.5
United Kingdom													
United States													
Uruguay	0.0	−31.2	0.0	−11.6	−1.7	59.1	9.4	0.0	17.7	0.3	0.7	1.5	44.2
Uzbekistan	0.0	9.0	0.0	86.0	0.0	49.2	0.0	0.0	3.1	0.6	3.4	1.3	152.6
Venezuela	0.0	−120.9	0.0	31.2	−1.3	−72.3	−5.6	0.0	6.3	0.4	2.0	2.6	−157.8
Vietnam	188.0	0.0	175.4	0.0	24.8	0.0	−0.8	12.6	11.8	4.5	12.1	4.8	433.1
West Bank and Gaza								3.1	47.1	0.7	3.0	120.1	174.0
Yemen, Rep.	86.4	0.0	0.0	122.0	0.0	0.0	20.6	9.0	7.1	3.2	3.7	8.6	260.5
Yugoslavia, FR (Serb./Mont.)	0.0	0.0	0.0	−1.1	0.0	0.0	0.0	139.8[a]	−0.7[a]	0.0[a]	4.4[a]	238.8[a]	381.2
Zambia	178.0	−47.1	0.0	0.0	14.4	−14.5	−3.9	3.9	6.7	1.7	8.4	5.3	152.9
Zimbabwe	11.0	−7.8	0.0	−8.6	0.6	38.5	3.6	0.0	4.6	1.3	6.1	4.5	54.0
World[b]	5,724 s	1,516 s	304 s	725 s	1,870 s	2,858 s	665 s	943 s	1,383 s	208 s	642 s	1,661 s	18,498 s
Low income	5,421	143	207	−1,014	1,676	1,063	171	642	511	113	454	607	9,993
Excluding China & India	3,959	−646	207	−41	1,676	−51	179	591	460	100	373	580	7,385
Middle income	303	1,374	97	1,739	195	1,795	494	289	689	56	177	644	7,850
Lower middle income	..	..	..	..	..	..	..	289	372	49	138	558	..
Upper middle income	..	..	..	..	..	..	..	0	317	7	39	86	..
Low & middle income	..	..	..	..	..	..	..	966	1,431	241	644	1,800	..
East Asia & Pacific	1,077	314	190	−308	234	−192	128	76	159	27	87	142	1,932
Europe and Central Asia	471	1,702	153	3,543	33	668	−580	163	53	7	50	365	6,682
Latin America & Caribbean	288	82	28	−2,001	311	1,651	193	53	633	24	90	111	1,488
Middle East & N. Africa	148	160	0	648	6	247	822	85	52	21	36	234	2,458
South Asia	1,295	−21	−217	−1,059	813	468	161	78	65	38	111	77	1,808
Sub-Saharan Africa	2,447	−721	151	−96	475	17	−59	477	327	65	261	77	3,876
High income	..	..	..	..	..	..	..	0	6	0	0	3	9

a. Includes net flows to states of the former Yugoslavia: Bosnia and Herzegovina, Croatia, Macedonia FYR, and Slovenia. b. Includes data for economies not specified elsewhere.

Net financial flows from multilateral institutions | 6.13

About the data

This table shows concessional and nonconcessional financial flows from the major multilateral institutions—the World Bank, the International Monetary Fund (IMF), regional development banks, United Nations agencies, and regional groups such as the Commission of the European Communities. Much of these data comes from the World Bank's Debtor Reporting System.

The multilateral development banks fund their non-concessional lending operations primarily by selling low-interest, highly rated bonds (the World Bank, for example, has a AAA rating) backed by prudent lending and financial policies and the strong financial backing of their members. These funds are then onlent at slightly higher interest rates, and with relatively long maturities (15–20 years), to developing countries. Lending terms vary with market conditions and the policies of the banks.

Concessional, or soft, lending by the World Bank Group is carried out through the International Development Association (IDA), although some loans by the International Bank for Reconstruction and Development (IBRD) are made on terms that may qualify as concessional under the Development Assistance Committee (DAC) definition. Eligibility for IDA lending is based on estimates of average GNP per capita, which

are revised annually. In 1997 countries with GNP per capita of $925 or less were eligible for IDA lending.

The IMF makes concessional funds available through its Enhanced Structural Adjustment Facility (ESAF), the successor to the Structural Adjustment Facility, and through the IMF Trust Fund. Low-income countries that face protracted balance of payments problems are eligible for ESAF funds.

Regional development banks also maintain concessional windows for funds. In *World Development Indicators 1997* loans from these institutions were classified using DAC definitions, under which concessional flows contain a grant element of at least 25 percent. (The grant element of loans is evaluated assuming a nominal, market interest rate of 10 percent. The grant element of a loan carrying a 10 percent interest rate is nil, and for a grant, which requires no repayment, it is 100 percent.) In some cases nonconcessional loans by these institutions may be on terms that meet DAC's definition of concessional; this year's *World Development Indicators* records loans from the major regional development banks—the African Development Bank (AfDB), Asian Development Bank (ADB), and the Inter-American Development Bank (IDB)—according to each institution's classification.

Definitions

• **Net financial flows** are disbursements of loans and credits less repayments of principal. • **IDA** is the International Development Association, the soft loan window of the World Bank Group. • **IBRD** is the International Bank for Reconstruction and Development, the founding and largest member of the World Bank Group. • **IMF** is the International Monetary Fund. Nonconcessional lending is the credit provided by the IMF to its members, principally to meet their balance of payments needs. Concessional assistance is provided through the Enhanced Structural Adjustment Facility. • **Regional development banks** include the African Development Bank (AfDB), based in Abidjan, Côte d'Ivoire, which lends to all of Africa, including North Africa; the Asian Development Bank (ADB), based in Manila, Philippines, which serves countries in South Asia and East Asia and the Pacific; and the Inter-American Development Bank (IDB), based in Washington, D.C., which is the principal development bank of the Americas. • **Others** is a residual category in the World Bank's Debtor Reporting System. It includes such institutions as the Caribbean Development Bank, European Investment Bank, and European Development Fund. • **United Nations** includes the World Food Programme (WFP), United Nations Development Programme (UNDP), United Nations Population Fund (UNFPA), United Nations Children's Fund (UNICEF), and other United Nations agencies such as the United Nations High Commissioner for Refugees, United Nations Relief and Works Agency for Palestine Refugees in the Near East, and United Nations Regular Program for Technical Assistance. • **Concessional financial flows** cover disbursements made through concessional lending facilities. • **Nonconcessional financial flows** cover all other disbursements.

Figure 6.13a

Maintaining financial flows from the World Bank to developing countries

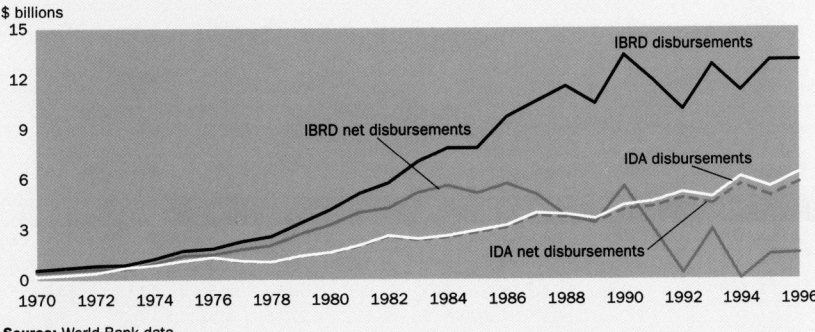

$ billions

IBRD disbursements
IBRD net disbursements
IDA disbursements
IDA net disbursements

1970 1972 1974 1976 1978 1980 1982 1984 1986 1988 1990 1992 1994 1996

Source: World Bank data.

As the World Bank's nonconcessional lending portfolio matures, repayments of principal have begun to balance out disbursements. The World Bank's concessional arm, the International Development Association (IDA), has maintained a steady flow of new funds to the world's poorest countries. In 1996 repayments to IDA were slightly over $500 million. While small relative to new IDA disbursements, repayments have helped IDA sustain its lending even as contributions from donors have declined.

Data sources

Data on net financial flows from international financial institutions come from the World Bank's Debtor Reporting System. These data are published annually in the World Bank's *Global Development Finance*. Data on aid from United Nations agencies come from the DAC chairman's report, *Development Co-operation*.

6.14 Foreign labor and population in OECD countries

	Foreign population[a] (thousands)		% of total population		Foreign labor force % of total labor force		Inflows of foreign population — Total thousands		Asylum seekers thousands	
	1990	1995	1990	1995	1990	1995	1990	1995	1990	1995
Austria	456 [b]	724 [b]	5.9	9.0	..	10.2	..	..	23 [c]	6 [c]
Belgium	905	910	9.1	9.0	7.5	8.1 [d]	51 [e]	53 [e]	13	12
Denmark	161	223	3.1	4.2	2.0	2.8 [d]	15	16 [d]	5	5
Finland	26	69	0.5	1.3	..	..	7	7	3	1
France	3,597 [f]	..	6.3	..	6.4	6.2	102	57	47	20
Germany	5,343 [g]	7,174	8.4 [g]	8.8	8.4	7.4	842	788	193	128
Ireland	80 [h]	96	2.3 [h]	2.7	2.6	3.0	..	..	..	..
Italy	781 [i]	991 [i]	1.4	1.7	..	1.9	..	..	5	2
Japan	1,075 [j]	1,362 [j]	0.9 [j]	1.1 [j]	..	0.9	224	210	..	..
Luxembourg	113	138	29.4	33.4	33.4	56.2	9	10	..	..
Netherlands	692	728 [k]	4.6	5.0 [k]	3.7	4.0	81 [e]	67 [e]	21	29
Norway	143 [l]	161 [l]	3.4 [l]	3.7 [l]	..	4.5	16	17	4	2
Portugal	108 [m]	168 [m]	1.1 [m]	1.7 [m]	..	1.7	..	..	0	1
Spain	279 [n]	500 [n]	0.7 [n]	1.2 [n]	..	0.6	..	..	9	6
Sweden	484	532	5.6	5.2	5.6	5.1 [d]	53	36	29	9
Switzerland	1,100 [o]	1,331 [o]	16.3 [o]	18.9 [o]	..	19.4	101	88	36	17
United Kingdom	1,723 [h]	2,060 [h]	3.2 [h]	3.4 [h]	3.5	3.6	52	56	38 [p]	55 [p]

	Foreign-born population[q] (thousands)		% of total population		Foreign-born labor force[r] % of total labor force		Inflows of foreign population — Total thousands		Asylum seekers thousands	
	1990	1995	1990	1995	1990	1995	1990	1995	1990	1995
Australia	4,125 [s]	..	22.7 [s]	..	25.8	24.0	121	99	4 [t]	5 [t]
Canada	4,343 [s]	..	15.6 [s]	..	18.4 [u]	18.5 [s]	214	212	37	26
United States	19,767	24,557 [v]	7.9	9.3 [v]	9.4	9.3	1,537	721	74 [w]	149 [w]

a. Except for France, Ireland, Portugal, and the United Kingdom, data are from population registers. Unless otherwise noted, they refer to the population on December 31 of the years indicated. b. Annual average. c. Data do not include de facto refugees from Bosnia and Herzegovina. d. Data refer to 1994. e. Includes some asylum seekers. f. Data are from the 1990 population census. g. Data refer to the Federal Republic of Germany before unification. h. Estimated from the annual labor force survey. i. Data are adjusted to take account of the regularizations in 1987–88 and 1990. Data for 1995 do not include permits delivered under the 1995–96 regularization program. j. Data refer to registered foreign nationals, who include foreigners staying in Japan for more than 90 days. k. Provisional data. l. Includes asylum seekers whose requests are being processed. m. Includes all foreigners who hold a valid residence permit. n. Data refer to foreigners with a residence permit. Those with permits for fewer than six months and students are excluded. o. Data refer to foreigners with an annual residence permit or with a settlement permit (permanent permit). p. Data adjusted to include dependents. q. Data are from the latest population census. r. Data are from labor force surveys except for Canada and the United States, for which data are from the latest population census. s. Data refer to 1991. t. Data refer to principal applicants and do not include dependents. u. Data refer to 1986. v. Data are from the United States Census Bureau March 1996 Population Survey and refer to 1996. w. Data refer to the fiscal year (October to September of years shown). Data do not include dependents.

Foreign labor and population in OECD countries | 6.14

The data in the table are based on national definitions and data collection practices and are not fully
comparable across countries. Japan and the
European members of the Organisation for Economic
Co-operation and Development (OECD) traditionally
have defined foreigners by nationality of descent.
Australia, Canada, and the United States use place
of birth, which is closer to the concept of the immigrant stock as defined by the United Nations. Few
countries, however, apply just one criterion in all circumstances. For this and other reasons data based
on the concept of foreign nationality and data based
on the concept of foreign-born cannot be completely
reconciled.

Data on the size of the foreign labor force are also
problematic. Countries use different permit systems
to gather information on immigrants. Some countries
issue a single permit for residence and work, while
others issue separate residence and work permits.
Differences in immigration laws across countries,
particularly with respect to immigrants' access to the
labor market, greatly affect the recording and measurement of migration and reduce the comparability
of raw data at the international level. The data
exclude temporary visitors and tourists (see table
6.15).

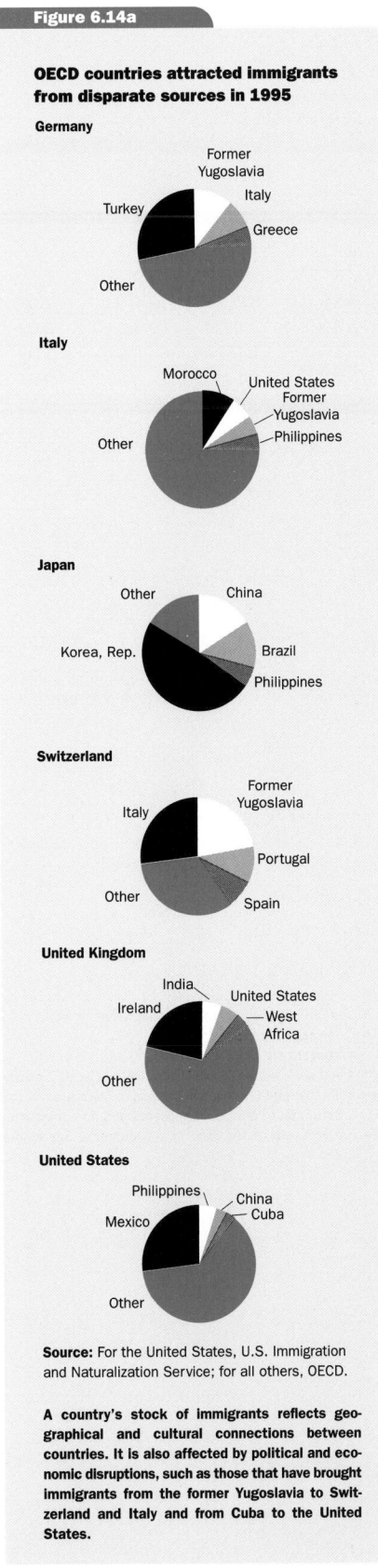

Figure 6.14a

OECD countries attracted immigrants from disparate sources in 1995

Source: For the United States, U.S. Immigration
and Naturalization Service; for all others, OECD.

A country's stock of immigrants reflects geographical and cultural connections between
countries. It is also affected by political and economic disruptions, such as those that have brought
immigrants from the former Yugoslavia to Switzerland and Italy and from Cuba to the United
States.

• **Foreign population** is the number of foreign or
foreign-born residents in a country. • **Foreign labor
force as a percentage of total labor force** is the
share of foreign or foreign-born workers in a country's
workforce. • **Inflows of foreign population** are the
gross arrivals of immigrants in the country shown.
The total does not include asylum seekers, except as
noted. • **Asylum seekers** are those who apply for
permission to remain in the country for humanitarian
reasons

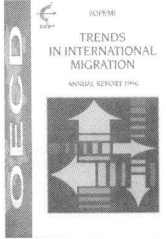

International migration data
are collected by the OECD
through information provided
by national correspondents to
the Continuous Reporting
System on Migration (SOPEMI)
network, which provides an
annual overview of trends and
policies. Data appear in the OECD's *Trends in
International Migration* (1996).

	International tourism				International tourism receipts				International tourism expenditures			
	Inbound tourists thousands		Outbound tourists thousands		$ millions		% of exports		$ millions		% of imports	
	1980	1996	1980	1996	1980	1996	1980	1996	1980	1996	1980	1996
Albania	4	56	..	16	..	11	..	2.9	..	7	..	0.6
Algeria	946	605	698	1,810	115	16	0.8	0.1	333	60	2.7	..
Angola	..	8	..	..	..	9	..	0.2	..	41	..	1.7
Argentina	1,120	4,286	..	3,550	345	4,572	3.5	16.9	1,791	2,340	13.6	8.4
Armenia	..	..	..	..	..	1	..	0.3	..	2	..	0.2
Australia	905	4,165	1,217	2,732	967	8,703	3.8	11.1	1,749	5,322	6.5	6.7
Austria	13,879	17,090	3,525	12,683	6,442	14,004	24.2	14.4	2,847	11,822	9.5	11.8
Azerbaijan	..	145	..	..	..	158	..	20.9	..	72	..	5.0
Bangladesh	57	166	..	935	15	32	1.7	0.7	16	200	0.6	2.6
Belarus	..	234	..	703	..	48	..	0.8	..	97	..	1.4
Belgium	3,777	5,829	9,565	5,645	1,810	5,893	2.6	3.1	3,272	9,895	4.4	5.5
Benin	39	147	..	415	7	29	3.1	5.3	4	6	1.0	1.3
Bolivia	155	375	..	258	40	160	3.9	12.5	52	155	6.2	8.8
Bosnia and Herzegovina	..	..	..	..	..	..	..	..	..	..	..	..
Botswana	236	707	..	460	22	178	3.4	6.8	17	149	2.1	7.2
Brazil	1,271	2,140	427	2,943	1,794	2,469	8.2	4.6	1,160	6,825	4.2	5.4
Bulgaria	1,933	2,795	759	3,006	260	450	2.8	7.4	..	197	..	3.4
Burkina Faso	38	136	..	..	6	23	2.9	6.4	32	32	5.5	4.8
Burundi	34	27	..	35	22	1	..	2.0	17	32	..	9.0
Cambodia	..	260	..	31	..	118	..	14.6	..	7	..	0.5
Cameroon	86	101	14	..	62	52	3.5	2.4	82	217	4.5	11.7
Canada	12,876	17,286	12,833	18,973	2,284	8,868	3.0	3.8	3,122	11,090	4.4	5.2
Central African Republic	7	29	..	..	3	5	1.5	2.6	18	39	5.5	17.6
Chad	7	8	..	11	3	10	4.2	3.2	14	24	17.6	6.3
Chile	420	1,450	379	1,070	166	918	2.8	4.9	195	801	2.8	4.0
China	3,500	22,765	..	5,061	617	10,200	3.6	5.9	66	4,000	0.3	2.6
Hong Kong, China	1,748	11,703	916	3,445	1,317	10,836	..	..	..	..	..	..
Colombia	553	1,254	781	1,073	357	909	6.7	6.3	250	856	4.6	5.1
Congo, Dem. Rep.	23	37	..	..	22	5	1.4	0.2	38	7	1.7	..
Congo, Rep.	48	27	..	..	10	4	1.0	0.3	29	38	2.8	1.8
Costa Rica	345	781	133	273	87	689	7.3	17.3	62	335	3.7	8.0
Côte d'Ivoire	194	237	..	5	79	89	2.2	1.7	270	164	6.5	4.1
Croatia	..	2,649	..	..	..	2,100	..	26.2	..	780	..	7.7
Cuba	101	999	7	55	40	1,350	..	..	..	..	..	..
Czech Republic	..	17,000	..	48,614	..	4,075	..	13.6	..	2,953	..	8.7
Denmark	950	1,794	..	4,955	1,337	3,425	5.6	5.3	1,560	4,142	5.8	7.4
Dominican Republic	383	1,926	257	175	168	1,755	13.2	44.6	166	92	8.7	2.0
Ecuador	243	500	..	279	91	281	3.2	4.9	228	219	7.7	4.9
Egypt, Arab Rep.	1,253	3,528	1,180	2,812	808	3,200	12.9	21.0	573	1,350	6.3	7.1
El Salvador	118	283	464	348	7	76	0.6	3.5	106	75	9.1	2.0
Eritrea	..	417	..	..	..	..	..	..	..	..	..	..
Estonia	..	600	..	217	..	470	..	14.8	..	98	..	2.6
Ethiopia	42	107	25	133	11	46	1.9	5.9	5	26	0.6	1.6
Finland	350	894	291	4,918	677	1,601	4.0	3.3	544	2,304	3.1	6.0
France	30,100	62,406	7,930	18,151	8,235	28,357	5.4	7.8	6,027	17,753	3.9	5.4
Gabon	17	136	..	..	17	4	0.7	0.1	96	115	6.5	6.1
Gambia, The	22	77	..	..	18	22	27.2	10.0	1	17	0.6	5.8
Georgia	..	..	..	..	..	..	..	..	..	..	..	..
Germany[a]	11,122	15,205	22,473	76,100	6,566	16,496	2.9	2.7	20,599	49,787	9.1	8.6
Ghana	40	298	..	..	1	239	0.1	13.8	27	22	2.3	0.9
Greece	4,796	8,987	1,374	1,620	1,734	3,660	21.3	24.0	190	1,400	1.7	5.5
Guatemala	466	520	178	333	183	284	10.6	10.2	183	179	9.3	5.1
Guinea	..	94	..	..	..	1	..	0.1	..	22	..	2.3
Guinea-Bissau	..	..	..	..	..	..	..	..	..	..	..	..
Haiti	138	150	..	..	65	81	21.3	42.3	41	32	8.5	4.1
Honduras	122	257	..	150	27	81	2.9	4.6	31	58	2.7	3.1

Travel and tourism | 6.15

	International tourism				International tourism receipts				International tourism expenditures			
	Inbound tourists thousands		Outbound tourists thousands		$ millions		% of exports		$ millions		% of imports	
	1980	1996	1980	1996	1980	1996	1980	1996	1980	1996	1980	1996
Hungary	9,413	20,674	5,164	12,064	160	2,246	2.5	11.7	88	958	0.9	4.7
India	1,194	2,288	1,017	3,056	1,150	3,027	10.2	6.9	113	415	0.7	0.8
Indonesia	527	5,034	635	1,782	246	6,087	1.2	10.8	375	2,300	3.0	4.1
Iran, Islamic Rep.	156	465	428	1,000	54	165	0.4	0.8	1,700	579	10.6	3.8
Iraq	1,222	345	443	200	170	13	..	..	..	..	..	..
Ireland	2,258	5,282	669	2,000	472	3,003	4.9	5.6	742	2,222	6.2	4.8
Israel	1,116	2,097	513	2,259	903	2,800	10.4	9.9	533	3,360	4.6	8.7
Italy	22,087	32,853	23,994	15,991	8,213	28,673	8.4	8.9	1,907	15,488	1.7	6.0
Jamaica	395	1,162	..	..	242	1,092	17.7	33.3	12	159	0.9	4.1
Japan	844	2,114	5,224	16,695	644	4,078	0.4	0.9	4,593	37,040	2.9	8.3
Jordan	393	1,103	720	1,141	431	744	36.5	20.3	301	381	12.5	7.0
Kazakhstan	..	..	..	..	..	..	..	..	..	..	..	..
Kenya	372	717	..	295	220	474	11.0	15.7	33	142	1.2	4.1
Korea, Dem. Rep.	..	127	..	..	..	..	..	..	..	..	..	..
Korea, Rep.	976	3,684	436	4,649	369	5,430	1.7	3.5	350	6,963	1.4	4.0
Kuwait	108	33	230	..	377	109	1.7	0.7	1,339	2,500	13.6	19.6
Kyrgyz Republic	..	13	..	42	..	5	..	0.9	..	7	..	0.7
Lao PDR	..	93	..	..	..	50	..	10.9	..	31	..	4.5
Latvia	..	97	..	1,798	..	182	..	6.9	..	25	..	0.8
Lebanon	..	420	..	..	..	715	..	50.6	..	..	..	..
Lesotho	73	108	..	..	12	19	13.3	9.3	8	8	1.7	0.8
Libya	126	88	95	185	10	6	0.0	..	470	215	3.7	..
Lithuania	..	832	..	2,864	..	345	..	8.2	..	270	..	5.4
Macedonia, FYR	..	..	..	..	..	..	..	..	..	..	..	..
Madagascar	13	83	..	38	5	65	1.0	8.1	31	52	2.9	5.2
Malawi	46	232	..	..	9	7	2.9	1.5	10	17	2.1	1.7
Malaysia	2,105	7,138	1,738	20,642	265	3,926	1.9	4.3	470	1,815	3.5	2.1
Mali	27	50	..	..	15	20	5.7	3.7	20	58	3.8	7.2
Mauritania	..	..	..	..	7	11	2.8	2.0	17	20	3.8	3.7
Mauritius	123	487	33	120	45	466	7.8	17.3	27	163	3.9	5.9
Mexico	11,945	21,405	3,322	9,001	5,393	6,934	23.8	6.5	4,174	3,387	15.1	3.4
Moldova	..	33	..	71	..	59	..	6.5	..	57	..	4.6
Mongolia	195	153	..	..	..	21	..	4.4	..	21	..	3.8
Morocco	1,425	2,693	578	1,212	397	1,381	12.3	14.9	98	316	1.9	2.9
Mozambique	..	..	..	..	..	..	..	..	..	..	..	..
Myanmar	38	165	..	..	10	90	1.9	2.1	3	25	0.4	1.4
Namibia	..	405	..	..	..	265	..	16.7	..	85	..	4.5
Nepal	163	404	23	70	45	130	20.1	13.0	26	140	7.1	8.5
Netherlands	2,784	6,580	6,749	10,261	1,668	6,256	1.8	2.8	4,664	11,370	5.1	5.6
New Zealand	465	1,529	454	920	211	2,444	3.3	12.9	534	1,382	7.7	7.4
Nicaragua	..	303	..	282	22	54	4.4	6.7	..	60	..	4.6
Niger	20	17	..	10	3	17	0.5	5.4	18	23	1.9	4.6
Nigeria	86	822	..	50	48	85	0.2	0.6	780	155	3.9	1.6
Norway	1,252	2,746	246	3,085	751	2,404	2.8	3.8	1,310	4,509	5.5	9.1
Oman	60	435	..	..	..	99	..	1.3	..	47	..	0.9
Pakistan	299	369	104	..	154	146	5.2	1.4	90	900	1.6	5.9
Panama	392	362	113	188	167	343	4.9	4.6	56	136	1.7	1.8
Papua New Guinea	40	56	..	51	12	68	1.2	2.3	18	77	1.4	3.4
Paraguay	302	425	..	418	91	236	13.0	6.0	35	229	2.7	10.0
Peru	373	515	127	508	208	535	4.5	7.4	107	350	2.7	3.5
Philippines	1,008	2,049	461	1,400	320	2,701	4.4	7.9	105	450	1.1	1.3
Poland	5,664	19,410	6,852	44,713	282	8,400	1.8	22.5	357	6,240	2.0	15.1
Portugal	2,730	9,730	..	2,358	1,147	4,265	17.2	12.8	290	2,353	2.9	5.6
Puerto Rico	1,639	3,065	2,758	1,237	619	1,898	..	..	400	895	..	..
Romania	..	136	1,711	5,737	..	20	..	0.2	..	26	..	0.2
Russian Federation	..	14,587	..	21,331	..	5,542	..	5.4	..	10,597	..	12.3

	International tourism				International tourism receipts				International tourism expenditures			
	Inbound tourists thousands		Outbound tourists thousands		$ millions		% of exports		$ millions		% of imports	
	1980	1996	1980	1996	1980	1996	1980	1996	1980	1996	1980	1996
Rwanda	30	1	..	..	4	1	2.4	1.2	11	17	3.4	4.7
Saudi Arabia	2,475	3,458	..	..	1,344	1,308	1.3	2.2	2,453	..	4.4	..
Senegal	186	263	..	..	68	147	8.4	9.3	45	77	3.7	4.1
Sierra Leone	46	46	..	..	10	10	3.6	9.0	8	2	1.7	1.4
Singapore	2,562	6,608	..	3,305	1,433	7,916	5.9	5.1	322	6,104	1.3	4.3
Slovak Republic	..	951	..	318	..	673	..	6.2	..	483	..	3.7
Slovenia	..	832	..	..	..	1,210	..	11.5	..	452	..	4.2
South Africa	700	4,944	572	2,775	652	1,995	2.3	6.0	756	2,100	3.4	6.4
Spain	22,388	41,295	18,022	12,644	6,968	27,414	21.7	18.7	1,229	4,921	3.2	3.5
Sri Lanka	322	302	138	494	111	168	8.6	3.5	34	176	1.5	2.9
Sudan	25	65	..	..	52	7	6.4	1.0	74	45	4.1	3.1
Sweden	1,366	2,376	2,941	6,582	962	3,683	2.5	3.6	1,235	6,285	3.1	7.4
Switzerland	8,873	10,600	4,451	10,860	3,149	8,891	6.5	7.3	2,357	7,479	4.5	6.9
Syrian Arab Republic	1,239	888	1,189	2,485	156	1,478	6.3	24.1	177	405	3.9	6.7
Tajikistan	..	..	..	..	..	..	..	..	..	..	..	..
Tanzania	84	310	..	148	20	322	3.0	23.6	20	504	1.6	23.1
Thailand	1,859	7,192	497	1,845	867	8,664	10.9	12.1	244	4,171	2.4	5.0
Togo	92	58	..	..	13	8	2.4	1.6	22	26	3.2	5.2
Trinidad and Tobago	199	282	206	261	151	74	4.8	2.6	140	82	5.8	3.7
Tunisia	1,602	3,885	478	1,778	601	1,436	18.4	17.6	55	268	1.5	3.1
Turkey	921	7,966	1,795	4,261	327	5,962	9.0	13.1	115	1,265	1.4	2.6
Turkmenistan	..	..	..	..	..	..	..	..	..	..	..	..
Uganda	36	205	..	..	5	100	1.5	13.8	18	110	4.1	6.9
Ukraine	..	814	..	..	..	202	..	1.0	..	250	..	1.2
United Arab Emirates	300	1,768	..	..	..	..	..	..	..	..	..	..
United Kingdom	12,420	25,293	15,507	41,873	6,932	19,296	4.7	5.7	6,893	25,445	5.1	7.3
United States	22,500	46,325	22,721	50,763	10,058	64,373	3.7	7.6	10,385	52,563	3.6	5.5
Uruguay	1,067	2,152	640	..	298	599	19.5	15.8	203	164	9.5	4.1
Uzbekistan	..	..	..	..	..	..	..	..	..	..	..	..
Venezuela	215	759	747	534	243	846	1.2	3.3	1,880	1,900	12.4	12.8
Vietnam	..	1,607	..	..	..	87	..	0.9	..	..	..	..
West Bank and Gaza	..	..	..	..	..	..	..	..	..	..	..	..
Yemen, Rep.	39	74	..	..	24	42	..	1.7	53	77	..	2.5
Yugoslavia, FR (Serb./Mont.)	..	162	..	..	..	43	..	..	..	..	..	..
Zambia	87	264	..	..	20	60	1.2	4.6	57	59	3.2	..
Zimbabwe	243	1,743	326	256	38	219	2.4	7.1	140	117	8.1	3.8

World	260,891 s	587,348 s	158,652 s	335,918 s	101,016 s	421,783 s	4.6 w	6.4 w	102,066 s	379,253 s	4.8 w	6.3 w
Low income	8,179	36,429	1,624	7,142	3,055	16,722	3.5	5.4	2,187	8,436	2.7	2.7
Excl. China & India	3,485	11,376	607	2,081	1,288	3,495	2.2	3.7	2,074	4,021	3.3	4.4
Middle income	59,321	186,536	33,393	144,683	19,388	96,400	4.7	8.8	20,299	59,596	5.8	6.2
Lower middle income	19,370	69,131	12,186	23,955	6,915	49,865	5.4	9.4	7,331	28,316	5.5	6.0
Upper middle income	39,951	117,405	21,207	120,728	12,473	46,535	4.4	8.2	12,968	31,280	6.0	6.5
Low & middle income	67,500	222,965	35,017	151,825	22,443	113,122	4.5	8.0	22,486	68,032	5.2	5.2
East Asia & Pacific	9,570	47,206	3,339	6,973	2,480	32,450	5.1	7.3	1,234	12,980	2.4	3.4
Europe & Central Asia	18,664	91,040	15,522	118,754	1,358	32,820	4.0	10.2	603	25,026	..	7.6
Latin America & Carib.	22,766	47,155	9,907	14,747	11,262	27,993	9.2	8.1	11,338	19,466	8.7	4.9
Middle East & N. Africa	11,086	19,744	4,620	6,165	4,260	10,903	2.3	8.2	6,323	3,868	5.0	..
South Asia	2,086	3,877	1,259	1,461	1,485	3,774	8.5	5.8	283	1,865	1.0	2.2
Sub-Saharan Africa	3,328	13,943	370	3,725	1,598	5,182	1.9	5.4	2,705	4,827	3.7	5.4
High income	193,391	364,383	123,635	184,093	78,573	308,661	4.6	5.9	79,580	311,221	4.7	6.5

a. Data prior to 1990 refer to the Federal Republic of Germany before unification.

Travel and tourism | 6.15

About the data

The data in the table are from the World Tourism Organization's *Yearbook of Tourism Statistics*. They are obtained primarily from a questionnaire sent to government offices, supplemented with data published by official sources. Although the World Tourism Organization tries to ensure the international comparability of national data and definitions, this is a relatively new area of statistical activity, and much work remains to be done.

Data on international inbound and outbound tourists refer to the number of arrivals and departures of visitors within the reference period, not to the number of people traveling. Thus a person who makes several trips to a country during a given period is counted each time as a new arrival. Regional and income group aggregates are based on the World Bank's classification of countries and differ from what is shown in the *Yearbook of Tourism Statistics*.

Definitions

• **International inbound tourists** are the number of visitors who travel to a country other than that where they have their usual residence for a period not exceeding 12 months and whose main purpose in visiting is other than an activity remunerated from within the country visited. • **International outbound tourists** are the number of departures that people make from their country of usual residence to any other country for any purpose other than a remunerated activity in the country visited. • **International tourism receipts** are expenditures by international inbound visitors, including payments to national carriers for international transport. These receipts should include any other prepayment made for goods or services received in the destination country. They also may include receipts from same-day visitors, except in cases where these are so important as to justify a separate classification. Their share in exports is calculated as a ratio to exports of goods and services. • **International tourism expenditures** are expenditures of international outbound visitors in other countries, including payments to foreign carriers for international transport. These may include expenditures by residents traveling abroad as same-day visitors, except in cases where these are so important as to justify a separate classification. Their share in imports is calculated as a ratio to imports of goods and services.

Figure 6.15a

Large, wealthy economies generated the most tourists in 1996

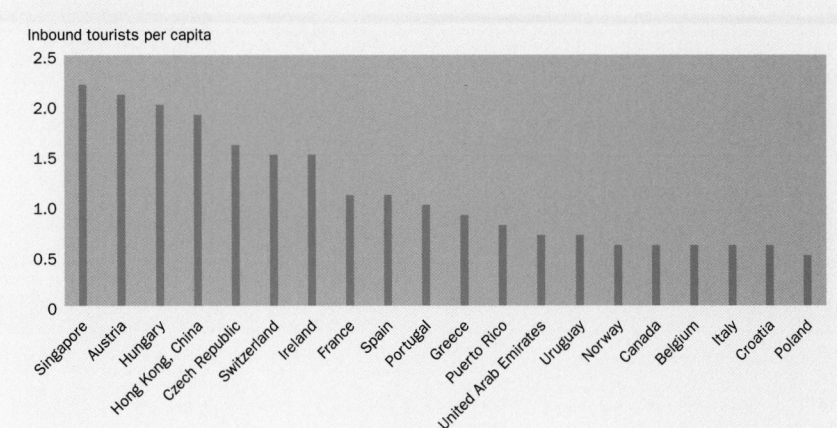

Czechs were the world's most active travelers in 1996

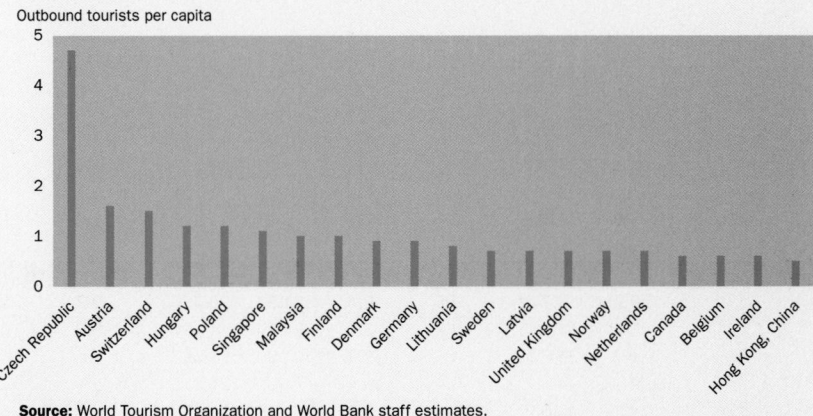

Source: World Tourism Organization and World Bank staff estimates.

Data sources

The visitor and expenditure data come from the World Tourism Organization's *Yearbook of Tourism Statistics*. Export and import data are from the International Monetary Fund's *International Financial Statistics* and World Bank staff estimates.

This section describes some of the statistical procedures used in preparing the *World Development Indicators*. It covers the methods employed for calculating regional and income group aggregates and for calculating growth rates, and it describes the World Bank's Atlas method for deriving the conversion factor used to estimate GNP and GNP per capita in U.S. dollars. Other statistical procedures and calculations are described in the *About the data* sections that follow each table.

Aggregation rules

Aggregates based on the World Bank's regional and income classification of economies appear at the end of most tables. The World Bank's regional and income classifications are shown on the front- and back-cover end flaps of the book. Specialized classifications, such as high-income OECD countries, are documented in *About the data* of the tables in which they appear.

Because of missing data, aggregations for groups of economies should be treated as approximations for unknown totals or average values. Regional and income group aggregates are based on the largest available set of data, including values for the 148 economies shown in the main tables, other economies shown in table 1.6, and Taiwan, China. The aggregation rules are intended to yield estimates for a consistent set of economies from one period to the next and for all indicators. Small differences between the values of subgroup aggregates and overall totals and averages may occur because of the approximations used. In addition, compilation errors and data reporting practices may cause discrepancies in theoretically identical aggregates such as world exports and world imports.

There are four principal methods of aggregation. For group and world totals noted in the tables by a *t*, missing data are imputed using a suitable proxy variable in a benchmark year, usually 1987. The imputed value is calculated so that it (or its proxy) bears the same relationship to the total of available data as it did in the benchmark year. Proxy variables are selected from a set of variables for which complete data are available for 1987. Imputed values are not calculated if missing data account for more than one-third of the total in the benchmark year. The variables used as proxies are GNP in U.S. dollars, GNP per capita in U.S. dollars, total population, exports and imports of goods and services in U.S. dollars, and value added in agriculture, industry, manufacturing, and services in local currency.

Aggregates marked by an *s* are sums of available data. Missing values are not imputed. Sums are not computed if more than one-third of the observations in the series or a proxy for the series are missing in a given year.

Aggregates of ratios are generally calculated as weighted averages of the ratios (indicated by *w*) using the value of the denominator or, in some cases, another indicator as a weight. The aggregate ratios are based on available data, including data for economies not shown in the main tables. Missing values are assumed to have the same average value as the available data. No aggregate is calculated if missing data account for more than one-third of the value of weights in the benchmark year. In a few cases the aggregate ratio may be computed as the ratio of group totals after imputing values for missing data according to the above rules for computing totals.

Aggregate growth rates are also generally calculated as a weighted average (and indicated by a *w*) of growth rates. In a few cases growth rates may be computed from time series of group totals. Growth rates are not calculated if more than one-third of the observations in a period are missing. For further discussion on methods of computing growth rates, see below.

Aggregates noted with an *m* are medians of the values shown in the table. No value is shown if more than one-third of the observations are missing.

Exceptions to the rules occur throughout the book. Depending on the judgment of World Bank analysts, the aggregates may be based on as little as 50 percent of the available data. In other cases, where missing or excluded values are judged to be small, irrelevant, or randomly distributed aggregates are based only on the data shown in the tables.

Growth rates

Growth rates are calculated as annual averages and represented as percentages. Except where noted, growth rates of values are computed from constant-price or real-value series. Three principal methods are used to calculate growth rates: the least-squares method, the exponential endpoint, and the geometric endpoint. Rates of change from one period to the next are calculated as proportional changes from the earlier period

Least-squares growth rate. Least-squares growth rates are used wherever there is a sufficiently long time series to permit a reliable calculation. No growth rate is calculated if more than half of the observations in a period are missing.

The least-squares growth rate, *r*, is estimated by fitting a linear regression trendline to the logarithmic annual values of the variable in the relevant period. The regression equation takes the form

$$\ln X_t = a + bt,$$

which is equivalent to the logarithmic transformation of the compound growth equation,

$$X_t = X_o (1 + r)^t.$$

In this equation, X is the variable, t is time, and $a = \log X_o$ and $b = \ln (1 + r)$ are parameters to be estimated. If $b*$ is the least-squares estimate of b, then the average annual growth rate, r, is obtained as $[\exp(b*)-1]$ and is multiplied by 100 to express it as a percentage.

The calculated growth rate is an average rate that is representative of the available observations over the entire period. It does not necessarily match the actual growth rate between any two periods.

Statistical methods

Exponential growth rate. The growth rate between two points in time for certain demographic data, notably labor force and population, is calculated from the equation:

$$r = \ln(p_n/p_1)/n,$$

where p_n and p_1 are the last and first observations in the period, n is the number of years in the period, and ln is the natural logarithm operator.

This growth rate is based on a model of continuous, exponential growth between two points in time. It does not take into account the intermediate values of the series. Note also that the exponential growth rate does not correspond to the annual rate of change measured at a one-year interval, which is given by $(p_n - p_{n-1})/p_{n-1}$.

Geometric growth rate. The geometric growth rate is applicable to compound growth over discrete periods, such as the payment and reinvestment of interest or dividends. Although continuous growth, as modeled by the exponential growth rate, may be more realistic, most economic phenomena are measured only at intervals for which the compound growth model is appropriate. The average growth rate over n periods is calculated as

$$r = \exp[\ln(p_n/p_1)/n] - 1.$$

Like the exponential growth rate, it does not take into account intermediate values of the series.

World Bank Atlas method

In calculating GNP in U.S. dollars and GNP per capita for certain operational purposes, the World Bank uses a synthetic exchange rate commonly called the Atlas conversion factor. The purpose of the Atlas conversion factor is to reduce the impact of exchange rate fluctuations in the cross-country comparison of national incomes.

The Atlas conversion factor for any year is the average of a country's exchange rate (or alternative conversion factor) for that year and its exchange rates for the two preceding years, after adjusting for differences between the rate of inflation in the country and the G-5 countries (France, Germany, Japan, the United Kingdom, and the United States). A country's inflation rate is measured by its GNP deflator. The inflation rate for G-5 countries is measured by changes in the SDR deflator. (Special drawing rights, or SDRs, are the International Monetary Fund's unit of account.) The SDR deflator is calculated as a weighted average of the G-5 countries' GDP deflators in SDR terms. The weights are determined by the amount of each currency included in one SDR unit. Weights vary over time because the IMF changes the composition of both the SDR and the SDR exchange rate for each currency changes. The SDR deflator is first calculated in SDR terms and then converted to U.S. dollars using the SDR to dollar Atlas conversion factor.

This three-year averaging smooths annual fluctuations in prices and exchange rates for each country. The Atlas conversion factor is then applied to a country's GNP. The resulting GNP in U.S. dollars is divided by the midyear population for the latest of the three years to derive GNP per capita. When official exchange rates are deemed to be unreliable or unrepresentative of the effective exchange rate during a period, an alternative estimate of the exchange rate is used in the Atlas formula (see below).

The following formulas describes the computation of the Atlas conversion factor for year t:

$$e_t^* = \frac{1}{3}\left[e_{t-2}\left(\frac{p_t}{p_{t-2}} \bigg/ \frac{p_t^{S\$}}{p_{t-2}^{S\$}} \right) + e_{t-1}\left(\frac{p_t}{p_{t-1}} \bigg/ \frac{p_t^{S\$}}{p_{t-1}^{S\$}} \right) + e_t \right]$$

and for calculating GNP per capita in U.S. dollars for year t:

$$Y_t^\$ = (Y_t/N_t)/e_t^*,$$

where e_t^* is the *Atlas* conversion factor (national currency to the U.S. dollar) for year t, e_t is the average annual exchange rate (national currency to the U.S. dollar) for year t, p_t is the GNP deflator for year t, $p_t^{S\$}$ is the SDR deflator in U.S. dollar terms for year t, $Y_t^\$$ is the Atlas GNP in U.S. dollars in year t, Y_t is current GNP (local currency) for year t, and N_t is midyear population for year t.

Alternative conversion factors

The World Bank systematically assesses the appropriateness of official exchange rates as conversion factors. An alternative conversion factor is used when the official exchange rate is judged to diverge by an exceptionally large margin from the rate effectively applied to domestic transactions of foreign currencies and traded products. This applies to only a small number of countries as shown in the *Primary data documentation*. Alternative conversion factors are used in the Atlas methodology and elsewhere in the *World Development Indicators* as single-year conversion factors.

Primary data documentation

The World Bank is not a primary data collection agency for most areas other than living standards surveys and external debt. As a major user of socioeconomic data, however, the World Bank places particular emphasis on data documentation to inform users of data in economic analysis and policymaking. The tables in this section provide information on the sources, treatment, and timeliness of the main demographic, economic, and environmental indicators in the *World Development Indicators*.

Differences in the methods and conventions used by the primary data collectors—usually national statistical agencies, central banks, and customs services—may give rise to significant discrepancies over time both among and within countries. Delays in reporting data and the use of old surveys as the base for current estimates may sometimes compromise the quality of national data.

Although data quality is improving in some countries, many developing countries lack the resources to train and maintain the skilled staff and obtain the equipment needed to measure and report demographic, economic, and environmental trends in an accurate and timely way. The World Bank recognizes the need for reliable data to measure living standards, track and evaluate economic trends, and plan and monitor development projects. Thus, working with bilateral and other multilateral agencies, it continues to fund and participate in technical assistance projects to improve statistical organization and basic data methods, collection, and dissemination.

The World Bank is working at several levels to meet the challenge of improving the quality of the data that it collates and disseminates. At the country level the Bank is carrying out technical assistance, training, and survey activities—with a view to strengthening national capacity—in the following areas:

- Poverty assessments in most borrower member countries.
- Living standards measurement and other household and farm surveys with country partner statistical agencies.
- National accounts and inflation.
- Price and expenditure surveys for the International Comparison Programme.
- Statistical improvement projects in the countries of the former Soviet Union.
- External debt management.
- Environmental and economic accounting.

At the international level, the World Bank is working closely with other agencies to achieve a common language in concepts, standards, classifications, nomenclatures, and definitions with a view to providing greater consistency and comparability to data from different countries and time periods.

Primary data documentation

	National currency	Fiscal year end	Reporting period	Base year	SNA price valuation	Alternative conversion factor	PPP survey year	Balance of Payments Manual in use	External debt	System of trade	Accounting concept	IMF special data dissemination
Albania	Albanian lek	Dec. 31	CY	1993	VAP			BPM5	Actual	G		
Algeria	Algerian dinar	Dec. 31	CY	1980	VAB			BPM5	Actual	S		
Angola	Angolan adjusted kwanza	Dec. 31	CY	1987	VAB	1993-96		BPM5	Actual	S		
Argentina	Argentine peso	Dec. 31	CY	1986	VAP	1972–81	1980	BPM5	Actual	S	C	S*
Armenia	Armenian dram	Dec. 31	CY	1993	VAP	1990–95	1990	BPM5	Preliminary			
Australia	Australian dollar	Jun. 30	CY	1989	VAP		1993	BPM5		G	C	S
Austria	Austrian schilling	Dec. 31	CY	1990	VAP		1993	BPM5		S	C	S*
Azerbaijan	Azeri manat	Dec. 31	CY	1987	VAP	1990–95	1990	BPM4	Preliminary			
Bangladesh	Bangladesh taka	Jun. 30	FY	1985	VAP		1993	BPM4	Actual	G		
Belarus	Belarussian ruble	Dec. 31	CY	1990	VAB	1990–96	1993	BPM4	Actual		C	
Belgium	Belgian franc	Dec. 31	CY	1985	VAP		1993	BPM5		S	C	S*
Benin	CFA franc	Dec. 31	CY	1985	VAP	1992	1993	BPM5	Actual	S		
Bolivia	Boliviano	Dec. 31	CY	1980	VAP	1974–85	1980	BPM5	Actual	S	C	
Bosnia and Herzegovina	Bosnian dinar	Dec. 31	CY	1987	VAP			BPM5	Actual			
Botswana	Botswana pula	Mar. 31	CY	1986	VAP		1993	BPM5	Actual	G	B	
Brazil	Brazilian real	Dec. 31	CY	1984	VAB		1980	BPM5	Preliminary	S	C	
Bulgaria	Bulgarian leva	Dec. 31	CY	1990	VAP	1991–93	1993	BPM5	Preliminary	G	C	
Burkina Faso	CFA franc	Dec. 31	CY	1985	VAB	1992–93		BPM5	Actual	S	C	
Burundi	Burundi franc	Dec. 31	CY	1980	VAB			BPM5	Actual	S		
Cambodia	Cambodian riel	Dec. 31	CY	1989	VAP			BPM5	Preliminary	S		
Cameroon	CFA franc	Jun. 30	FY	1980	VAP	1993	1993	BPM4	Preliminary		C	
Canada	Canadian dollar	Mar. 31	CY	1986	VAB		1993	BPM5		G	C	S*
Central African Republic	CFA franc	Dec. 31	CY	1987	VAB			BPM5	Actual	S		
Chad	CFA franc	Dec. 31	CY	1977	VAB			BPM5	Actual	S	C	
Chile	Chilean peso	Dec. 31	CY	1986	VAP		1980	BPM5	Actual	S	C	S*
China	Chinese yuan	Dec. 31	CY	1990	VAP		1986	BPM5	Preliminary	G	B	
Hong Kong, China	Hong Kong dollar	Dec. 31	CY	1990	VAB		1993	BPM4		G		S*
Colombia	Colombian peso	Dec. 31	CY	1975	VAP	1993–95	1980	BPM5	Preliminary	S	C	S*
Congo, Dem. Rep.	New zaire	Dec. 31	CY	1987	VAP	1993		BPM5	Preliminary	S	C	
Congo, Rep.	CFA franc	Dec. 31	CY	1978	VAP		1993	BPM5	Estimate	S		
Costa Rica	Costa Rican colon	Dec. 31	CY	1987	VAP		1980	BPM5	Actual		C	
Côte d'Ivoire	CFA franc	Dec. 31	CY	1986	VAP		1993	BPM5	Preliminary	S	C	
Croatia	Croatian kuna	Dec. 31	CY	1994	VAB		1993	BPM5	Preliminary		C	S*
Cuba	Cuban peso	Dec. 31	CY	..	..					S		
Czech Republic	Czech koruna	Dec. 31	CY	1984	VAP		1993	BPM5	Preliminary	G	C	
Denmark	Danish krone	Dec. 31	CY	1980	VAB		1993	BPM5		G	C	S*
Dominican Republic	Dominican peso	Dec. 31	CY	1970	VAP	1992–95	1980	BPM5	Actual	G	C	
Ecuador	Ecuadorian sucre	Dec. 31	CY	1975	VAP		1980	BPM5	Actual	G	B	
Egypt, Arab Rep.	Egyptian pound	Jun. 30	FY	1992	VAB		1993	BPM4	Actual		C	
El Salvador	Salvadoran colón	Dec. 31	CY	1962	VAP	1990–95	1980	BPM5	Actual	S	B	
Eritrea	Ethiopian birr	Dec. 31	CY	1992	VAB			BPM4	Actual			
Estonia	Estonian kroon	Dec. 31	CY	1993	VAB	1990–95	1993	BPM5	Actual		C	
Ethiopia	Ethiopian birr	Jul. 7	FY	1981	VAB	1991–92	1985	BPM4	Actual	G	B	
Finland	Finnish markka	Dec. 31	CY	1990	VAB		1993	BPM5		G	C	S*
France	French franc	Dec. 31	CY	1980	VAB		1993	BPM5		S	C	S*
Gabon	CFA franc	Dec. 31	CY	1989	VAP	1993	1993	BPM5	Actual	S	B	
Gambia, The	Gambian dalasi	Jun. 30	FY	1976	VAB			BPM5	Actual	G	B	
Georgia	Georgian lari	Dec. 31	CY	1987	VAB	1990–95	1990	BPM4	Preliminary			
Germany	Deutsche mark	Dec. 31	CY	1990	VAP		1993	BPM5		S	C	S*
Ghana	Ghanaian cedi	Dec. 31	CY	1975	VAP	1994		BPM5	Actual	G	C	
Greece	Greek drachma	Dec. 31	CY	1970	VAB		1993	BPM5		S	C	
Guatemala	Guatemalan quetzal	Dec. 31	CY	1958	VAP		1980	BPM5	Actual	S	B	
Guinea	Guinean franc	Dec. 31	CY	1989	VAP		1993	BPM5	Preliminary	S	C	
Guinea-Bissau	Guinea-Bissau peso	Dec. 31	CY	1986	VAB	1972–86		BPM5	Actual	S		
Haiti	Haitian gourde	Sep. 30	FY	1976	VAP			BPM5	Preliminary	G		
Honduras	Honduran lempira	Dec. 31	CY	1978	VAB		1980	BPM5	Actual	S		

	Latest population census	Latest household or demographic survey	Vital registration complete	Latest agricultural census	Latest industrial data	Latest water withdrawal data	Latest survey of scientists and engineers engaged in R&D	Latest survey of expenditure for R&D
Albania	1989	LSMS, 1996	✓		1993	1970		
Algeria	1987	PAPCHILD, 1992	✓		1993	1990		
Angola	1970					1987		
Argentina	1991		✓	1988	1993	1976	1988	1992
Armenia	1989		✓		1991	1994		
Australia	1996		✓	1990	1992	1985	1990	1990
Austria	1991		✓	1990	1994	1991	1993	1993
Azerbaijan	1989		✓			1995		
Bangladesh	1991	DHS, 1996–97			1992	1987		
Belarus	1989		✓			1990	1992	1992
Belgium	1991		✓	1990	1994	1980	1990	1990
Benin	1992	DHS, 1996		1992	1981	1994	1989	1989
Bolivia	1992	DHS, 1994			1994	1987	1991	1991
Bosnia and Herzegovina	1991		✓		1991			
Botswana	1991	DHS, 1988			1994	1992		
Brazil	1991	DHS, 1996			1992	1990	1995	1994
Bulgaria	1992	LSMS, 1995	✓		1994	1988	1992	1992
Burkina Faso	1996	SDA, 1995			1983	1992		
Burundi	1990				1991	1987	1989	1989
Cambodia	1962					1987		
Cameroon	1987	DHS, 1991			1994	1987		
Canada	1991		✓	1991	1993	1991	1991	1992
Central African Republic	1988	DHS, 1994–95			1992	1987	1990	1984
Chad	1993	DHS, 1996–97				1987		
Chile	1992		✓		1993	1975	1988	1994
China	1990	Population, 1995			1994	1980	1993	1993
Hong Kong, China	1991		✓		1993			
Colombia	1993	DHS, 1995		1988	1992	1987	1982	1982
Congo, Dem. Rep.	1984			1990		1990		
Congo, Rep.	1984				1998	1987	1984	1984
Costa Rica	1984	CDC, 1993	✓		1994	1970	1992	1986
Côte d'Ivoire	1988	DHS, 1994			1993	1987		
Croatia	1991		✓		1992		1992	1992
Cuba	1981		✓		1989	1975	1992	1992
Czech Republic	1991	CDC, 1993	✓	1990	1993	1991	1994	1994
Denmark	1991		✓	1989	1991	1990	1993	1993
Dominican Republic	1993	DHS, 1996			1985	1987		
Ecuador	1990	LSMS, 1995			1994	1987	1990	1990
Egypt, Arab Rep.	1996	DHS, 1995–96	✓		1992	1993	1991	1991
El Salvador	1992	CDC, 1994			1994	1975	1992	1992
Eritrea	1984							
Estonia	1989		✓			1995	1994	1994
Ethiopia	1994	Family and fertility, 1990		1989–92	1990	1987		
Finland	1990		✓		1994	1991	1993	1993
France	1990	Income, 1989		1988	1994	1990	1993	1993
Gabon	1993				1982		1987	1986
Gambia, The	1993				1982	1982		
Georgia	1989		✓		1992	1990	1991	1991
Germany			✓		1993	1991	1989	1989
Ghana	1984	DHS, 1993			1987	1970		
Greece	1991		✓		1993	1980	1993	1993
Guatemala	1994	DHS, 1995	✓		1990	1970	1988	1988
Guinea	1996	SDA, 1994–95		1989		1987	1984	1984
Guinea-Bissau	1991	SDA, 1991		1988		1991		
Haiti	1982	DHS, 1994–95				1987		
Honduras	1988	DHS, 1994		1993	1994	1992		

	National currency	Fiscal year end	National accounts					Balance of payments and trade			Government finance	IMF special data dissemi- nation
			Reporting period	Base year	SNA price valuation	Alternative conversion factor	PPP survey year	Balance of Payments Manual in use	External debt	System of trade	Accounting concept	
Hungary	Hungarian forint	Dec. 31	CY	1991	VAB		1993	BPM5	Actual	G	C	S*
India	Indian rupee	Mar. 31	FY	1980	VAB		1985	BPM4	Actual	G	C	S
Indonesia	Indonesian rupiah	Mar. 31	CY	1993	VAP		1993	BPM5	Preliminary	S	C	S*
Iran, Islamic Rep.	Iranian rial	Mar. 20	FY	1982	VAB		1993	BPM5	Estimate	S	C	
Iraq	Iraqi dinar	Dec. 31	CY	1969	VAB					S		
Ireland	Irish pound	Dec. 31	CY	1985	VAB		1993	BPM5		G	C	S*
Israel	Israeli new shekel	Dec. 31	CY	1990	VAB		1980	BPM5		S	C	S*
Italy	Italian lira	Dec. 31	CY	1985	VAP		1993	BPM5		S	C	S*
Jamaica	Jamaica dollar	Dec. 31	CY	1986	VAP	1995–96	1993	BPM5	Preliminary	G		
Japan	Japanese yen	Mar. 31	CY	1985	VAP		1993	BPM5		G	C	S*
Jordan	Jordan dinar	Dec. 31	CY	1985	VAB		1993	BPM4	Actual	G	B	
Kazakhstan	Kazakh tenge	Dec. 31	CY	1994	VAB	1990–95	1990	BPM4	Actual			
Kenya	Kenya shilling	Jun. 30	CY	1982	VAB		1993	BPM5	Actual	G	B	
Korea, Dem. Rep.	Democratic Republic of Korea won	Dec. 31	CY	1990	VAP			BPM5				S
Korea, Rep.	Korean won	Dec. 31	CY	1990	VAP		1993	BPM5		S	C	
Kuwait	Kuwaiti dinar	Jun. 30	CY	1984	VAP			BPM5		S	C	
Kyrgyz Republic	Kyrgyz som	Dec. 31	CY	1993	VAB	1990–95	1990	BPM4	Actual			
Lao PDR	Lao kip	Dec. 31	CY	1990	VAP			BPM5	Preliminary			
Latvia	Latvian lat	Dec. 31	CY	1993	VAB	1990–95	1993	BPM5	Actual	G	C	S
Lebanon	Lebanese pound	Dec. 31	CY	1994	VAB		1993	BPM4	Actual	G		
Lesotho	Lesotho loti	Mar. 31	CY	1980	VAB			BPM5	Actual	G	C	
Libya	Libyan dinar	Dec. 31	CY	1975	VAB		1993	BPM5		G		
Lithuania	Lithuanian litas	Dec. 31	CY	1993	VAB	1990–95		BPM5	Preliminary		C	S*
Macedonia, FYR	Macedonian denar	Dec. 31	CY	1990	VAP		1993		Actual			
Madagascar	Malagasy franc	Dec. 31	CY	1984	VAP		1993	BPM5	Actual	S	C	
Malawi	Malawi kwacha	Mar. 31	CY	1978	VAB		1985	BPM5	Preliminary	G	B	
Malaysia	Malaysian ringgit	Dec. 31	CY	1978	VAB		1993	BPM5	Preliminary	G	C	S*
Mali	CFA franc	Dec. 31	CY	1987	VAB		1985	BPM5	Actual	S		
Mauritania	Mauritanian ouguiya	Dec. 31	CY	1985	VAB		1993	BPM5	Preliminary	S		
Mauritius	Mauritian rupee	Jun. 30	CY	1992	VAB		1985	BPM5	Actual	G	C	
Mexico	Mexican new peso	Dec. 31	CY	1993	VAB		1993	BPM5	Actual	G	C	S*
Moldova	Moldovan leu	Dec. 31	CY	1995	VAB	1990–95	1993	BPM5	Actual			
Mongolia	Mongolian tugrik	Dec. 31	CY	1986	VAB	1993	1993	BPM5	Actual		C	
Morocco	Moroccan dirham	Dec. 31	CY	1980	VAP		1985	BPM5	Preliminary	S	C	
Mozambique	Mozambican metical	Dec. 31	CY	1995	VAB			BPM5	Actual	S		
Myanmar	Myanmar kyat	Mar. 31	FY	1985	VAP			BPM4	Actual	G	C	
Namibia	Namibia dollar	Mar. 31	CY	1990	VAB			BPM5			C	
Nepal	Nepalese rupee	Jul. 14	FY	1985	VAP		1993	BPM4	Actual	G	C	
Netherlands	Netherlands guilder	Dec. 31	CY	1990	VAP		1993	BPM5		S	C	S*
New Zealand	New Zealand dollar	Jun. 30	CY	1982	VAP		1993	BPM5		G	B	
Nicaragua	Nicaraguan gold cordoba	Dec. 31	CY	1980	VAP			BPM5	Actual	G	C	
Niger	CFA franc	Dec. 31	CY	1987	VAP	1993	1993	BPM5	Actual	S		
Nigeria	Nigerian naira	Dec. 31	CY	1987	VAB	1974–96	1993	BPM5	Estimate	G		
Norway	Norwegian krone	Dec. 31	CY	1990	VAP		1993	BPM5		G	C	S*
Oman	Rial Omani	Dec. 31	CY	1978	VAP		1993	BPM5	Actual	G	B	
Pakistan	Pakistan rupee	Jun. 30	FY	1981	VAB		1980	BPM4	Actual	G	C	
Panama	Panamanian balboa	Dec. 31	CY	1992	VAB			BPM5	Actual	S	C	
Papua New Guinea	Papua New Guinea kina	Dec. 31	CY	1983	VAP		1980	BPM5	Actual	G	B	
Paraguay	Paraguayan guaraní	Dec. 31	CY	1982	VAP		1980	BPM4	Actual	S	C	
Peru	Peruvian new sol	Dec. 31	CY	1979	VAP	1987–91	1993	BPM5	Actual	S	C	S*
Philippines	Philippine peso	Dec. 31	CY	1985	VAP		1993	BPM5	Preliminary	G	B	S*
Poland	Polish zloty	Dec. 31	CY	1990	VAP		1993	BPM5	Actual	G	C	S*
Portugal	Portuguese escudo	Dec. 31	CY	1985	VAP		1993	BPM5		S	C	S
Puerto Rico	U.S. dollar	Dec. 31	CY				1993					
Romania	Romanian leu	Dec. 31	CY	1993	VAB	1992	1993	BPM5	Preliminary	G	C	
Russian Federation	Russian ruble	Dec. 31	CY	1996	VAB	1990–94	1993	BPM4	Preliminary		C	

	Latest population census	Latest household or demographic survey	Vital registration complete	Latest agricultural census	Latest industrial data	Latest water withdrawal data	Latest survey of scientists and engineers engaged in R&D	Latest survey of expenditure for R&D
Hungary	1990	Income, 1995	✓		1994	1991	1993	1993
India	1991	National family health, 1992–93		1986	1992	1975	1990	1990
Indonesia	1990	DHS, 1994			1994	1987	1988	1988
Iran, Islamic Rep.	1991	Demographic, 1995		1988	1993	1993	1985	1985
Iraq	1997				1992	1970	1993	1993
Ireland	1996		✓	1991	1994	1980	1988	1988
Israel	1995		✓		1993	1989	1984	1992
Italy	1991		✓	1990	1991	1990	1993	1993
Jamaica	1991	LSMS, 1994	✓		1992	1975	1986	1986
Japan	1995		✓	1990	1993	1990	1992	1991
Jordan	1994	DHS, 1990			1994	1993	1986	1986
Kazakhstan	1989	DHS, 1995	✓			1993		
Kenya	1989	DHS, 1993			1994	1990		
Korea, Dem. Rep.	1993					1987		
Korea, Rep.	1995				1994	1994	1994	1994
Kuwait	1995		✓		1993	1974	1984	1984
Kyrgyz Republic	1989	LSMS, 1993	✓			1994		
Lao PDR	1995					1987		
Latvia	1989		✓		1994	1994	1994	1994
Lebanon	1970					1994	1980	1980
Lesotho	1996	DHS, 1991			1994	1987		
Libya	1995	PAPCHILD, 1995		1987	1989	1994	1980	1980
Lithuania	1989		✓		1994	1993	1992	1992
Macedonia, FYR	1994				1994		1991	1991
Madagascar	1993	DHS, 1992; SDA, 1993			1988	1984	1989	1988
Malawi	1987	DHS, 1996			1989	1994		
Malaysia	1991		✓		1993	1975	1992	1992
Mali	1987	DHS, 1995–96			1981	1987		
Mauritania	1988	PAPCHILD, 1990				1985		
Mauritius	1990	CDC, 1991	✓		1993	1974	1992	1992
Mexico	1990	Population, 1995			1991	1991	1993	1993
Moldova	1989		✓			1992		
Mongolia	1989				1994	1987		
Morocco	1994	DHS, 1995			1994	1992		
Mozambique	1997					1992		
Myanmar	1983			1993	1993	1987		
Namibia	1991	DHS, 1992				1991		
Nepal	1991	DHS, 1996		1992	1993	1987	1980	1980
Netherlands	1971		✓	1989	1993	1991	1991	1991
New Zealand	1996		✓	1990	1992	1991	1993	1993
Nicaragua	1995	LSMS, 1993			1985	1975	1987	1987
Niger	1988	Household budget and consumption, 1993			1982	1988		
Nigeria	1991	Consumption and expenditure, 1992			1990	1987	1987	1987
Norway	1990		✓	1989	1992	1985	1993	1993
Oman	1993	Child health, 1989		1992–93	1993	1991		
Pakistan	1981	LSMS, 1991		1990	1990	1991	1990	1987
Panama	1990			1990	1992	1975		
Papua New Guinea	1989				1989	1987		
Paraguay	1992	DHS, 1990; CDC, 1992		1991	1981	1987		
Peru	1993	DHS, 1996			1992	1987	1981	1984
Philippines	1995	DHS, 1993			1992	1975	1984	1984
Poland	1988		✓	1990	1994	1991	1992	1992
Portugal	1991		✓	1989	1994	1990	1990	1990
Puerto Rico	1990		✓	1987	1994			
Romania	1992	LSMS, 1994–95	✓		1993	1994	1994	1994
Russian Federation	1989	LSMS, 1992	✓		1994	1994	1993	1993

	National currency	Fiscal year end	National accounts					Balance of payments and trade			Government finance	IMF special data dissemi- nation
			Reporting period	Base year	SNA price valuation	Alternative conversion factor	PPP survey year	Balance of Payments Manual in use	External debt	System of trade	Accounting concept	
Rwanda	Rwanda franc	Dec. 31	CY	1985	VAB		1985	BPM5	Actual	G	C	
Saudi Arabia	Saudi Arabian riyal	Hijri year	Hijri year	1970	VAP		1993	BPM5		S		
Senegal	CFA franc	Dec. 31	CY	1987	VAB	1992–93	1993	BPM5	Preliminary	S		
Sierra Leone	Sierra Leonean leone	Jun. 30	CY	1990	VAB		1993	BPM5	Actual	G	B	
Singapore	Singapore dollar	Mar. 31	CY	1990	VAP		1993	BPM5		G	C	S*
Slovak Republic	Slovak koruna	Dec. 31	CY	1993	VAP		1993	BPM5	Preliminary			S
Slovenia	Slovenian tolar	Dec. 31	CY	1993	VAB		1993	BPM5	Actual			S*
South Africa	South African rand	Mar. 31	CY	1990	VAB		1993	BPM5	Preliminary		C	S*
Spain	Spanish peseta	Dec. 31	CY	1996	VAP		1993	BPM5		S	C	S
Sri Lanka	Sri Lanka rupee	Dec. 31	CY	1982	VAB		1985	BPM5	Actual	G	C	
Sudan	Sudanese dinar	Jun. 30	FY	1982	VAB			BPM4	Estimate	G		
Sweden	Swedish krona	Jun. 30	CY	1990	VAB		1993	BPM5		G	C	S*
Switzerland	Swiss franc	Dec. 31	CY	1990	VAP		1993	BPM5		S	C	S*
Syrian Arab Republic	Syrian pound	Dec. 31	CY	1985	VAP	1993–96	1993	BPM5	Estimate	S	C	
Tajikistan	Tajik ruble	Dec. 31	CY	1993	VAP	1990–96	1990	BPM4	Estimate			
Tanzania	Tanzania shilling	Jun. 30	FY	1992	VAB		1993	BPM5	Actual	G		
Thailand	Thai baht	Sep. 30	CY	1988	VAP		1993	BPM5	Preliminary	G	C	S*
Togo	CFA franc	Dec. 31	CY	1978	VAP			BPM5	Actual	S		
Trinidad and Tobago	Trinidad and Tobago dollar	Dec. 31	CY	1985	VAB		1993	BPM5	Preliminary	S		
Tunisia	Tunisian dinar	Dec. 31	CY	1990	VAP		1991	BPM5	Actual	G	C	
Turkey	Turkish lira	Dec. 31	CY	1993	VAB		1993	BPM5	Actual	S	C	S*
Turkmenistan	Turkmen manat	Dec. 31	CY	1987	VAP	1990–96	1990	BPM4	Actual			
Uganda	Uganda shilling	Jun. 30	FY	1991	VAB			BPM4	Actual	G		
Ukraine	Ukraine hrivnya	Dec. 31	CY	1995	VAB	1990–95	1993	BPM5	Actual			
United Arab Emirates	U.A.E dirham	Dec. 31	CY	1985	VAB			BPM4		G	B	
United Kingdom	Pound sterling	Dec. 31	CY	1990			1993	BPM5		G	C	S*
United States	U.S. dollar	Sep. 30	CY	1985	VAP		1993	BPM5		G	C	S*
Uruguay	Uruguayan peso	Dec. 31	CY	1983	VAP		1980	BPM5	Actual	S	C	
Uzbekistan	Uzbek sum	Dec. 31	CY	1987	VAB	1990–96	1990	BPM4	Actual			
Venezuela	Venezuelan bolívar	Dec. 31	CY	1984	VAP		1980	BPM5	Actual	G	C	
Vietnam	Vietnamese dong	Dec. 31	CY	1989	VAP		1993	BPM4	Preliminary			
West Bank and Gaza	Israeli new shekel	Dec. 31	CY		VAP							
Yemen, Rep.	Yemen rial	Dec. 31	CY	1990	VAB	1990–96	1993	BPM4	Actual	G	C	
Yugoslavia, FR (Serb./Mont.)	Yugoslav new dinar	Dec. 31	CY	1984	VAP		1985		Estimate	S		
Zambia	Zambian kwacha	Dec. 31	CY	1977	VAP		1993	BPM5	Preliminary	G	C	
Zimbabwe	Zimbabwe dollar	Jun. 30	CY	1990	VAB		1993	BPM5	Actual	G	C	

Note: For explanation of the abbreviations used in the table see the notes.

	Latest population census	Latest household or demographic survey	Vital registration complete	Latest agricultural census	Latest industrial data	Latest water withdrawal data	Latest survey of scientists and engineers engaged in R&D	Latest survey of expenditure for R&D
Rwanda	1991	DHS, 1992			1986	1993	1985	1985
Saudi Arabia	1992	Maternal and child health, 1993			1989	1992		
Senegal	1988	DHS, 1992–93			1994	1987	1981	1981
Sierra Leone	1985	SHEHEA, 1989–90			1993	1987		
Singapore	1990	General household, 1995	✓		1994	1975	1994	1994
Slovak Republic	1991		✓		1994	1991		
Slovenia	1991		✓		1994		1992	1992
South Africa	1996	LSMS, 1993			1992	1990	1991	1991
Spain	1991		✓	1989	1992	1991	1993	1993
Sri Lanka	1993	DHS, 1993	✓		1993	1970	1985	1984
Sudan	1993	DHS, 1989–90				1995		
Sweden	1990		✓		1994	1991	1993	1993
Switzerland	1990		✓	1990	1994	1991	1992	1992
Syrian Arab Republic	1994	PAPCHILD, 1995			1992	1993		
Tajikistan	1989		✓			1994		
Tanzania	1988	DHS, 1996			1988	1994		
Thailand	1990	DHS, 1987		1988	1991	1987	1991	1991
Togo	1981	DHS, 1988			1984	1987		
Trinidad and Tobago	1990	DHS, 1987	✓		1993	1975	1984	1984
Tunisia	1994	PAPCHILD, 1994–95			1993	1990	1992	1992
Turkey	1997	Population and health, 1983		1991	1993	1992	1991	1991
Turkmenistan	1995		✓			1994		
Uganda	1991	DHS, 1995		1991	1989	1970		
Ukraine	1991		✓			1992	1989	1989
United Arab Emirates	1980				1985	1995		
United Kingdom	1991		✓	1993	1994	1991	1993	1993
United States	1990	Current population, 1996	✓	1987	1994	1990	1993	1993
Uruguay	1996			1990	1993	1965		
Uzbekistan	1989	DHS, 1996	✓			1994	1992	1992
Venezuela	1990	LSMS, 1993	✓		1993	1970	1992	1992
Vietnam	1989	Intercensal demographic, 1995				1992	1985	1985
West Bank and Gaza		Demographic, 1995						
Yemen, Rep.	1994	DHS, 1991–92				1990		
Yugoslavia, FR (Serb./Mont.)	1991		✓		1994		1992	1992
Zambia	1990	DHS, 1996		1990	1994	1994		
Zimbabwe	1992	DHS, 1994			1993	1987		

Primary data documentation notes

• **Fiscal year end** is the date of the end of the fiscal year for the central government. Fiscal years for other levels of government and the reporting years for statistical surveys may differ, but if a country is designated as a fiscal year reporter in the following column, the date shown is the end of its national accounts reporting period. • **Reporting period** for national accounts and balance of payments data is designated as either calendar year basis (CY) or fiscal year (FY). Most economies report their national accounts and balance of payments data using calendar years, but some use fiscal years, which straddle two calendar years. In the *World Development Indicators* fiscal year data are assigned to the calendar year that contains the larger share of the fiscal year. Saudi Arabia follows a lunar year whose starting and ending dates change with respect to the solar year. Because the International Monetary Fund (IMF) reports most balance of payments data on a calendar year basis, balance of payments data for fiscal year reporters in the *World Development Indicators* are based on fiscal year estimates provided by World Bank staff. These estimates may differ from IMF data but allow consistent comparisons between national accounts and balance of payments data. • **Base year** is the year used as the base period for constant price calculations in the country's national accounts. Price indexes derived from national accounts aggregates, such as the GDP deflator, express prices in current years relative to prices in the base year. Constant price data reported in the *World Development Indicators* are partially rebased to a common 1987 base year. See the notes to table 4.1 for further discussion. • **SNA price valuation** shows whether value added in the national accounts is reported at basic or producers' prices (VAB) or at purchasers' prices (VAP). Purchasers' prices include the value of taxes levied on output and value added that are collected from consumers and thus tend to overstate the actual value added in production. See the notes to table 4.2 for further discussion of national accounts valuation. • **Alternative conversion factor** identifies the countries and years for which a World Bank–estimated conversion factor has been used in place of the official (IFS line rf) exchange rate. See *Statistical methods* for further discussion of the use of alternative conversion factors. • **PPP survey year** refers to the latest available survey year for the International Comparison Programme's estimates of purchasing power parities (PPPs). See the notes to tables 4.10, 4.11, and 5.6 for further details. • **Balance of Payments Manual in use** refers to the classification system used for compiling and reporting

data on balance of payments items in table 4.16. BPM4 refers to the fourth edition of the IMF's *Balance of Payments Manual* (1977), and BPM5 to the fifth edition (1993). Since 1995 the IMF has adjusted all balance of payments data to BPM5 conventions, but some countries continue to report using the older system. • **External debt** shows debt reporting status for 1996 data. "Actual" indicates data are as reported, "preliminary" indicates data are preliminary and include an element of staff estimation, and "estimate" indicates data are staff estimates. • **System of trade** refers to the general trade system (G) or the special trade system (S). For imports under the general trade system, both goods entering directly for domestic consumption and goods entered into customs storage are recorded, at the time of their first arrival, as imports. Under the special trade system goods are recorded as imports when declared for domestic consumption whether at time of entry or on withdrawal from customs storage. Exports under the general system comprise outward-moving goods: (a) national goods wholly or partly produced in the country; (b) foreign goods, neither transformed nor declared for domestic consumption in the country, that move outward from customs storage; and (c) nationalized goods that have been declared from domestic consumption and move outward without having been transformed. Under the special system of trade exports comprise categories (a) and (c). In some compilations categories (b) and (c) are classified as re-exports. Direct transit trade, consisting of goods entering or leaving for transport purposes only, is excluded from both import and export statistics. See the notes to tables 4.4 and 4.5 for further discussion. • **Government finance accounting concept** describes the accounting basis for reporting central government financial data. For most countries government finance data have been consolidated (C) into one set of accounts capturing all the central government's fiscal activities. Budgetary central government accounts (B) exclude central government units. See the notes to table 4.12 for further details. • **IMF special data dissemination** shows the countries that subscribe to the International Monetary Fund's (IMF) Special Data Dissemination Standard (SDDS). "S" refers to countries that subscribe; "S*" indicates subscribers that have posted data on the Internet. (Posted data can be reached through the IMF Dissemination Standard Bulletin Board at http://dsbb.imf.org/). The IMF established the SDDS to guide member countries that have or that are seeking access to international capital markets in providing economic and financial data to the public. The SDDS

is expected to enhance the availability of timely and comprehensive data and therefore contribute to the pursuit of sound macroeconomic policies; it is also expected to contribute to the improved functioning of financial markets. Although subscription is voluntary, it commits the subscriber to observing the standard and to providing information to the IMF about its practices in disseminating economic and financial data. A related General Data Dissemination System (GDDS), recently introduced by the IMF, is designed to encourage good practice in data production and dissemination. • **Latest population census** shows the most recent year in which a census was conducted. • **Latest household or demographic survey** gives information on the surveys used in compiling the household and demographic data presented in section 2. LSMS is Living Standards Measurement Study, PAPCHILD is the Pan Arab Project for Child Development, DHS is Demographic and Health Survey, SDA is Social Dimensions of Adjustment, CDC is Centers for Disease Control and Prevention, and SHEHEA is Survey of Household Expenditure and Household Economic Activities. • **Vital registration complete** identifies countries judged to have complete registries of vital (birth and death) statistics by the United Nations Department of Economic and Social Information and Policy Analysis, Statistical Division, and reported in *Population and Vital Statistics Reports*. Countries with complete vital statistics registries may have more accurate and more timely demographic indicators. • **Latest agricultural census** shows the most recent year in which an agricultural census was conducted and reported to the Food and Agriculture Organization. • **Latest industrial data** refer to the most recent year for which manufacturing value added data at the three-digit level of the International Standard Industrial Classification (rev. 2 or rev. 3) are available in the UNIDO database. • **Latest water withdrawal data** refer to the most recent year for which data have been compiled from a variety of sources. See the notes to table 3.5 for more information. • **Latest surveys of scientists and engineers engaged in R&D** and **expenditure for R&D** refer to the most recent year for which data are available from a data collection effort by UNESCO in science and technology and research and development (R&D). See the notes to table 5.12 for more information.

Acronyms and abbreviations

Technical terms

bbl	barrel
BOD	biochemical oxygen demand
btu	British thermal units
CFC	chlorofluorocarbon
c.i.f.	cost, insurance, and freight
CITES	Convention on International Trade in Endangered Species of Wild Flora and Fauna
CO$_2$	carbon dioxide
CPI	consumer price index
cu. m	cubic meter
DHS	demographic and health survey
DMTU	dry metric ton unit
DPT	diphtheria, pertussis, and tetanus
ESAF	Enhanced Structural Adjustment Facility
FDI	foreign direct investment
f.o.b.	free on board
FYR	former Yugoslav Republic
G-5	France, Germany, Japan, United Kingdom, and United States
G-7	G-5 plus Canada and Italy
GDP	gross domestic product
GEMS	Global Environment Monitoring System
GIS	Geographic Information System
GNI	gross national income
GNP	gross national product
ha	hectare
HIV	human immunodeficiency virus
ICD	international classification of diseases
ICRG	International Country Risk Guide
ICSE	International Classification of Status in Employment
IMR	infant mortality rate
ISCED	International Standard Classification of Education
ISIC	International Standard Industrial Classification
kg	kilogram
km	kilometer
kwh	kilowatt-hour
LIBOR	London interbank offered rate
M0	currency and coins (monetary base)
M1	narrow money (currency and demand deposits)
M2	money plus quasi money
M3	broad money or liquid liabilities
MFA	Multifiber Arrangement
mmbtu	millions of British thermal units
mt	metric ton
MUV	manufactures unit value
NEAP	national environmental action plan
NGO	nongovernmental organization
NTBs	nontariff barriers
ODA	official development assistance
P/E	price-earnings ratio
PPP	purchasing power parity
SAF	Structural Adjustment Facility
SDR	Special Drawing Right
SIC	Standard Industrial Classification
SITC	Standard International Trade Classification
SNA	U.N. System of National Accounts
SOPEMI	Continuous Reporting System on Migration
SO$_2$	sulfur dioxide
sq. km	square kilometer
STD	sexually transmitted disease
TB	tuberculosis
TFP	total factor productivity
ton-km	metric ton-kilometers
TSP	total suspended particulates
UN5MR	child (under-5) mortality rate

Organizations

ACDA	Arms Control and Disarmament Agency		**IRF**	International Road Federation
ADB	Asian Development Bank		**ITU**	International Telecommunication Union
AfDB	African Development Bank		**IUCN**	World Conservation Union
APEC	Asia-Pacific Economic Cooperation		**LAC**	Latin America and the Caribbean
CDC	Centers for Disease Control and Prevention		**LME**	London Metals Exchange
CDIAC	Carbon Dioxide Information Analysis Center		**MIGA**	Multilateral Investment Guarantee Agency
CEC	Commission of the European Community		**MNA**	Middle East and North Africa
DAC	Development Assistance Committee		**NAFTA**	North American Free Trade Agreement
EAP	East Asia and the Pacific		**NATO**	North Atlantic Treaty Organization
EBRD	European Bank for Reconstruction and Development		**OECD**	Organisation for Economic Co-operation and Development
ECA	Europe and Central Asia		**PAHO**	Pan American Health Organization
EDF	European Development Fund		**SAS**	South Asia
EFTA	European Free Trade Area		**SSA**	Sub-Saharan Africa
EIB	European Investment Bank		**UN**	United Nations
EU	European Union		**UNAIDS**	Joint United Nations Programme on HIV/AIDS
EUROSTAT	Statistical Office of the European Community		**UNCED**	United Nations Conference on Environment and Development
FAO	Food and Agriculture Organization		**UNCTAD**	United Nations Conference on Trade and Development
GATT	General Agreement on Tariffs and Trade		**UNDP**	United Nations Development Programme
GEF	Global Environment Facility		**UNECE**	United Nations Economic Commission for Europe
HIC	High-income countries		**UNEP**	United Nations Environment Programme
IBRD	International Bank for Reconstruction and Development		**UNESCO**	United Nations Educational, Scientific, and Cultural Organization
ICAO	International Civil Aviation Organization		**UNFPA**	United Nations Population Fund
ICCO	International Cocoa Organization		**UNICEF**	United Nations Children's Fund
ICO	International Coffee Organization		**UNIDO**	United Nations Industrial Development Organization
ICSE	International Classification of Status in Employment		**UNRISD**	United Nations Research Institute for Social Development
ICP	International Comparison Programme		**UNSO**	United Nations Statistical Office
IDA	International Development Association		**USAID**	U. S. Agency for International Development
IDB	Inter-American Development Bank		**WCMC**	World Conservation Monitoring Centre
IEA	International Energy Agency		**WFP**	World Food Programme
IFC	International Finance Corporation		**WHO**	World Health Organization
IFCI	International Finance Corporation Investable		**WIPO**	World Intellectual Property Organization
ILO	International Labour Organization		**WTO**	World Trade Organization
IMF	International Monetary Fund		**WWF**	World Wide Fund for Nature

Credits

This book has drawn on a wide range of World Bank reports and numerous external sources. These are listed in the bibliography that follows this section. Many people inside and outside the World Bank helped in writing and producing the *World Development Indicators*. This note identifies those who made specific contributions. Numerous others, too many to acknowledge here, helped in many ways for which the team is extremely grateful.

1. World view

was prepared by the members of the WDI team. K. Sarwar Lateef conceived the plan for the section and wrote the introduction with contributions from Eric Swanson and Sulekha Patel. The introduction drew heavily on Demery and Walton (1997). Crucial ideas and suggestions were provided by Lionel Demery, Paul Glewwe, Paul Isenman, Martin Ravallion, and Michael Walton. Colin Bradford and staff at USAID participated in early discussions on measuring development progress. Bernard Wood and Brian Hammond of the OECD made numerous helpful suggestions. Substantial assistance in preparing the data for this section was received from Jong-goo Park (GNP per capita) and Sultan Ahmad and Yonas Biru (GNP in PPP terms).

2. People

was prepared by Sulekha Patel in partnership with the World Bank's Human Development Network and Development Research Group. Cindy Alexis, Aelim Chi, and Endang Satyowati helped prepare the data for this section. Consultations were held with counterparts from the ILO, WHO, UNICEF, UNESCO, and UNAIDS, with Human Development Network staff serving as liaisons. Sulekha Patel wrote the introduction to the section. Substantial inputs to the section were provided by Eduard Bos (demography and health), Amit Dar, Monica Fong, and Dena Ringold (labor force and employment), Shaohua Chen, and Martin Ravallion (poverty and income distribution), and Martha Ainsworth and Vivian Hon (health). Bernherd Schwartlander (UNAIDS) and Karen Stanecki (U.S. Bureau of the Census) provided recent information on HIV prevalence and projections for the special feature on AIDS. David de Ferranti of the Human Development Council facilitated the data review by the Bank's regional staff and Helen

Saxenian facilitated collaboration between the Development Data Group and the Human Development Network. Comments and suggestions were received from Martha Ainsworth, Eduard Bos, Vivian Hon, Edna Jonas, Elizabeth King, Mead Over, Lant Pritchett, Martin Ravallion, Jee-Peng Tan, Lianqin Wang, and Michael Ward at various stages from design to production.

3. Environment

was prepared by M. H. Saeed Ordoubadi in partnership with the World Bank's Environmentally and Socially Sustainable Development Network and in collaboration with the World Bank's Development Research Group. Eric Rodenburg and Robin White of the World Resources Institute and Augusto Curti, Orio Tampieri, and Sami Zara of the FAO made important contributions. Demet Kaya assisted with research and data preparation. John Dixon, Kirk Hamilton, and Michael Ward provided invaluable comments and guidance in all stages of the work, from design to production. Andrew Steer and Robert Watson provided considerable encouragement and support and helped ensure the substantial ownership the World Bank's Environment Department feels for this section. The Environment Department devoted substantial staff resources to the book, for which we are very grateful. John Dixon and Kirk Hamilton wrote the introduction to the section with contributions from Per Fredriksson and Lisa Segnestam, and inputs were provided by Derek Byerlee, Peter Jipp, and William Magrath (land use and agriculture), John Dixon and Stefano Pagiola (biodiversity), John Briscoe (water), Susmita Dasgupta, Muthukumara Mani, and David Wheeler (water pollution), Kirk Hamilton and Shane Streifel (energy), Kseniya Lvovsky and Robin White (air pollution), and Charles Di Leva and Junko Funahashi (government commitment). The team received valuable comments at various stages from Jean Aden, Harry Burt, Anders Ekbom, Kristalina Georgieva, Ernst Lutz, William Martin, Glenn Morgan, Susan Shen, and Alan Winters.

4. Economy

was prepared by K. M. Vijayalakshmi and Eric Swanson in close collaboration with the Macroeconomic Data Team of the Development Data Group, led by Robin Lynch. Caroline Doggart prepared the introduction to

this section with additional inputs from Eric Swanson. Special acknowledgment is made to the *Global Economic Prospects* team, led by Uri Dadush—especially Robert Lynn and Mick Riordan. Substantial contributions to the section were provided by Robin Lynch and Michael Ward (national accounts), Azita Amjadi and Alexander Yeats (trade), Sultan Ahmad and Yonas Biru (structure of consumption and relative prices in PPP terms), Jong-goo Park (balance of payments), and Punam Chuhan (external debt). The national accounts and balance of payments data for low- and middle-income economies are gathered from the World Bank's regional staff through the annual Unified Survey under the direction of Soong Sup Lee and Monica Singh. Boris Blazic-Metzner reviewed and prepared the time series from 1960–65 for the CD-ROM, assisted by Premi Rathan Raj. Maja Bresslauer, Raquel Fok, and Jong-goo Park worked on updating, estimating, and validating the databases for national accounts and the balance of payments. The national accounts data for OECD countries were processed by Abdel Stambouli and reviewed by Robert King and Mick Riordan. The external debt tables were prepared by the Financial Data Team, led by Punam Chuhan, and were reviewed by Ibrahim Levent and Gloria Reyes. Shelley Fu and Sup Lee provided systems support.

5. States and markets

was prepared by David Cieslikowski in partnership with Yann Burtin, Andrew Ewing, Carsten Fink, Timothy Irwin, and Catherine Kleynhoff in the World Bank's Finance, Private Sector, and Infrastructure Network, the Poverty Reduction and Economic Management Network, and the International Finance Corporation. Demet Kaya contributed substantially to the overall preparation of this section, and Amy Wilson was responsible for updates to the state enterprise database. Caroline Doggart drafted the introduction to this section. Substantial inputs were made by William Shaw (private capital flows), Mariusz Sumlinski (private investment), Graeme Littler and Yuko Onuma (stock markets), Luke Haggerty (state enterprises), Sultan Ahmad and Yonas Biru (relative prices and PPP conversion factors), Geoffrey Lamb and Michael Stevens (military expenditures), Asli Demirgüç-Kunt (financial sector), Maria Concetta Gasbarro and Michael Minges of the ITU (communications and information), and Christine Kessides and Louis Thompson (transport).

6. Global links

was prepared by Eric Swanson with assistance from Aelim Chi in collaboration with the World Bank's Development Prospects Group, headed by Uri Dadush, and with special assistance from the OECD's Development Cooperation Directorate. Dipak Dasgupta provided a stimulating critique that helped in rethinking this section. Milan Brahmbhatt contributed to the redesign, suggested new indicators, and wrote the background paper from which the *World Development Indicators* team has drawn the introductory essay. He was assisted by Kumiko Imai. Substantial help in preparing the data for this section came from Azita Amjadi, Mick Riordan, and Alexander Yeats (trade), Jerzy Rozanski (tariffs), Betty Dow (commodity prices), Shelly Fu, Ibrahim Levent, and Gloria Reyes (financial data), Malvina Pollock (finance and aid), and Celine Thoreau of the OECD (migration). We wish to acknowledge the considerable assistance of Jean-Louis Grolleau of the OECD, who provided data on aid flows.

Other parts

The maps on the inside covers were prepared by the World Bank's Map Design Unit. The *Partners* section was coordinated and edited by Eric Swanson. The *Users guide* was prepared by David Cieslikowski. *Primary data documentation* was coordinated and written by David Cieslikowski. *Statistical methods* was written by Eric Swanson. *Acronyms and abbreviations* was prepared by Estela Zamora. The index was collated by Eric Swanson and Amy Wilson.

Systems support

Mehdi Akhlaghi was responsible for database management and programming and overall systems support. In this he drew on the Development Data Group's Systems Upgrade Team—in particular on Reza Farivari and Tariqul Khan—led by Henry Burt.

Administrative assistance and office technology support

Estela Zamora provided administrative assistance. She was supported by Karen Adams and Premi Rathan Raj. Office technology support was provided by Nacer Megherbi and Shahin Outadi.

Design, production, and editing

David Cieslikowski coordinated all aspects of production with the Communications Development Incorporated team, led by Laurel Morais. Bruce Ross-Larson and other staff at Communications Development Incorporated did the editing, layout, and design. In particular, we would like to thank the design team of Peter Grundy and Tilly Northedge, the editing team of Paul Holtz and Alison Strong, the desktopping team of Laurel Morais, Damon Iacovelli, Suzanne Luft, Terra Lynch, Donna McGreevy, and Christian Perez, and the production team of Daphne Levitas and Jessica Moore.

Client services

The Development Data Group's Client Services Team, led by Elizabeth Crayford, contributed to the design and planning of the books and helped coordinate with the Office of the Publisher.

External Affairs

Stephanie Gerard and Joyce Gates in the Office of the Publisher helped with the production of the *World Development Indicators*, the *Atlas*, and the related CD-ROM. Geoffrey Bergen, Phillip Hay, and Paul Zwaga of the External Affairs Vice Presidency assisted with the development of a communications strategy.

The Atlas

Production was managed by David Cieslikowski. The preparation of data benefited from the work on corresponding sections in the *World Development Indicators*. William Prince assisted with systems support and production of tables and graphs. The World Bank's Map Design Unit prepared the maps.

World Development Indicators CD-ROM

Design, programming, and production were carried out by Reza Farivari (the project leader) and his team: Mehdi Akhlaghi, Azita Amjadi, Elizabeth Crayford, Yusri Harun, Vasantha Hevaganinge, Angelo Kostopoulos, André Léger, Patricia McComas, and William Prince.

Client feedback

We are also grateful to David Beckman of Bread for the World, Deborah Brautigan of American University, Eric Rodenberg of World Resources Institute, and Ditta Smith of *The Washington Post*, who, along with colleagues from the World Bank, gave their time to provide comments and suggestions after the last edition was published. They, and those who responded to our readers' survey, all contributed greatly to improvements and changes in this year's edition. We urge others to send us their views on this edition. We would also like to extend our appreciation to Frances Stewart, Director, Queen Elizabeth House, Oxford; Keith Bezanson, Director, Institute of Development Studies, Sussex; and Sir Timothy Lancaster, Director, School of Oriental and African Studies, London, for arranging opportunities to present the *World Development Indicators* to invited guests, faculty, and students in their institutions and for the feedback we received on the book and CD-ROM at those presentations. We are also grateful to the Overseas Development Institute for organizing a seminar on the *World Development Indicators*.

Bibliography

ACDA (Arms Control and Disarmament Agency). 1997. *World Military Expenditures and Arms Transfers 1996.* Washington, D.C.

Ahmad, Sultan. 1992. "Regression Estimates of Per Capita GDP Based on Purchasing Power Parities." Policy Research Working Paper 956. World Bank, International Economics Department, Washington, D.C.

———. 1994. "Improving Inter-Spatial and Inter-Temporal Comparability of National Accounts." *Journal of Development Economics* 44:53–75.

American Automobile Manufacturers Association. 1995. *World Motor Vehicle Data.* Detroit, Mich.

Armington, Paul, and Yuri Dikhanov. 1996. "Multivariate Normalization of Infrastructure (e.g. Roads) for Comparative Purposes." World Bank, International Economics Department, Washington, D.C.

Aturupane, Harsha, Paul Glewwe, and Paul Isenman. 1994. "Poverty, Human Development and Growth: An Emerging Consensus?" *American Economic Association Papers and Proceedings* 84(2):244–49.

Ball, Nicole. 1984. "Measuring Third World Security Expenditure: A Research Note." *World Development* 12(2):157–64.

Barro, Robert J., and Jong-Wha Lee. 1996. "International Measures of Schooling Years and Schooling Quality." *Economic Reform and Growth* 86(2):218–23.

Berhrman, Jere R., and Mark R. Rosenzweig. 1994. "Caveat Emptor: Cross-Country Data on Education and the Labor Force." *Journal of Development Economics* 44:147–71.

Bos, Eduard, My T. Vu, Ernest Massiah, and Rodolfo Bulatao. 1994. *World Population Projections 1994–95.* Baltimore, Md.: Johns Hopkins University Press.

Braga, C.A. Primo, and Alexander Yeats. 1992. "How Minilateral Trading Arrangements May Affect the Post-Uruguay Round World." World Bank, International Economics Department, Washington, D.C.

Brunetti, Aymo, Gregory Kisunko, and Beatrice Weder. 1997. "Institutional Obstacles to Doing Business: Region by Region Results from a Worldwide Survey of the Private Sector." Policy Research Working Paper 1759. World Bank, Washington, D.C.

Caiola, Marcello. 1995. *A Manual for Country Economists.* Training Series 1, vol. 1. Washington, D.C.: International Monetary Fund.

Cassen, Robert, and associates. 1986. *Does Aid Work?* Report to Intergovernmental Task Force on Concessional Flows. Oxford: Clarendon Press.

Centro Latinoamericano de Demografía. Various years. *Boletín Demográfico.* Santiago, Chile.

Chamie, Joseph. 1994. "Demography: Population Databases in Development Analysis." *Journal of Development Economics* 44:131–46.

Chellaraj, Gnanaraj, Olusoji Adeyi, Alexander S. Preker, and Ellen Goldstein. 1996. *Trends in Health Status, Services, and Finance: The Transition in Central and Eastern Europe.* vol. 2. World Bank Technical Paper 348. Washington, D.C.

Conly, Shanti R., and Joanne E. Epp. 1997. *Falling Short: The World Bank's Role in Population and Reproductive Health.* Washington, D.C.: Population Action International.

Council of Europe. Various years. *Recent Demographic Developments in Europe and North America.* Strasbourg: Council of Europe Press.

Currency Data & Intelligence, Inc. Various issues. *Global Currency Report.* Brooklyn, N.Y.

Dasgupta, Partha. 1993. *An Inquiry into Well-Being and Destitution.* Oxford: Clarendon Press.

Dasgupta, Partha, and Martin Weale. 1992. "On Measuring the Quality of Life." *World Development* 20:119–31.

Davis, Lester. 1982. *Technology Intensity of U.S. Output and Trade.* Washington, D.C.: U.S. Department of Commerce.

Deininger, Klaus, and Lyn Squire. 1996. "A New Data Set Measuring Income Inequality." *The World Bank Economic Review* 10(3):565–91.

Demery, Lionel, and Michael Walton. 1997. "Are Poverty and Social Targets for the 21st Century Attainable?" Paper prepared for a Development Assistance Committee–Development Center seminar, December 4–5, Paris.

Demirgüç-Kunt, Asli, and Ross Levine. 1996a. "Stock Markets, Corporate Finance, and Economic Growth: An Overview." *The World Bank Economic Review* 10(2):223–39.

———. 1996b. "Stock Market Development and Financial Intermediaries: Stylized Facts." *The World Bank Economic Review* 10(2):291–321.

Dixon, John, and Paul Sherman. 1990. *Economics of Protected Areas: A New Look at Benefits and Costs.* Washington, D.C.: Island Press.

Drucker, Peter F. 1994. "The Age of Social Transformation." *Atlantic Monthly* 274 (November).

Engel, Charles, and John H. Rogers. 1995. "Regional Patterns in the Law of One Price: The Roles of Geography vs. Currencies." NBER Working Paper 5395. National Bureau of Economic Research, Cambridge, Mass.

———. 1996. "How Wide is the Border?" *American Economic Review* 86:112–25.

Euromoney. 1997. September. London.

Eurostat (Statistical Office of the European Communities). Various years. *Demographic Statistics.* Luxembourg.

———. Various years. *Statistical Yearbook.* Luxembourg.

Evenson, Robert E., and Carl E. Pray. 1994. "Measuring Food Production (with reference to South Asia)." *Journal of Development Economics* 44:173–97.

Faiz, Asif, Christopher S. Weaver, and Michael P. Walsh. 1996. *Air Pollution from Motor Vehicles: Standards and Technologies for Controlling Emissions.* Washington, D.C.: World Bank.

FAO (Food and Agriculture Organization). 1986. "Inter-Country Comparisons of Agricultural Production Aggregates." Economic and Social Development Paper 61. Rome.

———. 1990. "Tobacco: Supply, Demand and Trade Projections, 1995 and 2000." Economic and Social Development Paper 86. Rome.

———. 1996. *Food Aid in Figures 1994.* vol. 12. Rome.

———. 1997. *State of the World's Forests 1997.* Rome.

———. Various years. *Fertilizer Yearbook.* FAO Statistics Series. Rome.

———. Various years. *Production Yearbook.* FAO Statistics Series. Rome.

———. Various years. *Trade Yearbook.* FAO Statistics Series. Rome.

Feldstein, Martin, and Charles Horioka. 1980. "Domestic Savings and International Capital Flows." *Economic Journal* 90(358):314–29.

Filmer, Dean, Elizabeth King, and Lant Pritchett. 1998. "Gender Disparity in South Asia." Policy Research Working Paper 1867. World Bank, Development Research Group, Washington, D.C.

Finger, J. Michael, Merlinda Ingco, and Ulrich Reincke. 1996. *The Uruguay Round: Statistics on Tariff Concessions Given and Received.* Washington, D.C.: World Bank.

Fischer, Stanley. 1993. "The Role of Macroeconomic Factors in Growth." *Journal of Monetary Economics* 32:485–512.

Fox, James W. 1995. "What Do We Know about World Poverty?" USAID Evaluation Special Study 74. U.S. Agency for International Development, Center for Development Information and Evaluation, Washington, D.C.

Frankel, Jeffrey. 1993. "Quantifying International Capital Mobility in the 1990s." In Jeffrey

Frankel, ed., *On Exchange Rates*. Cambridge, Mass.: MIT Press.

Fredricksen, Birger. 1991. "An Introduction to the Analysis of Student Enrollment and Flows Statistics." PHREE Background Paper 91/39. World Bank, Washington, D.C.

French, Kenneth, and James M. Poterba. 1991. "Investor Diversification and International Equity Markets." *American Economic Review* 81: 222–26.

Furstenberg, George Von. 1997. "Where to Look for International Financial Integration? A Review Essay." Paper presented at an International Monetary Fund Research Department seminar, July 29, Washington, D.C.

Gannon, Colin, and Zmarak Shalizi. 1995. "The Use of Sectoral and Project Performance Indicators in Bank-Financed Transport Operations." TWU Discussion Paper 21. World Bank, Transportation, Water, and Urban Development Department, Washington, D.C.

GATT (General Agreement on Tariffs and Trade). 1966. *International Trade 1965*. Geneva.

———. 1989. *International Trade 1988–89*. Geneva.

Goldfinger, Charles. 1994. *L'utile et le futile: L'économie de l'immatériel*. Paris: Editions Odile Jacob.

Goldstein, Ellen, Alexander S. Preker, Olusoji Adeyi, and Gnanaraj Chellaraj. 1996. *Trends in Health Status, Services, and Finance: The Transition in Central and Eastern Europe*. vol. 1. World Bank Technical Paper 341. Washington, D.C.

Golub, Stephen S. 1990. "International Capital Mobility: Net versus Gross Stocks and Flows." *Journal of International Money and Finance* 9: 424–39.

Gordon, Roger H., and A. Lans Bovenberg. 1996. "Why is Capital So Immobile Internationally? Possible Explanations and Implications for Capital Income Taxation." *American Economic Review* 86:1057–75.

Gould, David M. 1994. "Immigrant Links to the Home Country: Empirical Implications for US Bilateral Trade Flows." *Review of Economics and Statistics* 76:302–16.

Graham, Edward M., and Robert Z. Lawrence. 1996. "Measuring the International Contestability of Markets: A Conceptual Approach." OECD Trade Directorate TD/TC (97)7. Paris.

Haggarty, Luke, and Mary M. Shirley. 1997. "A New Data Base on State-Owned Enterprises." *The World Bank Economic Review* 11(3):491–513.

Happe, Nancy, and John Wakeman-Linn. 1994. "Military Expenditures and Arms Trade: Alternative

Data Sources." IMF Working Paper 94/69. International Monetary Fund, Policy Development and Review Department, Washington, D.C.

Hatter, Victoria L. 1985. *U.S. High-Technology Trade and Competitiveness*. Washington, D.C.: U.S. Department of Commerce.

Heck, W.W. 1989. "Assessment of Crop Losses from Air Pollutants in the U.S." In J.J. McKenzie and M.T. El Ashry, eds., *Air Pollution's Toll on Forests and Crops*. New Haven, Conn.: Yale University Press.

Heston, Alan. 1994. "A Brief Review of Some Problems in Using National Accounts Data in Level of Output Comparisons and Growth Studies." *Journal of Development Economics* 44:29–52.

Hettige, Hemamala, Muthukumara Mani, and David Wheeler. 1998. "Industrial Pollution in Economic Development: Kuznets Revisited." Policy Research Working Paper 1876. World Bank, Development Research Group, Washington, D.C.

Heyneman, Stephen P. 1996. "The Quality of Education in the Middle East and North Africa." EMT Working Paper 3. World Bank, Europe and Central Asia and Middle East and North Africa Technical Department, Washington, D.C.

Hill, M. Anne, and Elizabeth M. King. 1993. "Women's Education in Developing Countries: An Overview." In Elizabeth M. King and M. Anne Hill, eds., *Women's Education in Developing Countries*. Baltimore, Md.: Johns Hopkins University Press.

Human Development Network. 1997. *Sector Strategy: Health, Nutrition, and Population*. Washington, D.C.: World Bank.

ICAO (International Civil Aviation Organization). 1997. "Civil Aviation Statistics of the World: 1996." *ICAO Statistical Yearbook*. 22nd ed. Montreal.

IEA (International Energy Agency). 1996a. *Energy Statistics and Balances of Non-OECD Countries 1993–94*. Paris.

———. 1996b. *Energy Statistics of OECD Countries 1993–94*. Paris.

IFC (International Finance Corporation). 1997a. *Emerging Stock Markets Factbook 1997*. Washington, D.C.

———. 1997b. *Trends in Private Investment in Developing Countries 1997*. Washington, D.C.

ILO (International Labour Organization). 1990a. *ILO Manual on Concepts and Methods*. Geneva: International Labour Office.

———. 1990b. *Yearbook of Labour Statistics: Retrospective Edition of Population Censuses 1945–89*. Geneva: International Labour Office.

———. Various years. *Sources and Methods: Labour Statistics*. (Formerly *Statistical Sources and Methods*.) Geneva: International Labour Office.

———. Various years. *Yearbook of Labour Statistics*. Geneva: International Labour Office.

IMF (International Monetary Fund). 1977. *Balance of Payments Manual*. 4th ed. Washington, D.C.

———. 1986. *A Manual on Government Finance Statistics*. Washington, D.C.

———. 1993. *Balance of Payments Manual*. 5th ed. Washington, D.C.

———. 1995. *Balance of Payments Compilation Guide*. Washington, D.C.

———. 1996a. *Balance of Payments Textbook*. Washington, D.C.

———. 1996b. *Manual on Monetary and Financial Statistics*. Washington, D.C.

———. 1997a. "Staff Studies for the World Economic Outlook." Research Department, Washington, D.C.

———. 1997b. *World Economic Outlook* (October). Washington, D.C.

———. Various years. *Balance of Payments Statistics Yearbook*. Parts 1 and 2. Washington, D.C.

———. Various years. *Direction of Trade Statistics Yearbook*. Washington, D.C.

———. Various years. *Government Finance Statistics Yearbook*. Washington, D.C.

———. Various issues. *International Financial Statistics* (monthly). Washington, D.C.

———. Various years. *International Financial Statistics Yearbook*. Washington, D.C.

Institutional Investor. 1997. September. New York.

Intergovernmental Panel on Climate Change. 1996. *Climate Change 1995*. Cambridge: Cambridge University Press.

International Telecommunication Union. 1997. *World Telecommunication Development Report*. Geneva.

International Working Group of External Debt Compilers (Bank for International Settlements, International Monetary Fund, Organisation for Economic Co-operation and Development, and World Bank). 1987. *External Debt Definitions*. Washington, D.C.

Inter-Secretariat Working Group on National Accounts (Commission of the European Community, International Monetary Fund, Organisation for Economic Co-operation and Development, United Nations, and World Bank). 1993. *System of National Accounts*. Brussels, Luxembourg, New York, and Washington, D.C.

IRF (International Road Federation). 1995. *World Road Statistics 1990–1994*. Geneva.

Irwin, Douglas A. 1996. "The United States in a New Global Economy? A Century's Perspective."

Bibliography

Papers and Proceedings of the 108th Annual Meeting of the American Economic Association. *American Economic Review* (May).

Isard, Peter. 1995. *Exchange Rate Economics.* Cambridge: Cambridge University Press.

IUCN (World Conservation Union). Various years. *Red List of Threatened Animals.* Gland, Switzerland.

Journal of Development Economics. 1994. Special issue on database for development analysis. Edited by T.N. Srinivasan. Vol. 44, no. 1.

Kanbur, Ravi. 1997. "Income Distribution and Development." Cornell University, Ithaca, N.Y.

Kaufmann, Daniel. 1997. "Corruption: The Facts." *Foreign Policy* 107:114–31.

Klugman, Jeni, and George Schieber. 1996. *A Survey of Health Reform in Central Asia.* World Bank Technical Paper 344. Washington, D.C.

Knetter, Michael. 1994. "Why are Retail Prices in Japan So High? Evidence from German Export Prices." NBER Working Paper 4894. National Bureau of Economic Research, Cambridge, Mass.

Kravis, Irving B. 1970. "Trade as a Handmaiden of Growth." *Economic Journal* 80(323):850–72.

Krueger, Anne O., Constantine Michalopoulos, and Vernon W. Ruttan. 1989. *Aid and Development.* Baltimore, Md.: Johns Hopkins University Press.

Krugman, Paul. 1991. *Geography and Trade.* Cambridge, Mass.: MIT Press.

Leamer, Edward. 1987. "Measures of Openness." Working Paper 447. University of California at Los Angeles, Department of Economics.

Levine, Ross, and Sara Zervos. 1996. "Stock Market Development and Long-Run Growth." *The World Bank Economic Review* 10(2):323–40.

Lewis, Karen K. 1995. "Puzzles in International Financial Markets." In Gene Grossman and Kenneth Rogoff, eds., *Handbook of International Economics.* vol. 3. Amsterdam: North Holland.

Lewis, Stephen R., Jr. 1989. "Primary Exporting Countries." In Hollis Chenery and T.N. Srinivasan, eds., *Handbook of Development Economics.* vol. 2. Amsterdam: North Holland.

Lim, Lin Lean. 1996. *More and Better Jobs for Women: An Action Guide.* Geneva: International Labour Office.

Lockheed, Marlaine E., Adriaan M. Verspoor, and associates. 1991. *Improving Primary Education in Developing Countries.* New York: Oxford University Press.

Lovei, Magdolna. 1997. "Toward Effective Pollution Management." *Environment Matters* (fall): 52–53.

Low, Patrick, and Alexander Yeats. 1994. "Nontariff Measures and Developing Countries: Has the Uruguay Round Leveled the Playing Field?" Policy Research Working Paper 1353. World Bank, International Economics Department, Washington, D.C.

Mani, Muthukumara, and David Wheeler. 1997. "In Search of Pollution Havens? Dirty Industry in the World Economy, 1960–95." World Bank, Policy Research Department, Washington, D.C.

Mauro, Pablo. 1997. "The Effects of Corruption on Growth, Investment and Government Expenditure: A Cross-Country Analysis." In Kimberly Ann Elliott, ed., *Corruption and the Global Economy.* Washington, D.C.: Institute for International Economics.

Midgley, Peter. 1994. *Urban Transport in Asia: An Operational Agenda for the 1990s.* World Bank Technical Paper 224. Washington, D.C.

Montiel, Peter. 1993. "Capital Mobility in Developing Countries." Policy Research Working Paper 1103. World Bank, International Economics Department, Washington, D.C.

Moody's Investors Service. 1998. *Sovereign, Subnational and Sovereign-Guaranteed Issuers.* January. New York.

Morgenstern, Oskar. 1963. *On the Accuracy of Economic Observations.* Princeton, N.J.: Princeton University Press.

Murray, Christopher, Ramesh Govindaraj, and Ganaraj Chellaraj. 1994. "Global Domestic Expenditures in Health." Background paper 13 to *World Development Report 1993.* World Bank, Washington, D.C.

Obstfeldt, Maurice. 1995. "International Capital Mobility in the 1990s." In P.B. Kenen, ed., *Understanding Interdependence: The Macroeconomics of the Open Economy.* Princeton, N.J.: Princeton University Press.

Obstfeldt, Maurice, and Kenneth Rogoff. 1996. *Foundations of International Macroeconomics.* Cambridge, Mass.: MIT Press.

OECD (Organisation for Economic Co-operation and Development). 1985. "Measuring Health Care 1960–1983: Expenditure, Costs, Performance." OECD Social Policy Studies 2. Paris.

———. 1989. "Health Care Expenditure and Other Data: An International Compendium from the OECD." *In Health Care Financing Review.* Annual supplement. Paris.

———. 1996a. *Development Assistance: Efforts and Policies of the Members of the Development Assistance Committee.* Paris.

———. 1996b. *Geographical Distribution of Financial Flows to Aid Recipients:*

Disbursements, Commitments, Country Indicators, 1990–1994. Paris.

———. 1996c. *National Accounts 1960–1994.* vol. 1, Main Aggregates. Paris.

———. 1996d. *National Accounts 1960–1994.* vol. 2, Detailed Tables. Paris.

———. 1996e. "Shaping the 21st Century: The Contribution of Development Cooperation." Paris.

———. 1997a. *Development Co-operation: 1996 Report.* Paris.

———. 1997b. *Employment Outlook.* Paris.

———. 1997c. *OECD Environmental Data: Compendium 1997.* Paris

———. 1997d. *Trends in International Migration: Continuous Reporting System on Migration.* Paris.

Political Risk Services. 1997. *International Country Risk Guide.* December. East Syracuse, N.Y.

Pritchett, Lant. 1996. "Measuring Outward Orientation in Developing Countries: Can It Be Done?" Policy Research Working Paper 566. World Bank, Washington, D.C.

Pritchett, Lant, and Geeta Sethi. 1994. "Tariff Rates, Tariff Revenue, and Tariff Reform—Some New Facts." *The World Bank Economic Review* 8(1):1–16.

Price Waterhouse. 1996a. *Corporate Taxes: A Worldwide Summary.* New York.

———. 1996b. *Individual Taxes: A Worldwide Summary.* New York.

Psacharopoulos, George. 1994. "Returns to Investment in Education: A Global Update." *World Development* 22(9):1325–43.

———. 1995. *Building Human Capital for Better Lives.* Washington, D.C.: World Bank.

Ravallion, Martin. 1996. "Poverty and Growth: Lessons from 40 Years of Data on India's Poor." DECNote 20. World Bank, Development Economics Vice Presidency, Washington, D.C.

Ravallion, Martin, and Shaohua Chen. 1996. "What Can New Survey Data Tell Us about the Recent Changes in Living Standards in Developing and Transitional Economies?" World Bank, Policy Research Department, Washington, D.C.

———. 1997. "Can High-Inequality Developing Countries Escape Absolute Poverty?" *Economic Letters* 56:51–57.

———. Forthcoming. "What Can New Survey Data Tell Us about Recent Changes in Distribution and Poverty?" *The World Bank Economic Review.*

Rogoff, Kenneth. 1996. "The Purchasing Power Parity Puzzle." *Journal of Economic Literature* 34:647–68.

Ruggles, Robert. 1994. "Issues Relating to the UN System of National Accounts and Developing

Countries." *Journal of Development Economics* 44(1):87–102.

Schultz, T. Paul. 1993. "Returns to Women's Education." In Elizabeth M. King and M. Anne Hill, eds., *Women's Education in Developing Countries.* Baltimore, Md.: Johns Hopkins University Press.

Sen, Amartya. 1988. "The Concept of Development." In Hollis Chenery and T.N. Srinivasan, eds., *Handbook of Development Economics.* vol. 1. Amsterdam: North Holland.

Serageldin, Ismail. 1995. *Toward Sustainable Management of Water Resources.* A Directions in Development book. Washington D.C.: World Bank.

Shang-jin Wei. 1996. "Intra-National versus International Trade: How Stubborn are Nations in Global Integration?" NBER Working Paper 5531. National Bureau of Economic Research, Cambridge, Mass.

Shapiro, Harvey. 1996. "Restoration Drama." *Institutional Investor* 21 (September):9.

Shiklovanov, Igor. 1993. "World Fresh Water Resources." In Peter H. Gleick, ed., *Water in Crisis: A Guide to Fresh Water Resources.* New York: Oxford University Press.

South Pacific Commission. 1997. *Pacific Island Populations Data Sheet.* Noumea, New Caledonia.

Srinivasan, T.N. 1991. "Development Thought, Policy, and Strategy, Then and Now." Background paper to *World Development Report 1991.* World Bank, Washington, D.C.

———. 1994. "Database for Development Analysis: An Overview." *Journal of Development Economics* 44(1):3–28.

Standard & Poor's. 1998. *Credit Week* (January). New York.

Sykes, Alan O. 1995. *Product Standards for Internationally Integrated Goods Markets.* Washington, D.C.: Brookings Institution.

———. "Strategies for Increasing Market Access Under Regulatory Heterogeneity." OECD Trade Directorate TD/TC (96)8. Paris.

Syrquin, Moshe. 1988. "Patterns of Structural Change." In Hollis Chenery and T.N. Srinivasan, eds., *Handbook of Development Economics.* vol. 1. Amsterdam: North Holland.

Stephenson, Sherry M. 1996. "Standards and Conformity Assessments as Nontariff Barriers to Trade." Policy Research Working Paper 1826. World Bank, Development Research Group, Washington, D.C.

Taylor, Alan M. 1996a. "International Capital Mobility in History: Purchasing Power Parity in the Long Run." NBER Working Paper 5742.

———. 1996b. "International Capital Mobility in History: The Saving-Investment Relationship." NBER Working Paper 5743.

Tesar, Linda, and Ingrid Werner. 1992. "Home Bias and the Globalization of Securities Markets." NBER Working Paper 4218. National Bureau of Economic Research, Cambridge, Mass.

UNAIDS. 1996. "HIV and Infant Feeding: An Interim Statement." Geneva.

UNCTAD (United Nations Conference on Trade and Development). Various years. *Handbook of International Trade and Development Statistics.* Geneva.

UNEP (United Nations Environment Programme). 1991. *Urban Air Pollution.* Nairobi.

UNEP (United Nations Environment Programme) and WHO (World Health Organization). 1992. *Urban Air Pollution in Megacities of the World.* Cambridge, Mass.: Blackwell.

———. 1995. *City Air Quality Trends.* Nairobi.

UNESCO (United Nations Educational, Scientific, and Cultural Organization). 1995. *World Education Report.* Paris: Oxford University Press.

———. 1997. *Statistical Yearbook.* Paris: Oxford University Press.

———. Forthcoming. *World Education Report 1998.* Paris: Oxford University Press.

UNICEF (United Nations Children's Fund). 1997. *The Progress of Nations.* New York: Oxford University Press.

———. Various years. *The State of the World's Children.* New York: Oxford University Press.

UNIDO (United Nations Industrial Development Organization). 1996. *International Yearbook of Industrial Statistics 1997.* Vienna.

United Nations. 1947. *Measurement of National Income and the Construction of Social Accounts.* New York.

———. 1968. *A System of National Accounts: Studies and Methods.* series F, no. 2, rev. 3. New York.

———. 1985. *National Accounts Statistics: Compendium of Income Distribution Statistics.* New York.

———. 1990. *Assessing the Nutritional Status of Young Children.* National Household Survey Capability Programme. New York.

———. 1991. *The World's Women, 1970–90: Trends and Statistics.* New York.

———. 1993a. *International Trade Statistics Yearbook.* vol. 1. New York.

———. 1993b. *Report on the World Social Situation, 1993.* New York.

———. 1997. *World Urbanization Prospects: The 1996 Revision.* New York.

———. Various years. *Energy Statistics Yearbook.* New York.

———. Various issues. *Monthly Bulletin of Statistics.* New York.

———. Various years. *National Income Accounts.* Statistics Division. New York.

———. Various years. *Statistical Yearbook.* New York.

———. Various years. *Update on the Nutrition Situation.* Administrative Committee on Coordination, Subcommittee on Nutrition. Geneva.

United Nations Department of Economic and Social Information and Policy Analysis. 1996. *World Population Prospects: The 1996 Edition.* New York.

———. Various years. *Levels and Trends of Contraceptive Use.* New York.

———. Various years. *Population and Vital Statistics Report.* New York.

United Nations Economic and Social Commission for Western Asia. 1997. *Purchasing Power Parities: Volume and Price Level Comparisons for the Middle East, 1993.* E/ESCWA/STAT/1997/2. Amman.

UNRISD (United Nations Research Institute for Social Development). 1977. *Research Data Bank of Development Indicators.* vol. 4, Notes on the Indicators. Geneva.

———. 1993. *Monitoring Social Progress in the 1990s: Data Constraints, Concerns, and Priorities.* Avebury.

U.S. Bureau of the Census. 1996. *World Population Profile 1996.* Washington, D.C.: U.S. Government Printing Office.

———. 1998. HIV/AIDS Surveillance Database. Washington, D.C.

U.S. Environmental Protection Agency. 1995. *National Air Quality and Emissions Trends Report 1995.* Washington, D.C.

Walsh, Michael P. 1994. "Motor Vehicle Pollution Control: An Increasingly Critical Issue for Developing Countries." World Bank, Washington, D.C.

Walton, Michael. 1997. "Will Global Advance Include the World's Poor?" Paper prepared for the Aspen Institute Conference, "Persistence Poverty in Developing Countries: Determining the Causes and Closing the Gaps," December 1–4, Broadway, England.

WCMC (World Conservation Monitoring Centre). 1992. *Global Biodiversity: Status of the Earth's Living Resources.* London: Chapman and Hall.

———. 1994. *Biodiversity Data Sourcebook.* Cambridge: World Conservation Press.

Wellenius, Bjorn. 1997. "Telecommunications Reform—How to Succeed." In Suzanne Smith, ed., *The Private Sector in Infrastructure: Strategy, Regulation, and Risk.* Washington, D.C.: World Bank.

Bibliography

WHO (World Health Organization). 1977. *International Classification of Diseases.* ninth revision. Geneva.

————. 1990. *World Health Statistics Quarterly* 43(4).

————. 1991. *Maternal Mortality: A Global Factbook.* Geneva.

————. 1994. *Progress Towards Health for All: Statistics of Member States.* Geneva.

————. 1995. *World Health Statistics Quarterly* 48(3/4).

————. 1996a. *Evaluating the Implementation of the Strategy for Health for All by the Year 2000.* Geneva.

————. 1996b. *Water Supply and Sanitation Sector Monitoring Report 1996.* Geneva.

————. 1997a. *Global Tuberculosis Control Report 1997.* Geneva.

————. 1997b. *Monitoring Reproductive Health: Selecting a Short List of National and Global Indicators.* Geneva.

————. 1997c. *Tobacco or Health: A Global Status Report 1997.* Geneva.

————. 1997d. *World Health Report 1997.* Geneva.

————. Various years. *World Health Statistics Annual.* Geneva.

WHO (World Health Organization) and UNICEF (United Nations Children's Fund). 1996. *Revised 1990 Estimates on Maternal Mortality: A New Approach.* Geneva.

Windham, Douglas M. 1988. *Indicators of Educational Effectiveness and Efficiency.* Tallahassee, Fla.: Florida State University, Educational Efficiency Clearinghouse.

Wolf, Holger C. 1997. "Patterns of Intra- and Inter-State Trade." NBER Working Paper 5939. National Bureau of Economic Research, Cambridge, Mass.

Wolfensohn, James D. 1997. "The Challenge of Inclusion." Address to the Board of Governors, 23 September, Hong Kong, China.

World Bank. 1990. *World Development Report 1990: Poverty.* New York: Oxford University Press.

————. 1991a. *Developing the Private Sector: The World Bank's Experience and Approach.* Washington, D.C.

————. 1991b. *World Development Report 1991: The Challenge of Development.* New York: Oxford University Press.

————. 1992. *World Development Report 1992: Development and the Environment.* New York: Oxford University Press.

————. 1993a. *The Environmental Data Book: A Guide to Statistics on the Environment and Development.* Washington, D.C.

————. 1993b. *Purchasing Power of Currencies: Comparing National Incomes Using ICP Data.* Washington, D.C.

————. 1993c. *World Development Report 1993: Investing in Health.* New York: Oxford University Press.

————. 1994a. *Global Economic Prospects and the Developing Countries 1994.* Washington, D.C.

————. 1994b. *World Development Report 1994: Infrastructure for Development.* New York: Oxford University Press.

————. 1995a. *Advancing Social Development: A World Bank Contribution to the Social Summit.* Washington, D.C.

————. 1995b. *Bureaucrats in Business: The Economics and Politics of Government Ownership.* Washington, D.C.

————. 1995c. *Global Economic Prospects and the Developing Countries 1995.* Washington, D.C.

————. 1995d. *Priorities and Strategies for Education: A Review.* Washington, D.C.

————. 1995e. *Private Sector Development in Low-Income Countries.* Washington, D.C.

————. 1995f. *Toward Gender Equality: The Role of Public Policy—An Overview.* Washington, D.C.

————. 1995g. *World Development Report 1995: Workers in an Integrating World.* New York: Oxford University Press.

————. 1996a. *From Vision to Action in the Rural Sector.* Agriculture Department. Washington, D.C.

————. 1996b. *Global Economic Prospects and the Developing Countries 1996.* Washington, D.C.

————. 1996c. *Livable Cities for the 21st Century.* Washington, D.C.

————. 1996d. *National Environmental Strategies: Learning from Experience.* Environment Department. Washington, D.C.

————. 1996e. *Poverty Reduction and the World Bank: Progress and Challenges in the 1990s.* Washington, D.C.

————. 1997a. *Confronting AIDS: Public Priorities in a Global Epidemic.* A Policy Research Report. Washington, D.C.

————. 1997b. *Global Economic Prospects and the Developing Countries 1997.* Washington, D.C.

————. 1997c. *Helping Countries Combat Corruption.* Poverty Reduction and Economic Management Network. Washington, D.C.

————. 1997d. *Poverty Reduction and the World Bank: Progress in Fiscal 1996 and 1997.* Washington, D.C.

————. 1997e. *Private Capital Flows to Developing Countries: The Road to Financial Integration.* World Bank Policy Research Report.

————. 1997f. *World Development Indicators 1997.* Washington, D.C.

————. 1997g. *World Development Report 1997: The State in a Changing World.* New York: Oxford University Press.

————. 1998. *Global Development Finance 1998.* Washington, D.C.

————. Various issues. *Commodity Markets and the Developing Countries* (quarterly). Washington, D.C.

————. Various years. *World Debt Tables.* Washington, D.C.

World Resources Institute, International Institute for Environment and Development, and IUCN (World Conservation Union). Various years. *World Directory of Country Environmental Studies.* Washington, D.C.

World Resources Institute, UNEP (United Nations Environment Programme), and UNDP (United Nations Development Programme). 1994. *World Resources 1994–95: A Guide to the Global Environment.* New York: Oxford University Press.

World Resources Institute, UNEP (United Nations Environment Programme), UNDP (United Nations Development Programme), and World Bank. 1998. *World Resources 1998–99: A Guide to the Global Environment.* New York: Oxford University Press.

World Tourism Organization. 1997. *Yearbook of Tourism Statistics.* vols. 1 and 2, 49 ed. Madrid.

Zimmermann, Klaus F. 1995. "European Migration: Push and Pull." In Michael Bruno and Boris Pleskovic, eds., *Proceedings of the World Bank Annual Conference on Development Economics 1994.* Washington, D.C.: World Bank.

Index of indicators

References are to table numbers.

Index of indicators

Index of indicators

Index of indicators

Index of indicators

World Development Indicators

1998 | Readers survey

Special offer

If you complete and return this questionnaire before 30 June 1998, we will send you a free copy of next year's *World Bank Atlas* when it is published in 1999. Please be sure to complete the mailing information on the reverse side and mail or fax a copy of the questionnaire to us immediately. Even if you can't make the deadline, we would still like to hear from you.

We need your help!

We want to know what you think about the new *World Development Indicators* and how it compares with other sources of information you use. Your response will help us continue to improve the family of *World Development Indicators* products.

1. How do you use the *World Development Indicators*?
   (Check as many as apply)
   - ○ Analysis and research
   - ○ Background information for my work
   - ○ Policy formulation
   - ○ Supplement to the WDI CD-ROM
   - ○ Reference source
   - ○ Teaching or training tool
   - ○ Update on world issues
   - ○ Other

2. How did you obtain the WDI?
   - ○ Own purchase
   - ○ Employer purchase
   - ○ Borrowed
   - ○ Library
   - ○ Gift

3. How do you rate the overall usefulness of the WDI? (Check one)
   - ○ Very useful
   - ○ Useful
   - ○ Marginally useful
   - ○ Not at all useful

4. What does the WDI offer that is not provided by other sources?
   - ○ Information and statistics
   - ○ Background for topical economic events and trends
   - ○ Insights into the development process
   - ○ Analysis of statistical issues
   - ○ Other

6. What sections do you refer to most?
   - ○ World view
   - ○ People
   - ○ Environment
   - ○ Economy
   - ○ States and markets
   - ○ Global links

7. Should we add/subtract tables or indicators to/from any of the sections? Please list against each section.
   - ○ World view
   - ○ People
   - ○ Environment
   - ○ Economy
   - ○ States and markets
   - ○ Global links

8. What do you like most about the WDI's design?
   - ○ Section introductions
   - ○ Tabular design
   - ○ Commentary
   - ○ About the data
   - ○ Definitions and Data sources
   - ○ Other design features

9. What aspects of the WDI do you like least? What should be changed or improved?
   - ○ Section introductions
   - ○ Tabular design
   - ○ Commentary
   - ○ About the data
   - ○ Definitions and Data sources
   - ○ Other design features

5. How do you rate the WDI?
   (Check one box in each case)

	Excellent	Good	Adequate	Poor
Accuracy	○	○	○	○
Coverage	○	○	○	○
Objectivity	○	○	○	○
Presentation and readability	○	○	○	○
Technical information	○	○	○	○
Usefulness of indicators	○	○	○	○

10. When seeking information on economic development and related subjects, how useful do you find the following sources?

	Very	Moderately	Slightly
Books	○	○	○
CD-ROM or diskette products	○	○	○
Colleagues	○	○	○
Courses and seminars	○	○	○
Internet or online services	○	○	○
Journals	○	○	○
Newspapers	○	○	○
Other	○	○	○

11. Please name one other source of information on economic development that you use regularly:

12. Do you also use the WDI CD-ROM or the *World Bank Atlas*?

	Yes	No
CD-ROM	○	○
Atlas	○	○

13. How do you classify the organization where you work?
- ○ Central bank
- ○ Finance ministry
- ○ Planning agency
- ○ Other government agency or public enterprise
- ○ International or regional organization
- ○ Commercial bank or financial organization
- ○ News media outlet
- ○ Other private enterprise
- ○ Nongovernmental organization
- ○ Policy/research institution
- ○ University or college
- ○ Primary or secondary school
- ○ Library
- ○ Other

14. What are your areas of specialization?
- ○ Administration/management
- ○ Economics
- ○ Engineering
- ○ Environmental sciences
- ○ Finance/banking
- ○ Health
- ○ Information management
- ○ Law
- ○ Natural sciences
- ○ Politics
- ○ Public affairs
- ○ Other social sciences
- ○ Teaching
- ○ Other (please state)

15. How large is your organization worldwide?
- ○ 99 or fewer employees
- ○ 100–999 employees
- ○ 1,000–9,999 employees
- ○ 10,000 or more employees

16. How would you categorize your position in the organization where you work?
- ○ Senior management
- ○ Middle management
- ○ Professional staff or faculty
- ○ Consultant
- ○ Student
- ○ Other

17. What is your age?
- ○ Under 25 years
- ○ 25–34 years
- ○ 35–44 years
- ○ 45–54 years
- ○ 55–64 years
- ○ 65 years or older

18. What is your highest level of educational attainment?
- ○ Secondary education or upper level
- ○ University level
- ○ Postgraduate work
- ○ Other

19. Finally, please tell us the country in which you are currently residing.

Thank you for completing this survey. Please use the space below or a separate sheet of paper to add any comments about this survey or to elaborate on any of your responses.

Your information:
(should you wish to provide it)

Name

Organization

Street address

City

State/Province

Country, Postal code

Return a copy of this form to:

Development Data Center
The World Bank
1818 H Street, N.W., Room MC2-812, Washington, D.C. 20433 USA
Fax: 202 522 1498

For further information, contact the Development Data Center
Tel: 800 590 1906 or 202 473 7824
Fax: 202 522 1498
Email: info@worldbank.org

Distributors of World Bank Publications

Prices and credit terms vary from country to country. Consult your local distributor before placing an order.

ARGENTINA
Oficina del Libro Internacional
Av. Cordoba 1877
1120 Buenos Aires
Tel: (54 1) 815-8354
Fax: (54 1) 815-8156
E-mail: olilibro@satlink.com

AUSTRALIA, FIJI, PAPUA NEW GUINEA, SOLOMON ISLANDS, VANUATU, AND WESTERN SAMOA
D.A. Information Services
648 Whitehorse Road
Mitcham 3132
Victoria
Tel: (61) 3 9210 7777
Fax: (61) 3 9210 7788
E-mail: service@dadirect.com.au
URL: http://www.dadirect.com.au

AUSTRIA
Gerold and Co.
Weihburggasse 26
A-1011 Wien
Tel: (43 1) 512-47-31-0
Fax: (43 1) 512-47-31-29
URL: http://www.gerold.co/at.online

BANGLADESH
Micro Industries Development
Assistance Society (MIDAS)
House 5, Road 16
Dhanmondi R/Area
Dhaka 1209
Tel: (880 2) 326427
Fax: (880 2) 811188

BELGIUM
Jean De Lannoy
Av. du Roi 202
1060 Brussels
Tel: (32 2) 538-5169
Fax: (32 2) 538-0841

BRAZIL
Publicações Tecnicas Internacionais Ltda.
Rua Peixoto Gomide, 209
01409 Sao Paulo, SP.
Tel: (55 11) 259-6644
Fax: (55 11) 258-6990
E-mail: postmaster@pti.uol.br
URL: http://www.uol.br

CANADA
Renouf Publishing Co. Ltd.
5369 Canotek Road
Ottawa, Ontario K1J 9J3
Tel: (613) 745-2665
Fax: (613) 745-7660
E-mail: order.dept@renoufbooks.com
URL: http://www.renoufbooks.com

CHINA
China Financial & Economic
Publishing House
8, Da Fo Si Dong Jie
Beijing
Tel: (86 10) 6333-8257
Fax: (86 10) 6401-7365

China Book Import Centre
P.O. Box 2825
Beijing

COLOMBIA
Infoenlace Ltda.
Carrera 6 No. 51-21
Apartado Aereo 34270
Santafé de Bogotá, D.C.
Tel: (57 1) 285-2798
Fax: (57 1) 285-2798

COTE D'IVOIRE
Center d'Edition et de Diffusion Africaines
(CEDA)
04 B.P. 541
Abidjan 04
Tel: (225) 24 6510;24 6511
Fax: (225) 25 0567

CYPRUS
Center for Applied Research
Cyprus College
6, Diogenes Street, Engomi
P.O. Box 2006
Nicosia
Tel: (357 2) 44-1730
Fax: (357 2) 46-2051

CZECH REPUBLIC
USIS, NIS Prodejna
Havelkova 22
130 00 Prague 3
Tel: (420 2) 2423 1486
Fax: (420 2) 2423 1114
URL: http://www.nis.cz/

DENMARK
SamfundsLitteratur
Rosenoerns Allé 11
DK-1970 Frederiksberg C
Tel: (45 31) 351942
Fax: (45 31) 357822
URL: http://www.sl.cbs.dk

ECUADOR
Libri Mundi
Libreria Internacional
P.O. Box 17-01-3029
Juan Leon Mera 851
Quito
Tel: (593 2) 521-606; (593 2) 544-185
Fax: (593 2) 504-209
E-mail: librimu1@librimundi.com.ec
E-mail: librimu2@librimundi.com.ec

CODEU
Ruiz de Castilla 763, Edif. Expocolor
Primer piso, Of. #2
Quito
Tel/Fax: (593 2) 507-383; 253-091
E-mail: codeu@impsat.net.ec

EGYPT, ARAB REPUBLIC OF
Al Ahram Distribution Agency
Al Galaa Street
Cairo
Tel: (20 2) 578-6083
Fax: (20 2) 578-6833

The Middle East Observer
41, Sherif Street
Cairo
Tel: (20 2) 393-9732
Fax: (20 2) 393-9732

FINLAND
Akateeminen Kirjakauppa
P.O. Box 128
FIN-00101 Helsinki
Tel: (358 0) 121 4418
Fax: (358 0) 121-4435
E-mail: akatilaus@stockmann.fi
URL: http://www.akateeminen.com/

FRANCE
World Bank Publications
66, avenue d'Iéna
75116 Paris
Tel: (33 1) 40-69-30-56/57
Fax: (33 1) 40-69-30-68

GERMANY
UNO-Verlag
Poppelsdorfer Allee 55
53115 Bonn
Tel: (49 228) 949020
Fax: (49 228) 217492
URL: http://www.uno-verlag.de
E-mail: unoverlag@aol.com

GHANA
Epp Books Services
P.O. Box 44
TUC
Accra

GREECE
Papasotiriou S.A.
35, Stournara Str.
106 82 Athens
Tel: (30 1) 364-1826
Fax: (30 1) 364-8254

HAITI
Culture Diffusion
5, Rue Capois
C.P. 257
Port-au-Prince
Tel: (509) 23 9260
Fax: (509) 23 4858

HONG KONG, MACAO
Asia 2000 Ltd.
Sales & Circulation Department
Seabird House, unit 1101-02
22-28 Wyndham Street, Central
Hong Kong
Tel: (852) 2530-1409
Fax: (852) 2526-1107
E-mail: sales@asia2000.com.hk
URL: http://www.asia2000.com.hk

HUNGARY
Euro Info Service
Margitszigeti Europa Haz
H-1138 Budapest
Tel: (36 1) 350 80 24, 350 80 25
Fax: (36 1) 350 90 32
E-mail: euroinfo@mail.matav.hu

INDIA
Allied Publishers Ltd.
751 Mount Road
Madras - 600 002
Tel: (44) 852-3938
Fax: (44) 852-0649

INDONESIA
Pt. Indira Limited
Jalan Borobudur 20
P.O. Box 181
Jakarta 10320
Tel: (62 21) 390-4290
Fax: (62 21) 390-4289

IRAN
Ketab Sara Co. Publishers
Khaled Eslamboli Ave., 6th Street
Delafrooz Alley No. 8
P.O. Box 15745-733
Tehran 15117
Tel: (98 21) 8717819; 8716104
Fax: (98 21) 8712479
E-mail: ketab-sara@neda.net.ir

Kowkab Publishers
P.O. Box 19575-511
Tehran
Tel: (98 21) 258-3723
Fax: (98 21) 258-3723

IRELAND
Government Supplies Agency
Oifig an tSoláthair
4-5 Harcourt Road
Dublin 2
Tel: (353 1) 661-3111
Fax: (353 1) 475-2670

ISRAEL
Yozmot Literature Ltd.
P.O. Box 56055
3 Yohanan Hasandlar Street
Tel Aviv 61560
Tel: (972 3) 5285-397
Fax: (972 3) 5285-397

R.O.Y. International
PO Box 13056
Tel Aviv 61130
Tel: (972 3) 5461423
Fax: (972 3) 5461442
E-mail: royil@netvision.net.il

Palestinian Authority/Middle East
Index Information Services
P.O.B. 19502 Jerusalem
Tel: (972 2) 6271219
Fax: (972 2) 6271634

ITALY
Licosa Commissionaria Sansoni SPA
Via Duca Di Calabria, 1/1
Casella Postale 552
50125 Firenze
Tel: (55) 645-415
Fax: (55) 641-257
E-mail: licosa@ftbcc.it
URL: http://www.ftbcc.it/licosa

JAMAICA
Ian Randle Publishers Ltd.
206 Old Hope Road, Kingston 6
Tel: 876-927-2085
Fax: 876-977-0243
E-mail: irpl@colis.com

JAPAN
Eastern Book Service
3-13 Hongo 3-chome, Bunkyo-ku
Tokyo 113
Tel: (81 3) 3818-0861
Fax: (81 3) 3818-0864
E-mail: orders@svt-ebs.co.jp
URL: http://www.bekkoame.or.jp/~svt-ebs

KENYA
Africa Book Service (E.A.) Ltd.
Quaran House, Mfangano Street
Nairobi
Tel: (254 2) 223 641
Fax: (254 2) 330 272

KOREA, REPUBLIC OF
Daejon Trading Co. Ltd.
P.O. Box 34, Youida, 706 Seoun Bldg
44-6 Youido-Dong, Yeongchengpo-Ku
Seoul
Tel: (82 2) 785-1631/4
Fax: (82 2) 784-0315

LEBANON
Librairie du Liban
P.O. Box 11-9232
Beirut
Tel: (961 9) 217 944
Fax: (961 9) 217 434

MALAYSIA
University of Malaya Cooperative
Bookshop, Limited
P.O. Box 1127
Jalan Pantai Baru
59700 Kuala Lumpur
Tel: (60 3) 756-5000
Fax: (60 3) 755-4424
E-mail: umkoop@tm.net.my

MEXICO
INFOTEC
Av. San Fernando No. 37
Col. Toriello Guerra
14050 Mexico, D.F.
Tel: (52 5) 624-2800
Fax: (52 5) 624-2822
E-mail: infotec@rtn.net.mx
URL: http://rtn.net.mx

Mundi-Prensa Mexico S.A. de C.V.
c/Rio Panuco, 141-Colonia Cuauhtemoc
06500 Mexico, D.F.
Tel: (52 5) 533-5658
Fax: (52 5) 514-6799

NEPAL
Everest Media International Services (P) Ltd.
GPO Box 5443
Kathmandu
Tel: (977 1) 472 152
Fax: (977 1) 224 431

NETHERLANDS
De Lindeboom/InOr-Publikaties
P.O. Box 202, 7480 AE Haaksbergen
Tel: (31 53) 574-0004
Fax: (31 53) 572-9296
E-mail: lindeboo@worldonline.nl
URL: http://www.worldonline.nl/~lindeboo

NEW ZEALAND
EBSCO NZ Ltd.
Private Mail Bag 99914
New Market
Auckland
Tel: (64 9) 524-8119
Fax: (64 9) 524-8067

NIGERIA
University Press Limited
Three Crowns Building Jericho
Private Mail Bag 5095
Ibadan
Tel: (234 22) 41-1356
Fax: (234 22) 41-2056

NORWAY
NIC Info A/S
Book Department, Postboks 6512 Etterstad
N-0606 Oslo
Tel: (47 22) 97-4500
Fax: (47 22) 97-4545

PAKISTAN
Mirza Book Agency
65, Shahrah-e-Quaid-e-Azam
Lahore 54000
Tel: (92 42) 735 3601
Fax: (92 42) 576 3714

Oxford University Press
5 Bangalore Town
Sharae Faisal
PO Box 13033
Karachi-75350
Tel: (92 21) 446307
Fax: (92 21) 4547640
E-mail: ouppak@TheOffice.net

Pak Book Corporation
Aziz Chambers 21, Queen's Road
Lahore
Tel: (92 42) 636 3222; 636 0885
Fax: (92 42) 636 2328
E-mail: pbc@brain.net.pk

PERU
Editorial Desarrollo SA
Apartado 3824, Lima 1
Tel: (51 14) 285380
Fax: (51 14) 286628

PHILIPPINES
International Booksource Center Inc.
1127-A Antipolo St, Barangay, Venezuela
Makati City
Tel: (63 2) 896 6501; 6505; 6507
Fax: (63 2) 896 1741

POLAND
International Publishing Service
Ul. Piekna 31/37
00-677 Warzawa

Tel: (48 2) 628-6089
Fax: (48 2) 621-7255
E-mail: books%ips@ikp.atm.com.pl
URL: http://www.ipscg.waw.pl/ips/export/

PORTUGAL
Livraria Portugal
Apartado 2681, Rua Do Carmo 70-74
1200 Lisbon
Tel: (1) 347-4982
Fax: (1) 347-0264

ROMANIA
Compani De Librarii Bucuresti S.A.
Str. Lipscani no. 26, sector 3
Bucharest
Tel: (40 1) 613 9645
Fax: (40 1) 312 4000

RUSSIAN FEDERATION
Isdatelstvo <Ves Mir>
9a, Kolpachniy Pereulok
Moscow 101831
Tel: (7 095) 917 87 49
Fax: (7 095) 917 92 59

SINGAPORE, TAIWAN, MYANMAR, BRUNEI
Ashgate Publishing Asia Pacific Pte. Ltd.
41 Kallang Pudding Road #04-03
Golden Wheel Building
Singapore 349316
Tel: (65) 741-5166
Fax: (65) 742-9356
E-mail: ashgate@asianconnect.com

SLOVENIA
Gospodarski Vestnik Publishing Group
Dunajska cesta 5
1000 Ljubljana
Tel: (386 61) 133 83 47; 132 12 30
Fax: (386 61) 133 80 30
E-mail: repansekj@gvestnik.si

SOUTH AFRICA, BOTSWANA
For single titles:
Oxford University Press Southern Africa
Vasco Boulevard, Goodwood
P.O. Box 12119, N1 City 7463
Cape Town
Tel: (27 21) 595 4400
Fax: (27 21) 595 4430
E-mail: oxford@oup.co.za

For subscription orders:
International Subscription Service
P.O. Box 41095
Craighall
Johannesburg 2024
Tel: (27 11) 880-1448
Fax: (27 11) 880-6248
E-mail: iss@is.co.za

SPAIN
Mundi-Prensa Libros, S.A.
Castello 37
28001 Madrid
Tel: (34 1) 431-3399
Fax: (34 1) 575-3998
E-mail: libreria@mundiprensa.es
URL: http://www.mundiprensa.es/

Mundi-Prensa Barcelona
Consell de Cent, 391
08009 Barcelona
Tel: (34 3) 488-3492
Fax: (34 3) 487-7659
E-mail: barcelona@mundiprensa.es

SRI LANKA, THE MALDIVES
Lake House Bookshop
100, Sir Chittampalam Gardiner Mawatha
Colombo 2
Tel: (94 1) 32105
Fax: (94 1) 432104
E-mail: LHL@sri.lanka.net

SWEDEN
Wennergren-Williams AB
P.O. Box 1305
S-171 25 Solna
Tel: (46 8) 705-97-50
Fax: (46 8) 27-00-71
E-mail: mail@wwi.se

SWITZERLAND
Librairie Payot Service Institutionnel
Côtes-de-Montbenon 30
1002 Lausanne
Tel: (41 21) 341-3229
Fax: (41 21) 341-3235

ADECO Van Diermen Editions Techniques
Ch. de Lacuez 41
CH1807 Blonay
Tel: (41 21) 943 2673
Fax: (41 21) 943 3605

THAILAND
Central Books Distribution
306 Silom Road
Bangkok 10500
Tel: (66 2) 235-5400
Fax: (66 2) 237-8321

TRINIDAD & TOBAGO AND THE CARRIBBEAN
Systematics Studies Ltd.
St. Augustine Shopping Center
Eastern Main Road, St. Augustine
Trinidad & Tobago, West Indies
Tel: (868) 645-8466
Fax: (868) 645-8467
E-mail: tobe@trinidad.net

UGANDA
Gustro Ltd.
PO Box 9997, Madhvani Building
Plot 16/4 Jinja Rd.
Kampala
Tel: (256 41) 251 467
Fax: (256 41) 251 468

UNITED KINGDOM
Microinfo Ltd.
P.O. Box 3, Alton, Hampshire GU34 2PG
England
Tel: (44 1420) 86848
Fax: (44 1420) 89889
E-mail: wbank@ukminfo.demon.co.uk
URL: http://www.microinfo.co.uk

The Stationery Office
51 Nine Elms Lane
London SW8 5DR
Tel: (44 171) 873-8400
Fax: (44 171) 873-8242

VENEZUELA
Tecni-Ciencia Libros, S.A.
Centro Cuidad Comercial Tamanco
Nivel C2, Caracas
Tel: (58 2) 959 5547; 5035; 0016
Fax: (58 2) 959 5636

ZAMBIA
University Bookshop, University of Zambia
Great East Road Campus
P.O. Box 32379
Lusaka
Tel: (260 1) 252 576
Fax: (260 1) 253 952

ZIMBABWE
Academic and Baobab Books (Pvt.) Ltd.
4 Conald Road, Graniteside
P.O. Box 567
Harare
Tel: 263 4 755035
Fax: 263 4 781913

2/25/98